WIRELESS COMMUNICATIONS

Second Edition

Andreas F. Molisch, *Fellow, IEEE*
University of Southern California, USA

A John Wiley and Sons, Ltd., Publication

Library of Congress Cataloguing-in-Publication Data

Molisch, Andreas F.
   Wireless communications / Andreas F. Molisch. – 2nd ed.
      p. cm.
   Includes bibliographical references and index.
   ISBN 978-0-470-74187-0 (cloth) – ISBN 978-0-470-74186-3 (pbk.)
   1. Wireless communication systems–Textbooks. I. Title.
   TK5103.2.M65 2011
   621.3845'6–dc22
                                                    2010017177

A catalogue record for this book is available from the British Library.

Print ISBN: 9780470741870 (H/B)
Print ISBN: 9780470741863 (P/B)
ePDF ISBN: 9780470666692

Typeset in 9/11 Times by Laserwords Private Limited, Chennai, India.
Printed and bound in Singapore by Markono Print Media Pte Ltd.

Contents

30   Exercises   751

**Peter Almers, Ove Edfors, Hao Feng, Fredrik Floren, Anders Johanson,
Johan Karedal, Buon Kiong Lau, Christian Mehlführer, Andreas F. Molisch,
Jan Plasberg, Barbara Resch, Jonas Samuelson, Junyang Shen, Andre Stranne,
Fredrik Tufvesson, Anthony Vetro and Shurjeel Wyne**

Preface and Acknowledgements to the Second Edition

Since the first edition of this book appeared in 2005, wireless communications research and technology continued its inexorable progress. This fact, together with the positive response to the first edition, motivated a second edition that would cover the topics that have emerged in the last years. Thus, the present edition aims to bring the book again in line with the breadth of topics relevant to state-of-the-art wireless communications engineering.

There are more than 150 pages of new material, covering

- cognitive radio (new Chapter 21);
- cooperative communications, relays, and ad hoc networks (new Chapter 22);
- video coding (new Chapter 23);
- 3GPP Long-Term Evolution (new Chapter 27);
- WiMAX (new Chapter 28).

There are furthermore significant extensions and additions on the following:

- MIMO (in Chapter 20), in particular a new section on multi-user MIMO (Section 20.3).
- IEEE 802.11n (high-throughput WiFi) in Section 29.3.
- Coding (bit-interleaved coded modulation) in Section 14.5.
- Introduction to information theory in Sections 14.1, 14.9, and Appendix 17.
- Channel models: updates of standardized channel models in Appendix 7.
- A number of minor modifications and reformulations, partly based on feedback from instructors and readers of the book.

These extensions are important for students (as well as researchers) to learn "up-to-date" skills, Most of the additional material might be best suited for a graduate course on advanced wireless concepts and techniques. However, the material on LTE (or WiMAX) is also well suited as an example for standardized systems in a more elementary course (replacing, e.g., discussions of GSM or WCDMA systems).

As for the first edition, presentation slides and a solutions manual are available *for instructors that adopt the textbook for their course*. This material can also be obtained from the publisher or from a new website, wides.usc.edu/teaching/textbook. This site will also contain important resources for all readers of the book, including an "errata list," updates, additional references, and similar material.

The writing of the new material was a major endeavor, and was greatly helped by the support of Sandy Sawchuk, Chair of the Department of Electrical Engineering at the University of Southern California. Particular thanks to Anthony Vetro, who wrote the new chapter on videocoding (Chapter 23). I am also grateful to the experts that kindly agreed to review the new material, namely

Honggang Zhang and Natasha Devroye (Chapter 21), Gerhard Kramer, Mike Neely, Bhaskar Krishnamachari (Chapter 22), Erik Dahlman (Chapter 27), Yang-Seok Choi Hujun Jin, and V. Shashidar (Chapter 28), Guiseppe Caire (new material in Chapters 14 and 17), Robert Heath and Claude Oestges (new material of Chapter 20), and Eldad Perahia (Section 29.3). As always, responsibility for any residual errors lies with me.

I also thank the students from my classes at USC, as well as readers and students from all over the world, who provided suggestions for corrections and improvements. Exercises for the new chapters were created by Junyang Shen, Hao Fang, and Christian Mehlfuehrer. Thanks to Neelesh B. Mehta for providing me with several figures on LTE.

As for the first edition, Mark Hammond from J. Wiley acted as acquisition editor; Sarah Tilley was the production editor. Special thanks to Dhanya Ramesh of Laserwords for her expert typesetting.

Preface to the First Edition

When, in 1994, I wrote the very first draft of this book in the form of lecture notes for a wireless course, the preface started by justifying the need for giving such a course at all. I explained at length why it is important that communications engineers understand wireless systems, especially digital cellular systems. Now, more than 10 years later, such a justification seems slightly quaint and outdated. Wireless industry has become the fastest growing sector of the telecommunications industry, and there is hardly anybody in the world who is not a user of some form of wireless technology. From the ubiquitous cellphones, to wireless LANs, to wireless sensors that are proliferating – we are surrounded by wireless communications devices.

One of the key challenges in studying wireless communications is the amazing breadth of topics that impacts this field. Traditionally, communications engineers have concentrated on, for example, digital modulation and coding theory, while the world of antennas and propagation studies was completely separate – "and never the twain shall meet." However, such an approach does not work for wireless communications. We need an understanding of *all* aspects that impact the performance of systems, and make the whole system work. This book is an attempt to provide such an overview, concentrating as it does on the physical layer of wireless communications.

Another challenge is that not only practical wireless systems, but also the science on which they are based is constantly changing. It is often claimed that while wireless systems rapidly change, the scientific basis of wireless communications stays the same, and thus engineers can rely on knowledge acquired at a given time to get them through many cycles of system changes, with just minor adjustments to their skill sets. This thought is comforting – and unfortunately false. For example, 10 years ago, topics like multiple-antenna systems, OFDM, turbo codes and LDPC codes, and multiuser detection, were mostly academic curiosities, and would at best be treated in PhD-level courses; today, they dominate not only mainstream research and system development, but represent vital, basic knowledge for students and practicing engineers. I hope that, by treating both new aspects as well as more "classical" topics, my book will give today's students and researchers knowledge and tools that will prove useful for them in the future.

The book is written for advanced undergraduate and graduate students, as well as for practicing engineers and researchers. Readers are assumed to have an understanding of elementary communication theory, like modulation/demodulation as well as of basic aspects of electromagnetic theory, though a brief review of these fields is given at the beginning of the corresponding chapters of the book. The core material of this book tries to get students to a stage where they can read more advanced monographs, and even research papers; for all those readers who want to dig deeper, the majority of chapters include a "further reading" section that cites the most important references. The text includes both mathematical formulations, and intuitive explanations. I firmly believe that such a dual approach leads to the deepest understanding of the material. In addition to being a textbook, the text is also intended to serve as a reference tool for researchers and practitioners. For this reason, I have tried to make it easier to read isolated chapters. All acronyms are explained the first time they occur in each chapter (not just at their first occurrence in the book); a list of symbols (see p. xlvii) explains the meaning of symbols used in the equations. Also, frequent cross-references should help for this purpose.

Synopsis

The book is divided into five parts. The first part, the introduction, gives a high-level overview of wireless communications. Chapter 1 first gives a taxonomy of different wireless services, and then describes the requirements for data rate, range, energy consumption, etc., that the various applications impose. This chapter also contains a brief history, and a discussion of the economic and social aspects of wireless communications. Chapter 2 describes the basic challenges of wireless communications, like multipath propagation and limited spectrum resources. Chapter 3 then discusses how noise and interference limit the capabilities of wireless systems, and how link budgets can serve as simple system-planning tools that give a first idea about the achievable range and performance.

The second part describes the various aspects of wireless propagation channels and antennas. As the propagation channel is the medium over which communication happens, understanding it is vital to understanding the remainder of the book. Chapter 4 describes the basic propagation processes: free space propagation, reflection, transmission, diffraction, diffuse scattering, and waveguiding. We find that the signal can get from the transmitter to the receiver via many different propagation paths that involve one or more of these processes, giving rise to many multipath components. It is often convenient to give a statistical description of the effects of multipath propagation. Chapter 5 gives a statistical formulation for narrowband systems, explaining both small-scale (Rayleigh) and large-scale fading. Chapter 6 then discusses formulations for wideband systems, and systems that can distinguish the directions of multipath components at the transmitter and receiver. Chapter 7 then gives specific models for propagation channels in different environments, covering path loss as well as wideband and directional models. Since all realistic channel models have to be based on (or confirmed by) measurements, Chapter 8 summarizes techniques that are used for measuring channel impulse responses. Finally, Chapter 9 briefly discusses antennas for wireless applications, especially with respect to different restrictions at base stations and mobile stations.

The third part of the book deals with the structure and theory of wireless transceivers. After a short summary of the components of a RF transceiver in Chapter 10, Chapter 11 then describes the different modulation formats that are used for wireless applications. The discussion not only includes mathematical formulations and signal space representations, but also an assessment of their advantages and disadvantages for various purposes. The performance of all these modems in flat-fading as well as frequency-selective channels is then the topic of Chapter 12. One critical observation we make here is the fact that fading leads to a drastic increase in error probability, and that increasing the transmit power is not a suitable way of improving performance. This motivates the next two Chapters, which deal with diversity and channel coding, respectively. We find that both these measures are very effective in reducing error probabilities in a fading channel. The coding chapter also includes a discussion of near-Shannon-limit-achieving codes (turbo codes and low-density parity check codes), which have gained great popularity in recent years. Since voice communication is still the most important application for cellphones and similar devices, Chapter 15 discusses the various ways of digitizing speech, and compressing information so that it can be transmitted over wireless channels in an efficient way. Chapter 16 finally discusses equalizers, which can be used to reduce the detrimental effect of frequency selectivity of wideband wireless channels. All the chapters in this part deal with a single link – i.e., the link between one transmitter and one receiver.

The fourth part then takes into account our desire to operate a number of wireless links simultaneously in a given area. This so-called *multiple-access* problem has a number of different solutions. Chapter 17 discusses frequency domain multiple access (FDMA) and time domain multiple access (TDMA), as well as packet radio, which has gained increasing importance for data transmission. This chapter also discusses the cellular principle, and the concept of frequency reuse that forms the basis not only for cellular, but also many other high-capacity wireless systems. Chapter 18 then describes spread spectrum techniques, in particular CDMA, where different users can be distinguished by different spreading sequences. This chapter also discusses multiuser detection, a very advanced receiver scheme that can greatly decrease the impact of multiple-access interference. Another topic of Part IV is "advanced transceiver techniques." Chapter 19 describes OFDM (orthogonal frequency domain multiplexing), which is a modulation method that can sustain very high data rates in channels with large delay spread. Chapter 20 finally discusses multiple-antenna techniques: "smart antennas," typically placed at the base station, are multiple-antenna elements with sophisticated signal processing that can (among other benefits) reduce interference and thus increase the capacity of cellular systems. MIMO (multiple-input-multiple-output) systems go one step further, allowing the transmission of parallel data streams from multiple-antenna elements at the transmitter, which are then received and demodulated by multiple-antenna elements at the receiver. These systems achieve a dramatic capacity increase even for a single link.

The last part of the book describes standardized wireless systems. Standardization is critical so that devices from different manufacturers can work together, and also systems can work seamlessly across national borders. The book describes the most successful cellular wireless standards – namely, GSM (Global System for Mobile communications), IS-95 and its advanced form CDMA 2000, as well as Wideband CDMA (also known as UMTS) in Chapters 21, 22, and 23, respectively. Furthermore, Chapter 24 describes the most important standard for wireless LANs – namely, IEEE 802.11.

A companion website (*www.wiley.com/go/molisch*) contains some material that I deemed as useful, but which would have made the printed version of the book overly bulky. In particular, the appendices to the various chapters, as well as supplementary material on the DECT (Digital Enhanced Cordless Telecommunications) system, the most important cordless phone standard, can be found there.

Suggestions for Courses

The book contains more material than can be presented in a single-semester course, and spans the gamut from very elementary to quite advanced topics. This gives the instructor the freedom to tailor teaching to the level and the interests of students. The book contains worked examples in the main text, and a large number of homework exercises at the end of the book. Solutions to these exercises, as well as presentation slides, are available to instructors on the companion website of this book.

A few examples for possible courses include:

Introductory course:

- Introduction (Chapters 1–3).
- Basic channel aspects (Sections 4.1–4.3, 5.1–5.4, 6.1, 6.2, 7.1–7.3):
  - elementary signal processing (Chapters 10, 11, and Sections 12.1, 12.2.1, 12.3.1, 13.1, 13.2, 13.4, 14.1–14.3, 16.1–16.2);
  - multiple access and system design (Chapters 17, 22 and Sections 18.2, 18.3, 21.1-21.7).
- Wireless propagation:
  - introduction (Chapter 2);
  - basic propagation effects (Chapter 4);
  - statistical channel description (Chapters 5 and 6);
  - channel modeling and measurement (Chapters 7 and 8);
  - antennas (Chapter 9).
  This course can also be combined with more basic material on electromagnetic theory and antennas.
- Advanced topics in wireless communications:
  - introduction and refresher: should be chosen by the instructor according to audience;

- o CDMA and multiuser detection (Sections 18.2, 18.3, 18.4);
- o OFDM (Chapter 19);
- o ultrawideband communications (Sections 6.6, 18.5);
- o multiantenna systems (Sections 6.7, 7.4, 8.5, 13.5, 13.6, and Chapter 20);
- o advanced coding (Sections 14.5, 14.6).
- Current wireless systems:
  - o TDMA-based cellular systems (Chapter 21);
  - o CDMA-based cellular systems (Chapters 22 and 23);
  - o cordless systems (supplementary material on companion website);
  - o wireless LANs (Chapter 24); and
  - o selected material from previous chapters for the underlying theory, according to the knowledge of the audience.

Acknowledgments to the First Edition

This book is the outgrowth of many years of teaching and research in the field of wireless communications. During that time, I worked at two universities (Technical University Vienna, Austria and Lund University, Sweden) and three industrial research labs (FTW Research Center for Telecommunications Vienna, Austria; AT&T (Bell) Laboratories–Research, Middletown, NJ, U.S.A.; and Mitsubishi Electric Research Labs., Cambridge, MA, U.S.A.), and cooperated with my colleagues there, as well as with numerous researchers at other institutions in Europe, the U.S.A., and Japan. All of them had an influence on how I see wireless communications, and thus, by extension, on this book. To all of them I owe a debt of gratitude. First and foremost, I want to thank Ernst Bonek, the pioneer and doyen of wireless communications in Austria, who initiated this project, and with whom I had countless discussions on technical as well as didactic aspects of this book and the lecture notes that preceded it (these lecture notes served for a course that we gave jointly at TU Vienna). Without his advice and encouragement, this book would never have seen the light of day. I also want to thank my colleagues and students at TU Vienna, particularly Paulina Erätuuli, Josef Fuhl, Alexander Kuchar, Juha Laurila, Gottfried Magerl, Markus Mayer, Thomas Neubauer, Heinz Novak, Berhard P. Oehry, Mario Paier, Helmut Rauscha, Alexander Schneider, Gerhard Schultes, and Martin Steinbauer, for their help. At Lund University, my colleagues and students also greatly contributed to this book: Peter Almers, Ove Edfors, Fredrik Floren, Anders Johanson, Johan Karedal, Vincent Lau, Andre Stranne, Fredrik Tufvesson, and Shurjeel Wyne. They contributed not only stimulating suggestions on how to present the material but also figures and examples; in particular, most of the exercises and solutions were created by them and Section 19.5 is based on the ideas of Ove Edfors. A special thanks to Gernot Kubin from Graz University of Technology, who contributed Chapter 15 on speech coding. My colleagues and managers at FTW, AT&T, and MERL – namely, Markus Kommenda, Christoph Mecklenbraueker, Helmut Hofstetter, Jack Winters, Len Cimini, Moe Win, Martin Clark, Yang-Seok Choi, Justin Chuang, Jin Zhang, Kent Wittenburg, Richard Waters, Neelesh Mehta, Phil Orlik, Zafer Sahinoglu, Daqin Gu (who greatly contributed to Chapter 24), Giovanni Vanucci, Jonathan Yedidia, Yves-Paul Nakache, and Hongyuan Zhang, also greatly influenced this book. Special thanks and appreciation to Larry Greenstein, who (in addition to the many instances of help and advice) took an active interest in this book and provided invaluable suggestions.

A special thanks also to the reviewers of this book. The manuscript was critically read by anonymous experts selected by the publisher, as well as several of my friends and colleagues at various research institutions: John B. Anderson (Chapters 11–13), Anders Derneryd (Chapter 9), Larry Greenstein (Chapters 1–3, 7, 17–19), Steve Howard (Chapter 22), Thomas Kaiser (Chapter 20), Achilles Kogantis (Chapter 23), Gerhard Kristensson (Chapter 4), Thomas Kuerner (Chapter 21), Gerald Matz (Chapters 5–6), Neelesh B. Mehta (Chapter 20), Bob O'Hara (Chapter 24), Phil Orlik (Section 17.4), John Proakis (Chapter 16), Said Tatesh (Chapter 23), Reiner Thomae (Chapter 8),

Chintha Tellambura (Chapters 11–13), Giorgia Vitetta (Chapter 16), Jonathan Yedidia (Chapter 14). To all of them goes my deepest appreciation. Of course, the responsibility for any possible remaining errors rests with me.

Mark Hammond as publisher, Sarah Hinton as project editor, and Olivia Underhill as assistant editor, all from John Wiley & Sons, Ltd, guided the writing of the book with expert advice and considerable patience. Manuela Heigl and Katalin Stibli performed many typing and drawing tasks with great care and cheerful enthusiasm. Originator expertly typeset the manuscript.

Abbreviations

2G	Second Generation
3G	Third Generation
3GPP	Third Generation Partnership Project
3GPP2	Third Generation Partnership Project 2
3SQM	Single Sided Speech Quality Measure
A/D	Analog to Digital
AB	Access Burst
AC	Access Category
AC	Administration Center
AC	Alternate Current
ACCH	Associated Control CHannel
ACELP	Algebraic Code Excited Linear Prediction
ACF	AutoCorrelation Function
ACI	Adjacent Channel Interference
ACK	ACKnowledgment
ACLR	Adjacent Channel Leakage Ratio
ACM	Address Complete Message
AD	Access Domain
ADC	Analog to Digital Converter
ADDTS	ADD Traffic Stream
ADF	Average Duration of Fades
ADPCM	Adaptive Differential Pulse Code Modulation
ADPM	Adaptive Differential Pulse Modulation
ADPS	Angular Delay Power Spectrum
ADSL	Asymmetric Digital Subscriber Line
AF	Amplify-and-Forward
AGC	Automatic Gain Control
AGCH	Access Grant CHannel
AICH	Acquisition Indication CHannel
AIFS	Arbitration Inter Frame Spacing
ALOHA	random access packet radio system
AMPS	Advanced Mobile Phone System
AMR	Adaptive Multi Rate
AN	Access Network
ANSI	American National Standards Institute
AODV	Ad hoc On-Demand Distance Vector
AP	Access Point
APS	Angular Power Spectrum
ARFCN	Absolute Radio Frequency Channel Number

ARIB	Association of Radio Industries and Businesses (Japan)
ARQ	Automatic Repeat reQuest
ASIC	Application Specific Integrated Circuit
ASK	Amplitude Shift Keying
ATDPICH	Auxiliary forward Transmit Diversity PIlot CHannel
ATIS	Alliance for Telecommunications Industry Solutions
ATM	Asynchronous Transfer Mode
ATSC-M/H	Advanced Television Systems Committee – Mobile/Handheld
AUC	AUthentication Center
AV	Audio and Video
AVC	Advanced Video Coding
AWGN	Additive White Gaussian Noise
BAM	Binary Amplitude Modulation
BAN	Body Area Network
BCC	Base station Color Code
BCCH	Broadcast Control CHannel
BCF	Base Control Function
BCH	Bose–Chaudhuri–Hocquenghem (code)
BCH	Broadcast CHannel
BCJR	Initials of the authors of Bahl et al. [1974]
BEC	Backward Error Correction
BER	Bit Error Rate
BFI	Bad Frame Indicator
BFSK	Binary Frequency Shift Keying
BICM	Bit Interleaved Coded Modulation
BLAST	Bell labs LAyered Space Time
Bm	Traffic channel for full-rate voice coder
BN	Bit Number
BNHO	Barring all outgoing calls except those to Home PLMN
BPF	BandPass Filter
BPPM	Binary Pulse Position Modulation
BPSK	Binary Phase Shift Keying
BS	Base Station
BSC	Base Station Controller
BSI	Base Station Interface
BSIC	Base Station Identity Code
BSS	Base Station Subsystem
BSS	Basic Service Set
BSSAP	Base Station Application Part
BTS	Base Transceiver Station
BU	Bad Urban
BW	Bandwidth
CA	Cell Allocation
CAF	Compute and Forward
CAP	Controlled Access Period
CAZAC	Constant Amplitude Zero AutoCorrelation
CB	Citizens' Band
CBCH	Cell Broadcast CHannel
CC	Country Code
CCBS	Completion of Calls to Busy Subscribers

CCCH	Common Control CHannel
CCF	Cross Correlation Function
CCI	Co Channel Interference
CCITT	Commite' Consultatif International de Telegraphique et Telephonique
CCK	Complementary Code Keying
CCPCH	Common Control Physical CHannel
CCPE	Control Channel Protocol Entity
CCSA	China Communications Standards Association
CCTrCH	Coded Composite Traffic CHannel
cdf	cumulative distribution function
CDG	CDMA Development Group
CDMA	Code Division Multiple Access
CELP	Code Excited Linear Prediction
CEPT	European Conference of Postal and Telecommunications Administrations
CF	Compress-and-Forward
CF-Poll	Contention-Free Poll
CFB	Contention Free Burst
CFP	Contention Free Period
CI	Cell Identify
C/I	Carrier-to-Interference ratio
CM	Connection Management
CMA	Constant Modulus Algorithm
CMOS	Complementary Metal Oxide Semiconductor
CN	Core Network
CND	Core Network Domain
CNG	Comfort Noise Generation
CONP	Connect Number Identification Presentation
COST	European COoperation in the field of Scientific and Technical research
CP	Contention Period
CP	Cyclic Prefix
CPC	Cognitive Plot Channels
CPCH	Common Packet CHannel
CPFSK	Continuous Phase Frequency Shift Keying
CPICH	Common PIlot CHannel
CRC	Cyclic Redundancy Check
CRC	Cyclic Redundancy Code
CS-ACELP	Conjugate Structure-Algebraic Code Excited Linear Prediction
CSD	Cyclic Shift Diversity
CSI	Channel State Information
CSIR	Channel State Information at the Receiver
CSIT	Channel State Information at the Transmitter
CSMA	Carrier Sense Multiple Access
CSMA/CA	Carrier Sense Multiple Access with Collision Avoidance
CTS	Clear To Send
CTT	Cellular Text Telephony
CUG	Closed User Group
CW	Contention Window
D-BLAST	Diagonal BLAST
DAA	Detect and Avoid
DAB	Digital Audio Broadcasting

DAC	Digital to Analog Converter
DAF	Diversity-Amplify-and-Forward
DAM	Diagnostic Acceptability Measure
dB	Decibel
DB	Dummy Burst
DBPSK	Differential Binary Phase Shift Keying
DC	Direct Current
DCCH	Dedicated Control CHannel
DCF	Distributed Coordination Function
DCH	Dedicated (transport) CHannel
DCM	Directional Channel Model
DCS1800	Digital Cellular System at the 1800-MHz band
DCT	Discrete Cosine Transform
DDF	Diversity-Decode-and-Forward
DDIR	Double Directional Impulse Response
DDDPS	Double Directional Delay Power Spectrum
DECT	Digital Enhanced Cordless Telecommunications (ETSI)
DF	Decode-and-Forward
DFE	Decision Feedback Equalizer
DFT	Discrete Fourier Transform
DIFS	Distributed Inter Frame Space
DL	Downlink
DLL	Data Link Layer
DLP	Direct Link Protocol
DM	Delta Modulation
DMC	Discrete Memoryless Channel
DMT	Discrete Multi Tone
DNS	Domain Name Server
DOA	Direction Of Arrival
DOD	Direction Of Departure
DPCCH	Dedicated Physical Control CHannel
DPDCH	Dedicated Physical Data CHannel
DPSK	Differential Phase Shift Keying
DQPSK	Differential Quadrature-Phase Shift Keying
DRM	Discontinuous Reception Mechanisms
DRT	Diagnostic Rhyme Test
DRX	Discontinuous Reception
DS	Direct Sequence
DS-CDMA	Direct Sequence–Code Division Multiple Access
DS-SS	Direct Sequence–Spread Spectrum
DSA	Dynamic Spectrum Access
DSCH	Downlink Shared Channel
DSDF	Destination-Sequenced Distance Vector
DSI	Digital Speech Interpolation
DSL	Digital Subscriber Line
DSMA	Data Sense Multiple Access
DSP	Digital Signal Processor
DSR	Distributed Speech Recognition
DSR	Dynamic Source Routing
DTAP	Direct Transfer Application Part

DTE	Data Terminal Equipment
DtFT	Discrete-time Fourier Transform
DTMF	Dual Tone Multi Frequency (signalling)
DTX	Discontinuous Transmission
DUT	Device Under Test
DVB	Digital Video Broadcasting
DVB-H	Digital Video Broadcasting – Handheld
DxF	Diversity xF
EC	European Commission
ECL	Emitter Coupled Logic
EDCA	Enhanced Distributed Channel Access
EDCSD	Enhanced Data rate Circuit Switched Data
EDGE	Enhanced Data rates for GSM Evolution
EDPRS	Enhanced Data rate GPRS
EFR	Enhanced Full Rate
EGC	Equal Gain Combining
EIA	Electronic Industries Alliance (U.S.A.)
EIFS	Extended Inter Frame Space
EIR	Equipment Identity Register
EIRP	Equivalent Isotropically Radiated Power
ELP	Equivalent Low Pass
EMS	Enhanced Messaging Service
EN	European Norm
ERLE	Echo Return Loss Enhancement
ESN	Electronic Serial Number
ESPRIT	Estimation of Signal Parameters by Rotational Invariance Techniques
ETS	European Telecommunication Standard
ETSI	European Telecommunications Standards Institute
ETX	Expected Number of Transmissions
EV-DO	EVolution-Data Optimized
EVD	Eigen Value Decomposition
EVM	Error Vector Magnitude
EVRC	Enhanced Variable Rate Coder
F-APICH	Forward dedicated Auxiliary PIlot CHannel
F-BCCH	Forward Broadcast Control CHannel
F-CACH	Forward Common Assignment CHannel
F-CCCH	Forward Common Control CHannel
F-CPCCH	Forward Common Power Control CHannel
F-DCCH	Forward Dedicated Control CHannel
F-PDCCH	Forward Packet Data Control CHannel
F-PDCH	Forward Packet Data CHannel
F-QPCH	Forward Quick Paging CHannel
F-SCH	Forward Supplemental CHannel
F-SYNC	Forward SYNChronization channel
F-TDPICH	Forward Transmit Diversity PIlot CHannel
F0	Fundamental frequency
FAC	Final Assembly Code
FACCH	Fast Associated Control CHannel
FACCH/F	Full-rate FACCH
FACCH/H	Half-rate FACCH

FACH	Forward Access CHannel
FB	Frequency correction Burst
FBI	Feed Back Information
FCC	Federal Communications Commission
FCCH	Frequency Correction CHannel
FCH	Fundamental CHannel
FCH	Frame Control Header
FCS	Frame Check Sequence
FDD	Frequency Domain Duplexing
FDMA	Frequency Division Multiple Access
FDTD	Finite Difference Time Domain
FEC	Forward Error Correction
FEM	Finite Element Method
FFT	Fast Fourier Transform
FH	Frequency Hopping
FHMA	Frequency Hopping Multiple Access
FIR	Finite Impulse Response
FM	Frequency Modulation
FN	Frame Number
FOMA	Japanese version of the UMTS standard
FQI	Frame Quality Indicator
FR	Full Rate
FS	Federal Standard
FSK	Frequency Shift Keying
FT	Fourier Transform
FTF	Fast Transversal Filter
FWA	Fixed Wireless Access
GF	Galois Field
GGSN	Gateway GPRS Support Node
GMSC	Gateway Mobile Services Switching Center
GMSK	Gaussian Minimum Shift Keying
GPRS	General Packet Radio Service
GPS	Global Positioning System
GSC	Generalized Selection Combining
GSCM	Geometry-based Stochastic Channel Model
GSM	Global System for Mobile communications
GSM PLMN	GSM Public Land Mobile Network
GSM 1800	Global System for Mobile communications at the 1800-MHz band
GTP	GPRS Tunneling Protocol
H-BLAST	Horizontal BLAST
H-S/MRC	Hybrid Selection/Maximum Ratio Combining
HC	Hybrid Coordinator
HCCA	HCF (Hybrid Coordination Function) Controlled Channel Access
HCF	Hybrid Coordination Function
HDLC	High Level Data Link Control
HF	High Frequency
HIPERLAN	HIgh PERformance Local Area Network
HLR	Home Location Register
HMSC	Home Mobile-services Switching Center
HNM	Harmonic + Noise Modeling

hostid	host address
HO	HandOver
HPA	High Power Amplifier
HR	Half Rate
HR/DS or HR/DSSS	High Rate Direct Sequence PHY
HRTF	Head Related Transfer Function
HSCSD	High Speed Circuit Switched Data
HSDPA	High Speed Downlink Packet Access
HSN	Hop Sequence Number
HSPA	High-Speed Packet Access
HT	High Throughput
HT	Hilly Terrain
HTTP	Hyper Text Transfer Protocol
IAF	InterSymbol Interference Amplify-and-Forward
IAM	Initial Address Message
ICB	Incoming Calls Barred
ICI	Inter Carrier Interference
ID	Identification
ID	Identifier
IDFT	Inverse Discrete Fourier Transform
I_e	Equipment impairment factor
IE	Information Element
IEC	International Electrotechnical Commission
IEEE	Institute of Electrical and Electronics Engineers
IETF	Internet Engineering Task Force
IF	Intermediate Frequency
IFFT	Inverse Fast Fourier Transformation
IFS	Inter Frame Space
iid	independent identically distributed
IIR	Infinite Impulse Response
ILBC	Internet Low Bit-rate Codec
IMBE	Improved Multi Band Excitation
IMEI	International Mobile station Equipment Identity
IMSI	International Mobile Subscriber Identity
IMT	International Mobile Telecommunications
IMT-2000	International Mobile Telecommunications 2000
INMARSAT	INternational MARitime SATellite System
IO	Interacting Object
I/O	Input/Output
IP	Internet Protocol
IPO	Initial Public Offering
IQ	In-Phase – Quadrature Phase
IR	Impulse Radio
IRIDIUM	Project
IRS	Intermediate Reference System
IS-95	Interim Standard 95 (the first CDMA system adopted by the American TIA)
ISDN	Integrated Services Digital Network
ISI	InterSymbol Interference
ISM	Industrial, Scientific, and Medical

ISO	International Standards Organization
ISPP	Interleaved Single Pulse Permutation
ITU	International Telecommunications Union
IWF	Inter Working Function
IWU	Inter Working Unit
IxF	Interference xF
JD-TCDMA	Joint Detection–Time/Code Division Multiple Access
JDC	Japanese Digital Cellular
JPEG	Joint Photographic Expert Group
JVT	Joint Video Team
Kc	Cipher Key
Ki	Key used to calculate SRES
Kl	Location Key
Ks	Session Key
KLT	Karhunen Loève Transform
LA	Location Area
LAC	Location Area Code
LAI	Location Area Identity
LAN	Local Area Network
LAP-Dm	Link Access Protocol on Dm Channel
LAR	Logarithmic Area Ratio
LBG	Linde–Buzo–Gray algorithm
LCR	Level Crossing Rate
LD-CELP	Low Delay–Code Excited Linear Prediction
LDPC	Low Density Parity Check
LEO	Low Earth Orbit
LFSR	Linear Feedback Shift Register
LLC	Logical Link Control
LLR	Log Likelihood Ratio
LMMSE	Linear Minimum Mean Square Error
LMS	Least Mean Square
LNA	Low Noise Amplifier
LO	Local Oscillator
LPF	LowPass Filter
LOS	Line Of Sight
LP	Linear Prediction
LP	Linear Predictor
LP	Linear Program
LPC	Linear Predictive Coding
LPC	Linear Predictive voCoder
LR	Location Register
LS	Least Squares
LSF	Line Spectral Frequency
LSP	Line Spectrum Pair
LTE	Long-Term Evolution
LTI	Linear Time Invariant
LTP	Long Term Prediction
LTP	Long Term Predictor
LTV	Linear Time Variant
M-QAM	M-ary Quadrature Amplitude Modulation

MA	Mobile Allocation
MA	Multiple Access
MAC	Medium Access Control
MACN	Mobile Allocation Channel Number
MAF	Mobile Additional Function
MAF	Multi-hop Amplify-and-Forward
MAHO	Mobile Assisted Hand Over
MAI	Multiple Access Interference
MAIO	Mobile Allocation Index Offset
MAN	Metropolitan Area Network
MAP	Maximum A Posteriori
MAP	Mobile Application Part
MB	Macroblock
MBE	Multi Band Excitation
MBOA	Multi Band OFDM Alliance
MC-CDMA	Multi Carrier Code Division Multiple Access
MCC	Mobile Country Code
MCS	Modulation and Coding Scheme
MDC	Multiple Description Coding
MDF	Multi-hop Decode-and-Forward
MDHO	Macro Diversity HandOver
ME	Maintenance Entity
ME	Mobile Equipment
MEA	Multiple Element Antenna
MEF	Maintenance Entity Function
MEG	Mean Effective Gain
MELP	Mixed Excitation Linear Prediction
MFEP	Matched Front End Processor
MIC	Mobile Interface Controller
MIME	Multipurpose Internet Mail Extensions
MIMO	Multiple Input Multiple Output system
MIPS	Million Instructions Per Second
ML	Maximum Likelihood
MLSE	Maximum Likelihood Sequence Estimators (or Estimation)
MMS	Multimedia Messaging Service
MMSE	Minimum Mean Square Error
MNC	Mobile Network Code
MNRU	Modulated Noise Reference Unit
MOS	Mean Opinion Score
MoU	Memorandum of Understanding
MP3	Motion Picture Experts Group-1 layer 3
MPC	Multi Path Component
MPDU	MAC Protocol Data Unit
MPEG	Motion Picture Experts Group
MPR	Multi-Point Relay
MPSK	M-ary Phase Shift Keying
MRC	Maximum Ratio Combining
MS	Mobile Station
MS ISDN	Mobile Station ISDN Number
MSC	Mobile Switching Center

MSCU	Mobile Station Control Unit
MSDU	MAC Service Data Unit
MSE	Mean Square Error
MSIN	Mobile Subscriber Identification Number
MSISDN	Mobile Station ISDN Number
MSK	Minimum Shift Keying
MSL	Main Signaling Link
MSRN	Mobile Station Roaming Number
MSS	Mobile Satellite Service
MT	Mobile Terminal
MT	Mobile Termination
MTP	Message Transfer Part
MUMS	Multi User Mobile Station
MUSIC	Multiple Signal Classification
MUX	Multiplexing
MVC	Multiview Video Coding
MVM	Minimum Variance Method
MxF	Multi-hop xF
NAV	Network Allocation Vector
NB	Narrow Band
NB	Normal Burst
NBIN	A parameter in the hopping sequence
NCELL	Neighboring (adjacent) Cell
NDC	National Destination Code
NDxF	Nonorthogonal Diversity xF
netid	network address
NF	Network Function
NLOS	Non Line Of Sight
NLP	Non Linear Processor
NM	Network Management
NMC	Network Management Centre
NMSI	National Mobile Station Identification number
NMT	Nordic Mobile Telephone
Node-B	Base station
NRZ	Non Return to Zero
NSAP	Network Service Access Point
NSS	Network and Switching Subsystem
NT	Network Termination
NTT	Nippon Telephone and Telegraph
O&M	Operations & Maintenance
OACSU	Off Air Call Set Up
OCB	Outgoing Calls Barred
ODC	Ornithine DeCarboxylase
OEM	Original Equipment Manufacturer
OFDM	Orthogonal Frequency Division Multiplexing
OFDMA	Orthogonal Frequency Division Multiple Access
OLSR	Optimized Link State Routing
OMC	Operations & Maintenance Center
OOK	On Off Keying
OPT	Operator Perturbation Technique

OQAM	Offset Quadrature Amplitude Modulation
OQPSK	Offset Quadrature Phase Shift Keying
OS	Operating Systems
OSI	Operator System Interface
OSS	Operation Support System
OTD	Orthogonal Transmit Diversity
OVSF	Orthogonal Variable Spreading Factor
P/S	Parallel to Serial (conversion)
PABX	Private Automatic Branch eXchange
PACCH	Packet Associated Control CHannel
PACS	Personal Access Communications System
PAD	Packet Assembly/Disassembly facility
PAGCH	Packet Access Grant CHannel
PAM	Pulse Amplitude Modulation
PAN	Personal Area Network
PAPR	Peak-to-Average Power Ratio
PAR	Peak-to-Average Ratio
PARCOR	PARtial CORrelation
PBCCH	Packet Broadcast Control CHannel
PC	Point Coordinator
PCCCH	Packet Common Control CHannel
PCF	Point Coordination Function
PCG	Power Control Group
PCH	Paging CHannel
PCM	Pulse Code Modulated
PCPCH	Physical Common Packet CHannel
PCS	Personal Communication System
PDA	Personal Digital Assistant
PDC	Pacific Digital Cellular (Japanese system)
PDCH	Packet Data CHannel
pdf	probability density function
PDN	Public Data Network
PDP	Power Delay Profile
PDSCH	Physical Downlink Shared CHannel
PDTCH	Packet Data Traffic CHannel
PDU	Packet Data Unit
PESQ	Perceptual Evaluation of Speech Quality
PHS	Personal Handyphone System
PHY	PHYsical layer
PIC	Parallel Interference Cancellation
PICH	Page Indication Channel
PIFA	Planar Inverted F Antenna
PIFS	Priority Inter Frame Space
PIN	Personal Identification Number
PLCP	Physical Layer Convergence Procedure
PLL	Physical Link Layer
PLMN	Public Land Mobile Network
PN	Pseudo Noise
PNCH	Packet Notification CHannel
POP	Peak to Off Peak

POTS	Plain Old Telephone Service
PPCH	Packet Paging CHannel
PPDU	Physical Layer Protocol Data Unit
PPM	Pulse Position Modulation
PRACH	Packet Random Access CHannel
PRACH	Physical Random Access CHannel
PRake	Partial Rake
PRB	Physical Resource Block
PRMA	Packet Reservation Multiple Access
PSD	Power Spectral Density
PSDU	Physical Layer Service Data Unit
PSK	Phase Shift Keying
PSMM	Pilot Strength Measurement Message
PSPDN	Packet Switched Public Data Network
PSQM	Perceptual Speech Quality Measurement
PSTN	Public Switched Telephone Network
PTCCH-D	Packet Timing advance Control CHannel-Downlink
PTCCH-U	Packet Timing advance Control CHannel-Uplink
PTM	Point To Multipoint
PTM-M	Point To Multipoint Multicast
PTM-SC	Point To Multipoint Service Center
PTO	Public Telecommunications Operators
PUSC	Partial Use of Subcarriers
PUK	Personal Unblocking Key
PWI	Prototype Waveform Interpolation
PWT	Personal Wireless Telephony
QAM	Quadrature Amplitude Modulation
QAP	QoS Access Point
QCELP	Qualcomm Code Excited Linear Prediction
QFGV	Quadratic Form Gaussian Variable
QOF	Quasi Orthogonal Function
QoS	Quality of Service
QPSK	Quadrature-Phase Shift Keying
QSTA	QoS STAtion
R-ACH	Reverse Access CHannel
R-ACKCH	Reverse ACKnowledgement CHannel
R-CCCH	Reverse Common Control CHannel
R-CQICH	Reverse Channel Quality Indicator CHannel
R-DCCH	Reverse Dedicated Control CHannel
R-EACH	Reverse Enhanced Access CHannel
R-FCH	Reverse Fundamental CHannel
R-PICH	Reverse PIlot CHannel
R-SCH	Reverse Supplemental CHannel
RA	Random Mode Request information field
RA	Routing Area
RA	Rural Area
RA	Random Access
RAB	Random Access Burst
RACH	Random Access CHannel
RAN	Radio Access Network

RC	Raised Cosine
RCDLA	Radiation Coupled Dual L Antenna
RE	Resource Element
RF	Radio Frequency
RFC	Radio Frequency Channel
RFC	Request For Comments
RFL	Radio Frequency subLayer
RFN	Reduced TDMA Frame Number
RLC	Radio Link Control
RLP	Radio Link Protocol
RLS	Recursive Least Squares
RNC	Radio Network Controller
RNS	Radio Network Subsystem
RNTABLE	Table of 128 integers in the hopping sequence
RPAR	Relay PAth Routing
RPE	Regular Pulse Excitation (Voice Codec)
RPE-LTP	Regular Pulse Excited with Long Term Prediction
RS	Reed–Solomon (code)
RS	Reference Signal
RS	Relay Station
RSC	Recursive Systematic Convolutional
RSC	Radio Spectrum Committee
RSSI	Received Signal Strength Indication
RTSP	Real Time Streaming Protocol
RTP	Real-time Transport Protocol
rv	random variable
RVLC	Reversible Variable Length Code
RX	Receiver
RXLEV	Received Signal Level
RXQUAL	Received Signal Quality
S-CCPCH	Secondary Common Control Physical CHannel
SABM	Set Asynchronous Balanced Mode
SACCH	Slow Associated Control CHannel
SAGE	Space Alternating Generalized Expectation – maximization
SAP	Service Access Point
SAPI	Service Access Point Identifier
SAPI	Service Access Points Indicator
SAR	Specific Absorption Rate
SB	Synchronization Burst
SC-CDMA	Single-Carrier CDMA
SC-FDMA	Single-Carrier FDMA
SCCP	Signalling Connection Control Part
SCH	Synchronisation CHannel
SCN	Sub Channel Number
SCxF	Split-Combine xF
SDCCH	Standalone Dedicated Control CHannel
SDCCH/4	Standalone Dedicated Control CHannel/4
SDCCH/8	Standalone Dedicated Control CHannel/8
SDMA	Space Division Multiple Access
SEGSNR	SEGmental Signal-to-Noise Ratio

SEP	Symbol Error Probability
SER	Symbol Error Rate
SFBC	Space Frequency Block Coding
SFIR	Spatial Filtering for Interference Reduction
SFN	System Frame Number
SGSN	Serving GPRS Support Node
SIC	Successive Interference Cancellation
SID	SIlence Descriptor
SIFS	Short Infer Frame Space
SIM	Subscriber Identity Module
SINR	Signal-to-Interference-and-Noise Ratio
SIR	Signal-to-Interference Ratio
SISO	Soft Input Soft Output
SISO	Single Input Single Output
SLNR	Signal-to-Leakage and Noise Ratio
SM	Spatial Multiplexing
SMS	Short Message Service
SMTP	Short Message Transfer Protocol
SN	Serial Number
SNDCP	Subnetwork Dependent Convergence Protocol
SNR	Signal-to-Noise Ratio
SOLT	Short Open Loss Termination
SON	Self Organizing Network
SP	Shortest Path
S/P	Serial to Parallel (conversion)
SQNR	Signal-to-Quantization Noise Ratio
SR	Spatial Reference
SRake	Selective Rake
SRMA	Split-channel Reservation Multiple Access
SSA	Small Scale Averaged
SSF	Small-Scale Fading
ST	Space – Time
STA	STAtion
STBC	Space Time Block Code
STC	Sinusoidal Transform Coder
STDCC	Swept Time Delay Cross Correlator
STP	Short Term Prediction
STP	Short Term Predictor
STS	Space Time Spreading
STTC	Space Time Trellis Code
SV	Saleh–Valenzuela model
SVD	Singular Value Decomposition
TA	Terminal Adapter
TAC	Type Approval Code
TAF	Terminal Adapter Function
TBF	Temporary Block Flow
TC	Traffic Category
TC	Topology Control
TCH	Traffic CHannel
TCH/F	Full-rate Traffic CHannels

TCH/H	Half-rate Traffic CHannels
TCM	Trellis Coded Modulation
TCP	Transmission Control Protocol
TD-SCDMA	Time Division-Synchronous Code Division Multiple Access
TDD	Time Domain Duplexing
TDMA	Time Division Multiple Access
TE	Temporal Reference
TE	Terminal Equipment
TE	Transmitted Reference
TE	Transversal Electric
TETRA	TErrestrial Trunked RAdio
TFCI	Transmit Format Combination Indicator
TFI	Transport Format Indicator
TH-IR	Time Hopping Impulse Radio
TIA	Telecommunications Industry Association (U.S.)
TM	Transversal Magnetic
TMSI	Temporary Mobile Subscriber Identity
TPC	Transmit Power Control
TR	Technical Report (ETSI)
TR	Temporal reference
TR	Transmitted reference
TS	Technical Specification
TS	Time Slot
TSPEC	Traffic SPECifications
TTA	Telecommunications Technology Association of Korea
TTC	Telecommunications Technology Committee
TTS	Text To Speech synthesis
TU	Typical Urban
TX	Transmitter
TXOP	Transmission OPportunity
U-NII	Unlicensed National Information Infrastructure
UARFCN	UTRA Absolute Radio Frequency Channel Number
UCPCH	Uplink Common Packet CHannel
UDP	User Datagram Protocol
UE	User Equipment
UE-ID	User Equipment in-band IDentification
UED	User Equipment Domain
UL	Uplink
ULA	Uniform Linear Array
UMTS	Universal Mobile Telecommunications System
UP	User Priority
US	Uncorrelated Scatterer
USB	Universal Serial Bus
USF	Uplink Status Flag
USIM	User Service Identity Module
UTRA	UMTS Terrestrial Radio Access
UTRAN	UMTS Terrestrial Radio Access Network
UWB	Ultra Wide Bandwidth
UWC	Universal Wireless Communications
VAD	Voice Activity Detection/Detector

VCDA	Virtual Cell Deployment Area
VCEG	Video Coding Expert Group
VCO	Voltage Controlled Oscillator
VLC	Variable Length Coding
VLR	Visitor Location Register
VoIP	Voice over Internet Protocol
VRB	Virtual Resource Block
VQ	Vector Quantization/Quantizer
VSELP	Vector Sum Excited Linear Prediction
WAP	Wireless Application Protocol
WB	Wide Band
WCDMA	Wideband Code Division Multiple Access
WF	Whitening Filter
WG	Working Group
WH	Walsh-Hadamard
WI	Waveform Interpolation
WiFi	Wireless Fidelity
WLAN	Wireless Local Area Network
WLL	Wireless Local Loop
WM	Wireless Medium
WSS	Wide Sense Stationary
WSSUS	Wide Sense Stationary Uncorrelated Scatterer
ZF	Zero-Forcing

Symbols

This list gives a brief overview of the use of variables in the text. Due to the large number of quantities occurring, the same letter might be used in different chapters for different quantities. For this reason, this will need checking as chapter numbers have changed in which the variable is primarily used, though they can occur in other chapters as well. Those variables that are used only locally, and explained directly at their occurrence, are not mentioned here.

Lowercase symbols:

a_p, a_m	auxiliary variables	4
a_1	amplitudes of the MPCs	5, 6, 7, 8
$a(h_m)$	auxiliary function	7
$\mathbf{a}(\phi)$	steering vector	8
$a_{n,m}$	amplitudes of components from direction n, delay m	13
b_m	mth bit	11, 12, 13, 14, 16, 17
cdf	cumulative distribution function	5
$c_{i,k}$	amplitudes of resolvable MPCs; tap weights for tapped delay lines	7
c_0	speed of light	5, 13, 19
c_m	complex transmit symbols	11, 12, 13, 16
d	distance BS–MS	4
d	distance in signal space diagram	12, 13
d_R	Rayleigh distance	4
d_{break}	distance BS–breakpoint	4
d_{layer}	thickness of layer	4
d_{direct}	direct pathlength	4
d_{refl}	length of reflected path	4
d_p	distance to previous screen	4
d_n	distance to subsequent screen	4
d_0	distance to reference point	5
d_a	distance between antenna elements	8, 13
d_w	distance between turns of helix antenna	8
d_{km}	euclidean distance between signal points with index k and m	11
$d(\vec{x}, \vec{y})$	distance of codewords	14
d_H	Hamming distance	14
d_{cov}	coverage distance	3
d_{div}	diversity order	20
e	basis of natural logarithm	

$e(t)$	impulse response of equalizer	16
f	frequency	5
$f(\ )$	function	
f_c	carrier frequency	5, 7, 8
f_{rep}	repetition frequency	8
f_{slip}	frequency slip	8
f_{inst}	instantaneous frequency	12
f_k	impulse response of discrete-time channel	16
f_n	carrier frequencies in OFDM	19
f_D	modulation frequency in FSK	11
$\mathbf{g}$	network encoding vector	22
$g(\ )$	function	
$g(t)$	basis pulse	11, 12, 19
$g_R(t)$	rectangular basis pulse	11
$\tilde{g}(t)$	phase pulse	11
g_m	discrete impulse response of channel plus feedforward filter	16
g_N	Nyquist raised cosine pulse	11
g_{NR}	root Nyquist raised cosine pulse	11
$h(t,\tau)$	channel impulse response	2, 6, 7, 8, 12, 13, 18, 19, 20
h_{TX}	height of TX	4
h_{RX}	height of RX	4
h_s	height of screen	4
h_b	height of BS	7
h_m	height of MS	7
$h_{r,d}$	complex channel gain from relay to destination	22
h_{roof}	height of rooftop	7
$h_{meas}(t_i,\tau)$	measured impulse response	8
$h_{s,d}$	complex channel gain from source to destination	22
$h_{s,r}$	complex channel gain from source to relay	22
h_w	height of helix antenna	9
$h(t,\tau,\phi)$	directionally resolved impulse response	7
h_{mod}	modulation index of CPFSK signal	11
$\mathbf{h}_d$	vector of desired impulse responses	13, 20
i	index counter	
j	index, imaginary unit	4
k, k_0	wavenumber	4, 13
k	index counter	8, 11, 13, 19, 20, 22, 28
k_{scale}	scaling factor for STDCC	8
k_B	Boltzmann constant	3
l	index counter	8, 19, 20
m	Nakagami m-factor	5, 13
m	counter	11, 12, 13, 14, 16, 17
m	index for parity check bits	14
n	propagation exponent	4, 7

n_1	refraction index for medium	4
$n(t)$	noise signal	8, 12, 13, 14, 16, 18
$n_{LP}(t)$	low-pass noise	12
$n_{BP}(t)$	bandpass noise	12
n	index counter	11
n_m	sampled noise values	16, 19
n_n	sampled noise values	14, 21
$\mathbf{n}$	vector of noise samples	20
$\tilde{n}_m$	sample values of colored noise	
p	transition probability	14
pdf	probability density function	5
$p(t)$	modulated pulse	8
$p(t)$	pulse sequence	11
q_m	impulse response of channel + equalizer	16
$\mathbf{r}$	position vector	4
r	absolute value of fieldstrength	5
r	spectral efficiency	14
$r_{LP}(t)$	low-pass representation of received signal	12
$\mathbf{r}$	received signal vector	12
$r(t)$	received signal	14, 15, 16
s	subcarrier channel index	28
$s(t)$	sounding signal	8
$s_1(t)$	auxiliary signal	8
$s_{LP,BP}(t)$	low-pass (bandpass) signal	11
$\mathbf{s}_{LP,BP}$	signal vector in low-pass (bandpass)	11
$\mathbf{s}_{synd}$	syndrome vector	14
$\mathbf{s}$	vector of signals at antenna array	8, 20, 27, 28
$\mathbf{s}$	transmit signal vector	14
t	absolute time	2, 11, 12, 13, 16, 17
$\mathbf{t}$	precoding vector	20
t_0	start time	6
t_s	sampling time	12
u	auxiliary variable	11
u_m	sequence of sample values at equalizer input	16
$\mathbf{u}$	vector of information symbols	14
v	velocity	5
$\mathbf{v}$	singlar vector	20
w_l	antenna weights	8, 13, 20
x	x-coordinate	4
x	general variable	5
x	transmit signal	22
$x(t)$	input signal	6
$\mathbf{x}$	code vector	14
$\mathbf{x}$	sequence of transmit signals	14
y	y-coordinate	4
y	decision variable	21
y	received signal	22
$\mathbf{y}$	sequence of receive signals	14

$y(t)$	output signal	6
z	z-coordinate	

Uppercase symbols:

$\mathbf{A}$	steering matrix	8
$\mathbf{A}$	antenna mapping matrix	27
A_{RX}	antenna area of receiver	4
$A(d_{TX}, d_{RX})$	amplitude factors for diffraction	4
ADF	average duration of fades	
A	amplitude of dominant component	7
A	state in the trellis diagram	14
$B(\nu f)$	Doppler-variant transfer function	6
B_{coh}	coherence bandwidth	6
B	bandwidth	11
BER	bit error probability	12
B_n	noise bandwidth	12
B_r	receiver bandwidth	12
B	state in the trellis diagram	14
B_G	bandwidth of Gaussian filter	11
C	capacity	14, 17, 20
$\mathbf{C}$	covariance matrix	
C_{crest}	crest factor	8
C	proportionality constant	
C	state in the trellis diagram	14
D	diffraction coefficient	5
D_W	diameter of helix antenna	8
D	quadratic form	12
D	maximum distortion	16
D	state in the trellis diagram	14
D	unit delay	27
D_{leav}	interleaver separation	14
D	antenna directivity	
E	electric fieldstrength	4
E_{diff}	fieldstrength of diffracted field	4
E_{inc}	fieldstrength of incident field	4
$E\{\}$	expectation	4, 13, 14, 18
$E1, E2$	fieldstrength of multipath components	5
E_0	normalization fieldstrength	13
E_S	Symbol energy	11, 12, 13
E_B	bit energy	11, 12, 13
E_C	chip energy	18
$E_{s,k}$	energy of kth signal	11, 18
$E(f)$	transfer function of equalizer	16
$F(\nu_F)$	Fresnel integral	4
$\tilde{F}$	modified Fresnel integral	4
F	local mean of fieldstrength	5
$F(z)$	factorization of the transfer function of the equivalent time discrete channel	16
F	noise figure	3
G_{RX}	antenna gain of receive antenna	3, 4

G_{TX}	antenna gain transmit antenna	4
$G(D)$	code polynomial	27
$G(\gamma), G(\varphi, \theta)$	antenna pattern	5
$G(v, v_1, v_2)$	Gaussian function	7
G_{max}	maximum gain	
G_R	spectrum of rectangular pulse	11
G_N	spectrum of Nyquist pulse	11
G_{NR}	spectrum of root Nyquist pulse	11
$\mathbf{G}$	generator matrix	14
G_{code}	code gain	14
$\mathbf{G}_G$	matrix with iid Gaussian entries	20
G	gain of an amplifier stage	3
H	transfer function of the channel	5, 6, 19, 20
$H(X)$	entropy	14
$H_D(X)$	entropy of binary symmetric channel	14
$H_R(f)$	transfer function of receive filter	12, 18
$\mathbf{H}$	parity check matrix	14
$\mathbf{H}_{had}$	Hadamard matrix	18, 19
$I(t)$	in-phase component	5
$I(t)$	link control action	22
$I(x, y)$	mutual information	14
I_0	modified Bessel function	5, 12
$|J|$	Jacobi determinant	5
J_0	Bessel function	7
K_r	Rice factor	5
$K(t, \tau)$	kernel function	6
K	number of resolvable directions	8
K_I	system margin	3
K	number of bits in a symbol	11
K	number of information symbols in a codeword	14
$2K + 1$	number of equalizer taps	16
K	scaling constant for STDCC	8
K	number of users	20
K	number of relays	22
L	number of clusters	7
L_{msd}	multiscreenloss	
L_{rts}	diffraction loss	7
L_{ori}	street orientation loss	7
L_a	antenna dimension	4
L_i	attenuation at the ith screen	
L_c	correlation length	4
L	duration of the impulse response of the equivalent time discrete channel f	16
L	number of cells in convolutional encoder	14
L	number of data streams	20
$\tilde{L}$	dimension of space-time code	20
L_{Tr}	truncation depth	14
L_{symb}	number of symbols where two possible sequences differ	14

L_f	losses in feeder	3
L	number of RF chains in HS-MRC	
$M(\phi, \theta)$	array factor	8
M	number of elements in the alphabet	11
$M(s)$	moment-generating function	12, 13
N	number of screens	4
N	number of MPCs	5, 7, 8
N	size of the set of expansion functions	12
N	total number of symbols in the code	14
N	number of mod-2 adders in the convolutional encoder	14
N_0	noise power-spectral density	12
$N(f)$	noise spectrum	6, 12
N_{symb}	number of information bits/symbols in TCM	14
$\tilde{N}$	number of bits for convolutional encoder in TCM	14
N	number of users in MA	17
N_{reg}	length of shift register	14
N_r	number of receive antennas	8, 13, 20
$N_{subchannel}$	number of subchannels	28
N_t	number of transmit antennas	13, 20
N_s	number of significant scatterers	20
N_{BS}	number of BS antennas	20
N_{MS}	number of MS antennas	20
N_R	level crossing rate	5
P_{TX}	transmit power	4
P_{RX}	receive power	4
P_m	average power	5
$P_{h,S,B}$	cross-power spectral densities	6
$P_h(\tau)$	PDP	6
$P(t, \tau)$	instantaneous PDP	7
P_n	noise power	
P_{pair}	pairwise error probability	12
P_{inst}	instantaneous received power	12
P_{max}	maximum TX power	20
P_f	false alarm probability	21
P_{md}	missed detection probability	21
P_s	transmit power of source	22
P_r	transmit power of relay	22
Pr	probability	
Pr_{out}	outage probability	
PL	pathloss	4, 5
P_s	signal power	3, 11
$Q(x)$	Q-function	12, 13
Q	antenna quality	9
Q	codebook size	20
Q	queue backlog	22
Q	quantization function	23
$Q(t)$	quadrature component	5
Q_T	interference quotient	6

Q_M	Marcum's Q-function	12
$Q(z)$	transfer function of equivalent channel and equalizer	16
R	radius of circle	5
R	cell size	17
R	transmission rate	14
R_{th}	threshold transmission rate	22
$\mathbf{R}_{\text{TX}}$	transmit correlation matrix	6, 7, 20
$\mathbf{R}_{\text{RX}}$	receive correlation matrix	6, 7, 20
R_{xx}	autocorrelation function of x	11
$\mathbf{R}_{xx}$	correlation matrix of x	8
R_{rad}	radiation resistance	8
R_{S}	symbol rate	11
R_{B}	bit rate	11
R_{c}	code rate	14
R_{e}	rank of error matrix	17
R_h	impulse response correlation function	6
$\mathbf{R}_{\text{ni}}$	noise and interference correlation matrix	20
$\tilde{R}_{yy}(t, t')$	autocovariance signal of received signal	7
SIR	signal-to-interference ratio	5
$S(f)$	power spectrum	5, 6, 12
$S(t)$	topology state	22
$S(\nu, \tau)$	spreading function	6
S_τ	delay spread	6
$S_{\text{D}}(\nu, \tau)$	Doppler spectrum	7
$S_{\text{LP,BP}}(f)$	power spectrum of LP (BP) signal	11
SER	symbol error probability	12
S_{N}	noise power-spectral density	
S_ϕ	angular spread	6, 7, 13
T_{B}	bit duration	
T	transmission factor	4
T_{m}	mean delay	6
$T_{\text{m}}(t)$	instantaneous mean delay	6
T_{rep}	repetition time of pulse signal	8
TB	time bandwidth product	8
T_{slip}	slip period	8
$\mathbf{T}$	auxiliary matrix	8
$\mathbf{T}$	transmit beamforming matrix	20
T	duration (general)	11
T_{per}	periodicity	11
T_{s}	sampling time	
T_{S}	symbol duration	11
T_{p}	packet duration	17
T_{cp}	duration of cyclic prefix	19
T_{C}	chip duration	18
T_{e}	temperature of environment	3
T_{d}	delay of pulse in PPM	11
T_{coh}	coherence time	6
T_{g}	group delay	12
$\mathbf{U}$	unitary matrix	8, 20

W	correlation spectrum	4
W_a	delay window	6
W	system bandwidth	
$\mathbf{W}$	unitary matrix	19
X	complex Gaussian random variable	12
$X(x)$	code polynomial	14
Y	complex Gaussian random variable	12
Z	complex Gaussian random variable	12
Z	virtual queue backlog	22

Lowercase Greek:

α	dielectric length	4
α	complex channel gain	12
α	rolloff factor	11
α	steering vector	8
β	decay time constant	7
β	amplification at relay	22
γ	SNR	
$\overline{\gamma}$	mean SNR	
γ_{MRC}	SNR at output of maximum ratio combiner	13
γ_{EGC}	SNR at output of equal gain combiner	13
γ	angle for Doppler shift	5
γ_S	E_S/N_0	12
γ_B	E_B/N_0	12, 13, 16, 17
δ	complex dielectric constant	4
δ_{ik}	Kronecker delta	13, 16, 19
$\delta(\tau)$	Dirac function	12, 13, 16, 18
$\in$	dielectric constant	4
$\in_r$	relative dielectric constant	4
$\in_{eff}$	effective relative dielectric constant	4
ε_m	error signal	16
ε	error vector of a code symbol	14
φ	orientation of a street	7
φ	phase of an MPC	5
$\tilde{\varphi}$	deterministic phaseshift	7
$\varphi_m(t)$	base functions for expansion	11
ϕ	azimuth angle of arrival	6, 7
$\eta(t)$	$g(t) * h(t)$	16
κ	auxiliary variable	13
λ, λ_0	wavelength	
λ_p	packet transmission rate	17
λ_i	ith eigenvalue	
μ	metric	12
μ	stepwidth of LMS	16
$\mu(t)$	transmission matrix	22
ν	Doppler shift	6
ν_{max}	maximum Doppler shift	7
ν_m	mean Doppler shift	5
ν_F	Fresnel parameter	5
ω	angular frequency	4

ρ	position vector	
ρ_{km}	correlation coefficients between signals	11
σ_c	conductivity	4
σ_h	standard deviation of height	4
σ	standard deviation	5
σ_F	standard deviation of local mean	5
σ_G	standard deviation of Gaussian pulse	11
σ_n	noise standard deviation	
σ_S^2	power in symbol sequence	11
τ	delay	4, 5
τ_{Gr}	group delay	5, 12
τ_i	delay of the ith MPC	7
τ_{max}	maximum excess delay	
$\chi_i(t)$	distortion of the i-th pulse	13
ζ	ACF of $\eta(t)$	16
ξ	SNR loss for discrete precoding	20
$\xi_n(t)$	noise correlation function	12
$\tilde{\xi}_s(t)$	FT of the normalized Doppler spectrum	12
$\xi_s(\nu, \tau)$	scattering function	12
$\xi_h(t, \tau)$	FT of the scattering function	12

Uppercase Greek:

Δh_b	$= h_b - h_{roof}$	7
Δx_s	distance between measurement points	8
$\Delta\tau_{min}$	minimum resolvable τ	8
Δf_{chip}	difference in chip frequency	8
$\Delta\varphi$	angle difference of paths	4
$\Delta\tau$	runtime difference	5
$\Delta\nu$	Doppler shift	5
Δ	phaseshift between two antenna elements	8
Δ	Lyapunov drift	22
Δ_C	determinant of $\mathbf{C}$	13
$\Delta\phi$	angular range	13
Φ_H	phase of the channel transfer function	5, 12
$\Phi_{CPFSK}(t)$	phase of transmit signal for CPFSK signal	11
Λ	matrix of eigenvalues	
$\Phi_{TX}(t)$	phase of transmit signal	
Ω	mean quadratic power Nakagami	5
Ω	direction of departure	
Ω_n	nth moment of Doppler spectrum	5, 13
$\Theta(t)$	queue backlog	22
Θ_e	angle of incidence	4
Θ_r	angle of reflection	4
Θ_t	angle of transmission	4
Θ_n	transmission phase of nth bit	
Θ_d	diffraction angle	
ϕ_{TX}	angle TX wedge	4
ϕ_{RX}	angle RX wedge	4
ϕ	azimuth	7
ϕ_0	nominal DOA	7, 13

Part I

Introduction

In the first part of this book, we give an introduction to the basic applications of wireless communications, as well as the technical problems inherent in this communication paradigm. After a brief history of wireless, Chapter 1 describes the different types of wireless services, and works out their fundamental differences. The subsequent Section 1.3 looks at the same problem from a different angle: what data rates, ranges, etc., occur in practical systems, and especially, what combination of performance measures are demanded (e.g., what data rates need to be transmitted over short distances; what data rates are required over long distances?) Chapter 2 then describes the technical challenges of communicating without wires, putting special emphasis on fading and co-channel interference. Chapter 3 describes the most elementary problem of designing a wireless system, namely to set up a link budget in either a noise-limited or an interference-limited system.

After studying this part of the book, the reader should have an overview of different types of wireless services, and understand the technical challenges involved in each of them. The solutions to those challenges are described in the later parts of this book.

Part I

1

Applications and Requirements of Wireless Services

Wireless communications is one of the big engineering success stories of the last 25 years – not only from a scientific point of view, where the progress has been phenomenal but also in terms of market size and impact on society. Companies that were completely unknown 25 years ago are now household names all over the world, due to their wireless products, and in several countries the wireless industry is dominating the whole economy. Working habits, and even more generally the ways we all communicate, have been changed by the possibility of talking "anywhere, anytime."

For a long time, wireless communications has been associated with cellular telephony, as this is the biggest market segment, and has had the highest impact on everyday lives. In recent times, wireless computer networks have also led to a significant change in working habits and mobility of workers – answering emails in a coffee shop has become an everyday occurrence. But besides these widely publicized cases, a large number of less obvious applications have been developed, and are starting to change our lives. Wireless sensor networks monitor factories, wireless links replace the cables between computers and keyboards, and wireless positioning systems monitor the location of trucks that have goods identified by wireless Radio Frequency (RF) tags. This variety of new applications causes the technical challenges for the wireless engineers to become bigger with each day. This book aims to give an overview of the solution methods for current as well as future challenges.

Quite generally, there are two paths to developing new technical solutions: engineering driven and market driven. In the first case, the engineers come up with a brilliant scientific idea – without having an immediate application in mind. As time progresses, the market finds applications enabled by this idea.[1] In the other approach, the market demands a specific product and the engineers try to develop a technical solution that fulfills this demand. In this chapter, we describe these market demands. We start out with a brief history of wireless communications, in order to convey a feeling of how the science, as well as the market, has developed in the past 100 years. Then follows a description of the types of services that constitute the majority of the wireless market today. Each of these services makes specific demands in terms of data rate, range, number of users, energy consumption, mobility, and so on. We discuss all these aspects in Section 1.3. We wrap up this section with a description of the interaction between the engineering of wireless devices and the behavioral changes induced by them in society.

---

[1] The second chapter gives a summary of the main technical challenges in wireless communications – i.e., the basis for the engineering-driven solutions. Chapters 3–23 discuss the technical details of these challenges and the scientific basis, while Chapters 24–29 expound specific systems that have been developed in recent years.

Wireless Communications, Second Edition   Andreas F. Molisch
© 2011 John Wiley & Sons, Ltd

1.1 History

1.1.1 How It All Started

When looking at the history of communications, we find that wireless communications is actually the oldest form – shouts and jungle drums did not require any wires or cables to function. Even the oldest "electromagnetic" (optical) communications are wireless: smoke signals are based on propagation of optical signals along a line-of-sight connection. However, wireless communications as we know it started only with the work of Maxwell and Hertz, who laid the basis for our understanding of the transmission of electromagnetic waves. It was not long after their groundbreaking work that Tesla demonstrated the transmission of information via these waves – in essence, the first wireless communications system. In 1898, Marconi made his well-publicized demonstration of wireless communications from a boat to the Isle of Wight in the English Channel. It is noteworthy that while Tesla was the first to succeed in this important endeavor, Marconi had the better public relations, and is widely cited as the inventor of wireless communications, receiving a Nobel prize in 1909.[2]

In the subsequent years, the use of radio (and later television) became widespread throughout the world. While in the "normal" language, we usually do not think of radio or TV as "wireless communications," they certainly are, in a scientific sense, information transmission from one place to another by means of electromagnetic waves. They can even constitute "mobile communications," as evidenced by car radios. A lot of basic research – especially concerning wireless propagation channels – was done for entertainment broadcasting. By the late 1930s, a wide network of wireless information transmission – though unidirectional – was in place.

1.1.2 The First Systems

At the same time, the need for bidirectional mobile communications emerged. Police departments and the military had obvious applications for such two-way communications, and were the first to use wireless systems with closed user groups. Military applications drove a lot of the research during, and shortly after, the Second World War. This was also the time when much of the theoretical foundations for communications in general were laid. Claude Shannon's [1948] groundbreaking work *A Mathematical Theory of Communications* appeared during that time, and established the possibility of error-free transmission under restrictions for the data rate and the Signal-to-Noise Ratio (SNR). Some of the suggestions in that work, like the use of optimum power allocation in frequency-selective channels, are only now being introduced into wireless systems.

The 1940s and 1950s saw several important developments: the use of *Citizens' Band* (CB) radios became widespread, establishing a new way of communicating between cars on the road. Communicating with these systems was useful for transferring vital traffic information and related aspects within the closed community of the drivers owning such devices, but it lacked an interface to the public telephone system, and the range was limited to some 100 km, depending on the power of the (mobile) transmitters. In 1946, the first mobile telephone system was installed in the U.S.A. (St. Louis). This system did have an interface to the *Public Switched Telephone Network* (PSTN), the landline phone system, though this interface was not automated, but rather consisted of human telephone operators. However, with a total of six speech channels for the whole city, the system soon met its limits. This motivated investigations of how the number of users could be increased, even though the allocated spectrum would remain limited. Researchers at AT&T's Bell Labs found the answer: the cellular principle, where the geographical area is divided into cells; different cells might use the same frequencies. To this day, this principle forms the basis for the majority of wireless communications.

---

[2] Marconi's patents were actually overturned in the 1940s.

Despite the theoretical breakthrough, cellular telephony did not experience significant growth during the 1960s. However, there were exciting developments on a different front: in 1957, the Soviet Union launched the first satellite (*Sputnik*) and the U.S.A. soon followed. This development fostered research in the new area of satellite communications.[3] Many basic questions had to be solved, including the effects of propagation through the atmosphere, the impact of solar storms, the design of solar panels and other long-lasting energy sources for the satellites, and so on. To this day, satellite communications is an important area of wireless communications (though not one that we address specifically in this book). The most widespread application lies in satellite TV transmission.

1.1.3 Analog Cellular Systems

The 1970s saw a revived interest in cellular communications. In scientific research, these years saw the formulation of models for path loss, Doppler spectra, fading statistics, and other quantities that determine the performance of analog telephone systems. A highlight of that work was Jakes' book *Microwave Mobile Radio* that summed up the state of the art in this area [Jakes 1974]. The 1960s and 1970s also saw a lot of basic research that was originally intended for landline communications, but later also proved to be instrumental for wireless communications. For example, the basics of adaptive equalizers, as well as multicarrier communications, were developed during that time.

For the practical use of wireless telephony, the progress in device miniaturization made the vision of "portable" devices more realistic. Companies like Motorola and AT&T vied for leadership in this area and made vital contributions. Nippon Telephone and Telegraph (NTT) established a commercial cellphone system in Tokyo in 1979. However, it was a Swedish company that built up the first system with large coverage and automated switching: up to that point, Ericsson AB had been mostly known for telephone switches while radio communications was of limited interest to them. However, it was just that expertise in switching technology and the (for that time, daring) decision to use digital switching technology that allowed them to combine different cells in a large area into a single network, and establish the *Nordic Mobile Telephone* (NMT) system [Meurling and Jeans 1994]. Note that while the switching technology was digital, the radio transmission technology was still analog, and the systems became therefore known as *analog* systems. Subsequently, other countries developed their own analog phone standards. The system in the U.S.A., e.g., was called *Advanced Mobile Phone System* (AMPS).

An investigation of NMT also established an interesting method for estimating market size: business consultants equated the possible number of mobile phone users with the number of Mercedes 600 (the top-of-the-line luxury car at that time) in Sweden. Obviously, mobile telephony could never become a mass market, could it? Similar thoughts must have occurred to the management of the inventor of cellular telephony, AT&T. Upon advice from a consulting company, they decided that mobile telephony could never attract a significant number of participants and stopped business activities in cellular communications.[4]

The analog systems paved the way for the wireless revolution. During the 1980s, they grew at a frenetic pace and reached market penetrations of up to 10% in Europe, though their impact was somewhat less in the U.S.A. In the beginning of the 1980s, the phones were "portable," but definitely not handheld. In most languages, they were just called "carphones," because the battery and transmitter were stored in the trunk of the car and were too heavy to be carried around. But at the end of the 1980s, handheld phones with good speech quality and quite acceptable battery

---

[3] Satellite communications – specifically by geostationary satellites – had already been suggested by science fiction writer Arthur C. Clark in the 1940s.

[4] These activities were restarted in the early 1990s, when the folly of the original decision became clear. AT&T then paid more than 10 billion dollars to acquire McCaw, which it renamed "AT&T Wireless."

lifetime abounded. The quality had become so good that in some markets digital phones had difficulty establishing themselves – there just did not seem to be a need for further improvements.

1.1.4   GSM and the Worldwide Cellular Revolution

Even though the public did not see a need for changing from analog to digital, the network operators knew better. Analog phones have a bad spectral efficiency (we will see why in Chapter 3), and due to the rapid growth of the cellular market, operators had a high interest in making room for more customers. Also, research in communications had started its inexorable turn to digital communications, and that included digital wireless communications as well. In the late 1970s and the 1980s, research into spectrally efficient modulation formats, the impact of channel distortions, and temporal variations on digital signals, as well as multiple access schemes and much more, were explored in research labs throughout the world. It thus became clear to the cognoscenti that the real-world systems would soon follow the research.

Again, it was Europe that led the way. The *European Telecommunications Standards Institute* (ETSI) group started the development of a digital cellular standard that would become mandatory throughout Europe and was later adopted in most parts of the world: *Global System for Mobile communications* (GSM). The system was developed throughout the 1980s; deployment started in the early 1990s and user acceptance was swift. Due to additional features, better speech quality, and the possibility for secure communications, GSM-based services overtook analog services typically within 2 years of their introduction. In the U.S.A., the change to digital systems was somewhat slower, but by the end of the 1990s, this country also was overwhelmingly digital.

Digital phones turned cellular communications, which was already on the road to success, into a blockbuster. By the year 2000, market penetration in Western Europe and Japan had exceeded 50%, and though the U.S.A. showed a somewhat delayed development, growth rates were spectacular as well.

The development of wireless systems also made clear the necessity of standards. Devices can only communicate if they are compatible, and each receiver can "understand" each transmitter – i.e., if they follow the same standard. But how should these standards be set? Different countries developed different approaches. The approach in the U.S.A. is "hands-off": allow a wide variety of standards and let the market establish the winner (or several winners). When frequencies for digital cellular communications were auctioned off in the 1990s, the buyers of the spectrum licences could choose the system standard they would use. For this reason, three different standards are now being used in the U.S.A. A similar approach was used by Japan, where two different systems fought for the market of Second Generation (2G) cellular systems. In both Japan and the U.S.A., the networks based on different standards work in the *same* geographical regions, allowing consumers to choose between different technical standards.

The situation was different in Europe. When digital communications were introduced, usually only one operator *per country* (typically, the incumbent public telephone operators) existed. If each of these operators would adopt a different standard, the result would be high market fragmentation (i.e., a small market for each standard), without the benefit of competition between operators. Furthermore, roaming from country to country, which for obvious geographical regions is much more frequent in Europe than in the U.S.A. or Japan, would be impossible. It was thus logical to establish a single common standard for all of Europe. This decision proved to be beneficial for wireless communications in general, as it provided the economy of scales that decreased cost and thus increased the popularity of the new services.

1.1.5 New Wireless Systems and the Burst of the Bubble

Though cellular communications defined the picture of wireless communications in the general population, a whole range of new services was introduced in the 1990s. Cordless telephones started to replace the "normal" telephones in many homes. The first versions of these phones used analog technology; however, also for this application, digital technology proved to be superior. Among other aspects, the possibility of listening in to analog conversations, and the possibility for neighbors to "highjack" an analog cordless Base Station (BS) and make calls at other people's expense, led to a shift to digital communications. While cordless phones never achieved the spectacular market size of cellphones, they constitute a solid market.

Another market that seemed to have great promise in the 1990s was *fixed wireless access* and *Wireless Local Loop* (WLL) – in other words, replacing the copper lines to the homes of the users by wireless links, but without the specific benefit of mobility. A number of technical solutions were developed, but all of them ultimately failed. The reasons were as much economical and political as they were technical. The original motivation for WLL was to give access to customers for alternative providers of phone services, bypassing the copper lines that belonged to the incumbents. However, regulators throughout the world ruled in the mid-1990s that the incumbents *have* to lease their lines to the alternative providers, often at favorable prices. This eliminated much of the economic basis for WLL. Similarly, fixed wireless access was touted as the scheme to provide broadband data access at competitive prices. However, the price war between Digital Subscriber Line (DSL) technology and cable TV has greatly dimmed the economic attractiveness of this approach.

The biggest treasure thus seemed to lie in a further development of cellular systems, establishing the *Third Generation* (3G) (after the analog systems and 2G systems like GSM) [Bi et al. 2001]. 2G systems were essentially pure voice transmission systems (though some simple data services, like the *Short Message Service* – SMS – were included as well). The new 3G systems were to provide data transmission at rates comparable with the ill-fated *Integrated Services Digital Network* (ISDN) (144 kbit/s), and even up to 2 Mbit/s, at speeds of up to 500 km/h. After long deliberations, two standards were established: *Third Generation Partnership Project* (3GPP) (supported by Europe, Japan, and some American companies) and 3GPP2 (supported by another faction of American companies). The new standards also required a new spectrum allocation in most of the world, and the selling of this spectrum became a bonanza for the treasuries of several countries.

The development of 3GPP, and the earlier introduction of the IS-95 CDMA (*Code Division Multiple Access*) system in the U.S.A., sparked a lot of research into CDMA and other spread spectrum techniques (see Chapter 18) for wireless communications; by the end of the decade, multicarrier techniques (Chapter 19) had also gained a strong footing in the research community. Multiuser detection – i.e., the fact that the effect of interference can be greatly mitigated by exploiting its structure – was another area that many researchers concentrated on, particularly in the early 1990s. Finally, the field of multiantenna systems (Chapter 20) saw an enormous growth since 1995, and for some time accounted for almost half of all published research in the area of the physical layer design of wireless communications.

The spectrum sales for 3G cellular systems and the Initial Public Offerings (IPOs) of some wireless start-up companies represented the peak of the "telecom bubble" of the 1990s. In 2000/2001, the market collapsed with a spectacular crash. Developments on many of the new wireless systems (like fixed wireless) were stopped as their proponents went bankrupt, while the deployment of other systems, including 3G cellular systems, was slowed down considerably. Most worrisome, many companies slowed or completely stopped research, and the general economic malaise led to decreased funding of academic research as well.

1.1.6 Wireless Revival

Since 2003, several developments have led to a renewed interest in wireless communications. The first one is a continued growth of 2G and 2.5G cellular communications, stimulated by new markets as well as new applications. To give just one example, in 2008, China had more than 500 million cellphone users – even before the first 3G networks became operative. Worldwide, about 3.5 billion cellphones were in use in 2008, most of them based on 2G and 2.5G standards.

Furthermore, 3G networks have become widely available and popular – especially in Japan, Europe, and the U.S.A. – (in 2008, overall cellphone market penetration of cellphones in Western Europe was more than 100% and was approaching 90% in the U.S.A.). Data transmission speeds comparable to cable (5 Mbit/s) are available. This development has, in turn, spurred the proliferation of devices that not only allow voice calls but also Internet browsing and reception of streaming audio and video. One such device, called *iPhone*, received enormous attention among the general public when first introduced, but there exist actually dozens of cellphones with similar capabilities – these so-called "smartphones" account for 20% of the cellphone market in the U.S.A. As a consequence of all these developments, transmission of data to and from cellphones has become a large market.

Even while 3G networks are still being deployed, the next generation (sometimes called *4G* or *3.9G*) has been developed. Most infrastructure manufacturers are concentrating on the Long-Term Evolution (LTE) of the dominating 3G standard. An alternative standard, whose roots are based in fixed wireless access systems, is also under deployment. In addition, access to TV programming (either live TV or prerecorded episodes) from cellphones is becoming more and more popular. For 4G networks as well as TV transmission, Multiple Input Multiple Output system-Orthogonal Frequency Division Multiplexing (MIMO-OFDM) (see Chapters 19 and 20) is the modulation method of choice, which has spurred research in this area.

A second important development was the unexpected success of wireless computer networks (wireless Local Area Networks (LANs)). Devices following the Institute of Electrical and Electronics Engineers (IEEE) 802.11 standard (Chapter 29) have enabled computers to be used in a way that is almost as versatile and mobile as cellphones. The standardization process had already started in the mid-1990s, but it took several versions, and the impact of intense competition from manufacturers, to turn this into a mass product. Currently, wireless access points are widely available not only at homes and offices but also at airports, coffee shops, and similar locations. As a consequence, many people who depend on laptops and Internet connections to do their work now have more freedom to choose when and where to work.

Thirdly, wireless sensor networks offer new possibilities of monitoring and controlling factories and even homes from remote sites, and also find applications for military and surveillance purposes. The interest in sensor networks has also spurred a wave of research into ad hoc and peer-to-peer networks. Such networks do not use a dedicated infrastructure. If the distance between source and destination is too large, other nodes of the network help in forwarding the information to the final destination. Since the structure of those networks is significantly different from the traditional cellular networks, a lot of new research is required.

Summarizing, the "wireless revival" is based on three tendencies: (i) a much broader range of products, (ii) data transmission with a higher rate for already existing products, and (iii) higher user densities. These trends determine the directions of research in the field and provide a motivation for many of the more recent scientific developments.

1.2 Types of Services

1.2.1 Broadcast

The first wireless service was broadcast radio. In this application, information is transmitted to different, possibly mobile, users (see Figure 1.1). Four properties differentiate broadcast radio

Broadcast

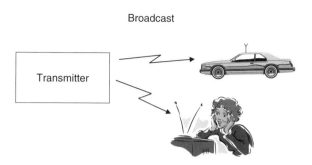

Figure 1.1   Principle of broadcast transmission.

from, e.g., cellular telephony:

1. The information is only sent in one direction. It is only the broadcast station that sends information to the radio or TV receivers; the listeners (or viewers) do not transmit any information back to the broadcast station.
2. The transmitted information is the same for all users.
3. The information is transmitted continuously.
4. In many cases, multiple transmitters send the same information. This is especially true in Europe, where national broadcast networks cover a whole country and broadcast the same program in every part of that country.[5]

The above properties led to a great many simplifications in the design of broadcast radio networks. The transmitter does not need to have any knowledge or consideration about the receivers. There is no requirement to provide for duplex channels (i.e., for bringing information from the receiver to the transmitter). The number of possible users of the service does not influence the transmitter structure either – irrespective of whether there are millions of users, or just a single one, the transmitter sends out the same information.

The above description has been mainly true for traditional broadcast TV and radio. Satellite TV and radio differ in the fact that often the transmissions are intended only for a subset of all possible users (pay-TV or pay-per-view customers), and therefore, encryption of the content is required in order to prevent unauthorized viewing. Note, however, that this "privacy" problem is different from regular cellphones: for pay-TV, the content should be accessible to all members of the authorized user group, ("multicast") while for cellphones, each call should be accessible only for the single person it is intended for ("unicast") and not to all customers of a network provider.

Despite their undisputed economic importance, broadcast networks are not at the center of interest for this book – space restrictions prevent a more detailed discussion. Still, it is useful to keep in mind that they are a specific case of wireless information transmission, and recent developments, like simulcast digital TV, interactive TV, and especially streaming TV to computers and cellphones, tend to obscure the distinction from cellular telephony even more.

1.2.2 Paging

Similar to broadcast, paging systems are unidirectional wireless communications systems. They are characterized by the following properties (see also Figure 1.2):

---

[5] The situation is slightly different in the U.S.A., where a "local station" usually covers only a single metropolitan area, often with a single transmitter.

1. The user can only receive information, but cannot transmit. Consequently, a "call" (message) can only be initiated by the call center, not by the user.
2. The information is intended for, and received by, only a single user.
3. The amount of transmitted information is very small. Originally, the received information consisted of a single bit of information, which indicated to the user that "somebody has sent you a message." The user then had to make a phone call (usually from a payphone) to the call center, where a human operator repeated the content of the waiting message. Later, paging systems became more sophisticated, allowing the transmission of short messages (e.g., a different phone number that should be called, or the nature of an emergency). Still, the amount of information was rather limited.

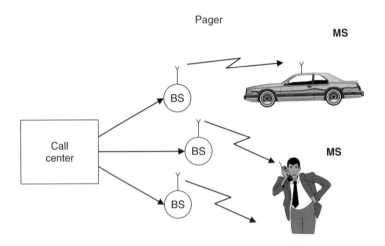

Figure 1.2   Principle of a pager.

Due to the unidirectional nature of the communications, and the small amount of information, the bandwidth required for this service is small. This in turn allows the service to operate at lower carrier frequencies – e.g., 150 MHz – where only small amounts of spectrum are available. As we will see later on, such lower carrier frequencies make it much easier to achieve good coverage of a large area with just a few transmitters.

Pagers were very popular during the 1980s and early 1990s. For some professional groups, like doctors, they were essential tools of the trade, allowing them to react to emergencies in shorter time. However, the success of cellular telephony has considerably reduced their appeal. Cellphones allow provision of all the services of a pager, plus many other features as well. The main appeal of paging systems, after the year 2000, lies in the better area coverage that they can achieve.

1.2.3   Cellular Telephony

Cellular telephony is the economically most important form of wireless communications. It is characterized by the following properties:

1. The information flow is bidirectional. A user can transmit and receive information at the same time.

2. The user can be anywhere within a (nationwide or international) network. Neither (s)he nor the calling party need to know the user's location; it is the network that has to take the mobility of the user into account.
3. A call can originate from either the network or the user. In other words, a cellular customer can be called or can initiate a call.
4. A call is intended only for a single user; other users of the network should not be able to listen in.
5. High mobility of the users. The location of a user can change significantly during a call.

Figure 1.3 shows a block diagram of a cellular system. A mobile user is communicating with a BS that has a good radio connection with that user. The BSs are connected to a mobile switching center, which is connected to the public telephone system.

Figure 1.3 Principle of a cellular system. *In this figure*: PSTN, Public Switched Telephone Network; MSC, Mobile Switching Center; BS, Base Station; MS, Mobile Station.

Since each user wants to transmit or receive different information, the number of active users in a network is limited. The available bandwidth must be shared between the different users; this can be done via different "multiple access" schemes (see also Chapters 17–20). This is an important difference from broadcast systems, where the number of users (receivers) is unlimited, since they all receive the same information.

In order to increase the number of possible users, the *cellular principle* is used: the area served by a network provider is divided into a number of subareas called *cells*. Within each cell, different users have to share the available bandwidth: let us consider in the following the case, that each user occupies a different carrier frequency. Even users in neighboring cells have to use different frequencies, in order to keep co-channel interference low. However, for cells that are sufficiently far apart, the same frequencies can be used, because the signals get weaker with increasing distance from their transmitter. Thus, within one country, there can be hundreds or thousands of cells that are using the same frequencies.

Another important aspect of cellular telephony is the unlimited mobility. The user can be anywhere within the coverage area of the *network* (i.e., the user is not limited to a specific cell), in order to be able to communicate. Also, (s)he can move from one cell to the other during one call. The cellular network interfaces with the PSTN as well as with other wireless systems.

As mentioned in our brief wireless history, cellphones started to become popular in the 1980s and are now a dominant form of communications, with more than three billion users worldwide.

Due to this reason, this book often draws its examples from cellular telephony, even though the general principles are applicable to other wireless systems as well. Chapters 24–28 give a detailed description of the most popular cellular systems.

1.2.4 Trunking Radio

Trunking radio systems are an important variant of cellular phones, where there is no connection between the wireless system and the PSTN; therefore, it allows the communications of closed user groups. Obvious applications include police departments, fire departments, taxis, and similar services. The closed user group allows implementation of several technical innovations that are not possible (or more difficult) in normal cellular systems:

1. *Group calls*: a communication can be sent to several users simultaneously, or several users can set up a conference call between multiple users of the system.
2. *Call priorities*: a normal cellular system operates on a "first-come, first-serve" basis. Once a call is established, it cannot be interrupted.[6] This is reasonable for cellphone systems, where the network operator cannot ascertain the importance or urgency of a call. However, for the trunk radio system of, e.g., a fire department, this is not an acceptable procedure. Notifications of emergencies have to go through to the affected parties, even if that means interrupting an existing, lower priority call. A trunking radio system thus has to enable the prioritization of calls and has to allow dropping a low-priority call in favor of a high-priority one.
3. *Relay networks*: the range of the network can be extended by using each *Mobile Station* (MS) as a relay station for other MSs. Thus, an MS that is out of the coverage region of the BS might send its information to another MS that is within the coverage region, and that MS will forward the message to the BS; the system can even use multiple relays to finally reach the BS. Such an approach increases the effective coverage area and the reliability of the network. However, it can only be used in a trunking radio system and not in a cellular system – normal cellular users would not want to have to spend "their" battery power on relaying messages for other users.

1.2.5 Cordless Telephony

Cordless telephony describes a wireless link between a handset and a BS that is directly connected to the public telephone system. The main difference from a cellphone is that the cordless telephone is associated with, and can communicate with, only a single BS (see Figure 1.4). There is thus no *mobile switching center*; rather, the BS is directly connected to the PSTN. This has several important consequences:

1. The BS does not need to have any network functionality. When a call is coming in from the PSTN, there is no need to find out the location of the MS. Similarly, there is no need to provide for handover between different BSs.
2. There is no central system. A user typically has one BS for his/her apartment or business under control, but no influence on any other BSs. For that reason, there is no need for (and no possibility for) frequency planning.
3. The fact that the cordless phone is under the control of the user also implies a different pricing structure: there are no network operators that can charge fees for connections from the MS to the BS; rather, the only occurring fees are the fees from the BS into the PSTN.

---

[6] Except for interrupts due to technical problems, like the user moving outside the coverage region.

Cordless phone

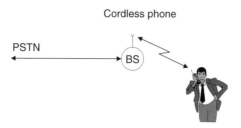

Figure 1.4 Principle of a simple cordless phone.

Wireless PABX

Figure 1.5 Principle of a wireless private automatic branch exchange.

In many other respects, the cordless phone is similar to the cellular phone: it allows mobility *within* the cell area; the information flow is bidirectional; calls can originate from either the PSTN or the mobile user, and there have to be provisions such that calls cannot be intercepted or listened to by unauthorized users and no unauthorized calls can be made.

Cordless systems have also evolved into wireless *Private Automatic Branch eXchanges* (PABXs) (see Figure 1.5). In its most simple form, a PABX has a single BS that can serve several handsets simultaneously – either connecting them to the PSTN or establishing a connection between them (for calls within the same company or house). In its more advanced form, the PABX contains several BSs that are connected to a central control station. Such a system has essentially the same functionality as a cellular system; it is only the size of the coverage area that distinguishes such a full functionality wireless PABX from a cellular network.

The first cordless phone systems were analog systems that just established a simple wireless link between a handset and a BS; often, they did not even provide rudimentary security (i.e., stopping unauthorized calls). Current systems are digital and provide more sophisticated functionality. In Europe, the Digital Enhanced Cordless Telecommunications (DECT) system (see the companion website at *www.wiley.com/go/molisch*) is the dominant standard; Japan has a similar system called the Personal Handyphone System (PHS) that provides both the possibility for cordless telephony and an alternative cellular system (a full functionality PABX system that covers most of Japan and provides the possibility of public access). Both systems operate in the 1,800-MHz band, using a spectrum specifically dedicated to cordless applications. In the U.S.A., digital cordless phones mainly operate in the 2.45-GHz *Industrial, Scientific, and Medical* (ISM) band, which they share with many other wireless services.

1.2.6   Wireless Local Area Networks

The functionality of Wireless Local Area Networks (WLANs) is very similar to that of cordless phones – connecting a single mobile user device to a public landline system. The "mobile user device" in this case is usually a laptop computer and the public landline system is the Internet. As in the cordless phone case, the main advantage is convenience for the user, allowing mobility. Wireless LANs can even be useful for connecting fixed-location computers (desktops) to the Internet, as they save the costs for laying cables to the desired location of the computer.

A major difference between wireless LANs and cordless phones is the required data rate. While cordless phones need to transmit (digitized) speech, which requires at most 64 kbit/s, wireless LANs should be at least as fast as the Internet that they are connected to. For consumer (home) applications, this means between 700 kbit/s (the speed of DSLs in the U.S.A.) and 3–5 Mbit/s (speed of cable providers in the U.S.A. and Europe) to $\geq$20 Mbit/s (speed of DSLs in Japan). For companies that have faster Internet connections, the requirements are proportionately higher. In order to satisfy the need for these high data rates, a number of standards have been developed, all of which carry the identifier IEEE 802.11. The original IEEE 802.11 standard enabled transmission with 1 Mbit/s, the very popular 802.11b standard (also known under the name WiFi) allows up to 11 Mbit/s and the 802.11a standard extends that to 55 Mbit/s. Even higher rates are realized by the 802.11n standard that was introduced in 2008/2009.

WLAN devices can, in principle, connect to any BS (access point) that uses the same standard. However, the owner of the access point can restrict the access – e.g., by appropriate security settings.

1.2.7   Personal Area Networks

When the coverage area becomes even smaller than that of WLANs, we speak of *Personal Area Networks* (PANs). Such networks are mostly intended for simple "cable replacement" duties. For example, devices following the *Bluetooth* standard allow to connect a hands-free headset to a phone without requiring a cable; in that case, the distance between the two devices is less than a meter. In such applications, data rates are fairly low (<1 Mbit/s). Recently, wireless communications between components in an entertainment system (DVD player to TV), between computer and peripheral devices (printer, mouse), and similar applications have gained importance, and a number of standards for PANs have been developed by the IEEE 802.15 group. For these applications, data rates in excess of 100 Mbit/s are used.

Networks for even smaller distances are called *Body Area Networks* (BANs), which enable communications between devices located on various parts of a user's body. Such BANs play an increasingly important role in the monitoring of patients' health and of medical devices (e.g., pacemakers).

We note finally that PANs and BANs can either have a network structure similar to a cellular approach or they can be ad hoc networks as discussed in Section 1.2.9.

1.2.8   Fixed Wireless Access

Fixed wireless access systems can also be considered as a derivative of cordless phones or WLANs, essentially replacing a dedicated cable connection between the user and the public landline system. The main difference from a cordless system is that (i) there is no mobility of the user devices and (ii) the BS almost always serves multiple users. Furthermore, the distances bridged by fixed wireless access devices are much larger (between 100 m and several tens of kilometers) than those bridged by cordless telephones.

The purpose of fixed wireless access lies in providing users with telephone and data connections without having to lay cables from a central switching office to the office or apartment the user is in. Considering the high cost of labor for the cable-laying operations, this can be an economical approach. However, it is worth keeping in mind that most buildings, especially in the urban areas of developed countries, are already supplied by some form of cable – regular telephone cable, cable TV, or even optical fiber. Rulings of the telecom regulators in various countries have stressed that incumbent operators (owners of these lines) have to allow competing companies to use these lines. As a consequence, fixed wireless access has its main market for covering rural areas, and for establishing connections in developing countries that do not have any wired infrastructure in place. In general, the business cases for fixed wireless has been disappointing (see "Burst of the Bubble" in Section 1.1.5). The IEEE 802.16 (WiMAX) standard tries to alleviate that problem by allowing some limited mobility in the system, and thus blurs the distinction from cellular telephony.

1.2.9 Ad hoc Networks and Sensor Networks

Up to now, we have dealt with "infrastructure-based" wireless communications, where certain components (base stations, TV transmitters, etc.) are intended by design to be in a fixed location, to exercise control over the network and interface with other networks. The size of the networks may differ (from LANs covering just one apartment to cellular networks covering whole countries), but the central principle of distinguishing between "infrastructure" and "user equipment" is common to them all. There is, however, an alternative in which there is only one type of equipment, and those devices, all of which may be mobile, organize themselves into a network according to their location and according to necessity. Such networks are called *ad hoc networks* (see Figure 1.6). There can still be "controllers" in an ad hoc network, but the choice of which device acts as master and which as slave is done opportunistically whenever a network is formed. There are also ad hoc networks without any hierarchy. While the actual transmission of the data (i.e., physical layer communication) is almost identical to that of the infrastructure-based networks, the medium access and the networking functionalities are very different.

Figure 1.6 Principle of an ad hoc network.

The advantages of ad hoc networks lie in their low costs (because no infrastructure is required) and high flexibility. The drawbacks include reduced efficiency, smaller communication range, and restrictions on the number of devices that can be included in a network. Ad hoc networks play a major role in the recent proliferation of sensor networks, which allow communications between machines for the purpose of building control (controlling air conditioning, lighting, etc., based on sensor data), factory automation, surveillance, etc. Ad hoc networks also play a role in emergency communications (when infrastructure was destroyed, e.g., by an earthquake) as well as military communications.

1.2.10 Satellite Cellular Communications

Besides TV, which creates the biggest revenues in the satellite market, cellular communications are a second important application of satellites. Satellite cellular communications mostly have the same operating principles as land-based cellular communications. However, there are some key differences.

The distance between the "BS" (i.e., the satellite) and the MS is *much* larger: for geostationary satellites, that distance is 36,000 km; for *Low Earth Orbit* (LEO) satellites, it is several hundred kilometers. Consequently, the transmit powers need to be larger, high-gain antennas need to be used on the satellite (and in many cases also on the MS), and communications from within buildings is almost impossible.

Another important difference from the land-based cellular system lies in the cell size: due to the large distance between the satellite and the Earth, it is impossible to have cells with diameters less than 100 km even with LEO satellites; for geostationary satellites, the cell areas are even larger. This large cell size is the biggest advantage as well as the biggest drawback of the satellite systems. On the positive side, it makes it easy to have good coverage even of large, sparsely populated areas – a single cell might cover most of the Sahara region. On the other hand, the area spectral efficiency is very low, which means that (given the limited spectrum assigned to this service) only a few people can communicate at the same time.

The costs of setting up a "BS" – i.e., a satellite – are much higher than for a land-based system. Not only is the launching of a communications satellite very expensive but it is also necessary to build up an appropriate infrastructure of ground stations for linking the satellites to the PSTN.

As a consequence of all these issues, the business case for satellite communications systems is quite different: it is based on supplying a small number of users with vital communications at a much higher price. Emergency workers and journalists in disaster and war areas, ship-based communications, and workers on offshore oil drilling platforms are typical users for such systems. The INMARSAT system is the leading provider for such communications. In the late 1990s, the IRIDIUM project attempted to provide lower priced satellite communications services by means of some 60 LEO satellites, but ended in bankruptcy.

1.3 Requirements for the Services

A key to understanding wireless design is to realize that different applications have different requirements in terms of data rate, range, mobility, energy consumption, and so on. It is *not* necessary to design a system that can sustain gigabit per second data rates over a 100-km range when the user is moving at 500 km/h. We stress this fact because there is a tendency among engineers to design a system that "does everything but wash the dishes"; while appealing from a scientific point of view, such systems tend to have a high price and low spectral efficiency. In the following, we list the range of requirements encountered in system design and we enumerate which requirements occur in which applications.

1.3.1 Data Rate

Data rates for wireless services span the gamut from a few bits per second to several gigabit per second, depending on the application:

- *Sensor networks* usually require data rates from a few bits per second to about 1 kbit/s. Typically, a sensor measures some critical parameter, like temperature, speed, etc., and transmits the current value (which corresponds to just a few bits) at intervals that can range from milliseconds to several hours. Higher data rates are often required for the central nodes of sensor networks that collect the information from a large number of sensors and forward it for further processing. In that case, data rates of up to 10 Mbit/s can be required. These "central nodes" show more similarity to WLANs or fixed wireless access.

- *Speech communications* usually require between 5 and 64 kbit/s depending on the required quality and the amount of compression. For cellular systems, which require higher spectral efficiency, source data rates of about 10 kbit/s are standard. For cordless systems, less elaborate compression and therefore higher data rates (32 kbit/s) are used.

- *Elementary data services* require between 10 and 100 kbit/s. One category of these services uses the display of the cellphone to provide Internet-like information. Since the displays are smaller, the required data rates are often smaller than for conventional Internet applications. Another type of data service provides a wireless mobile connection to laptop computers. In this case, speeds that are at least comparable with dial-up (around 50 kbit/s) are demanded by most users, though elementary services with 10 kbit/s (exploiting the same type of communications channels foreseen for speech) are sometimes used as well. Elementary data services are mostly replaced by high-speed data services in the U.S.A., Europe, and Japan, but still play an important role in other parts of the world.

- *Communications between computer peripherals and similar devices*: for the replacement of cables that link computer peripherals, like mouse and keyboard, to the computer (or similarly for cellphones), wireless links with data rates around 1 Mbit/s are used. The functionality of these links is similar to the previously popular infrared links, but usually provides higher reliability.

- *High-speed data services*: WLANs and 3G cellular systems are used to provide fast Internet access, with speeds that range from 0.5 to 100 Mbit/s (currently under development).

- *Personal Area Networks* (PANs) is a newly coined term that refers mostly to the range of a wireless network (up to 10 m), but often also has the connotation of high data rates (over 100 Mbit/s), mostly for linking the components of consumer entertainment systems (streaming video from computer or DVD player to a TV) or high-speed computer connections (wireless Universal Serial Bus (USB)).

1.3.2   Range and Number of Users

Another distinction among the different networks is the range and the number of users that they serve. By "range," we mean here the distance between one transmitter and receiver. The coverage area of a system can be made almost independent of the range, by just combining a larger number of BSs into one big network.

- *Body Area Networks* (BANs) cover the communication between different devices attached to one body – e.g., from a cellphone in a hip holster to a headset attached to the ear. The range is thus on the order of 1 m. BANs are often subsumed into PANs.

- *Personal Area Networks* include networks that achieve distances of up to or about 10 m, covering the "personal space" of one user. Examples are networks linking components of computers and home entertainment systems. Due to the small range, the number of devices within a PAN is small, and all are associated with a single "owner." Also, the number of overlapping PANs (i.e., sharing the same space or room) is small – usually less than five. That makes cell planning and multiple access much simpler.

- *WLANs*, as well as cordless telephones cover still larger ranges of up to 100 m. The number of users is usually limited to about 10. When much larger numbers occur (e.g., at conferences or meetings), the data rates for each user decrease. Similarly, cordless phones have a range of up to 300 m and the number of users connected to one BS is of the same order as for WLANs. Note, however, that wireless PABXs can have much larger ranges and user numbers – as mentioned before, they can be seen essentially as small private cellular systems.

- *Cellular systems* have a range that is larger than, e.g., the range of WLANs. Microcells typically cover cells with 500 m radius, while macrocells can have a radius of 10 or even 30 km. Depending on the available bandwidth and the multiple access scheme, the number of active users in a cell is usually between 5 and 50. If the system is providing high-speed data services to one user, the number of active users usually shrinks.
- *Fixed wireless access services* cover a range that is similar to that of cellphones – namely, between 100 m and several tens of kilometers. Also, the number of users is of a similar order as for cellular systems.
- *Satellite systems* provide even larger cell sizes, often covering whole countries and even continents. Cell size depends critically on the orbit of the satellite: geostationary satellites provide larger cell sizes (1,000-km radius) than LEOs.

Figure 1.7 gives a graphical representation of the link between data rate and range. Obviously, higher data rates are easier to achieve if the required range is smaller. One exception is fixed wireless access, which demands a high data rate at rather large distances.

Figure 1.7 Data rate versus range for various applications.

1.3.3 Mobility

Wireless systems also differ in the amount of mobility that they have to allow for the users. The ability to move around while communicating is one of the main charms of wireless communication for the user. Still, within that requirement of mobility, different grades exist:

- *Fixed devices* are placed only once, and after that time communicate with their BS, or with each other, always from the same location. The main motivation for using wireless transmission techniques for such devices lies in avoiding the laying of cables. Even though the devices are not mobile, the propagation channel they transmit over can change with time: both due to people walking by and due to changes in the environment (rearranging of machinery, furniture, etc.). Fixed wireless access is a typical case in point. Note also that all wired communications (e.g., the PSTN) fall into this category.

- *Nomadic devices*: nomadic devices are placed at a certain location for a limited duration of time (minutes to hours) and then moved to a different location. This means that during one "drop" (placing of the device), the device is similar to a fixed device. However, from one drop to the next, the environment can change radically. Laptops are typical examples: people do not operate their laptops while walking around, but place them on a desk to work with them. Minutes or hours later, they might bring them to a different location and operate them there.
- *Low mobility*: many communications devices are operated at pedestrian speeds. Cordless phones, as well as cellphones operated by walking human users are typical examples. The effect of the low mobility is a channel that changes rather slowly, and – in a system with multiple BSs – handover from one cell to another is a rare event.
- *High mobility* usually describes speed ranges from about 30 to 150 km/h. Cellphones operated by people in moving cars are one typical example.
- *Extremely high mobility* is represented by high-speed trains and planes, which cover speeds between 300 and 1000 km/h. These speeds pose unique challenges both for the design of the physical layer (Doppler shift, see Chapter 5) and for the handover between cells.

Figure 1.8 shows the relationship between mobility and data rate.

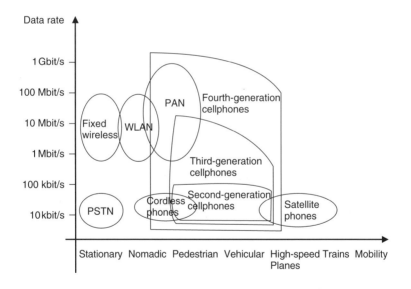

Figure 1.8   Data rate versus mobility for various applications.

1.3.4 Energy Consumption

Energy consumption is a critical aspect for wireless devices. Most wireless devices use (one-way or rechargeable) batteries, as they should be free of *any* wires – both the ones used for communication and the ones providing the power supply.

- *Rechargeable batteries*: nomadic and mobile devices, like laptops, cellphones, and cordless phones, are usually operated with rechargeable batteries. Standby times as well as operating times are one of the determining factors for customer satisfaction. Energy consumption is determined on one hand by the distance over which the data have to be transmitted (remember that a minimum SNR has to be maintained), and on the other hand, by the amount of data that are to be transmitted (the SNR is proportional to the energy per bit). The energy density of batteries has increased slowly over the past 100 years, so that the main improvements in terms of operating and standby time stem from reduced energy consumption of the devices. For cellphones, talk times of more than 2 hours and standby times of more than 48 hours are minimum requirements. For laptops, power consumption is not mainly determined by the wireless transmitter, but rather by other factors like hard drive usage and processor speed. For smartphones, the energy consumption of the processor and of the wireless connection is of the same order, and both have to be considered for maximizing battery lifetime.

- *One-way batteries*: sensor network nodes often use one-way batteries, which offer higher energy density at lower prices. Furthermore, changing the battery is often not an option; rather, the sensor including the battery and the wireless transceiver is often discarded after the battery has run out. It is obvious that in this case energy-efficient operation is even more important than for devices with rechargeable batteries.
- *Power mains*: BSs and other fixed devices can be connected to the power mains. Therefore, energy efficiency is not a major concern for them. It is thus desirable, if possible, to shift as much functionality (and thus energy consumption) from the MS to the BS.

User requirements concerning batteries are also important sales issues, especially in the market for cellular handsets:

- The weight of an MS is determined mostly (70–80%) by the battery. Weight and size of a handset are critical sales issues. It was in the mid-1980s that cellphones were commonly called "carphones," because the MS could only be transported in the trunk of a car and was powered by the car battery. By the end of the 1980s, the weight and dimensions of the batteries had decreased to about 2 kg, so that it could be carried by the user in a backpack. By the year 2000, the battery weight had decreased to about 200 g. Part of this improvement stems from more efficient battery technology, but to a large part, it is caused by the decrease of the power consumption of the handsets.
- Also, the costs of a cellphone (raw materials) are determined to a considerable degree by the battery.
- Users require standby times of several days, as well as talk times of at least 2 hours before recharging.

These "commercial" aspects determine the maximum size (and thus energy content) of the battery, and consequently, the admissible energy consumption of the phone during standby and talk operation.

1.3.5 Use of Spectrum

Spectrum can be assigned on an exclusive basis, or on a shared basis. That determines to a large degree the multiple access scheme and the interference resistance that the system has to provide:

- *Spectrum dedicated to service and operator*: in this case, a certain part of the electromagnetic spectrum is assigned, on an exclusive basis, to a service provider. A prime point in case is cellular telephony, where the network operators buy or lease the spectrum on an exclusive basis (often for a very high price). Due to this arrangement, the operator has control over the spectrum and can plan the use of different parts of this spectrum in different geographical regions, in order to minimize interference.

- *Spectrum allowing multiple operators*:
  - ○ *Spectrum dedicated to a service*: in this case, the spectrum can be used only for a certain service (e.g., cordless telephones in Europe and Japan), but is not assigned to a specific operator. Rather, users can set up qualified equipment without a license. Such an approach does not require (or allow) interference planning. Rather, the system must be designed in such a way that it avoids interfering with other users in the same region. Since the only interference *can* come from equipment of the same type, coordination between different devices is relatively simple. Limits on transmit power (identical for all users) are a key component of this approach – without them, each user would just increase the transmit power to drown out interferers, leading essentially to an "arms race" between users.
  - ○ *Free spectrum*: is assigned for different services as well as for different operators. The ISM band at 2.45 GHz is the best known example – it is allowed to operate microwave ovens, WiFi LANs, and Bluetooth wireless links, among others, in this band. Also for this case, each user has to adhere to strict emission limits, in order not to interfere too much with other systems and users. However, coordination between users (in order to minimize interference) becomes almost impossible – different systems cannot exchange coordination messages with each other, and often even have problems determining the exact characteristics (bandwidth, duty cycle) of the interferers.

After 2000, two new approaches have been promulgated, but are not yet in widespread use:

- *Ultra Wide Bandwidth systems* (UWB) spread their information over a very large bandwidth, while at the same time keeping a very low-power spectral density. Therefore, the transmit band can include frequency bands that have already been assigned to other services, without creating significant interference. UWB is discussed in more detail in Chapter 17.
- *Adaptive spectral usage*: another approach relies on first determining the current spectrum usage at a certain location and then employing unused parts of the spectrum. This approach, also known as *cognitive radio*, is described in detail in Chapter 21.

1.3.6 Direction of Transmission

Not all wireless services need to convey information in both directions.

- *Simplex systems* send the information only in one direction – e.g., broadcast systems and pagers.
- *Semi-duplex systems* can transmit information in both directions. However, only one direction is allowed at any time. Walkie-talkies, which require the user to push a button in order to talk, are a typical example. Note that one user must signify (e.g., by using the word "over") that (s)he has finished his/her transmission; then the other user knows that now (s)he can transmit.
- *Full-duplex systems* allow simultaneous transmission in both directions – e.g., cellphones and cordless phones.
- *Asymmetric duplex systems*: for data transmission, we often find that the required data rate in one direction (usually the downlink) is higher than in the other direction. However, even in this case, full duplex capability is maintained.

1.3.7 Service Quality

The requirements for service quality also differ vastly for different wireless services. The first main indicator for service quality is *speech quality* for speech services and *file transfer speed* for data services. Speech quality is usually measured by the *Mean Opinion Score* (MOS). It represents the average of a large number of (subjective) human judgments (on a scale from 1 to 5) about the quality of received speech (see also Chapter 15). The speed of data transmission is simply measured in bit/s – obviously, a higher speed is better.

An even more important factor is the availability of a service. For cellphones and other speech services, the *service quality* is often computed as the complement of "fraction of blocked calls[7] plus 10 times the fraction of dropped calls." This formula takes into account that the dropping of an active call is more annoying to the user than the inability to make a call at all. For cellular systems in Europe and Japan, this service quality measure usually exceeds 95%; in the U.S.A., the rate is considerably lower.[8]

For emergency services and military applications, service quality is better measured as the complement of "fraction of blocked calls plus fraction of dropped calls." In emergency situations, the inability to make a call is as annoying as the situation of having a call interrupted. Also, the systems must be planned in a much more robust way, as service qualities better than 99% are required. "Ultrareliable systems," which are required, e.g., for factory automation systems, require service quality in excess of 99.99%.

A related aspect is the *admissible delay (latency)* of the communication. For voice communications, the delay between the time when one person speaks and the other hears the message must not be larger than about 100 ms. For streaming video and music, delays can be larger, as buffering of the streams (up to several tens of seconds) is deemed acceptable by most users. In both voice and streaming video communications, it is important that the data transmitted first are also the ones made available to the receiving user first. For data files, the acceptable delays can be usually larger and the sequence with which the data arrive at the receiver is not critical (e.g., when downloading email from a server, it is not important whether the first or the seventh of the emails is the first to arrive). However, there are some data applications where small latency is vital – e.g., for control applications, security and safety monitoring, etc.

1.4 Economic and Social Aspects

1.4.1 Economic Requirements for Building Wireless Communications Systems

The design of wireless systems not only aims to optimize performance for specific applications but also to do that at a reasonable cost. As economic factors impact the design, scientists and engineers have to have at least a basic understanding of the constraints imposed by marketing and sales divisions. Some of the guidelines for the design of wireless *devices* are as follows:

- Move as much functionality as possible from the (more expensive) analog components to digital circuitry. The costs for digital circuits decrease much faster with time than those of analog components.
- For mass-market applications, try to integrate as many components onto one chip as possible. Most systems strive to use only two chips; one for analog RF circuitry and one for digital (baseband) processing. Further integration into a single chip (system on a chip) is desirable. Exceptions are niche market products, which typically try to use general-purpose processors, Application Specific Integrated Circuits (ASICs), or off-the-shelf components, as the number of sold units does not justify the cost of designing more highly integrated chips.
- As human labor is very expensive, any circuit that requires human intervention (e.g., tuning of RF elements) is to be avoided. Again, this aspect is more important for mass-market products.

---

[7] Here, "blocked calls" encompasses all failed call attempts, including those that are caused by insufficient signal strength, as well as insufficient network capacity.

[8] The reason for this discrepancy is partly historical and economical, and partly geographical.

- In order to increase the efficiency of the development process and production, the same chips should be used in as many systems as possible.

When it comes to the design of wireless *systems and services*, we have to distinguish between two different categories:

- Systems where the mobility is of value of itself – e.g., in cellular telephony. Such services can charge a premium to the customer – i.e., be more expensive than equivalent, wired systems. Cellular telephony is a case in point: the per-minute price has been higher than that of landline telephony in the past, and is expected to remain so (especially when compared, e.g., with Voice over Internet Protocol (VoIP) telephone services). Despite this fact, the services might compete (and ultimately edge out) traditional wired services if the price difference is not too large. The years since 1990 have certainly seen such a trend, with many consumers (and even companies) canceling wired services and relying on cellular telephony alone.
- Services where wireless access is only intended as a cheap cable replacement, without enabling additional features – e.g., fixed wireless access. Such systems have to be especially cost-conscious, as the buildup of the infrastructure has to remain cheaper than the laying of new wired connections, or buying access to existing ones.

1.4.2   The Market for Wireless Communications

Cellphones are a highly dynamic market that has grown tremendously. Still, different countries show different market penetrations. Some of the factors influencing this penetration are:

- *Price of the offered services*: the price of the services is in turn influenced by the amount of competition, the willingness of the operators to accept losses in order to gain greater market penetration, and the external costs of the operators (especially, the cost of spectrum licenses). However, the price of the services is not always the decisive factor for market penetration: Scandinavia, with its relatively high prices, has the highest market penetration in the world.
- *Price of the MSs*: the MSs are usually subsidized by the operators and are either free or sold at a nominal price, if the consumer agrees to a long-term contract. Exceptions are "prepaid" services, where a user buys a certain number of minutes of service usage (in that case, the handsets are sold to the consumer at full cost); at the other end of the market spectrum, high-end devices usually require a significant co-payment by the consumer.
- *Attractiveness of the offered services*: in many markets, the price of the services offered by different network operators is almost identical. Operators try to distinguish themselves by different features, like better coverage, text and picture message service, etc. The offering of these improved features also helps to increase the market size in general, as it allows customers to find services tailored to their needs.
- *General economic situation*: obviously, a good general economic situation allows the general population to spend more money on such "non-essential" things as mobile communications services. In countries where a very large percentage of the income goes to basic necessities like food and housing, the market for cellular telephony is obviously more limited.
- *Existing telecom infrastructure*: in countries or areas with a bad existing landline-based telecom infrastructure, cellular telephony and other wireless services can be the only way of communicating. This would enable high market penetration. Unfortunately, these areas are usually also the ones that have the bad economic situation mentioned above (large percentage of income goes to basic necessities). This fact has hindered especially the development of fixed wireless services.[9]

---

[9] To put it succinctly: "the market for this product is the people who cannot afford to pay for it."

- *Predisposition of the population*: there are several social factors that can increase the cellular market: (i) people have a positive attitude to new technology (gadgets) – e.g., Japan and Scandinavia; (ii) people consider communication as an essential part of their lives – e.g., China; and (iii) high mobility of the population, with people being absent from their offices or homes for a significant part of the day – e.g., the U.S.A.

Wireless communications has become such a huge market that most of the companies in this business are not even known to most of the consumers. Consumers tend to know network operators and handset manufacturers. However, component suppliers and other auxiliary industries abound:

- *Infrastructure manufacturers for cellular telephony*: most of the major handset manufacturers also provide infrastructure (BSs, switches, etc.) to the network operators.
- *Component manufacturers*: most handset manufacturers buy chips, batteries, antennas, etc., from external suppliers. This trend was accelerated by the fact that many manufacturers and system integrators spun off their semiconductor divisions. There are even handset companies that do not manufacture anything, but are just design and marketing operations.
- *Software suppliers*: software and applications are becoming an increasingly important part of the market. For example, ringtones for cellphones have become a multibillion dollar (Euro) market. Similarly, operating systems and applications software for cellphones have become increasingly important with the proliferation of smartphones, i.e., as cellphones acquire more and more of the functionalities of *Personal Digital Assistants* (PDAs).
- *Systems integrators*: WLANs and sensor networks need to be integrated either into larger networks or combined with other hardware (e.g., sensor networks have to be integrated into a factory automation system). This offers new fields of business for OEMs and system integrators.

1.4.3 Behavioral Impact

Engineering does not happen in a vacuum – the demands of people change what the engineers develop and the products of their labor influences how people behave. Cellphones have enabled us to communicate anytime – something that most people think of as desirable. But we have to be aware that it changes our lifestyle. In former times, one did not call a person, but rather a location. That meant a rather clean separation between professional or personal life. Due to the cellphone, anybody can be reached at any time – somebody from work can call in the evening; a private acquaintance can call in the middle of a meeting; in other words, professional and private lives get intermingled. On the positive side, this also allows new and more convenient forms of working and increased flexibility.

Another important behavioral impact is the development of (or lack of) *cellphone etiquette*. Most people tend to agree that hearing a cellphone ring during an opera performance is exasperating – and still there is a significant number of people who are not willing to turn their phones to the "silent" mode during such occasions. There also seems to be an innate reluctance in humans to just ignore a ringing cellphone. People are willing to interrupt whatever they are doing in order to answer a ringing phone. Caller identification, automatic callback features, etc., are solutions that engineers can provide to alleviate these problems.

On a more serious note, wireless devices, and especially cellphones, can be a matter of life and death. Being able to call for help in the middle of the wilderness after a mountaineering accident is definitely a lifesaving feature. Location devices for victims of avalanches have a similar beneficial role. On the downside, drivers who are distracted by their cellphone conversations or text messages constitute a serious hazard on the roads. Recent studies at Virginia Technological University showed that talking on a cellphone while driving – even when using hands-free equipment – constitutes a hazard similar to being drunk. Text messaging while driving is even worse! While the author of this book hopes that the readers will remember many of the technical facts presented throughout the text, the by far most important message to remember from this book is: **Do not text or phone while driving !!!** The problem is, again, not purely a technical one, as a multitude of solutions (including the "off" button) have already been developed to solve this issue. Rather, it is a matter of behavioral changes by the users and the question of what the engineers can do to further these changes.

For updates and errata for this chapter, see wides.usc.edu/teaching/textbook

2

Technical Challenges of Wireless Communications

In the previous chapter, we described the requirements for wireless communications systems stemming from the applications and user demands. In this chapter, we give a high-level description of the technical challenges to wireless communications systems. Most notably, they are as follows:

- multipath propagation: i.e., the fact that a transmit signal can reach the receiver via different paths (e.g., reflections from different houses or mountains);
- spectrum limitations;
- energy limitations;
- user mobility.

This sets the stage for the rest of the book, where these challenges, as well as remedies, are discussed in more detail.

As a first step, it is useful to investigate the differences between wired and wireless communications. Let us first repeat some important properties of wired and wireless systems, as summarized in Table 2.1.

2.1 Multipath Propagation

For wireless communications, the transmission medium is the radio channel between transmitter TX and receiver RX. The signal can get from the TX to the RX via a number of different propagation paths. In some cases, a Line Of Sight (LOS) connection might exist between TX and RX. Furthermore, the signal can get from the TX to the RX by being reflected at or diffracted by different *Interacting Objects* (IOs) in the environment: houses, mountains (for outdoor environments), windows, walls, etc. The number of these possible propagation paths is very large. As shown in Figure 2.1, each of the paths has a distinct amplitude, delay (runtime of the signal), direction of departure from the TX, and direction of arrival; most importantly, the components have different *phase shifts* with respect to each other. In the following, we discuss some implications of the multipath propagation for system design.

2.1.1 Fading

A simple RX cannot distinguish between the different *Multi Path Components* (MPCs); it just adds them up, so that they interfere with each other. The interference between them can be constructive

Wireless Communications, Second Edition   Andreas F. Molisch
© 2011 John Wiley & Sons, Ltd

Table 2.1 Wired and wireless communications

Wired communications	Wireless communications
The communication takes place over a more or less stable medium like copper wires or optical fibers. The properties of the medium are well defined and time-invariant.	Due to user mobility as well as multipath propagation, the transmission medium varies strongly with time.
Increasing the transmission capacity can be achieved by using a different frequency on an existing cable, and/or by stringing new cables.	Increasing the transmit capacity must be achieved by more sophisticated transceiver concepts and smaller cell sizes (in cellular systems), as the amount of available spectrum is limited.
The range over which communications can be performed without repeater stations is mostly limited by attenuation by the medium (and thus noise); for optical fibers, the distortion of transmitted pulses can also limit the speed of data transmission.	The range that can be covered is limited both by the transmission medium (attenuation, fading, and signal distortion) and by the requirements of spectral efficiency (cell size).
Interference and crosstalk from other users either do not happen or the properties of the interference are stationary.	Interference and crosstalk from other users are inherent in the principle of cellular communications. Due to the mobility of the users, they also are time-variant.
The delay in the transmission process is also constant, determined by the length of the cable and the group delay of possible repeater amplifiers.	The delay of the transmission depends partly on the distance between base station and Mobile Station (MS), and is thus time-variant.
The *Bit Error Rate* (BER) decreases strongly (approximately exponentially) with increasing *Signal-to-Noise Ratio* (SNR). This means that a relatively small increase in transmit power can greatly decrease the error rate.	For simple systems, the average BER decreases only slowly (linearly) with increasing average SNR. Increasing the transmit power usually does not lead to a significant reduction in BER. However, more sophisticated signal processing helps.
Due to the well-behaved transmission medium, the quality of wired transmission is generally high.	Due to the difficult medium, transmission quality is generally low unless special measures are used.
Jamming and interception of dedicated links with wired transmission is almost impossible without consent by the network operator.[a]	Jamming a wireless link is straightforward, unless special measures are taken. Interception of the on-air signal is possible. Encryption is therefore necessary to prevent unauthorized use of the information.
Establishing a link is *location* based. In other words, a link is established from one outlet to another, independent of which *person* is connected to the outlet.	Establishing a connection is based on the (mobile) equipment, usually associated with a specific person. The connection is not associated with a fixed location.
Power is either provided through the communications network itself (e.g., for traditional landline telephones), or from traditional power mains (e.g., fax). In neither case is energy consumption a major concern for the designer of the device.	MSs use rechargeable or one-way batteries. Energy efficiency is thus a major concern.

[a]Note, though, that interception of wired Internet communication is simple due to the design of this particular communication protocol.

or destructive, depending on the phases of the MPCs, (see Figure 2.2). The phases, in turn, depend mostly on the run length of the MPC, and thus on the position of the Mobile Station (MS) and the IOs. For this reason, the interference, and thus the amplitude of the total signal, changes with time if either TX, RX, or IOs is moving. This effect – namely, the changing of the total signal amplitude due to interference of the different MPCs – is called *small-scale fading*. At 2-GHz carrier frequency, a movement by less than 10 cm can already effect a change from constructive to

Figure 2.1 Multipath propagation.

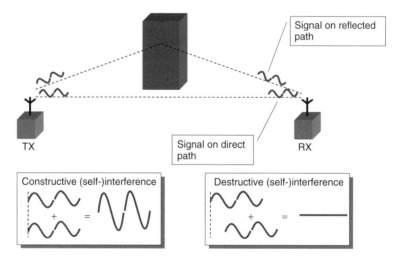

Figure 2.2 Principle of small-scale fading.

destructive interference and vice versa. In other words, even a small movement can result in a large change in signal amplitude. A similar effect is known to all owners of car radios – moving the car by less than 1 m (e.g., in stop-and-go traffic) can greatly affect the quality of the received signal. For cellphones, it can often be sufficient to move one step in order to improve signal quality.

As an additional effect, the amplitudes of each separate MPC change with time (or with location). Obstacles can lead to a shadowing of one or several MPCs. Imagine, e.g., the MS in Figure 2.3 that at first (at position A) has LOS to the Base Station (BS). As the MS moves behind the high-rise building (at position B), the amplitude of the component that propagates along the direct connection (LOS) between BS and MS greatly decreases. This is due to the fact that the MS is now in the radio shadow of the high-rise building, and any wave going through or around that building is greatly attenuated – an effect called *shadowing*. Of course, shadowing can occur not only for an LOS component but also for *any* MPC. Note also that obstacles do not throw "sharp" shadows: the transition from the "light" (i.e., LOS) zone to the "dark" (shadowed) zone is gradual.[1] The MS

---

[1] This is due to (i) diffraction effects, as also explained in more detail in Chapter 4 and (ii) the fact that secondary radiation sources like houses are spatially extended (compare how a long fluorescent tube never throws a sharp shadow).

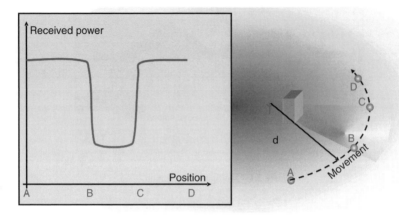

Figure 2.3   The principle of shadowing.

has to move over large distances (from a few meters up to several hundreds of meters) to move from the light to the dark zone. For this reason, shadowing gives rise to *large-scale fading*.

Large-scale and small-scale fading overlap, so that the received signal amplitude can look like the one depicted in Figure 2.4. Obviously, the transmission quality is low at the times (or places) with low signal amplitude. This can lead to bad speech quality (for voice telephony), high Bit Error Rate (BER) and low data rate (for data transmission), and – if the quality is too low for an extended period of time – to termination of the connection.

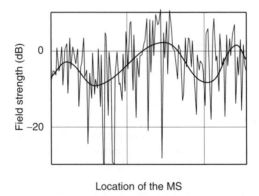

Location of the MS

Figure 2.4   Typical example of fading. The thin line is the (normalized) instantaneous field strength; the thick line is the average over a 1-m distance.

It is well known from conventional digital communications that for nonfading communications links, the BER decreases approximately exponentially with increasing Signal-to-Noise Ratio (SNR) if no special measures are taken. However, in a fading channel, the SNR is not constant; rather, the probability that the link is in a *fading dip* (i.e., location with low SNR) dominates the behavior of the BER. For this reason, the average BER decreases only *linearly* with increasing average SNR. Consequently, improving the BER often cannot be achieved by simply increasing the transmit power. Rather more sophisticated transmission and reception schemes have to be used. Most of the third and fourth parts of this book (Chapters 13, 14, 16, 18–22) are devoted to such techniques.

Due to fading, it is almost impossible to exactly predict the received signal amplitude at specific locations. For many aspects of system development and deployment, it is considered sufficient to predict the mean amplitude and the *statistics* of fluctuations around that mean. Completely deterministic predictions of the signal amplitude – e.g., by solving approximations to Maxwell's equations[2] in a given environment – usually show errors of between 3 and 10 dB (for the total amplitude), and are even less reliable for the properties of individual MPCs. More details on fading can be found in Chapters 5 to 7.

2.1.2 Intersymbol Interference

The runtimes for different MPCs are different. We have already mentioned above that this can lead to different phases of MPCs, which lead to interference in narrowband systems. In a system with large bandwidth, and thus good resolution in the time domain,[3] the major consequence is signal dispersion: in other words, the impulse response of the channel is not a single delta pulse but rather a sequence of pulses (corresponding to different MPCs), each of which has a distinct arrival time in addition to having a different amplitude and phase (see Figure 2.5). This signal dispersion leads to InterSymbol Interference (ISI) at the RX. MPCs with long runtimes, carrying information from bit k, and MPCs with short runtimes, carrying contributions from bit $k + 1$ arrive at the RX at the same time, and interfere with each other (see Figure 2.6). Assuming that no special measures[4] are taken, this ISI leads to errors that cannot be eliminated by simply increasing the transmit power, and are therefore often called *irreducible errors*.

Multi path components with different runtimes   Channel impulse response

Figure 2.5 Multipath propagation and resulting impulse response.

---

[2] The most popular of these deterministic prediction tools are "ray tracing" and "ray launching," which are discussed in Chapter 7.

[3] Strictly speaking, we refer here to resolution in the delay domain. An explanation for the difference between the time domain and delay domain is given in Chapters 5 and 6.

[4] Special measures include equalizers (Chapter 16), Rake receivers (Chapter 18), and Orthogonal Frequency Division Multiplexing (OFDM) (Chapter 19).

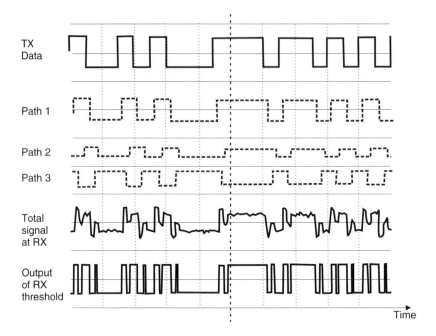

Figure 2.6   Intersymbol interference.

ISI is essentially determined by the ratio between symbol duration and the duration of the impulse response of the channel. This implies that ISI is not only more important for higher data rates but also for multiple access methods that lead to an increase in transmitted *peak* data rate (e.g., time division multiple access, see Chapter 17). Finally, it is also noteworthy that ISI can even play a role when the duration of the impulse response is *shorter* (but not *much* shorter) than bit duration (see Chapters 12 and 16).

2.2   Spectrum Limitations

The spectrum available for wireless communications services is limited, and regulated by international agreements. For this reason, the spectrum has to be used in a highly efficient manner. Two approaches are used: regulated spectrum usage, where a single network operator has control over the usage of the spectrum, and unregulated spectrum, where each user can transmit without additional control, as long as (s)he complies with certain restrictions on the emission power and bandwidth. In the following, we first review the frequency ranges assigned to different communications services. We then discuss the basic principle of frequency reuse for both regulated and unregulated access.

2.2.1   Assigned Frequencies

The frequency assignment for different wireless services is regulated by the *International Telecommunications Union* (ITU), a suborganization of the United Nations. In its tri-annual conferences (*World Radio Conferences*), it establishes worldwide guidelines for the usage of spectrum in different regions and countries. Further regulations are issued by the frequency regulators of individual countries, including the *Federal Communications Commission* (FCC) in the U.S.A., the *Association of Radio Industries and Businesses* (ARIB) in Japan, and the *European Conference of Postal and Telecommunications Administrations* (CEPT) in Europe. While the exact frequency assignments differ, similar services tend to use the same frequency ranges all over the world:

- Below 100 MHz: at these frequencies, we find Citizens' Band (CB) radio, pagers, and analog cordless phones.
- 100–800 MHz: these frequencies are mainly used for broadcast (radio and TV) applications.
- 400–500 MHz: a number of cellular and trunking radio systems make use of this band. It is mostly systems that need good coverage, but show low user density.
- 800–1000 MHz: several cellular systems use this band (analog systems as well as second-generation cellular). Also some emergency communications systems (trunking radio) make use of this band.
- 1.8–2.1 GHz: this is the main frequency band for cellular communications. The current (second-generation) cellular systems operate in this band, as do most of the third-generation systems. Many cordless systems also operate in this band.
- 2.4–2.5 GHz: the Industrial, Scientific, and Medical (ISM) band. Cordless phones, Wireless Local Area Networks (WLANs) and wireless Personal Area Networks (PANs) operate in this band; they share it with many other devices, including microwave ovens.
- 3.3–3.8 GHz: is envisioned for fixed wireless access systems.
- 4.8–5.8 GHz: in this range, most WLANs can be found. Also, the frequency range between 5.7 and 5.8 GHz can be used for fixed wireless access, complementing the 3-GHz band. Also car-to-car communications are working in this band.
- 11–15 GHz: in this range we can find the most popular satellite TV services, which use 14.0–14.5 GHz for the uplink, and 11.7–12.2 GHz for the downlink.

More details about the exact frequencies for specific services can be obtained from the national frequency regulators, as well as from the ITU.

Different frequency ranges are optimum for different applications. Low carrier frequencies usually propagate more easily (see also Chapter 4), so that a single BS can cover a large area. On the other hand, absolute bandwidths are smaller, and also the frequency reuse is not as efficient as it is at higher frequencies.[5] For this reason, low frequency bands are best for services that require good coverage, but have a small aggregate rate of information that has to be exchanged. Typical cases in point are paging services and television; paging is suitable because the amount of information transmitted to each user is small, while in the latter case, only a single information stream is sent to *all* users. For cellular systems, low carrier frequencies are ideal for covering large regions with low user density (rural areas in the Midwest of the U.S.A. and in Russia, Northern Scandinavia, Alpine regions, etc.). For cellular systems with high user densities, as well as for WLANs, higher carrier frequencies are usually more desirable.[6]

The amount of spectrum assigned to the different services does not always follow technical necessities, but rather historical developments. For example, for many years, the amount of precious low-frequency spectrum assigned to TV stations was much higher than would be justified by technical requirements. Using appropriate frequency planning and different transmission techniques (including simulcast), a considerable part of the spectrum below 1 GHz could be freed up for alternative usage – a process that took place in U.S.A. and Europe around 2008/2009, but might take longer in other countries. Broadcast stations tend to fight such a development, as it would

---

[5] As we will see in the next subsection, frequency reuse requires that a signal is attenuated strongly outside the cell it is assigned to. However, low carrier frequency results in good propagation so that the signal can remain strong far outside its assigned cell.

[6] However, the carrier frequency should not become *too* high: at extremely high frequencies, it becomes difficult to cover even small areas.

require modifications in their transmitters. As these stations have a considerable influence on pub-lic opinion as well as lobbying power, frequency regulators are hesitant to enforce appropriate rule changes.

It is also noticeable that the financial terms on which spectrum is assigned to different services differ vastly – from country to country, from service to service, and even depending on the time at which the spectrum is assigned. Obviously, spectrum is assigned to public safety services (police, fire department, military) without monetary compensation. Even television stations usually get the spectrum assigned for free. In the 1980s, spectrum for cellular telephony was often assigned for a rather small fee, in order to encourage the development of this then-new service. In the mid- and late-1990s, spectrum auctions were seen by some countries as a method to increase the country's revenues (consider the frequency auctions for the PCS band in the U.S.A. in 1995, and the auctions of the Universal Mobile Telecommunications System (UMTS) bands in the United Kingdom and Germany around 2000). Other countries chose to assign spectrum based on a "beauty contest," where the applicant had to guarantee a certain service quality, coverage etc., in order to obtain a license. Unregulated services, like WLANs, are assigned spectrum without fees.

2.2.2   Frequency Reuse in Regulated Spectrum

Since spectrum is limited, the *same* spectrum has to be used for *different* wireless connections in *different* locations. To simplify the discussion, let us consider in the following a cellular system where different connections (different users) are distinguished by the frequency channel (band around a certain carrier frequency) that they employ. If an area is served by a single BS, then the available spectrum can be divided into N frequency channels that can serve N users simultaneously. If more than N users are to be served, multiple BSs are required, and frequency channels have to be reused in different locations.

For this purpose, we divide the area (a region, a country, or a whole continent) into a number of *cells*; we also divide the available frequency channels into several groups. The channel groups are now assigned to the cells. The important thing is that channel groups can be used in multiple cells. The only requirement is that cells that use the same frequency group do not interfere with each other *significantly*.[7] It is fairly obvious that the same carrier frequency can be used for different connections in, say, Rome and Stockholm, at the same time. The large distance between the two cities makes sure that a signal from the MS in Stockholm does not reach the BS in Rome, and can therefore not cause any interference at all. But in order to achieve high efficiency, frequencies must actually be reused much more often – typically, several times within each city. Consequently, *intercell interference* (also known as *co-channel interference*) becomes a dominant factor that limits transmission quality. More details on co-channel interference can be found in Part IV.

Spectral efficiency describes the effectiveness of reuse – i.e., the traffic density that can be achieved per unit bandwidth and unit area. It is therefore given in units of Erland/(Hz m^2) for voice traffic and bit/(s Hz m^2) for data. Since the area covered by a network provider, as well as the bandwidth that it can use, are fixed, increasing the spectral efficiency is the only way to increase the number of customers that can be served, and thus revenue. Methods for increasing this spectral efficiency are thus at the center of wireless communications research.

Since a network operator buys a license for a spectrum, it can use that spectrum according to its own planning – i.e., network planning can make sure that the users in different cells do not interfere with each other significantly. The network operator is allowed to use as much transmit power as it desires; it can also dictate limits on the emission power of the MSs of different

---

[7] The threshold for *significant* interference (i.e., the admissible signal-to-interference ratio) is determined by mod-ulation and reception schemes, as well as by propagation conditions.

users.[8] The operator can also be sure that the only interference in the network is created by its own network and users.

2.2.3 Frequency Reuse in Unregulated Spectrum

In contrast to regulated spectrum, several services use frequency bands that are available to the general public. For example, some WLANs operate in the 2.45-GHz band, which has been assigned to "ISM" services. Anybody is allowed to transmit in these bands, as long as they (i) limit the emission power to a prescribed value, (ii) follow certain rules for the signal shape and band-width, and (iii) use the band according to the (rather broadly defined) purposes stipulated by the frequency regulators.

As a consequence, a WLAN receiver can be faced with a large amount of interference. This interference can either stem from other WLAN transmitters or from microwave ovens, cordless phones, and other devices that operate in the ISM band. For this reason, a WLAN link must have the capability to deal with interference. That can be achieved by selecting a frequency band within the ISM band at which there is little interference, by using spread spectrum techniques (see Chapter 18), or some other appropriate technique.

There are also cases where the spectrum is assigned to a specific service (e.g., DECT), but not to a specific operator. In that case, receivers might still have to deal with strong interference, but the structure of this interference is known. This allows the use of special interference mitigation techniques like dynamic frequency assignment, see the material on DECT on the companion website (*www.wiley.com/go/molisch*).

Dynamic frequency assignment can be seen as a special case of *cognitive radio* (see Chapter 21), where a transmitter senses which part of the spectrum is currently unused in the location of interest, and dynamically moves the transmit frequency accordingly.

2.3 Limited Energy

Truly wireless communications requires not only that the information is sent over the air (not via cables) but also that the MS is powered by one-way or rechargeable batteries. Otherwise, an MS would be tied to the "wire" of the power supply, batteries in turn impose restrictions on the power consumption of the devices. The requirement for small energy consumption results in several technical imperatives:

- The power amplifiers in the transmitter have to have high efficiency. As power amplifiers account for a considerable fraction of the power consumption in an MS, mainly amplifiers with an efficiency above 50% should be used in MSs. Such amplifiers – specifically, class-C or class-F amplifiers – are highly nonlinear.[9] As a consequence, wireless communications tend to use modulation formats that are insensitive to nonlinear distortions. For example, constant envelope signals are preferred (see Chapter 11).
- Signal processing must be done in an energy-saving manner. This implies that the digital logic should be implemented using power-saving semiconductor technology like Complementary Metal Oxide Semiconductor (CMOS), while the faster but more power-hungry approaches like Emitter Coupled Logic (ECL) do not seem suitable for MSs. This restriction has important consequences for the algorithms that can be used for interference suppression, combating of ISI, etc.

---

[8] There are some exceptions to that rule – e.g., emission limits dictated by health concerns, as well as limits imposed by the standard of the system used by the operator (e.g., GSM).
[9] Linear amplifiers, like class A, class B, or class AB, have efficiencies of less than 30%.

- The RX (especially at the BS) needs to have high sensitivity. For example, Global System for Mobile Communications (GSM) is specified so that even a received signal power of $-100\,\mathrm{dBm}$ leads to an acceptable transmission quality. Such an RX is several orders of magnitude more sensitive than a TV RX. If the GSM standard had defined $-80\,\mathrm{dBm}$ instead, then the transmit power would have to be higher by a factor of 100 in order to achieve the same coverage. This in turn would mean that – for identical talktime – the battery would have to be 100 times as large – i.e., 20 kg instead of the current 200 g. But the high requirements on RX sensitivity have important consequences for the construction of the RX (low-noise amplifiers, sophisticated signal processing to fully exploit the received signal) as well as for network planning.
- Maximum transmit power should be used only when required. In other words, transmit power should be adapted to the channel state, which in turn depends on the distance between TX and RX (*power control*). If the MS is close to the BS, and thus the channel has only a small attenuation, transmit power should be kept low. Furthermore, for voice transmission, the MS should only transmit if the user at the MS actually talks, which is the case only about 50% of the time (*Discontinuous Voice Transmission* (DTX)).
- For cellular phones, and even more so for sensor networks, an energy-efficient "standby" or "sleep" mode has to be defined.

Several of the mentioned requirements are contradictory. For example, the requirement to build an RX with high sensitivity (and thus, sophisticated signal processing) is in contrast to the requirement of having energy-saving (and thus slow) signal processing. Engineering tradeoffs are thus called for.

2.4 User Mobility

Mobility is an inherent feature of most wireless systems, and has important consequences for system design. Fading was already discussed in Section 2.1.1. A second important effect is particular to mobile users in cellular systems: the system has to know at any time which cell a user is in:[10]

- If there is an incoming call for a certain MS (user), the network has to know in which cell the user is located. The first requirement is that an MS emits a signal at regular intervals, informing nearby BSs that it is "in the neighborhood." Two databanks then employ this information: the *Home Location Register* (HLR) and the *Visitor Location Register* (VLR). The HLR is a central database that keeps track of the location a user is currently at; the VLR is a database associated with a certain BS that notes all the users who are currently within the coverage area of this specific BS. Consider user A, who is registered in San Francisco, but is currently located in Los Angeles. It informs the nearest BS (in Los Angeles) that it is now within its coverage area; the BS enters that information into its VLR. At the same time, the information is forwarded to the central HLR (located, e.g., in New York). If now somebody calls user A, an enquiry is sent to the HLR to find out the current location of the user. After receiving the answer, the call is rerouted to Los Angeles. For the Los Angeles BS, user A is just a "regular" user, whose data are all stored in the VLR.
- If an MS moves across a cell boundary, a different BS becomes the *serving BS*; in other words, the MS is *handed over* from one BS to another. Such a handover has to be performed without interrupting the call; as a matter of fact, it should not be noticeable at all to the user. This requires complicated signaling. Different forms of handover are described in Chapter 18 for code-division-multiple-access-based systems, and in Chapter 24 for GSM.

For updates and errata for this chapter, see wides.usc.edu/teaching/textbook

---

[10] This effect is not relevant, e.g., for simple cordless systems: either a user is within the coverage region of the (one and only) BS, or (s)he is not.

3

Noise- and Interference-Limited Systems

3.1 Introduction

This chapter explains the principles of link budgets, and the planning of wireless systems with one or multiple users. In Section 3.2, we set up link budgets for noise-limited systems and compute the minimum transmit power (or maximum range) that can be achieved in the absence of interference. Such computations give a first insight into the basic capabilities of wireless systems and also have practical applications. For example, Wireless Local Area Networks (WLANs) and cordless phones often operate in a noise-limited mode, if no other Base Station (BS) is in the vicinity. Even cellular systems sometimes operate in that mode if the user density is low (this happens, e.g., during the build-up phase of a network).

In Section 3.3, we discuss interference-limited systems. As we described in the first two chapters, the unregulated use of spectrum leads to interference that cannot be controlled by the user. When the spectrum is regulated, the network operator can determine the location of BSs, and thus impact the Signal-to-Interference Ratio (SIR). For either case, it is important to set up the link budgets that take the presence of interference into account; Section 3.3 describes these link budgets. In Chapter 17, we then see how these calculations are related to the cellular principle and the reuse of frequencies in different cells.

3.2 Noise-Limited Systems

Wireless systems are required to provide a certain minimum transmission quality (see Section 1.3). This transmission quality in turn requires a minimum *Signal-to-Noise Ratio* (SNR) at the receiver (RX). Consider now a situation where only a single BS transmits, and a Mobile Station (MS) receives; thus, the performance of the system is determined only by the strength of the (useful) signal and the noise. As the MS moves further away from the BS, the received signal power decreases, and at a certain distance, the SNR does not achieve the required threshold for reliable communications. Therefore, the range of the system is noise limited; equivalently, we can call it *signal power limited*. Depending on the interpretation, it is too much noise or too little signal power that leads to bad link quality.

Let us assume for the moment that the received power decreases with d^2, the square of the distance between BS and MS. More precisely, let the received power P_{RX} be

$$P_{RX} = P_{TX} G_{RX} G_{TX} \left(\frac{\lambda}{4\pi d} \right)^2 \tag{3.1}$$

where G_{RX} and G_{TX} are the gains of the receive and transmit antennas, respectively,[1] λ is the wavelength, and P_{TX} is the transmit power (see Chapter 4 for a derivation of this equation and for more details).

The noise that disturbs the signal can consist of several components, as follows:

1. *Thermal noise*: The power spectral density of thermal noise depends on the environmental temperature T_e that the antenna "sees." The temperature of the Earth is around $300\,K$, while the temperature of the (cold) sky is approximately $T_e \approx 4\,K$ (the temperature in the direction of the Sun is of course much higher). As a first approximation, it is usually assumed that the environmental temperature is isotropically $300\,K$. Noise power spectral density is then

$$N_0 = k_B T_e \tag{3.2}$$

where k_B is Boltzmann's constant, $k_B = 1.38 \cdot 10^{-23}\,J/K$, and the noise power is

$$P_n = N_0 B \tag{3.3}$$

where B is RX bandwidth (in units of Hz). It is common to write Eq. (3.2) using logarithmic units (power P expressed in units of dBm is $10 \log_{10} (P/1\,mW)$):

$$N_0 = -174\,dBm/Hz \tag{3.4}$$

This means that the noise power contained in a 1-Hz bandwidth is $-174\,dBm$. The noise power contained in bandwidth B is

$$-174 + 10 \log_{10}(B)\,dBm \tag{3.5}$$

The logarithm of bandwidth B, specifically $10 \log_{10}(B)$, has the units dBHz.

2. *Man-made noise*: We can distinguish two types of man-made noise:
   (a) *Spurious emissions*: Many electrical appliances as well as radio transmitters (TXs) designed for other frequency bands have spurious emissions over a large bandwidth that includes the frequency range in which wireless communications systems operate. For urban outdoor environments, car ignitions and other impulse sources are especially significant sources of noise. In contrast to thermal noise, the noise created by impulse sources decreases with frequency (see Figure 3.1). At 150 MHz, it can be 20 dB stronger than thermal noise; at 900 MHz, it is typically 10 dB stronger. At Universal Mobile Telecommunications System (UMTS) frequencies, Neubauer et al. [2001] measured 5-dB noise enhancement by man-made noise in urban environments and about 1 dB in rural environments. Note that frequency regulators in most countries impose limits on "spurious" or "out-of-band" emissions for *all* electrical devices. Furthermore, for communications operating in licensed bands, such spurious emissions are the only source of man-made noise. It lies in the nature of the license (for which the license holder usually has paid) that no other intentional emitters are allowed to operate in this band. In contrast to thermal noise, man-made noise is not necessarily Gaussian distributed. However, as a matter of convenience, most system-planning tools, as well as theoretical designs, assume *Gaussianity* anyway.

---

[1] Roughly speaking, "receive antenna gain" is a measure of how much more power we can receive (from a certain direction) by using a specific antenna, compared with the use of an isotropic antenna; the definition for transmit antennas is similar. See Chapter 9 and/or Stutzman and Thiele [1997] for details.

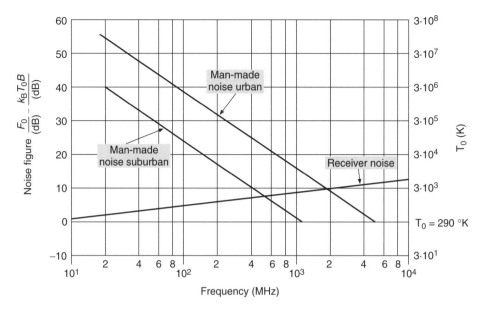

Figure 3.1 Noise as a function of frequency.
Reproduced with permission from Jakes [1974] © IEEE.

(b) *Other intentional emission sources*: Several wireless communications systems operate in unlicensed bands. In these bands, everybody is allowed to operate (emit electromagnetic radiation) as long as certain restrictions with respect to transmit power, etc. are fulfilled. The most important of these bands is the 2.45-GHz Industrial, Scientific, and Medical (ISM) band. The amount of interference in these bands can be considerable.

3. *Receiver noise*: The amplifiers and mixers in the RX are noisy, and thus increase the total noise power. This effect is described by the noise figure F, which is defined as the SNR at the RX input (typically after downconversion to baseband) divided by the SNR at the RX output. As the amplifiers have gain, noise added in the later stages does not have as much of an impact as noise added in the first stage of the RX. Mathematically, the total noise figure F_{eq} of a cascade of components is

$$F_{eq} = F_1 + \frac{F_2 - 1}{G_1} + \frac{F_3 - 1}{G_1 G_2} + \cdots \qquad (3.6)$$

where F_i and G_i are noise figures and noise gains of the individual stages in absolute units (not in decibels (dB)). Note that for this equation, passive components, like attenuators with gain $m < 1$, can be interpreted as *either* having a noise figure of $F = 1/m$ and unit gain of $G = 1$, *or* unit noise figure $F = 1$, and gain $G = m$.

For a digital system, the transmission quality is often described in terms of the *Bit Error Rate* (BER) probability. Depending on the modulation scheme, coding, and a range of other factors (discussed in Part III of this book), there is a relationship between SNR and BER for each digital communications systems. A minimum transmission quality can thus be linked to the minimum SNR, SNR_{min}, by this mapping (see Figure 3.2). Thus, the planning methods of all analog and digital links in noise-limited environments are the same; the goal is to determine the minimum

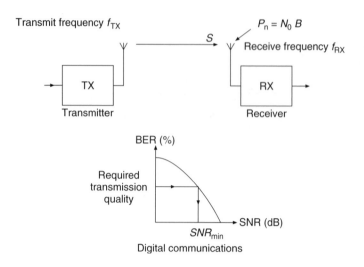

Figure 3.2 Noise-limited systems.
Reproduced with permission from Oehrvik [1994] © Ericsson AB.

signal power P_S:

$$P_S = SNR_{\min} + P_n \tag{3.7}$$

where all quantities are in dB. However, note that the actual *values* will be different for different systems.

3.2.1 Link Budget

A link budget is the clearest and most intuitive way of computing the required TX power. It tabulates all equations that connect the TX power to the received SNR. As most factors influencing the SNR enter in a multiplicative way, it is convenient to write all the equations in a logarithmic form – specifically, in dB. It has to be noted, however, that the link budget gives only an approximation (often a worst case estimate) for the total SNR, because some interactions between different effects are not taken into account.

Before showing some examples, the following points should be stressed:

- Chapters 4 and 7 provide extensive discussions of path loss, i.e., the attenuation due to propagation effects, between TX and RX. For the purpose of this chapter, we use a simple model, the so-called "breakpoint" model. For distances $d < d_{\text{break}}$, the received power is proportional to d^{-2}, according to Eq. (3.1). Beyond that point, the power is proportional to d^{-n}, where n typically lies between 3.5 and 4.5. The received power is thus

$$P_{RX}(d) = P_{RX}(d_{\text{break}}) \left(\frac{d}{d_{\text{break}}} \right)^{-n} \quad \text{for } d > d_{\text{break}} \tag{3.8}$$

- Wireless systems, especially mobile systems, suffer from temporal and spatial variations of the transmission channel (*fading*) (see Section 2.1). In other words, even if the distance is approximately constant, the received power can change significantly with small movements of the

Figure 3.3 Fading margin to guarantee a certain outage probability.

TX and/or RX. The power computed from Eq. (3.8) is only a *mean* value; the ratio of the transmit power to this mean received power is also known as the *path loss* (inverse of the path gain).

If the mean received power is used as the basis for the link budget, then the transmission quality will be above the threshold only in approximately 50% of the times and locations.[2] This is completely unacceptable quality of service. Therefore, we have to add a *fading margin*, which makes sure that the minimum received power is exceeded in at least, e.g., 90% of all cases (see Figure 3.3). The value of the fading margin depends on the amplitude statistics of the fading and is discussed in more detail in Chapter 5.

- Uplink (MS to BS) and downlink (BS to MS) are reciprocal, in the sense that the voltage and currents at the antenna ports are reciprocal (as long as uplink and downlink use the same carrier frequency). However, the noise figures of BSs and MSs are typically quite different. As MSs have to be produced in quantity, it is desirable to use low-cost components, which typically have higher noise figures. Furthermore, battery lifetime considerations dictate that BSs can emit more power than MSs. Finally, BSs and MSs differ with respect to antenna diversity, how close they are to interferers, etc. Thus, the link budgets of uplinks and downlinks are different.

Example 3.1 *Link budget*

Consider the downlink of a GSM system (see also Chapter 24). The carrier frequency is 950 MHz and the RX sensitivity is (according to GSM specifications) -102 dBm. The output power of the TX amplifier is 30 W. The antenna gain of the TX antenna is 10 dB and the aggregate attenuation of connectors, combiners, etc. is 5 dB. The fading margin is 12 dB and the breakpoint d_{break} is at a distance of 100 m. What distance can be covered?

TX side:

TX power	P_{TX}	30 W	45 dBm
Antenna gain	G_{TX}	10	10 dB
Losses (combiner, connector, etc.)	L_f		-5 dB

EIRP (Equivalent Isotropically Radiated Power)			50 dBm

RX side:

RX sensitivity	P_{min}		-102 dBm
Fading margin			12 dB
Minimum RX power (mean)			-90 dBm

Admissible path loss (difference EIRP and min. RX power)			140 dB
Path loss at $d_{\text{break}} = 100$ m	$[\lambda/(4\pi d)]^2$		72 dB
Path loss beyond breakpoint	$\propto d^{-n}$		68 dB

---

[2] It would lie above the threshold in exactly 50% of the cases if Eq. (3.8) represented the *median* power.

Depending on the path loss exponent,

$$n = 1.5 \ldots 2.5 \text{ (line-of-sight)}^3$$

$$n = 3.5 \ldots 4.5 \text{ (non-line-of-sight)}$$

we obtain the coverage distance,

$$d_{\text{cov}} = 100 \cdot 10^{68/(10n)} \text{m} \qquad (3.9)$$

If, e.g., $n = 3.5$, then the coverage distance is 8.8 km.

This example was particularly easy, because RX sensitivity was prescribed by the system specifications. If it is not available, the computations at the RX become more complicated, as shown in the next example.

Example 3.2 *Link budget*

Consider a mobile radio system at 900-MHz carrier frequency, and with 25-kHz bandwidth, that is affected only by thermal noise (temperature of the environment $T_e = 300$ K). Antenna gains at the TX and RX sides are 8 dB and -2 dB,[4] respectively. Losses in cables, combiners, etc. at the TX are 2 dB. The noise figure of the RX is 7 dB and the 3-dB bandwidth of the signal is 25 kHz. The required operating SNR is 18 dB and the desired range of coverage is 2 km. The breakpoint is at 10-m distance; beyond that point, the path loss exponent is 3.8, and the fading margin is 10 dB. What is the minimum TX power?

The way this problem is formulated makes working our way backward from the RX to the TX advantageous.

Noise spectral density	$k_B T_e$	-174 dBm/Hz
Bandwidth		44 dBHz
⋮		
Thermal noise power at the RX		-130 dBm
RX excess noise		7 dB
Required SNR		18 dB
⋮		
Required RX power		-105 dBm
Path loss from 10 m to 2-km distance	$(200^{3.8})$	87 dB
Path loss from TX to breakpoint at 10 m	$[\lambda/(4\pi d)]^2$	52 dB
Antenna gain at the MS G_{RX}	(2-dB loss)	$-(-2)$ dB
Fading margin		10 dB
Required EIRP		46 dBm

---

[3] Note that the Line-Of-Sight (LOS) cannot exist beyond a certain distance even in environments that have no buildings or hills. The curvature of the earth cuts off the LOS at a distance that depends on the heights of the BS and the MS.

[4] In most link budgets, the antenna gain for the MS is assumed to be 0 dB. However, recent measurements have shown that absorption and reflection by the head and body of the user reduce the antenna gain, leading to losses up to 10 dB. This is discussed in more detail in Chapter 9.

TX antenna gain G_{TX}	(8-dB gain)	−8 dB
Losses in cables, combiners, etc. at TX L_f		2 dB
Required TX power (amplifier output)		40 dBm

The required TX power is thus 40 dBm, or 10 W. The link budget is also represented in Figure 3.4.

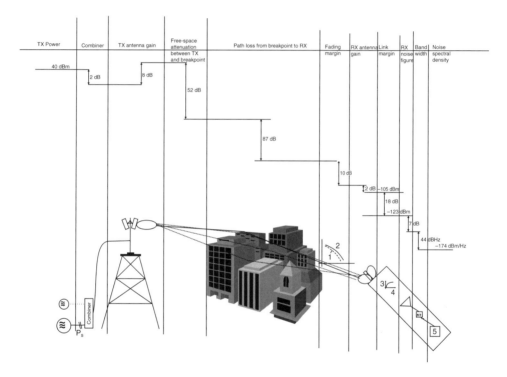

Figure 3.4 Link budget of Example 3.2. 1 = 10% decile; 2 = median; 3 = MOS; 4 = SNR; 5 = detector.

3.3 Interference-Limited Systems

Consider now the case that the interference is so strong that it completely dominates the performance, so that the noise can be neglected. Let a BS cover an area (cell) that is approximately described by a circle with radius R and center at the location of the BS. Furthermore, there is an interfering TX at distance D from the "desired" BS, which operates at the same frequency, and with the same transmit power. How large does D have to be in order to guarantee satisfactory transmission quality 90% of the time, assuming that the MS is at the cell boundary (worst case)? The computations follow the link budget computations of the previous section. As a first approximation, we treat the interference as Gaussian. This allows us to treat the interference as equivalent noise, and the minimum SIR, $SIR_{\min}$, takes on the same values as $SNR_{\min}$ in the noise-limited case.

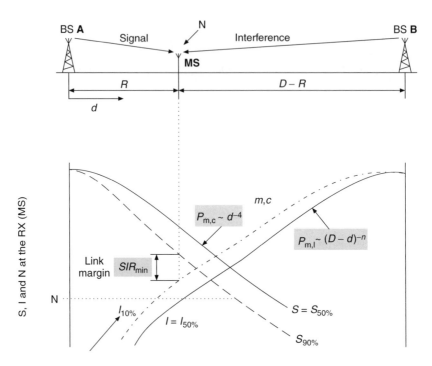

Figure 3.5 Relationship between cell radius and reuse distance. Solid lines: median values. Dashed lines: 90% decile of the desired signal. Dash–dotted lines: 10%-decile of the interfering signal.

Reproduced with permission from Oehrvik [1994] © Ericsson AB.

One difference between interference and noise lies in the fact that interference suffers from fading, while the noise power is typically constant (averaged over a short time interval). For determination of the fading margin, we thus have to account for the fact that (i) the desired signal is weaker than its median value during 50% of the time and (ii) the interfering signal is stronger than its median value 50% of the time. Mathematically speaking, the cumulative distribution function of the SIR is the probability that the ratio of two random variables is larger than a certain value in x% of all cases (where x is the percentage of locations in which transmission quality is satisfactory), see Chapter 5. As a first approximation, we can add the fading margin for the desired signal (i.e., the additional power we have to transmit to make sure that the desired signal level exceeds a certain value, x%, of the time, instead of 50%) and the fading margin of the interference –i.e., the power *reduction* to make sure that the interference exceeds a certain value only $(100 - x)$% of the time, instead of 50% of the time (see Figure 3.5). This results in an overestimation of the true fading margin. Therefore, if we use that value in system planning, we are on the safe side.

For updates and errata for this chapter, see wides.usc.edu/teaching/textbook

Part II

Wireless Propagation Channels

A wireless propagation channel is the medium linking the transmitter and the receiver. Its properties determine the information-theoretic capacity, i.e., the ultimate performance limit of wireless communications, and also determine how specific wireless systems behave. It is thus essential that we know and understand the wireless propagation channel and apply this knowledge to system design. Part II of this book, consisting of Chapters 4–9, is intended to provide such an understanding.

Wireless channels differ from wired channels by *multipath propagation* (see also Chapter 2), i.e., the existence of a multitude of propagation paths from transmitter and receiver, where the signal can be reflected, diffracted, or scattered along its way. One way to understand the channel is to consider all those different *propagation phenomena*, and how they impact each *Multi Path Component* (MPC). The propagation phenomena will be at the center of Chapter 4; Section 7.5 will explain how to apply this knowledge to deterministic channel models and prediction (ray tracing).

The alternative is a more phenomenological view. We consider important channel parameters, like received power, and analyze their statistics. In other words, we do not care how the channel looks in a specific location, or how it is influenced by specific MPCs. Rather, we are interested in describing the *probability* that a channel parameter attains a certain value. The parameter of highest interest is the received power or field strength; it determines the behavior of narrowband systems in both noise- and interference-limited systems (see Chapter 3). We will find that the received power *on average* decreases with distance. The physical reasons for this decrease are described in Sections 4.1 and 4.2; models are detailed in Section 7.1. However, there are variations around that mean that can be modeled stochastically. These stochastic variations are described in detail in Chapter 5.

The interference of the different MPCs creates not only fading (i.e., variations of the received power with time and/or place), but also delay dispersion. Delay dispersion means that if we transmit a signal of duration T, the received signal has a *longer* duration T'. Naturally, this leads to *Inter Symbol Interference* (ISI). While this effect was not relevant for earlier, analog, systems, it is very important for digital systems like Global System for Mobile communications (GSM). Third- and fourth-generation cellular systems and wireless LANs are even more influenced by the delay dispersion, and possibly also by angular dispersion, of the wireless channel. This requires the introduction of new parameters and description methods to quantify those characteristics, as described in Chapter 6. Sections 7.3 and 7.4 describe typical values for those parameters in outdoor and indoor environments.

For both the understanding of the propagation phenomena, and for the stochastic description of the channel, we need measurements. Chapter 8 describes the measurement equipment, and how the output from that equipment needs to be processed. While measuring the received power is fairly straightforward, finding the delay dispersion and angular characteristics of the channel is much more involved. Finally, Chapter 9 describes antennas for wireless channels. The antennas represent the interface between the transceivers and the propagation channel, and determine how the signal is sent out into the propagation channel, and collected from it.

4

Propagation Mechanisms

In this chapter, we describe the basic mechanisms that govern the propagation of electromagnetic waves, emphasizing those aspects that are relevant for wireless communications. The simplest case is free space propagation – in other words, one transmit and one receive antenna in free space. In a more realistic scenario, there are dielectric and conducting obstacles (*Interacting Objects* (IOs)). If these IOs have a *smooth* surface, waves are *reflected* and a part of the energy penetrates the IO (*transmission*). If the surfaces are *rough*, the waves are diffusely *scattered*. Finally, waves can also be *diffracted* at the edges of the IOs. These effects will now be discussed one after the other.[1]

4.1 Free Space Attenuation

We start with the simplest possible scenario: a transmit and a receive antenna in free space, and derive the received power as a function of distance.

Energy conservation dictates that the integral of the power density over any closed surface surrounding the transmit antenna must be equal to the transmitted power. If the closed surface is a sphere of radius d, centered at the transmitter (TX) antenna, and if the TX antenna radiates isotropically, then the power density on the surface is $P_{TX}/(4\pi d^2)$. The receiver (RX) antenna has an "effective area" A_{RX}. We can envision that all power impinging on that area is collected by the RX antenna [Stutzman and Thiele 1997]. Then the received power is given by:

$$P_{RX}(d) = P_{TX}\frac{1}{4\pi d^2}A_{RX}$$

If the transmit antenna is not isotropic, then the energy density has to be multiplied with the antenna gain G_{TX} in the direction of the receive antenna.[2]

$$P_{RX}(d) = P_{TX}G_{TX}\frac{1}{4\pi d^2}A_{RX} \tag{4.1}$$

The product of transmit power and gain in the considered direction is also known as *Equivalent Isotropically Radiated Power* (EIRP).

The effective antenna area is proportional to the power that can be extracted from the antenna connectors for a given energy density. For a parabolic antenna, e.g., the effective antenna area is roughly the geometrical area of the surface. However, antennas with very small geometrical area – e.g., dipole antennas – can also have a considerable effective area.

---

[1] In the literature, the IOs are often called "scatterers," even if the interaction process is not scattering.
[2] If not stated otherwise, this book always defines the antenna gain as the gain over an isotropic radiator. For further discussions of antenna properties, see Chapter 9.

Wireless Communications, Second Edition   Andreas F. Molisch
© 2011 John Wiley & Sons, Ltd

It can be shown that there is a simple relationship between effective area and antenna gain [Stutzman and Thiele 1997]:

$$G_{\mathrm{RX}} = \frac{4\pi}{\lambda^2} A_{\mathrm{RX}} \qquad (4.2)$$

Most noteworthy in this equation is the fact that – for a fixed antenna area – the antenna gain increases with frequency. This is also intuitive, as the directivity of an antenna is determined by its size in terms of wavelengths.

Substituting Eq. (4.2) into (4.1) gives the received power P_{RX} as a function of the distance d in free space, also known as *Friis' law*:

$$P_{\mathrm{RX}}(d) = P_{\mathrm{TX}} G_{\mathrm{TX}} G_{\mathrm{RX}} \left(\frac{\lambda}{4\pi d} \right)^2 \qquad (4.3)$$

The factor $(\lambda/4\pi d)^2$ is also known as the *free space loss factor*.

Friis' law seems to indicate that the "attenuation" in free space increases with frequency. This is counterintuitive, as the energy is not lost, but rather redistributed over a sphere surface of area $4\pi d^2$. This mechanism has to be independent of the wavelength. This seeming contradiction is caused by the fact that we assume the *antenna gain* to be independent of the wavelength. If we assume, on the other hand, that the effective *antenna area* of the RX antenna is independent of frequency, then the received power becomes independent of the frequency (see Eq. (4.1)). For wireless systems, it is often useful to assume constant gain, as different systems (e.g., operating at 900 and 1800 MHz) use the same antenna *type* (e.g., $\lambda/2$ dipole or monopole), and not the same antenna *size*.

The validity of Friis' law is restricted to the far field of the antenna – i.e., the TX and RX antennas have to be at least one *Rayleigh distance* apart. The Rayleigh distance (also known as the *Fraunhofer distance*) is defined as:

$$d_{\mathrm{R}} = \frac{2L_{\mathrm{a}}^2}{\lambda} \qquad (4.4)$$

where L_{a} is the largest dimension of the antenna; furthermore, the far field requires $d \gg \lambda$ and $d \gg L_{\mathrm{a}}$.

Example 4.1   *Compute the Rayleigh distance of a square antenna with 20-dB gain.*

The gain is 100 on a linear scale. In that case, the effective area is approximately

$$A_{\mathrm{RX}} = \frac{\lambda^2}{4\pi} G_{\mathrm{RX}} = 8\lambda^2 \qquad (4.5)$$

For a square-shaped antenna with $A_{\mathrm{RX}} = L_{\mathrm{a}}^2$, the Rayleigh distance is given by:

$$d_{\mathrm{R}} = \frac{2 \cdot 8\lambda^2}{\lambda} = 16\lambda \qquad (4.6)$$

For setting up link budgets, it is advantageous to write Friis' law on a logarithmic scale. Equation (4.3) then reads

$$P_{\mathrm{RX}}|_{\mathrm{dBm}} = P_{\mathrm{TX}}|_{\mathrm{dBm}} + G_{\mathrm{TX}}|_{\mathrm{dB}} + G_{\mathrm{RX}}|_{\mathrm{dB}} + 20\log\left(\frac{\lambda}{4\pi d} \right) \qquad (4.7)$$

where $|_{dB}$ means "in units of dB." In order to better point out the distance dependence, it is advantageous to first compute the received power at 1-m distance:

$$P_{RX}(1m) = P_{TX}|_{dBm} + G_{TX}|_{dB} + G_{RX}|_{dB} + 20 \log \left(\frac{\lambda|_m}{4\pi \cdot 1} \right) \qquad (4.8)$$

The last term on the r.h.s. of Eq. (4.8) is about -32 dB at 900 MHz and -38 dB at 1800 MHz. The actual received power at a distance d (in meters) is then:

$$P_{RX}(d) = P_{RX}(1m) - 20 \log(d|_m) \qquad (4.9)$$

4.2   Reflection and Transmission

4.2.1   Snell's Law

Electromagnetic waves are often reflected at one or more IOs before arriving at the RX. The reflection coefficient of the IO, as well as the direction into which this reflection occurs, determines the power that arrives at the RX position. In this section, we deal with *specular reflections*. This type of reflection occurs when waves are incident onto smooth, large (compared with the wavelength) objects. A related mechanism is the *transmission* of waves – i.e., the penetration of waves into and through an IO. Transmission is especially important for wave propagation inside buildings. If the Base Station (BS) is either outside the building, or in a different room, then the waves have to penetrate a wall (dielectric layer) in order to get to the RX.

We now derive the reflection and transmission coefficients of a homogeneous plane wave incident onto a dielectric halfspace. The dielectric material is characterized by its dielectric constant $\varepsilon = \varepsilon_0 \varepsilon_r$ (where ε_0 is the vacuum dielectric constant $8.854 \cdot 10^{-12}$ Farad/m, and ε_r is the relative dielectric constant of the material) and conductivity σ_e. Furthermore, we assume that the material is isotropic and has a relative permeability $\mu_r = 1$.[3] The dielectric constant and conductivity can be merged into a single parameter, the *complex* dielectric constant:

$$\delta = \varepsilon_0 \delta_r = \varepsilon - j \frac{\sigma_e}{2\pi f_c} \qquad (4.10)$$

where f_c is the carrier frequency, and j is the imaginary unit. Though this definition is strictly valid only for a single frequency, it can actually be used for all *narrowband* systems, where the bandwidth is much smaller than the carrier frequency, as well as much smaller than the bandwidth over which the quantities σ_e and ε vary significantly.[4]

The plane wave is incident on the halfspace at an angle Θ_e, which is defined as the angle between the wave vector $\mathbf{k}$ and the unit vector that is orthogonal to the dielectric boundary. We have to distinguish between the *Transversal Magnetic* (TM) case, where the magnetic field component is parallel to the boundary between the two dielectrics, and the *Transversal Electric* (TE) case, where the electric field component is parallel (see Figure 4.1).

The reflection and transmission coefficients can now be computed from postulating incident, reflected, and transmitted plane waves, and enforcing the continuity conditions at the boundary – see, e.g., Ramo et al. [1967]. From these considerations, we obtain Snell's law: the angle of incidence is the same as the reflected angle:

$$\Theta_r = \Theta_e \qquad (4.11)$$

---

[3] This is approximately true for most materials influencing mobile radio wave propagation.
[4] Note that this means "narrowband" in the Radio Frequency (RF) sense. We will later encounter a different definition for narrowband that is related to delay dispersion in wireless channels.

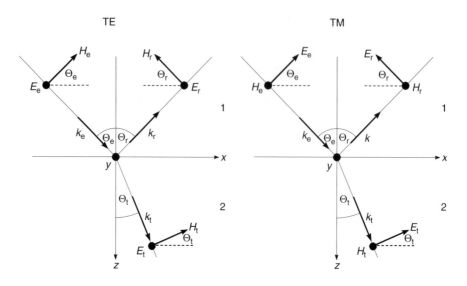

Figure 4.1 Reflection and transmission.

and the angle of the transmitted wave is given by:

$$\frac{\sin \Theta_t}{\sin \Theta_e} = \frac{\sqrt{\delta_1}}{\sqrt{\delta_2}} \tag{4.12}$$

where subscripts 1 and 2 index the considered medium.

The reflection and transmission coefficients are different for TE and for TM waves. For TM polarization:

$$\rho_{TM} = \frac{\sqrt{\delta_2}\cos\Theta_e - \sqrt{\delta_1}\cos(\Theta_t)}{\sqrt{\delta_2}\cos\Theta_e + \sqrt{\delta_1}\cos(\Theta_t)} \tag{4.13}$$

$$T_{TM} = \frac{2\sqrt{\delta_1}\cos(\Theta_e)}{\sqrt{\delta_2}\cos\Theta_e + \sqrt{\delta_1}\cos(\Theta_t)} \tag{4.14}$$

and for TE polarization:

$$\rho_{TE} = \frac{\sqrt{\delta_1}\cos(\Theta_e) - \sqrt{\delta_2}\cos(\Theta_t)}{\sqrt{\delta_1}\cos(\Theta_e) + \sqrt{\delta_2}\cos(\Theta_t)} \tag{4.15}$$

$$T_{TE} = \frac{2\sqrt{\delta_1}\cos(\Theta_e)}{\sqrt{\delta_1}\cos(\Theta_e) + \sqrt{\delta_2}\cos(\Theta_t)} \tag{4.16}$$

where $\rho_{TM} = E_r/E_e$, and $T_{TM} = E_t/E_e$ (and similarly for ρ_{TE} and T_{TE}). Note that the reflection coefficient has both an amplitude and a phase. Figure 4.2 shows both for a dielectric material with complex dielectric constant $\delta = 4 - 0.25j$. It is noteworthy that for both TE and TM waves, the reflection coefficient becomes -1 (magnitude 1, phase shift of $180°$) at grazing incidence ($\Theta_e \to 90°$). This is the same reflection coefficient that would occur for reflection on an ideally conducting surface. We will see later on that this has important consequences for the impact of ground-reflected waves in wireless systems.

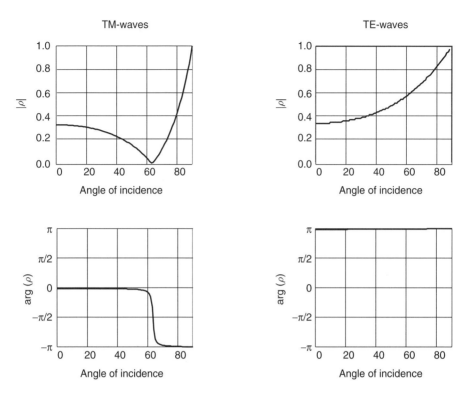

Figure 4.2 Reflection coefficient for a dielectric material with complex dielectricity constant $\delta = (4 - 0.25j)\varepsilon_0$.

In highly lossy materials, the transmitted wave is not a homogeneous plane wave, so that Snell's law is not applicable anymore. Instead, there is a guided wave at the dielectric boundary. However, these considerations have more theoretical than practical use.

4.2.2 Reflection and Transmission for Layered Dielectric Structures

The previous section discussed the reflection and transmission in a dielectric halfspace. This is of interest, e.g., for ground reflections and reflections by terrain features, like mountains. A related problem is the problem of *transmission through* a dielectric layer. It occurs when a user inside a building is communicating with an outdoor BS, or in a picocell where the Mobile Station (MS) and the BS are in different rooms. In that case, we are interested in the attenuation and phase shift of a wave transmitted through a wall. Fortunately, the basic problem of dielectric layers is well known from other areas of electrical engineering – e.g., optical thin film technology [Heavens 1965], and the results can be easily applied to wireless communications.

The most simple, and practically most important, case occurs when a dielectric layer is surrounded on both sides by air. The reflection and transmission coefficients can be determined by summation of the partial waves, resulting in a total transmission coefficient:

$$T = \frac{T_1 T_2 e^{-j\alpha}}{1 + \rho_1 \rho_2 e^{-2j\alpha}} \tag{4.17}$$

and a reflection coefficient:

$$\rho = \frac{\rho_1 + \rho_2 e^{-j2\alpha}}{1 + \rho_1 \rho_2 e^{-2j\alpha}} \tag{4.18}$$

where T_1 is the transmission coefficient of a wave from air into a dielectric halfspace (with the same dielectric properties as the considered layer) and T_2 is the transmission coefficient from dielectric into air; they can be computed from the results of Section 4.2.1. The quantity α is the electrical length of the dielectric as seen by waves that are at an angle Θ_t with the layer:

$$\alpha = \frac{2\pi}{\lambda} \sqrt{\varepsilon_{r,2}} d_{\text{layer}} \cos(\Theta_t) \tag{4.19}$$

where d_{layer} is the geometrical length of the layer. Note also that there is a waveguiding effect in lossy materials (see discussion in the previous section), so that the results of this section are not strictly applicable for dielectrics with losses.

In *multilayer* structures, the problem becomes considerably more complicated [Heavens 1965]. However, in practice, even multilayer structures are described by "effective" dielectric constants or reflection/transmission constants. These are measured directly for the composite structure. The alternative of measuring the dielectric properties for each layer separately and computing the resulting effective dielectric constant is prone to errors, as the measurement errors for the different layers add up.

Example 4.2 *Compute the effective ρ and T for a 50-cm-thick brick wall at 4-GHz carrier frequency for perpendicularly incident waves.*

Since $\Theta_e = 0$, Eqs. (4.11) and (4.12) imply that also, $\Theta_r = \Theta_t = 0$. At $f = 4\,\text{GHz}$ ($\lambda = 7.5\,\text{cm}$), brick has a relative permittivity ε_r of 4.44 [Rappaport 1996]; we neglect the conductivity. With the air having $\varepsilon_{\text{air}} = \varepsilon_1 = \varepsilon_0$, Eqs. (4.13)–(4.16) give the reflection and transmission coefficients for the surface between air and brick as:

$$\left. \begin{aligned}
\rho_{1,\text{TM}} &= \frac{\sqrt{\varepsilon_2} - \sqrt{\varepsilon_1}}{\sqrt{\varepsilon_2} + \sqrt{\varepsilon_1}} = 0.36 \\[2ex]
\rho_{1,\text{TE}} &= \frac{\sqrt{\varepsilon_1} - \sqrt{\varepsilon_2}}{\sqrt{\varepsilon_2} + \sqrt{\varepsilon_1}} = -0.36 \\[2ex]
T_{1,\text{TM}} &= \frac{2\sqrt{\varepsilon_1}}{\sqrt{\varepsilon_2} + \sqrt{\varepsilon_1}} = 0.64 \\[2ex]
T_{1,\text{TE}} &= \frac{2\sqrt{\varepsilon_1}}{\sqrt{\varepsilon_2} + \sqrt{\varepsilon_1}} = 0.64
\end{aligned} \right\} \tag{4.20}$$

and between brick and air as:

$$\left. \begin{aligned}
\rho_{2,\text{TM}} &= \rho_{1,\text{TE}} \\[1ex]
\rho_{2,\text{TE}} &= \rho_{1,\text{TM}} \\[1ex]
T_{2,\text{TM}} &= T_{1,\text{TE}} \cdot \sqrt{\varepsilon_2/\varepsilon_1} = 1.36 \\[1ex]
T_{2,\text{TE}} &= T_{1,\text{TM}} \cdot \sqrt{\varepsilon_2/\varepsilon_1} = 1.36
\end{aligned} \right\} \tag{4.21}$$

Note that the transmission coefficient can become larger than unity (e.g., $T_{2,\text{TE}}$). This is not a violation of energy conservation: the transmission coefficient is defined as the ratio of the amplitudes of the incident and reflected field. Energy conservation only dictates that the flux (energy) density of a reflected and transmitted field equals that of the incident field.

The electrical length of the wall, α, is determined using Eq. (4.19) as:

$$\alpha = \frac{2\pi}{\lambda}\sqrt{\varepsilon_r}\, d = \frac{2\pi}{0.075}\sqrt{4.44}\cdot 0.5 = 88.26 \tag{4.22}$$

Finally, the total reflection and transmission coefficients can be determined using Eqs. (4.17), (4.18), which gives:

$$
\left.
\begin{aligned}
T_{\text{TM}} &= \frac{T_{1,\text{TM}}T_{2,\text{TM}}e^{-j\alpha}}{1+\rho_{1,\text{TM}}\rho_{2,\text{TM}}e^{-2j\alpha}} = \frac{0.64\cdot 1.356e^{-j88.26}}{1-0.36^2 e^{-j176.53}} = 0.90-0.36j \\[2mm]
\rho_{\text{TM}} &= \frac{\rho_{1,\text{TM}}+\rho_{2,\text{TM}}e^{-2j\alpha}}{1+\rho_{1,\text{TM}}\rho_{2,\text{TM}}e^{-2j\alpha}} = \frac{0.36(1-e^{-j176.53})}{1-0.36^2 e^{-j176.53}} = 0.086+0.22j \\[2mm]
T_{\text{TE}} &= \frac{T_{1,\text{TE}}T_{2,\text{TE}}e^{-j\alpha}}{1+\rho_{1,\text{TE}}\rho_{2,\text{TE}}e^{-2j\alpha}} = \frac{T_{2,\text{TM}}T_{1,\text{TM}}e^{-j\alpha}}{1+\rho_{2,\text{TM}}\rho_{1,\text{TM}}e^{-2j\alpha}} = T_{\text{TM}} \\[2mm]
\rho_{\text{TE}} &= \frac{\rho_{1,\text{TE}}+\rho_{2,\text{TE}}e^{-2j\alpha}}{1+\rho_{1,\text{TE}}\rho_{2,\text{TE}}e^{-2j\alpha}} = \frac{\rho_{2,\text{TM}}+\rho_{1,\text{TM}}e^{-2j\alpha}}{1+\rho_{2,\text{TM}}\rho_{1,\text{TM}}e^{-2j\alpha}} = \frac{-\rho_{1,\text{TM}}-\rho_{2,\text{TM}}e^{-2j\alpha}}{1+\rho_{2,\text{TM}}\rho_{1,\text{TM}}e^{-2j\alpha}} = -\rho_{\text{TM}}
\end{aligned}
\right\}
\tag{4.23}
$$

It is easily verified that in both cases $|\rho|^2 + |T|^2 = 1$.

4.2.3 The d^{-4} Power Law

One of the "folk laws" of wireless communications says that the received signal power is inversely proportional to the *fourth* power of the distance between TX and RX. This law is often justified by computing the received power for the case that only a direct (Line Of Sight, LOS) wave, plus a ground-reflected wave, exists. For this specific case, the following equation is derived in Appendix 4.A (see *www.wiley.com/go/molisch*):

$$P_{\text{RX}}(d) \approx P_{\text{TX}} G_{\text{TX}} G_{\text{RX}} \left(\frac{h_{\text{TX}} h_{\text{RX}}}{d^2}\right)^2 \tag{4.24}$$

where h_{TX} and h_{RX} are the height of the transmit and the receive antenna, respectively; it is valid for distances larger than:

$$d_{\text{break}} \gtrsim \frac{4 h_{\text{TX}} h_{\text{RX}}}{\lambda} \tag{4.25}$$

This equation, which replaces the standard Friis' law, implies that the received power becomes independent of frequency. Furthermore, it follows from Eq. (4.24) that the received power increases with the square of the height of both BS and MS. For distances $d < d_{\text{break}}$, Friis' law remains approximately valid.

For the link budget, it is useful to rewrite the power law on a logarithmic scale. Assuming that the power decays as d^{-2} until a breakpoint d_{break}, and from there with d^{-n}, then the received power is given by (see also Chapter 3):

$$P_{RX}(d) = P_{RX}(1 \text{ m}) - 20 \log(d_{break}|m) - n10 \log(d/d_{break}) \qquad (4.26)$$

Figure 4.3 shows the received power when there is a direct wave and a ground-reflected wave, both from the exact formulation (see Appendix 4.A) and from Eq. (4.24). We find that the transition between the attenuation coefficient $n = 2$ and $n = 4$ is actually not a sharp breakpoint, but rather smooth. According to Eq. (4.25), the breakpoint is at $d = 90$ m; this seems to be approximated reasonably well in the plot.

The equations derived above (and in Appendix 4.A) are self-consistent, but it has to be emphasized that they are not a *universal* description of wireless channels. After all, propagation does not always happen over a flat Earth. Rather, multiple propagation paths are possible, LOS connections are often shadowed by IOs, etc. Thus the derived theory does not agree with the measurement results in realistic channels in several respects:

1. $n = 4$ is not a universally valid decay exponent at larger distances. Rather, values between $1.5 < n < 5.5$ have been measured, and the actual value strongly depends on the surrounding environment. The value of $n = 4$ is, at best, a mean value of various environments.
2. The transition between $n = 2$ and $n = 4$ almost never occurs at d_{break} predicted by Eq. (4.25).
3. Measurements also show that there is a second breakpoint, beyond which an $n > 6$ exponent is valid. This effect is not predicted at all by the above model. For some situations, the effect can be explained by the radio horizon (i.e., the curvature of the Earth), which is not included in the model above [Parsons 1992].

Figure 4.3 Propagation over an ideally reflecting ground. Height of BS: 5 m. Height of MS: 1.5 m.

4.3   Diffraction

All the equations derived up to now deal with infinitely extended IOs. However, real IOs like buildings, cars, etc. have a finite extent. A finite-sized object does not create sharp shadows (the way geometrical optics would have it), but rather there is diffraction due to the wave nature of electromagnetic radiation. Only in the limit of very small wavelength (large frequency) does geometrical optics become exact.

In the following, we first treat two canonical diffraction problems: diffraction of a homogeneous plane wave (i) by a knife edge or screen and (ii) by a wedge, and derive the diffraction coefficients that tell us how much power can be received in the shadow region behind an obstacle. Subsequently, we consider the effect of a concatenation of several screens.

4.3.1 Diffraction by a Single Screen or Wedge

The Diffraction Coefficient

The simplest diffraction problem is the diffraction of a homogeneous plane wave by a semi-infinite screen, as sketched in Figure 4.4. Diffraction can be understood from Huygen's principle that each point of a wavefront can be considered the source of a spherical wave. For a homogeneous plane wave, the superposition of these spherical waves results in another homogeneous plane wave, see transition from plane A' to B'. If, however, the screen eliminates parts of the point sources (and their associated spherical waves), the resulting wavefront is not plane anymore, see the transition from plane B' to C'. Constructive and destructive interferences occur in different directions.[5]

Figure 4.4 Huygen's principle.

The electric field at any point to the right of the screen ($x \geq 0$) can be expressed in a form that involves only a standard integral, the *Fresnel integral*. With the incident field represented as $\exp(-jk_0 x)$, the total field becomes [Vaughan and Andersen 2003]:

$$E_{\text{total}} = \exp(-jk_0 x)\left(\frac{1}{2} - \frac{\exp(j\pi/4)}{\sqrt{2}} F(v_{\text{F}})\right) = \exp(-jk_0 x)\tilde{F}(v_{\text{F}}) \qquad (4.27)$$

---

[5] For more accurate considerations, it is noteworthy that Huygen's principle is not exact. A derivation from Maxwell's equations, which also includes a discussion of the necessary assumptions, is given, e.g., in Marcuse [1991].

where $\nu_F = -2y/\sqrt{\lambda x}$ and the Fresnel integral $F(\nu_F)$ is defined as:

$$F(\nu_F) = \int_0^{\nu_F} \exp\left(-j\pi \frac{t^2}{2}\right) dt \tag{4.28}$$

Figure 4.5 plots this function. It is interesting that $\tilde{F}(\nu_F)$ can become larger than unity for some values of ν_F. This implies that the received power at a specific location can actually be *increased* by the presence of screen. Huygen's principle again provides the explanation: some spherical waves that would normally interfere destructively in a specific location are blocked off. However, note that the *total* energy (integrated over the whole wavefront) *cannot* be increased by the screen.

Consider now the more general geometry of Figure 4.6. The TX is at height h_{TX}, the RX at h_{RX}, and the screen extends from $-\infty$ to h_s. The diffraction angle θ_d is thus:

$$\theta_d = \arctan\left(\frac{h_s - h_{TX}}{d_{TX}}\right) + \arctan\left(\frac{h_s - h_{RX}}{d_{RX}}\right) \tag{4.29}$$

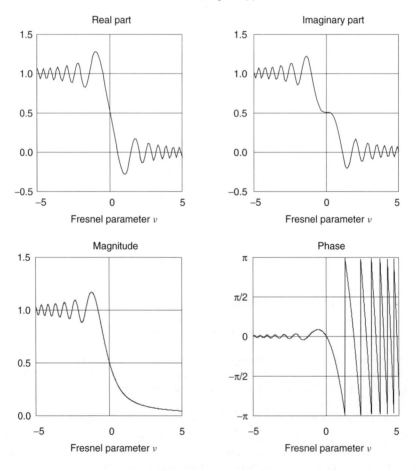

Fresnel integral $\tilde{F}(\nu)$

Figure 4.5   Fresnel integral.

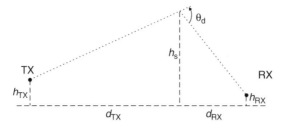

Figure 4.6 Geometry for the computation of the Fresnel parameters.

and the Fresnel parameter v_F can be obtained from θ_d as:

$$v_F = \theta_d \sqrt{\frac{2d_{TX}d_{RX}}{\lambda(d_{TX} + d_{RX})}} \qquad (4.30)$$

The field strength can again be computed from Eq. (4.27), just using the Fresnel parameter from Eq. (4.30).

Note that the result given above is approximate in the sense that it neglects the polarization of the incident field. More accurate equations for both the TE and the TM case can be found in Bowman et al. [1987].

Example 4.3 *Consider diffraction by a screen with $d_{TX} = 200\,m$, $d_{RX} = 50\,m$, $h_{TX} = 20\,m$, $h_{RX} = 1.5\,m$, $h_s = 40\,m$, at a center frequency of 900 MHz. Compute the diffraction coefficient.*

A center frequency of 900 MHz implies a wavelength $\lambda = 1/3$ m. Computing the diffraction angle θ_d from Eq. (4.29) gives:

$$\theta_d = \arctan\left(\frac{h_s - h_{TX}}{d_{TX}}\right) + \arctan\left(\frac{h_s - h_{RX}}{d_{RX}}\right)$$

$$= \arctan\left(\frac{40 - 20}{200}\right) + \arctan\left(\frac{40 - 1.5}{50}\right) = 0.756\,\text{rad} \qquad (4.31)$$

Then, the Fresnel parameter is given by Eq. (4.30) as:

$$v_F = \theta_d\sqrt{\frac{2d_{TX}d_{RX}}{\lambda(d_{TX} + d_{RX})}} = 0.756\sqrt{\frac{2 \cdot 200 \cdot 50}{1/3 \cdot (200 + 50)}} = 11.71 \qquad (4.32)$$

Evaluation of Eq. (4.28), with MATLAB or Abramowitz and Stegun [1965], yields

$$F(11.71) = \int_0^{v_F} \exp\left(-j\pi\frac{t^2}{2}\right)dt \approx 0.527 - j0.505 \qquad (4.33)$$

Finally, Eq. (4.27) gives the total received field as:

$$E_{\text{total}} = \exp(-jk_0x)\left(\frac{1}{2} - \frac{\exp(j\pi/4)}{\sqrt{2}}F(11.71)\right)$$

$$= \exp(-jk_0x)\left(\frac{1}{2} - \frac{\exp(j\pi/4)}{\sqrt{2}}(0.527 - j0.505)\right)$$

$$= (-0.016 - j0.011)\exp(-jk_0x) \qquad (4.34)$$

Fresnel Zones

The impact of an obstacle can also be assessed qualitatively, and intuitively, by the concept of *Fresnel zones*. Figure 4.7 shows the basic principle. Draw an ellipsoid whose foci are the BS and the MS locations. According to the definition of an ellipsoid, all rays that are reflected at points on this ellipsoid have the same run length (equivalent to runtime). The eccentricity of the ellipsoid determines the extra run length compared with the LOS – i.e., the direct connection between the two foci. Ellipsoids where this extra distance is an integer multiple of $\lambda/2$ are called "Fresnel ellipsoids." Now extra run length also leads to an additional phase shift, so that the ellipsoids can be described by the phase shift that they cause. More specifically, the ith Fresnel ellipsoid is the one that results in a phase shift of $i \cdot \pi$.

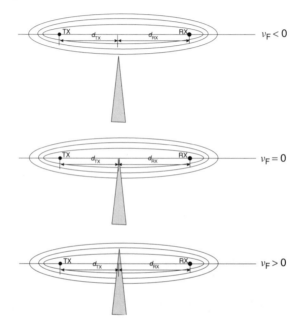

Figure 4.7   The principle of Fresnel ellipsoids.

Fresnel zones can also be used for explanation of the d^{-4} law. The propagation follows a free space law up to the distance where the first Fresnel ellipsoid touches the ground. At this distance, which is the breakpoint distance, the phase difference between the direct and the reflected ray becomes π.

Diffraction by a Wedge

The semi-infinite absorbing screen is a useful tool for the explanation of diffraction, since it is the simplest possible configuration. However, many obstacles especially in urban environments are much better represented by a wedge structure, as sketched in Figure 4.8. The problem of diffraction by a wedge has been treated for some 100 years, and is still an area of active research. Depending on the boundary conditions, solutions can be derived that are either valid at arbitrary observation points or approximate solutions that are only valid in the far field (i.e., far away from the wedge). These latter solutions are usually much simpler, and will thus be the only ones considered here.

The part of the field that is created by diffraction can be written as the product of the incident field with a phase factor $\exp(-jk_0 d_{RX})$, a geometry factor $A(d_{TX}, d_{RX})$ that depends only on the distance of TX and RX from the wedge, and the diffraction coefficient $D(\phi_{TX}, \phi_{RX})$ that depends on the diffraction angles [Vaughan and Andersen 2003]:

$$E_{\text{diff}} = E_{\text{inc},0} D(\phi_{TX}, \phi_{RX}) A(d_{TX}, d_{RX}) \exp(-jk_0 d_{RX}) \qquad (4.35)$$

The diffracted field has to be added to the field as computed by geometrical optics.[6]

The definition of the geometry parameters is shown in Figure 4.8. The geometry factor is given by:

$$A(d_{TX}, d_{RX}) = \sqrt{\frac{d_{TX}}{d_{RX}(d_{TX} + d_{RX})}} \qquad (4.36)$$

The diffraction coefficient D depends on the boundary conditions – namely, the reflection coefficients ρ_{TX} and ρ_{RX}. Explicit equations are given in Appendix 4.B (see *www.wiley.com/go/molisch*).

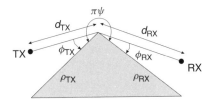

Figure 4.8   Geometry for wedge diffraction.

4.3.2   *Diffraction by Multiple Screens*

Diffraction by a single screen is a problem that has been widely studied, because it is amenable to closed-form mathematical treatment, and forms the basis for the treatment of more complex problems. However, in practice, we usually encounter situations where *multiple IOs* are located between TX and RX. Such a situation occurs, e.g., for propagation over the rooftops of an urban environment. As we see in Figure 4.9, such a situation can be well approximated by diffraction by multiple screens. Unfortunately, diffraction by multiple screens is an extremely challenging mathematical problem, and – except for a few special cases – no exact solutions are available. Still, a wealth of approximate methods has been proposed in the literature, of which we give an overview in the remainder of this section.

Bullington's Method

Bullington's method replaces the multiple screens by a single, "equivalent" screen. This equivalent screen is derived in the following way: put a tangential straight line from the TX to the real obstacles, and select the steepest one (i.e., the one with the largest elevation angle), so that all obstacles either touch this tangent, or lie below it. Similarly, take the tangents from the RX to the obstacles, and select the steepest one. The equivalent screen is then determined by the intersection of the steepest TX tangent and the steepest RX tangent (see Figure 4.10). The field resulting from diffraction at this single screen can be computed according to Section 4.3.1.

---

[6] If the field incident on the wedge can be written as $E_{\text{inc},0} = E_0 \exp(-jk_0 d_{TX})/d_{TX}$, the above equations become completely symmetrical with respect to d_{TX} and d_{RX}.

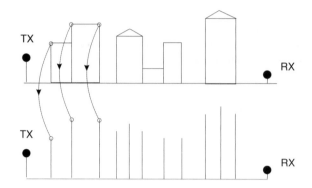

Figure 4.9 Approximation of multiple buildings by a series of screens.

Equivalent screen

Figure 4.10 Equivalent screen after Bullington.
Reproduced with permission from Parsons [1992] © J. Wiley & Sons, Ltd.

The major attraction of Bullington's method is its simplicity. However, this simplicity also leads to considerable inaccuracies. Most of the physically existing screens do not impact the location of the equivalent screen. Even the highest obstacle might not have an impact. Consider Figure 4.10: if the highest obstacle lies between screens 01 and 02, it could lie below the tangential lines, and thus not influence the "equivalent" screen, even though it is higher than either screen 01 or screen 02. In reality, these high obstacles *do* have an effect on propagation loss, and cause an additional attenuation. The Bullington method thus tends to give optimistic predictions of the received power.

The Epstein–Petersen Method

The low accuracy of the Bullington method is due to the fact that only two obstacles determine the equivalent screen, and thus the total diffraction coefficient. This problem can be somewhat mitigated by the Epstein–Petersen method [Epstein and Peterson 1953]. This approach computes the diffraction losses for each screen separately. The attenuation of a specific screen is computed by putting a virtual "TX" and "RX" on the tips of the screens to the left and right of this considered screen (see Figure 4.11). The diffraction coefficient, and the attenuation, of this one screen can be

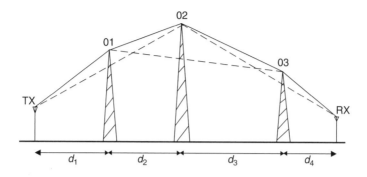

Figure 4.11 The Epstein–Petersen method.
Reproduced with permission from Parsons [1992] © J. Wiley & Sons, Ltd.

easily computed from the principles of Section 4.3.1. Attenuations by the different screens are then added up (on a logarithmic scale). The method thus includes the effects of *all* screens.

Despite this more refined modeling, the method is still only approximate. It uses the diffraction attenuation (Eq. 4.27) that is based on the assumption that the RX is in the far field of the screen. If, however, two of the screens are close together, this assumption is violated, and significant errors can occur.

The inaccuracies caused by this "far-field assumption" can be reduced considerably by the *slope diffraction method*. In this approach, the field is expanded into a Taylor series. In addition to the zeroth-order term (far field), which enforces continuity of the electrical field at the screen, also the first-order term is taken into account, and used to enforce continuity of the first derivative of the field. This results in modified coefficients A and D, which are determined by recursion equations [Andersen 1997].

Deygout's Method

The philosophy of Deygout's method is similar to that of the Epstein–Petersen method, as it also adds up the attenuations caused by each screen [Deygout 1966]. However, the diffraction angles are defined in the Deygout method by a different algorithm:

- In the first step, determine the attenuation between TX and RX if only the ith screen is present (for all i).
- The screen that causes the largest attenuation is defined as the "main screen" – its index is defined as i_{ms}.
- Compute the attenuation between the TX and the tip of the main screen caused by the jth screen (with j running now from 1 to i_{ms}). The screen resulting in the largest attenuation is called the "subsidiary main screen." Similarly, compute the attenuation between the main screen and the RX, caused by the jth screen ($j > i_{ms} + 1$).
- Optionally, repeat that procedure to create "subsidiary screens," etc.
- Add up the losses (in dB) from all considered screens.

The Deygout method works well if there is actually one dominant screen that creates most of the losses. Otherwise, it can create considerable errors.

Example 4.4 *There are three screens, 20 m apart from each other and 30, 40, and 25 m high.*
The first screen is 30 m from the TX, the last screen is 100 m from the RX. The TX is 1.5 m high, the
RX is 30 m high. Compute the attenuation due to diffraction at 900 MHz by the Deygout method.

The attenuation L caused by a certain screen is given as:

$$L = -20 \log \tilde{F}(\nu_F) \tag{4.37}$$

where $\tilde{F}(\nu_F)$ as defined in Eq. (4.27) is given by:

$$\tilde{F}(\nu_F) = \frac{1}{2} - \frac{\exp(j\pi/4)}{\sqrt{2}} F(\nu_F) \tag{4.38}$$

First, determine the attenuation caused by screen 1. The diffraction angle θ_d, Fresnel parameter
ν_F, and attenuation L are given by:

$$
\left.
\begin{aligned}
\theta_d &= \arctan\left(\frac{30-1.5}{30}\right) + \arctan\left(\frac{30-30}{140}\right) = 0.760\,\text{rad} \\[2mm]
\nu_F &= 0.760\sqrt{\frac{2 \cdot 30 \cdot 140}{1/3 \cdot (30+140)}} = 9.25 \\[2mm]
L_1 &= -20 \log\left(\left|\frac{1}{2} - \frac{\exp(j\pi/4)}{\sqrt{2}} F(9.25)\right|\right) \\[2mm]
&\approx -20 \log\left(\left|\frac{1}{2} - \frac{\exp(j\pi/4)}{\sqrt{2}} \cdot (0.522 - j0.527)\right|\right) = 32.28\,\text{dB}
\end{aligned}
\right\} \tag{4.39}
$$

where the Fresnel integral $F(\nu_F)$ is numerically evaluated. Similarly, the attenuation caused by
screens 2 and 3 is $L_2 = 33.59\,\text{dB}$ and $L_3 = 25.64\,\text{dB}$, respectively. Hence, the main attenuation
is caused by screen 2, which therefore becomes the "main screen."

Next, the attenuation from the TX to the tip of screen 2, as caused by screen 1, is determined.
The diffraction angle θ_d, Fresnel parameter ν_F, and attenuation L become:

$$
\left.
\begin{aligned}
\theta_d &= \arctan\left(\frac{30-1.5}{30}\right) + \arctan\left(\frac{30-40}{20}\right) = 0.296\,\text{rad} \\[2mm]
\nu_F &= 0.296\sqrt{\frac{2 \cdot 30 \cdot 20}{1/3 \cdot (30+20)}} = 2.51 \\[2mm]
L_4 &= -20 \log\left(\frac{1}{2} - \frac{\exp(-j\pi/4)}{\sqrt{2}} F(2.51)\right) \\[2mm]
&\approx -20 \log\left(\frac{1}{2} - \frac{\exp(-j\pi/4)}{\sqrt{2}} \cdot (0.446 - j0.614)\right) = 21.01\,\text{dB}
\end{aligned}
\right\} \tag{4.40}
$$

Similarly, the attenuation from the tip of screen 2 to the RX, as caused by screen 3, is
$L_5 = 0.17\,\text{dB}$. The total attenuation caused by diffraction is then determined as the sum of all
attenuations – i.e.:

$$
\begin{aligned}
L_{total} &= L_2 + L_4 + L_5 \\
&= 33.59 + 21.01 + 0.17 = 54.77\,\text{dB}
\end{aligned}
$$

Empirical Models

The *International Telecommunications Union* (ITU) proposed an extremely simple, semi-empirical model for diffraction losses (i.e., losses in addition to free space attenuation):

$$L_{\text{total}} = \sum_{i=1}^{N} L_i + 20 \log C_N \tag{4.41}$$

where L_i is the diffraction loss from each separate screen (in dB), and C_N is defined as:

$$C_N = \sqrt{\frac{P_a}{P_b}} \tag{4.42}$$

where

$$\left.\begin{aligned} P_a &= d_{p1} \prod_{i=1}^{N} d_{ni} \left(d_{p1} + \sum_{j=1}^{N} d_{nj} \right) \\ P_b &= d_{p1} d_{nN} \prod_{i=1}^{N} (d_{pi} + d_{ni}) \end{aligned}\right\} \tag{4.43}$$

Here, d_{pi} is the (geometrical) distance to the preceding screen tip, and d_{ni} is the distance to the following one. Li et al. [1997] showed that this equation leads to large errors and proposed a modified definition:

$$C_N = \frac{P_a}{P_b} \tag{4.44}$$

Comparison of the Different Methods

The only special case where an exact solution can be easily computed is the case when all screens have the same height, and are at the same height as the TX and RX antennas. In that case, diffraction loss (on a linear scale!) is proportional to the number of screens, $1/(N_{\text{screen}} + 1)$. Let us now check whether the above approximation methods give this result (see Figure 4.12):

- The Bullington method is independent of the number of screens, and thus obviously gives a wrong functional dependence.
- The Epstein–Petersen method adds the attenuations *on a logarithmic scale* and thus leads to an exponential increase of the total attenuation on a linear scale.
- Similarly, the Deygout method and the ITU-R method predict an exponential increase of the total attenuation as the number of screens increases.
- The slope diffraction method (up to 15 screens) and the *modified* ITU method lead to a linear increase in total attenuation, and thus predict the trend correctly.

However, note that the above comparison considers a specific, limiting case. For a small number of screens of different height, both the Deygout and the Epstein–Petersen method can be used successfully.

4.4 Scattering by Rough Surfaces

Scattering on rough surfaces (Figure 4.13) is a process that is very important for wireless communications. Scattering theory usually assumes roughness to be *random*. However, in wireless

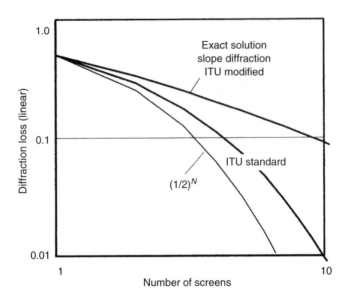

Figure 4.12 Comparison of different computation methods for multiple-screen diffraction.

Figure 4.13 Scattering by a rough surface.

communications it is common to also describe *deterministic*, possibly periodic, structures (e.g., bookshelves or windowsills) as rough. For ray-tracing predictions (see Section 7.5), "roughness" thus describes all (physically present) objects that are not included in the used maps and building plans. The justifications for this approach are rather heuristic: (i) the errors made are smaller than some other error sources in ray-tracing predictions and (ii) there is no better alternative.

That being said, the remainder of this section will consider the mathematical treatment of genuinely rough surfaces. This area has been investigated extensively in the last 30 years, mostly due to its great importance in radar technology. Two main theories have evolved: the Kirchhoff theory and the perturbation theory.

4.4.1 The Kirchhoff Theory

The Kirchhoff theory is conceptually very simple and requires only a small amount of information – namely, the probability density function of surface amplitude (height). The theory assumes that height variations are so small that different *scattering points* on the surface do not influence each other – in other words, that one point of the surface does not "cast a shadow" onto other points of the surface. This assumption is actually not fulfilled very well in wireless communications.

Assuming that the above condition is actually fulfilled, surface roughness leads to a reduction in power of the specularly reflected ray, as radiation is also scattered in other directions (see r.h.s. of Figure 4.13). This power reduction can be described by an *effective* reflection coefficient ρ_{rough}. In the case of Gaussian height distribution, this reflection factor becomes:

$$\rho_{\mathrm{rough}} = \rho_{\mathrm{smooth}} \exp\left[-2(k_0 \sigma_{\mathrm{h}} \sin \psi)^2\right] \tag{4.45}$$

where σ_{h} is the standard deviation of the height distribution, k_0 is the wavenumber $2\pi/\lambda$, and ψ is the angle of incidence (defined as the angle between the wave vector and the surface). The term $2k_0\sigma_{\mathrm{h}}\sin\psi$ is also known as Rayleigh roughness. Note that for grazing incidence ($\psi \approx 0$), the effect of the roughness vanishes, and the reflection becomes specular again.

4.4.2 Perturbation Theory

The perturbation theory generalizes the Kirchhoff theory, using not only the probability density function of the surface height but also its spatial correlation function. In other words, it takes into account the question "how fast does the height vary if we move a certain distance along the surface?" (see Figure 4.14).

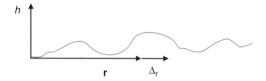

Figure 4.14   Geometry for perturbation theory of rough scattering.

Mathematically, the spatial correlation function is defined as:

$$\sigma_{\mathrm{h}}^2 W(\Delta_{\mathrm{r}}) = E_{\mathbf{r}}\{h(\mathbf{r})h(\mathbf{r} + \Delta_{\mathrm{r}})\} \tag{4.46}$$

where $\mathbf{r}$ and Δ_{r} are (two-dimensional) location vectors, and $E_{\mathbf{r}}$ is expectation with respect to $\mathbf{r}$. We need this information to find whether one point on the surface can "cast a shadow" onto another point of the surface. If extremely fast amplitude variations are allowed, shadowing situations are much more common. The above definition enforces spatial statistical stationarity – i.e., the correlation is independent of the absolute location $\mathbf{r}$. The correlation length L_{c} is defined as the distance so that $W(L_{\mathrm{c}}) = 0.5 \cdot W(0)$.[7]

The effect of surface roughness on the amplitude of a specularly reflected wave can be described by an "effective" (complex) dielectric constant δ_{eff}, which in turn gives rise to an "effective" reflection coefficient, as computed from Snell's law.[8] For vertical polarization, the δ_{eff} is given by

---

[7] A more extensive discussion of expectation values and correlation functions will be given in Chapter 6.
[8] We assume here that the material is not conductive, $\sigma_{\mathrm{e}} = 0$. The imaginary contribution arises from surface roughness.

Vaughan and Andersen [2003]:

$$\frac{1}{\sqrt{\delta_{\mathrm{r,eff}}}} = \begin{cases} \dfrac{1}{\sqrt{\varepsilon_{\mathrm{r}}}} + j\dfrac{k_0\sigma_{\mathrm{h}}^2 \sin(2\psi)}{2L_{\mathrm{c}}} \displaystyle\int_0^\infty \dfrac{1}{x}\dfrac{d\hat{W}(x)}{dx}\,dx, & k_0 L_{\mathrm{c}} \ll 1 \\[2ex] \dfrac{1}{\sqrt{\varepsilon_{\mathrm{r}}}} + (k_0\sigma_{\mathrm{h}})^2 (\sin\psi)^3, & k_0 L_{\mathrm{c}} \gg 1, \psi \gg \dfrac{1}{\sqrt{k_0 L_{\mathrm{c}}}} \\[2ex] \dfrac{1}{\sqrt{\varepsilon_{\mathrm{r}}}} - \dfrac{\sigma_{\mathrm{h}}^2}{2L_{\mathrm{c}}}\dfrac{\sqrt{k_0 L_{\mathrm{c}}}}{\sqrt{2\pi}}\exp(j\pi/4) \displaystyle\int_0^\infty \dfrac{1}{x\sqrt{x}}\dfrac{d\hat{W}(x)}{dx}\,dx, & k_0 L_{\mathrm{c}} \gg 1, \psi \ll \dfrac{1}{\sqrt{k_0 L_{\mathrm{c}}}} \end{cases}$$

$$(4.47)$$

where $\hat{W}(x) = W(x/L_{\mathrm{c}})$, while for horizontal polarization, it is

$$\frac{1}{\sqrt{\delta_{\mathrm{r,eff}}}} = \begin{cases} \dfrac{1}{\sqrt{\varepsilon_{\mathrm{r}}}} + j\dfrac{k_0\sigma_{\mathrm{h}}^2}{2L_{\mathrm{c}}} \displaystyle\int_0^\infty \dfrac{1}{x}\dfrac{d\hat{W}(x)}{dx}\,dx, & k_0 L_{\mathrm{c}} \ll 1 \\[2ex] \dfrac{1}{\sqrt{\varepsilon_{\mathrm{r}}}} + (k_0\sigma_{\mathrm{h}})^2 \sin\psi, & k_0 L_{\mathrm{c}} \gg 1, \psi \gg \dfrac{1}{\sqrt{k_0 L_{\mathrm{c}}}} \\[2ex] \dfrac{1}{\sqrt{\varepsilon_{\mathrm{r}}}} - \dfrac{(k_0\sigma_{\mathrm{h}})^2}{\sqrt{k_0 L_{\mathrm{c}}}}\dfrac{2}{\sqrt{2\pi}}\exp(-j\pi/4) \displaystyle\int_0^\infty \dfrac{1}{\sqrt{x}}\dfrac{d\hat{W}(x)}{dx}\,dx, & \sqrt{k_0 L_{\mathrm{c}}} \gg 1, \psi \ll \dfrac{1}{\sqrt{k_0 L_{\mathrm{c}}}} \end{cases}$$

$$(4.48)$$

Comparing these results to the Kirchhoff theory, we find that there is good agreement in the case $k_0 L_{\mathrm{c}} \gg 1$, $\psi \gg 1/\sqrt{k_0 L_{\mathrm{c}}}$. This agrees with our above discussion of the limits of the Kirchhoff theory: by assuming that the coherence length is long compared with the wavelength, there cannot be diffraction by a sudden "spike" in the surface. And by fulfilling $\psi \gg 1/\sqrt{k_0 L_{\mathrm{c}}}$, it is assured that a wave incident under angle ψ cannot cast a shadow onto other points of the surface.

4.5 Waveguiding

Another important process is propagation in (dielectric) waveguides. This process models propagation in street canyons, corridors, and tunnels. The basic equations of dielectric waveguides are well established [Collin 1991], [Marcuse 1991]. However, those waveguides occurring in wireless communications deviate from the idealized assumptions of theoretical work:

- The materials are lossy.
- Street canyons (and most corridors) do not have continuous walls, but are interrupted at more or less regular intervals by cross-streets. Furthermore, street canyons lack the "upper" wall of the waveguide.
- The surfaces are rough (window sills, etc.).
- The waveguides are not empty, but filled with metallic (cars) and dielectric (pedestrian) IOs.

Propagation prediction can be done either by computing the waveguide modes, or by a geometric optics approximation. If the waveguide cross-section as well as the IOs in it are much larger than the wavelength, the latter method gives good results [Klemenschits and Bonek 1994].

Conventional waveguide theory predicts a propagation loss that increases exponentially with distance. Some measurements in corridors observed a similar behavior. The majority of measurements, however, fitted a d^{-n} law, where n varies between 1.5 and 5. Note that a loss exponent smaller than 2 does not contradict energy conservation or any other laws of physics. The d^{-2} law in free space propagation just stems from the fact that energy is spread out over a larger surface as distance is increased. If energy is guided, even d^0 becomes theoretically possible.

4.6 Appendices

Please go to *www.wiley.com/go/molisch*

Further Reading

Several books on wireless propagation processes give an overview of all the phenomena discussed in this chapter: [Bertoni 2000, Blaunstein 1999, Parsons 1992, Vaughan and Andersen 2003, Haslett 2008, and Barclay 2002]; MATLAB programs for simulating them are described in Fontan and Espineira [2008]. The basic propagation processes, especially reflection and transmission, are described in any number of classic textbooks on electromagnetics – e.g., Ramo et al. [1967], which is a personal favorite – but also in many others. More involved results can be found in Bowman et al. [1987] and Felsen and Marcuvitz [1973]. The description of transmission through multilayer dielectric films can be read in Heavens [1965]. The geometric theory of diffraction was originally invented by Keller [1962], extended to the uniform theory of diffraction in Kouyoumjian and Pathak [1974], and summarized in McNamara et al. [1990]. The different theories for multiple knife edge diffraction are nicely summarized in Parsons [1992]; however, continuous research is being done on that topic, and further solutions are constantly being published (e.g., Bergljung [1994]). Scattering by rough surfaces, mostly inspired by radar problems, is described in Bass and Fuks [1979], de Santo and Brown [1986], Ogilvie [1991]; an excellent summary is given in Vaughan and Andersen [2003]. The theory of dielectric waveguides is described in Collin [1991], Marcuse [1991], and Ramo et al. [1967].

For updates and errata for this chapter, see wides.usc.edu/teaching/textbook

5

Statistical Description of the Wireless Channel

5.1 Introduction

In many circumstances, it is too complicated to describe all reflection, diffraction, and scattering processes that determine the different Multi Path Components (MPCs). Rather, it is preferable to describe the *probability* that a channel parameter attains a certain value. The most important parameter is channel gain, as it determines received power or field strength; it will be at the center of our interest in this chapter.[1]

In order to arrive at a better understanding of the channel, let us first look at a typical plot depicting received power as a function of distance (Figure 5.1). The first thing we notice is that received power can vary strongly – by 100 dB or more. We also see that variations happen on different spatial scales:

- On a very-short-distance scale, power fluctuates around a (local) mean value, as can be seen from the insert in Figure 5.1. These fluctuations happen on a scale that is comparable with one wavelength, and are therefore called *small-scale fading*. The reason for these fluctuations is interference between different MPCs, as mentioned in Chapter 2. Fluctuations in field strength can be well described statistically – namely, by the (local) mean value of the power and the statistics of the fluctuations around this mean.
- Mean power, averaged over about 10 wavelengths, itself shows fluctuations. These fluctuations occur on a larger scale – typically a few hundred wavelengths. These variations can be seen most clearly when moving on a circle around the transmitter (TX) (Figure 5.2). The reason for these variations is shadowing by large objects (Chapter 2), and is thus fundamentally different from the interference that causes small-scale fading. However, this *large-scale fading* can also be described by a mean and the statistics of fluctuations around this mean.
- The large-scale mean itself depends monotonically on the distance between TX and receiver (RX). This effect is related to free space path loss or some variation thereof. This effect is usually described in a deterministic manner, and was already treated in Chapter 4 (further details and models can be found in Chapter 7).

[1] Note that we talk of channel gain (and not attenuation) because it is directly proportional to receive power (receive power is transmit power times channel gain); of course, this gain is smaller than unity. In the following, we often assume unit transmit power and use "receive power" and "channel gain" interchangeably.

Wireless Communications, Second Edition   Andreas F. Molisch
© 2011 John Wiley & Sons, Ltd

Figure 5.1 Received power as a function of distance from the transmitter.

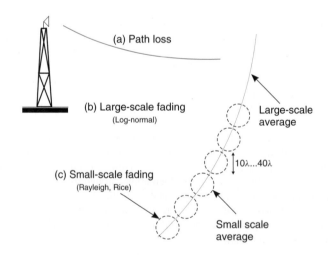

Figure 5.2 Types of received power variations.

Chapter 5 only concentrates on the statistical variations of the channel gain; delay dispersion and other effects will be treated later. It is thus sufficient for this chapter to consider channel gain for an unmodulated (sinusoidal) carrier signal, though the considerations are also valid for *narrowband* systems (as defined in Section 6.1). Sections 5.2 and 5.3 explain the *two-path model* – it is the most simple model explaining small-scale fading effects. Sections 5.4 and 5.5 generalize these considerations to more realistic channel models, and give statistics of the received field strength for various scenarios. The next two sections describe the statistics of fading dips (instances of very low received power), the frequency of their occurrence, and their average duration. Finally, the statistical variations due to shadowing effects are described in Section 5.8.

5.2 The Time-Invariant Two-Path Model

As an introduction to the rather involved subject of multipath propagation and fading, we consider the simplest possible case – time-invariant propagation along two paths. We transmit a sinusoidal waveform and determine the (complex) transfer function at the location of the RX.

First, consider a single wave. Let the transmit signal be a sinusoidal wave:

$$E_{\text{TX}}(t) \propto \cos(2\pi f_c t) \tag{5.1}$$

Let the received signal be approximated as a homogeneous plane wave. If the run length between TX and RX is d, the received signal can be described as:

$$E(t) = E_0 \cdot \cos(2\pi f_c t - k_0 d) \tag{5.2}$$

where k_0 is the wavenumber $2\pi/\lambda$. Using complex baseband notation,[2] this reads

$$E = E_0 \exp(-jk_0 d) \tag{5.3}$$

Note that the real part of the field in complex representation, $\text{Re}\{E\}$ is equal to the instantaneous value of the field strength at time $t = 0$.

Now consider the case that the transmit signal gets to the RX via two different propagation paths, created by two different Interacting Objects (IOs, see Figure 5.3). These paths have different runtimes:

$$\tau_1 = d_1/c_0, \text{ and } \tau_2 = d_2/c_0 \tag{5.4}$$

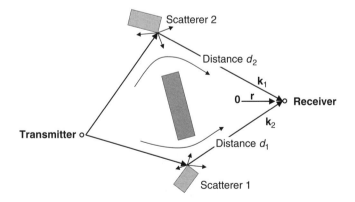

Figure 5.3 Geometry of the time-invariant two-path model.

---

[2] The bandpass signal – i.e., the physically existing signal – is related to the complex baseband (lowpass) representation as $s_{\text{BP}}(t) = \text{Re}\{s_{\text{LP}}(t)\exp[j2\pi f_c t]\}$ (see also Section 11.1).

The RX is in the far field of the IOs, so that the arriving waves are homogeneous planewaves. We assume furthermore that both waves are vertically polarized, and have amplitudes $E1$ and $E2$ at the reference position (origin of the coordinate system) $\mathbf{r} = 0$. We get the following expression for the superposition of two planewaves:

$$E(\mathbf{r}) = E1 \, \exp(-j\mathbf{k}_1\mathbf{r}) + E2 \, \exp(-j\mathbf{k}_2\mathbf{r}) \qquad (5.5)$$

where $\mathbf{k}_1$ is the vector-valued wavenumber (i.e., has the absolute magnitude k_0, and is pointing into the direction of wave 1).

We assume here that the two waves arriving at the RX position $\mathbf{r}$ are two plane waves whose absolute amplitudes do not vary as a function of RX position (we vary RX positions only within an area that has less than about 10 λ diameter).

The upper part of Figure 5.4 depicts the real part of $E(\mathbf{r})$, and the lower part shows the magnitude, which is proportional to the square root of received power. We see locations of both constructive and destructive interference – i.e., location-dependent fading. There are fading dips where the phase differences due to the different runtimes are exactly 180°. If the RX is at a point of destructive interference, it sees a signal whose total amplitude is the difference of the amplitudes of the constituting waves. If the amplitudes of the constituting waves are equal, destructive interference can be complete. In the points of constructive interference, the total amplitude is the sum of the amplitudes of the constituting waves. This phenomenon can be seen in the lower right part of Figure 5.4.

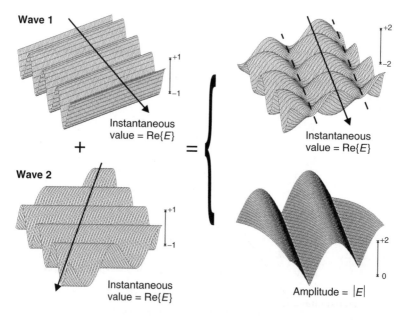

Figure 5.4 Interference of two planewaves with $E1 = E2 = 1$ and arg $(\mathbf{k}_1, \mathbf{k}_2) = 30°$.

5.3 The Time-Variant Two-Path Model

In general, the runtime (path length) difference between the different propagation paths changes with time. This change can be due to movements of the TX, the RX, the IOs, or any combination thereof; to simplify discussion, we henceforth assume only movement of the RX. The RX then "sees" a time-varying interference pattern; we can imagine that the RX moves through the "mountains and valleys" of the field strength plot. Spatially varying fading thus becomes time-varying fading. Since fading dips are approximately half a wavelength apart (corresponding to 16 cm at a carrier frequency of 900 MHz), this fading is called *small-scale fading*, also known as *short-term fading*, or *fast fading*.[3] The fading rate (number of fading dips per second) depends on the speed of the RX.

The movement of the RX also leads to a shift of the received frequency, called the Doppler shift. In order to explain this phenomenon, let us first revert to the case of a *single* sinusoidal wave reaching the RX, and also revert to real passband notation. If the RX moves away from the TX with speed v, then the distance d between TX and RX increases with that speed. Thus:

$$E(t) = E_0 \cdot \cos(2\pi f_c t - k_0[d_0 + vt])$$

$$= E_0 \cdot \cos\left(2\pi t\left[f_c - \frac{v}{\lambda}\right] - k_0 d_0\right) \tag{5.6}$$

where d_0 is the distance at time $t = 0$. The frequency of the received oscillation is thus decreased by v/λ – in other words, the Doppler shift is given by:

$$\nu = -\frac{v}{\lambda} = -f_c \cdot \frac{v}{c_0} \tag{5.7}$$

Note that the Doppler shift is negative when the TX and RX move away from each other. Since the speed of the movement is always small compared with the speed of light, the Doppler shifts are relatively small.

In the above example, we had assumed that the direction of RX movement is aligned with the direction of wave propagation. If that is not the case, the Doppler shift is determined by the speed of movement *in the direction of wave propagation*, v $\cos(\gamma)$ (see Figure 5.5). The Doppler shift is then:

$$\nu = -\frac{v}{\lambda}\cos(\gamma) = -f_c \cdot \frac{v}{c_0}\cos(\gamma) = -\nu_{max}\cos(\gamma) \tag{5.8}$$

The maximum Doppler shift ν_{max} typically lies between 1 Hz and 1 kHz. Note that in general, the relationship $\nu_{max} = f_c \cdot v/c_0$ is based on several assumptions – e.g., static IOs, no double reflections on moving objects, etc.

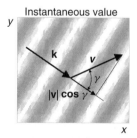

Figure 5.5   Projection of velocity vector $|v|$ onto the direction of propagation **k**.

---

[3] Unfortunately, the expression *fast* fading is often used for two completely different phenomena. On one hand, it is used as a synonym for small-scale fading, independent of the actual *temporal* scale of the fading (which depends on the movement speed of the TX, RX, and IOs). On the other hand, it is used to denote channel variations within the duration of a symbol length (in contrast to the "quasi-static" channel). It is thus preferable to call the fading due to interference effects *small-scale fading*, as this is unambiguous.

Since the Doppler shifts are so small, it seems natural to ask whether they have a significant influence on the radio link. If all constituent waves were Doppler shifted by the same amount – e.g., 100 Hz – the effect on radio link performance would really be negligible – the local oscillator in the RX could easily compensate for such a shift. The important point is, however, that the different MPCs have *different* Doppler shifts. The superposition of several Doppler-shifted waves creates the sequence of fading dips. Again, this can be demonstrated using the two-path model. As the RX moves, it receives two waves that are each Doppler shifted, but by different amounts. By Fourier transformation to the time domain, we get the well-known effect of *beating* of two oscillations with slightly different frequencies (see Figure 5.6). The frequency of this beating envelope is equal to the frequency difference between the two carriers – i.e., the difference of the two Doppler shifts. The RX thus sees a signal whose amplitude undergoes periodic variations – this is exactly the fast fading that is created by traversing the "mountains and valleys" of the field strength plot of Figure 5.4.

Figure 5.6 Superposition of two carriers with different frequencies (beating).

Summarizing, we can obtain the fading rate in the two-path model by two equivalent considerations:

1. We superimpose two incident waves, plot the resulting interference pattern (field strength "mountains and valleys"), and count the number of fading dips per second that an RX sees when moving through that pattern.
2. Alternatively, we can think of superimposing two signals with different Doppler shifts at the receive antenna, and determine the fading rate from the beat frequency – i.e., the difference of the Doppler shifts of the two waves.

Doppler frequency is thus an important parameter of the channel, even though it is so small.

- Doppler frequency is a measure for the rate of change of the channel, as we have discussed above.
- Furthermore, the superposition of many slightly Doppler-shifted signals leads to phase shifts of the total received signal that can impair the reception of angle-modulated signals (Chapters 11 and 12). These phase shifts lead to a random Frequency Modulation (FM) of the received signal (see Section 5.7), and are especially important for signals with low bit rates.

5.4 Small-Scale Fading without a Dominant Component

Following these basic considerations utilizing the two-path model, we now investigate a more general case of multipath propagation. We consider a radio channel with many IOs and a moving RX. Due to the large number of IOs, a deterministic description of the radio channel is not efficient any more, which is why we take refuge in stochastic description methods. This stochastic description is essential for the whole field of wireless communications, and is thus explained in considerable detail. We start out with a computer experiment in Section 5.4.1, followed by a more general mathematical derivation in Section 5.4.2.

5.4.1 A Computer Experiment

Consider the following simple computer experiment. The signals from several IOs are incident onto an RX that moves over a small area. The IOs are distributed approximately uniformly around the receiving area. They are also assumed to be sufficiently far away so that all received waves are homogeneous plane waves, and that movements of the RX within the considered area do not change the amplitudes of these waves. The different distances and strength of the interactions are taken into account by assigning a random phase and a random amplitude to each wave. We are then creating eight constituting waves E_i with absolute amplitudes $|a_i|$, angle of incidence (with respect to the x-axis) ϕ_i and phase φ_i:

	$|a_i|$	ϕ_i	φ_i
$E_1(x, y) = 1.0 \exp[-jk_0(x \cos(169°) + y \sin(169°))] \exp(j311°)$	1.0	169°	311°
$E_2(x, y) = 0.8 \exp[-jk_0(x \cos(213°) + y \sin(213°))] \exp(j32°)$	0.8	213°	32°
$E_3(x, y) = 1.1 \exp[-jk_0(x \cos(87°) + y \sin(87°))] \exp(j161°)$	1.1	87°	161°
$E_4(x, y) = 1.3 \exp[-jk_0(x \cos(256°) + y \sin(256°))] \exp(j356°)$	1.3	256°	356°
$E_5(x, y) = 0.9 \exp[-jk_0(x \cos(17°) + y \sin(17°))] \exp(j191°)$	0.9	17°	191°
$E_6(x, y) = 0.5 \exp[-jk_0(x \cos(126°) + y \sin(126°))] \exp(j56°)$	0.5	126°	56°
$E_7(x, y) = 0.7 \exp[-jk_0(x \cos(343°) + y \sin(343°))] \exp(j268°)$	0.7	343°	268°
$E_8(x, y) = 0.9 \exp[-jk_0(x \cos(297°) + y \sin(297°))] \exp(j131°)$	0.9	297°	131°

We now superimpose the constituting waves, using the complex baseband notation again. The total complex field strength E thus results from the sum of the complex field strengths of the constituting waves. We can also interpret this as adding up complex random phasors. Figure 5.7 shows the instantaneous value of the total field strength at time $t = 0$ – i.e., Re$\{E\}$, in an area of size $5\lambda \cdot 5\lambda$.

Let us now consider the statistics of the field strengths occurring in that area. As can be seen from the histogram (Figure 5.8), the values Re$\{E\}$ follow, to a good approximation, a zero-mean Gaussian distribution. This is a consequence of the central limit theorem: when superimposing N statistically independent random variables, none of which is dominant, the associated *probability density function* (pdf) approaches a normal distribution for $N \to \infty$ (see Appendix 5.A – at www.wiley.com/go/molisch – for a more exact formulation of this statement). The conditions for the validity of the central limit theorem are fulfilled approximately: the eight constituting waves have random angles of incidence and phases, and none of the amplitudes a_i is dominant. Figures 5.9 and 5.10 show that the imaginary part of the field strength, Im$\{E\}$, is also normally distributed.[4]

The behavior of most RXs is determined by the received (absolute) *amplitude (magnitude)*. We thus need to investigate the distribution of the envelope of the received signal, corresponding to the magnitude of the (complex) field strength phasor. Figure 5.11 shows the field strength seen by an RX that moves along the y-axis of Figure 5.7 or 5.8. The left diagram of Figure 5.12 shows the complex field strength phasor, and the right side the absolute amplitude of the received signal.

Figures 5.11 and 5.13 show a three-dimensional representation of the amplitudes and the statistics of the amplitude over that area, respectively. Figure 5.13 also exhibits a plot of a Rayleigh pdf.

---

[4] The imaginary part represents the instantaneous value of the field strength at a time that corresponds to a quarter period of the radio frequency oscillation, $\omega t = \pi/2$.

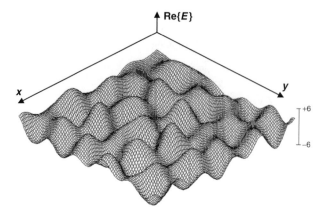

Figure 5.7 Instantaneous value of the field strength at time $t = 0$ – i.e., Re$\{E\}$. Superposition of the eight constituting waves in the area $0 < x < 5\lambda, 0 < y < 5\lambda$.

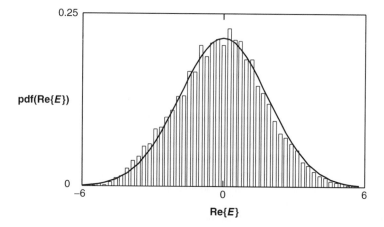

Figure 5.8 Histogram of the field strength of Figure 5.7. A Gaussian pdf is shown for comparison.

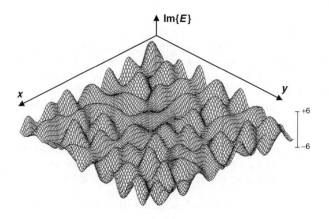

Figure 5.9 Imaginary part of E.

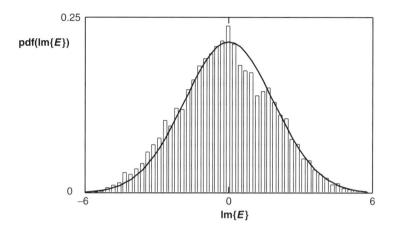

Figure 5.10 Histogram of the field strength of Figure 5.9. A Gaussian pdf is shown for comparison.

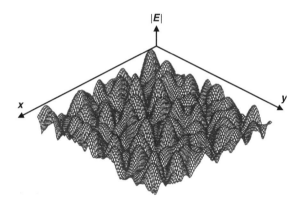

Figure 5.11 Amplitude of the field strength.

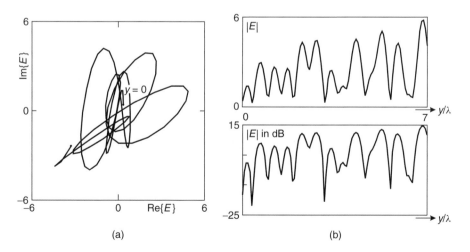

Figure 5.12 Complex phasor of the field strength (a) and received amplitude $|E|$ (b) for a receiver moving along the y-axis.

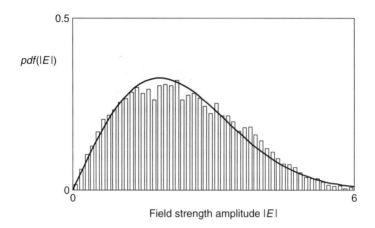

Figure 5.13   Pdf of the received amplitude.

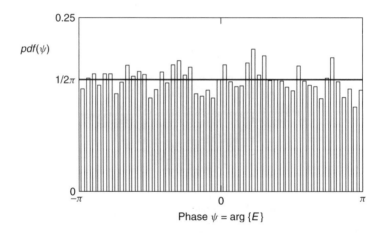

Figure 5.14   Pdf of the received phase.

A Rayleigh distribution describes the magnitude of a complex stochastic variable whose real and imaginary parts are independent and normally distributed; more details are given below. Figure 5.14 shows that the *phase* is approximately uniformly distributed.

5.4.2   Mathematical Derivation of the Statistics of Amplitude and Phase

After these pseudo-experimental considerations, we now turn to a more detailed, and more mathematically sound, derivation of Rayleigh distribution. Consider a scenario where N homogeneous plane waves (MPCs) have been created by reflection/scattering from different IOs. The IOs and the TX do not move, and the RX moves with a velocity v. As above, we assume that the absolute amplitudes of the MPCs do not change over the region of observation. The sum of the squared amplitudes is thus:

$$\sum_{i=1}^{N} |a_i|^2 = C_{\mathrm{P}} \tag{5.9}$$

where C_{P} is a constant. However, the phases φ_i vary strongly, and are thus approximated as random variables that are uniformly distributed in the range $[0, 2\pi]$. The real part of the received field strength due to the ith MPC is thus $|a_i| \cos(\varphi_i)$, the imaginary part is $|a_i| \sin(\varphi_i)$.

Furthermore, we need to consider the Doppler shift for computation of the total field strength $E(t)$. If we look at an unmodulated carrier, we get (in real passband notation):

$$E(t) = \sum_{i=1}^{N} |a_i| \cos[2\pi f_c t - 2\pi \nu_{\max} \cos(\gamma_i) t + \varphi_i] \tag{5.10}$$

Rewriting this in terms of in-phase and quadrature-phase components in real passband notation, we obtain

$$E_{\mathrm{BP}}(t) = I(t) \cdot \cos(2\pi f_c t) - Q(t) \cdot \sin(2\pi f_c t) \tag{5.11}$$

with

$$I(t) = \sum_{i=1}^{N} |a_i| \cos[-2\pi \nu_{\max} \cos(\gamma_i) t + \varphi_i] \tag{5.12}$$

$$Q(t) = \sum_{i=1}^{N} |a_i| \sin[-2\pi \nu_{\max} \cos(\gamma_i) t + \varphi_i] \tag{5.13}$$

Both the in-phase and the quadrature-phase component are the sum of many random variables, none of which dominate (i.e., $|a_i| \ll C_p$). It follows from the central limit theorem that the pdf of such a sum is a normal (Gaussian) distribution, regardless of the exact pdf of the constituent amplitudes – i.e., we do not need knowledge of the a_i or their distributions!!! (see Appendix 5.A – at *www.wiley.com/go/molisch* – for more details). A zero-mean Gaussian random variable has the pdf:

$$pdf_x(x) = \frac{1}{\sqrt{2\pi}\sigma} \exp\left(-\frac{x^2}{2\sigma^2}\right) \tag{5.14}$$

where σ^2 denotes the variance.

Starting with the statistics of the real and imaginary parts, Appendix 5.B derives the statistics of amplitude and phase of the received signal. The pdf is a product of a pdf for ψ – namely, a uniform distribution:

$$pdf_\psi(\psi) = \frac{1}{2\pi} \tag{5.15}$$

and a pdf for r – namely, a Rayleigh distribution:

$$pdf_r(r) = \frac{r}{\sigma^2} \cdot \exp\left[-\frac{r^2}{2\sigma^2}\right] \quad 0 \le r < \infty \tag{5.16}$$

For $r < 0$ the pdf is zero, as absolute amplitudes are by definition positive.

5.4.3 Properties of the Rayleigh Distribution

A Rayleigh distribution has the following properties, which are also shown in Figure 5.15:

$$
\left.
\begin{array}{rl}
\text{Mean value } & \bar{r} = \sigma\sqrt{\tfrac{\pi}{2}} \\
\text{Mean square value } & \overline{r^2} = 2\sigma^2 \\
\text{Variance } & \overline{r^2} - (\bar{r})^2 = 2\sigma^2 - \sigma^2\tfrac{\pi}{2} = 0.429\sigma^2 \\
\text{Median value } & r_{50} = \sigma\sqrt{2\cdot ln2} = 1.18\sigma \\
\text{Location of maximum } & \max\{pdf(r)\} \text{ occurs at } r = \sigma
\end{array}
\right\}
\qquad (5.17)
$$

where the bar denotes expected value (we deviate here from our usual notation $E\{\}$ in order to avoid confusion with the field strength E).

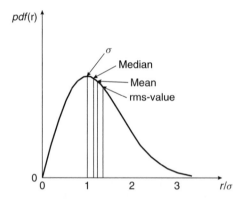

Figure 5.15 Pdf of a Rayleigh distribution.

The *cumulative distribution function*, cdf (x), is defined as the probability that the realization of the random variable has a value smaller than x. The cdf is thus the integral of the pdf:

$$
\text{cdf}(r) = \int_{-\infty}^{r} pdf(u)\, du
\qquad (5.18)
$$

Applying this equation to the Rayleigh pdf, we get

$$
\text{cdf}(r) = 1 - \exp\left(-\frac{r^2}{2\sigma^2}\right)
\qquad (5.19)
$$

For small values of r this can be approximated as:

$$
\text{cdf}(r) \approx \frac{r^2}{2\sigma^2}
\qquad (5.20)
$$

It is straightforward to check whether a measured ensemble of field strength values follows a Rayleigh distribution: its empirical cdf is plotted on the so-called Weibull paper (see Figure 5.16). The cdf of the Rayleigh distribution is a straight line on it. For small values of r, an increase of

Figure 5.16   Measured cdf of the (normalized) receive power of an indoor non-line of sight scenario [Gahleitner 1993].

r by 10 dB has to increase the value of the cdf by 10 dB. By transformation of variables, we can easily see that the squared amplitude, and by extension the power, has an exponential distribution $pdf_{\mathrm{P}}(P) = \frac{1}{\overline{\Omega}}exp(-P/\overline{\Omega})$, where $\overline{\Omega}$ is the mean power.[5]

The Rayleigh distribution is widely used in wireless communications. This is due to several reasons:

- It is an *excellent approximation* in a large number of practical scenarios, as confirmed by a multitude of measurements. However, it is noteworthy that there *are* scenarios where it is not valid. These can occur, e.g., in *Line Of Sight* (LOS) scenarios, some indoor scenarios, and in (ultra) wideband scenarios (see Chapters 6 and 7).
- It describes a *worst case scenario* in the sense that there is no dominant signal component, and thus there is a large number of fading dips. Such a worst case assumption is useful for the design of robust systems.[6]
- It depends only on a *single parameter*, the mean received power – once this parameter is known, the complete signal statistics are known. It is easier, and less error-prone, to obtain this single parameter either from measurements or deterministic prediction methods than to obtain the multiple parameters of more involved channel models.
- *Mathematical convenience*: computations of error probabilities and other parameters can often be done in closed form when the field strength distribution is Rayleigh.

---

[5] To simplify discussion, we do not distinguish in this chapter between the squared absolute amplitude and the power, though they are actually related by a proportionality constant.
[6] As we will see later on, there are some fading distributions that show a larger number of fading dips – e.g., Nakagami distributions with $m < 1$; furthermore, the large number of MPCs can also be an advantage for specific systems, see Section 20.2.

5.4.4   Fading Margin for Rayleigh-Distributed Field Strength

Knowledge of the fading statistics is extremely important for the design of wireless systems. We saw in Chapter 3 that for noise-limited systems the received field strength determines the performance. As field strength is a random variable, even a large mean field strength does not guarantee successful communications at *all* times. Rather, the field strength exceeds a minimum value only in a certain *percentage* of situations. The task is therefore to answer the following question: "Given a minimum receive power or field strength required for successful communications, how large does the mean power have to be in order to ensure that communication is successful in $x\%$ of all situations?" In other words, how large does the *fading margin* have to be?

The cdf gives by definition the probability that a certain field strength level is not exceeded. In order to achieve an $x\%$ outage probability, it follows that:

$$x = cdf(r_{min}) \approx \frac{r_{min}^2}{2\sigma^2} \tag{5.21}$$

where the r.h.s follows from Eq. (5.20). From this, we can immediately compute the mean square value of field strength $2\sigma^2$ as $2\sigma^2 = r_{min}^2/x$.

Example 5.1   *For a signal with Rayleigh-distributed amplitude, what is the probability that the received signal power is at least 20, 6, 3 dB below the mean power. Compare the exact result and the result from the approximate formulation of Eq. (5.20).*

From a Rayleigh-distributed signal envelope r:

$$\overline{r^2} = 2\sigma^2 \tag{5.22}$$

A power level 20 dB below the mean power corresponds to $\frac{r_{min}^2}{2\sigma^2} = \frac{1}{100}$:

$$\Pr\{r < r_{min}\} = 1 - \exp\left(-\frac{1}{100}\right) = 9.95 \cdot 10^{-3} \tag{5.23}$$

Similarly, the exact results for 6 and 3 dB are 0.221 and 0.393, respectively.

The approximate formulation of Eq. (5.20) gives $\frac{r_{min}^2}{2\sigma^2} = 0.01$, 0.25, and 0.5, respectively. It is thus reasonably accurate for power levels 6 dB below the mean power, but breaks down for higher values of r_{min}.

For the interference-limited case, the situation is somewhat more complicated: not only does the desired signal fade but so do the interferers. For the computation of the statistics of the amplitude ratio of signal and interference, we note that both the desired signal and the interference are Rayleigh-fading; we thus need the pdf of the ratio of two random variables, each of which is Rayleigh-distributed [Molisch et al. 1996]:

$$pdf(r) = \frac{2\tilde{\sigma}^2 r}{(\tilde{\sigma}^2 + r^2)^2} \tag{5.24}$$

where $\tilde{\sigma}^2 = \sigma_1^2/\sigma_2^2$ is the ratio of mean signal power to mean interference power. The associated cdf is given by:

$$cdf(r) = 1 - \frac{\tilde{\sigma}^2}{(\tilde{\sigma}^2 + r^2)} \tag{5.25}$$

This formulation is essential for computation of the reuse distance (see Chapters 3 and 17).

5.5 Small-Scale Fading with a Dominant Component

5.5.1 A Computer Experiment

Fading statistics change when a dominant MPC – e.g., an LOS component or a dominant specular component – is present. We can gain some insights by repeating the computer experiment of Section 5.4.1, but now adding an additional wave with the (dominant) amplitude 5:

	$|a_9$	ϕ_9	φ_9
$E_9(x, y) = 5.0 \exp[-jk_0(x \cos(0°) + y \sin(0°))] \exp(j0°)$	5.0	0°	0°

Figure 5.17 shows the real part of E, and the contribution from the dominant component is visible; while Figure 5.18 shows the absolute value. The histogram of the absolute value of the field strength is shown in Figure 5.19. It is clear that the probability of deep fades is much smaller than in the Rayleigh-fading case.

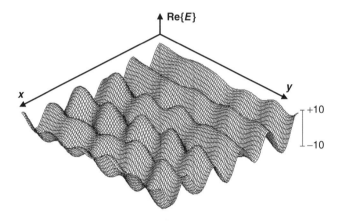

Figure 5.17 Re(E) – i.e., the instantaneous value at $t = 0$, in the presence of a dominant MPC.

5.5.2 Derivation of the Amplitude and Phase Distribution

The pdf of the amplitude can be computed in a way that is similar to our derivation of the Rayleigh distribution (Appendix 5.B – see *www.wiley.com/go/molisch*). Without restriction of generality, we assume that the LOS component has zero phase, so that it is purely real. The real part thus has a non-zero-mean Gaussian distribution, while the imaginary part has a zero-mean Gaussian distribution. Performing the variable transformation as in Appendix 5.B, we get the joint pdf of amplitude r and phase ψ [Rice 1947]:

$$pdf_{r,\psi}(r, \psi) = \frac{r}{2\pi\sigma^2} \exp\left(-\frac{r^2 + A^2 - 2rA\cos(\psi)}{2\sigma^2}\right) \qquad (5.26)$$

where A is the amplitude of the dominant component. In contrast to the Rayleigh case, this distribution is not separable. Rather, we have to integrate over the phases to get the amplitude pdf, and vice versa.

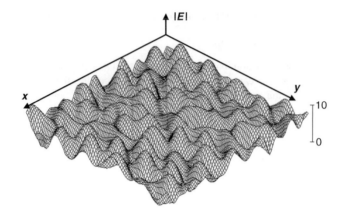

Figure 5.18 Magnitude of the electric field strength, $|E|$, in an example area, in the presence of a dominant MPC.

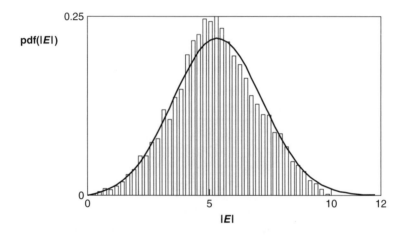

Figure 5.19   Histogram of the amplitudes in the presence of a dominant MPC.

The pdf of the amplitude is given by the *Rice distribution* (solid line in Figure 5.19):

$$pdf_r(r) = \frac{r}{\sigma^2} \cdot \exp\left[-\frac{r^2 + A^2}{2\sigma^2}\right] \cdot I_0\left(\frac{rA}{\sigma^2}\right) \quad 0 \leq r < \infty \tag{5.27}$$

$I_0(x)$ is the modified Bessel function of the first kind, zero order [Abramowitz and Stegun 1965]. The mean square value of a Rice-distributed random variable r is given by:

$$\overline{r^2} = 2\sigma^2 + A^2 \tag{5.28}$$

The ratio of the power in the LOS component to the power in the diffuse component, $A^2/(2\sigma^2)$, is called the Rice factor K_r.

Figure 5.20 shows the Rice distribution for three different values of the Rice factor. The stronger the LOS component, the rarer the occurrence of deep fades. For $K_r \rightarrow 0$, the Rice distribution

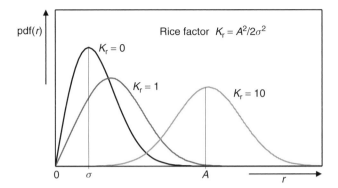

Figure 5.20   Rice distribution for three different values of K_r – i.e., the ratio between the power of the LOS component and the diffuse components.

becomes a Rayleigh distribution, while for large K_r it approximates a Gaussian distribution with mean value A.

Example 5.2   *Compute the fading margin for a Rice distribution with $K_r = 0.3, 3$ and $20\,dB$ so that the outage probability is less than 5%.*

Recall that the outage probability can be expressed in terms of the *cdf* of the Rician envelope:

$$P_{\text{out}} = \text{cdf}(r_{\min}) \tag{5.29}$$

For the Rician distribution the cdf is given as:

$$\text{cdf}(r_{\min}) = \int_0^{r_{\min}} \frac{r}{\sigma^2} \cdot \exp\left[-\frac{r^2 + A^2}{2\sigma^2}\right] \cdot I_0\left(\frac{rA}{\sigma^2}\right) dr \quad 0 \le r < \infty$$

$$= 1 - Q_{\text{M}}\left(\frac{A}{\sigma}, \frac{r_{\min}}{\sigma}\right) \tag{5.30}$$

where $Q_{\text{M}}(a, b)$ is Marcum's Q-function (see also Chapter 12) given by:

$$Q_{\text{M}}(a, b) = e^{-(a^2 + b^2)/2} \sum_{n=0}^{\infty} \left(\frac{a}{b}\right)^n I_n(ab) \tag{5.31}$$

$I_n(\cdot)$ is the modified Bessel function of the first kind, order n. The fading margin is given by:

$$\frac{\overline{r^2}}{r_{\min}^2} = \frac{2\sigma^2(1 + K_r)}{r_{\min}^2} \tag{5.32}$$

The Rice power *cdf* is plotted in Figure 5.21. The required fading margins at different K_r can be found from that figure: they are 11.5, 9.7, and 1.1 dB, for Rice factors of 0.3, 3, and 20 dB, respectively.

The presence of a dominant component also changes the *phase distribution*. This becomes intuitively clear by recalling that for a very strong dominant component, the phase of the total signal must be very close to the phase of the dominant component – in other words, the phase

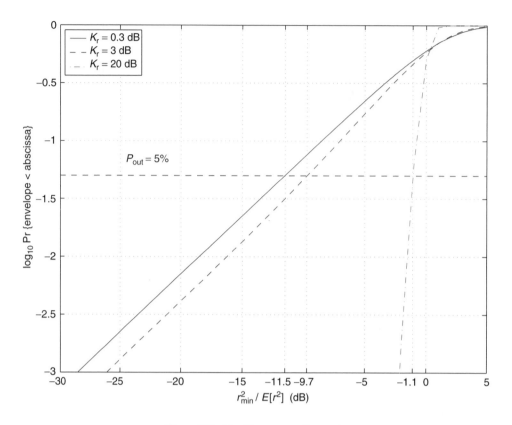

Figure 5.21 The Rice power *cdf*, $\sigma = 1$.

distribution converges to a delta function. For the general case (remember that we define the phase of the LOS component as $\psi = 0$), the pdf of the phase can be computed from the joint pdf of r and ψ and becomes [Lustmann and Porrat 2010]:

$$\text{pdf}(\psi) = \frac{1 + \sqrt{\pi K_r} e^{K_r \cos^2(\psi)} \cos(\psi) \left(1 + \text{erf}\left[\sqrt{K_r} \cos(\psi)\right]\right)}{2\pi e^{K_r}} \tag{5.33}$$

where $erf(x)$ is the error function [Abramowitz and Stegun 1965]:

$$\text{erf}(x) = (2/\sqrt{\pi}) \int_0^x \exp(-t^2)\, dt$$

Figure 5.22 shows the phase distribution for $\sigma = 1$ and different values of A.

The pdf of the power is given as

$$pdf_P(P) = \frac{1 + K_r}{\overline{\Omega}} \exp\left(-K_r - \frac{(K_r + 1)P}{\overline{\Omega}}\right) I_0\left(2\sqrt{\frac{K_r(K_r + 1)P}{\overline{\Omega}}}\right) \quad \text{for } P \geq 0 \tag{5.34}$$

From a historical perspective, it is interesting to note that all the work about Rice distributions was performed without the slightest regard for wireless channels. The classical paper of Rice [Rice 1947] considered the problem of a sinusoidal wave in additive white Gaussian noise. However,

Figure 5.22 Pdf of the phase of a non-zero-mean complex Gaussian distribution, with $\sigma = 1$, $A = 0, 1, 3, 10$.

from a mathematical point of view, this is just the problem of a deterministic phasor (giving rise to a non-zero-mean) added to a zero-mean complex Gaussian distribution – exactly the same problem as in the field strength computation. Existing results thus just had to be reinterpreted by wireless engineers. This fact is so interesting because there are probably other wireless problems that can be solved by such "reinterpretation" methods.

5.5.3 Nakagami Distribution

Another probability distribution for field strength that is in widespread use is the Nakagami m-distribution. The pdf is given as:

$$pdf_r(r) = \frac{2}{\Gamma(m)} \left(\frac{m}{\overline{\Omega}} \right)^m r^{2m-1} \exp\left(-\frac{m}{\overline{\Omega}} r^2 \right) \tag{5.35}$$

for $r \geq 0$ and $m \geq 1/2$; $\Gamma(m)$ is Euler's Gamma function [Abramowitz and Stegun 1965]. The parameter $\overline{\Omega}$ is the mean square value $\overline{\Omega} = \overline{r^2}$, and the parameter m is

$$m = \frac{\overline{\Omega^2}}{\overline{(r^2 - \overline{\Omega})}^2} \tag{5.36}$$

It is straightforward to extract these parameters from measured values. If the amplitude is Nakagami-fading, then the power follows a Gamma distribution:

$$pdf_P(P) = \frac{m}{\overline{\Omega}\Gamma(m)} \left(\frac{mP}{\overline{\Omega}} \right)^{m-1} \exp\left(-\frac{mP}{\overline{\Omega}} \right) \tag{5.37}$$

Nakagami and Rice distribution have a quite similar shape, and one can be used to approximate the other. For $m > 1$ the m-factor can be computed from K_r by Stueber [1996]:

$$m = \frac{(K_r + 1)^2}{(2K_r + 1)} \tag{5.38}$$

while

$$K_r = \frac{\sqrt{m^2 - m}}{m - \sqrt{m^2 - m}} \tag{5.39}$$

While Nakagami and Rice pdfs show good "general" agreement, they have different slopes close to $r = 0$. This in turn has an important impact on the achievable diversity order (see Chapter 13).

The main difference between the two pdfs is that the Rice distribution gives the *exact* distribution of the amplitude of a non-zero-mean complex Gaussian distribution – this implies the presence of one dominant component, and a large number of non-dominant components. The Nakagami distribution describes *in an approximate way* the amplitude distribution of a vector process where the central limit theorem is not necessarily valid (e.g., ultrawideband channels – see Chapter 7).

5.6 Doppler Spectra and Temporal Channel Variations

5.6.1 Temporal Variations for Moving MS

Section 5.3 showed us the physical interpretation of the frequency shift by movement – i.e., the Doppler effect. If the Mobile Station (MS) is moving, then different directions of the MPCs arriving at the MS give rise to different frequency shifts. This leads to a broadening of the received spectrum. The goal of this section is to derive this spectrum, assuming that the transmit signal is a sinusoidal signal (i.e., the narrowband case). The more general case of a wideband transmit signal is treated in Chapter 6.

Let us first repeat the expressions for Doppler shift when a wave only comes from a single direction. Let γ denote the angle between the velocity vector v of the MS and the direction of the wave at the location of the MS. As shown in Eq. (5.8), the Doppler effect leads to a shift of the received frequency f by the amount ν, so that the received frequency is given by:

$$f = f_c \left[1 - \frac{\text{v}}{c_0} \cos(\gamma) \right] = f_c - \nu \tag{5.40}$$

where $\text{v} = |\mathbf{v}|$. Obviously, the frequency shift depends on the direction of the wave, and must lie in the range $f_c - \nu_{\max} \ldots f_c + \nu_{\max}$, where $\nu_{\max} = f_c \text{v}/c_0$.

If there are multiple MPCs, we need to know the distribution of power of the incident waves as a function of γ. As we are interested in the *statistical* distribution of the received signal, we consider the pdf of the received power; in a slight abuse of notation, we call it the pdf of the incident waves $pdf_\gamma(\gamma)$. The MPCs arriving at the RX are also weighted by the antenna pattern of the MS; therefore, an MPC arriving in the direction γ has to be multiplied by the pattern $G(\gamma)$. The received power spectrum as a function of direction is thus:

$$S(\gamma) = \overline{\Omega} \left[pdf_\gamma(\gamma) G(\gamma) + pdf_\gamma(-\gamma) G(-\gamma) \right] \tag{5.41}$$

where $\overline{\Omega}$ is the mean power of the arriving field. In Eq. (5.41), we have also exploited the fact that waves from the direction γ and $-\gamma$ lead to the same Doppler shift, and thus need not be distinguished for the purpose of deriving a Doppler spectrum.

In a final step, we have to perform the variable transformation $\gamma \to \nu$. The Jacobian can be determined as:

$$\left| \frac{d\gamma}{d\nu} \right| = \left| \frac{1}{\frac{d\nu}{d\gamma}} \right| = \frac{1}{\left| \frac{\nu}{c_0} f_c \sin(\gamma) \right|} = \frac{1}{\sqrt{\left(f_c \frac{\nu}{c_0} \right)^2 - (f - f_c)^2}} = \frac{1}{\sqrt{\nu_{max}^2 - \nu^2}} \tag{5.42}$$

so that the Doppler spectrum becomes:

$$S_D(\nu) = \begin{cases} \overline{\Omega}[pdf_\gamma(\gamma)G(\gamma) + pdf_\gamma(-\gamma)G(-\gamma)]\frac{1}{\sqrt{\nu_{max}^2 - \nu^2}} & \text{for} - \nu_{max} \le \nu \le \nu_{max} \\ 0 & \text{otherwise} \end{cases} \tag{5.43}$$

For further use, we also define

$$\Omega_n = (2\pi)^n \int_{-\nu_{max}}^{\nu_{max}} S_D(\nu)\nu^n \, d\nu \tag{5.44}$$

as the nth moment of the Doppler spectrum.

More specific equations can be obtained for specific angular distributions and antenna patterns. A very popular model for the angular spectrum at the MS is that the waves are incident uniformly from all azimuthal directions, and all arrive in the horizontal plane, so that:

$$pdf_\gamma(\gamma) = \frac{1}{2\pi} \tag{5.45}$$

This situation corresponds to the case when there is no LOS connection, and a large number of IOs are distributed uniformly around the MS (see also Chapter 7). Assuming furthermore that the antenna is a vertical dipole, with an antenna pattern $G(\gamma) = 1.5$ (see Chapter 9), the Doppler spectrum becomes

$$S_D(\nu) = \frac{1.5\overline{\Omega}}{\pi \sqrt{\nu_{max}^2 - \nu^2}} \tag{5.46}$$

This spectrum, known as the *classical* or *Jakes* spectrum, is depicted in Figure 5.23. It has the characteristic "bathtub" shape – i.e., (integrable) singularities at the minimum and maximum Doppler frequencies $\nu = \pm\nu_{max} = \pm f_c \nu/c_0$. These singularities thus correspond to the direction of

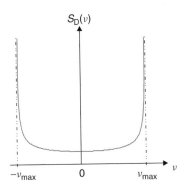

Figure 5.23 Classical Doppler spectrum.

the movement of the MS, or its opposite. It is remarkable that a uniform azimuthal distribution can lead to a highly *non*uniform Doppler spectrum.

Naturally, *measured* Doppler spectra do not show singularities; even if the model and underlying assumptions would be strictly valid, it would require an infinite number of measurement samples to arrive at a singularity. Despite this fact, the classical Doppler spectrum is the most widely used model. Alternative models include the following:

- Aulin's model [see Parsons 1992], which limits the amplitude of the Jakes spectrum at and near its singularities;
- Gaussian spectrum;
- uniform spectrum (this corresponds to the case when all waves are incident uniformly in all three dimensions, and the antenna has an isotropic pattern).

As already mentioned in Section 5.2, the Doppler spectrum has two important interpretations:

1. It describes *frequency dispersion*. For narrowband systems, as well as Orthogonal Frequency Division Multiplexing (OFDM), such frequency dispersion can lead to transmission errors. This is discussed in more detail in Chapters 12 and 19. It has, however, no *direct* impact on most other wideband systems (like single-carrier Time Division Multiple Access (TDMA) or Code Division Multiple Access (CDMA) systems).
2. It is a measure for the *temporal variability* of the channel. As such, it is important for *all* systems.

The temporal dependence of fading is best described by the autocorrelation function of fading. The normalized correlation between the in-phase component at time t, and the in-phase component at time $t + \Delta t$ can be shown to be:

$$\frac{\overline{I(t)I(t + \Delta t)}}{\overline{I(t)^2}} = J_0(2\pi v_{\max} \Delta t) \tag{5.47}$$

which is proportional to the inverse Fourier transform of the Doppler spectrum $S_D(v)$, while the correlation $\overline{I(t)Q(t + \Delta t)} = 0$ for all values of Δt. The normalized covariance function of the envelope is thus:

$$\frac{\overline{r(t)r(t + \Delta t)} - \overline{r(t)}^2}{\overline{r(t)^2}} = J_0^2(2\pi v_{\max} \Delta t) \tag{5.48}$$

More details about the autocorrelation function can be found in Chapters 6 and 13.

Example 5.3 *Assume that an MS is located in a fading dip. On average, what minimum distance should the MS move so that it is no longer influenced by this fading dip?*

As a first step, we plot the envelope correlation function (Eq. 5.48) in Figure 5.24. If we now define "no longer influenced" as "envelope correlation coefficient equals 0.5," then we see that (on average) the RX has to move through 0.18 λ. If we want complete decorrelation from the fading dip, then moving through 0.38 λ is required.

5.6.2 Temporal Variations in Fixed Wireless Systems

In a fixed wireless system (i.e., a system where the MS does not move, see Anderson [2003]), temporal variations can arise only from movement of the IOs. The resulting fading leads to a

Figure 5.24 Amplitude correlation as a function of displacement of the receiver.

different way of thinking about, and modeling of, the temporal channel changes. In a mobile link, an MS tracks the spatial fading (i.e., the fading "landscape" of Figure 5.11) as it moves. The spatial and temporal fading can thus be mapped onto each other, the scaling factor being the mobile speed v. In a fixed wireless link, the variation is due instead to the fact that some scattering objects are moving, a prime example being cars, or windblown leaves.

Measurements have shown that in many cases, the amplitude statistics of the fading (taken over the ensemble of values observed at different times) are Rician. However, the Rice factor now represents the ratio of the power in the *time-invariant* MPCs vs. the power in the *time-variant* MPCs. In other words, the temporal Rice factor has nothing to do with LOS (or other dominant) components: in a pure NLOS situation with many equally strong components the temporal Rice factor can approach infinity if all MPCs are time invariant.

The Doppler spectrum of such a system consists of a delta pulse at $\nu = 0$, and a continuous spectrum that depends on the movement and location of the moving IOs. Various measurement campaigns have shown that the shape of this diffuse spectrum is approximately Gaussian.

5.7 Temporal Dependence of Fading

5.7.1 Level Crossing Rate

The Doppler spectrum is a complete characterization of the temporal statistics of fading. However, it is often desirable to have a different formulation that allows more direct insights into system behavior. A quantity that allows immediate interpretation is the occurrence rate of fading dips – this occurrence rate is known as the *Level Crossing Rate* (LCR). Obviously, it depends on which level we are considering (i.e., how a fading dip is defined): falling below a level that is 30 dB below the mean happens more rarely than falling 3 dB below this mean. As the admissible depth of fading dips depends on the mean field strength, as well as on the considered system, we want to derive the LCR for arbitrary levels (i.e., depth of fading dips).

Providing a mathematical formulation, the LCR is defined as the expected value of the rate at which the received field strength crosses a certain level r in the positive direction. This can also be written as:

$$N_R(r) = \int_0^\infty \dot{r} \cdot pdf_{r,\dot{r}}(r, \dot{r}) \, d\dot{r} \quad \text{for } r \geq 0 \tag{5.49}$$

where $\dot{r} = dr/dt$ is the temporal derivative, and $pdf_{r,\dot{r}}$ is the joint pdf of r and $\dot{r}$.

In Appendix 5.C (see *www.wiley.com/go/molisch*), we derive the LCR as:

$$N_R(r) = \sqrt{\frac{\Omega_2}{\pi \Omega_0}} \frac{r}{\sqrt{2\Omega_0}} \exp\left(-\frac{r^2}{2\Omega_0}\right) \tag{5.50}$$

Note that $\sqrt{2\Omega_0}$ is the root-mean-square value of amplitude. Figure 5.25 shows the LCR for a Rayleigh-fading amplitude and Jakes Doppler spectrum.

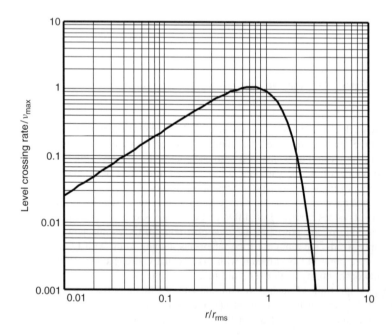

Figure 5.25 Level crossing rate normalized to the maximum Doppler frequency as a function of the normalized level r/r_{rms}, for a Rayleigh-fading amplitude and Jakes spectrum.

5.7.2 Average Duration of Fades

Another parameter of interest is the *Average Duration of Fades* (ADF). In the previous sections, we have already derived the rate at which the field strength goes below the considered threshold (i.e., the LCR), and the total percentage of time the field strength is lower than this threshold (i.e., the cdf of the field strength). The ADFs can be simply computed as the quotient of these two quantities:

$$ADF(r) = \frac{cdf_r(r)}{N_R(r)} \tag{5.51}$$

A plot of the ADF is shown in Figure 5.26.

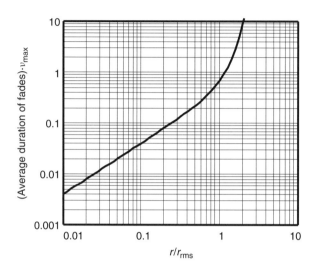

Figure 5.26 Average duration of fades normalized to the maximum Doppler frequency as a function of the normalized level r/r_{rms}, for a Rayliegh-fading amplitude and Jakes spectrum.

Example 5.4 *Assume a multipath environment where the received signal has a Rayleigh distribution and the Doppler spectrum has the classical bathtub (Jakes) shape. Compute the LCR and the ADF for a maximum Doppler frequency $\upsilon_{max} = 50\,Hz$, and amplitude thresholds:*

$$r_{min} = \frac{\sqrt{2\Omega_0}}{10}, \frac{\sqrt{2\Omega_0}}{2}, \sqrt{2\Omega_0}$$

The LCR is computed from Eq. (5.50). For a Jakes scenario the second moment of the Doppler spectrum is given as:

$$\Omega_2 = \frac{1}{2}\Omega_0(2\pi\upsilon_{max})^2 \tag{5.52}$$

therefore,

$$N_R(r_{min}) = \sqrt{2\pi} \cdot \upsilon_{max} \cdot \frac{r_{min}}{\sqrt{2\Omega_0}} \exp\left(-\frac{r_{min}^2}{2\Omega_0}\right) \tag{5.53}$$

The ADF is computed from Eq. (5.51) where:

$$cdf(r_{min}) = 1 - \exp\left(-\left(\frac{r_{min}}{\sqrt{2\Omega_0}}\right)^2\right) \tag{5.54}$$

These expressions are evaluated for different values of the threshold, r_{min}. The results are tabulated in Table 5.1.

Table 5.1 Effect of threshold on ADF and LCR

r_{min}	$N_R(r_{min})$	$cdf(r_{min})$	ADF (msec)
$\dfrac{\sqrt{2\Omega_0}}{10}$	12.4	0.01	0.8
$\dfrac{\sqrt{2\Omega_0}}{2}$	48.8	0.22	4.5
$\sqrt{2\Omega_0}$	46.1	0.63	13.7

5.7.3 Random Frequency Modulation

A random channel leads to a random phase shift of the received signal; in a time-variant channel, these phase shifts are time variant as well. By definition, a temporally varying phase shift is an FM. In this section, we compute this *random FM*.

The pdf of the instantaneous frequency $\dot{\psi}$ can be computed from the joint pdf $pdf_{r,\dot{r},\psi,\dot{\psi}}$ of Appendix 5.C (see *www.wiley.com/go/molisch*). We now have to integrate over the variables r, $\dot{r}$, and ψ. This results in:

$$pdf_{\dot{\psi}}(\dot{\psi}) = \frac{1}{2}\sqrt{\frac{\Omega_0}{\Omega_2}}\left(1 + \frac{\Omega_0}{\Omega_2}\dot{\psi}^2\right)^{-3/2} \tag{5.55}$$

See Figure 5.27.

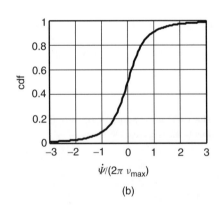

(a)   (b)

Figure 5.27 Normalized pdf (a) and cdf (b) of a random FM.

This is a "student's *t*-distribution" with two degrees of freedom [Mardia et al. 1979]. The cumulative distribution is given as:

$$cdf_{\dot{\psi}}(\dot{\psi}) = \frac{1}{2}\left[1 + \sqrt{\frac{\Omega_0}{\Omega_2}}\dot{\psi}\left(1 + \frac{\Omega_0}{\Omega_2}\dot{\psi}^2\right)^{-1/2}\right] \tag{5.56}$$

It is remarkable that all instantaneous frequencies in a range $-\infty$ to ∞ can occur – they are not restricted to the range of possible Doppler frequencies!

Strong FM most likely occurs in fading dips: if the signal level is low, very strong relative changes can occur. This intuitively pleasing result can be shown mathematically by considering the pdf of the instantaneous frequency conditioned on the amplitude level r_0 [Jakes 1974]:

$$pdf(\dot{\psi}|r_0) = \frac{r_0}{\sqrt{2\pi\,\Omega_2}}\exp\left(-r_0^2\frac{\dot{\psi}^2}{2\Omega_2}\right) \tag{5.57}$$

This is a zero-mean Gaussian distribution with variance Ω_2/r_0^2. The variance is the smaller the larger the signal level is.

We thus see that fading dips can create errors in two ways: on one hand, the lower signal level leads to higher susceptibility to noise. On the other hand, they increase the probability of strong random FMs, which introduces errors in any system that conveys information by means of the *phase* of the transmitted signal. In addition, we find in Chapter 12 that fading dips are also related to intersymbol interference.

5.8   Large-Scale Fading

Small-scale fading, created by the superposition of different MPCs, changes rapidly over the spatial scale of a few wavelengths. If the field strength is averaged over a small area (e.g., ten by ten wavelengths), we obtain the *Small Scale Averaged* (SSA) field strength.[7] In the previous sections, we treated the SSA field strength as a constant. However, as explained in the introduction, it varies when considered on a larger spatial scale, due to shadowing of the MPCs by IOs.

Many experimental investigations have shown that the SSA field strength F, *plotted on a logarithmic scale*, shows a Gaussian distribution around a mean μ. Such a distribution is known as lognormal, and its pdf is given by:

$$pdf_{\rm F}(F) = \frac{20/\ln(10)}{F\sigma_{\rm F}\sqrt{2\pi}} \cdot \exp\left[-\frac{(20\log_{10}(F) - \mu_{\rm dB})^2}{2\cdot\sigma_{\rm F}^2}\right] \tag{5.58}$$

where $\sigma_{\rm F}$ is the standard deviation of F, and $\mu_{\rm dB}$ is the mean of the values of F expressed in dB. Also, power is distributed lognormally. However, there is an important fine point: when fitting the logarithm of SSA *power* to a normal distribution, we find that the median value of this distribution, $\mu_{\rm P, dB}$, is related to the median of the field strength distribution $\mu_{\rm dB}$ as $\mu_{\rm P, dB} = \mu_{\rm dB} + 10\log(4/\pi)$ dB. The pdf for the power thus reads

$$pdf_{\rm P}(F) = \frac{10/\ln(10)}{P\sigma_{\rm P}\sqrt{2\pi}} \cdot \exp\left[-\frac{(10\log_{10}(P) - \mu_{\rm P, dB}))^2}{2\cdot\sigma_{\rm P}^2}\right] \tag{5.59}$$

where $\sigma_{\rm P} = \sigma_{\rm F}$. Typical values of $\sigma_{\rm F}$ are 4 to 10 dB.

Another interesting property of the lognormal distribution is that the sum of lognormally distributed values is also approximately lognormally distributed. A variety of methods exist for computing the mean and variance of the sum from the parameters of the composite distributions. Methods suitable for various special cases are described in Stueber [1996]; a general, flexible, method was recently derived in Mehta et al. [2007].

The lognormal variations of F have commonly been attributed to shadowing effects. Consider the situation in Figure 5.28. If the MS moves, it changes the angle γ, and thus the diffraction

---

[7] Strictly speaking, this is only an approximation to the SSA field strength, as there can only be a finite number of statistically independent sample values within the finite area over which we average.

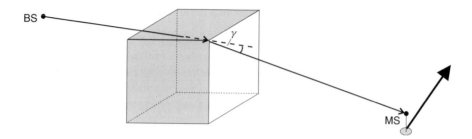

Figure 5.28 Shadowing by a building.

parameter ν_F. If the distance between the edge and the MS is now large, the MS must move over a large distance (e.g., several tens of wavelengths) for the field strength to change noticeably. Note that this effect changes the absolute amplitude of the MPC, $|a|$, and has nothing to do with an interference effect. Consider now the situation where an MPC undergoes several of these or similar processes on its way from the TX and RX, each of which contributes a certain attenuation. The received field strength then depends on the *product* of the attenuations; on a logarithmic scale, these attenuations add up. We can thus model the effect as the sum of random variables *on a dB scale* – i.e., a lognormal distribution. However, the mechanism just described is not necessarily valid in all physical situations.[8]

Consider now the statistics of the field strength based on samples taken from a large area (i.e., including both lognormal fading and interference effects). The pdf of these samples is given by the so-called Suzuki distribution, which follows in a straightforward manner from the laws of conditional statistics. The mean value of the field strength, taken over a small area, is $\bar{r} = \sigma\sqrt{\pi/2}$, and the distribution of the local value of the field strength conditioned on σ is

$$pdf(r|\bar{r}) = \frac{\pi r}{2\bar{r}^2}\exp\left(-\frac{\pi r^2}{4\bar{r}^2}\right) \tag{5.60}$$

The local mean (i.e., the expected value used above) is distributed according to lognormal statistics. Unconditioning results in the pdf of the field strength:

$$pdf_r(r) = \int_0^\infty \frac{\pi r}{2\bar{r}^2}\exp\left(-\frac{\pi r^2}{4\bar{r}^2}\right)\frac{2 \cdot 10\ln(10)}{\bar{r}\sigma_F\sqrt{2\pi}}\exp\left(-\frac{(20\log_{10}(\bar{r}) - \mu_{dB})^2}{2\sigma_F^2}\right)d\bar{r} \tag{5.61}$$

This pdf is also known as the Suzuki distribution, an example is shown in Figure 5.29. A similar function can be defined if the small-scale statistics are Rician or Nakagami.

The pdf for the power reads

$$pdf_P(P) = \int_0^\infty \frac{1}{\Omega}\exp\left(-\frac{P}{\Omega}\right)\frac{10/\ln(10)}{\Omega\sigma_P\sqrt{2\pi}}\exp\left(-\frac{(10\log_{10}(\Omega) - \mu_{P,\,dB})^2}{2\sigma_P^2}\right)d\Omega. \tag{5.62}$$

Since both large-scale and small-scale fading occurs in practical situations, the fading margin must account for the combination of the two effects (see also Chapter 3). One possibility is to just add up the fading margin for the Rayleigh distribution and the fading margin for a lognormal

---

[8] An alternative explanation, based on double-scattering processes, was proposed by Andersen [2002]. But, independently of the mechanism that leads to its creation, many measurements have confirmed that the pdf of F is well-approximated by a lognormal function.

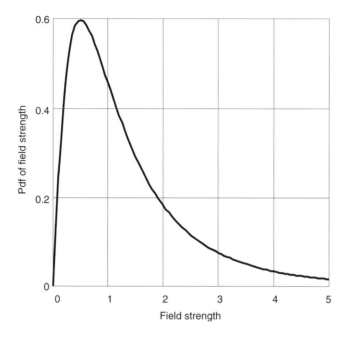

Figure 5.29   Suzuki distribution with $\sigma_F = 5.6\,\text{dB}$, $\mu = 1.4\,\text{dB}$.

distribution. This method is commonly used because of its simplicity, but overestimates the required fading margin. The more accurate method is based on the cdf of the Suzuki distribution, which can be obtained by integrating Eq. (5.61) from $-\infty$ to x. This then allows computation of the necessary mean field strength r_0 for a given admissible outage probability (e.g., 0.05), as shown in the following example.

Example 5.5   *Consider a channel with $\sigma_F = 6\,dB$ and $\mu = 0\,dB$. Compute the fading margin for a Suzuki distribution relative to the mean-dB value of the shadowing so that the outage probability is smaller than 5%. Also compute the fading margin for Rayleigh fading and shadow fading separately.*

For the Suzuki distribution:

$$
P_{\text{out}} = cdf(r_{\min}) = \int_0^{r_{\min}} pdf_r(r)\, dr
$$

$$
= \int_0^\infty \left(1 - \exp\left(-\frac{\pi r_{\min}^2}{4\bar{r}^2}\right)\right) \frac{20/\ln(10)}{\bar{r}\sigma_F\sqrt{2\pi}} \exp\left(-\frac{(20\log_{10}\bar{r} - \mu_{\text{dB}})^2}{2\sigma_F^2}\right) d\bar{r} \quad (5.63)
$$

Inserting $\sigma_F = 6\,\text{dB}$ and $\mu_{\text{dB}} = 0$, the *cdf* is expressed as:

$$
cdf(r_{\min}) = \frac{20/\ln 10}{\sqrt{2\pi} \cdot 6} \int_0^\infty \frac{1}{\bar{r}} \exp\left(-\frac{(20\log\bar{r} - 0)^2}{2 \cdot 36}\right) \left(1 - \exp\left(-\frac{\pi}{4} \cdot \frac{r_{\min}^2}{\bar{r}^2}\right)\right) d\bar{r} \quad (5.64)
$$

The fading margin is defined as:

$$
M = \frac{\mu^2}{r_{\min}^2} \quad (5.65)
$$

The *cdf* plot is obtained by evaluating Eq. (5.65) for different values of r_{min}, the result is shown in Figure 5.30. A fading margin of 15.5 dB is required to get an outage probability of 5%.

The outage probability for Rayleigh fading only is evaluated as:

$$P_{out} = cdf(r_{min}) = 1 - \exp\left(-\frac{r_{min}^2}{2\sigma^2}\right) \tag{5.66}$$

$$= 1 - \exp(-1/M)$$

After some manipulation, we get:

$$M_{Rayleigh, dB} = -10\log_{10}(-\ln(1 - P_{out})) \tag{5.67}$$

$$= 12.9 \, dB$$

For shadow fading, the pdf of the field strength values in dB is a standard Gaussian distribution. The complementary cdf (i.e., unity minus the cdf) is thus given by a Q-function defined as:

$$Q(a) = \frac{1}{\sqrt{2\pi}} \int_a^\infty \exp\left(-\frac{x^2}{2}\right) dx$$

The outage probability is given as:

$$P_{out} = Q\left(\frac{M_{large\text{-}scale_{dB}}}{\sigma_F}\right) \tag{5.68}$$

Inserting $\sigma_F = 6 \, dB$ and $P_{out} = 0.05$ into Eq. (5.68) we get:

$$M_{large\text{-}scale_{dB}} = 6 \cdot Q^{-1}(0.05) \tag{5.69}$$

$$= 9.9 \, dB$$

When computing the fading margin as the sum of the margin for Rayleigh fading and shadowing, we obtain

$$M_{dB} = 12.9 + 9.9 \, dB = 22.8 \, dB \tag{5.70}$$

Compared with the fading margin obtained from the Suzuki distribution, this is a more conservative estimate.

It turns out that a Suzuki distribution can be approximated reasonably well by a lognormal distribution, where the parameters of approximating lognormal distribution are related to the parameters of the shadowing alone (i.e., without the Rayleigh fading) by

$$\mu_{approx, dB} = \mu_{shadow, dB} - 1.5 \tag{5.71}$$

$$\sigma_{dB}^2 = \sigma_{shadow, dB}^2 + 13 \tag{5.72}$$

We finally turn to the spatial correlation of the shadowing alone (without small-scale fading). The value of the shadowing changes only slowly with the location of the MS. Thus, realizations of the shadowing field strength (or power) that are measured at a distance Δx apart are correlated. The most common model is that of an exponential correlation, i.e., $E_x\{F(x)F(x + \Delta x)\} = \exp(-\Delta x/\bar{x})$, where the decorrelation distance $\bar{x}$ is typically on the order of 5–50 m, depending on the environment.

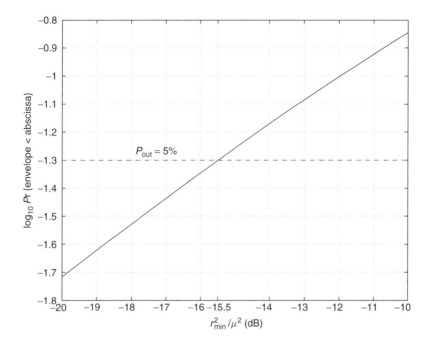

Figure 5.30 The Suzuki *cdf*, $\sigma_F = 6\,dB$ and $\mu_{dB} = 0$.

5.9 Appendices

Please see companion website *www.wiley.com/go/molisch*

Further Reading

The mathematical basis for the derivations in this chapter is described in a number of standard textbooks on statistics and random processes, especially the classical book of Papoulis [1991]. The statistical model for the amplitude distribution and the Doppler spectrum was first described in Clarke [1968]. A comprehensive exposition of the statistics of the channel, derivation of the Doppler spectrum, LCR, and ADFs, for the Rayleigh case, can be found in Jakes [1974]. Since Rayleigh fading is based on Gaussian fading of the I- and Q-component, the rich literature on Gaussian multivariate analysis is applicable [Muirhead 1982]. Derivation of the Nakagami distribution, and many of its statistical properties, can be found in Nakagami [1960]; a physical interpretation is given in Braun and Dersch [1991]; the Rice distribution is derived in Rice [1947]. The Suzuki distribution is derived in Suzuki [1977]. More details about the lognormal distribution, especially the summing of several lognormally distributed variables, can be found in Stueber [1996], Cardieri and Rappaport [2001], and Mehta et al. [2007]. The combination of Nakagami small-scale fading and lognormal shadowing is treated in Thjung and Chai [1999]. The fading statistics, including ADF and LCR, of the Nakagami and Rice distributions are summarized in Abdi et al. [2000]. Another important aspect of statistical channel descriptions is the generation of random variables according to a prescribed Doppler spectrum. An extensive description of this area can be found in Paetzold [2002].

For updates and errata for this chapter, see wides.usc.edu/teaching/textbook

6

Wideband and Directional Channel Characterization

6.1 Introduction

In the previous chapter, we considered the effect of multipath propagation on the received field strength and the temporal variations if the transmit signal is a pure sinusoid. These considerations are also valid for all systems where the bandwidth of the transmitted signal is "very small" (see below for a more precise definition). However, most current and future wireless systems use a large bandwidth, either because they are intended for high data rates or because of their multiple access scheme (see Chapters 17–19). We therefore have to describe variations of the channel over a large bandwidth as well. These description methods are the topic of the current chapter.

The impact of multipath propagation in wideband systems can be interpreted in two different ways: (i) the transfer function of the channel varies over the bandwidth of interest (this is called the *frequency selectivity* of the channel) or (ii) the impulse response of the channel is not a delta function; in other words, the arriving signal has a longer duration than the transmitted signal (this is called *delay dispersion*). These two interpretations are equivalent, as can be shown by performing Fourier transformations between the delay (time) domain and the frequency domain.

In this chapter, we first explain the basic concepts of wideband channels using again the most simple channel – namely, the two-path channel. We then formulate the most general statistical description methods for wideband, time-variant channels (Section 6.3), and discuss its most common special form, the WSSUS – Wide Sense Stationary Uncorrelated Scatterer – model (Section 6.4). Since these description methods are rather complicated, condensed parameters are often used for a more compact description (Section 6.5). Section 6.6 considers the case when the channel is not only wide enough to show appreciable variations of the transfer function but even so wide that the bandwidth becomes comparable with the carrier frequency – this case is called "ultra wideband."

Systems operating in wideband channels have some important properties:

- They suffer from InterSymbol Interference (ISI). This can be most easily understood from the interpretation of delay dispersion. If we transmit a symbol of length T_S, the arriving signal corresponding to that symbol has a longer duration, and therefore interferes with the subsequent symbol (Chapter 2). Section 12.3 describes the effect of this ISI on the Bit Error Rate (BER) if

Wireless Communications, Second Edition   Andreas F. Molisch
© 2011 John Wiley & Sons, Ltd

no further measures are taken; Chapter 16 describes equalizer structures that can actively combat the detrimental effect of the ISI.

- They can reduce the detrimental effect of fading. This effect can be most easily understood in the frequency domain: even if some part of the transmit spectrum is strongly attenuated, there are other frequencies that do not suffer from attenuation. Appropriate coding and signal processing can exploit this fact, as explained in Chapters 16, 18, and 19.

The properties of the channel can vary not only depending on the frequency at which we consider it but also depending on the location. This latter effect is related to the directional properties of the channel – i.e., the directions from which the Multi Path Components (MPCs) are incident. Section 6.7 discusses the stochastic description methods for these directional properties; they are especially important for antenna diversity (Chapter 13) and multielement antennas (Chapter 20).

6.2   The Causes of Delay Dispersion

6.2.1   The Two-Path Model

Why does a channel exhibit delay dispersion – or, equivalently, why are there variations of the channel over a given frequency range? The most simple picture arises again from the two-path model, as introduced in the beginning of Chapter 5. The transmit signal gets to the receiver (RX) via two different propagation paths with different runtimes:

$$\tau_1 = d_1/c_0 \quad \text{and} \quad \tau_2 = d_2/c_0 \tag{6.1}$$

We assume now that runtimes do not change with time (this occurs when neither transmitter (TX), RX, nor Interacting Objects (IOs) move). Consequently, the channel is linear and time invariant, and has an impulse response:

$$h(\tau) = a_1\delta(\tau - \tau_1) + a_2\delta(\tau - \tau_2) \tag{6.2}$$

where again the complex amplitude $a = |a| \exp(j\varphi)$. Clearly, such a channel exhibits delay dispersion; the support (duration) of the impulse response has a finite extent, namely $\tau_2 - \tau_1$.

A Fourier transformation of the impulse response gives the transfer function $H(j\omega)$:

$$H(f) = \int_{-\infty}^{\infty} h(\tau) \exp[-j2\pi f\tau]\,d\tau = a_1 \exp[-j2\pi f\tau_1] + a_2 \exp[-j2\pi f\tau_2] \tag{6.3}$$

The magnitude of the transfer function is

$$|H(f)| = \sqrt{|a_1|^2 + |a_2|^2 + 2|a_1||a_2| \cos(2\pi f \cdot \Delta\tau - \Delta\varphi)}$$

$$\text{with } \Delta\tau = \tau_2 - \tau_1 \text{ and } \Delta\varphi = \varphi_2 - \varphi_1 \tag{6.4}$$

Figure 6.1 shows the transfer function for a typical case. We observe first that the transfer function depends on the frequency, so that we have *frequency-selective fading*. We also see that there are dips (*notches*) in the transfer function at the so-called *notch frequencies*. In the two-path model, the notch frequencies are those frequencies where the phase difference of the two arriving waves becomes 180°. The frequency difference between two adjacent notch frequencies is

$$\Delta f_{\text{Notch}} = \frac{1}{\Delta\tau} \tag{6.5}$$

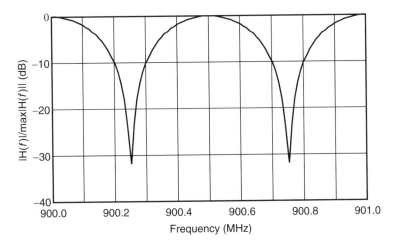

Figure 6.1 Normalized transfer function for $|a_1| = 1.0$, $|a_2| = 0.95$, $\Delta\varphi = 0$, $\tau_1 = 4\,\mu s$, $\tau_2 = 6\,\mu s$ at the 900-MHz carrier frequency.

The destructive interference between the two waves is stronger the more similar the amplitudes of the two waves are.

Channels with fading dips distort not only the amplitude but also the phase of the signal. This can be best seen by considering the group delay, which is defined as the derivative of the phase of the channel transfer function $\phi_H = \arg(H(f))$:

$$\tau_{Gr} = -\frac{1}{2\pi}\frac{d\phi_H}{df} \tag{6.6}$$

As can be seen in Figure 6.2, group delay can become very large in fading dips. As we will see later, this group delay can be related to ISI.

Figure 6.2 Group delay as a function of frequency (same parameters as in Figure 6.1).

6.2.2   The General Case

After the simple two-path model, we now progress to the more general case where IOs can be at any place in the plane. Again, the scenario is static, so that neither TX, RX, nor IOs move. We now draw the ellipses that are defined by their focal points – TX and RX – and the eccentricity determining the runtime.[1] All rays that undergo a single interaction with an object on a specific ellipse arrive at the RX at the same time. Signals that interact with objects on different ellipses arrive at different times. Thus, the channels are *delay-dispersive* if the IOs in the environment are not all located on a single ellipse.

It is immediately obvious that in a realistic environment, IOs never lie exactly on a single ellipse. The next question is thus: How strict must this "single ellipse" condition be fulfilled so that the channel is still "effectively" nondispersive? The answer depends on the system bandwidth. An RX with bandwidth W cannot distinguish between echoes arriving at τ and $\tau + \Delta\tau$, if $\Delta\tau \ll 1/W$ (for many qualitative considerations, it is sufficient to consider the above condition with $\Delta\tau = 1/W$). Thus echoes that are reflected in the donut-shaped region corresponding to runtimes between τ and $(\tau + \Delta\tau)$ arrive at "effectively" the same time (see Figure 6.3).

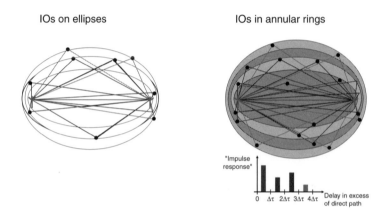

Figure 6.3   Scatterers located on the same ellipses lead to the same delays.

A time-discrete approximation to the impulse response of a wideband channel can thus be obtained by dividing the impulse response into bins of width $\Delta\tau$ and then computing the sum of echoes within each bin. If enough nondominant IOs are in each donut-shaped region, then the MPCs falling into each delay bin fulfill the central limit theorem. In that case, the amplitude of each bin can be described statistically, and the probability density function (pdf) of this amplitude is Rayleigh or Rician. Thus, all the equations of Chapter 5 are still valid; but now they apply for the field strength *within* one delay bin. We furthermore define the minimum delay as the runtime of the direct path between the Base Station (BS) and the Mobile Station (MS) d/c_0 and we define the maximum delay as the runtime from the BS to the MS via the farthest "significant" IO – i.e., the farthest IO that gives a measurable contribution to the impulse response.[2] The *maximum excess delay* τ_{max} is then defined as the difference between minimum and maximum delay.

---

[1] These ellipses are thus quite similar to the Fresnel ellipses described in Chapter 4. The difference is that in Chapter 4 we were interested in excess runtimes that introduce a phase shift of $i \cdot \pi$, while here we are interested in delays that are typically much larger.

[2] We see from this definition that the maximum delay is a quantity that is extremely difficult to measure, and depends on the measurement system.

The above considerations also lead us to a mathematical formulation for *narrowband* and *wideband* from a time domain point of view: a system is narrowband if the inverse of the system bandwidth $1/W$ is much larger than the maximum excess delay τ_{max}. In that case, all echoes fall into a single delay bin, and the amplitude of this delay bin is $\alpha(t)$. A system is wideband in all other cases. In a wideband system, the *shape* and duration of the arriving signal is different from the shape of the transmitted signal; in a narrowband system, they stay the same.

If the impulse response has a finite extent in the delay domain, it follows from the theory of Fourier transforms (FTs) that the transfer function $\mathcal{F}\{h(\tau)\} = H(f)$ is frequency dependent. Delay dispersion is thus equivalent to *frequency selectivity*. A frequency-selective channel cannot be described by a simple attenuation coefficient, but rather the details of the transfer function must be modeled. Note that any real channel is frequency selective if analyzed over a large enough bandwidth; in practice, the question is whether this is true over the bandwidth of the considered system. This is equivalent to comparing the maximum excess delay of the channel impulse response with the inverse system bandwidth. Figure 6.4 sketches these relationships, demonstrating the variations of wideband systems in the delay and frequency domain.

We stress that the definition of a wideband wireless system is fundamentally different from the definition of "wideband" in the usage of Radio Frequency (RF) engineers. The RF definition

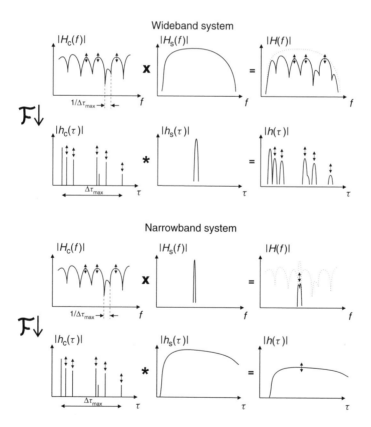

Figure 6.4   Narrowband and wideband systems. $H_C(f)$, channel transfer function; $h_C(\tau)$, channel impulse response.

of wideband implies that the system bandwidth becomes comparable with carrier frequency.[3] In wireless communications, on the other hand, we compare the properties of the *channel* with the properties of the system. It is thus possible that the same system is wideband in one channel, but narrowband for another.

6.3   System-Theoretic Description of Wireless Channels

As we have seen in the previous section, a wireless channel can be described by an impulse response; it thus can be interpreted as a linear filter. If the BS, MS, and IOs are all static, then the channel is time invariant, with an impulse response $h(\tau)$. In that case, the well-known theory of *Linear Time Invariant* (LTI) systems [Oppenheim and Schafer 2009] is applicable. In general, however, wireless channels are time variant, with an impulse response $h(t, \tau)$ that changes with time; we have to distinguish between the absolute time t and the delay τ. Thus, the theory of the *Linear Time Variant* (LTV) system must be used. This is not just a trivial extension of the LTI theory, but gives rise to considerable theoretical challenges and causes the breakdown of many intuitive concepts. Fortunately, most wireless channels can be classified as *slowly* time-variant systems, also known as *quasi-static*. In that case, many of the concepts of LTI systems can be retained with only minor modifications.

6.3.1   Characterization of Deterministic Linear Time Variant Systems

As the impulse response of a time-variant system, $h(t, \tau)$, depends on two variables, τ and t, we can perform Fourier transformations with respect to either (or both) of them. This results in four different, but equivalent, representations. In this section, we investigate these representations, their advantages, and drawbacks.

From a system-theoretic point of view, it is most straightforward to write the relationship between the system input (transmit signal) $x(t)$ and the system output (received signal) $y(t)$ as:

$$y(t) = \int_{-\infty}^{\infty} x(\tau) K(t, \tau) \, d\tau \tag{6.7}$$

where $K(t, \tau)$ is the *kernel* of the integral equation, which can be related to the impulse response. For LTI systems, the well-known relationship $K(t, \tau) = h(t - \tau)$ holds. Generally, we define the time-variant impulse response as:

$$h(t, \tau) = K(t, t - \tau) \tag{6.8}$$

so that

$$y(t) = \int_{-\infty}^{\infty} x(t - \tau) h(t, \tau) \, d\tau \tag{6.9}$$

An intuitive interpretation is possible if the impulse response changes only slowly with time – more exactly, the duration of the impulse response (and the signal) should be much shorter than the time over which the channel changes significantly. Then we can consider the behavior of the system at one time t like that of an LTI system. The variable t can thus be viewed as "absolute" time that

---

[3] In wireless communications, it has become common to denote systems whose bandwidth is larger than 20% of the carrier frequency as *Ultra Wide Bandwidth* (UWB) systems (see Section 6.6). This definition is similar to the RF definition of wideband.

parameterizes the impulse response, i.e., tells us which (out of a large ensemble) impulse response $h(\tau)$ is currently valid. Such a system is also called *quasi-static*.

Fourier transforming the impulse response with respect to the variable τ results in the *time-variant transfer function $H(t, f)$*:

$$H(t, f) = \int_{-\infty}^{\infty} h(t, \tau) \exp(-j2\pi f \tau) \, d\tau \tag{6.10}$$

The input–output relationship is given by:

$$y(t) = \int_{-\infty}^{\infty} X(f) H(t, f) \exp(j2\pi f t) \, df \tag{6.11}$$

The interpretation is straightforward for the case of the quasi-static system – the spectrum of the input signal is multiplied by the spectrum of the "currently valid" transfer function, to give the spectrum of the output signal. If, however, the channel is quickly time varying, then Eq. (6.11) is a purely mathematical relationship. The spectrum of the output signal is given by a double integral

$$Y(\tilde{f}) = \int_{-\infty}^{\infty} \int_{-\infty}^{\infty} X(f) H(t, f) \exp(j2\pi f t) \exp(-j2\pi \tilde{f} t) \, df \, dt \tag{6.12}$$

which does *not* reduce to $Y(f) = H(f)X(f)$ [Matz and Hlawatsch 1998].

A Fourier transformation of the impulse response with respect to t results in a different representation – namely, the Doppler-variant impulse response, better known as *spreading function $s(\nu, \tau)$*:

$$s(\nu, \tau) = \int_{-\infty}^{\infty} h(t, \tau) \exp(-j2\pi \nu t) \, dt \tag{6.13}$$

This function describes the spreading of the input signal in the delay and Doppler domains.

Finally, the function $s(\nu, \tau)$ can be transformed with respect to the variable τ, resulting in the *Doppler-variant transfer function $B(\nu, f)$*:

$$B(\nu, f) = \int_{-\infty}^{\infty} s(\nu, \tau) \exp(-j2\pi f \tau) \, d\tau \tag{6.14}$$

A summary of the interrelations between the system functions is given in Figure 6.5. Figure 6.6 shows an example of a measured impulse response; Figure 6.7 shows the spreading function computed from it.

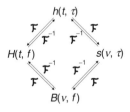

Figure 6.5   Interrelation between deterministic system functions.

6.3.2 Stochastic System Functions

We now return to the stochastic description of wireless channels. Interpreting them as time-variant stochastic systems, a complete description requires the multidimensional pdf of the impulse

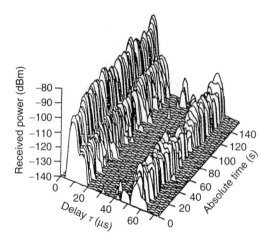

Figure 6.6 Squared magnitude of the impulse response $|h(t, \tau)|^2$ measured in hilly terrain near Darmstadt, Germany. Measurement duration 140 s; center frequency 900 MHz. τ denotes the excess delay.
Reproduced with permission from Liebenow and Kuhlmann [1993] © U. Liebenow.

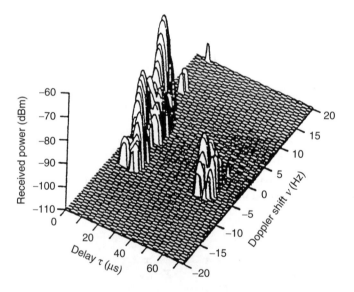

Figure 6.7 Spreading function computed from the data of Figure 6.6.
Reproduced with permission from U. Liebenow.

response – i.e., the joint pdf of the complex amplitudes at all possible values of delay and time. However, this is usually much too complicated in practice. Instead, we restrict our attention to a second-order description – namely, the *AutoCorrelation Function* (ACF).

Let us first repeat some facts about the ACFs of one-dimensional stochastic processes (i.e., processes that depend on a single parameter t). The ACF of a stochastic process y is defined as:

$$R_{yy}(t, t') = E\{y^*(t)y(t')\} \tag{6.15}$$

where the expectation is taken over the *ensemble of possible realizations* (for the definition of this ensemble see Appendix 6.A at *www.wiley.com/go/molisch*). The ACF describes the relationship between the second-order moments of the amplitude pdf of the signal y at different times. If the pdf is zero-mean Gaussian, then the second-order description contains all the required information. If the pdf is non-zero-mean Gaussian, the mean

$$\overline{y}(t) = E\{y(t)\} \tag{6.16}$$

together with the auto covariance function

$$\tilde{R}_{yy}(t, t') = E\{[y(t) - \overline{y}(t)]^*[y(t') - \overline{y}(t')]\} \tag{6.17}$$

constitutes a complete description. If the channel is non-Gaussian, then the first- and second-order statistics are not a complete description of the channel. In the following, we mainly concentrate on zero-mean Gaussian channels.

Let us now revert to the problem of giving a stochastic description of the channel. Inserting the input–output relationship into Eq. (6.15), we obtain the following expression for the ACF of the received signal:

$$R_{yy}(t, t') = E\left\{\int_{-\infty}^{\infty} x^*(t - \tau)h^*(t, \tau)\,d\tau \int_{-\infty}^{\infty} x(t' - \tau')h(t', \tau')\,d\tau'\right\} \tag{6.18}$$

The system is linear, so that expectation can be interchanged with integration. Furthermore, the transmit signal can be interpreted as a stochastic process that is independent of the channel, so that expectations over the transmit signal and over the channel can be performed independently. Thus, the ACF of the received signal is given by:

$$R_{yy}(t, t') = \int_{-\infty}^{\infty} \int_{-\infty}^{\infty} E\{x^*(t - \tau)x(t' - \tau')\}E\{h^*(t, \tau)h(t', \tau')\}\,d\tau\,d\tau'$$

$$= \int_{-\infty}^{\infty} \int_{-\infty}^{\infty} R_{xx}(t - \tau, t' - \tau')R_{h}(t, t', \tau, \tau')\,d\tau\,d\tau' \tag{6.19}$$

i.e., a combination of the ACF of the transmit signal and the ACF of the channel:

$$R_{h}(t, t', \tau, \tau') = E\{h^*(t, \tau)h(t', \tau')\} \tag{6.20}$$

Note that the ACF of the channel depends on four variables since the underlying stochastic process is two dimensional.

We observe a formal similarity of the channel ACF to the impulse response of a deterministic channel: we can form stochastic system functions by Fourier transformations. In contrast to the deterministic case, we now have to perform a *double Fourier transformation*, with respect to the pair of variables t, t' and/or τ, τ'. From that, we obtain in an elementary way the relationships between the different formulations of the ACFs of input and output – e.g., $R_{s}(\nu, \nu', \tau, \tau') = E\{s^*(\nu, \tau)s(\nu', \tau')\} = \int\int R_{h}(t, t', \tau, \tau') \cdot \exp(+j2\pi\nu t)\exp(-j2\pi\nu't')dt\,dt'$.

6.4   The WSSUS Model

The correlation functions depend on four variables, and are thus a rather complicated form for the characterization of the channel. Further assumptions about the physics of the channel can lead to a simplification of the correlation function. The most frequently used assumptions are the so-called *Wide-Sense Stationary* (WSS) assumption and the *Uncorrelated Scatterers* (US) assumption. A model using both assumptions simultaneously is called a *WSSUS* model.

6.4.1 Wide-Sense Stationarity

The mathematical definition of *wide-sense stationarity* is that the ACF depends not on the two variables t, t' separately, but only on their difference $t - t'$. Consequently, *second-order amplitude statistics* do not change with time.[4] We can thus write

$$R_\mathrm{h}(t, t', \tau, \tau') = R_\mathrm{h}(t, t + \Delta t, \tau, \tau') = R_\mathrm{h}(\Delta t, \tau, \tau') \tag{6.21}$$

Physically speaking, WSS means that the *statistical properties* of the channel do not change with time. This must not be confused with a static channel, where fading *realizations* do not change with time. For the simple case of a flat Rayleigh-fading channel, WSS means that the mean power and the Doppler spectrum do not change with time, while the instantaneous amplitude can change.

According to the mathematical definition, WSS has to be fulfilled for any arbitrary time, t. In practice, this is not possible: as the MS moves over larger distances, the mean received power changes because of shadowing and variations in path loss. Rather, WSS is typically fulfilled over an area of about 10λ diameter (compare also Section 5.1). We can thus define quasi-stationarity over a finite time interval (associated with a movement distance of the MS), over which statistics do not change noticeably.

WSS also implies that components with different Doppler shifts undergo independent fading. This can be shown by considering the Doppler-variant impulse response $s(\nu, \tau)$. Inserting Eq. (6.21) into the definition of R_s, we get

$$R_\mathrm{s}(\nu, \nu', \tau, \tau') = \int_{-\infty}^{\infty} \int_{-\infty}^{\infty} R_\mathrm{h}(t, t + \Delta t, \tau, \tau') \exp[2\pi j(\nu t - \nu'(t + \Delta t))]\, dt\, dt' \tag{6.22}$$

which can be rewritten as:

$$R_\mathrm{s}(\nu, \nu', \tau, \tau') = \int_{-\infty}^{\infty} \exp[2\pi j t(\nu - \nu')]\, dt \int_{-\infty}^{\infty} R_h(\Delta t, \tau, \tau') \exp[-2\pi j\nu'\Delta t]\, d\Delta t \tag{6.23}$$

The first integral is an integral representation of the delta function $\delta(\nu - \nu')$. Thus, R_s can be factored as:

$$R_\mathrm{s}(\nu, \nu', \tau, \tau') = \overset{\approx}{P}_\mathrm{s}(\nu, \tau, \tau')\delta(\nu - \nu') \tag{6.24}$$

This implies that contributions undergo uncorrelated fading if they have different Doppler shifts. The function $\overset{\approx}{P}_\mathrm{s}()$, which is implicitly defined by Eq. (6.24), is discussed below in more detail.

Analogously, we can write R_B as:

$$R_\mathrm{B}(\nu, \nu', f, f') = P_\mathrm{B}(\nu, f, f')\delta(\nu - \nu') \tag{6.25}$$

6.4.2 Uncorrelated Scatterers

The US assumption is defined as "contributions with different delays are uncorrelated," which is written mathematically as:

$$R_\mathrm{h}(t, t', \tau, \tau') = P_\mathrm{h}(t, t', \tau)\delta(\tau - \tau') \tag{6.26}$$

or for R_s as:

$$R_\mathrm{s}(\nu, \nu', \tau, \tau') = \tilde{P}_\mathrm{s}(\nu, \nu', \tau)\delta(\tau - \tau') \tag{6.27}$$

---

[4] Strict sense stationarity means that fading statistics of arbitrary order do not change with time. For Gaussian channels, WSS implies strict sense stationarity.

The US condition is fulfilled if the phase of an MPC does not contain any information about the phase of another MPC with a different delay. If scatterers are distributed randomly in space, phases change in an uncorrelated way even when the MS moves only a small distance.

For the transfer function, the US condition means that R_H does not depend on the absolute frequency, but only on the frequency difference:[5]

$$R_H(t, t', f, f + \Delta f) = R_H(t, t', \Delta f) \qquad (6.28)$$

6.4.3 WSSUS Assumption

The US and WSS assumptions are duals: US defines contributions with different delays as uncorrelated, while WSS defines contributions with different Doppler shifts as uncorrelated. Alternatively, we can state that US means that R_H depends only on the frequency difference, while WSS means that R_H depends only on the time difference.

It is thus natural to combine these two definitions in the WSSUS condition, so that the ACF has to fulfill the following conditions:

$$R_h(t, t + \Delta t, \tau, \tau') = P_h(\Delta t, \tau)\, \delta(\tau - \tau') \qquad (6.29)$$

$$R_H(t, t + \Delta t, f, f + \Delta f) = R_H(\Delta t, \Delta f) \qquad (6.30)$$

$$R_s(\nu, \nu', \tau, \tau') = P_s(\nu, \tau)\, \delta(\nu - \nu')\, \delta(\tau - \tau') \qquad (6.31)$$

$$R_B(\nu, \nu', f, f + \Delta f) = P_B(\nu, \Delta f)\, \delta(\nu - \nu') \qquad (6.32)$$

In contrast to the ACFs, which depend on four variables, the P-functions on the r.h.s. depend only on *two* variables. This greatly simplifies their formal description, parameterization, and application in further derivations. Because of their importance, they have been given distinct names. Following Kattenbach [1997], we define

- $P_h(\Delta t, \tau)$ as *delay cross power spectral density*;
- $R_H(\Delta t, \Delta f)$ as *time frequency correlation function*;
- $P_s(\nu, \tau)$ as *scattering function*;
- $P_B(\nu, \Delta f)$ as *Doppler cross power spectral density*.

The scattering function has special importance because it can be easily interpreted physically. If only single interactions occur, then each differential element of the scattering function corresponds to a physically existing IO. From the Doppler shift, we can determine the Direction Of Arrival (DOA); the delay determines the radii of the ellipse on which the scatterer lies.

The WSSUS assumption is very popular, but not always fulfilled in practice. Appendix 6.A (see *www.wiley.com/go/molisch*) gives a more detailed discussion of the assumptions and their validity.

6.4.4 Tapped Delay Line Models

A WSSUS channel can be represented as a tapped delay line, where the coefficients multiplying the output from each tap vary with time. The impulse response is then written as:

$$h(t, \tau) = \sum_{i=1}^{N} c_i(t)\delta(\tau - \tau_i) \qquad (6.33)$$

where N is the number of taps, $c_i(t)$ are the time-dependent complex coefficients for the taps, and τ_i is the delay of the ith tap. For each tap, a Doppler spectrum determines the changes of the

---

[5] Proof is left as an exercise for the reader.

coefficients with time. This spectrum can be different for each tap, though many models assume the same spectrum for each tap, see also Chapter 7.

One interpretation of a tapped delay line is as a physical representation of the multipath propagation in the channel. Each of the N components corresponds to one group of closely spaced MPCs: the model would be purely deterministic only if the arriving signals consisted of *completely resolvable* echoes from discrete IOs. However, in most practical cases, the resolution of the RX is not sufficient to resolve all MPCs. We thus write the impulse response as:

$$h(t, \tau) = \sum_{i=1}^{N} \sum_{k} a_{i,k}(t) \delta(\tau - \tau_i) = \sum_{i=1}^{N} c_i(t) \delta(\tau - \tau_i) \tag{6.34}$$

Note that the second part of this equation makes sense only in a band-limited system. In that case, each complex amplitude $c_i(t)$ represents the sum of several MPCs, which fades. WSSUS implies that all the taps are fading independently, and that their average power does not depend on time.

Another interpretation of the tapped delay line is based on the sampling theorem. Any wireless system, and thus the channel we are interested in, is band limited. Therefore, the impulse response can be represented by a sampled version of the continuous impulse response $\tilde{h}_{bl}(t, \tau) = \sum A_\ell(t) \delta(\tau - \tau_\ell)$; similarly, the scattering function, correlation functions, etc., can be represented by their sampled versions. Commonly, the samples are spaced equidistantly, $\tau_\ell = \ell \cdot \Delta\tilde{\tau}$, where the distance between the taps $\Delta\tilde{\tau}$ is determined by the Nyquist theorem. The continuous version of the impulse response can be recovered by interpolation:

$$h_{bl}(t, \tau) = \sum_{\ell} A_\ell(t) \text{ sinc } (W(\tau - \tau_\ell)) \tag{6.35}$$

where W is the bandwidth. Note that if the *physical* IOs fulfill the WSSUS condition, but are not equidistantly spaced, then the tap weights $A_\ell(t)$ are *not* necessarily WSSUS.

Many of the standard models for wireless channels (see Chapter 7) were developed with a specific system and thus a specific system bandwidth in mind. It is often necessary to adjust the tap locations to a different sampling grid for a discrete simulation: in other words, a discrete simulation requires a channel representation $h(t, \tau) = \sum A_\ell(t) \delta(\tau - \ell T_s)$, but T_ℓ/T_s is a non-integer. The following methods are in widespread use:

1. *Rounding to the nearest integer*: This method leads to errors, which are smaller the higher the new sampling rate is.
2. *Splitting the tap energy*: The energy is divided between the two adjacent taps $kT_s < \tau_\ell < (k + 1)T_s$, possibly weighted by the distance to the original tap.
3. *Resampling*: This can be done by using the interpolation formula – i.e., resampling $h_{bl}(t, \tau)$ in Eq. (6.35) at the desired rate. Alternatively, we can describe the channel in the frequency domain and transform it back (with a discrete Fourier transform) with the desired tap spacing.

6.5   Condensed Parameters

The correlation functions are a somewhat cumbersome way of describing wireless channels. Even when the WSSUS assumption is used, they are still *functions* of two variables. A preferable representation would be a function of one variable, or even better, just a single parameter. Obviously, such a representation implies a serious loss of information, but this is a sometimes acceptable price for a compact representation.

6.5.1 Integrals of the Correlation Functions

A straightforward way of getting from two variables to one is to integrate over one of them. Integrating the scattering function over the Doppler shift v gives the *delay power spectral density* $P_h(\tau)$, more popularly known as the *Power Delay Profile* (PDP). The PDP contains information about how much power (from a transmitted delta pulse with unit energy) arrives at the RX with a delay between $(\tau, \tau + d\tau)$, irrespective of a possible Doppler shift. The PDP can be obtained from the complex impulse responses $h(t, \tau)$ as:

$$P_h(\tau) = \lim_{T \to \infty} \frac{1}{2T} \int_{-T}^{T} |h(t, \tau)|^2 \, dt \qquad (6.36)$$

if ergodicity holds. Note that in practice the integral will not extend over infinite time, but rather over the time span during which quasi-stationarity is valid (see above).

Analogously, integrating the scattering function over τ results in the *Doppler power spectral density* $P_B(v)$.

The *frequency correlation function* can be obtained from the time frequency correlation function by setting $\Delta t = 0$ – i.e., $R_H(\Delta f) = R_H(0, \Delta f)$. It is noteworthy that this frequency correlation function is the Fourier transform of the PDP. The *temporal correlation function* $R_H(\Delta t) = R_H(\Delta t, 0)$ is the inverse Fourier transform of the Doppler power spectral density.

6.5.2 Moments of the Power Delay Profile

The PDP is a *function*, but for obtaining a quick overview of measurement results, it is preferable to have each measurement campaign described by a single *parameter*. While there are a large number of possible parameters, normalized moments of the PDP are the most popular.

We start out by computing the zeroth-order moment – i.e., time-integrated power:

$$P_m = \int_{-\infty}^{\infty} P_h(\tau) \, d\tau \qquad (6.37)$$

The normalized first-order moment, the *mean delay*, is given by:

$$T_m = \frac{\int_{-\infty}^{\infty} P_h(\tau)\tau \, d\tau}{P_m} \qquad (6.38)$$

The normalized second-order central moment is known as *rms delay spread* and is defined as:

$$S_\tau = \sqrt{\frac{\int_{-\infty}^{\infty} P_h(\tau)\tau^2 \, d\tau}{P_m} - T_m^2} \qquad (6.39)$$

The rms delay spread has obtained a special stature among all parameters. It has been shown that under some specific circumstances, the error probability due to delay dispersion is proportional to the rms delay spread only (see Chapter 12), while the actual *shape* of the PDP does not have a significant influence. In that case, S_τ is all we need to know about the channel. It cannot be stressed enough, however, that this is true only under specific circumstances, and that the rms delay spread is not a "solve-it-all." It is also noteworthy that S_τ does not attain finite values for all physically reasonable signals. A channel with $P_h(\tau) \propto 1/(1 + \tau^2)$ is physically possible, and does not contradict energy conservation, but $\int_{-\infty}^{\infty} P_h(\tau)\tau^2 d\tau$ does not converge.

Example 6.1 *Compute the rms delay spread of a two-spike profile*

$$P_h(\tau) = \delta(\tau - 10\,\mu s) + 0.3\delta(\tau - 17\,\mu s).$$

The time-integrated power is given by Eq. (6.37):

$$P_m = \int_{-\infty}^{\infty} (\delta(\tau - 10^{-5}) + 0.3\delta(\tau - 1.7 \cdot 10^{-5}))\,d\tau$$

$$= 1.30 \tag{6.40}$$

and the mean delay is given by Eq. (6.38):

$$T_m = \int_{-\infty}^{\infty} (\delta(\tau - 10^{-5}) + 0.3\delta(\tau - 1.7 \cdot 10^{-5}))\tau\,d\tau/P_m$$

$$= (10^{-5} + 0.3 \cdot 1.7 \cdot 10^{-5})/1.3 = 1.16 \cdot 10^{-5}\,s \tag{6.41}$$

Finally, the rms delay spread is computed according to Eq. (6.39):

$$S_\tau = \sqrt{\frac{\int_{-\infty}^{\infty}(\delta(\tau - 10^{-5}) + 0.3\delta(\tau - 1.7 \cdot 10^{-5}))\tau^2\,d\tau}{P_m} - T_m^2}$$

$$= \sqrt{\frac{(10^{-5})^2 + 0.3(1.7 \cdot 10^{-5})^2}{1.3} - (1.16 \cdot 10^{-5})^2} = 3\,\mu s \tag{6.42}$$

6.5.3 Moments of the Doppler Spectra

Moments of the Doppler spectra can be computed in complete analogy to the moments of the PDP. The integrated power is

$$P_{B,m} = \int_{-\infty}^{\infty} P_B(\nu)\,d\nu \tag{6.43}$$

where obviously $P_{B,m} = P_m$. The mean Doppler shift is

$$\nu_m = \frac{\int_{-\infty}^{\infty} P_B(\nu)\nu\,d\nu}{P_{B,m}} \tag{6.44}$$

The rms Doppler spread is

$$S_\nu = \sqrt{\frac{\int_{-\infty}^{\infty} P_B(\nu)\nu^2\,d\nu}{P_{B,m}} - \nu_m^2} \tag{6.45}$$

6.5.4 Coherence Bandwidth and Coherence Time

In a frequency-selective channel, different frequency components fade differently. Obviously, the correlation between fading at two different frequencies is the smaller the more these two frequencies are apart. The *coherence bandwidth* B_{coh} defines the frequency difference that is required so that the correlation coefficient is smaller than a given threshold.

A mathematically exact definition starts out with the frequency correlation function $R_H(0, \Delta f)$, assuming WSSUS (see Figure 6.8a for an example). The coherence bandwidth can then be defined as:

$$B_{\text{coh}} = \frac{1}{2} \left[\arg \max_{\Delta f > 0} \left(\frac{|R_H(0, \Delta f)|}{R_H(0, 0)} = 0.5 \right) - \arg \min_{\Delta f < 0} \left(\frac{|R_H(0, \Delta f)|}{R_H(0, 0)} = 0.5 \right) \right] \qquad (6.46)$$

Figure 6.8 Typical frequency correlation function.

Reproduced with permission from Kattenbach [1997] © Shaker Verlag.

This is essentially the half-width half-maximum bandwidth of the correlation function. The somewhat complicated formulation stems from the fact that the correlation function need not decay *monotonically*. Rather, there can be local maxima that exceed the threshold. A precise definition thus uses the bandwidth that encompasses all parts of the correlation function exceeding the threshold.[6]

The rms delay spread S_τ and the coherence bandwidth B_{coh} are obviously related: S_τ is derived from the PDP $P_h(\tau)$ while B_{coh} is obtained from the frequency correlation function, which is the Fourier transform of the PDP. Based on this insight, Fleury [1996] derived an "uncertainty relationship:"

$$B_{coh} \gtrsim \frac{1}{2\pi S_\tau} \tag{6.47}$$

Equation (6.47) is an inequality and therefore *does not* offer the possibility to obtain one parameter from the other. The question thus arises whether B_{coh} or S_τ better reflects the channel properties. An answer to that question can only be given for a specific system. For a Frequency Division Multiple Access (FDMA) or Time Division Multiple Access (TDMA) system without an equalizer, the rms delay spread is the quantity of interest, as it is related to the BER (see Chapter 12), though generally it overemphasizes long-delayed echoes. For Orthogonal Frequency Division Multiplexing (OFDM) systems (Chapter 19), where the information is transmitted on many parallel carriers, the coherence bandwidth is obviously a better measure.

The temporal correlation function is a measure of how fast a channel changes. The definition of the coherence time T_{coh} is thus analogous to the coherence bandwidth; it also has an uncertainty relationship with the rms Doppler spread.

Figure 6.9 summarizes the relationships between system functions, correlation functions, and special parameters. Ergodicity is assumed throughout this figure.

6.5.5   Window Parameters

Another useful set of parameters are the so-called *window parameters* [de Weck 1992], more precisely the *interference quotient* Q_T and the *delay window* W_Q. They are a measure for the percentage of energy of the average PDP arriving within a certain delay interval. In contrast to the delay spread and coherence bandwidth, the window parameters need to be defined in the context of specific systems.

The *interference quotient* Q_T is the ratio between the signal power arriving within a time window of duration T, relative to the power arriving outside that window. The delay window characterizes the self-interference due to delay dispersion. If, e.g., a system has an equalizer that can process MPCs with a delay up to T, then every MCP within the window is "useful," while energy outside the window creates interference – these components carry information about bits that cannot be processed, and thus act as independent interferers.[7] For the Global System for Mobile communications (GSM) (see Chapter 24), an equalizer length of four-symbol duration, corresponding to a 16-μs delay, is required by the specifications. It is thus common to define a parameter Q_{16} – i.e., the interference quotient for $T = 16\,\mu s$.

---

[6] An alternative definition would define the coherence bandwidth as the second central moment of the correlation function; this would circumvent all problems with local maxima. Unfortunately, this second moment becomes infinite in the practically important case that the squared correlation function is Lorentzian $1/(1 + \Delta f^2)$, which corresponds to an exponential PDP.

[7] The interpretation is only an approximate one. There is no sharp "jump" from "useful" to "interference" when the delay of an MPC exceeds the equalizer length. Rather, there is a smooth transition, similar to the effect of delay dispersion in unequalized systems, see Chapter 12.

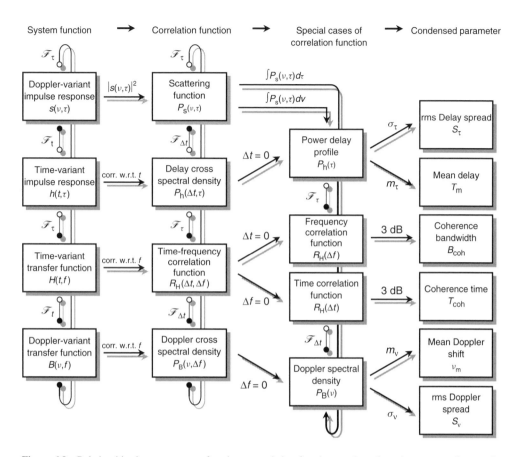

Figure 6.9 Relationships between system functions, correlation functions, and condensed parameters for ergodic channel impulse responses.

A mathematical definition of the interference quotient is given as:

$$Q_T = \frac{\int_{t_0}^{t_0+T} P_h(\tau)\, d\tau}{P_m - \int_{t_0}^{t_0+T} P_h(\tau)\, d\tau} \tag{6.48}$$

This quotient depends not only on the PDP and the duration T but also on the starting delay of the window t_0. This latter dependence is often eliminated by either setting the starting delay to the minimum excess delay (i.e., the first MPC determines the start of the window) $t_0 = \tau_{\min}$. Alternatively, the t_0 can be chosen to maximize Q_T:

$$Q_T = \max_{t_0} \left\{ \frac{\int_{t_0}^{t_0+T} P_h(\tau)\, d\tau}{P_m - \int_{t_0}^{t_0+T} P_h(\tau)\, d\tau} \right\} \tag{6.49}$$

This definition makes sense because an RX can often adapt equalizer timing to optimize performance.

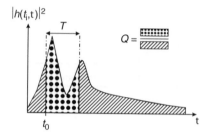

Figure 6.10   Definition of window parameters.
Reproduced with permission from Molisch [2000] © Prentice Hall.

A related parameter is the delay window W_Q (see Figure 6.10). This defines how long a window has to be so that the power within that window is a factor of Q larger than the power outside the window. The defining equations are the same as for the interference quotient. The difference is just that now T is considered as variable, and Q as fixed.

Example 6.2   *For an exponential PDP, $P_h(\tau) = \exp(-\tau/2\,\mu s)$, compute the delay window so that the interference quotient becomes 10 and 20 dB, respectively – i.e., so that 91% and 99% of the energy are contained in the window. Do the same of the two-spike profile $\delta(\tau - 10\,\mu s) + 0.3\delta(\tau - 17\,\mu s)$.*

Let the starting delay be equal to the minimum excess delay. For an exponential PDP, the interference quotient, given by Eq. (6.48), is

$$Q_T = \frac{\int_0^T e^{-\tau/2\cdot10^{-6}}\,d\tau}{\int_0^\infty e^{-\tau/2\cdot10^{-6}}\,d\tau - \int_0^T e^{-\tau/2\cdot10^{-6}}\,d\tau}$$

Solving for T yields

$$T = 2\cdot10^{-6}\ln(Q_T + 1)$$

and the 91% and 99% windows are thus $T_{91\%} = 4.8\,\mu s$ and $T_{99\%} = 9.2\,\mu s$, respectively. For the two-spike profile, the starting delay is $t_0 = 10^{-5}$ and the energy within the window is

$$\int_{10^{-5}}^{T+10^{-5}} (\delta(\tau - 10^{-5}) + 0.3\delta(\tau - 1.7\cdot10^{-5}))\,d\tau = \begin{cases} 1, & 0 < T < 7\,\mu s \\ 1.3, & T > 7\,\mu s \\ 0, & \text{otherwise} \end{cases}$$

Hence, the interference quotient is greater than 10 dB and/or 20 dB for $T > 7\,\mu s$.

6.6   Ultra Wideband Channels

6.6.1   UWB Signals with Large Relative Bandwidth

The above models are wideband in the sense that they model the delay dispersion caused by multipath propagation. However, they are still based on the following two assumptions.

1. The reflection, transmission, and diffraction coefficients of the IOs are constant over the considered bandwidth.
2. The relative bandwidth of the system (bandwidth divided by carrier frequency) is *much* smaller than unity.

Note that these conditions are met for the bandwidth of most currently used wireless systems. However, in recent years, a technique called *Ultra Wide Band* (UWB) transmission (see also Chapter 18) has gained increased interest. UWB systems have a relative bandwidth of more than 20%. In that case, the different frequency components contained in the transmitted signal "see" different propagation environments. For example, the diffraction coefficient of a building corner is different at 100 MHz compared with 1 GHz; similarly, the reflection coefficients of walls and furniture can vary over the bandwidth of interest. Channel impulse realization is then given by:

$$h(\tau) = \sum_{i=1}^{N} a_i \chi_i(\tau) \otimes \delta(\tau - \tau_i) \qquad (6.50)$$

where $\chi_i(\tau)$ denotes the distortion of the ith MPC by the frequency selectivity of IOs. Expressions for these distortions are given, e.g., in Molisch [2005] and Qiu [2002]; one example for a distortion of a short pulse by diffraction by a screen is shown in Figure 6.11.

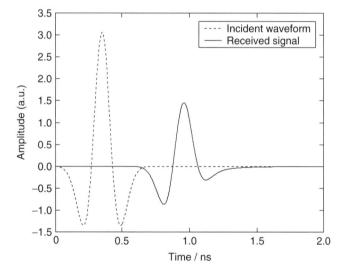

Figure 6.11 UWB pulse diffracted by a semi-infinite screen.
Reproduced with permission from Qiu [2002] © IEEE.

For UWB systems, propagation effects can also show frequency dependence. As explained in Chapter 4, path loss is a function of frequency if the antennas have constant gain. Similarly, diffraction and reflection are frequency dependent. Thus, the higher frequency components of the transmitted signal are usually attenuated more strongly by the combination of antenna and channel. Also, this effect leads to a distortion of individual MPCs since *any* frequency dependence of the transfer function leads to delay dispersion, and thus distortion of an MPC.

As a consequence of the distortion of the frequency dependence, statistical channel models also change. First of all, the path loss has to be redefined according to $G_{\mathrm{pr}}(d, f) =$

$E\{\int_{f-\Delta f/2}^{f+\Delta f/2} |H(f, d)|^2 df\}$, where the expectation is taken over both small-scale and large-scale and Δ is a bandwidth that is sufficiently small so that all propagation effects stay constant within it. Furthermore, when representing the impulse response as a tapped delay line, the fading at the different taps becomes correlated. The MPC distortion makes each multipath contribution influence several subsequent taps. From that, it follows that the fading of one component influences the amplitudes of several taps, and therefore causes correlation.

6.6.2 UWB Channels with Large Absolute Bandwidth

Another definition of UWB signals is that they have more than 500 MHz absolute bandwidth. Despite the high temporal resolution of UWB systems, there is still an appreciable probability that several MPCs fall into one resolvable delay bin, and add up there; in other words, there is fading even in UWB. The difference to conventional system lies mainly in the number of MPCs that fall into one bin. This number is influenced by the environment: the more objects are in the environments, the more MPCs can occur. For example, residential environments tend to have fewer MPCs than industrial environments. Furthermore, the delay of the considered bin plays a role: for larger excess delays, there are more feasible paths causing this particular delay. Thus, fading depth increases with increasing delay. Depending on these factors, a Rayleigh distribution of the amplitudes might or might not be suitable. Nakagami, Rice, or lognormal distributions have been suggested.

At a high absolute bandwidth, not every resolvable delay bin contains MPCs, so that delay bins containing MPCs are interspersed with "empty" delay bins, i.e., not containing any discrete (plane wave) MPCs. The resulting PDP is called "sparse." The bandwidth required for these phenomena to occur depends on the environment. Such impulse responses are often described by the Saleh–Valenzuela model that will be described in detail in Section 7.3.3.

6.7 Directional Description

We now turn to channel descriptions that take the *directions* of the MPCs into account (in addition to their amplitude and delay). Such a directional description is useful for two reasons.

- The directional properties are important for spatial diversity (Chapter 13) and multielement antennas (Chapter 20).
- It allows to separate the propagation effects from the impact of the antenna. Note that in a conventional wideband representation, as used in the previous sections, the impulse response contains the sum of the weighted MPCs, where the weighting depends on the specific antenna used; consequently, changing the antenna changes the impulse response, even though the true propagation channel remains unchanged.

For these reasons, it is useful to employ the *Double-Directional Impulse Response* (DDIR)[8], which consists of the sum of contributions from MPCs:

$$h(t, \tau, \Omega, \Psi) = \sum_{\ell=1}^{N(t)} h_\ell(t, \tau, \Omega, \Psi) \tag{6.51}$$

---

[8] To be completely general, we would have to include a description of polarization as well. To avoid the cumbersome matrix notation, we omit this case here and only briefly discuss it in Section 7.4.4.

The DDIR depends on the time t, delay τ, the *Direction Of Departure* (DOD) Ω, the *DOA* Ψ, and the number of MPCs, $N(t)$, for the specific time; dependence on location is not written explicitly. The $h_\ell(t, \tau, \Omega, \Psi)$ is the contribution of the ℓth MPC, modeled as:

$$h_\ell(t, \tau, \Omega, \Psi) = |a_\ell|e^{j\varphi_\ell}\delta(\tau - \tau_\ell)\delta(\Omega - \Omega_\ell)\delta(\Psi - \Psi_\ell) \tag{6.52}$$

which essentially just adds the directional properties to the tapped delay line model. Besides the absolute amplitude $|a|$ and the delay, the DOA and DOD also vary slowly (over many wavelengths), while again phase φ varies quickly.

The single-directional impulse response can be obtained by integrating the DDIR (weighted by the transmit antenna pattern) over the DODs. Integrating the single-directional impulse response (weighted by the RX antenna pattern) over the DOAs results in the conventional impulse response.

The stochastic description of directional channels is analogous to the nondirectional case. The ACF of the impulse response can be generalized to include directional dependence so that it depends on six or eight variables. We can also introduce a "generalized WSSUS condition" so that contributions coming from different directions fade independently. Note that the directions of the MPCs at the MS, on one hand, and Doppler spreading, on the other hand, are linked, and thus ν and Ψ are not independent variables anymore (we assume in the following that Ψ are the directions at the MS).

Analogously to the nondirectional case, we can then define condensed descriptions of the wireless channel. We first define

$$E\{s^*(\Omega, \Psi, \tau, \nu)s(\Omega', \Psi', \tau', \nu')\} = P_s(\Omega, \Psi, \tau, \nu)\delta(\Omega - \Omega')\delta(\Psi - \Psi')\delta(\tau - \tau')\delta(\nu - \nu') \tag{6.53}$$

from which the *Double Directional Delay Power Spectrum* (DDDPS) is derived as:

$$DDDPS(\Omega, \Psi, \tau) = \int P_s(\Psi, \Omega, \tau, \nu)\,d\nu \tag{6.54}$$

From this, we can establish the *Angular Delay Power Spectrum* (ADPS) as seen from the BS antenna:

$$ADPS(\Omega, \tau) = \int DDDPS(\Psi, \Omega, \tau)G_{MS}(\Psi)\,d\Psi \tag{6.55}$$

where G_{MS} is the antenna power pattern of the MS. The ADPS is usually normalized as:

$$\int\int ADPS(\tau, \Omega)\,d\tau\,d\Omega = 1 \tag{6.56}$$

The *Angular Power Spectrum* (APS) is given by:

$$APS(\Omega) = \int APDS(\Omega, \tau)\,d\tau \tag{6.57}$$

Note also that an integration of the ADPS over Ω recovers the PDP.

The *azimuthal spread* is defined as the second central moment of the APS if all MPCs are incident in the horizontal plane, so that $\Omega = \phi$. In many papers, it is defined in a form analogous to Eq. (6.39) – namely:

$$S_\phi = \sqrt{\frac{\int APS(\phi)\phi^2\,d\phi}{\int APS(\phi)\,d\phi} - \left(\frac{\int APS(\phi)\phi\,d\phi}{\int APS(\phi)\,d\phi}\right)^2} \tag{6.58}$$

However, this definition is ambiguous because of the periodicity of the azimuthal angle: by this definition. $APS = \delta(\phi - \pi/10) + \delta(\phi - 19\pi/10)$ would have a different angular spread from $APS = \delta(\phi - 3\pi/10) + \delta(\phi - \pi/10)$, even though physical intuition tells us that they should be the same, since the two APSs differ just by a constant offset. A better definition is given in Fleury [2000]:

$$S_\phi = \sqrt{\frac{\int |\exp(j\phi) - \mu_\phi|^2 APS(\phi)\, d\phi}{\int APS(\phi)\, d\phi}} \tag{6.59}$$

with

$$\mu_\phi = \frac{\int \exp(j\phi) APS(\phi)\, d\phi}{\int APS(\phi)\, d\phi} \tag{6.60}$$

Example 6.3 *Consider the APS defined as $APS = 1$ for $0° < \varphi < 90°$ and $340° < \varphi < 360°$, compute the angular spread according to the definitions of Eqs. (6.58) and (6.59), respectively.*

According to Eq. (6.58), we have

$$S_\phi = \sqrt{\frac{\int_0^{\pi/2} \phi^2\, d\phi + \int_{17\pi/9}^{2\pi} \phi^2\, d\phi}{\int_0^{\pi/2} d\phi + \int_{17\pi/9}^{2\pi} d\phi} - \left(\frac{\int_0^{\pi/2} \phi\, d\phi + \int_{17\pi/9}^{2\pi} \phi\, d\phi}{\int_0^{\pi/2} d\phi + \int_{17\pi/9}^{2\pi} d\phi}\right)^2}$$

$$= 2.09\,\mathrm{rad} = 119.7° \tag{6.61}$$

In contrast, Eqs. (6.59) and (6.60) yield

$$\left.\begin{aligned}
\mu_\phi &= \frac{18}{11\pi} \int_{-\pi/9}^{\pi/2} \exp(j\phi)\, d\phi = 0.7 + 0.49 j \\[2mm]
S_\phi &= \sqrt{\frac{18}{11\pi} \int_{-\pi/9}^{\pi/2} (\cos(\phi) - Re(\mu_\phi))^2 + (\sin(\phi) - Im(\mu_\phi))^2)\, d\phi} \\[2mm]
&= 0.521\,\mathrm{rad} = 29.9°
\end{aligned}\right\} \tag{6.62}$$

The values obtained from the two methods differ radically. We can easily see that the second value – namely, 29.9°, makes more sense: the APS extends continuously from $-20°$ to $90°$. The angular spread should thus be the same as for an APS that extends from $0°$ to $111°$. Inserting this modified APS in Eq. (6.58), we obtain an angular spread of $32°$, while the value from Eq. (6.59) remains at 29.9°. It is also interesting that this value is close to $(\phi_{max} - \phi_{min})/(2/\sqrt{3})$ – and remember that for a rectangular PDP, the relationship between rms delay spread and maximum excess delay is also $S_\tau = (\tau_{max} - \tau_{min})/(2\sqrt{3})$.

Similar to delay spread, angular spread also is only a partial description of angular dispersion. It has been shown that the correlation of signals at the elements of a uniform linear array depends only on the rms angular spread and not on the shape of the APS; however, this is valid only under some very specific assumptions.

Directional channel descriptions are especially valuable in the context of multiantenna systems (Chapter 20). In that case, we are often interested in obtaining joint impulse responses at the different antenna elements. The impulse response thus becomes a matrix if we have antenna arrays at both link ends, and a vector if there is an array at one link end. We denote the transmit and receive element coordinates as $\mathbf{r}_{TX}^{(1)}, \mathbf{r}_{TX}^{(2)}, \ldots, \mathbf{r}_{TX}^{(N_t)}$, and $\mathbf{r}_{RX}^{(1)}, \mathbf{r}_{RX}^{(2)}, \ldots, \mathbf{r}_{RX}^{(N_r)}$, respectively, so that the impulse response from the jth transmit to the ith receive element becomes

$$
\begin{aligned}
h_{ij} &= h\left(\mathbf{r}_{TX}^{(j)}, \mathbf{r}_{RX}^{(i)}\right) \\
&= \sum_\ell h_\ell\left(\mathbf{r}_{TX}^{(1)}, \mathbf{r}_{RX}^{(1)}, \tau, \Omega_\ell, \Psi_\ell\right) \tilde{G}_{TX}(\Omega_\ell)\tilde{G}_{RX}(\Psi_\ell) \exp\left(j\langle \mathbf{k}(\Omega_\ell), (\mathbf{r}_{TX}^{(j)} - \mathbf{r}_{TX}^{(1)})\rangle\right) \\
&\quad \times \exp\left(-j\langle \mathbf{k}(\Psi_\ell), (\mathbf{r}_{RX}^{(i)} - \mathbf{r}_{RX}^{(1)})\rangle\right)
\end{aligned}
\tag{6.63}
$$

Here we have explicitly written the dependence of the impulse responses on the location, and exploited the fact that the contribution from the ℓ-th MPC at location $\mathbf{r}_{TX}^{(j)}$ is related to the impulse response at the reference antenna element location $\mathbf{r}_{TX}^{(1)}$ by a phase shift. Furthermore, we let h_ℓ depend on the location of the reference antenna elements $\mathbf{r}_{TX}^{(1)}$ and $\mathbf{r}_{RX}^{(1)}$ instead of on absolute time t. Here $\tilde{G}_{TX}$ and $\tilde{G}_{RX}$ are the complex (amplitude) patterns of the transmit and receive antenna elements, respectively, $\{\mathbf{k}\}$ is the unit wave vector in the direction of the ℓth DOD or DOA, and $\langle .,. \rangle$ denotes the dot product. We thus see that it is always possible to obtain the impulse response matrix from a DDIR.

If the receive arrays are uniform linear arrays, we can write Eq. (6.63) as:

$$
\mathbf{H} = \int\int h(\tau, \Omega, \Psi)\tilde{G}_{TX}(\Omega)\tilde{G}_{RX}(\Psi)\boldsymbol{\alpha}_{RX}(\Psi)\boldsymbol{\alpha}_{TX}^\dagger(\Omega)d\Psi d\Omega
\tag{6.64}
$$

where we used the *steering vectors*:

$$
\boldsymbol{\alpha}_{TX}(\Omega) = \frac{1}{\sqrt{N_t}}\left[1, \exp(-j2\pi\frac{d_a}{\lambda}\sin(\Omega)), \ldots, \exp(-j2\pi(N_t-1)\frac{d_a}{\lambda}\sin(\Omega))\right]^T
$$

and analogously defined $\boldsymbol{\alpha}_{RX}(\Psi)$. Ω and Ψ are measured from the antenna broadside.

6.8 Appendices

Please see companion website *www.wiley.com/go/molisch*

Further Reading

The theory of linear time-variant systems is described in the classical paper of Bello [1963]; further details are discussed in Kozek [1997] and Matz and Hlawatsch [1998]. The theory of WSSUS systems was established in Bello [1963], and further investigated in Hoeher [1992], Kattenbach [1997], Molnar et al. [1996], Paetzold [2002]; more considerations about the validity of WSSUS in wireless communications can be found in Fleury [1990], Kattenbach [1997], Kozek [1997], Molisch and Steinbauer [1999]. A method for characterizing non-WSSUS channels is described in Matz [2003]. An overview of condensed parameters, including delay spread, is given in Molisch and Steinbauer [1999]. Generic descriptions of UWB channels can be found in Molisch [2005], Molisch [2009], Qiu [2004]. Description methods for spatial channels are discussed in Durgin [2003], Ertel et al. [1998], Molisch [2002], Yu and Ottersten [2002]. Generalizations of the WSSUS approach to directional models are discussed in Fleury [2000], Kattenbach [2002].

For updates and errata for this chapter, see wides.usc.edu/teaching/textbook

7

Channel Models

7.1 Introduction

For the design, simulation, and planning of wireless systems, we need *models* for the propagation channels. In the previous chapters, we have discussed some basic properties of wireless channels, and how they can be described mathematically – amplitude-fading statistics, scattering function, delay spread, etc. In this chapter, we discuss in a more concrete way how these mathematical description methods can be converted into generic simulation models, and how to parameterize these models.

There are two main applications for channel models:

1. For the design, testing, and type approval of *wireless systems*, we need simple channel models that reflect the important properties of propagation channels – i.e., properties that have an impact on system performance. This is usually achieved by simplified channel models that describe the statistics of the impulse response in parametric form. The number of parameters is small and *independent of specific locations*. Such models sometimes lead to insights due to closed-form relationships between channel parameters and system performance. Furthermore, they can easily be implemented by system designers for testing purposes.
2. The designers of *wireless networks* are interested in optimizing a given system in a certain geographical region. Locations of Base Stations (BSs) and other network design parameters should be optimized on the computer, and not by field tests, and trial and error. For such applications, *location-specific channel models* that make good use of available geographical and morphological information are desirable. However, the models should be robust with respect to small errors in geographical databases.

The following three modeling methods are in use for these applications:

1. *Stored channel impulse responses*: a channel sounder (see Chapter 8) measures, digitizes, and stores impulse responses $h(t, \tau)$. The main advantage of this approach is that the resulting impulse responses are realistic. Furthermore, system simulations using the stored data are reproducible, as the data remain available and can be reused indefinitely, even for simulations of different systems. This is an important distinction from field trials of whole systems, where there can be no guarantee that the impulse response remains constant over time. The disadvantages of using stored impulse responses are (i) the large effort in acquiring and storing the data and (ii) the fact that the data characterize only a certain area, and need not be typical for a propagation environment.

Wireless Communications, Second Edition   Andreas F. Molisch
© 2011 John Wiley & Sons, Ltd

2. *Deterministic channel models*: these models use the geographical and morphological information from a database for a deterministic solution of Maxwell's equation or some approximation thereof. The basic philosophy is the same as for stored impulse responses: determining the impulse response in a certain geographic location. Both of these methods are therefore often subsumed as *site-specific models*. The drawbacks of deterministic (computed) channel models compared with stored (measured) impulse responses are (i) the large computational effort and (ii) the fact that the results are inherently less accurate, due to inaccuracies in the underlying databases and the approximate nature of numerical computation methods. The main advantage is that computer simulations are easier to perform than measurement campaigns. Furthermore, certain types of computation methods (e.g., ray tracing, see Section 7.5) allow the effects of different propagation mechanisms to be isolated.

3. *Stochastic channel models*: these model the probability density function (pdf) of the channel impulse response (or equivalent functions). These methods do not attempt to correctly predict the impulse response in one specific location, but rather to predict the pdf over a large area. The simplest example of this approach is the Rayleigh-fading model: it does not attempt to correctly predict the field strength at each location, but rather attempts to correctly describe the pdf of the field strength over a large area. Stochastic wideband models can be created in the same spirit.

Generally speaking, stochastic models are used more for the design and comparison of systems, while site-specific models are preferable for network planning and system deployment. Furthermore, deterministic and stochastic approaches can be combined to enhance the efficiency of a model: e.g., large-scale averaged power can be obtained from deterministic models, while the variations within an averaging area are modeled stochastically.

It is obvious that none of the above models can achieve perfect accuracy. Establishing a criterion for "satisfactory accuracy" is thus important:

- From a purely scientific point of view, any inaccuracy is unsatisfactory. From an engineering point of view, however, there is no point in increasing modeling accuracy (and thus effort) beyond a certain point.[1]
- For deterministic modeling methods, inaccuracies in the underlying databases lead to unavoidable errors. For stochastic models that are derived from measurements, the finite number of underlying measurement points, as well as measurement errors, limit the possible accuracy. Ideally, errors due to a specific modeling method should be smaller than errors due to these unavoidable inaccuracies.
- Requirements on modeling accuracy can be relaxed even more by the following pragmatic criterion: the inaccuracies in the model should not "significantly" alter a system design or deployment plan. For this definition, the system designer has to determine what "significant" is.

7.2   Narrowband Models

7.2.1   Modeling of Small-Scale and Large-Scale Fading

For a narrowband channel, the impulse response is a delta function with a time-varying attenuation, so that for slowly time-varying channels:

$$h(t, \tau) = \alpha(t)\delta(\tau) \tag{7.1}$$

---

[1] Many research papers use the words "satisfactory accuracy" as a rhetorical tool to emphasize the value of a new modeling method. If a method decreases the deviation between theory and measurement from 12 to 9 dB, it will consider 9 dB as "satisfactory." A subsequent paper that decreases the error to 6 dB will consider the same 9 dB as "unsatisfactory."

As mentioned in Chapter 5, the variations in amplitude over a small area are typically modeled as a random process, with an autocorrelation function that is determined by the Doppler spectrum. The complex amplitude is modeled as a zero-mean, circularly symmetric complex Gaussian random variable. As this gives rise to a Rayleigh distribution of the absolute amplitude, we henceforth refer to this case simply as "Rayleigh fading."

When considering variations in a somewhat larger area, the small-scale averaged amplitude F obeys a lognormal distribution, with standard deviation σ_F; typically, values of σ_F are 4 to 10 dB. The spatial autocorrelation function of lognormal shadowing is usually assumed to be a double-sided exponential, with correlation distances between 5 and 100 m, depending on the environment.

7.2.2   Path Loss Models

Next, we consider models for the received field strength, averaged over both small-scale and the large-scale fading. This quantity is modeled completely deterministically. The most simple models of that kind are the free space path loss model, and the "breakpoint" model (with $n = 2$ valid for distances up to $d < d_{break}$, and $n = 4$ beyond that, as described in Chapter 4). In more sophisticated models, described below, path loss depends not only on distance but also on some additional external parameters like building height, measurement environment (e.g., suburban environment), etc.

The Okumura–Hata Model

The Okumura–Hata model is by far the most popular model in that category. Path loss (in dB) is written as

$$PL = A + B \log(d) + C \tag{7.2}$$

where A, B, and C are factors that depend on frequency and antenna height. Factor A increases with carrier frequency and decreases with increasing height of the BS and Mobile Station (MS). Also, the path loss exponent (proportional to B) decreases with increasing height of the BS. Appendix 7.A – see *www.wiley.com/go/molisch* – gives details of these correction factors. The model is only intended for large cells, with the BS being placed higher than the surrounding rooftops.

The COST[2] 231–Walfish–Ikegami Model

The COST 231–Walfish–Ikegami model is also suitable for microcells and small macrocells, as it has fewer restrictions on the distance between the BS and MS and the antenna height.

In this model, total path loss consists of the free space path loss PL_0, *multiscreen loss* L_{msd} along the propagation path, and attenuation from the last roof-edge to the MS, L_{rts} (*rooftop-to-street diffraction and scatter loss*) (Figure 7.1). Free space loss depends on carrier frequency and distance, while the rooftop-to-street diffraction loss depends on frequency, the width of the street, and the height of the MS, as well as on the orientation of the street with respect to the connection line BS–MS. Multiscreen loss depends on the distance between buildings and the distance between the BS and MS, as well as on carrier frequency, BS height, and rooftop height. The model assumes a Manhatten street grid (streets intersecting at right angles), constant building height, and flat terrain. Furthermore, the model does not include the effect of waveguiding through street canyons, which can lead to an underestimation of the received field strength. Details of the model can be found in Appendix 7.B (see companion website *www.wiley.com/go/molisch*).

---

[2] European COoperation in the field of Scientific and Technical research.

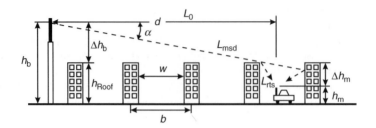

Figure 7.1 Parameters in the COST 231–Walfish–Ikegami model.
Reproduced with permission from Damosso and Correia [1999] © European Union.

The Motley–Keenan Model

For indoor environments, wall attenuation plays an important role. Based on this consideration, the Motley–Keenan model suggests that path loss (expressed in decibel (dB)) can be written as [Motley and Keenan 1988]

$$PL = PL_0 + 10n \log(d/d_0) + F_{\text{wall}} + F_{\text{floor}}$$

where F_{wall} is the sum of attenuations by the walls that a Multi Path Component (MPC) has to penetrate on its way from the transmitter (TX) to the receiver (RX); similarly, F_{floor} describes the summed-up attenuation of the floors that are located between the BS and MS. Depending on the building material, attenuation by one wall can lie between 1 and 20 dB in the 300 MHz–5 GHz range, and can be much higher at higher frequencies.

The Motley–Keenan model is a site-specific model, in the sense that it requires knowledge of the location of the BS and MS, and the building plan. It is, however, not very accurate, as it neglects propagation paths that "go around" the walls. For example, propagation between two widely separated offices can occur either through many walls (quasi-Line Of Sight – LOS), or through a corridor (signal leaves the office, propagates down a corridor, and enters from there into the office of the RX). The latter type of propagation path can often be more efficient, but is not taken into account by the Motley–Keenan model.

7.3 Wideband Models

7.3.1 Tapped Delay Line Models

The most commonly used wideband model is an N-tap Rayleigh-fading model. This is a fairly generic structure, and is basically just the tapped delay line structure of Chapter 6, with the added restriction that the amplitudes of all taps are subject to Rayleigh fading. Adding an LOS component does not pose any difficulties; the impulse response then just becomes

$$h(t, \tau) = a_0 \delta(\tau - \tau_0) + \sum_{i=1}^{N} c_i(t) \delta(\tau - \tau_i) \tag{7.3}$$

where the LOS component a_0 does *not* vary with time, while the $c_i(t)$ are zero-mean complex Gaussian random processes, whose autocorrelation function is determined by their associated Doppler spectra (e.g., Jakes spectra). In most cases, $\tau_0 = \tau_1$, so the amplitude distribution of the first tap is Rician.

The model is further simplified when the number of taps is limited to $N = 2$, and no LOS component is allowed. This is the simplest stochastic fading channel exhibiting delay dispersion,[3] and thus very popular for theoretical analysis. it is alternatively called the *two-path channel, two-delay channel*, or *two-spike channel*.

Another popular channel model consists of a purely deterministic LOS component plus *one* fading tap ($N = 1$) whose delay τ_0 can differ from τ_1. This model is widely used for satellite channels – in these channels, there is almost always an LOS connection, and the reflections from buildings near the RX give rise to a delayed fading component. The channel reduces to a flat-fading Rician channel when $\tau_0 = \tau_1$.

7.3.2 Models for the Power Delay Profile

It has been observed in many measurements that the Power Delay Profile (PDP) can be approximated by a one-sided exponential function:

$$P_{\mathrm{h}}(\tau) = P_{\mathrm{sc}}(\tau) = \begin{cases} \exp(-\tau/S_\tau) & \tau \geq 0 \\ 0 & \text{otherwise} \end{cases} \tag{7.4}$$

In a more general model (see also Section 7.3.3), the PDP is the sum of several delayed exponential functions, corresponding to multiple *clusters* of Interacting Objects (IOs):

$$P_{\mathrm{h}}(\tau) = \sum_{\ell} \frac{P_l^{\mathrm{c}}}{S_{\tau,l}^{\mathrm{c}}} P_{\mathrm{sc}}(\tau - \tau_{0,l}^{\mathrm{c}}) \tag{7.5}$$

where p_l^{c}, $\tau_{0,l}^{\mathrm{c}}$, $S_{\tau,l}^{\mathrm{c}}$ are the power, delay, and delay spread of the *l*th cluster, respectively. The sum of all cluster powers has to add up to the narrowband power described in Section 7.2.

For a PDP in the form of Eq. (7.4), the rms delay spread characterizes delay dispersion. In the case of multiple clusters, Eq. (7.5), the rms delay spread is defined mathematically, but often has a limited physical meaning. Still, the vast majority of measurement campaigns available in the literature use just this parameter for characterization of delay dispersion.

Typical values of the delay spread for different environments are (see Molisch and Tufvesson [2004] for more details and extensive references) as follows:

- *Indoor residential buildings*: 5–10 ns are typical; but up to 30 ns have been measured.
- *Indoor office environments*: these show typical delay spreads of between 10 and 100 ns, but even 300 ns have been measured. Room size has a clear influence on delay spread. Building size and shape have an impact as well.
- *Factories and airport halls*: these have delay spreads that range from 50 to 200 ns.
- *Microcells*: in microcells, delay spreads range from around 5–100 ns (for LOS situations) to 100–500 ns (for non-LOS).
- *Tunnels and mines*: empty tunnels typically show a very small delay spread (on the order of 20 ns), while car-filled tunnels exhibit larger values (up to 100 ns).
- *Typical urban and suburban environments*: these show delay spreads between 100 and 800 ns, although values up to 3 μs have also been observed.
- *Bad Urban (BU) and Hilly Terrain (HT) environments*: these show clear examples of multiple clusters that lead to much larger delay spreads. Delay spreads up to 18 μs, with cluster delays of up to 50 μs, have been measured in various European cities, while American cities show somewhat smaller values. Cluster delays of up to 100 μs occur in mountainous terrain.

---

[3] Note that each of the taps of this channel exhibits Rayleigh fading; the channel is thus different from the two-path model used in Chapters 5 and 6 for purely didactic reasons.

The delay spread is a function of the distance BS–MS, increasing with distance approximately as d^ε, where $\varepsilon = 0.5$ in urban and suburban environments, and $\varepsilon = 1$ in mountainous regions. The delay spread also shows considerable large-scale variations. Several papers find that the delay spread has a lognormal distribution with a variance of typically 2–3 dB in suburban and urban environments. A comprehensive model containing all these effects was first proposed by Greenstein et al. [1997].

7.3.3 Models for the Arrival Times of Rays and Clusters

In the previous section, we described models for the PDP. The modeled PDPs were continuous functions of the delay; this implies that the RX bandwidth was so small that different discrete MPCs could not be resolved, and were "smeared" into a continuous PDP. For systems with higher bandwidth, MPCs can be resolved. In that case, it is advantageous to describe the PDP by the arrival times of the MPCs, plus an "envelope" function that describes the power of the MPCs as a function of delay.

In order to statistically model the arrival times of MPCs, a first-order approximation assumes that objects that cause reflections in an urban area are located randomly in space, giving rise to a *Poisson distribution* for excess delays. However, measurements have shown that MPCs tend to arrive in groups ("clusters"). Two models have been developed to reflect this fact: the $\Delta - K$ model, and the Saleh–Valenzuela (SV) model.

The $\Delta - K$ *model* has two states: S_1, where the mean arrival rate is $\lambda_0(t)$, and S_2, where the mean arrival rate is $K\lambda_0(t)$. The process starts in S_1. If an MPC arrives at time t, a transition is made to S_2 for the interval $[t, t + \Delta]$. If no further paths arrive in this interval, a transition is made back to S_1 at the end of the interval. Note that for $K = 1$ or $\Delta = 0$, the above-mentioned process reverts to a standard Poisson process.

The SV model takes a slightly different approach. It assumes a priori the existence of cluster. *Within* each cluster, the MPCs are arriving according to a Poisson distribution, and the arrival times of the *clusters themselves* are Poisson distributed (but with a different interarrival time constant). Furthermore, the powers of the MPCs within a cluster decrease exponentially with delay, and the power of the clusters follows a (different) exponential distribution (see Figure 7.2).

Mathematically, the following discrete time impulse response is used:

$$h(\tau) = \sum_{l=0}^{L} \sum_{k=0}^{K} c_{k,l}(\tau)\delta(\tau - T_l - \tau_{k,l})$$

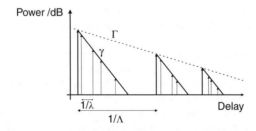

Figure 7.2 The Saleh–Valenzuela model.

where the distribution of cluster arrival time and the ray arrival time is described (with a slight abuse of notation) as

$$\text{pdf}(T_l | T_{l-1}) = \Lambda \exp[-\Lambda(T_l - T_{l-1})], \qquad l > 0$$

$$\text{pdf}(\tau_{k,l} | \tau_{(k-1),l}) = \lambda \exp[-\lambda(\tau_{k,l} - \tau_{(k-1),l})], \qquad k > 0$$

where T_l is the arrival time of the first path of the lth cluster, $\tau_{k,l}$ is the delay of the kth path within the lth cluster relative to the first path arrival time of this (by definition, $\tau_{0,l} = 0$), Λ is the cluster arrival rate, and λ is the ray arrival rate – i.e., the arrival rate of paths within each cluster. All dependences of the parameters on absolute time have been suppressed in the above equations.

The PDP within each cluster is

$$E\{|c_{k,l}|^2\} \propto P_l^c \exp(-\tau_{k,l}/\gamma) \qquad (7.6)$$

where P_l^c is the energy of the lth cluster, and γ is the intracluster decay time constant. Cluster power decreases exponentially as well:

$$P_l^c \propto \exp(-T_l / \Gamma) \qquad (7.7)$$

7.3.4   Standardized Channel Model

A special case of the tapped delay line model is the COST 207 model, which specifies the PDPs or tap weights and Doppler spectra for four typical environments. These PDPs were derived from numerous measurement campaigns in Europe. The model distinguishes between four different types of macrocellular environments – namely, *typical urban (TU)*, *bad urban (BU)*, *rural area (RA)*, and *hilly terrain (HT)*. Depending on the environment, the PDP has a single-exponential decay, or it consists of two single-exponential functions (clusters) that are delayed with respect to each other (see Figure 7.3). The second cluster corresponds to groups of faraway high-rise buildings or mountains that act as efficient IOs, and thus give rise to a group of delayed MPCs with considerable power. More details can be found in Appendix 7.C (see companion website *www.wiley.com/go/molisch*).

The COST 207 models are based on measurements with a rather low bandwidth, and are applicable only for systems with 200-kHz bandwidth or less. For simulation of third-generation cellular systems, which have a bandwidth of 5 MHz, the *International Telecommunications Union* (ITU) specified another set of models that accounts for the larger bandwidth. This model distinguishes between pedestrian, vehicular, and indoor environments. Details can be found in Appendix 7.D (see *www.wiley.com/go/molisch*). Additional tapped delay line models were also derived for indoor wireless Local Area Network (LAN) systems and Personal Area Networks (PANs); for an overview, see Molisch and Tufvesson [2004].

7.4   Directional Models

7.4.1   General Model Structure and Factorization

As discussed in Chapter 6, a fairly general model is based on the Double Directional Delay Power Spectrum (DDDPS), which depends on the three variables Direction Of Departure (DOD), Direction Of Arrival (DOA), and delay. An important simplification is obtained if the DDDPS can be factored

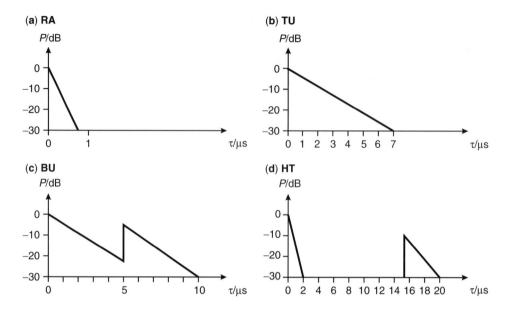

Figure 7.3   COST 207 power delay profiles.

Reprinted with permission from Molisch [2000] © Prentice Hall.

into three functions, each of which depends on just a single parameter:

$$DDDPS(\Omega, \Psi, \tau) = APS^{\mathrm{BS}}(\Omega)APS^{\mathrm{MS}}(\Psi)P_h(\tau) \tag{7.8}$$

This implies that the Angular Power Spectrum (APS) at the BS is independent of delay, as is the APS at the MS. Furthermore, the APS at the MS is independent of the direction in which the BS transmits, and vice versa.

Such a factorization greatly simplifies theoretical computations, and also the parameterization of channel models. However, it does not always correspond to physical reality. A more general model assumes that the DDDPS consists of several *clusters*, each of which has a separable DDDPS:

$$DDDPS(\Omega, \Psi, \tau) = \sum_l P_l^c APS_l^{c,\mathrm{BS}}(\Omega)APS_l^{c,\mathrm{MS}}(\Psi)P_{h,l}^c(\tau) \tag{7.9}$$

where superscript c stands for "cluster" and l indexes the clusters. Obviously, this model reduces to Eq. (7.8) only if a single cluster exists.

In the remainder of this section, we assume that factorization is possible and describe only models for the angular spectra $APS_l^{c,\mathrm{BS}}$ and $APS_l^{c,\mathrm{MS}}(\Psi)$ – i.e., the components in this factorization (the PDP has already been discussed in Section 7.3).

7.4.2   Angular Dispersion at the Base Station

The most common model for the APS at the BS is a Laplacian distribution in *azimuth* [Pedersen et al. 1997]:

$$APS(\phi) \propto \exp\left[-\sqrt{2}\frac{|\phi - \phi_0|}{S_\phi}\right] \tag{7.10}$$

where ϕ_0 is the mean azimuthal angle. The *elevation* spectrum is usually modeled as a delta function (i.e., all radiation is incident in the horizontal plane) so that $\Omega = \phi$; alternatively, it has been modeled as a Laplacian function.

The following range of rms angular spreads (see Section 6.7) and cluster angular spreads can be considered typical [Molisch and Tufvesson 2004]:

- *Indoor office environments*: rms cluster angular spreads between $10°$ and $20°$ for non-LOS situations, and typically around $5°$ for LOS.
- *Industrial environments*: rms angular spreads between $20°$ and $30°$ for non-LOS situations.
- *Microcells*: rms angular spreads between $5°$ and $20°$ for LOS, and $10°-40°$ for non-LOS.
- *Typical urban and suburban environments*: measured rms angular spreads on the order of $3°-20°$ in dense urban environments. In suburban environments, the angular spread is smaller, due to the frequent occurrence of LOS.
- *Bad urban and hilly terrain environments*: rms angular spreads of $20°$ or larger, due to the existence of multiple clusters.
- *Rural environments*: rms angular spreads between $1°$ and $5°$ have been observed.

In outdoor environments, the distribution of the angular spread over large areas has also been found to be lognormal and correlated with the delay spread. This permits the logarithms of the spreads to be treated as correlated Gaussian random variables. The dependence of angular spread on distance is still a matter of discussion.

7.4.3 Angular Dispersion at the Mobile Station

For outdoor environments, it is commonly assumed that radiation is incident from all azimuthal directions onto the MS, because the MS is surrounded by "local IOs" (cars, people, houses, etc.). This model dates back to the 1970s. However, recent studies indicate that the azimuthal spread can be considerably smaller, especially in street canyons. The APS is then again approximated as Laplacian; cluster angular spreads on the order of $20°$ have been suggested. Furthermore, the angular distribution is a function of delay. For MSs located in street canyons without LOS, small delays are related to over-the-rooftop propagation, which results in large angular spreads, while later components are waveguided through the streets and thus confined to a smaller angular range. In indoor environments with (quasi-) LOS, early components have a very small angular spread, while components with larger delay have an almost uniform APS.

For the outdoor elevation spectrum, MPCs that propagate over the rooftops have an elevation distribution that is uniform between 0 and the angle under which the rooftops are seen; later-arriving components, which have propagated through the street canyons, show a Laplacian elevation distribution.

7.4.4 Polarization

Most channel models analyze only the propagation of vertical polarization, corresponding to transmission and reception using vertically polarized antennas. However, there is increased interest in polarization diversity – i.e., antennas that are colocated, but receive waves with different polarizations. In order to simulate such systems, models for the propagation of dual-polarized radiation are required.

Transmission from a vertically polarized antenna will undergo interactions that result in energy being leaked into the horizontal polarization component before reaching the RX antenna (and vice versa). The fading coefficients for MPCs thus have to be written as a polarimetric 2×2 matrix, so that the complex amplitude $\mathbf{a}_\ell$ becomes

$$\mathbf{a}_\ell = \begin{pmatrix} a_\ell^{VV} & a_\ell^{VH} \\ a_\ell^{HV} & a_\ell^{HH} \end{pmatrix} \tag{7.11}$$

where V and H denote vertical and horizontal polarization, respectively.

The most common polarimetric channel model assumes that the entries in the matrix are statistically independent, complex Gaussian fading variables. The *mean* powers of the *VV* and the *HH* components are assumed to be identical; similarly, the *mean* powers of the *VH* and *HV* components are the same. The cross-polarization ratio, *XPD*, which is the ratio (expressed in dB) of the mean powers in *VV* and *VH*, is modeled as a Gaussian random variable. The mean and variance of the XPD can depend on the propagation environment, and even on the delay of the considered components. Typical values for the mean of the XPD lie between 0 and 12 dB; for the variance, around 3–6 dB [Shafi et al. 2006].

7.4.5 Model Implementations

The above sections have discussed a continuous model for the angular spectra. For a computer implementation, we usually need a discretized version. One way of implementing a Directional Channel Model (DCM) is a generalized tapped delay line. In this approach, the DDDPS is discretized, according to the same principles as described in Section 6.4.

An alternative is the so-called *Geometry-based Stochastic Channel Model* (GSCM). In this approach, it is not the strength and direction of the MPCs that is modeled stochastically but rather the location of IOs and the strength of the interaction processes (see Figure 7.4). Additionally, it is assumed that only single interaction processes can occur. The directionally resolved impulse response is then obtained in two steps:

1. Assign locations to the IOs, according to the pdf of their position.
2. Based on the assumption of single interaction only, determine the contributions of the IOs to the double-directional impulse response. Each MPC (corresponding to one IO) has a unique DOA, DOD, amplitude delay, and phase shift.

The simplest model is based on the assumption that all relevant IOs are close to the MS. This case occurs, e.g., in macrocells with regular building structures like suburban environments. In this

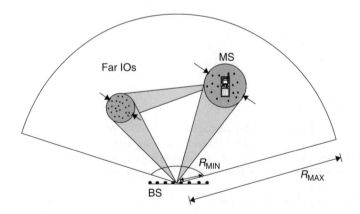

Figure 7.4 Principle of geometry-based stochastic channel model.

case, radiation from the MS interacts with IOs around the MS, but can proceed without further interaction from those objects to the BS. Different models exist for the distribution of the IOs around the MS:

- Some papers place all IOs on a circle around the MS, see Lee [1973].
- Other papers suggest a uniform distribution within a disk. When the MS moves, the disk around the MS also moves. Some IOs thus "fall out" of the IO disk, while new IOs enter (see Figure 7.5). This corresponds to the physical reality that IOs that are far away from the MS do not make significant contributions (although naturally they still "exist" physically).
- A one-sided Gaussian distribution $pdf(r) = \exp(-r^2/2\sigma^2)$, $r \geq 0$, has also been suggested. Computing the PDP and the APS from this distribution gives the results shown in Figures 7.6 and 7.7. We see that these results are fairly similar to an exponential PDP and Laplacian APS.

The case when all IOs are close to the MS is the "single-cluster" case, with an Angular Delay Power Spectrum (ADPS) that is approximately given as

$$ADPS(\tau, \phi) \propto \exp(-\tau/S_\tau) \exp(-\sqrt{2}|\phi - \phi_0|/S_\phi)$$

The generalization includes the so-called *far IOs* (also known as far scatterers), which correspond to high-rise buildings or mountains. Such a far IO can be modeled either as a single specular reflector (corresponding, e.g., to a high-rise building with a smooth glass front) or a cluster of IOs. In contrast to the IOs around the MS, the location of far IOs stays constant during a whole simulation process.

Geometric channel models have advantages especially when movement is to be simulated. Whenever the MS moves, adjustments to the parameters of the MPCs are automatically made. Thus, the correct fading correlation results automatically from movement; also the correlation between the DOAs at the MS and the Doppler shift is taken into account. Any changes in the mean DOAs, DODs, and delays due to large-scale movement of the MS are automatically included, while they would be difficult to model in tapped delay line models.

Figure 7.5 "Vanishing" and "appearing" of IOs when the MS moves. It is assumed that all the IOs are in a disk around the MS. Scatterers that are active only at time $t_1(t_2)$ are shown as black (empty) circles; scatterers that are active at both time instants are shown in grey.

Figure 7.6 PDP for different distributions of scatterers.
Reproduced with permission from Laurila et al. [1998] © IEEE.

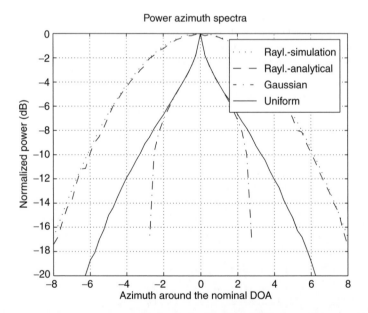

Figure 7.7 Angular power spectrum for different distributions of scatterers.
Reproduced with permission from Laurila et al. [1998] © IEEE.

7.4.6 Standardized Directional Models

The European research initiative COST 259 developed a DCM that has gained widespread acceptance. It is very realistic, incorporating a wealth of effects and their interplay, for a number of different environments. As the model is rather involved, this section only points out some basic features. A more detailed description of the first version of the model is described in Steinbauer and Molisch [2001], and a full account is found in Asplund et al. [2006] and Molisch et al. [2006].

The COST 259 DCM includes small-scale as well as continuous large-scale changes of the channel. This is achieved efficiently by distinguishing between three different layers:

1. At the top layer, there is a distinction between different *Radio Environments* (REs) – i.e., environmental classes with similar propagation characteristics (e.g., "TU"). All in all, there are 13 REs: four macrocellular REs (i.e., BS height above rooftop), four microcellular REs (outdoor, BS height below rooftop), and five picocellular REs (indoor).
2. *Large-scale effects* are described by their pdfs, whose parameters differ for different REs. For example, delay spread, angular spread, shadowing, and the Rice factor change as the MS moves over large distances. Each realization of large-scale fading parameters determines a DDDPS.
3. On a third layer, double-directional impulse responses are realizations of the DDDPS, created by the *small-scale fading*.

Large-scale effects are described in a mixed geometrical–stochastic fashion, applying the concept of IO clusters as described above. At the beginning of a simulation, IO clusters (one local cluster around the MS and several far IO clusters) are distributed at random in the coverage area; this is the stochastic component. During the simulation, the delays and angles between the clusters are obtained deterministically from their position and the positions of the BS and MS; this is the geometrical component. Each of the clusters has a small-scale averaged DDDPS that is exponential in delay, Laplacian in azimuth and elevation at the BS, and uniform or Laplacian in azimuth and elevation at the MS. Double-directional complex impulse responses are then obtained from the average ADPS either directly, or by mapping it onto an IO distribution and obtaining impulse responses in a geometrical way.

In macrocells the positions of clusters are random. In micro- and picocells, the positions are deterministic, using the concept of *Virtual Cell Deployment Areas* (VCDAs). A VCDA is a map of a virtual town or office building, with the route of the MS prescribed in it. This approach is similar to the ray-tracing approach but differs in two important respects: (i) the "city maps" need not reflect an actual city and can thus be made to be "typical" for many cities; (ii) only the cluster *positions* are determined by ray tracing, while the behavior *within* one cluster is treated stochastically.

Other standardized models are described in the "Further Reading" section and the appendices.

7.4.7 Multiple-Input Multiple-Output Matrix Models

The previous sections have described models that include the directional information of MPCs. An alternative concept that is popular in the context of multiantenna systems is to stochastically model the impulse response matrix (see Section 6.7) of a Multiple Input Multiple Output (MIMO) channel. In this case, the channel is characterized not only by the amplitude statistics of each matrix entry (which is usually Rayleigh or Rician) but also by the *correlation* between these entries. The correlation matrix (for each delay tap) is defined by first "stacking" all the entries of the channel matrix in one vector $\mathbf{h}_{\mathrm{stack}} = [h_{1,1}, h_{2,1}, \ldots, h_{N_{r,1}}, h_{1,2}, \ldots, h_{N_r N_t}]^T$ and then computing the correlation matrix as $\mathbf{R} = E\{\mathbf{h}_{\mathrm{stack}}\mathbf{h}_{\mathrm{stack}}^{\dagger}\}$, where superscript † denotes the Hermitian transpose. One popular simplified model assumes that the correlation matrix can be written as a Kronecker product $\mathbf{R} = \mathbf{R}_{\mathrm{TX}} \otimes \mathbf{R}_{\mathrm{RX}}$, where $\mathbf{R}_{\mathrm{TX}} = E\{\mathbf{H}^{\dagger}\mathbf{H}\}$ and $\mathbf{R}_{\mathrm{RX}} = E\{\mathbf{H}\mathbf{H}^{\dagger}\}$. This model implies that the correlation matrix at the RX is independent of the direction of transmission; this is equivalent to assuming that the DDDPS can be factored into independent APSs at the BS and at the MS. In that case, the channel transfer function matrix can be generated as

$$\mathbf{H} = \frac{1}{\mathrm{E}\{\mathrm{tr}(\mathbf{H}\mathbf{H}^\dagger)\}} \mathbf{R}_{\mathrm{RX}}^{1/2} \mathbf{G}_{\mathrm{G}} \mathbf{R}_{\mathrm{TX}}^{1/2} \tag{7.12}$$

where $\mathbf{G}_\mathrm{G}$ is a matrix with independent identically distributed (iid) complex Gaussian entries.

7.5 Deterministic Channel-Modeling Methods

In principle, a wireless propagation channel can be viewed as a deterministic channel. Maxwell's equations, together with electromagnetic boundary conditions (location, shape, and dielectric and conductive properties of all objects in the environment), allow determination of the field strength at all points and times. For outdoor environments, such purely deterministic channel models have to take into account all the geographical and morphological features of a propagation environment; for indoor environments, the building structure, wall properties, and even furniture should be taken into consideration. In this section, we outline the basic principles of channel models based on such a deterministic point of view.

To make deterministic modeling a viable option, two major challenges had to be overcome: (i) the high amount of required computer time and (ii) the need for exact knowledge of boundary conditions.

- *Computer time and storage* were prohibitive up to about 1990. However, this has changed since then. On one hand, computers have become so much faster that tasks that seemed unfeasible even with supercomputers in the 1990s are realistic options on a personal computer nowadays. On the other hand, the development of more efficient deterministic algorithms has improved the situation as well.
- *Exact knowledge of boundary conditions* is required for the successful application of deterministic models. This implies that the position and the electromagnetic properties of the whole "relevant" environment have to be known (we will discuss later what "relevant" means in this context). The creation of digital terrain maps and city plans, based on satellite images or building plans, has also made considerable progress in the last few years.

The most accurate solution (given an environment database) is a "brute force" solution of Maxwell's equations, employing either integral or differential equation formulations. Integral equations are most often variations of the well-known *Method of Moments*, where the unknown currents induced in the IOs are represented by a set of basis functions. In their most simple form, basis functions are rectangular functions, extending over a fraction of a wavelength. Differential equation formulations include the *Finite Element Method* (FEM) or the increasingly popular *Finite Difference Time Domain* (FDTD) method.

All these methods are highly accurate, but the computational requirements are prohibitive in most environments. It is thus much more common to use approximations to Maxwell's equations as a basis for solution. The most widespread approximation is the *high-frequency* approximation (also known as *ray approximation*).[4] In this approximation, electromagnetic waves are modeled as rays that follow the laws of geometrical optics (Snell's laws for reflection and transmission); further refinements allow inclusion of diffraction and diffuse scattering in an approximate way. In the remainder of this section, we will concentrate on various implementations of ray-based schemes.

---

[4] In the literature, such methods are often generally known as *ray tracing*. However, the expression "ray tracing" is also used for one specific implementation method (described below). We will thus stick with the name "high-frequency approximation" for the general class of algorithms.

7.5.1 Ray Launching

In the *ray-launching* approach, the transmit antenna sends out (launches) rays into different directions. Typically, the total spatial angle 4π is divided into N units of equal magnitude, and each ray
is sent in the direction of the center of one such unit (i.e., uniform sampling of the spatial angle)
(see Figure 7.8). The number of launched rays is a tradeoff between accuracy of the method and
computation time.

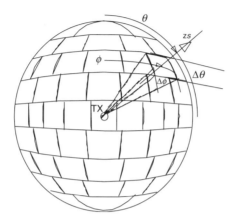

Figure 7.8 Principle of ray launching.
Reproduced with permission from Damosso and Correia [1999] © European Union.

The algorithm follows the propagation of each ray until it either hits the RX or becomes too
weak to be significant (e.g., drops below the noise level). When following a ray, a number of effects
have to be taken into account:

- *Free space attenuation*: as each ray represents a certain spatial angle, the energy *per unit area*
  decreases d^{-2} along the path of the ray.
- *Reflections* change the direction of a ray and cause an additional attenuation. Reflection coefficients can be computed from Snell's laws (see Chapter 4) depending on the angle of incidence
  and possibly the polarization of the incident ray.
- *Diffraction and diffuse scattering* are included in more advanced models. In those cases, a ray that
  is incident on an IO gives rise to several new rays. The amplitudes of diffracted rays are usually
  computed from the geometrical or uniform theory of diffraction, as discussed in Chapter 4.

The *ray-splitting* algorithm is an important improvement in the accuracy of the method. The
algorithm is based on the premise that the effective cross-section of the ray should never exceed
a certain size (e.g., the size of a typical IO). Thus, if a ray has propagated too far from the TX,
it is subdivided into two rays. Let us explain that principle in more detail using Figure 7.9. To
simplify the discussion, we consider only the two-dimensional case. Each ray represents not only
a certain angle but rather an angular *range* of width ϕ – corresponding to the angle between two
launched rays. The intersection of such an angular range with a circle of radius d has a length of
approximately ϕd (for the three-dimensional case, think "cross-section" instead of "length"). Thus,
the farther we get away from the TX, the larger the length that is covered by the ray. In order to
maintain high accuracy of the simulation, this length should not become too large. As soon as it
reaches a length $\overline{L}$, the ray is split (thus reducing the length to $\overline{L}/2$). The resulting subrays (which
again represent a whole angular range of width ϕ) then propagate until they reach a length $\overline{L}$, etc.

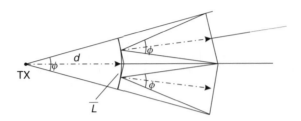

Figure 7.9 Principle of ray splitting.

Ray launching gives the channel characteristics in the whole environment – i.e., for many different RX positions and a given TX position. In other words, once we have decided on a BS location, we can compute coverage, delay spread, and other channel characteristics in the whole envisioned cell area. Furthermore, a preprocessing scheme allows the inclusion of multiple TX locations. The environment (the IOs) is subdivided into "tiles" (areas of finite size, typically the same size as the maximum effective area of a ray) and the interaction between all tiles is computed. Then, for each TX position, only the interaction between the TX and the tiles that can act as first IOs has to be computed [Hoppe et al. 2003].

7.5.2 Ray Tracing

Classical *ray tracing* determines all rays that can go from *one* TX location to *one* RX location. The method operates in two steps:

1. First, all rays that can transfer energy from the TX location to the RX location are determined. This is usually done by means of the image principle. Rays that can get to the RX via a reflection show the same behavior as rays from a virtual source that is located where an image of the original source (with respect to the reflecting surface) would be located (see Figure 7.10).
2. In a second step, attenuations (due to free space propagation and finite reflection coefficients) are computed, thus providing the parameters of all MPCs.

Ray tracing allows fast computation of single- and double-reflection processes, and also does not require ray splitting. On the downside, effort increases exponentially with the order of reflections that are included in the simulation. Also, the inclusion of diffuse scattering and diffraction is nontrivial. Finally, the method is less efficient than ray launching for the computation of channel characteristics over a wide area.

7.5.3 Efficiency Considerations

Both for ray launching and ray tracing, it is almost impossible to correctly predict the *phases* of arriving rays. Such a prediction would require a geographical and building database that is accurate to within a fraction of a wavelength. It is thus preferable to assume that all rays have uniformly distributed random phases. In this case, it is only possible to deterministically predict the small-scale *statistics* of channel characteristics; realizations of the impulse responses are obtained by ascribing random phases to MPCs. This is another form of the mixed deterministic–stochastic approach mentioned at the beginning of this chapter.

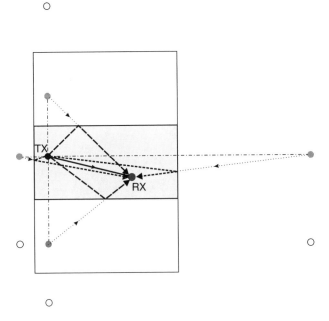

Figure 7.10 The image principle. Grey circles: virtual sources corresponding to a single reflection. White circles: virtual sources corresponding to double reflections. Dotted lines: rays from the virtual sources to the RX. Dashed lines: actual reflections. Solid lines: line of sight.

A further method to reduce the computational effort is to perform ray tracing not in all three dimensions but rather only in two dimensions. It depends on the propagation environment whether this simplification is admissible:

- *Indoor*: indoor environments practically always require three-dimensional considerations. Even when the BS and the MS are on the same floor, reflections at floors and ceilings represent important propagation paths.
- *Macrocells*: by definition, the BS antenna is considerably above the rooftops. Propagation thus occurs mostly over the rooftops to points that are close to the MS. From these points, they then reach the MS, possibly via a diffraction or a reflection from the wall of the house opposite. Ray tracing in the vertical plane alone can thus be sufficient for *some* cases. This is especially true when ray tracing should only predict the received power and delay spread. On the other hand, such a purely vertical ray tracing will not correctly predict the directions of the rays at the MS.
- *Microcells, small distance BS–MS*: as both BS and MS antennas are below the rooftop, the diffraction loss of over-the-rooftop propagation is large. Propagation in the horizontal plane – i.e., through street canyons – can be a much more efficient process. Under these conditions, ray tracing in just the horizontal plane can be sufficient.
- *Microcells, large distance BS–MS*: in this case, the relative power of rays propagating in the horizontal plane (compared with over-the-rooftop components) is smaller. Horizontal components undergo multiple diffraction and reflection processes, while losses from over-the-rooftop components are mostly determined by diffraction losses near the BS and the MS, and thus depend less on distance. In this case, a so-called *2.5-dimensional model* can be used: only propagation in

the horizontal plane, on one hand, and only in the vertical plane, on the other hand, is simulated, and these two contributions are added together.

2.5-dimensional modeling can also be used for macrocells. However, both in macrocells and in microcells, with the BS antenna close to the rooftop height, there are propagation processes that cannot be correctly modeled by 2.5-dimensional ray tracing. For example, reflections at a far IO cluster (high-rise building) are not accounted for in this approach (see Figure 7.11).

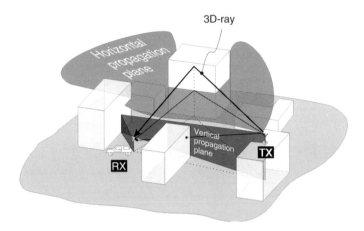

Figure 7.11   Two- and three-dimensional modeling.

7.5.4   Geographical Databases

The foundation of all deterministic methods is the information about the geography and morphology of the environment. The accuracy of that information determines the achievable accuracy of any deterministic channel model.

For *indoor environments*, that information can usually be obtained from building plans, which nowadays are often available in digital form.

In *RAs*, geographical databases are available with a resolution of 10–100 m. These databases are often created by means of satellite observations. In many countries, morphological information (land usage) is also available; however, obtaining this information in an automated and consistent way can be quite challenging.

In *urban areas*, digital databases use two different types of data: vector data and pixel data. For vector data, the actual location of building endpoints is stored. For pixel data, a regular grid of points is superimposed on the area, and for each pixel it is stated whether it falls on "free space" (streets, parks, etc.) or is covered by a building. In both cases, building heights and materials might be included in the database.

7.6   Appendices

Please see companion website *www.wiley.com/go/molisch*

Further Reading

There is a rich literature on channel models. Besides the original papers already mentioned in the main text, Andersen et al. [1995] and Molisch and Tufvesson [2004] give overviews of different models. For implementation of the tapped delay line model, we recommend Paetzold [2002]. Generalizations of the tapped delay line approach to the MIMO case were suggested by Xu et al. [2002]. Poisson approximations for arrival times were developed by Turin et al. [1972] and later improved and extended by Suzuki [1977] and Hashemi [1979]. The Saleh–Valenzuela model was first proposed in Saleh and Valenzuela [1987]. The Delta-K model was described in Hashemi [1993]. Parameterizations of various models are reviewed in Molisch and Tufvesson [2004]; parameters for MIMO channel models in Almers et al. [2007]; parameters for models of car-to-car propagation channels in Molisch et al. [2009].

The Okumura–Hata model is based on the extensive measurements of Okumura et al. [1968] in Japan, and was brought into a form suitable for computer simulations by Hata [1980]. Extensions exploiting the terrain profile are discussed in Badsberg et al. [1995].

The COST 231–Walfish–Ikegami model was developed by the research and standardization group COST 231 [Damosso and Correia 1999] based on the work of Walfish and Bertoni [1988] and Ikegami et al. [1984].

The GSCM was proposed in one form or the other in Blanz and Jung [1998], Fuhl et al. [1998], Norklit and Andersen [1998], and Petrus et al. [2002], see also Liberti and Rappaport [1996]. More details about efficient implementations of a GSCM can be found in Molisch et al. [2003], while a generalization to multiple-interaction processes is described in Molisch [2004] and [Molisch and Hofstetter 2006]. The Laplacian structure of the power azimuthal spectrum was first suggested in Pedersen et al. [1997], and though there has been some discussion about its validity it is now in widespread use.

The description of the channels for MIMO systems by transfer function matrices stems from the classical work of Foschini and Gans [1998] and Winters [1987]. The Kronecker assumption was proposed in Kermoal et al. [2002]; a more general model encompassing correlations between DOAs and DODs was introduced by Weichselberger et al. [2006], other generalized models were proposed by Gesbert et al. [2002] and Sayeed [2002]. Directional channel modeling methods, as well as typical parameterizations, are reviewed in Almers et al. [2007].

The Method of Moments is described in the classical book by Harrington [1993]. Special methods that increase the efficiency of the method include *natural basis sets* [Moroney and Cullen 1995], the *fast multipole method* [Rokhlin 1990], and the *tabulated interaction method* [Brennan and Cullen 1998]. The FEM is described in Zienkiewicz and Taylor [2000], while the FDTD method is described in Kunz and Luebbers [1993].

Ray tracing originally comes from the field of computer graphics, and a number of books (e.g., Glassner [1989]) are available from that perspective; its application to wireless is described, e.g., in Valenzuela [1993]. Descriptions of the ray-launching algorithm can be found in Lawton and McGeehan [1994]. Ray splitting was introduced in Kreuzgruber et al. [1993]. There is also a considerable number of commercial software programs for radio channel prediction that use ray tracing.

As far as standardized models are concerned, a number of models that include directional information or have a larger bandwidth have been developed recently. The COST 259 model is described in [Molisch et al. 2006b] and [Asplund et al. 2006], while the IEEE 802.11n model covers spatial models for indoor environments [Erceg et al. 2004]. The IEEE 802.15.3a and 4a channel models [Molisch et al. 2003a, 2006a] describe channel models for the ultrawideband case.

Another double-direct channel model was standardized by the Third Generation Partnership Project (3GPP) and 3GPP2, the standardization organizations for third-generation cellular systems (see Chapter 26). This model, which is similar to the COST 259 model, is described in detail in the Appendix and in Calcev et al. [2007]. Further double-directional models were published by COST 273 [Molisch and Hofstetter 2006], the European WINNER project [Winner 2007], and the International Telecommunications Union [ITU 2008] (see Appendix).

For updates and errata for this chapter, see wides.usc.edu/teaching/textbook

8

Channel Sounding

8.1 Introduction

8.1.1 Requirements for Channel Sounding

Measurement of the properties (impulse responses) of wireless channels, better known as *channel sounding*, is a fundamental task for wireless communications engineering because any channel model is based on measurement data. For stochastic channel models, parameter values have to be obtained from extensive measurement campaigns, while for deterministic models the quality of the prediction has to be checked by comparisons with measured data.

As the systems, and the required channel models, become more complex, so do the tasks of channel sounding. Measurement devices in the 1960s only had to measure the received field strength. The transition to wideband systems necessitated the development of a new class of channel sounders that could measure impulse responses – i.e., delay dispersion. The focus on directional propagation properties that arose in the 1990s, caused by the interest in multiantenna systems, also affects channel sounders. These sounders now have to be able to measure double-directional impulse responses.

In addition to the changes in measured quantities, the *environments* in which the measurements are done are changing. Up to 1990, measurement campaigns were usually performed in macrocells. Since then, microcells and especially indoor propagation has become the focus of interest.

In the following, we discuss the most important channel-sounding approaches. After a discussion of the basic requirements of wideband measurements, different types of sounders are described. An outline of spatially resolved channel sounding concludes the chapter.

8.1.2 Generic Sounder Structure

The word *channel sounder* gives a graphic description of the functionality of such a measurement device. A transmitter (TX) sends out a signal that excites – i.e., "sounds" – the channel. The output of the channel is observed ("listened to") by the receiver (RX), and stored. From the knowledge of the transmit and the receive signal, the time-variant impulse response or one of the other (deterministic) system functions is obtained.

Figure 8.1 shows a block diagram of the channel sounder that is conceptually most simple.

The TX sends out a signal $s(t)$ that consists of periodically repeated pulses $p(t)$:

$$s(t) = \sum_{i=0}^{N-1} p(t - iT_{\text{rep}}) \qquad (8.1)$$

Wireless Communications, Second Edition   Andreas F. Molisch
© 2011 John Wiley & Sons, Ltd

Figure 8.1 Principle of a channel sounder. Correct synchronization between transmitter and receiver is especially important. *In this figure*: AGC, Automatic Gain Control.

where T_{rep} is the repetition interval of the transmitted pulses. One *measurement run* consists of N pulses that are transmitted at fixed intervals. The pulses are the convolution of a basis pulse $\tilde{s}(t)$ created by a pulse generator and a transmit filter:

$$p(t) = \tilde{s}(t) * g(t) \tag{8.2}$$

where $g(t)$ is the impulse response of the transmit filter. The waveform used for $p(t)$ depends on the type of sounder, and can greatly differ for sounders working in the time domain, compared with sounders working in the frequency domain, as discussed in Sections 8.2 and 8.3.

This block diagram is generic; the properties of the sounder are mostly determined by the choice of the sounding signal. In order to perform efficient measurements, the following requirements should be fulfilled by the sounding signal:

- *Large bandwidth*: the bandwidth is inversely proportional to the shortest temporal changes in the sounding signal and thus determines the achievable *delay resolution*.
- *Large time bandwidth product*: it is often advantageous if the sounding signal has a duration that is longer than the inverse of the bandwidth – i.e., a time bandwidth product TW larger than unity. For many systems, the transmit *power* is limited. In this case, a large TW allows the transmission of high *energy* in the sounding signal, and thus obtains a higher Signal-to-Noise Ratio (SNR) at the RX. Sounding schemes with large TW are related to spread spectrum systems (Chapter 18). At the RX, special signal processing (despreading) is required in order to exploit the benefits of large TW.
- *Signal duration*: the effective signal duration must be adapted to channel properties. On one hand, a long sounding signal can give a large time bandwidth product, which is beneficial (see above). On the other hand, the sounding signal should not be longer than the coherence time of the channel – i.e., the time during which the channel can be considered to be approximately constant. For practical reasons, the pulse repetition time T_{rep} should be larger than the duration of the constituent pulse $p(t)$ and the maximum excess delay of the channel.
- *Power-spectral density*: the power-spectral density of the sounding signal, $|P_{\text{TX}}(j\omega)|^2$, should be uniform across the bandwidth of interest. This allows us to have the same quality of the channel estimate at all frequencies. Due to efficiency considerations, little energy should be transmitted outside the bandwidth of interest.
- *Low crest factor*: signals with a low crest factor

$$C_{\text{crest}} = \frac{\text{Peak amplitude}}{\text{rms amplitude}} = \frac{\max\{s(t)\}}{\sqrt{\overline{s^2(t)}}} \tag{8.3}$$

allow efficient use of the transmit power amplifier. A first estimate for the crest factor of *any* signal can be obtained from Felhauer et al. [1993]:

$$1 < C_{\text{crest}} \leq \sqrt{TW} \tag{8.4}$$

- *Good correlation properties*: correlation-based channel sounders require signals whose Auto-Correlation Function (ACF) has a high *Peak to Off Peak* (POP) ratio, and a zero mean. The latter property allows *unbiased estimates* (see Section 8.4.2). Correlation properties are critical for channel estimates that directly use the correlation function while it is less important for parameter-based estimation techniques (see Section 8.5).

The design of optimum, digitally synthesized sounding signals thus proceeds in the following steps:

1. Choose the duration of the sounding signal according to the channel coherence time and the required time bandwidth product.
2. For a constant power-spectral density, all frequency components need to have the same absolute value.
3. Now the only remaining free parameters are the phases of the frequency components. These can be adjusted to yield a low crest factor.

8.1.3 Identifiability of Wireless Channels

The temporal variability of wireless channels has an impact on whether the channel can be identified (measured) in a unique way.

A band-limited *time-invariant* channel can always be identified by appropriate measurement methods, the only requirement being that the RX fulfills the Nyquist theorem [Proakis 2005] in the delay domain – i.e., samples the received signal sufficiently fast.

In a *time-variant* system, the repetition period T_{rep} of the sounding pulse $p(t)$ is of fundamental importance. The channel response to any excitation pulse $p(t)$ can be seen as one "snapshot" (sample) of the channel (see Figure 8.2). In order to track changes in the channel, these snapshots need to be taken sufficiently often. Intuitively, T_{rep} must be smaller than the time over which the channel changes. This notion can be formalized by establishing a sampling theorem in the time domain. Just as there is a minimum sampling rate to identify a signal with a band-limited spectrum, so is there a minimum temporal sampling rate to identify a time-variant process with a band-limited Doppler spectrum. Thus, the temporal sampling frequency must be twice the maximum Doppler frequency ν_{max}:

$$f_{\text{rep}} \geq 2\nu_{\text{max}} \tag{8.5}$$

Rewriting Eq. (8.5), and using the relationship between the movement speed of the MS and the Doppler frequency $\nu_{\text{max}} = f_0 v_{\text{max}}/c_0$ (see Eq. 5.8), the repetition frequency for the pulses can be written as

$$T_{\text{rep}} \leq \frac{c_0}{2 f_c v_{\text{max}}} \tag{8.6}$$

In that case,

$$\frac{v}{\Delta x_s} \geq 2 \frac{v_{\text{max}}}{\lambda_c} \tag{8.7}$$

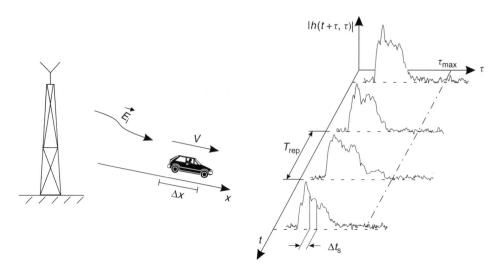

Figure 8.2 Time-variant impulse response of the channel and channel identifiability. A new snapshot can only be taken after the impulse response of the previous excitation has died down.

holds, so that the distance Δx_s between the locations at which the sounding has to take place is upper bounded as

$$\Delta x_s \leq \frac{v}{v_{max}} \frac{\lambda}{2} \leq \frac{\lambda}{2} \qquad (8.8)$$

Equation (8.8) thus tells us that for an aliasing-free measurement at least two snapshots per wavelength are required.

Strongly time-varying channels can be fundamentally unidentifiable because requirements for the design of sounding signals can become contradictory. On one hand, the repetition frequency T_{rep} has to be larger than the maximum excess delay of the channel τ_{max}; otherwise the impulse responses from the different excitation pulses start to overlap. On the other hand, we have just shown that the repetition frequency has to fulfill $T_{rep} \leq 1/2v_{max}$. Thus, channels can be identified in an unambiguous way only if

$$2\tau_{max}v_{max} \leq 1 \qquad (8.9)$$

This equation is also known as the *two-dimensional Nyquist criterion*. A channel that fulfills these requirements is known as *underspread*. If Eq. (8.9) is not fulfilled, then the channel can only be identified by making specific assumptions – e.g., a certain parametric model. Fortunately, the overwhelming majority of wireless channels are underspread; in many cases, even $2\tau_{max}v_{max} \ll 1$ is fulfilled. We will assume that this is fulfilled in the following. This also implies that the channel is *slowly time variant* (see Chapter 6), so that $h(t, \tau)$ can be interpreted as the impulse response $h(\tau)$ that is valid at a certain (fixed) time instant t.[1]

---

[1] Strictly speaking, use of the "slow time variance" concept also requires that the sounding signal has a duration that is much smaller than the coherence time of a channel. We will assume this in the following.

Example 8.1 *A channel sounder is in a car that moves along a street at 36 km/h. It measures the channel impulse response at a carrier frequency of 2 GHz. At what intervals does it have to measure? What is the maximum excess delay the channel can have so as to still remain underspread?*

$$v = 36\,\text{km/h} = 10\,\text{m/s} \left.\begin{matrix} \\ \\ \end{matrix}\right\}$$
$$\lambda_c = \frac{c_0}{f_c} = \frac{3 \cdot 10^8}{2 \cdot 10^9} = 0.15\,\text{m} \tag{8.10}$$

The channel must be sampled in the time domain at a rate that is, at minimum, twice the maximum Doppler shift. Using Eq. (8.5),

$$f_{\text{rep}} = 2 \cdot \upsilon_{\text{max}} = 2 \cdot \frac{v}{\lambda_c} \tag{8.11}$$

The sampling interval T_{rep} is given as

$$T_{\text{rep}} = \frac{1}{f_{\text{rep}}} = \frac{\lambda_c}{2 \cdot v} = 7.5\,\text{ms} \tag{8.12}$$

At a mobile speed of 36 km/h, this corresponds to a channel snapshot taken every 75 mm. To calculate the maximum excess delay τ_{max}, we make use of the fact that the channel must be underspread in order to be identifiable. Hence from Eq. (8.9),

$$2 \cdot \tau_{\text{max}} \cdot \upsilon_{\text{max}} = 1 \left.\begin{matrix} \\ \\ \end{matrix}\right.$$
$$\tau_{\text{max}} = \frac{1}{2 \cdot \upsilon_{\text{max}}} = T_{\text{rep}} = 7.5\,\text{ms} \tag{8.13}$$

This is orders of magnitude larger than the maximum excess delays that occur in typical wireless channels (see Chapter 7). However, note that this is only the *theoretical* maximum delay spread that still guarantees identifiability. A correlative channel sounder would show a significant degradation in estimation quality for this high τ_{max}.

If τ_{max} is smaller, then the repetition period T_{rep} of the sounding pulse should be increased; this would allow averaging of the snapshots and thus improvement of the SNR.

8.1.4 Influence on Measurement Data

When performing the measurements, we have to be aware of the fact that measured impulse responses carry undesired contributions as well. These are mainly:

- interference from other (independent) signal sources that also use the channel;
- additive white Gaussian noise.

Interference is created especially when measurements are done in an environment where other wireless systems are already active in the same frequency range. Wideband measurements in the 2-GHz range, e.g., become quite difficult, as these bands are heavily used by various systems. If the number of interferers is large, the resulting interference can usually be approximated as equivalent Gaussian noise. This equivalent noise raises the noise floor, and thus decreases the dynamic range.

8.2   Time-Domain Measurements

A time-domain measurement directly measures the (time-variant) impulse response. Assuming that the channel is slowly time variant, the measured impulse response is the convolution of the *true channel impulse response* with the *impulse response of the sounder*:

$$h_{\text{meas}}(t_i, \tau) = \tilde{p}(\tau) * h(t_i, \tau) \tag{8.14}$$

where the effective *sounder impulse response* $\tilde{p}(\tau)$ is the convolution of the transmitted pulse shape and the RX filter impulse response:

$$\tilde{p}(\tau) = p_{\text{TX}}(\tau) * p_{\text{RX}}(\tau) \tag{8.15}$$

if the channel and the transceiver are linear.[2] The sounder impulse response should be as close to an ideal delta (Dirac) function as possible.[3] This minimizes the impact of the measurement system on the results. If the impulse response of the sounder is not a delta function, it has to be eliminated from the measured impulse response by a deconvolution procedure, which leads to noise enhancement and other additional errors.

8.2.1   *Impulse Sounder*

This type of channel sounder, which is comparable with an impulse radar, sends out a sequence of short pulses $p_{\text{TX}}(\tau)$. These pulses should be as short as possible, in order to achieve good spatial resolution, but also contain as much energy as possible, in order to obtain a good SNR. Figure 8.3 shows a rough sketch of a transmit pulse, and the received signal after this pulse has propagated through the channel.

Figure 8.3   Principle of pulse-based measurements. (a) shows one sample of a (periodically repeated) transmit pulse. (b) shows the impulse response of the channel. (c) shows the output from the channel, as measured by the RX.

The receive filter is a bandpass filter – i.e., has a constant-magnitude spectrum in the frequency range of interest. Ideally, $p_{\text{RX}}(\tau)$ should not have an impact, so that

$$p(\tau) = p_{\text{TX}}(\tau) \tag{8.16}$$

---

[2] Strictly speaking, a sounder impulse response is also time variant, due to second-order effects like temperature drift. However, it is more common to recalibrate the sounder, and consider it as time invariant until the next calibration.

[3] This corresponds to a spectrum that is flat over all frequencies. For practical purposes, it is sufficient that the spectrum is flat over the bandwidth of interest.

When comparing the sounding signal with the requirements of Section 8.1.2, we find that the signal has a small time bandwidth product, a short signal duration, and a high crest factor. The requirement of high pulse energy and short duration for transmitted pulses implies that the pulses have a very high peak power. Amplifiers and other Radio Frequency (RF) components that are designed for such high peak powers are expensive, or show other grave disadvantages (e.g., non-linearities). A further disadvantage of an impulse sounder is its low resistance to interference. As the sounder interprets the received signal directly as the impulse response of the channel, any interfering signal – e.g., from a cellphone active in the band of interest – is interpreted as part of the channel impulse response.

8.2.2 Correlative Sounders

The time bandwidth product can be increased by using correlative channel sounders. Equations (8.14) and (8.15) show that it is not the transmit pulse shape alone that determines the impact of the measurement system on the observed impulse response. Rather, it is the convolution of $p_{TX}(\tau)$ and $p_{RX}(\tau)$. This offers additional degrees of freedom for designing transmit signals that result in high delay resolution but low crest factors.

The first step is to establish a general relationship between the desired $p_{TX}(\tau)$ and $p_{RX}(\tau)$. As is well known from digital communications theory, the SNR of the RX filter output is maximized if the receive filter is the *matched filter* with respect to the transmit waveform [Barry et al. 2003, Proakis 2005].[4] Concatenation of the transmit and receive filters thus has an impulse response that is identical to the ACF of the transmit filter:

$$\tilde{p}(\tau) = p_{TX}(\tau) * p_{RX}(\tau) = R_{p_{TX}}(\tau) \tag{8.17}$$

The sounding pulses thus should have an ACF that is a good approximation of a delta function; in other words, a high autocorrelation peak $R_{p_{TX}}(0)$, as well as low ACF sidelobes. The ratio between the height of the autocorrelation peak and the largest sidelobe is called the *POP* ratio and is an important quantity for characterization of correlative sounding signals. Figure 8.4 shows an example of an ACF, and the delay resolution that can be achieved with such a signal [de Weck 1992].

Figure 8.4 Definition of the *peak to off peak ratio* and the delay resolution $\Delta\tau_{min} \cdot (\Delta = 6\,dB)$.

In practice, *Pseudo Noise* (PN) *sequences* or linearly frequency modulated signals (chirp signals) have become the prevalent sounding sequences. Maximum-length PN sequences (*m-sequences*), which can be created by means of a shift register with feedback, are especially popular. Such sequences are well known from Code Division Multiple Access (CDMA) systems, and have been

---

[4] Strictly speaking, the SNR at the output of the RX-matched filter is maximized if the receive filter is matched to the signal that is actually received at the RX antenna connector. However, that would require knowledge of the channel impulse response – the very quantity we are trying to measure. Thus, matching the filter to the transmit signal is the best we can do.

extensively studied both in the mathematical and the communications engineering literature (see Chapter 18 for more details). The ACF of an m-sequence with periodicity M_c has only a single peak of height M_c, and a POP of M_c. Following the CDMA literature, each of the M_c elements of such a sequence is called a *chip*.

For constant chip duration and increasing length of the m-sequence, the POP, as well as the time bandwidth product, increases: signal duration increases linearly with M_c, while bandwidth, which is approximately the inverse of chip duration, stays constant. The increased time bandwidth product improves immunity to noise and interference. More exactly, noise and interference are suppressed by a factor M_c. The reason for this is discussed in more detail in Chapter 18, as the principle is identical to that of direct-sequence CDMA.

The interpretation of measurements with correlative channel sounders in time-varying channels requires some extra care. The basic principle of correlative channel sounders is that $p_{TX}(\tau) *$ $h(t, \tau) * p_{RX}(\tau)$ is identical to $[p_{TX}(\tau) * p_{RX}(\tau)] * h(t, \tau)$. In other words, we require that the channel at the beginning of the PN sequence is the same as the one at the end of the PN sequence. This is a good approximation for slowly time-variant channels. However, if this condition is not fulfilled, correction procedures need to be used [Matz et al. 2002].

8.3   Frequency Domain Analysis

The techniques described in the previous section directly estimate the impulse response of the channel in the time domain. Alternatively, we can try to directly estimate the transfer function – i.e., measure in the frequency domain. The fundamental relationship (Eq. 8.1) still holds. However, the shape of the waveform $p(t)$ is now different. The main criterion for its design is that it has a power spectrum $|P(j\omega)|^2$ that is approximately constant in the bandwidth of interest, and that it allows interpretation of the measurement result directly in the frequency domain.

One method of frequency domain analysis is based on chirping. The transmit waveform is given as

$$p(t) = \exp\left[2\pi j \left(f_0 t + \Delta f \frac{t^2}{2T_{\text{chirp}}} \right)\right] \quad \text{for } 0 \leq t \leq T_{\text{chirp}} \tag{8.18}$$

Consequently, the instantaneous frequency is

$$f_0 + \Delta f \frac{t}{T_{\text{chirp}}} \tag{8.19}$$

and thus changes linearly with time, covering the whole range Δf of interest. The receive filter is again a matched filter. Intuitively, the chirp filter "sweeps" through the different frequencies, measuring different frequencies at different times.

Alternatively, we can sound the channel on different frequencies at the same time. The conceptually most simple way is to generate different, sinusoidal sounding signals with different weights, phases, and frequencies and transmit them all from the TX antenna simultaneously:

$$p(t) = \sum_{i=1}^{N_{\text{tones}}} a_i \cdot \exp[2\pi j t (f_0 + i \Delta f / N_{\text{tones}}) + j\varphi_i] \quad \text{for } 0 \leq t \leq T_{\text{ss}} \tag{8.20}$$

Due to hardware costs, calibration issues, etc., analog generation of $p(t)$ using multiple oscillators to generate multiple frequencies is not practical. However, it is possible to generate $p(t)$ digitally, similar to the principles of Orthogonal Frequency Division Multiplexing (OFDM) described in Chapter 19, and then use just a single oscillator to upconvert the signal to the desired passband (and similarly at the RX).

8.4   Modified Measurement Methods

8.4.1   Swept Time Delay Cross Correlator (STDCC)

The STDCC is a modification of correlative channel sounders that aims to reduce the sampling rate at the RX. Normal correlative channel sounders require sampling at the Nyquist rate. In contrast, the STDCC samples at rate T_{rep}, i.e., uses just a single sample value for each m-sequence – namely, at the maximum of the ACF. The position of this maximum is changed for each repetition of the m-sequence, by shifting the time base of the RX with respect to the TX. Thus, K_{scal} transmissions of the m-sequence give the sampled values of a single impulse response $h(\tau_i), i = 1, \ldots, K_{scal}$. The delay resolution is thus better, by a factor K_{scal}, than the inverse sampling rate. This drastically reduces the sampling rate and the requirements for subsequent processing and storing of the impulse response. On the downside, the duration of each measurement is increased by the same factor K_{scal}.

In an STDCC, shifting of the maximum of the ACF for subsequent repetitions of the sounding signal is achieved by using different time bases in the TX and RX. In particular, the delayed time base of the RX correlator (compared with the TX sequence) is achieved by using a chipping frequency (inverse of the chip duration) that is smaller by Δf. This results in a slow relative shift of the TX and RX sequences. During each repetition of the sequence, the correlation maximum corresponds to a different delay. After a duration,

$$T_{slip} = \frac{1}{f_{TX} - f_{RX}} = \frac{1}{\Delta f} \tag{8.21}$$

the TX and RX signals are fully aligned again – i.e., the ACF maximum again occurs at delay $\tau = 0$. This means that (for a static channel) the output from the sampler is periodic with the so-called *slip rate*:

$$f_{slip} = f_{TX} - f_{RX} \tag{8.22}$$

The ratio:

$$K_{scal} = \frac{f_{TX}}{f_{slip}} \gg 1 \tag{8.23}$$

is the *scaling factor* K_{scal} of the impulse response. The actual impulse response can be obtained from the measured sample values as

$$\hat{h}(t_i, k\Delta\tau) = C \cdot h_{STDCC}(t_i, k\Delta\tau K_{scal}) \tag{8.24}$$

where C is a proportionality constant.

The drawback of the measurement method is increased measurement duration. Remember that a channel is identifiable only if it is underspread – i.e., $2\nu_{max}\tau_{max} < 1$, which is usually fulfilled in wireless channels. For an STDCC, this requirement changes to $2K_{scal}\nu_{max}\tau_{max} < 1$, which is *not* fulfilled for many outdoor channels and typical values of K_{scal}.

Example 8.2   *Consider an STDCC that performs measurements in an environment with 500-Hz maximum Doppler frequency and maximum excess delay of 1 μs. The sounder can sample at most with 1 Msample/s. What is the maximum delay resolution (inverse bandwidth) that the sounder can achieve?*

In order for the channel to remain identifiable (underspread) when measured with an STDCC, the following condition has to hold:

$$2 \cdot K_{\text{scal}} \cdot \tau_{\max} \cdot \nu_{\max} = 1$$

$$K_{\text{scal}} = 1/(2 \cdot \tau_{\max} \cdot \nu_{\max}) = 1000$$

The sounder can take one sample for each repetition of the sounding pulse – i.e., one sample per µs. Hence, the STDCC sounder can resolve Multi Path Components (MPCs) separated by a delay of $1\,\mu s/1000 = 1\,$ns.

8.4.2 Inverse Filtering

In some cases, it is advantageous to use a receive filter that optimizes the POP ratio but is *not* ideally matched to the transmit signal. At first glance, it sounds paradoxical to use such a filter, which results in a worse SNR. However, there can be good practical reasons for this approach. Small variations of the SNR are usually less important than the sidelobes of the ACF: while sidelobes can be eliminated by appropriate deconvolution procedures, they can give rise to additional errors. It is thus meaningful to optimize the receive filters with respect to the POP ratio, not with respect to the SNR.

Let us in particular consider *inverse filtering*, and compare it to matched filtering. For the matched filter, the receive filter transfer function is chosen as $P_{\text{TX}}^*(f)$, so that the total filter transfer function $P_{\text{MF}}(f)$ (concatenation of transmit and receive filter) is given as

$$P_{\text{MF}}(f) = P_{\text{TX}}(f) \cdot P_{\text{TX}}^*(f) \tag{8.25}$$

For inverse filtering, the receive filter transfer function is chosen as $1/P_{\text{TX}}(f)$ in the bandwidth of interest, so that the total transfer function is made as close to unity as possible:

$$P_{\text{IF}}(f) = P_{\text{TX}}(f) \cdot \frac{1}{P_{\text{TX}}(f)} \approx 1 \tag{8.26}$$

The inverse filter is thus essentially a *zero-forcing equalizer* (see Chapter 16) for compensation of distortions by the transmit filter. It is important that the transmit spectrum P_{TX} does not have any nulls in the bandwidth of interest. The inverse filter leads to noise enhancement, and thus to a worse SNR than a matched filter. On the positive side, the inverse filter is *unbiased*, so that the estimation error is zero-mean.

8.4.3 Averaging

It is common to average over several, subsequently recorded, impulse responses of transfer functions. Assuming that the channel does not change during the whole measurement time, and that the noise is statistically independent for the different measurements, then the averaging of M profiles results in an enhancement of the SNR by $10 \cdot \log_{10} M$ dB. However, note that the maximum measurable Doppler frequency decreases by a factor of M.

Averaging over different realizations of the channel is used to obtain, e.g., the Small Scale Averaged (SSA) power.

Example 8.3 *A Mobile Station (MS) moves along a straight line, and can measure a statistically independent sample of the impulse response every 15 cm. Measurements are taken over a distance* *of 1.5 m, so that shadowing can be considered constant over this distance. The goal is estimation of channel gain (attenuation), averaged over small-scale fading. Assume first that the measurements are only taken at a single frequency. What is the standard deviation of the estimate of channel gain? What is the probability that the estimator is more than 20% off?*

Since measurement is only taken at a single frequency, we assume a flat Rayleigh-fading channel with mean power $\overline{P}$. Each sample of the power of the impulse response is then an exponentially distributed random variable with mean power $\overline{P}$. A reasonable estimate of mean power is

$$\hat{\overline{P}} = \frac{1}{N} \sum_{k=1}^{N} P_k$$

where P_k is the kth sample of the power of the impulse response and N is the number of samples. As a measure of the error we choose the normalized standard deviation, $\sigma_{\hat{\overline{P}}}/\overline{P}$. Since the P_k are identically distributed and independent, we have

$$\frac{\sigma_{\hat{\overline{P}}}}{\overline{P}} = \frac{1}{\overline{P}} \sqrt{\mathrm{Var}\left(\frac{1}{N}\sum_{k=1}^{N} P_k\right)} = \frac{1}{\overline{P}}\sqrt{\frac{1}{N^2}N\overline{P}^2} = \frac{1}{\sqrt{N}}$$

Using 11 samples, the relative standard deviation is 0.3. Approximating the probability density function (pdf) of the estimator to be Gaussian with mean $\overline{P}$ and variance $\overline{P}^2/N$, the probability that the estimate is more than 20% off is then

$$1 - \mathrm{Pr}(0.8\overline{P} < \hat{\overline{P}} < 1.2\overline{P}) = 2\,Q(0.2\sqrt{N}) \qquad (8.27)$$

For $N = 11$ this probability becomes 0.5.

Example 8.4 *Consider now the case of wideband measurements, where measurements are done at ten independently fading frequencies. How do the results change?*

Again aiming to estimate the narrowband channel attenuation averaged over small-scale fading, usage of wideband measurements just implies that $N = 110$ measurements are now available. Modifying Eq. (8.27), we find that the probability for more than 20% error has decreased to 0.036.

8.4.4 Synchronization

The synchronization of TX and RX is a key problem for wireless channel sounding. It is required to establish synchronization in frequency and time at a TX and RX that can be separated by distances up to several kilometers. This task is made more difficult by the presence of multipath propagation and time variations of the channel. Several different approaches are in use:

1. In indoor environments, *synchronization by cables* is possible. For distances up to about 10 m, coaxial cables are useful; for larger distances, fiber-optic cables are preferable. In either case, the synchronization signal is transmitted on a known and well-defined medium from the TX to the RX.

2. For many outdoor environments, the *Global Positioning System* (GPS) offers a way of establishing common time and frequency references. The reference signals required by channel sounders are an integral part of the signals that GPS satellites transmit. An additional benefit lies in the

fact that the measurement location is automatically recorded as well. The disadvantage is that this method requires that both TX and RX have line-of-sight connection to GPS satellites; the latter condition is rarely fulfilled in microcellular and indoor scenarios.

3. *Rubidium clocks* at the TX and RX are an alternative to GPS signals. They can be synchronized at the beginning of a measurement campaign; as they are extremely stable (relative drifts of 10^{-11} are typical), they retain synchronization for several hours.

4. *Measurements without synchronization*: it is possible to synchronize via the wireless link itself – i.e., the received signal self-triggers the recording at the RX by exceeding a certain threshold. The advantage of this technique is its simplicity. However, the drawbacks are twofold: (i) noise or interference can erroneously trigger the RX and (ii) it is not possible to determine absolute delays.

8.4.5   Vector Network Analyzer Measurements

The measurement techniques described in Sections 8.2 and 8.3, including the frequency domain techniques, require dedicated equipment that can be quite expensive, due especially to the high-speed components required for the generation of short pulses or sequences with short chip duration. An alternative measurement technique is based on a slow sweep in the frequency domain. In the following, we discuss measurements by means of a vector network analyzer, as these devices are present in many RF labs.

A vector network analyzer measures the S-parameters of a *Device Under Test* (DUT). The DUT can be a wireless channel, in which case the parameter S_{21} is the channel transfer function at the frequency that is used to excite the channel. By having the excitation signal sweep or step through the frequency band of interest, we obtain a continuous or sampled version of the transfer function $H(t, f)$.

In order to reduce the impact of the network analyzer itself, back-to-back calibration has to be performed. The necessity to do a calibration is common to all sorts of channel sounders, and is indeed a basic principle of a good measurement procedure. What is specific to network analyzers is the type of calibration, which is commonly *SOLT calibration* (Short Open Loss Termination). This calibration establishes the reference planes and measures the frequency response of the network analyzer. During subsequent measurement, the network analyzer compensates for this frequency response, so that it measures only the frequency response of the DUT. Note that calibration does not include the antennas. This is not a problem if the antennas are to be considered a part of the channel. If, however, antenna effects are to be eliminated, a separate calibration of the antennas has to be performed, and taken into account during evaluation of the measurements.[5]

Measurements using a vector network analyzer are usually accurate, and can be performed in a straightforward way. However, there are also important disadvantages:

- Such measurements are slow, so that repetition rates typically cannot exceed a few Hz (!). Since we require that the channel does not change significantly during one measurement, network analyzer measurements are limited to static environments.
- The TX and RX are often placed in the same casing. This puts an upper limit on the distance that TX and RX antennas can be spaced apart.

From these restrictions it follows that network analyzers are mainly suitable for indoor measurements.

---

[5] Note that corrections for antenna pattern are possible only if the directions of the MPCs are known (see Section 8.5).

8.5   Directionally Resolved Measurements

Directionally resolved channel measurements, and models based on those measurements, are important for the design and simulation of multiantenna systems (see Chapters 6 and 7). In the first three subsections of this section, we discuss how to make measurements that are directionally resolved at just the RX. These concepts are then generalized to Multiple Input Multiple Output (MIMO) measurements (directionally resolved at both link ends) at the end of this section.

Fortunately, it is not necessary to devise directional channel sounders from scratch. Rather, a clever combination of existing devices can be used to allow directional measurements. We can distinguish two basic approaches: *measurements with directional antennas* and *array measurements*.

- *Measurements with directional antennas*: a highly directive antenna is installed at the RX. This antenna is then connected to the RX of a "regular" channel sounder. The output of the RX is thus the impulse response of the combination of the channel and the antenna pointing in a specific direction, i.e.,

$$h(t, \tau, \phi_i) = \int h(t, \tau, \phi)\tilde{G}_{\mathrm{RX}}(\phi - \phi_i)\, d\phi \tag{8.28}$$

where ϕ_i is the direction in which the maximum of the receive antenna pattern is pointing. By stepping through different values of ϕ_i, we can obtain an approximation of the directionally resolved impulse response. One requirement for this measurement is that the channel stays constant during the *total* measurement duration, which encompasses the measurements of all the different ϕ_i. As the antenna has to be rotated mechanically in order to point to a new direction, the total measurement duration can be several seconds or even minutes. The better the directional resolution, the longer the measurement duration.

- *Measurement with an antenna array*: an antenna array consists of a number of antenna elements, each of which has low (or no) directivity, which are spaced apart at a distance d_{a} that is on the order of one wavelength. The impulse response is measured at all these antenna elements (quasi-) simultaneously. The resulting vector of impulse responses is either useful by itself (e.g., for the prediction of diversity performance) or the directional impulse response can be extracted from it by appropriate signal-processing techniques (*array processing*).

Measurement of the impulse response at the different antenna elements can be done by means of three different approaches (see Figure 8.5):

  ○ *real arrays*: in this case one demodulator chain exists for each receive antenna element. Measurement of the impulse response thus truly occurs at all antenna elements simultaneously. The drawbacks include high costs, as well as the necessity to calibrate multiple demodulator chains.

  ○ *multiplexed arrays*: in this technique, multiple antenna elements, but only one demodulator chain, exist. The different antenna elements are connected to a demodulator chain (conventional channel sounder) via a fast RF switch [Thomae et al. 2000]. The RX thus first measures the impulse response at the first antenna element, then it connects the switch to the second element, measures its impulse response, and so on.

  ○ *virtual array*: in this technique, there is only a single antenna element, which is moved mechanically from one position to the next, measuring the impulse responses at the different antenna elements.

A basic assumption for evaluation is again that the environment does not change during the measurement procedure. "Virtual arrays" (which – due to the need of mechanically moving an antenna – require a few seconds or even minutes for one measurement run) can thus only be used in static environments. This precludes scenarios where cars or moving persons are significant

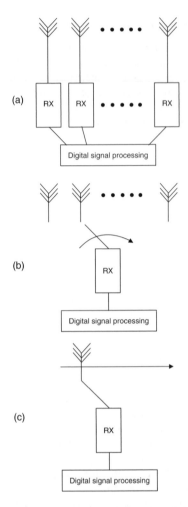

Figure 8.5 Types of arrays for channel sounding: real array (a), switched array (b), virtual array (c).
Reproduced with permission from Molisch and Tufvesson [2005] © Hindawi.

Interacting Objects (IOs). In nonstatic environments, multiplexed arrays are usually the best compromise between measurement speed and hardware effort. A related aspect is the impact of frequency drift and loss of synchronization of the TX/RX. The longer a measurement lasts, the higher the impact of these impairments.

We now turn to the question of how to extract directional information from array measurements. Section 8.5.1 describes the fundamental data model; Sections 8.5.2 and 8.5.3 discuss various types of signal-processing methods.

8.5.1   Data Model for Receive Arrays

Let us establish a mathematical model for the array and incoming signals. We analyze the case of a *Uniform Linear Array* (ULA) consisting of N_r elements, where the signal $r_n(t)$ is detected at

Figure 8.6   Plane waves incident at angle $\phi_1, \ldots, \phi_N$ on a uniform linear array.

the nth element. To simplify the discussion, we will assume that all waves are propagating in the horizontal plane.

Consider now the case when plane waves from N different directions are incident on the array (Figure 8.6), where each wave is described by its Direction Of Arrival (DOA) ϕ_i. The relationship between incident signals s_i and the signal it creates at the first antenna element is simply:

$$r_1(t) = \sum_i a_{i,1} s_i(t - \tau_{i,1}) + n_1(t) \tag{8.29}$$

where $\tau_{i,1}$ is the runtime between the source of the ith signal and the first antenna element, $a_{i,1}$ is the (complex) amplitude of the signal, and $n_1(t)$ is the noise at the first antenna element. Consider now the second antenna element. Here the received signal is

$$r_2(t) = \sum_i a_{i,2} s_i(t - \tau_{i,2}) + n_2(t) \tag{8.30}$$

If the ith source is in the far field, then $|a_{i,2}| = |a_{i,1}|$, and

$$s_i(t - \tau_{i,2}) = s_i(t - \tau_{i,1}) \exp(-j(\tau_{i,2} - \tau_{i,1}) 2\pi f_c) \tag{8.31}$$

This last equation assumes that the signal is narrowband in the RF sense – i.e., the bandwidth of the signal is much smaller than the carrier frequency. The physical interpretation of this fact is that the only influence of the antenna position is a phase shift due to the additional runtime. This runtime difference is

$$\tau_{i,2} - \tau_{i,1} = (d_a/c_0) \cos(\phi_i) \tag{8.32}$$

The relationship between r_2 and s is thus

$$r_1(t) = \sum_i \tilde{s}_i(t) + n_1(t) \tag{8.33}$$

$$r_2(t) = \sum_i \tilde{s}_i(t) \exp(-j2\pi d_a \cos(\phi_i)/\lambda_0) + n_2(t) \tag{8.34}$$

where $\tilde{s}_i(t) = a_{i,1} s_i(t - \tau_{i,1})$. For the next antenna element, we get

$$r_3(t) = \sum_i \tilde{s}_i(t) \exp(-j2\pi 2 d_a \cos(\phi_i)/\lambda_0) + n_3(t) \tag{8.35}$$

From this, we can conclude the general relationship between r and s:

$$\mathbf{r}(t) = \mathbf{A}\mathbf{s}(t) + \mathbf{n}(t) \tag{8.36}$$

where $\mathbf{r}(t) = [r_1(t), r_2(t), \ldots, r_{N_r}(t)]^T$, $\mathbf{s}(t) = [s_1(t), s_2(t), \ldots, s_N(t)]^T$, $\mathbf{n}(t) = [n_1(t), n_2(t), \ldots, n_{N_r}(t)]^T$, and

$$
\mathbf{A} = \begin{pmatrix}
1 & 1 & \cdots & 1 \\
\exp(-jk_0 d_a\cos(\phi_1)) & \exp(-jk_0 d_a\cos(\phi_2)) & \cdots & \exp(-jk_0 d_a\cos(\phi_N)) \\
\exp(-j2k_0 d_a\cos(\phi_1)) & \exp(-j2k_0 d_a\cos(\phi_2)) & \cdots & \exp(-j2k_0 d_a\cos(\phi_N)) \\
\vdots & \vdots & \vdots & \vdots \\
\exp(-j(N_r-1)k_0 d_a\cos(\phi_1)) & \exp(-j(N_r-1)k_0 d_a\cos(\phi_2)) & \cdots & \exp(-j(N_r-1)k_0 d_a\cos(\phi_N))
\end{pmatrix}
\tag{8.37}
$$

is the *steering matrix*.

Additionally, we assume that the noise at the different antenna elements is independent (spatially white), so that the correlation matrix of the noise is a diagonal matrix with entries σ_n^2 on the main diagonal.

8.5.2 Beamforming

The most simple determination of the angle of incidence can be obtained by a Fourier transform of the signal vector $\mathbf{r}$. This gives the directions of arrival ϕ_i with an angular resolution that is determined by the size of the array, approximately $2\pi/N_r$. The advantage of this method is its simple implementability (requiring only a Fast Fourier Transform (FFT)); the drawback is its small resolution.

More exactly, the angular spectrum $P_{\mathrm{BF}}(\phi)$ is given as

$$
P_{\mathrm{BF}}(\phi) = \frac{\boldsymbol{\alpha}^\dagger(\phi)\mathbf{R}_{\mathrm{rr}}\boldsymbol{\alpha}(\phi)}{\boldsymbol{\alpha}^\dagger(\phi)\boldsymbol{\alpha}(\phi)}
\tag{8.38}
$$

where $\mathbf{R}_{\mathrm{rr}}$ is the correlation matrix of the incident signal and

$$
\boldsymbol{\alpha}_{\mathrm{RX}}(\phi) = \begin{pmatrix}
1 \\
\exp(-jk_0 d_a\cos(\phi)) \\
\exp(-j2k_0 d_a\cos(\phi)) \\
\vdots \\
\exp(-j(N_r-1)k_0 d_a\cos(\phi))
\end{pmatrix}
\tag{8.39}
$$

is the steering vector into direction ϕ (compare also Eq. 8.37).

8.5.3 High-Resolution Algorithms

The problem of low resolution can be eliminated by means of so-called *high-resolution methods*. The resolution of these methods is not limited by the size of the antenna array, but only by modeling errors and noise. This advantage is paid for by high computational complexity. Furthermore, there is often a limit on the *number* of MPCs whose directions can be estimated.

High-resolution methods include the following:

- ESPRIT: it determines the signal subspace, and extracts the directions of arrival in closed form. A description of this algorithm, which is mainly suitable for ULAs, is given in Appendix 8.A (see *www.wiley.com/go/molisch*).

- MUltiple SIgnal Classification (MUSIC): this algorithm also requires determination of the signal and noise subspaces, but then uses a spectral search to find the directions of arrival.
- Minimum Variance Method (**MVM**: Capon's beamformer): this method is a pure spectral search method, determining an angular spectrum such that for each considered direction the sum of the noise and the interference from other directions is minimized. The modified spectrum is easy to compute, namely,

$$P_{\mathrm{MVM}}(\phi) = \frac{1}{\alpha^{\dagger}(\phi)\mathbf{R}_{\mathrm{rr}}^{-1}\alpha(\phi)} \qquad (8.40)$$

- Maximum-likelihood estimation of the parameters of incident waves: the problem of maximum-likelihood parameter extraction is its high computational complexity. An efficient, iterative implementation is the Space Alternating Generalized Expectation (SAGE) maximization algorithm. Since 2000, it has become the most popular method for channel-sounding evaluations. The drawback is that the iterations might converge to a local optimum, not the global one.

One problem that is common to all the algorithms is array calibration. Most of the algorithms use certain assumptions about the array: the antenna patterns of all elements are identical, no *mutual coupling* between antenna elements, and the distance between all antenna elements is identical. If the actual array does not fulfill the assumptions, calibration is required, so that appropriate corrections can be applied. Such calibrations have to be done repeatedly, as temperature drift, aging of components, etc., tend to destroy the calibration.

For many high-resolution algorithms (including subspace-based algorithms), it is required that the correlation matrix does not become singular. Such singular $\mathbf{R}_{\mathrm{rr}}$ typically occur if the sources of the waves from the different directions are correlated. For channel sounding, all signals typically come from the same source, so that they are completely correlated. In that case, subarray averaging ("spatial smoothing" or "forward–backward" averaging) has to be used to obtain the correct correlation matrix [Haardt and Nossek 1995]. The drawback with subarray averaging is that it decreases the effective size of the array.

Example 8.5 *Three independent signals with amplitudes 1, 0.8, and 0.2 are incident from directions 10°, 45°, and 72°, respectively. The noise level is such that the SNR of the first signal is 15 dB. Compute first the correlation matrix, and steering vectors for a five-element linear array with λ/2 spacing of the antenna elements. Then plot $P_{\mathrm{BF}}(\phi)$ and $P_{\mathrm{MVM}}(\phi)$.*

Let us first assume that the three signals arrive at the linear array with zero initial phase. Thus we have the following parameters for the three MPCs:

$$a_1 = 1 \cdot e^{j.0}, \quad \phi_1 = 10\pi/180\,\mathrm{rad}$$
$$a_2 = 0.8 \cdot e^{j.0}, \quad \phi_2 = 45\pi/180\,\mathrm{rad} \qquad (8.41)$$
$$a_3 = 0.2 \cdot e^{j.0}, \quad \phi_3 = 72\pi/180\,\mathrm{rad}$$

We assume that measurement noise has a complex Gaussian distribution, and is spatially white. The common variance σ_n^2 of the noise samples is given as

$$\sigma_n^2 = \frac{1}{10^{\frac{15}{10}}} = 0.032 \qquad (8.42)$$

According to Eq. (8.37), the steering matrix for the five-element linear array with element spacing $d = \dfrac{\lambda}{2}$ is given as

$$
\mathbf{A} = \begin{bmatrix}
1 & 1 & 1 \\
\exp(-j \cdot \pi \cdot \cos(\phi_1)) & \exp(-j \cdot \pi \cdot \cos(\phi_2)) & \exp(-j \cdot \pi \cdot \cos(\phi_3)) \\
\exp(-j \cdot 2 \cdot \pi \cdot \cos(\phi_1)) & \exp(-j \cdot 2 \cdot \pi \cdot \cos(\phi_2)) & \exp(-j \cdot 2 \cdot \pi \cdot \cos(\phi_3)) \\
\exp(-j \cdot 3 \cdot \pi \cdot \cos(\phi_1)) & \exp(-j \cdot 3 \cdot \pi \cdot \cos(\phi_2)) & \exp(-j \cdot 3 \cdot \pi \cdot \cos(\phi_3)) \\
\exp(-j \cdot 4 \cdot \pi \cdot \cos(\phi_1)) & \exp(-j \cdot 4 \cdot \pi \cdot \cos(\phi_2)) & \exp(-j \cdot 4 \cdot \pi \cdot \cos(\phi_3))
\end{bmatrix} \quad (8.43)
$$

We note that each column of this matrix is a steering vector corresponding to one MPC. Inserting values for the directions of arrival gives

$$
\mathbf{A} = \begin{bmatrix}
1 & 1 & 1 \\
-0.9989 - 0.0477j & -0.6057 - 0.7957j & 0.5646 - 0.8253j \\
0.9954 + 0.0953j & -0.2663 + 0.9639j & -0.3624 - 0.9320j \\
-0.9898 - 0.1427j & 0.9282 - 0.3720j & -0.9739 - 0.2272j \\
0.9818 + 0.1898j & -0.8582 - 0.5133j & -0.7374 + 0.6755j
\end{bmatrix} \quad (8.44)
$$

The three incident signals result in an observed array response given by

$$
\mathbf{r}(t) = \mathbf{A} \cdot \mathbf{s}(t) + \mathbf{n}(t) \quad (8.45)
$$

where $\mathbf{s}(t) = [1 \ 0.8 \ 0.2]^T$ and $\mathbf{n}(t)$ is the vector of additive noise samples. We evaluate the correlation matrix $\mathbf{R}_{rr} = E[\mathbf{r}(t)\mathbf{r}^\dagger(t)]$ for 10,000 realizations (where the columns of the steering vectors have different phases φ_i, corresponding to independent realizations[6]) resulting in:

$$
\mathbf{R}_{rr} = \begin{bmatrix}
1.7295 & -1.3776 + 0.5941j & 0.8098 - 0.6763j & -0.4378 + 0.3755j & 0.4118 + 0.1286j \\
-1.3776 - 0.5941j & 1.7244 & -1.3621 + 0.5957j & 0.7993 - 0.6642j & -0.4275 + 0.3666j \\
0.8098 + 0.6763j & -1.3621 - 0.5957j & 1.7080 & -1.3493 + 0.5850j & 0.7932 - 0.6599j \\
-0.4378 - 0.3755j & 0.7993 + 0.6642j & -1.3493 - 0.5850j & 1.6903 & -1.3439 + 0.5780j \\
0.4118 - 0.1286j & -0.4275 - 0.3666j & 0.7932 + 0.6599j & -1.3439 - 0.5780j & 1.6875
\end{bmatrix}
$$

$$
(8.46)
$$

The angular spectrum for the conventional beamformer is given by Eq. (8.38), and it is plotted as the dashed line in Figure 8.7. We see that the conventional beamformer fails to identify the three incident signals. The angular spectrum for the MVM (Capon's beamformer) is given by Eq. (8.40); its result is plotted as a solid line in Figure 8.7. Three peaks can be identified in the vicinity of the true angles of arrival $10°$, $45°$, and $72°$.

8.5.4 Multiple Input Multiple Output Measurements

The methods described above are intended for getting the directions of arrival at one link end. They can be easily generalized to double-directional or MIMO measurements. Antenna arrays can be used at *both* link ends. In this case, it is necessary to use transmit signals in such a way that the RX can determine which antenna they were transmitted from. This can be done, e.g., by sending signals at different times, on different frequencies, or modulated with different codes (see Figure 8.8). Of these methods, using different times requires the least hardware effort, and is thus in widespread use in commercial MIMO channel sounders. The signal-processing techniques for determination

---

[6] We *assume* here that we have different realizations available. As discussed above, many channel-sounding applications require additional measures like subarray averaging to obtain those realizations.

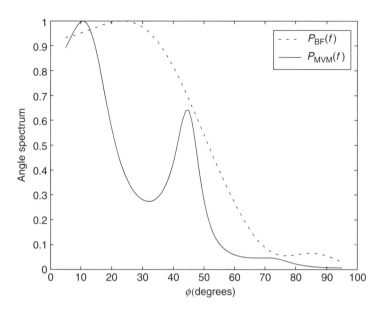

Figure 8.7 Comparison of conventional beamformer and the MVM angle spectrum for a five-element linear array. The true angles are 10°, 45°, and 72°.

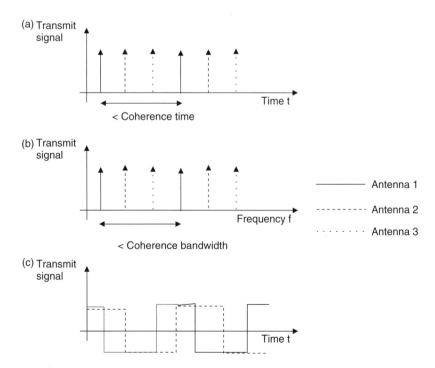

Figure 8.8 Transmission of sounding signals from different antennas: signals orthogonal in time (a), frequency (b), or code (c).

of the directions at the two link ends are also fairly similar to the processing techniques for the one-dimensional case.

8.6 Appendix

Please see companion website *www.wiley.com/go/molisch*

Further Reading

Overviews of different channel-sounding techniques are given in Parsons [1992] and Parsons et al. [1991]. Cullen et al. [1993] concentrates on correlative channel sounders; the STDCC is also described in detail in Cox [1972], where it was first introduced. Matz et al. [2002] discuss the impact of time variations of the channel on measurement results. Measurement procedures are also discussed by most papers presenting measurement results (especially back-to-back calibration and deconvolution). For the measurement of directional properties, the alternative method of using a rotating directional antenna is discussed, e.g., in Pajusco [1998].

The ESPRIT algorithm is described in Haardt and Nossek [1995] and Roy et al. [1986]; MUSIC in Schmidt [1986]; for the MVM (Capon's beamformer), see Krim and Viberg [1996]; SAGE is described in Fleury et al. [1999]. An elegant combination of the SAGE algorithm with the gradient method is described in Thomae et al. [2005]. The incorporation of diffuse radiation into the extraction is discussed in Richter [2006]. Other high-resolution algorithms include the CLEAN algorithm that is especially suitable for the analysis of ultrawideband signals [Cramer et al. 2002], the JADE algorithm [Vanderveen et al. 1997], and many others. Tracking of multi path components obtained from a sequence of measurements is discussed in Salmi et al. [2009].

For updates and errata for this chapter, see wides.usc.edu/teaching/textbook

9

Antennas

9.1 Introduction

9.1.1 Integration of Antennas into Systems

Antennas are the interface between the wireless propagation channel and the transmitter (TX) and receiver (RX), and thus have a major impact on the performance of wireless systems. This chapter therefore discusses some important aspects of antennas for wireless systems, both at the Base Station (BS) and at the Mobile Station (MS). We concentrate mainly on the antenna aspects that are specific to practical wireless systems.

Antenna design for wireless communications is influenced by two factors: (i) performance considerations and (ii) size and cost considerations. The latter aspect is especially important for antennas on MSs. These antennas must show not only good electromagnetic performance but must also be small, mechanically robust, and easy to produce. Furthermore, the performance of these antennas is influenced by the casing on which they are mounted, and by the person operating the handset. Antennas for BSs, on the other hand, are more similar to "conventional" antennas. Both size and cost are less restricted, and – at least for outdoor applications – the immediate surroundings of the antennas are clear of obstacles (some exceptions to this rule are also discussed in this chapter).

The remainder of this chapter thus distinguishes between MS antennas and BS antennas. Different types of antennas are discussed, as is the impact of the environment on antenna performance.

9.1.2 Characteristic Antenna Quantities

Directivity

The *directivity* D of an antenna is a measure of how much a transmit antenna concentrates the emitted radiation to a certain direction, or how much a receive antenna emphasizes radiation from a certain direction. More precisely, it is defined as [Vaughan and Andersen 2003]

$$D(\Omega) = \frac{\text{Total power radiated per unit solid angle in a direction } \Omega}{\text{Average power radiated per unit solid angle}} \tag{9.1}$$

Due to the principle of reciprocity, directivity is the same in the transmit and in the receive case. It is related to the far-field antenna power pattern $G(\Omega)$.[1] It is worth keeping in mind that the antenna

---

[1] Note that $G(\Omega)$ refers to antenna power, while we define $\tilde{G}(\Omega)$ as the complex amplitude gain, so that $G(\Omega) = |\tilde{G}(\Omega)|^2$. Both quantities are defined for the far field.

Wireless Communications, Second Edition   Andreas F. Molisch
© 2011 John Wiley & Sons, Ltd

power pattern is normalized so that

$$\frac{1}{4\pi} \int G(\Omega) \, d\Omega = 1 \tag{9.2}$$

In many cases, antennas have different patterns for different polarizations. In such cases,

$$D(\phi_0, \theta_0) = \frac{G_\phi(\phi_0, \theta_0) + G_\theta(\phi_0, \theta_0)}{\frac{1}{4\pi} \int \int (G_\phi(\phi, \theta) + G_\theta(\phi, \theta)) \sin(\theta) \, d\theta \, d\phi} \tag{9.3}$$

where θ and ϕ are elevation and azimuth, respectively, and G_ϕ and G_θ are the power gains for radiation polarized in ϕ and θ, respectively.

The *gain* of an antenna in a certain direction is related to directivity. However, the gain also has to account for losses – e.g., ohmic losses. Those losses are described by antenna efficiency, which is discussed in the next subsection.

Efficiency

Losses in the antenna can be caused by several different phenomena. First, ohmic losses (i.e., due to the finite conductivity of the antenna material) can occur. Second, polarizations between the receive antenna and the incident radiated field can be misaligned (see below). Finally, losses can occur because of imperfect matching. The efficiency can thus be written as

$$\eta = \frac{R_{\mathrm{rad}}}{R_{\mathrm{rad}} + R_{\mathrm{ohmic}} + R_{\mathrm{match}}} \tag{9.4}$$

where R_{rad} is the radiation resistance; it is defined as the resistance of an equivalent network element so that the radiated power P_{rad} can be written as $0.5|I_0|^2 R_{\mathrm{rad}}$, where I_0 is the excitation current magnitude. For example, the input impedance of a half-wavelength dipole is $Z_0 = 73 + j42\,\Omega$, so that the radiation resistance is $R_{\mathrm{rad}} = 73\,\Omega$.

For antennas that are to operate on just a single frequency, perfect matching can be achieved. However, there are limits to how well an antenna can be matched over a larger band. The so-called Fano bound for the case that the antenna impedance can be modeled as a resistor and a capacitor in series reads

$$\int \frac{1}{(2\pi f_{\mathrm{c}})^2} \ln\left(\frac{1}{|\rho(f)|}\right) df < \pi RC \tag{9.5}$$

where ρ is the reflection coefficient, and R and C are resistance and capacitance, respectively.

High efficiency is a key criterion for any wireless antenna. From the point of view of the transmit antenna, high efficiency reduces the required power of the amplifier to achieve a given field strength. From the point of view of the receive antenna, the achievable Signal-to-Noise Ratio (SNR) is directly proportional to antenna efficiency. Antenna efficiency thus enters the battery lifetime of the MS, as well as the transmission quality of a link. It is difficult, however, to define an absolute goal for efficiency. Ideally, $\eta = 1$ should be obtained over the whole bandwidth of interest. However, it must be noted that in recent years the antenna efficiency of MS antennas has *decreased*. It has been sacrificed mainly for cosmetic reasons – namely, to decrease the size of the antennas.

Another factor influencing radiation efficiency is the presence of dielectric and/or conducting material – namely, the user – in the vicinity of an MS. When computing antenna gain that can be used in a link budget, this material has to be taken into account, as it leads to strong distortions of

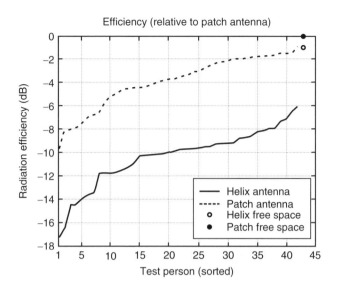

Figure 9.1 Efficiency of mobile station antennas close to a human body, relative to the efficiency of the patch antenna in free space. Different users and positions of the MS were used in the ensemble.
Reproduced with permission from Pedersen et al. [1998] © IEEE.

the antenna pattern and absorption of energy. Therefore, radiation efficiency is decreased when these effects are taken into account. The effective gain should be analyzed for a large number of users in order to eliminate the specific properties of any one user (at which angle with respect to the head is the MS being held?, is the user right-handed or left-handed?, what are the dielectric properties of the hand holding the device?, etc.). Figure 9.1 shows an exemplary cumulative distribution function for a patch antenna and a helical antenna. Note that the difference between effective antenna gain and theoretical values can exceed 10 dB. This has a large impact on network planning.

Q-Factor

A fundamental quantity of antennas is their Q-factor, defined as [Vaughan and Andersen 2003]

$$Q = 2\pi \frac{\text{Energy stored}}{\text{Energy dissipated per cycle}} \tag{9.6}$$

The Q-factor can be related to input impedance as

$$Q = \frac{f_c}{2R} \frac{\partial X}{\partial f} \tag{9.7}$$

where input impedance $Z = R + jX$. For an antenna contained in a sphere with diameter L_a, the Q-factor is given as

$$Q \geq \frac{1}{(k_0 L_a/2)^3} + \frac{1}{k_0 L_a/2} \tag{9.8}$$

where k_0 is again $2\pi/\lambda$.

Mean Effective Gain

The antenna pattern, and thus antenna directivity, is defined for the far field – i.e., assumes the existence of homogeneous plane waves – and is a function of the considered direction. It is measured in an anechoic chamber, where all obstacles and distorting objects are removed from the vicinity of the antenna. The patterns and gains measured this way are very close to the numbers found in any book on antenna theory – e.g., a Hertzian dipole has a gain of 1.5. When the antenna is operated in a random scattering environment, it becomes meaningful to investigate the *Mean Effective Gain* (MEG), which is an average of the gain over different directions when the directions of incident radiation are determined by a random environment [Andersen and Hansen 1977, Taga 1990]. The MEG is defined as the ratio of the average power received at the mobile antenna and the sum of the average power of the vertically and horizontally polarized waves received by isotropic antennas.

Polarization

In some wireless systems (e.g., satellite TV), the polarizations of the radiation and the receive antenna should be carefully aligned. This alignment can be measured by the Poincaré sphere. Each polarization state is associated with a point on the sphere: right-hand circular and left-hand circular polarizations are the north and south poles, respectively, while the different linear polarization states are on the equator; all other points correspond to elliptical polarizations. The angle between two points on the sphere is a measure of their mismatch.

For non-line-of-sight scenarios, requirements for a specific polarization of the antennas are typically not very stringent. Even if the transmit antenna sends mainly with a single polarization, propagation through the wireless channel leads to depolarization (see Chapter 7), so that cross-polarization at the *RX* is rarely higher than 10 dB. This fact is advantageous in many practical situations: as the orientation of MS antennas cannot be predicted (different users hold handsets in different ways), low sensitivity of the antenna to polarization of the incident radiation and/or uniform polarization of the incident radiation is advantageous.

One situation where good cross-polarization discrimination of antennas is required is the case of polarization diversity (see Chapter 13). In that case, the absolute polarization of the antennas is not critical; however, cross-polarization discrimination of two antenna elements plays an important role.

Bandwidth

Antenna bandwidth is defined as the bandwidth over which antenna characteristics (reflection coefficient, antenna gain, etc.) fulfill the specifications. Most wireless systems have a relative bandwidth of approximately 10%. This number already accounts for the fact that most systems use Frequency Domain Duplexing ((FDD), see Chapter 17) – i.e., transmit and receive on different frequencies. For example, in a GSM1800[2] system (Chapter 24), bandwidth is about 200 MHz, while the carrier frequency is around 1.8 GHz. As it is desirable to have only a single antenna for TX and RX cases, the antenna bandwidth must be large enough to include both the TX and RX bands. As we have mentioned above, a large bandwidth implies that the matching circuits can no longer be perfect. This effect is the stronger, the smaller the physical dimensions of the antenna are.

Another interesting special case involves dual- or multimode devices. For example, most GSM phones have to be able to operate at both the 900- and the 1,800-MHz carrier frequencies. Antennas for such devices can be constructed by requiring good performance across the whole 0.9–2-GHz band. However, using such an antenna with 50% relative bandwidth would be "overkill," as the frequency range from 1 GHz to 1.7 GHz need not be covered. It is therefore better to just design

---

[2] Global System for Mobile communications in the 1,800-MHz band.

antennas that have good performance in the required bands; this goal is easier to fulfill if the different carrier frequencies are integer multiples of each other. Still, the design of antennas that cover two or more bands with high efficiency is a very challenging task. For BS antennas, it is thus often preferable from a technical point of view to use different antennas for different frequency bands. However, multiband antennas are to be preferred for esthetical reasons and reduced visual impact.

9.2 Antennas for Mobile Stations

9.2.1 Monopole and Dipole Antennas

Linear antennas are the "classical" antennas for MSs, and have for long determined the typical "look" of these devices. The most common ones are electric monopoles, located above a conducting plane (the casing), and dipoles. The antenna pattern of a short (Hertzian) dipole oriented along the z-axis is uniform in azimuth, and sine-shaped in polar angle θ (measured from the z-axis):

$$\tilde{G}(\varphi, \theta) \propto \sin(\theta) \tag{9.9}$$

with a maximum gain:

$$G_{\max} = 1.5 \tag{9.10}$$

A $\lambda/2$ dipole has the following properties (see Figure 9.2):

$$\tilde{G}(\varphi, \theta) \propto \frac{\cos\left(\frac{\pi}{2}\cos(\theta)\right)}{\sin(\theta)} \tag{9.11}$$

and a maximum gain:

$$G_{\max} = 1.64 \tag{9.12}$$

Figure 9.2 Shape and radiation pattern of a $\lambda/2$ dipole antenna.

We can thus see that the patterns of the $\lambda/2$ dipole and the Hertzian dipole do not differ dramatically. However, the radiation resistance can be quite different. For a dipole with *uniform current distribution*, the radiation resistance is

$$R_{\text{rad}}^{\text{uniform}} = 80\pi^2(L_a/\lambda)^2 \qquad (9.13)$$

For dipoles with a tapered current distribution (maximum at the feed, and linear decrease towards the end), the radiation resistance is $0.25R_{\text{rad}}^{\text{uniform}}$.

From the image principle it follows that the radiation pattern of a monopole located above a conducting plane is identical to that of a dipole antenna in the upper half-plane. Since the energy transmitted into the upper half-space $0 \leq \theta \leq \pi/2$ is twice that of the dipole, the maximum gain is twice and the radiation resistance half that of the dipole. Note, however, that the image principle is valid only if the conducting plane extends infinitely – this is not the case for MS casings. We can thus also anticipate considerable radiation in the lower half-space even for monopole antennas.

The reduction in radiation resistance is often undesirable, since it makes matching more difficult, and leads to a reduction in efficiency due to ohmic losses. One way of increasing efficiency without increasing the physical size of the antenna is to use *folded* dipoles. A folded dipole consists of a pair of half-wavelength wires that are joined at the non-feed end [Vaughan and Andersen 2003]; these increase the input impedance.

The biggest plus of monopole and dipole antennas is that they can be produced easily and cheaply. The relative bandwidth is sufficient for most applications in single-antenna systems. The disadvantage is the fact that a relatively long metal stick must be attached to the MS casing. In the 900-MHz band, a $\lambda/4$ monopole is 8 cm long – often longer than the MS itself. Even when realized as a retractable element, it is easily damaged. For this reason, shorter and/or integrated antennas, which are less efficient, are becoming increasingly widespread. This does not pose a significant problem in Europe and Japan, where coverage is usually good, and most systems are interference-limited anyway. However, in the U.S.A. and other countries where coverage is somewhat haphazard in many regions, this can have considerable influence on performance.

9.2.2   Helical Antennas

The geometry of helical antennas is outlined in Figure 9.3. A helical antenna can be seen as a combination of a loop antenna and a linear antenna, and thus has two modes of operation. The dimensions of the antenna determine which mode it is operating in. If the dimensions of the helix are much smaller than a wavelength, then the antenna is operating in normal mode. It behaves similar to a linear antenna, and has a pattern that is shaped mainly in the radial direction. This is the operating condition used in MS antennas. In general, the polarization is elliptical, though it can

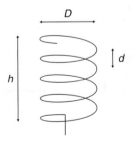

Figure 9.3   Geometry of a helical antenna.

become circular if the ratio $2\lambda d/(\pi D)^2$ becomes unity [Balanis 2005]. The polarization becomes vertical if the helical antenna is arranged over a conducting plane, as the horizontal components of the actual antenna and its image cancel out. When the circumference of the helix is on the order of one wavelength, the antenna pattern has its maximum along the axis of the helix, and polarization is almost circular.

Since the helical antenna in normal mode is similar to a linear antenna, the number of turns the antenna makes does not influence the antenna pattern. However, the bandwidth, efficiency, and radiation resistance increase with increasing h. In general, a helical antenna has lower bandwidth and a smaller input impedance than a monopole antenna; however, a relative bandwidth of 10% can be achieved by appropriate matching circuits. The main advantage of the helical antenna is its smaller size; this has made it (together with linear antennas) the most widely used external antenna for MSs.

9.2.3 Microstrip Antennas

A microstrip antenna (patch antenna) consists of a thin dielectric substrate, which is covered on one side by a thin layer of conducting material (ground plane), while on the other side there is a *patch* of conducting material. The configuration is outlined in Figure 9.4 (see also Fujimoto [2008] and Fujimoto et al. [1987]).

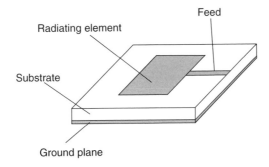

Figure 9.4 Geometry of a microstrip antenna.

The properties of a microstrip antenna are determined by the shape and dimension of the metallic patch, as well as by the dielectric properties of the used substrate. Essentially, the patch is a resonator whose dimensions have to be multiples of the effective dielectric wavelength. Thus, a high dielectric constant of the substrate allows the construction of small antennas. The most commonly used patch shapes are rectangular, circular, and triangular.

The patch is usually fed either by a coaxial cable or a microstrip line. It is also possible to feed the patch via electromagnetic coupling. This latter case uses a substrate where the ground plane is sandwiched between two layers of dielectric material. On the top of one material is the patch, while the feedline is at the bottom of the other dielectric layer. Coupling is effected through a slot (*aperture*) in the ground plane. These antennas are thus called aperture-coupled patch antennas. This design has the advantage that the dielectric properties of the two layers can be chosen differently, depending on the requirements for the patch and the feedline. Furthermore, this design shows a larger bandwidth than conventional patch antennas.

As mentioned above, the size and efficiency of the microstrip antenna are determined by the parameters of the dielectric substrate. A large ε_r reduces the size. This follows immediately from the fact that in a resonator the length of one side must be

$$L = 0.5\lambda_{\text{substrate}} \tag{9.14}$$

where

$$\lambda_{\text{substrate}} = \lambda_0/\sqrt{\varepsilon_r} \tag{9.15}$$

Unfortunately, a reduction in physical size also leads to a smaller bandwidth, which is usually undesirable. For this reason, substrates used in practice usually have a very low ε_r – even air is used quite frequently. A further possibility for reducing the size of patch antennas is the use of short-circuited resonators, which reduces the required size of a resonator from $\lambda/2$ to $\lambda/4$ (see Figure 9.5).

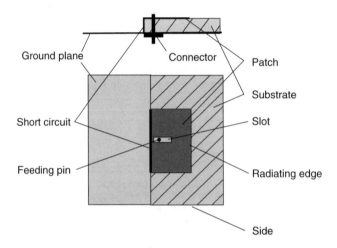

Figure 9.5   Short-circuited $\lambda/4$ patch antenna.

The bandwidth of microstrip antennas can be increased by various measures. The most straightforward one is an increase in antenna volume – i.e., the use of thicker substrates with a lower ε_r. Alternatives are the use of matching circuits and the use of parasitic elements.

Microstrip antennas have several important advantages for wireless applications:

- They are small and can be manufactured cheaply.
- The feedlines can be manufactured on the same substrate as the antenna.
- They can be integrated into the MS, without sticking out from the casing.

However, they also have serious weaknesses:

- They have a low bandwidth (usually just a few percent of the carrier frequency).
- They have low efficiency.

9.2.4   Planar Inverted F Antenna

Some of the problems of microstrip antennas can be alleviated by a *Planar Inverted F Antenna* (PIFA). The shape of the PIFA is similar to that of a $\lambda/4$ short-circuited microstrip antenna (see

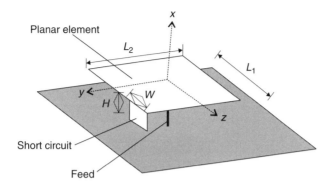

Figure 9.6 Planar inverted-F antenna.

Figure 9.6). A planar, radiating element is located parallel to a ground plane. This element is short-circuited over a distance W. If W is chosen equal to the length of the edge L, then we obtain a short-circuited $\lambda/4$ microstrip antenna. If W is chosen smaller, then the resonance length increases, and the current distribution on the radiating element changes.

9.2.5 Radiation Coupled Dual L Antenna

A further improvement is achieved by the so-called *Radiation Coupled Dual L Antenna* (RCDLA) (see Figure 9.7 and Rasinger et al. [1990]). It consists of two L-shaped angular structures only one of which is fed directly (conductively). The other L-shaped structure is fed by the first L by means of radiation coupling. This increases the bandwidth of the total arrangement. By optimally placing the antenna on the casing, relative bandwidths of up to 10% can be achieved, which is about twice the bandwidth of a PIFA.

Figure 9.7 Radiation-coupled dual-L antenna.

9.2.6 Multiband Antennas

Modern cellular handsets are anticipated to be able to handle different frequencies for communications. As discussed in Section 9.2.2 a GSM handset, e.g., needs to be able to deal at least with 900 and 1,800 MHz foreseen in the specifications for most countries. As an added difficulty, many handsets should also be able to cope with the 1,900-MHz mode used in the U.S.A., as well as 2.4 GHz if Bluetooth connections (e.g., to a wireless headset) are required. The situation becomes even more complicated for dual-mode devices that can handle both GSM and Wideband Code Division Multiple Access (WCDMA) (see Chapter 26). The design of internal multiband antennas is very complicated, and few rules for a closed-form design are available. Figure 9.8 shows an example of a microstrip multiband antenna.

Figure 9.8   Integrated multiband antenna.

Reproduced with permission from Ying and Anderson [2003] © Z. Ying.

9.2.7   Antenna Mounting on the Mobile Station

Antennas do not operate in empty space but are placed on top of the casing, which can be considered to be part of the radiator. Furthermore, antenna characteristics are influenced by the hand and the head of the user; this influence also depends on the mounting of the antenna on the MS. It is therefore important to investigate different options for placing the antenna, and to see how this placement influences performance.

Linear and helical antennas are usually placed on the upper, narrow side of the casing – i.e., they stick out from that part of the casing. This has mainly ergonomic reasons – if they were sticking out from the lower side, they would feel uncomfortable to the user, and the hand of the user would often cover the antenna, leading to additional attenuation.

For microstrip antennas, PIFAs, and RCDLAs, there are more options for placements. These antennas are usually used as internal antennas – i.e., integrated into the casing or enclosed within the casing. This greatly reduces the danger of mechanical damage. However, there is an increased probability that users will place their hands over the antenna, which increases absorption of the electromagnetic energy and thus worsens link performance [Erätuuli and Bonek 1997]. Examples for the positioning of RCDLAs can be found in Figure 9.9; other types of microstrip antennas can be placed in a similar way.

The impact of the human body on antenna patterns is discussed in Section 9.3.4.

Figure 9.9   Placement of radiation-coupled dual-L antennas on the casing of a mobile station.

9.3 Antennas for Base Stations

9.3.1 Types of Antennas

The design requirements for BS antennas are different from those of MS antennas. Cost has a smaller impact, as BSs are much more expensive in the first place. Also, size restrictions are less stringent: for macrocells, it is only required that (i) the mechanical stress on the antenna mast, especially due to wind forces, must remain reasonable and (ii) the "cosmetic" impact on the surroundings must be small. For micro- and picocells, antennas need to be considerably smaller, as they are mounted on building surfaces, street lanterns, or on office walls. The desired antenna pattern is quite different for BS antennas compared with MS antennas. As the physical placement and orientation of the BS antenna is known, patterns should be shaped in such a way that no energy is wasted (see also Section 9.3.3). Such pattern shaping can be most easily implemented by multielement antennas.

For these reasons, macrocellular antennas typically are antenna arrays or Yagi antennas whose elements are linear antennas. For micro- and picocells, antenna arrays consisting of patch antennas are common. All considerations from Section 9.2 about bandwidth, efficiency, etc., remain valid, except for the size requirements which are relaxed quite a bit.

9.3.2 Array Antennas

Array antennas are often used for BS antennas. They result in an antenna pattern that can be more easily shaped. This shaping can either be done in a predetermined way (see Section 9.3.3) or adaptively (see Chapters 13 and 20). The pattern of an antenna array can be written as the product of the pattern of a single element and the array factor $M(\phi, \theta)$. In the plane of the array antenna, the array factor of a uniform linear array is (compare Section 8.5.1) [Stutzman and Thiele 1997]

$$M(\phi) = \sum_{n=0}^{N_r-1} w_n \exp[-j \cos(\phi) 2\pi d_a n/\lambda] \tag{9.16}$$

where d_a is the distance between the antenna elements,[3] and w_n are the complex weights of element excitations. For the case that all $|w_n| = 1$ and $\arg(w_n) = n\Delta$ the array factor becomes

$$|M(\phi)| = \left| \frac{\sin\left[\dfrac{N_r}{2}\left(\dfrac{2\pi}{\lambda}d_a\cos\phi - \Delta\right)\right]}{\sin\left[\dfrac{1}{2}\left(\dfrac{2\pi}{\lambda}d_a\cos\phi - \Delta\right)\right]} \right| \tag{9.17}$$

The phase shift Δ of the feed currents determines the direction of the antenna main lobe. This principle is well known from the theory of phased array antennas ([Hansen 1998], see also Chapter 8). By imposing $\varphi_n = n\Delta$, the degrees of freedom have reduced to one; while the main lobe can be put into an arbitrary direction, the placement of the sidelobes and the nulls follows uniquely from that direction.

---

[3] Distance is usually chosen as $d_a = \lambda/2 (d_a \leq \lambda/2$ is necessary to avoid spatial aliasing – i.e., periodicities in the antenna pattern). In the following, we assume that the elements of the linear array are on the x-axis. Obviously, the same principles are valid when antenna elements are stacked vertically, and the elevation pattern of the array should be shaped.

Example 9.1 *Consider a BS antenna mounted at a height of 51 m, providing coverage for a cell with a radius of 1 km. The antenna consists of eight vertically stacked short dipoles, separated by λ/2. How much should the phase shift Δ be so that the maximum points at the cell edge at an MS height of 1 m?*

In order to get constructive interference of the contributions from the different antennas in the direction θ_0, the angle Δ should fulfill

$$\Delta = \frac{2\pi}{\lambda} d_a \cos \theta_0 \tag{9.18}$$

Obviously, if $\theta_0 = \pi/2$, then $\Delta = 0$ – i.e., no phase shift is necessary. In our example, we are looking for a tilt angle:

$$\theta_0 = \frac{\pi}{2} + \arctan\left(\frac{51-1}{1000}\right) = \frac{\pi}{2} + 0.05 \tag{9.19}$$

The phase shift thus has to be

$$\Delta = \frac{2\pi}{\lambda} d_a \cos\left(\frac{\pi}{2} + 0.05\right) = -0.05\pi \tag{9.20}$$

The impact of the element pattern can be neglected.

9.3.3 Modifying the Antenna Pattern

It is undesirable that a BS antenna radiates isotropically. Radiation emitted in a direction that has a large elevation angle (i.e., into the sky) is not only wasted but actually increases interference with other systems. The optimum antenna pattern is achieved when the received power is constant within the whole cell area, and vanishes outside. The problem is then to synthesize such an elevation pattern. This can be achieved approximately by means of array antennas, where the complex weights are chosen in such a way as to minimize the mean-square error of the pattern. An even simpler approach is to use an antenna array with $w_n = 1$, but tilt the main lobe down by about $5°$. This downtilt can be achieved either mechanically (by tilting the whole antenna array) or electronically.

The desired azimuthal antenna pattern is either (i) omnidirectional – i.e., uniform in $[0, 2\pi)$ – or (ii) uniform within a sector, and zero outside. The angular range of a sector is usually $60°$ or $120°$. For omnidirectional antennas, which are mostly used in rural macrocells, linear antennas are used. The antennas are arranged in a linear array that extends only along the vertical axis. For sector antennas, either linear antennas with appropriately shaped reflectors or microstrip antennas (which have inherently nonuniform antenna patterns) can be used.

9.3.4 Impact of the Environment on Antenna Pattern

The antenna pattern of BS antennas is usually defined for the case when an antenna is in free space, or above an ideally conducting plane. This is what is measured in an anechoic chamber, and is also the easiest to compute. However, at its location of operation, a BS antenna is surrounded by different objects of finite extent and conductivity. The antenna patterns in such surroundings can deviate significantly from theoretical patterns. The antenna mast, which is usually metallic, and thus highly conductive, can lead to distortions. Similarly, roofs made out of certain building materials, like reinforced concrete, can distort antenna patterns (see Figure 9.10).

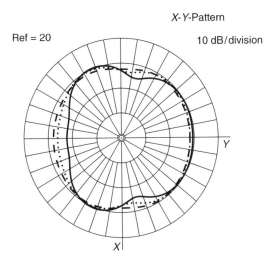

Figure 9.10 Antenna pattern of an omnidirectional antenna close to an antenna mast. Distance from the antenna mast 30 cm. Diameter of the mast: very small (solid), 5 cm (dashed), 10 cm (dotted).

Reproduced with permission from Molisch et al. [1995] © European Microwave Association.

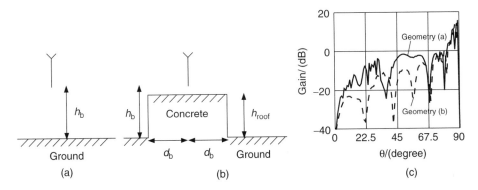

Figure 9.11 Distortion in the vertical antenna pattern. (a) Idealized geometry. (b) Real geometry, including the roof. (c) Distortions in antenna patterns. Properties of all materials: relative dielectric constant $\varepsilon_r = 2$: conductivity 0.01 S/m; $d_b = 20$ m.

Reproduced with permission from Molisch et al. [1995] © European Microwave Association.

Distortions in the vertical pattern are plotted in Figure 9.11. They arise from two effects: the fact that the (finite extent) roof is much closer to the antenna than the ground and that the dielectric properties of the roof are different from those of the ground.

The presence of a human body also distorts antenna patterns. One way of taking this into account is to consider the antenna and the human body as a "superantenna," whose characteristics (efficiency, radiation pattern) can be measured and characterized in the same manner as "regular" antennas. Figure 9.12 shows example measurements of antenna patterns by a human head and body.

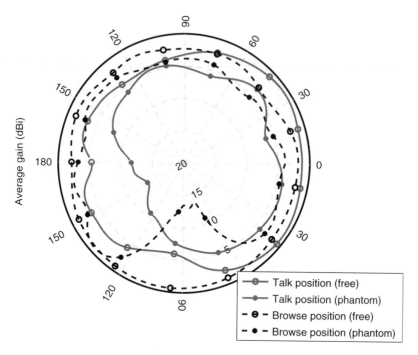

Figure 9.12   Pattern of antenna distortion by human body in talk and data-browsing position.
Reproduced with permission from Harryson et al. [2010]. Copyright IEEE.

Further Reading

For a general introduction to antenna theory, we just refer to the many excellent books on antenna theory: Balanis [2005], Kraus and Marhefka [2002], Ramo et al. [1967], and Stutzman and Thiele [1997], as well as the somewhat more advanced text of Collin [1985]; the latter also devotes a separate chapter to the properties of receiving antennas. For more details on antenna specifically for wireless communications, see Godara [2001] and Vaughan and Andersen [2003]. Antennas for the MS are discussed in Fujimoto [2008] and Hirasawa and Haneishi [1991]. Antenna design for BSs is surveyed in the monograph of Chen and Luk [2009] and in the conference paper by Beckman and Lindmark [2007]. Finally, phased array antennas are treated in detail in Hansen [1998] and Mailloux [1994]. Discussions on beam tilting can be found in Manholm et al. [2003]. The impact of the human head and body on antenna characteristics is discussed, e.g., in Ogawa and Matsuyoshi [2001], Kivekaes et al. [2004] and [Harryson et al. 2010].

For updates and errata for this chapter, see wides.usc.edu/teaching/textbook

Part III

Transceivers and Signal Processing

The ultimate performance limits of wireless systems are determined by the wireless propagation channels that we have investigated in the previous parts. The task of practical transceiver design now involves finding suitable modulation schemes, codes, and signal processing algorithms so that these performance limits can be approximated "as closely as possible." This task always involves a tradeoff between the performance and the hardware and software effort. As technology progresses, more and more complicated schemes can be implemented. For example, a third-generation cellphone has a computation power that is comparable to a (year 2000) personal computer, and can thus implement signal processing algorithms whose practical use was unthinkable in the mid-1990s. For this reason, this part of the book will not pay too much attention to algorithm complexity – what is too complex at the current time might well be a standard solution a few years down the road.

The part starts with a description of the general structure of a transceiver in Chapter 10. It describes the various blocks in a transmitter and receiver, as well as simplified models that can be used for system simulations and design. Next, we discuss the various *modulation formats*, and their specific advantages and disadvantages for their use in a wireless context. For example, we find that constant-modulus modulation methods are especially useful for battery-powered transmitters, since they allow the use of high-efficiency amplifiers. Building on the formal mathematical description of those modulation formats, Chapter 12 then describes how to evaluate their performance in terms of *bit error probability* in different types of fading channels. We find that the performance of such systems is mostly limited by two effects: fading and delay dispersion. The effect of fading can be greatly mitigated by *diversity*, i.e., by transmitting the same signal via different paths. Chapter 13 describes the different methods of obtaining such different paths, e.g., by implementing multiple antennas, by repeating the signal at different frequencies, or at different times. The chapter also discusses the effect that the diversity has on the performance of the different modulation schemes. Negative effects of the delay dispersion can also be combated by diversity; however, it is more effective to use equalization. *Equalizers* do not only combat intersymbol interference created by delayed echoes of the original signal, but they make use of them, exploiting the energy contained in such echoes. They can thus lead to a considerable improvement of performance, especially in systems with high data rates and/or systems operating in channels with large delay spreads. Chapter 16 describes different equalizer structures, from the simple linear equalizers to the optimum (but highly complex) maximum-likelihood sequence detectors.

Diversity and equalizers are not always a sufficient or effective way of improving the error probability. In many cases, *coding* can greatly enhance the performance, and provide the transmission quality required by a specific application. Chapter 14 thus gives an overview of the different coding schemes that are most popular for wireless communications, including the near-optimum turbo codes and Low Density Parity Check (LDPC) codes that have drawn great attention since the early 1990s. These coding schemes are intended for correcting the errors introduced on the propagation channel. The chapter also describes the fundamentals of information theory, which establishes the ultimate performance limits that can be achieved with "ideal" codes. A different type of coding is source coding, which translates the information from the source into a bitstream that can be transmitted most efficiently over the wireless channel. Chapter 15 gives an overview of *speech coding*, which is the most important type of source coding for wireless applications. Finally, Chapter 16 describes equalization, i.e., methods for compensating for delay dispersion of the channel

Modulation, coding, and equalization for wireless communications are, of course, strongly related to digital communications in general. This part of the book is *not* intended as a textbook of digital communications, but rather assumes that the reader is already familiar with the topic from either previous courses, or one of the many excellent textbooks (e.g., [Proakis 2005], [Barry et al. 2003], [Sklar 2001], [Anderson 2005]). While the text gives summaries of the most salient facts, they are rather terse, and only intended as a reminder to the reader.

10

Structure of a Wireless Communication Link

10.1 Transceiver Block Structure

In this section, we describe a block diagram of a wireless communication link, and give a brief description of the different blocks. More detailed considerations are left for the later chapters of Parts III and IV. We start out with a rough overview that concentrates on the *functionality* of the different blocks. Subsequently, we describe a block diagram that concentrates more on the different *hardware* elements.

Figure 10.1 shows a functional block diagram of a communications link. In most cases, the goal of a wireless link is the transmission of information from an analog information source (microphone, videocamera) via an analog wireless propagation channel to an analog information sink (loudspeaker, TV screen); the digitizing of information is done only in order to increase the reliability of the link. Chapter 15 describes speech coding, which represents the most common form of digitizing analog information; Chapter 23 describes video coding. For other transmissions – e.g., file transfer – information is already digital.

The transmitter (TX) can then add redundancy in the form of a forward error correction code, in order to make it more resistant to errors introduced by the channel (note that such encoding is done for most, but not all, wireless systems). The encoded data are then used as input to a modulator, which maps the data to output waveforms that can be transmitted. By transmitting these symbols on specific frequencies or at specific times, different users can be distinguished.[1] The signal is then sent through the propagation channel, which attenuates and distorts it, and adds noise, as discussed in Part II.

At the receiver (RX), the signal is received by one or more antennas (see Chapter 13 for a discussion on how to combine the signals from multiple antennas). The different users are separated (e.g., by receiving signals only at a single frequency). If the channel is delay dispersive, then an equalizer can be used to reverse that dispersion, and eliminate intersymbol interference. Afterwards, the signal is demodulated, and a channel decoder eliminates (most of) the errors that are present in the resulting bitstream. A source decoder finally maps this bitstream to an analog information stream that goes to the information sink (loudspeaker, TV monitor, etc.); in the case when the information was originally digital, this last stage is omitted.

The above description of the blocks is of course oversimplified, and – especially in the RX – the separation of blocks need not be that clear cut. An optimum RX would use as its input the

---

[1] Alternative multiple-access methods are described in Chapters 18–22.

Wireless Communications, Second Edition   Andreas F. Molisch
© 2011 John Wiley & Sons, Ltd

Figure 10.1   Block diagram of a transmitter and receiver, denoting which blocks are discussed in which chapter.

sampled received signal, and compute from it the signal that has been transmitted with the largest likelihood. While this is still too computationally intensive for most applications, joint decoding and demodulation and similar schemes have already been investigated in the literature.

Figures 10.2 and 10.3 show a more detailed block diagram of a digital TX and RX that concentrate on the hardware aspects and the interfaces between analog and digital components:

- The *information source* provides an analog source signal and feeds it into the *source ADC* (Analog to Digital Converter). This ADC first band limits the signal from the analog information source (if necessary), and then converts the signal into a stream of digital data at a certain sampling rate and resolution (number of bits per sample). For example, speech would typically be sampled at 8 ksamples/s, with 8-bit resolution, resulting in a datastream at 64 kbit/s. For the transmission of digital data, these steps can be omitted, and the digital source directly provides the input to interface "G" in Figure 10.2.
- The *source coder* uses a priori information on the properties of the source data in order to reduce redundancy in the source signal. This reduces the amount of source data to be transmitted, and thus the required transmission time and/or bandwidth. For example, the Global System for Mobile communications (GSM) speech coder reduces the source data rate from 64 kbit/s mentioned above to 13 kbit/s. Similar reductions are possible for music and video (MPEG standards). Also, fax information can be compressed significantly. One thousand subsequent symbols "00" (representing "white" color), which have to be represented by 2,000 bits, can be replaced by the statement: "what follows now are 1,000 symbols 00," which requires only 12 bits. For a typical fax, compression by a factor of 10 can be achieved. The source coder increases the entropy (information per bit) of the data at interface F; as a consequence, bit errors have greater impact. For some applications, source data are *encrypted* in order to prevent unauthorized listening in.
- The *channel coder* adds redundancy in order to protect data against transmission errors. This increases the data rate that has to be transmitted at interface E – e.g., GSM channel coding increases the data rate from 13 to 22.8 kbit/s. Channel coders often use information about the statistics of error sources in the channel (noise power, interference statistics) to design codes that are especially well suited for certain types of channels (e.g., Reed–Solomon codes protect especially well against burst errors). Data can be sorted according to importance; more important bits then get stronger protection. Furthermore, it is possible to use interleaving to break up error bursts; note that interleaving is mainly effective if it is combined with channel coding.
- *Signaling* adds control information for the establishing and ending of connections, for associating information with the correct users, synchronization, etc. Signaling information is usually strongly protected by error correction codes.
- The *multiplexer* combines user data and signaling information, and combines the data from multiple users.[2] If this is done by time multiplexing, the multiplexing requires some time compression.

---

[2] Actually, only the multiplexer at a Base Station (BS) really combines the data from multiple users for transmission. At a Mobile Station (MS), the multiplexer only makes sure that the RX at the BS can distinguish between the data streams from different users.

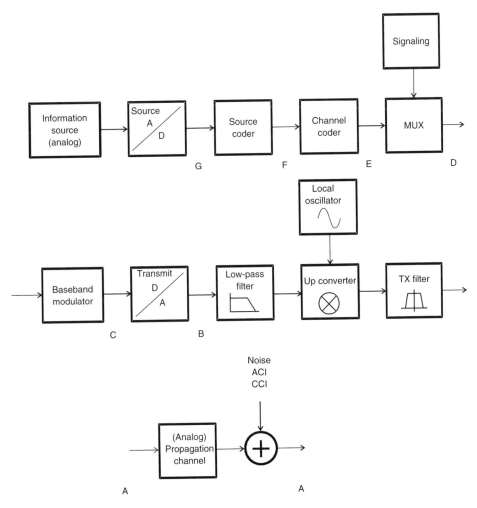

Figure 10.2 Block diagram of a radio link with digital transmitter and analog propagation channel. MUX, multiplexing; ACI, Adjacent Channel Interference; CCI, Co Channel Interference.

In GSM, multiaccess multiplexing increases the data rate from 22.8 to 182.4 kbit/s (8 · 22.8) for the standard case of eight participants. The addition of signaling information increases the data rate to 271 kbit/s.

- The *baseband modulator* assigns the gross data bits (user data and signaling at interface D) to complex transmit symbols in the baseband. Spectral properties, intersymbol interference, peak-to-average ratio, and other properties of the transmit signal are determined by this step. The output from the baseband modulator (interface C) provides the transmit symbols in oversampled form, discrete in time and amplitude.

  Oversampling and quantization determine the aliasing and quantization noise. Therefore, high resolution is desirable, and the data rate at the output of the baseband modulator should be much higher than at the input. For a GSM system, an oversampling factor of 16 and 8-bit amplitude resolution result in a data rate of about 70 Mbit/s.

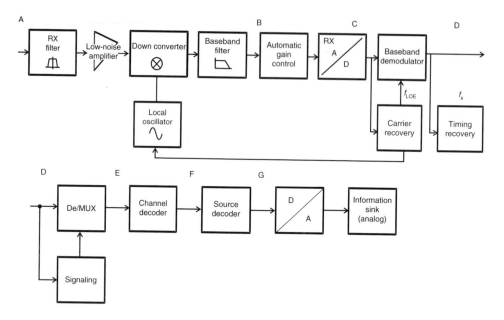

Figure 10.3 Block diagram of a digital receiver chain for mobile communications. MUX, multiplexing.

- The *TX Digital to Analog Converter* (DAC) generates a pair of analog, discrete amplitude voltages corresponding to the real and imaginary part of the transmit symbols, respectively.
- The *analog low-pass filter* in the TX eliminates the (inevitable) spectral components outside the desired transmission bandwidth. These components are created by the out-of-band emission of an (ideal) baseband modulator, which stem from the properties of the chosen modulation format. Furthermore, imperfections of the baseband modulator and imperfections of the DAC lead to additional spurious emissions that have to be suppressed by the TX filter.
- The *TX Local Oscillator* (LO) provides an unmodulated sinusoidal signal, corresponding to one of the admissible center frequencies of the considered system. The requirements for frequency stability, phase noise, and switching speed between different frequencies depend on the modulation and multiaccess method.
- The *upconverter* converts the analog, filtered baseband signal to a passband signal by mixing it with the LO signal. Upconversion can occur in a single step, or in several steps. Finally, amplification in the Radio Frequency (RF) domain is required.
- The *RF TX filter* eliminates out-of-band emissions in the RF domain. Even if the low-pass filter succeeded in eliminating all out-of-band emissions, upconversion can lead to the creation of additional out-of-band components. Especially, nonlinearities of mixers and amplifiers lead to intermodulation products and "spectral regrowth" – i.e., creation of additional out-of-band emissions.

- The *(analog) propagation channel* attenuates the signal, and leads to delay and frequency dispersion. Furthermore, the environment adds noise (Additive White Gaussian Noise – AWGN) and co-channel interference.
- The *RX filter* performs a rough selection of the received band. The bandwidth of the filter corresponds to the total bandwidth assigned to a specific service, and can thus cover multiple communications channels belonging to the same service.
- The *low-noise amplifier* amplifies the signal, so that the noise added by later components of the RX chain has less effect on the Signal-to-Noise Ratio (SNR). Further amplification occurs in the subsequent steps of downconversion.
- The *RX LO* provides sinusoidal signals corresponding to possible signals at the TX LO. The frequency of the LO can be fine-tuned by a carrier recovery algorithm (see below), to make sure that the LOs at the TX and the RX produce oscillations with the same frequency and phase.
- The *RX downconverter* converts the received signal (in one or several steps) into baseband. In baseband, the signal is thus available as a complex analog signal.
- The *RX low-pass filter* provides a selection of desired frequency bands for one specific user (in contrast to the RX bandpass filter that selects the frequency range in which the service operates). It eliminates adjacent channel interference as well as noise. The filter should influence the desired signal as little as possible.
- The *Automatic Gain Control* (AGC) amplifies the signal such that its level is well adjusted to the quantization at the subsequent ADC.
- The RX *ADC* converts the analog signal into values that are discrete in time and amplitude. The required resolution of the ADC is determined essentially by the dynamics of the subsequent signal processing. The sampling rate is of limited importance as long as the conditions of the sampling theorem are fulfilled. Oversampling increases the requirements for the ADC, but simplifies subsequent signal processing.
- *Carrier recovery* determines the frequency and phase of the carrier of the received signal, and uses it to adjust the RX LO.
- The *baseband demodulator* obtains *soft-decision* data from digitized baseband data, and hands them over to the decoder. The baseband demodulator can be an optimum, coherent demodulator, or a simpler differential or incoherent demodulator. This stage can also include further signal processing like equalization.
- If there are *multiple antennas*, then the RX either selects the signal from one of them for further processing or the signals from all of the antennas have to be processed (filtering, amplification, downconversion). In the latter case, those baseband signals are then either combined before being fed into a conventional baseband demodulator or they are fed directly into a "joint" demodulator that can make use of information from the different antenna elements.
- *Symbol-timing recovery* uses demodulated data to determine an estimate of the duration of symbols, and uses it to fine-tune sampling intervals.
- The *decoder* uses soft estimates from the demodulator to find the original (digital) source data. In the most simple case of an uncoded system, the decoder is just a hard-decision (threshold) device. For convolutional codes, *Maximum Likelihood Sequence Estimators* (MLSEs, such as the Viterbi decoder) are used. Recently, iterative RXs that perform joint demodulation and decoding have been proposed. Remaining errors are either taken care of by repetition of a data packet (*Automatic Repeat reQuest* – ARQ) or are ignored. The latter solution is usually resorted to for speech communications, where the delay entailed by retransmission is unacceptable.
- *Signaling recovery* identifies the parts of the data that represent signaling information and controls the subsequent demultiplexer.
- The *demultiplexer* separates the user data and signaling information and reverses possible time compression of the TX multiplexer. Note that the demultiplexer can also be placed earlier

in the transmission scheme; its optimum placement depends on the specific multiplexing and multiaccess scheme.

- The *source decoder* reconstructs the source signal from the rules of source coding. If the source data are digital, the output signal is transferred to the data sink. Otherwise, the data are transferred to the DAC, which converts the transmitted information into an analog signal, and hands it over to the information sink.

10.2   Simplified Models

It is often preferable to have simplified models for the link. Figure 10.4 shows a model that is suitable for the analysis of modulation methods. The parts of the TX between the information source and the output of the TX multiplexer are subsumed into a "black box" digital data source. The analog radio channel, together with the upconverters, downconverters, RF elements (filters, amplifiers), and all noise and interference signals, is subsumed into an equivalent time-discrete low-pass channel, characterized by a time-variant impulse response and the statistics of additive disturbances. The criterion for judging the quality of the modulation format is the bit error probability at the interfaces D–D.

Figure 10.4   Mathematical link model for the analysis of modulation formats.

Other simplified models use a digital representation of the channel (e.g., binary symmetric channel), and are mainly suitable for the analysis of coding schemes.

Further Reading

This chapter gave in a very brief form an overview of the structure of TXs and RXs. Many of the aspects of digital signal processing are discussed in subsequent chapters, so we refer the reader to them for details and further references.

An important aspect for power consumption as well as the manufacturing cost of high-speed wireless devices are the ADCs and DACs. van der Plassche [2003] gives many details for implementation in Complementary Metal Oxide Semiconductor (CMOS), which is the technology preferred by the majority of manufacturers. The RF hardware, including amplifiers, mixers, and synthesizers, is discussed in Pozar [2000], Razavi [1997], and Sayre [2001]. Amplifier design is discussed in Gonzalez [1984]. Another very important topic – synchronization – is discussed in Mengali and D'Andrea [1997], Meyr and Ascheid [1990], and Meyr et al. [1997].

For updates and errata for this chapter, see wides.usc.edu/teaching/textbook

11

Modulation Formats

11.1 Introduction

Digital modulation is the mapping of data bits to signal waveforms that can be transmitted over an (analog) channel. As we saw in the previous chapter, the data that we want to transmit over the wireless propagation channel are digital – either because we really want to transmit data files or because the source coder has rendered the source information into digital form. On the other hand, the wireless propagation channel is an analog medium, over which analog waveforms have to be sent. For this reason, the digital modulator at the transmitter (TX) has to convert the digital source data to analog waveforms. At the receiver (RX), the demodulator tries to recover the bits from the received waveform. This chapter gives a brief review of digital *modulation formats*, concentrating mostly on the *results*; for references with more details see the "Further Reading" at the end of this chapter. Chapter 12 then describes optimum and suboptimum demodulators, and their performance in wireless channels.

An analog waveform can represent either one bit or a group of bits, depending on the type of modulation. The most simple modulation is binary modulation, where a $+1$-bit value is mapped to one specific waveform, while a -1-bit value is mapped to a different waveform. More generally, a group of K bits can be subsumed into a symbol, which in turn is mapped to one out of a set of $M = 2^K$ waveforms; in that case we speak of *M-ary modulation, higher order modulation, multilevel modulation*, or modulation with *alphabet size M*. In any case, different modulation formats differ in the waveforms that are transmitted, and in the way the mapping from bit groups to waveforms is achieved. Typically, the waveform corresponding to one symbol is time limited to a time T_S, and waveforms corresponding to different symbols are transmitted one after the other. Obviously, the data (bit) rate is K times the transmitted symbol rate (signaling rate).

When choosing a modulation format in a wireless system, the ultimate goal is to transmit with a certain energy as much information as possible over a channel with a certain bandwidth, while assuring a certain transmission quality (Bit Error Rate – BER). From this basic requirement, some additional criteria follow logically:

- The *spectral efficiency* of the modulation format should be as high as possible. This can best be achieved by a higher order modulation format. This allows the transmission of many data bits with each symbol.
- *Adjacent channel interference* must be small. This entails that the power spectrum of the signal should show a strong roll-off outside the desired band. Furthermore, the signal must be filtered before transmission.

- The *sensitivity with respect to noise* should be very small. This can be best achieved with a low-order modulation format, where (assuming equal average power) the difference between the waveforms of the alphabet is largest.
- *Robustness with respect to delay and Doppler dispersion* should be as large as possible. Thus, the transmit signal should be filtered as little as possible, as filtering creates delay dispersion that makes the system more sensitive to channel-induced delay dispersion.
- Waveforms should be *easy to generate* with hardware that is easy to produce and highly energy efficient. This requirement stems from the practical requirements of wireless TXs. In order to be able to use efficient class-C (or class-E and -F) amplifiers, modulation formats with constant envelopes are preferable. If, on the other hand, the modulation format is sensitive to distortions of the envelope, then the TX has to use linear (class-A or -B) amplifiers. In the former case, power efficiency can be up to 80%, while in the latter case, it is below 40%. This has important consequences for battery lifetime.

The above outline shows that some of these requirements are contradictory. There is thus no "ideal" modulation format for wireless communications. Rather, the modulation format has to be selected according to the requirements of a specific system and application.

11.2   Basics

The remainder of this chapter uses an equivalent baseband representation for the description of modulation formats in equivalent baseband. The bandpass signal – i.e., the physically existing signal – is related to the complex baseband (low-pass) representation as:[1]

$$s_{BP}(t) = \text{Re}\{s_{LP}(t)\exp[j2\pi f_c t]\} \qquad (11.1)$$

11.2.1   Pulse Amplitude Modulation

Many modulation formats can be interpreted as Pulse Amplitude Modulation (PAM) formats, where a *basis pulse* $g(t)$ is multiplied with a modulation coefficient c_i:

$$s_{LP}(t) = \sum_{i=-\infty}^{\infty} c_i g(t - iT_S) \qquad (11.2)$$

where T_S is symbol duration. This means that the transmitted analog waveform s consists of a series of time-shifted *basis pulses*, each of which is linearly weighted by a (complex) scalar that is related to the symbol we want to transmit; the modulation of the various basis pulses is independent of each other.

Different PAM formats differ in how the data bits b are mapped to the modulation coefficients c. This is analyzed in more detail in subsequent sections. For now, let us turn to the possible

---

[1] Note that definition (11.1) results in an energy of the bandpass signal that is half the energy of the baseband signal. In order to achieve equal energy, the energy must be defined as $E = ||s_{BP}||^2 = 0.5||s_{LP}||^2$. Furthermore, it is required that the impulse response of a filter $h_{\text{filter,BP}}(t)$ in the passband has the following equivalent baseband representation $h_{\text{filter,BP}}(t) = 2\text{Re}\{h_{\text{filter,LP}}(t)\exp[j2\pi f_c t]\}$, in order to assure that the output of the filter has the same normalization as the input signal. The normalization used is thus not very logical, but is used here because it is the one that is most commonly used in the literature. Alternatively, some authors (e.g., Barry et al. [2003]) define $s_{BP}(t) = \sqrt{2}\text{Re}\{s_{LP}(t)\exp[j2\pi f_c t]\}$.

shapes of the basis pulse $g(t)$ and the impact on the spectrum. We will assume that basis pulses are normalized to unit average power, so that

$$\int_{-\infty}^{\infty} |g(t)|^2 \, dt = \int_{-\infty}^{\infty} |G(f)|^2 \, df = T \tag{11.3}$$

where the second equality follows from Parseval's relation.

Rectangular Basis Pulses

The most simple basis pulse is a rectangular pulse with duration T. Figure 11.1 shows the pulse as a function of time, Figure 11.2 the corresponding spectrum:

$$\left. \begin{aligned} g_R(t, T) &= \begin{cases} 1 \ldots \text{for } 0 \leq t \leq T \\ 0 \ldots \text{otherwise} \end{cases} \\ G_R(f, T) &= \mathcal{F}\{g_R(t, T)\} = T \operatorname{sinc}(\pi f T) \exp(-j\pi f T) \end{aligned} \right\} \tag{11.4}$$

where $\operatorname{sinc}(x) = \sin(x)/x$.

Nyquist Pulse

The rectangular pulse has a spectrum that extends over a large bandwidth; the first sidelobes are only 13 dB weaker than the maximum. This leads to large adjacent channel interference, which

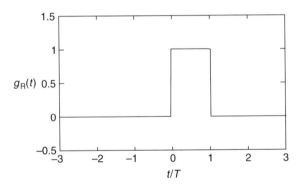

Figure 11.1   Rectangular basis pulse.

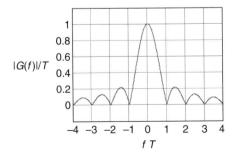

Figure 11.2   Spectrum of a rectangular basis pulse.

in turn decreases the spectral efficiency of a cellular system. It is thus often required to obtain a spectrum that shows a stronger roll-off in the frequency domain. The most common class of pulses with strong spectral roll-offs are Nyquist pulses – i.e., pulses that fulfill the Nyquist criterion, and thus do not lead to InterSymbol Interference (ISI).[2]

As an example, we present equations for a raised cosine pulse (see Figure 11.3). The spectrum of this pulse has a roll-off that follows a sinusoidal shape; the parameter determining the steepness of spectral decay is the roll-off factor α. Defining the functions,

$$
G_{N0}(f, \alpha, T) = \begin{cases} 1 & 0 \le |2\pi f| \le (1-\alpha)\dfrac{\pi}{T} \\[2ex] \dfrac{1}{2} \cdot \left(1 - \sin\left(\dfrac{T}{2\alpha}\left(|2\pi f| - \dfrac{\pi}{T}\right)\right)\right) & (1-\alpha)\dfrac{\pi}{T} \le |2\pi f| \le (1+\alpha)\dfrac{\pi}{T} \\[2ex] 0 & (1+\alpha)\dfrac{\pi}{T} \le |2\pi f| \end{cases} \tag{11.5}
$$

the spectrum of the raised cosine pulse is

$$
G_N(f, \alpha, T) = \frac{T}{\sqrt{1 - \dfrac{\alpha}{4}}} \cdot G_{N0}(f, \alpha, T)\exp(-j\pi f T_S) \tag{11.6}
$$

where the normalization factors $\frac{T}{\sqrt{1-\alpha/4}}$ follows from Eq. (11.3).

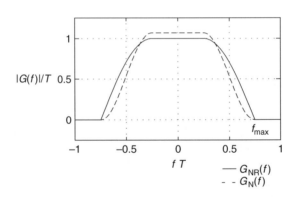

Figure 11.3 Spectrum of raised cosine pulse G_N and root-raised cosine pulse G_{NR}.

For many applications, it is the *concatenation* of the TX filter and RX filter that should result in a raised cosine shape, so that the pulse shape as observed after the RX filter fulfills the Nyquist criterion. Due to the requirements of matching the receive filter to the transmit waveform (see Chapter 12), both the TX pulse spectrum and the RX filter spectrum should be the square root of a raised cosine spectrum. Such a filter is known as a root-raised cosine filter, and henceforth denoted by subscript NR.

The impulse responses for raised cosine and root-raised cosine pulses follow from the inverse Fourier transformation (see Figure 11.4); close-form equations can be found in Chennakeshu and Saulnier [1993].

---

[2] More precisely, Nyquist pulses do not create ISIat the ideal sampling times, assuming that there are no other sources of delay dispersion in the transmission/channel/reception chain.

Figure 11.4   Raised cosine pulse and root-raised cosine pulse.

Example 11.1   *Compute the ratio of signal power to adjacent channel interference when using (i) raised cosine pulses and (ii) root-raised cosine pulses with $\alpha = 0.5$ when the two considered signals have center frequencies 0 and 1.25/T.*

The two signals overlap slightly in the roll-off region and with $\alpha = 0.5$ the desired signal has spectral components in the band $-0.75/T \leq f \leq 0.75/T$. There is interfering energy between $0.5/T \leq f \leq 0.75/T$. Without loss of generality we assume $T = 1$. In the case of raised cosine pulses and assuming a matched filter at the RX, signal energy is proportional to (dropping normalization factors occurring in both signal and interference)

$$S = 2 \int\limits_{f=0}^{0.75/T} |G_N(f, \alpha, T)|^2 \, df$$

$$= 2 \int\limits_{f=0}^{0.25} |1|^4 \, df + 2 \int\limits_{f=0.25}^{0.75} \left| \frac{1}{2} \cdot \left(1 - \sin\left(\frac{2\pi}{2\alpha} \left(|f| - \frac{1}{2} \right) \right) \right) \right|^4 \, df$$

$$= 0.77 \tag{11.7}$$

while the interfering signal energy (assuming that there is an interferer at both the lower and the upper adjacent channel) is proportional to

$$I = 2 \int\limits_{f=0.5}^{f=0.75} |G_N(f, \alpha, T) G_N(f - 1.25, \alpha, T)|^2 \, df$$

$$= 2 \int\limits_{f=0.5}^{f=0.75} \left| \frac{1}{2} \cdot \left(1 - \sin\left[2\pi \left(|f| - \frac{1}{2} \right) \right] \right) \frac{1}{2} \cdot \left(1 - \sin\left[2\pi \left(|f - 1.25| - \frac{1}{2} \right) \right] \right) \right|^2 \, df$$

$$= 0.9 \cdot 10^{-4} \tag{11.8}$$

so the signal-to-interference ratio is $10 \log_{10}\left(\dfrac{S}{I}\right) \approx 39\,\mathrm{dB}$. Using root-raised cosine pulses, the signal energy is given by

$$S = 2 \int\limits_{f=0}^{f=0.75/T} |G_\mathrm{N}(f, \alpha, T)|^2\, df = 0.875 \tag{11.9}$$

and the interfering energy is given by

$$I = 2 \int\limits_{f=0.5}^{f=0.75} |G_\mathrm{N}(f, \alpha, T) G_\mathrm{N}(f - 1.25, \alpha, T)|\, df$$

$$= 2 \int\limits_{f=0.5}^{f=0.75} \left| \left| \frac{1}{2} \cdot \left(1 - \sin\left[2\pi\left(|f| - \frac{1}{2}\right)\right]\right) \right| \left| \frac{1}{2} \cdot \left(1 - \sin\left[2\pi\left(|f - 1.25| - \frac{1}{2}\right)\right]\right)\right| \right|\, df$$

$$= 5.6 \cdot 10^{-3} \tag{11.10}$$

so the signal-to-interference ratio is $10 \log_{10}(S/I) \approx 22\,\mathrm{dB}$. Obviously, the root-raised cosine filter, which has a flatter decay in the frequency domain, leads to a worse signal-to-interference ratio.

11.2.2 Multipulse Modulation and Continuous Phase Modulation

PAM (Eq. 11.2) can be generalized to multipulse modulation, where the signal is composed of a set of basis pulses; the pulse to be transmitted depends on the modulation coefficient c_i:

$$s_\mathrm{LP}(t) = \sum_{i=-\infty}^{\infty} g_{c_i}(t - iT) \tag{11.11}$$

A typical example is M-ary Frequency Shift Keying (FSK): here the basis pulses have an offset from the carrier frequency $f_\mathrm{mod} = i\Delta f/2$, where $i = \pm 1, \pm 3, \ldots, \pm(M-1)$. It is common to choose a set of orthogonal or bi-orthogonal pulses, as this simplifies the detector.

In *Continuous Phase Frequency Shift Keying* (CPFSK), the transmit waveform at a given time depends not just on one specific symbol that we want to transmit but also on the *history* of the transmit signal. Specifically, the contributions associated with the modulation symbols follow each other in such a way that the phase of the total signal is continuous. The amplitude of the total signal is chosen as constant; the phase $\Phi(t)$ can be written as

$$\Phi_\mathrm{CPFSK}(t) = 2\pi h_\mathrm{mod} \sum_{i=-\infty}^{\infty} c_i \int_{-\infty}^{t} \tilde{g}(u - iT)\, du \tag{11.12}$$

where u is the integration variable, h_mod is the modulation index, and $\tilde{g}(t)$ is the *basis phase pulse*, which is normalized to

$$\int_{-\infty}^{\infty} \tilde{g}(t)\, dt = 1/2 \tag{11.13}$$

Note that the normalization of the basis phase pulse is fundamentally different from the basis pulse in PAM, and is not related to the energy of the signal.

Gaussian Basis Pulses

A Gaussian basis pulse is the convolution of a rectangular and a Gaussian function – in other words, the output of a filter with Gaussian impulse response that is excited by a rectangular waveform (Figure 11.5). Speaking mathematically, the rectangular waveform is given by Eq. (11.4), and the impulse response of the Gaussian filter is

$$\frac{1}{\sqrt{2\pi}\sigma_G T} \exp\left(-\frac{t^2}{2\sigma_G^2 T^2}\right) \tag{11.14}$$

where

$$\sigma_G = \frac{\sqrt{\ln(2)}}{2\pi B_G T} \tag{11.15}$$

and B_G is the 3-dB bandwidth of the Gaussian filter. The spectrum of the Gaussian filter is

$$\exp\left(-\frac{(2\pi f)^2 \sigma_G^2 T^2}{2}\right) \tag{11.16}$$

Using the normalization for phase basis pulse (11.13):

$$\tilde{g}_G(t) = \frac{1}{4T}\left[\operatorname{erfc}\left(\frac{2\pi}{\sqrt{2\ln(2)}} B_G T\left(-\frac{t}{T}\right)\right) - \operatorname{erfc}\left(\frac{2\pi}{\sqrt{2\ln(2)}} B_G T\left(1 - \frac{t}{T}\right)\right)\right] \tag{11.17}$$

where $\operatorname{erfc}(x)$ is the complementary error function:

$$\operatorname{erfc}(x) = \frac{2}{\sqrt{\pi}} \int_x^\infty \exp(-t^2)\, dt \tag{11.18}$$

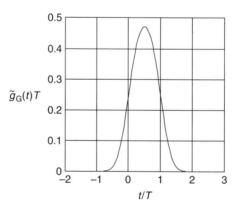

Figure 11.5 Shape of a Gaussian phase basis pulse with $B_G T = 0.5$.

11.2.3 Power Spectrum

Occupied Bandwidth

The bandwidth of the transmitted signal is a very important characteristic of a modulation format, especially in the context of a wireless signal, where spectrum is at a premium. Before going into a deeper discussion, however, we have to clarify what we mean by "bandwidth," since various definitions are possible:

- *The noise bandwidth* is defined as the bandwidth of a system with a rectangular transfer function $|H_{rect}(f)|$ (and identical peak amplitude $\max(|H(f)|)$) that receives as much noise as the system under consideration.
- The *3 dB bandwidth* is the bandwidth at which $|H(f)|^2$ has decreased to a value that is 3 dB below its maximum value.
- The 90% *energy bandwidth* is the bandwidth that contains 90% of total emitted energy; analogous definitions are possible for the 99% energy bandwidth or other percentages of the contained energy.
  Bandwidth efficiency is defined as the ratio of the data (bit) rate to the occupied bandwidth.

Power-Spectral Density of Cyclostationary Processes

A cyclostationary process $x(t)$ is defined as a stochastic process whose mean and autocorrelation functions are periodic with period T_{per}:

$$\left. \begin{array}{rcl} E\{x(t + T_{per})\} & = & E\{x(t)\} \\ R_{xx}(t + T_{per} + \tau, t + T_{per}) & = & R_{xx}(t + \tau, t) \end{array} \right\} \tag{11.19}$$

These properties are fulfilled by most modulation formats; periodicity T_{per} is the symbol duration T_S. The power-spectral density of a PAM signal can then be computed as the product of the power spectrum of a basis pulse $|S_G(f)|^2$ and the spectral density of the data σ_S^2:

$$S_{LP}(f) = \frac{1}{T_{per}} \cdot |S_G(f)|^2 \sigma_S^2(f) \tag{11.20}$$

The power-spectral density of the basis pulse is simply the squared magnitude of the Fourier transform $G(f)$ of the basis pulse $g(t)$:

$$|S_G(f)|^2 = |G(f)|^2 \tag{11.21}$$

The power-spectral density in passband is

$$S_{BP}(f) = \frac{1}{2}[S_{LP}(f - f_c) + S_{LP}(-f - f_c)] \tag{11.22}$$

We furthermore assume that the data symbols are zero-mean and uncorrelated, so that the spectral density of the data symbols is white $\sigma_S(f) = \sigma_S$.[3]

Note that CPFSK signals have memory and thus correlation between the symbols, so that the computation of power-spectral densities is much more complicated. Details can be found in Proakis [2005].

11.2.4 Signal Space Diagram

The signal space diagram represents the analog transmit symbols as vectors (points) in a finite-dimensional space. It is one of the most important tools in the analysis of modulation formats, providing a graphical representation of signals that allows an intuitive and uniform treatment of different modulation methods. In Chapter 12, we explain in more detail how the signal space

---

[3] Data that are uncoded are usually assumed to be uncorrelated. Encoding adds correlation between data (see Chapter 14). Many systems use scrambling (multiplication of the encoded data with pseudorandom data) to eliminate any correlation of the transmit symbol.

diagram can be used to compute bit error probabilities. For now, we mainly use it as a convenient shorthand and a method for representing different modulation formats.

To simplify explanations, we assume in the following rectangular basis pulses so that $g(t) = g_R(t, T)$. The signals $s(t)$ that can be present during the ith interval $iT_S < t < (i + 1)T_S$ form a finite set of functions (in the following, we assume $i = 0$ without loss of generality). The size of this set is M. This representation covers both PAM and multipulse modulation.

We now choose an orthogonal set (of size N) of expansion functions $\varphi_n(t)$.[4] The set of expansion functions should be complete, in the sense that all transmit signals can be represented as linear combinations of expansion functions. Such a complete set of expansion functions can be obtained by a Gram–Schmidt orthogonalization procedure (see, e.g., Wozencraft and Jacobs [1965]).

Given the set of expansion functions $\{\varphi_n(t)\}$, any complex (baseband) transmit signal $s_m(t)$ can be represented as a vector $\mathbf{s}_m = (s_{m,1}, s_{m,2}, \ldots, s_{m,N})$ where

$$s_{m,n} = \int_0^{T_S} s_m(t)\varphi_n^*(t)\, dt \tag{11.23}$$

where * denotes complex conjugation. The vector components of a *bandpass signal* are computed as

$$s_{BP,m,n} = \int_0^{T_S} s_{BP,m}(t)\varphi_{BP,n}(t)\, dt \tag{11.24}$$

Conversely, the actual transmit signal can be obtained from the vector components (both for passband and baseband) as

$$s_m(t) = \sum_{n=1}^{N} s_{m,n}\varphi_n(t) \tag{11.25}$$

As each signal is represented by a vector $\mathbf{s}_m$, we can plot these vectors in the so-called *signal space diagram*.[5] A graphical representation of these points is especially simple for $N = 2$, which fortunately covers most important modulation formats.

For the passband representation of PAM signals with rectangular basis pulses, the expansion functions are commonly

$$\left. \begin{aligned} \varphi_{BP,1}(t) &= \sqrt{\tfrac{2}{T_s}} \cos(2\pi f_c t) \\ \varphi_{BP,2}(t) &= \sqrt{\tfrac{2}{T_s}} \sin(2\pi f_c t) \end{aligned} \right\} \quad iT_S \le t < (i + 1)T_s; 0 \text{ otherwise} \tag{11.26}$$

Here it is assumed that $f_c \gg 1/T_S$; this implies that all products containing $\cos(2 \cdot 2\pi f_c t)$ are negligible and/or eliminated by a filter.

For example, for a binary antipodal signal (Binary Phase Shift Keying – BPSK, see Section 11.3.1), where the signal is

$$s_{BP,\frac{1}{2}} = \pm\sqrt{\frac{2E_S}{T_S}} \cos(2\pi f_c t) \tag{11.27}$$

the points in the signal space diagram are located at $\pm\sqrt{E_S}$, with E_S being the symbol energy.

---

[4] Expansion functions are often called "basis functions." However, we avoid this name in order to avoid confusion with "basis pulses." It is noteworthy that expansion functions are *by definition* orthogonal, while basis pulses can be orthogonal, but are not necessarily so.

[5] More precisely, the endpoints of the vectors starting in the origin of the coordinate system.

Generally, the *energy* contained in a symbol can be computed from the bandpass signal as

$$E_{\text{S},m} = \int_0^{T_\text{S}} s_{\text{BP},m}{}^2(t)\, dt = ||\mathbf{s}_{\text{BP},m}||^2 \tag{11.28}$$

where $||\mathbf{s}||$ denotes the L_2-norm (Euclidian norm) of $\mathbf{s}$. Note that for many modulation formats (e.g., BPSK), E_S is independent of m.

The *correlation coefficient* between $s_\text{m}(t)$ and $s_k(t)$ can be computed for the bandpass representation as

$$\text{Re}\{\rho_{k,m}\} = \frac{\mathbf{s}_{\text{BP},m}\mathbf{s}_{\text{BP},k}}{||\mathbf{s}_{\text{BP},m}||\,||\mathbf{s}_{\text{BP},k}||} \tag{11.29}$$

The squared *Euclidean distance* between the two signals is

$$d_{k,m}^2 = [E_{\text{S},m} + E_{\text{S},k} - 2\sqrt{E_{\text{S},m}E_{\text{S},k}}\text{Re}\{\rho_{k,m}\}] \tag{11.30}$$

We will see later on that this distance is important for computation of the bit error probability.

If starting from the *complex baseband representation*, the signal space diagram can be obtained with the expansion vectors:

$$\left.\begin{aligned}\varphi_1(t) &= \sqrt{\frac{1}{T_\text{S}}} \cdot 1 \\[2mm] \varphi_2(t) &= \sqrt{\frac{1}{T_\text{S}}} \cdot j\end{aligned}\right\} \quad iT_\text{S} \le t < (i+1)T_\text{S}; 0\,\text{otherwise} \tag{11.31}$$

It is important that the signal space diagrams that result from bandpass and low-pass representation differ by a factor of $\sqrt{2}$. Thus, the signal energy is given as

$$E_{\text{S},m} = \frac{1}{2}\int_0^{T_\text{S}} ||s_{\text{LP},m}(t)||^2\, dt \simeq \frac{1}{2}||\mathbf{s}_{\text{LP},m}||^2 \tag{11.32}$$

where $\mathbf{s}_{\text{LP},m}$ are the signal vectors obtained from the equivalent baseband representation.

The correlation coefficient is given as

$$\rho_{k,m} = \frac{\mathbf{s}_{\text{LP},m}(\mathbf{s}_{\text{LP},k})^*}{||\mathbf{s}_{\text{LP},m}||\,||\mathbf{s}_{\text{LP},k}||} \tag{11.33}$$

11.3 Important Modulation Formats

In this section, we summarize in a very concise form the most important properties of different digital modulation formats. We will give the following information for each modulation format:

- bandpass and baseband signal as a function of time;
- representation in terms of PAM or multipulse modulation;
- signal space diagram;
- spectral efficiency.

11.3.1 Binary Phase Shift Keying

BPSK modulation is the simplest modulation method: the carrier phase is shifted by $\pm\pi/2$, depending on whether a $+1$ or -1 is sent.[6] Despite this simplicity, two different interpretations of BPSK

---

[6] Strictly speaking, the reference phase φ also has to be given, which determines the phase at $t = 0$. Without restriction of generality, we assume in the following $\varphi_{\text{S},0} = 0$.

are possible. The first one is to see BPSK as a *phase modulation*, in which the data stream influences the phase of the transmit signal. Depending on the data bit b_i, the phase of the transmitted signal is $\pi/2$ or $-\pi/2$.

The second, and more popular interpretation, is to view BPSK as a *PAM*, where the basis pulses are rectangular pulses with amplitude 1, so that

$$s_{\mathrm{BP}}(t) = \sqrt{2E_{\mathrm{B}}/T_{\mathrm{B}}}\, p_{\mathrm{D}}(t) \cos\left(2\pi f_{\mathrm{c}} t + \frac{\pi}{2}\right)$$

where

$$p_{\mathrm{D}}(t) = \sum_{i=-\infty}^{\infty} b_i g(t - iT) = b_i * g(t) \tag{11.34}$$

and

$$g(t) = g_{\mathrm{R}}(t, T_{\mathrm{B}}) \tag{11.35}$$

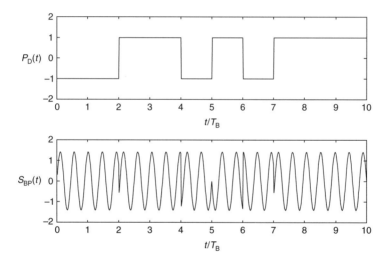

Figure 11.6 Binary phase shift keying signal as function of time.

Figure 11.6 shows the signal waveform and Figure 11.7 the signal space diagram. In equivalent baseband, the complex modulation symbols are $\pm j$:

$$c_i = j \cdot b_i \tag{11.36}$$

so that the real part of the signal is

$$\mathrm{Re}\{s_{\mathrm{LP}}(t)\} = 0 \tag{11.37}$$

and the imaginary part is

$$\mathrm{Im}\{s_{\mathrm{LP}}(t)\} = \sqrt{\frac{2E_{\mathrm{B}}}{T_{\mathrm{B}}}}\, p_{\mathrm{D}}(t) \tag{11.38}$$

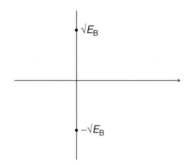

Figure 11.7   Signal space diagram for binary phase shift keying.

The envelope has constant amplitude, except at times $t = iT_B$. The spectrum shows a very slow roll-off, due to the use of unfiltered rectangular pulses as basis pulses. The bandwidth efficiency is 0.59 bit/s/Hz when we consider the bandwidth that contains 90% of the energy, but only 0.05 bit/s/Hz when considering the 99% energy bandwidth (see also Figure 11.8).

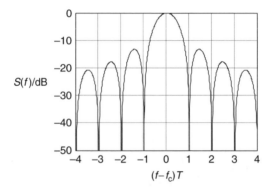

Figure 11.8   Normalized power-spectral density for binary phase shift keying.

Because of the low bandwidth efficiency of rectangular pulses, practical TXs often use Nyquist-shaped pulses as basis pulses.[7] Even the relatively mild filtering of $\alpha = 0.5$ leads to a dramatic increase in spectral efficiency: for 90% and 99% energy bandwidth, spectral efficiency becomes 1.02 and 0.79 bit/s/Hz, respectively. On the other hand, we see that the signal no longer has a constant envelope (see Figures 11.9 and 11.10).

An important variant is *Differential PSK* (DPSK). The basic idea is that the transmitted phase is not solely determined by the current symbol; rather, we transmit the phase of the previous symbol plus the phase corresponding to the current symbol. For BPSK, this reduces to a particularly simple form; we first encode the data bits according to

$$\tilde{b}_i = b_i b_{i-1} \tag{11.39}$$

and then use $\tilde{b}_i$ instead of b_i in Eq. (11.34).

---

[7] Note that there is a difference between a PAM with Nyquist-shaped basis pulses and a phase modulation with Nyquist-shaped phase pulses. In the following, we will consider only PAM with Nyquist basis pulses, and call it *Binary Amplitude Modulation* (BAM) for clarity.

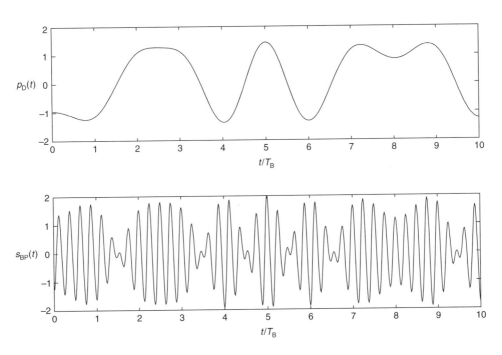

Figure 11.9   Binary amplitude modulation signal (with roll-off factor $\alpha = 0.5$) as a function of time.

Figure 11.10   Normalized power-spectral density of a binary amplitude modulation signal (with roll-off factor $\alpha = 0.5$).

The advantage of differential encoding is that it enables a differential decoder, which only needs to compare the phases of two subsequent symbols in order to demodulate received signals. This obviates the need to recover the absolute phase of the received signal, and thus allows simpler and cheaper RXs to be built.

11.3.2 Quadrature-Phase Shift Keying

A Quadrature-Phase Shift Keying (QPSK)-modulated signal is a PAM where the signal carries 1 bit per symbol interval on both the in-phase and quadrature-phase component. The original data stream is split into two streams, $b1_i$ and $b2_i$:

$$\left.\begin{array}{rcl} b1_i &=& b_{2i} \\ b2_i &=& b_{2i+1} \end{array}\right\} \tag{11.40}$$

each of which has a data rate that is half that of the original data stream:

$$R_S = 1/T_S = R_B/2 = 1/(2T_B) \tag{11.41}$$

Let us first consider the situation where basis pulses are rectangular pulses, $g(t) = g_R(t, T_S)$. Then we can give an interpretation of QPSK as either a phase modulation or as a PAM. We first define two sequences of pulses (see Figure 11.11):

$$\left.\begin{array}{rcl} p1_D(t) &=& \displaystyle\sum_{i=-\infty}^{\infty} b1_i g(t - iT_S) = b1_i * g(t) \\[3mm] p2_D(t) &=& \displaystyle\sum_{i=-\infty}^{\infty} b2_i g(t - iT_S) = b2_i * g(t) \end{array}\right\} \tag{11.42}$$

When interpreting QPSK as a *PAM*, the bandpass signal reads

$$s_{BP}(t) = \sqrt{E_B/T_B}[p1_D(t)\cos(2\pi f_c t) - p2_D(t)\sin(2\pi f_c t)] \tag{11.43}$$

Normalization is done in such a way that the energy within one symbol interval is $\int_0^{T_S} s_{BP}(t)^2 dt = 2E_B$, where E_B is the energy expended on transmission of a bit. Figure 11.12 shows the signal

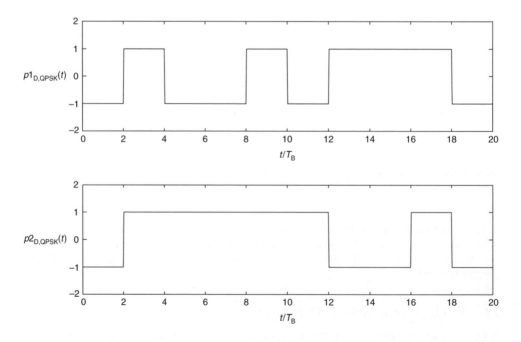

Figure 11.11 Data streams of in-phase and quadrature-phase components in quadrature-phase shift keying.

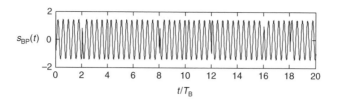

Figure 11.12   Quadrature-phase shift keying signal as a function of time.

space diagram. The baseband signal is

$$s_{LP}(t) = [p1_D(t) + jp2_D(t)]\sqrt{E_B/T_B} \qquad (11.44)$$

When interpreting QPSK as a *phase modulation*, the low-pass signal can be written as $\sqrt{2E_B/T_B} \exp(j\Phi_S(t))$ with:

$$\Phi_S(t) = \pi \cdot \left[\frac{1}{2} \cdot p2_D(t) - \frac{1}{4} \cdot p1_D(t) \cdot p2_D(t) \right] \qquad (11.45)$$

It is obvious from this representation that the signal is constant envelope, except for the transitions at $t = iT_S$ (see Figure 11.13).

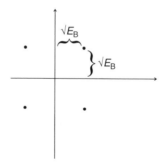

Figure 11.13   Signal space diagram of quadrature-phase shift keying.

The spectral efficiency of QPSK is twice the efficiency of BPSK, since both the in-phase and the quadrature-phase components are exploited for the transmission of information. This means that when considering the 90% energy bandwidth, the efficiency is 1.1 bit/s/Hz, while for the 99% energy bandwidth, it is 0.1 bit/s/Hz (see Figure 11.14). The slow spectral roll-off motivates (similarly to the BPSK) the use of raised cosine basis pulses (see Figure 11.15); we will refer to the resulting modulation format as quadrature amplitude modulation (QAM) in the following. The spectral efficiency increases to 2.04 and 1.58 bit/s/Hz, respectively (see Figure 11.16). On the other hand, the signal shows strong envelope fluctuations (Figure 11.17).

11.3.3   π/4-Differential Quadrature-Phase Shift Keying

Even though QPSK is nominally a constant envelope format, it has amplitude dips at bit transitions; this can also be seen by the fact that the trajectories in the I–Q diagram pass through the origin for

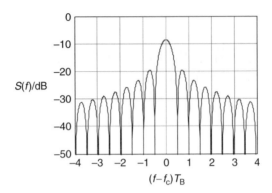

Figure 11.14   Normalized power-spectral density of quadrature-phase shift keying.

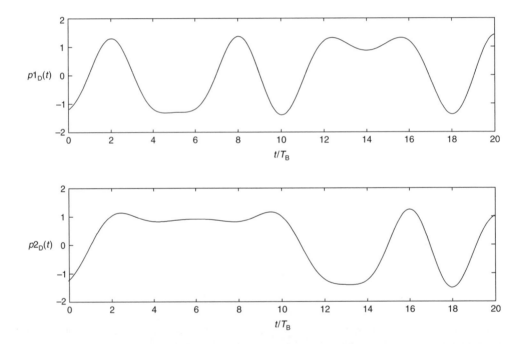

Figure 11.15   Quadrature amplitude modulation pulse sequence.

some of the bit transitions. The duration of the dips is longer when non-rectangular basis pulses are used. Such variations of the signal envelope are undesirable, because they make the design of suitable amplifiers more difficult. One possibility for reducing these problems lies in the use of $\pi/4$-DQPSK ($\pi/4$ differential quadrature-phase shift keying). This modulation format had great importance for second-generation cellphones – it was used in several American standards (IS-54, IS-136, PWT), as well as the Japanese cellphone (JDC) and cordless (PHS) standards, and the European trunk radio standard (TETRA).

The principle of $\pi/4$-DQPSK can be understood from the signal space diagram of DQPSK (see Figure 11.18). There exist *two* sets of signal constellations: (0, 90, 180, 270°) and (45, 135, 225,

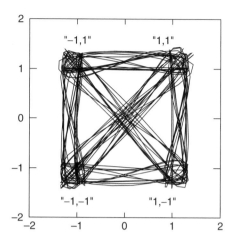

Figure 11.16 I–Q diagram of quadrature amplitude modulation with raised cosine basis pulses. Also shown are the four normalized points of the normalized signal space diagram, $(1, 1), (1, -1), (-1, -1), (-1, 1)$.

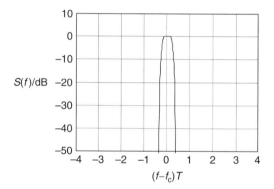

Figure 11.17 Normalized power-spectral density of quadrature amplitude modulation with raised cosine filters with $\alpha = 0.5$.

$315°$). All symbols with an even temporal index i are chosen from the first set, while all symbols with odd index are chosen from the second set. In other words: whenever t is an integer multiple of the symbol duration, the transmit phase is increased by $\pi/4$, in addition to the change of phase due to the transmit symbol. Therefore, transitions between subsequent signal constellations can never pass through the origin (see Figure 11.19); in physical terms, this means smaller fluctuations of the envelope.

The signal phase is given by

$$\Phi_s(t) = \pi \left[\frac{1}{2} p 2_\mathrm{D}(t) - \frac{1}{4} p 1_\mathrm{D}(t) p 2_\mathrm{D}(t) + \frac{1}{4} \left\lfloor \frac{t}{T_\mathrm{S}} \right\rfloor \right] \tag{11.46}$$

where $\lfloor x \rfloor$ denotes the largest integer smaller or equal to x. Comparing this with Eq. (11.45), we can clearly see the change in phase at each integer multiple of T_S. Figure 11.20 shows the underlying

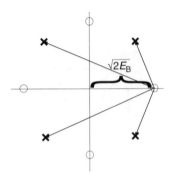

Figure 11.18   Allowed transitions in the signal space diagram of $\pi/4$ differential quadrature-phase shift keying.

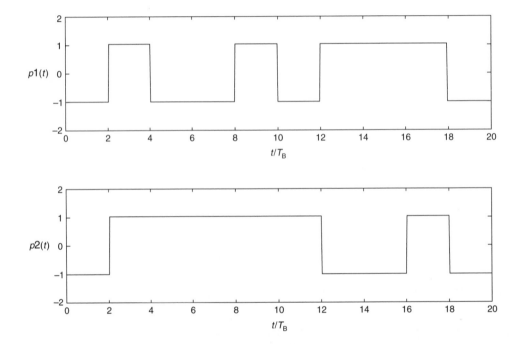

Figure 11.19   Sequence of basis pulses for $\pi/4$ differential quadrature-phase shift keying.

data sequences, and Figure 11.21 depicts the resulting bandpass signals when using rectangular or raised cosine basis pulses.

11.3.4   Offset Quadrature-Phase Shift Keying

Another way of improving the peak-to-average ratio in QPSK is to make sure that bit transitions for the in-phase and the quadrature-phase components occur at different time instants. This method is called OQPSK (offset QPSK). The bitstreams modulating the in-phase and quadrature-phase components are offset half a symbol duration with respect to each other (see Figure 11.22), so

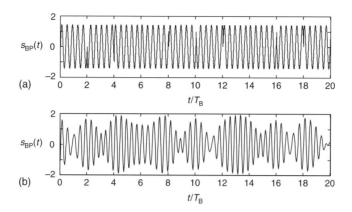

Figure 11.20 $\pi/4$ differential quadrature-phase shift keying signals as function of time for rectangular basis functions (a) and raised cosine basis pulses (b).

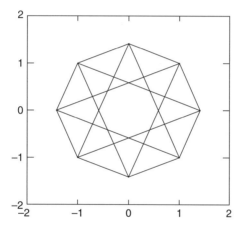

Figure 11.21 I–Q diagram of a $\pi/4$-differential quadrature-phase shift keying signal with rectangular basis functions.

that transitions for the in-phase component occur at integer multiples of the symbol duration (even integer multiples of the bit duration), while quadrature component transitions occur half a symbol duration (1-bit duration) later. Thus, the transmit pulse streams are

$$\left. \begin{array}{l} p1_{\mathrm{D}}(t) = \sum\limits_{i=-\infty}^{\infty} b1_i g(t - iT_{\mathrm{S}}) = b1_i * g(t) \\[2mm] p2_{\mathrm{D}}(t) = \sum\limits_{i=-\infty}^{\infty} b2_i g\left(t - \left(i + \tfrac{1}{2}\right)T_{\mathrm{S}}\right) = b2_i * g\left(t - \dfrac{T_{\mathrm{S}}}{2}\right) \end{array} \right\} \qquad (11.47)$$

These data streams can again be used for interpretation as PAM (Eq. 11.44) or as phase modulation, according to Eq. (11.45). The resulting bandpass signal is shown in Figure 11.23.

The representation in the I–Q diagram (Figure 11.24) makes clear that there are no transitions passing through the origin of the coordinate system; thus this modulation format takes care of envelope fluctuations as well.

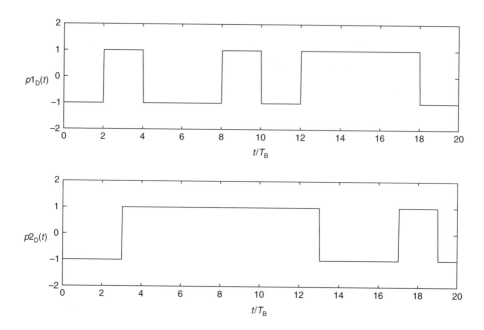

Figure 11.22   Sequence of basis pulses for offset quadrature-phase shift keying.

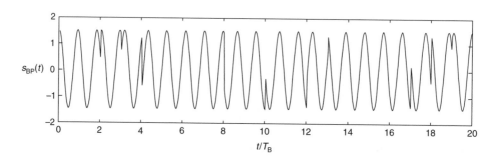

Figure 11.23   Offset quadrature-phase shift keying signal as a function of time.

As for regular QPSK, we can use smoother basis pulses, like raised cosine pulses, to improve spectral efficiency. Figure 11.25 shows the resulting I–Q diagram.

11.3.5   Higher Order Modulation

Up to now, we have treated modulation formats that transmit at most 2 bits per symbol. In this section, we mention how these schemes can be generalized to transmit more information in each symbol interval. Such schemes result in higher spectral efficiency, but consequently also in higher sensitivity to noise and interference. They were therefore not used as often in the early history of wireless communications. However, in recent times multilevel QAM has found adaptation in the standards for wireless Local Area Networks (LANs) (see Chapter 29) and fourth-generation cellular systems (Chapters 27 and 28).

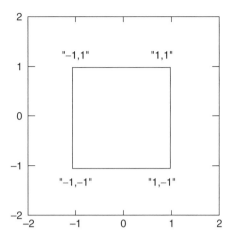

Figure 11.24 I–Q diagram for offset quadrature-phase shift keying with rectangular basis functions. Also shown are the four points of the normalized signal space diagram, $(1, 1), (1, -1), (-1, -1), (-1, 1)$.

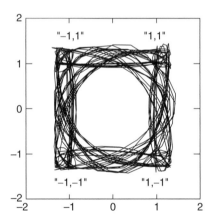

Figure 11.25 I–Q diagram for offset quadrature amplitude modulation with raised cosine basis pulses.

Higher Order QAM

Higher order QAM transmits multiple bits in both the in-phase and the quadrature-phase component. It does so by sending a signal with positive or negative polarity, *as well as multiple amplitude levels*, on each component. The mathematical representation is the same as for 4-QAM, just that in the pulse sequences of Eq. (11.42) we not only allow the levels ± 1 but $2m - 1 - \sqrt{M}$, with $m = 1, \ldots, \sqrt{M}$.

The signal space diagram for 16-QAM is shown in Figure 11.26. Larger constellations, including 64-QAM and 256-QAM, can be constructed according to similar principles. Naturally, the larger the peak-to-average ratio of the output signal, the larger the constellation is.

Example 11.2 *Relate the average energy of a 16-QAM signal to the distance between two adjacent points in the signal space diagram.*

Let us for simplicity consider the first quadrant of the signal space diagram. Assume that the signal points are located at $d + jd$, $d + j3d$, $3d + j$, and $3d + j3d$. The average energy is given by $E_s = \frac{d^2}{4}(2 + 10 + 10 + 18) = 10d^2$. The squared distance between two points is $4d^2$, so the ratio then becomes 2.5.

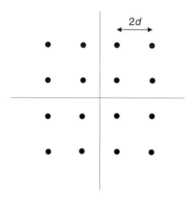

Figure 11.26   Signal space diagram for 16-QAM.

Higher Order Phase Modulation

The drawback of higher order QAM is the fact that the resulting signals show strong variations of output amplitude; this implies that linear amplifiers need to be used. An alternative is the use of higher order phase shift keying (PSK), where the transmit signal can be written as:

$$s_{\text{BP}}(t) = \sqrt{2E_S/T_S} \cos \left(2\pi f_c t + \frac{2\pi}{M}(m - 1) \right), \quad m = 1, 2, \ldots, M \qquad (11.48)$$

that is, the TX picks one of the M transmit phases (see Figure 11.27); note that we normalize energy here in terms of *symbol* energy and *symbol* duration. The equivalent low-pass signal is

$$s_{\text{LP}}(t) = \sqrt{2E_S/T_S} \exp \left(j\frac{2\pi}{M}(m - 1) \right) \qquad (11.49)$$

The correlation coefficient between two signals is

$$\rho_{km} = \exp \left(j\frac{2\pi}{M}(m - k) \right) \qquad (11.50)$$

11.3.6   Binary Frequency Shift Keying

In *FSK*, each symbol is represented by transmitting (for a time T_S) a sinusoidal signal whose frequency depends on the symbol to be transmitted. FSK cannot be represented as PAM. Rather, it is a form of multipulse modulation: depending on the bit to be transmitted, basis pulses with different center frequencies are transmitted (see also Figure 11.28):

$$g_m(t) = \cos[(2\pi f_c + b_m 2\pi f_{\text{mod}})t/T + \psi] \qquad (11.51)$$

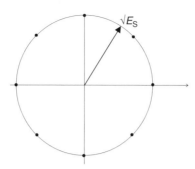

Figure 11.27   Envelope diagram of 8-PSK. Also shown are the eight points of the signal space diagram.

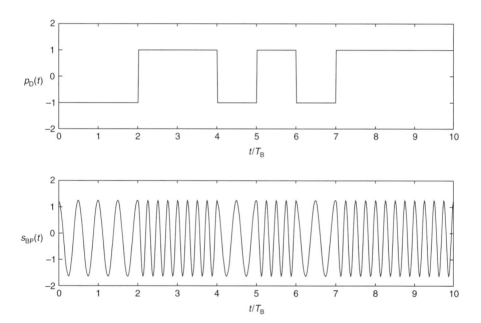

Figure 11.28   Frequency shift keying signal as a function of time.

Note that the phase of the transmit signal can jump at the bit transitions. This leads to undesirable spectral properties.

The power spectrum of FSK can be shown to consist of a continuous and a discrete (spectral lines) part:

$$S(f) = S_{\text{cont}}(f) + S_{\text{disc}}(f) \tag{11.52}$$

where (see Benedetto and Biglieri [1999]):

$$S_{\text{cont}}(f) = \frac{1}{2T} \left\{ \sum_{m=1}^{2} |G_m(f)|^2 - \frac{1}{2} \left| \sum_{m=1}^{2} G_m(f) \right|^2 \right\} \tag{11.53}$$

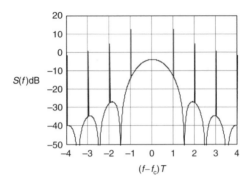

Figure 11.29 Power-spectral density of (noncontinuous phase) frequency shift keying with $h_{mod} = 1$.

and

$$S_{\text{disc}}(f) = \frac{1}{(2T)^2} \left| \sum_{m=1}^{2} G_m(f) \right|^2 \sum_{n} \delta \left(f - \frac{n}{T} \right) \tag{11.54}$$

where $G_m(f)$ is the Fourier transform of $g_m(t)$. An example is shown in Figure 11.29.

CPFSK enforces a smooth change of the phase at the bit transitions. The phase of the transmission signal is chosen as

$$\Phi_S(t) = 2\pi h_{\text{mod}} \int_{-\infty}^{t} \tilde{p}_{D,\text{FSK}}(\tau) \, d\tau \tag{11.55}$$

The resulting signal has a constant envelope, where h_{mod} is the modulation index. Using the normalization for phase pulses (Eq. 11.13) and assuming rectangular phase basis pulses:

$$\tilde{g}_{\text{FSK}}(t) = \frac{1}{2T_{\text{B}}} g_R(t, T_{\text{B}}) \tag{11.56}$$

the phase pulse sequence is

$$\tilde{p}_{D,\text{FSK}}(t) = \sum_{i=-\infty}^{\infty} b_i \tilde{g}_{\text{FSK}}(t - iT_{\text{B}}) = b_i * \tilde{g}_{\text{FSK}}(t) \tag{11.57}$$

The instantaneous frequency is given as

$$f(t) = f_c + b_i f_{\text{mod}}(t) = f_c + f_D(t) = f_c + \frac{1}{2\pi} \frac{d\Phi_S(t)}{dt} \tag{11.58}$$

The real and imaginary parts of the equivalent baseband signal are then

$$\text{Re}(s_{\text{LP}}(t)) = \sqrt{2E_B/T_B} \cos \left[2\pi h_{\text{mod}} \int_{-\infty}^{t} \tilde{p}_{D,\text{FSK}}(\tau) \, d\tau \right] \tag{11.59}$$

$$\text{Im}(s_{\text{LP}}(t)) = \sqrt{2E_B/T_B} \sin \left[2\pi h_{\text{mod}} \int_{-\infty}^{t} \tilde{p}_{D,\text{FSK}}(\tau) \, d\tau \right] \tag{11.60}$$

The resulting signal has a *memory*, as the signal at time t depends on all previously transmitted bits.

For the signal space diagram, we can find two different representations:

1. Use sinusoidal oscillations at the two possible signal frequencies $f_c \pm f_{mod}$. Expressing this in terms of phase pulses, the expansion functions (in the passband) read

$$\left.\begin{array}{l} \varphi_{BP,1}(t) = \sqrt{2/T_B} \cos(2\pi f_c t + 2\pi f_{mod} t) \\ \varphi_{BP,2}(t) = \sqrt{2/T_B} \cos(2\pi f_c t - 2\pi f_{mod} t) \end{array}\right\} \tag{11.61}$$

In this case, the signal space diagram consists of two points on the two orthogonal axes (see Figure 11.30). Note that we have made the implicit assumption that the two signals $\varphi_{BP,1}(t)$ and $\varphi_{BP,2}(t)$ are orthogonal to each other.
2. Use the in-phase and quadrature-phase component of the center frequency f_c. In this case, the signal shows up in the envelope diagram not as a discrete point but as a continuous trajectory – namely, a circle (see Figure 11.31). At any time instant, the transmit signal is represented by a different point in the diagram.

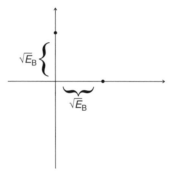

Figure 11.30   Signal space diagram of frequency shift keying when using $\cos[2\pi (f_c \pm f_D)t]$ as basis functions.

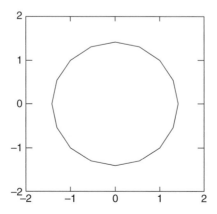

Figure 11.31   I–Q diagram of frequency shift keying. This is equivalent to a signal space diagram for FSK when using $\cos(2\pi f_c t)$ and $\sin(2\pi f_c t)$ as basis functions.

11.3.7 Minimum Shift Keying

Minimum Shift Keying (MSK) is one of the most important modulation formats for wireless communications. However, it can be interpreted in different ways, which leads to considerable confusion:

1. The first interpretation is as CPFSK with a modulation index:

$$h_{\mathrm{mod}} = 0.5, \quad f_{\mathrm{mod}} = 1/4T \tag{11.62}$$

   This implies that the phase changes by $\pm\pi/2$ during a 1-bit duration (see Figure 11.32). The bandpass signal is shown in Figure 11.33.

2. Alternatively, we can interpret MSK as *Offset QAM* (OQAM) with basis pulses that are sinusoidal half-waves extending over a duration of $2T_{\mathrm{B}}$ (see also Figure 11.34):

$$g(t) = \sin(2\pi f_{\mathrm{mod}}(t + T_{\mathrm{B}}))g_{\mathrm{R}}(t, 2T_{\mathrm{B}}) \tag{11.63}$$

   For proof, see Appendix 11.A at *www.wiley.com/go/molisch*.

   Due to the use of smoother basis functions, the spectrum decreases faster than that of "regular" OQPSK:

$$S(f) = \frac{16T_{\mathrm{B}}}{\pi^2}\left(\frac{\cos(2\pi f T_{\mathrm{B}})}{1 - 16f^2T_{\mathrm{B}}^2}\right)^2 \tag{11.64}$$

see also Figure 11.35. On the other hand, MSK is only a binary modulation format, while OQPSK transmits 2 bits per symbol duration. As a consequence, MSK has lower spectral efficiency when considering the 90% energy bandwidth (1.29 bit/s/H$_z$), but still performs reasonably well when considering the 99% energy bandwidth (0.85 bit/s/H$_z$).

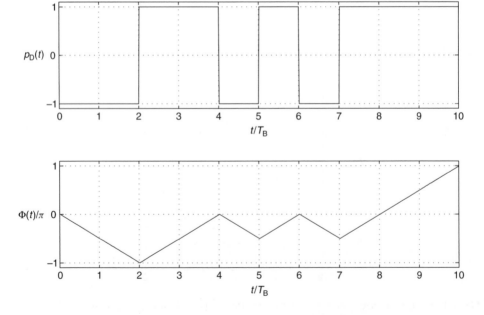

Figure 11.32   Phase pulse and phase as function of time for minimum shift keying signal.

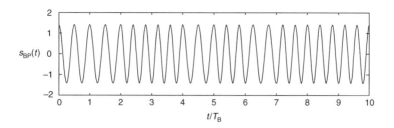

Figure 11.33 Minimum shift keying modulated signal.

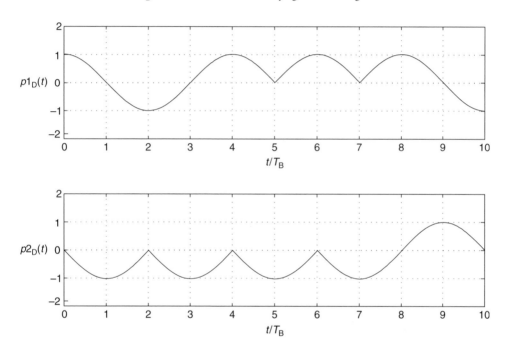

Figure 11.34 Composition of minimum shift keying from sinusoidal half-waves.

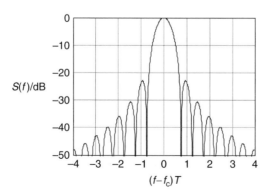

Figure 11.35 Power-spectral density of minimum shift keying.

Example 11.3 *Compare the spectral efficiency of MSK and QPSK with rectangular constituent pulses. Consider systems with equal bit duration. Compute the out-of-band energy at $1/T_B$, $2/T_B$, and $3/T_B$.*

The power-spectral density of MSK is given by

$$S_{\text{MSK}}(f) = \frac{16T_B}{\pi^2}\left(\frac{\cos(2\pi f T_B)}{1 - 16f^2 T_B^2}\right)^2 \tag{11.65}$$

whereas, the power-spectral density for QPSK with rectangular pulses is the same as for ordinary QAM given by (note that we normalize such that the integral over the power-spectral density becomes unity):

$$S_{\text{OQPSK}}(f) = (1/T_S)(T_S \text{ sinc}(\pi f T_S))^2 \tag{11.66}$$

where it must be noted that $T_S = 2T_B$ for QPSK. The out-of-band power is, for MSK and $T_B = 1$, given by

$$P_{\text{out}}(f_0) = 2\int\limits_{f=f_0}^{\infty} S(f)\,df = 2\int\limits_{1}^{\infty}\frac{16}{\pi^2}\left(\frac{\cos(2\pi f)}{1 - 16f^2}\right)^2 df$$

$$= \frac{32}{\pi^2}\int_1^{\infty}\frac{\cos^2 2\pi f}{256f^4 - 32f^2 + 1}\,df = \frac{32}{\pi^2}\int_1^{\infty}\frac{\frac{1}{2}\cos 4\pi f + \frac{1}{2}}{256f^4 - 32f^2 + 1}\,df \tag{11.67}$$

and for QPSK with $T_B = 1$ given by

$$P_{\text{out}}(f_0) = \int_1^{\infty}\left(2\frac{\sin(2\pi f)}{(2\pi f)}\right)^2 df \tag{11.68}$$

Solving these integrals numerically gives the following table:

	$1/T_B$	$2/T_B$	$3/T_B$
QPSK	0.050	0.025	0.017
MSK	0.0024	$2.8 * 10^{-4}$	$7.7 * 10^{-5}$

11.3.8 Demodulation of Minimum Shift Keying

The different interpretations of MSK are not just useful for gaining insights into the modulation scheme but also for building demodulators. Different demodulator structures correspond to different interpretations:

- *Frequency discriminator detection*: since MSK is a type of FSK, it is straightforward to check whether the instantaneous frequency is larger or smaller than the carrier frequency (larger or smaller than 0 when considering equivalent baseband). The instantaneous frequency can be sampled in the middle of the bit, or it can be integrated over (part of the) bit duration in order to reduce the effect of noise. This RX structure is simple, but suboptimum, since it does not exploit the continuity of the phase at bit transitions.
- *Differential detection*: the phase of the signal changes by $+\pi/2$ or $-\pi/2$ over a 1-bit duration, depending on the bit that was transmitted. An RX thus just needs to determine the phases at

times iT and $(i + 1)T$, in order to make a decision. It is remarkable that no differential encoding of the transmit signal is required; an erroneous estimate of the phase at one sampling time leads to two (but not more) bit errors.

- *Matched filter reception*: it is well known that matched filter reception is optimum (see also Chapter 12). This is true both when considering MSK as OQPSK, and when considering it as multipulse modulation. However, it has to be noted that MSK is a modulation format with memory. Thus, bit-by-bit detection is suboptimum: consider the signal space diagram: four constellation points (at $0°$, $90°$, $180°$, and $270°$) are possible. For a bit-by-bit decision, the decision boundaries are thus the first and second main diagonals. The distance between the signal constellations and the decision boundary are $\sqrt{E}/\sqrt{2}$, and thus worse by 3 dB compared with BPSK. However, such a decision method has thrown away the information arising from the memory of the system: if the previous constellation point had been at $0°$, the subsequent signal constellations can only be at either $90°$ or $270°$. The decision boundary should thus be the x-axis, and the distance from the signal constellations to the decision boundary is $\sqrt{E}$ – i.e., equal to BPSK. Memory can be exploited, e.g., by a maximum-likelihood sequence estimation,[8] (compare Section 14.3).

11.3.9   Gaussian Minimum Shift Keying

GMSK (Gaussian MSK) is CPFSK with modulation index $h_{\mathrm{mod}} = 0.5$ and *Gaussian* phase basis pulses:

$$\tilde{g}(t) = g_{\mathrm{G}}(t, T_{\mathrm{B}}, B_{\mathrm{G}}T) \tag{11.69}$$

Thus the sequence of transmit phase pulses is

$$p_{\mathrm{D}}(t) = \sum_{i=-\infty}^{\infty} b_i \tilde{g}(t - iT_{\mathrm{B}}) = b_i * \tilde{g}(t) \tag{11.70}$$

(see Figure 11.36). The spectrum is shown in Figure 11.37. We see that GMSK achieves better spectral efficiency than MSK because it uses the smoother Gaussian phase basis pulses as opposed to the rectangular ones of MSK.

GMSK is the modulation format most widely used in Europe. It is applied in the cellular Global System for Mobile communications (GSM) standard (with $B_{\mathrm{G}}T = 0.3$) and the cordless standard Digital Enhanced Cordless Telecommunications (DECT) (with $B_{\mathrm{G}}T = 0.5$) (see Chapter 24 and the Appendix on DECT, respectively). It is also used in the Bluetooth (IEEE 802.15.1) standard for wireless personal area networks.[9]

It is noteworthy that GMSK *cannot* be interpreted as PAM. However, Laurent [1986] derived equations that allow the interpretation of GMSK as PAM with finite memory.

11.3.10   Pulse Position Modulation

Pulse Position Modulation (PPM) is another form of multipulse modulation. Remember that for FSK we used pulses that had different center frequencies as basis pulses. For PPM, we use pulses

---

[8] It can be shown that for a maximum-likelihood sequence estimation, only three sampling values can influence the decision for a specific bit (see Benedetto and Biglieri [1999]).

[9] Strictly speaking, DECT and Bluetooth use GFSK with modulation index 0.5, which is equivalent to GMSK. However, deviations from the nominal modulation index are tolerated.

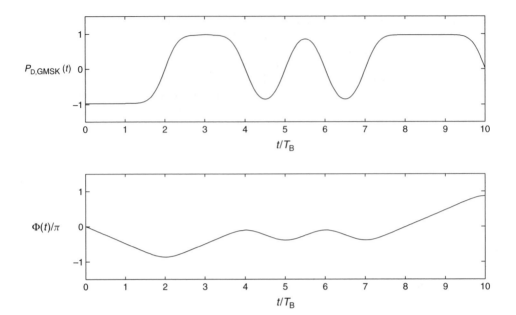

Figure 11.36 Pulse sequence and phase of Gaussian minimum shift keying signal.

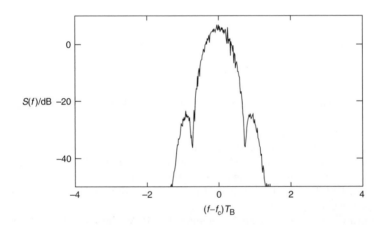

Figure 11.37 Gaussian minimum shift keying power-spectral density (from simulations).

that have different delays. In the following, we will consider M-ary PPM:

$$s_{\mathrm{LP}}(t) = \sqrt{\frac{2E_{\mathrm{B}}}{T_{\mathrm{B}}}} \sum_{i=-\infty}^{\infty} g_{c_i}(t - iT) \tag{11.71}$$

$$= \sqrt{\frac{2E_{\mathrm{B}}}{T_{\mathrm{B}}}} \sum_i g_{\mathrm{PPM}}(t - iT - c_i T_{\mathrm{d}}) \tag{11.72}$$

where T_d is the modulation delay (see Figure 11.38). Note that the modulation symbols are directly mapped to the delay of the pulses, and are thus real.

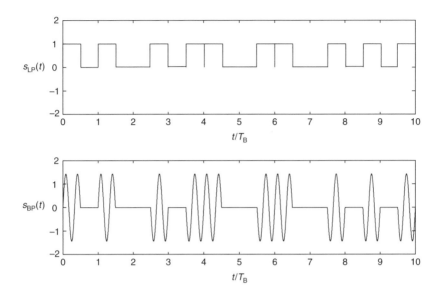

Figure 11.38   Temporal signal in the low-pass and bandpass domain for pulse position modulation.

The envelope shows strong fluctuations; however, it is still possible to use nonlinear amplifiers, since only two output amplitude levels are allowed.

Because PPM is a nonlinear modulation, the spectrum is not given by Eq. (11.20), but rather by the more complicated expression for multipulse modulation (see also Eq. 11.52):

$$
S(f) = \frac{1}{M^2 T_S^2} \sum_{i=-\infty}^{+\infty} \left(\left| \sum_{m=0}^{M-1} G_m(f) \right|^2 \delta\left(f - \frac{i}{T_S} \right) \right) + \frac{1}{T_S} \left(\sum_{m=0}^{M-1} \frac{1}{M} |G_m(f)|^2 - \left| \sum_{m=0}^{M-1} \frac{1}{M} G_m(f) \right|^2 \right)
$$

(11.73)

We note that the spectrum has a number of lines.

Up to now, we have assumed that the transmitted pulses are rectangular pulses, and offset with respect to each other by one pulse width. However, this is not spectrally efficient, as the pulses have a very broad spectrum. Other pulse shapes can be used as well. The performance of these different pulses depends on the detection method. When coherent detection is employed, then the key quantity determining performance is the correlation between the pulses representing $+1$ and -1 (assuming binary PPM):

$$
\rho = \frac{\int g(t) g^*(t - T_d)\, dt}{\int |g(t)|^2\, dt}
$$

(11.74)

If $\rho = 0$, modulation is orthogonal as is the case for rectangular pulses if $T_d > T_B$. It is actually possible to choose pulse shapes so that $\rho < 0$. In that case, we have not only better spectral efficiency but also better performance. However, when the RX uses *incoherent* (energy) detection, then it is best if the pulses do not have any overlap. The relevant correlation coefficient in this case is

$$\rho_{\text{env}} = \frac{\int |g(t)||g(t-T_{\text{d}})|\, dt}{\int |g(t)|^2\, dt} \tag{11.75}$$

PPM can be combined with other modulation formats. For example, binary PPM can be combined with BPSK, so that each symbol represents 2 bits −1 bit determined by the phase of the transmitted pulse, and 1 bit by its position.

Example 11.4 *Consider a PPM system with $g(t) = \sin(t/T)/t$. What is the correlation coefficient ρ and the correlation coefficient of the envelopes when T_{d} is (i) T, (ii) 5T?*

The correlation coefficient is given by

$$\rho = \frac{\int g(t)g(t-T_{\text{d}})\, dt}{\int |g(t)|^2\, dt} \tag{11.76}$$

Assuming $T = 1$, the denominator is

$$\int |g(t)|^2\, dt = \int_{-\infty}^{\infty} \frac{\sin^2(t)}{t^2}\, dt = \pi \tag{11.77}$$

For $T_{\text{d}} = 1$, using numerical integration, the numerator becomes

$$\int g(t)g(t-T_{\text{d}})\, dt = \int_{-\infty}^{\infty} \frac{\sin(t)}{t}\frac{\sin(t-1)}{t-1}\, dt \approx 2.63 \tag{11.78}$$

so that the correlation coefficient becomes $\rho \approx 0.84$. For $T_{\text{d}} = 5$, using numerical integration, the numerator becomes

$$\int g(t)g(t-T_{\text{d}})\, dt = \int_{-\infty}^{\infty} \frac{\sin(t)}{t}\frac{\sin(t-5)}{t-5}\, dt \approx -0.61 \tag{11.79}$$

so that the correlation coefficient becomes $\rho \approx -0.2$. For $T_{\text{d}} = 1$, the numerator for the envelope correlation is given by

$$\rho \int_{-\infty}^{\infty} \left|\frac{\sin(t)}{t}\right| \left|\frac{\sin(t-1)}{t-1}\right|\, dt \approx 2.73 \tag{11.80}$$

so that the envelope correlation becomes $\rho \approx 0.87$. For $T_{\text{d}} = 5$, the numerator for the envelop correlation is given by

$$\rho \int_{-\infty}^{\infty} \left|\frac{\sin(t)}{t}\right| \left|\frac{\sin(t-5)}{t-5}\right|\, dt \approx 1.04 \tag{11.81}$$

so that the envelope correlation become $\rho \approx 0.33$.

PPM is not used very often for wireless systems. This is due to its relatively low spectral efficiency, as well as to the effect of delay dispersion on a PPM system. Still, there are some emerging applications where PPM is used, especially impulse radio (see Section 18.5).

11.3.11 Summary of Spectral Efficiencies

Table 11.1 summarizes the spectral efficiencies of different modulation methods, assuming that the "occupied bandwidth" is defined as the bandwidth that contains 90 or 99%, respectively, of the overall energy.

Table 11.1 Spectral efficiency for different modulation schemes

Modulation method	Spectral efficiency for 90% of total energy (bit/s/Hz)	Spectral efficiency for 99% of total energy (bit/s/Hz)
BPSK	0.59	0.05
BAM ($\alpha = 0.5$)	1.02	0.79
QPSK, OQPSK	1.18	0.10
MSK	1.29	0.85
GMSK ($B_G = 0.5$)	1.45	0.97
QAM ($\alpha = 0.5$)	2.04	1.58

11.4 Appendix

Please go to *www.wiley.com/go/molisch*

Further Reading

The description of different modulation formats, and their representation in a signal space diagram, can be found in any of the numerous textbooks on digital communications – e.g., Anderson [2005], Barry et al. [2003], Proakis [2005], Sklar [2001], Wilson [1996], and Xiong [2006]. A book specifically dedicated to modulation for wireless communications is Burr [2001]. QAM is described in Hanzo et al. [2000]. A description of the multiple interpretations of MSK, and the resulting demodulation structures, can be found in Benedetto and Biglieri [1999]. GMSK was invented by Murota and Hirade [1981]. An authoritative description of different forms of continuous-phase modulation is Anderson et al. [1986].

For updates and errata for this chapter, see wides.usc.edu/teaching/textbook

12

Demodulation

In this chapter, we describe how to demodulate the received signal, and discuss the performance of different demodulation schemes in *Additive White Gaussian Noise* (AWGN) channels as well as flat-fading channels and dispersive channels.

12.1 Demodulator Structure and Error Probability in Additive White Gaussian Noise Channels

This section deals with basic demodulator structures for the modulation formats described in Chapter 11, and the computation of the Bit Error Rate (BER) and Symbol Error Rate (SER) in an AWGN channel. As in the previous chapter, we concentrate on the most important results; for more detailed derivations, we refer the reader to the monographs of Anderson [2005], Benedetto and Biglieri [1999], Proakis [2005], and Sklar [2001].

12.1.1 Model for Channel and Noise

The AWGN channel attenuates the transmit signal, causes phase rotation, and adds Gaussian-distributed noise. Attenuation and phase rotation are temporally constant, and are thus easily taken into account. Thus, the received signal (in complex baseband notation) is given by

$$r_{\text{LP}}(t) = \alpha s_{\text{LP}}(t) + n_{\text{LP}}(t) \tag{12.1}$$

where α is the (complex) channel gain $|\alpha| \exp(j\varphi)$, and $n(t)$ is a (complex) Gaussian noise process.

In order to derive the properties of noise, let us first consider noise in the bandpass system. We assume that over the bandwidth of interest the noise power-spectral density is constant. The value of the *two-sided* noise power-spectral density is $N_0/2$ (see Figure 12.1a). The (complex) equivalent low-pass noise has a power-spectral density (see Figure 12.1b):

$$S_{\text{n,LP}}(f) = \begin{cases} N_0 & |f| \leq B/2 \\ 0 & \text{otherwise} \end{cases} \tag{12.2}$$

Note that $S_{\text{n,LP}}(f)$ is symmetric with respect to f – i.e., $S_{\text{n,LP}}(f) = S_{\text{n,LP}}(-f)$.

When considering noise in the time domain, we find that it is described by its autocorrelation function $R_{\text{LP},nn}(\tau) = (1/2)E\{n_{\text{LP}}^*(t)n_{\text{LP}}(t + \tau)\}$.[1] For a system limited to the band $[-B/2, B/2]$,

---

[1] Note that this definition of the AutoCorrelation Function (ACF) differs by a factor of 2 from the one used in Chapter 6. The reason for using this modified definition here is that it is commonly used in communications theory for BER computations.

Wireless Communications, Second Edition   Andreas F. Molisch
© 2011 John Wiley & Sons, Ltd

Bandpass noise

Equivalent low-pass noise

Figure 12.1 Bandpass noise and equivalent low-pass noise power spectrum.

the ACF is

$$R_{\text{LP},nn}(\tau) = N_0 \frac{\sin(\pi B \tau)}{\pi \tau} \tag{12.3}$$

which becomes

$$R_{\text{LP},nn}(\tau) = N_0 \delta(\tau) \tag{12.4}$$

as the bandwidth B tends to infinity. We also note that the correlations of the in-phase and the quadrature-phase components of noise, respectively, are both $R_{\text{LP},nn}(\tau)$, while the cross-correlation between in-phase and quadrature-phase noise is zero.

The ACF of the bandpass signal is

$$R_{\text{BP},nn}(\tau) = \text{Re}\{R_{\text{LP},nn}(\tau) \exp(j 2\pi f_c \tau)\} \tag{12.5}$$

12.1.2   Signal Space Diagram and Optimum Receivers

We now derive the structure of the optimum receivers for digital modulation. In the process, we will also find an additional motivation for using signal space diagrams. For these derivations we make the following assumptions:

1. All transmit symbols are equally likely.
2. The modulation format does not have memory.
3. The channel is an AWGN channel, and both absolute channel gain and phase rotation are completely known. Without loss of generality, we assume henceforth that phase rotation has been compensated completely, so that the channel attenuation is real, so that $\alpha = |\alpha|$.

The ideal detector, called the *Maximum A Posteriori* (MAP) detector, aims to answer the following question: "If a signal $r(t)$ was received, then which symbol $s_m(t)$ was most likely transmitted?" In other words, which symbol maximizes m?:

$$\Pr[s_m(t)|r(t)] \tag{12.6}$$

Bayes' rules can be used to write this as (in other words, find the symbol m that achieves)

$$\max_m \Pr[n(t) = r(t) - \alpha s_m(t)] \Pr[s_m(t)] \tag{12.7}$$

Since we assume that all symbols are equiprobable, the MAP detector becomes identical to the *Maximum Likelihood* (ML) detector:

$$\max_m \Pr[n(t) = r(t) - \alpha s_m(t)] \tag{12.8}$$

In Chapter 11, we introduced the signal spaced diagram for the representation of modulated transmit signals. There, we treated the signal space diagram just as a convenient shorthand. Now we will show how a transmit and received signal can be related, and how the signal space diagram can be used to derive optimum receiver structures. In particular, we will derive that the ML detector finds in the signal space diagram the transmit symbol that has the smallest Euclidean distance to the receive signal.

Remember that the transmit signal can be represented in the form:[2]

$$s_m(t) = \sum_{n=1}^{N} s_{m,n} \varphi_n(t) \tag{12.9}$$

where

$$s_{m,n} = \int_0^{T_S} s_m(t) \varphi_n^*(t) \, dt \tag{12.10}$$

Now we find that the received signal can be represented by a similar expansion. Using the same expansion functions $\varphi_n(t)$ we obtain

$$r(t) = \sum_{n=1}^{\infty} r_n \varphi_n(t) \tag{12.11}$$

where

$$r_n = \int_0^{T_S} r(t) \varphi_n^*(t) \, dt \tag{12.12}$$

Since the received signal contains noise, it seems at first glance that we need more terms in the expansion – infinitely many, to be exact. However, we find it useful to split the series into two parts:

$$r(t) = \sum_{n=1}^{N} r_n \varphi_n(t) + \sum_{n=N+1}^{\infty} r_n \varphi_n(t) \tag{12.13}$$

and similarly:

$$n(t) = \sum_{n=1}^{N} n_n \varphi_n(t) + \sum_{n=N+1}^{\infty} n_n \varphi_n(t) \tag{12.14}$$

Using these expansions, the expression that the ML receiver aims to maximize can be written as

$$\max_m \Pr[\mathbf{n} = \mathbf{r} - \alpha \mathbf{s}_m] \tag{12.15}$$

---

[2] Note that these equations are valid for both the baseband and the bandpass representation – it is just a matter of inserting the correct expansion functions.

where the received signal vector $\mathbf{r}$ is simply $\mathbf{r} = (r_1, r_2, \ldots)^T$, and similarly for $\mathbf{n}$. Since the noise components are independent, the probability density function (pdf) of the received vector $\mathbf{r}$, assuming that $\mathbf{s}_m$ was transmitted, is given as

$$p(\mathbf{r}_{\text{LP}}|\alpha\mathbf{s}_{\text{LP},m}) \propto \exp\left\{-\frac{1}{2N_0}\|\mathbf{r} - \alpha\mathbf{s}_m\|^2\right\} \tag{12.16}$$

$$= \prod_{n=1}^{\infty} \exp\left\{-\frac{1}{2N_0}(r_n - \alpha s_{m,n})^2\right\} \tag{12.17}$$

Since $s_{m,n}$ is nonzero only for $n \leq N$, the ML detector aims to find:

$$\max_m \prod_{n=1}^{N} \exp\left\{-\frac{1}{2N_0}(r_n - \alpha s_{m,n})^2\right\} \prod_{n=N+1}^{\infty} \exp\left\{-\frac{1}{2N_0}(r_n)^2\right\} \tag{12.18}$$

A key realization is now that the second product in Eq. (12.18) is independent of s_m, and thus does not influence the decision. This is another way of saying that components of the noise (and thus of the received signal) that do not lie in the signal space of the transmit signal are irrelevant for the decision of the detector (Wozencraft's irrelevance theorem). Finally, as exp(.)is a monotonic function, we find that it is sufficient to minimize the metric:

$$\mu(\mathbf{s}_{\text{LP},m}) = \|\mathbf{r}_{\text{LP}} - \alpha\mathbf{s}_{\text{LP},m}\|^2 \tag{12.19}$$

Geometrically, this means that the ML receiver decides for symbol m which transmit vector $\mathbf{s}_{\text{LP},m}$ has the smallest Euclidean distance to the received vector $\mathbf{r}_{\text{LP}}$. We need to keep in mind that this is an optimum detection method only in memoryless, uncoded systems. Vector $\mathbf{r}$ contains "soft" information – i.e., how sure the receiver is about its decision. This soft information, which is lost in the decision process of finding the nearest neighbor, is irrelevant when one bit does not tell us anything about any other bit. However, for coded systems and systems with memory (Chapters 14 and 16), this information can be very helpful.

The metric can be rewritten as

$$\mu(\mathbf{s}_{\text{LP},m}) = \|\mathbf{r}_{\text{LP}}\|^2 + \|\alpha\mathbf{s}_{\text{LP},m}\|^2 - 2\alpha\text{Re}\{\mathbf{r}_{\text{LP}}\mathbf{s}_{\text{LP},m}^*\} \tag{12.20}$$

Since the term $\|\mathbf{r}_{\text{LP}}\|^2$ is independent of the considered $\mathbf{s}_{\text{LP},m}$, minimizing the metric is equivalent to maximizing:

$$\text{Re}\{\mathbf{r}_{\text{LP}}\mathbf{s}_{\text{LP},m}^*\} - \alpha E_m \tag{12.21}$$

(remember that $E_m = \|\mathbf{s}_{\text{LP},m}\|^2/2$, see Chapter 11).

One important consequence of this decision rule is that the receiver has to know the value of channel gain α. This can be difficult in wireless systems, especially if channel properties quickly change (see Part II). Thus, modulation and detection methods that do not require this information are preferred. Specifically, the magnitude of channel gain does not need to be known if all transmit signals have equal energy, $E_m = E$. The phase rotation of the channel (argument of alpha) can be ignored if either incoherent detection or differential detection is used.

The beauty of the above derivation is that it is independent of the actual modulation scheme. The transmit signal is represented in the signal space diagram, which gives all the *relevant* information. The receiver structure of Figure 12.2 is then valid for optimum reception for any modulation alphabet represented in a signal space diagram. The only prerequisite is that the conditions mentioned at the beginning of this chapter are fulfilled. This receiver can be interpreted as a correlator, or as a matched filter, matched to the different possible transmit waveforms.

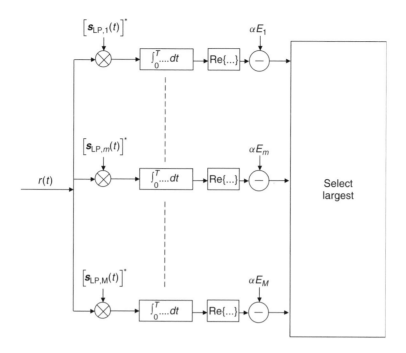

Figure 12.2 Structure of a generic optimum receiver.

In the following, we will find that the Euclidean distance of two points in the signal space diagram is a vital parameter for modulation formats. We find that in the bandpass representation for equal-energy signals, which we assumed in Chapter 11, the Euclidean distance is related to the energy of the signal component as

$$d_{12}^2 = 2E(1 - \text{Re}\{\rho_{12}\}) \tag{12.22}$$

where ρ_{jk} is the correlation coefficient as defined in Chapter 11:

$$\text{Re}\{\rho_{k,m}\} = \frac{\mathbf{s}_k \mathbf{s}_m}{|\mathbf{s}_k||\mathbf{s}_m|} \tag{12.23}$$

12.1.3 Methods for the Computation of Error Probability

In this subsection, we discuss how to compute the performance that can be achieved with optimum receivers. Before going into details, let us define some important expressions: the *bit error rate* is, as its name says, a rate. Thus, it describes the number of bit errors per unit time, and has dimension s^{-1}. The *bit error ratio* is the number of errors divided by the number of transmitted bits; it thus is dimensionless. For the case when infinitely many bits are transmitted, this becomes the *bit error probability*. In the literature, there is a tendency to mix up these three expressions; specifically, the expression BER is often used when the authors mean bit error probability. This misnomer has become so common that in the following we will use it as well.

Though we have treated a multitude of modulation formats in the previous chapter, they can all be classified easily with the framework of signal space diagrams:

1. Binary Phase Shift Keying (BPSK) signals are antipodal signals.

2. Binary Frequency Shift Keying (BFSK), and Binary Pulse Position Modulation (BPPM), are orthogonal signals.
3. Quadrature-Phase Shift Keying (QPSK), $\pi/4$-DQPSK (Differential Quadrature-Phase Shift Keying), and Offset Quadrature-Phase Shift Keying (OQPSK) are bi-orthogonal signals.

Error Probability for Coherent Receivers – General Case

As mentioned above, coherent receivers compensate for phase rotation of the channel by means of *carrier recovery*. Furthermore, the channel gain α is assumed to be known, and absorbed into the received signal, so that in the absence of noise, $\mathbf{r} = \mathbf{s}$ holds. The probability that symbol $\mathbf{s}_j$ is mistaken for symbol $\mathbf{s}_k$ that has Euclidean distance d_{jk} from $\mathbf{s}_j$ (*pairwise error probability*) is given as

$$\mathrm{Pr_{pair}}(\mathbf{s}_j, \mathbf{s}_k) = Q\left(\sqrt{\frac{d_{jk}^2}{2N_0}}\right) = Q\left(\sqrt{\frac{E}{N_0}(1 - \mathrm{Re}\{\rho_{jk}\})}\right) \tag{12.24}$$

where the Q-function is defined as

$$Q(x) = \frac{1}{\sqrt{2\pi}} \int_x^\infty \exp(-t^2/2)\, dt \tag{12.25}$$

This is related to the complementary error function:

$$Q(x) = \frac{1}{2}\mathrm{erfc}\left(\frac{x}{\sqrt{2}}\right) \tag{12.26}$$

and

$$\mathrm{erfc}(x) = 2Q(\sqrt{2x}) \tag{12.27}$$

Equation (12.24) can be found by computing the probability that the noise is large enough to make the received signal geometrically closer to the point $\mathbf{s}_k$ in the signal space diagram, even though the signal $\mathbf{s}_j$ was transmitted.

Error Probability for Coherent Receivers – Binary Orthogonal Signals

As we saw in Chapter 11, a number of important modulation formats can be viewed as binary orthogonal signals – most prominently, binary frequency shift keying (FSK) and binary Pulse Position Modulation (PPM). Figure 12.3 shows the signal space diagram for this case. The figure also shows the decision boundary: if a received signal point falls into the shaded region, then it is decided that a $+1$ was transmitted, otherwise a -1 was transmitted.

Defining the Signal-to-Noise Ratio (SNR) for one symbols as $\gamma_S = E_S/N_0$, we get

$$\mathrm{Pr_{pair}}(\mathbf{s}_j, \mathbf{s}_k) = Q\left(\sqrt{\gamma_S(1 - \mathrm{Re}\{\rho_{jk}\})}\right) \tag{12.28}$$

$$= Q(\sqrt{\gamma_S}) \tag{12.29}$$

Note that since we are considering binary signaling $\gamma_S = \gamma_B$.

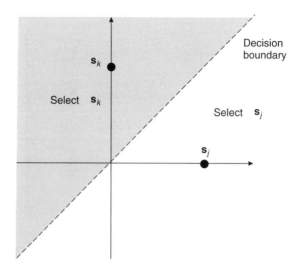

Figure 12.3 Decision boundary for the selection between s_k and s_j.

Error Probability for Coherent Receivers – Antipodal Signaling

For antipodal signals, the pairwise error probability is:

$$\text{Pr}_{\text{pair}}(s_j, s_k) = Q\left(\sqrt{\gamma_S(1 - \text{Re}\{\rho_{jk}\})}\right) \tag{12.30}$$

$$= Q(\sqrt{2\gamma_S}) \tag{12.31}$$

For binary signals with equal-probability transmit symbols, pairwise error probability is equal to symbol error probability, which in turn is equal to bit error probability. This is the case, e.g., for BPSK, as well as for MSK with ideal coherent detection (see Chapter 11), and the BER is given by Eq. (12.30). Note that MSK can be detected like FSK, but then does not exploit the continuity of the phase. In this case, it becomes an orthogonal modulation format, and the BER is given by Eq. (12.29). This means deterioration of the effective SNR by 3 dB.

Union Bound and Bi-Orthogonal Signaling

For M-ary modulation methods, exact computation of the BER is much more difficult; therefore, the BER is often upper-bounded by the *union bound* method. The principle of this bounding technique is outlined in Figure 12.4. The region of the signal space diagram that results in an erroneous decision consists of partial regions, each of which represents a pairwise error – i.e., confusing the correct symbol with another symbol. The symbol error probability is then written as the sum of these pairwise probabilities. Since the pairwise- error regions overlap, this represents an upper bound for true symbol error probabilities. The approximation improves as the SNR increases, because the SER is then mainly determined by regions close to the decision boundaries whereas overlap regions have little impact.

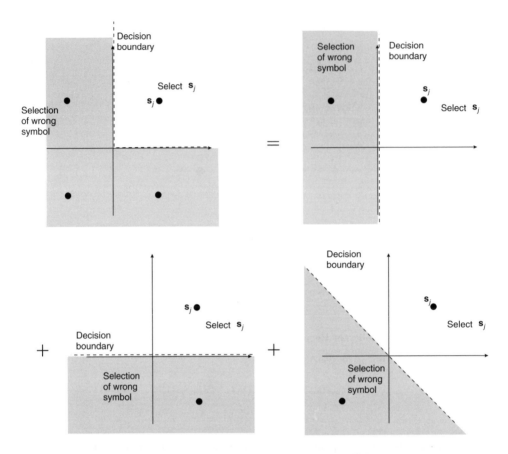

Figure 12.4   Union bound for symbol error probabilities.

Great care must be taken when using equations for the BER of higher order modulation formats from the literature. There are several possible pitfalls:

- Does the equation give the symbol error probability or the bit error probability? Take, e.g., the 4-Quadrature Amplitude Modulation (QAM) of Figure 12.4, and assume furthermore that the signal constellations are Gray-coded, so that the four points represent the bit combinations 00, 01, 11, and 10 (when read clockwise). There is a high probability of errors occurring between two neighboring signal constellations. One symbol error (in one transmitted symbol) then corresponds to one bit error (one error in *two* transmitted bits). Thus, bit error probability is only about half symbol error probability.[3]
- Does computation of the SNR use bit energy or symbol energy? A fair comparison between different modulation formats should be based on E_B/N_0.
- Some authors define the distance between the origin and (equal-energy) signal constellation points not as $\sqrt{E}$, but rather as $\sqrt{2E}$. This is related to a different normalization of the expansion

---

[3] This demonstrates the drawbacks of mixing up the expressions "bit error probability" and " bit error rate". In our example, the BER is the same as the symbol errror rate, while the bit error probability is only half the symbol error probability.

functions, which is compensated by different values for n_n. The final results do not change, but this makes the combination of intermediate results from different sources much more difficult.

As an example, we compute in the following the BER of 4-QAM. As it is a four-state modulation format (2 bit/symbol), we can invest twice the bit energy for each symbol. The signal points thus have squared Euclidean distance $d^2 = E_S = 2E_B$ from the origin; the points in the signal constellation diagram are thus at $\pm\sqrt{E_B}(\pm 1 \pm j)$, and the distance between two neighboring points is $d_{jk}^2 = 4E_B$.

We can now consider two types of union bounds:

1. For a "full" union bound, we compute the pairwise error probability with *all* possible signal constellation points, and add them up. This is shown in Figure 12.4.
2. We compute pairwise error probability using nothing more than neighboring points. In this case, we omit the last decision region in Figure 12.4 from our computations. As we can see, the union of the first two regions (for pairwise error with nearest neighbors) already covers the whole "erroneous decision region." In the following, we will use this type of union bound.

Example 12.1   *Compute the BER and SER of QPSK.*

From Eq. (12.24) it follows that the pairwise error probability is

$$Q\left(\sqrt{2\frac{E_B}{N_0}}\right) = Q(\sqrt{2\gamma_B}) \tag{12.32}$$

According to Figure 12.4, the symbol error probability as computed from the union bound is twice as large:

$$SER \approx 2Q(\sqrt{2\gamma_B}) \tag{12.33}$$

Now, as discussed above, the BER is half the SER:

$$BER = Q(\sqrt{2\gamma_B}) \tag{12.34}$$

This is the same BER as for BPSK.

For QPSK, it is also possible to compute the symbol error probability exactly: the probability of a correct decision (in Figure 12.4) is to stay in the right half plane (probability for that event is $1 - Q(\sqrt{2\gamma_B})$) times the probability of staying in the upper half-plane (the probability for this is independent of the probability of being in the right half plane, and also is $1 - Q(\sqrt{2\gamma_B})$). The overall symbol error probability in thus $1 - (1 - Q(\sqrt{2\gamma_B})^2$, i.e.,

$$SER = 2Q(\sqrt{2\gamma_B})\left[1 - \frac{1}{2}Q(\sqrt{2\gamma_B})\right] \tag{12.35}$$

which shows the magnitude of the error made by the union bound. The relative error is $0.5Q(\sqrt{2\gamma_B})$, which tends to 0 as γ_B tends to infinity.

The exact computation of the BER or SER of higher order modulation formats can be complicated. However, the union bound offers a simple and (especially for high SNRs) very accurate approximation.

Error Probability for Differential Detection

Carrier recovery can be a challenging problem (see Meyr et al. [1997] and Proakis [2005]), which makes coherent detection difficult for many situations. Differential detection is an attractive alternative to coherent detection, because it renders irrelevant the absolute phase of the detected signal. The receiver just compares the phases (and possibly amplitudes) of two subsequent symbols to recover the information in the symbol; this phase difference is independent of the absolute phase. If the phase rotation introduced by the channel is slowly time varying (and thus effectively the same for two subsequent symbols), it enters just the absolute phase, and thus need not be taken into account in the detection process.

For differential detection of Phase Shift Keying (PSK), the transmitter needs to provide differential encoding. For binary symmetric PSK, the transmit phase Φ_i of the ith bit is

$$\Phi_i = \Phi_{i-1} + \begin{cases} +\dfrac{\pi}{2} & \text{if } b_i = +1 \\[2mm] -\dfrac{\pi}{2} & \text{if } b_i = -1 \end{cases} \tag{12.36}$$

Comparison of the difference between phases on two subsequent sampling instances determines whether the transmitted bit b_i was $+1$ or -1.[4]

For Continuous Phase Frequency Shift Keying (CPFSK), such differential encoding can be avoided. Remember that in the case of MSK (without differential encoding), the phase rotation over a 1-bit duration is $\pm \pi/2$. It is thus possible to determine the phases at two subsequent sampling points, take the difference, and conclude which bit has been transmitted. This can also be interpreted by the fact that the phase of the transmit signal is an integral over the uncoded bit sequence. Computing the phase difference is a first approximation to taking the derivative (it is exact if the phase change is linear), and thus reverses the integration.

For binary orthogonal signals, the BER for differential detection is [Proakis 2005]

$$BER = \frac{1}{2} \exp(-\gamma_b) \tag{12.37}$$

For 4-PSK with Gray-coding, it is

$$BER = Q_\mathrm{M}(a, b) - \frac{1}{2} I_0(ab) \exp\left(-\frac{1}{2}(a^2 + b^2)\right) \tag{12.38}$$

where

$$a = \sqrt{2\gamma_\mathrm{B}\left(1 - \frac{1}{\sqrt{2}}\right)}, \quad b = \sqrt{2\gamma_\mathrm{B}\left(1 + \frac{1}{\sqrt{2}}\right)} \tag{12.39}$$

and $Q_\mathrm{M}(a, b)$ is Marcum's Q-function:

$$Q_\mathrm{M}(a, b) = \int_b^\infty x \exp\left[-\frac{a^2 + x^2}{2}\right] I_0(ax)\, dx \tag{12.40}$$

whose series representation is given in Eq. (5.31).

Error Probability for Noncoherent Detection

When the carrier phase is completely unknown, and differential detection is not an option, then noncoherent detection can be used. For equal-energy signals, the detector tries to maximize the metric:

$$|\mathbf{r}_\mathrm{LP}\mathbf{s}_{\mathrm{LP},m}^*| \tag{12.41}$$

so that the optimum receiver has a structure according to Figure 12.5.

---

[4] Theoretically, differential detection could also be performed on a non-encoded signal. However, one bit error would then lead to a whole chain of errors.

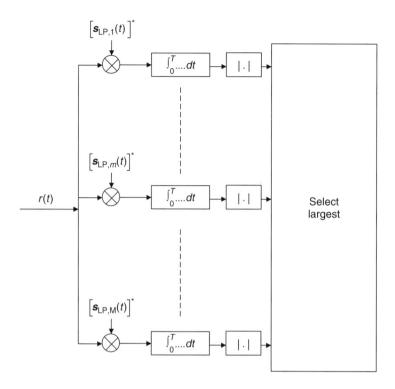

Figure 12.5 Optimum receiver structure for noncoherent detection.

An actual realization, where $r(t)$ is a bandpass signal, will in each branch of Figure 12.5 split the signal into two subbranches, in which it obtains and processes the I- and Q-branches of the signal separately; the outputs of the "absolute value" operation of the I- and Q-branches are then added up before the "select largest" operation.

In this case, the BER can be computed from Eq. (12.38), but with a different definition of the parameters a and b:

$$a = \sqrt{\frac{\gamma_B}{2}\left(1 - \sqrt{1 - |\rho|^2}\right)}, \quad b = \sqrt{\frac{\gamma_B}{2}\left(1 + \sqrt{1 - |\rho|^2}\right)} \qquad (12.42)$$

The optimum performance is achieved in this case if $\rho = 0$ – i.e., the signals are orthogonal. For the case when $|\rho| = 1$, which occurs for PSK signals, including BPSK and 4-QAM, the BER becomes 0.5.

Example 12.2 *Compute the BER of binary FSK in an AWGN channel with $\gamma_B = 5\,dB$, and compare it with Differential Binary Phase Shift Keying (DBPSK) and BPSK.*

For binary FSK in AWGN, the BER is given by Eq. (12.29) with $\gamma_S = \gamma_B$. Thus, with $\gamma_B = 5\,dB$, we have:

$$\mathrm{BER_{BFSK}} = Q(\sqrt{\gamma_B})$$

$$= 0.038 \qquad (12.43)$$

For BPSK, the BER is instead given by Eq. (12.37), and thus:

$$\text{BER}_{\text{BPSK}} = Q(\sqrt{2\gamma_B})$$

$$= 0.006 \qquad (12.44)$$

Finally, for differential BPSK, the BER is given by Eq. (12.37), which results in:

$$\text{BER}_{\text{DBPSK}} = (1/2)e^{-\gamma_B}$$

$$= 0.021 \qquad (12.45)$$

12.2 Error Probability in Flat-Fading Channels

12.2.1 Average BER – Classical Computation Method

In fading channels, the received signal power (and thus the SNR) is not constant but changes as the fading of the channel changes. In many cases, we are interested in the BER in a fading channel averaged over the different fading states. For a mathematical computation of the BER in such a channel, we have to proceed in three steps:

1. Determine the BER for any arbitrary SNR.
2. Determine the probability that a certain SNR occurs in the channel – in other words, determine the pdf of the power gain of the channel.
3. Average the BER over the distribution of SNRs.

In an AWGN channel, the BER decreases approximately exponentially as the SNR increases: for binary modulation formats, a 10-dB SNR is sufficient to give a BER on the order of 10^{-4}, for 15 dB the BER is below 10^{-8}. In contrast, we will see below that in a fading channel the BER decreases only linearly with the (average) SNR. At first glance, this is astonishing: sometimes fading leads to high SNRs, sometimes it leads to low SNRs, and it could be assumed that high and low values would compensate for each other. The important point here is that the relationship between (instantaneous) BER and (instantaneous) SNR is highly nonlinear, so that the cases of low SNR essentially determine the overall BER.

Example 12.3   *BER in a two-state fading channel.*

Consider the following simple example: a fading channel has an average SNR of 10 dB, where fading causes the SNR to be $-\infty$dB half of the time, while it is 13 dB the rest of the time. The BERs corresponding to the two channel states are 0.5 and 10^{-9} respectively (assuming antipodal modulation with differential detection). The mean BER is then $\overline{BER} = 0.5 \cdot 0.5 + 0.5 \cdot 10^{-9} = 0.25$. For an AWGN channel with a 10-dB SNR, the BER is $2 \cdot 10^{-5}$.

Following this intuitive explanation, we now turn to the mathematical details of the above-mentioned three-step procedure. Step 1, the determination of the BER of an arbitrary given SNR, was treated in Section 12.1. Point 2 requires computation of the SNR distribution. In Chapter 5, we mostly concentrated on the distribution of *amplitude* $r = |r(t)|$. This has to be converted to distribution of (instantaneous) received *power* $P_{\text{inst}} = r^2$. Such a transformation can be done by means of the Jacobian. As an example, we show the transformation for a Rayleigh distribution of the amplitude of the received signal,

$$pdf_r(r) = \frac{r}{\sigma^2} \exp\left(-\frac{r^2}{2\sigma^2}\right) \quad \text{for } r \geq 0 \tag{12.46}$$

The mean power is $P_m = 2\sigma^2$. Using the Jacobian $|dP_{inst}/dr| = 2r$, the pdf of the received power becomes

$$pdf_{P_{inst}}(P_{inst}) = \frac{1}{P_m} \exp\left(-\frac{P_{inst}}{P_m}\right) \quad \text{for } P_{inst} \geq 0 \tag{12.47}$$

Since the SNR is the received power scaled with the noise power, the pdf of the SNR is

$$pdf_{\gamma_B}(\gamma_B) = \frac{1}{\overline{\gamma_B}} \exp\left(-\frac{\gamma_B}{\overline{\gamma_B}}\right) \tag{12.48}$$

where $\overline{\gamma_B}$ is the mean SNR. For Rician-fading channels, a similar, but more tedious, computation gives [Rappaport 1996]:

$$pdf_{\gamma_B}(\gamma_B) = \frac{1+K_r}{\overline{\gamma_B}} \exp\left(-\frac{\gamma_B(1+K_r) + K_r\overline{\gamma_B}}{\overline{\gamma_B}}\right) I_0\left(\sqrt{\frac{4(1+K_r)K_r\gamma_B}{\overline{\gamma_B}}}\right) \tag{12.49}$$

where K_r is the Rice factor. Analogous computations are possible for other amplitude distributions.

In the last step, the BER has to be averaged over the distribution of the SNR:

$$\overline{BER} = \int pdf_{\gamma_B}(\gamma_B)BER(\gamma_B)d\gamma_B \tag{12.50}$$

For Rayleigh fading, a closed-form evaluation of Eq. (12.50) is possible for many modulation formats. Using

$$2\int_0^\infty Q\left(\sqrt{2x}\right) a \exp(-ax)\, dx = 1 - \sqrt{\frac{1}{1+a}} \tag{12.51}$$

the mean BER for coherent detection of binary antipodal signals is

$$\overline{BER} = \frac{1}{2}\left[1 - \sqrt{\frac{\overline{\gamma_B}}{1+\overline{\gamma_B}}}\right] \approx \frac{1}{4\overline{\gamma_B}} \tag{12.52}$$

and the mean BER for coherent detection of binary orthogonal signals is

$$\overline{BER} = \frac{1}{2}\left[1 - \sqrt{\frac{\overline{\gamma_B}}{2+\overline{\gamma_B}}}\right] \approx \frac{1}{2\overline{\gamma_B}} \tag{12.53}$$

These BERs are plotted in Figure 12.6.

For differential detection, the averaging process is even simpler, as the BER curve is an exponential function of the instantaneous SNR. For binary antipodal signals:

$$\overline{BER} = \frac{1}{2(1+\overline{\gamma_B})} \approx \frac{1}{2\overline{\gamma_B}} \tag{12.54}$$

and for binary orthogonal signals:

$$\overline{BER} = \frac{1}{2+\overline{\gamma_B}} \approx \frac{1}{\overline{\gamma_B}} \tag{12.55}$$

Figure 12.6 BER for binary signals with coherent detection: antipodal (solid) and orthogonal (dashed).

For differential detection, the BER in Rician channels can also be computed in closed form. For antipodal signals:

$$\overline{BER} = \frac{1 + K_r}{2(1 + K_r + \overline{\gamma_B})} \exp\left(-\frac{K_r\overline{\gamma_B}}{1 + K_r + \overline{\gamma_B}}\right) \tag{12.56}$$

and for orthogonal signals:

$$\overline{BER} = \frac{1 + K_r}{(2 + 2K_r + \overline{\gamma_B})} \exp\left(-\frac{K_r\overline{\gamma_B}}{2 + 2K_r + \overline{\gamma_B}}\right) \tag{12.57}$$

Example 12.4 *Compute the BER of DBPSK with $\overline{\gamma_B} = 12$ dB and $K_r = -3$ dB, 0 dB, 10 dB. The BER is given by Eq. (12.56), and hence, with $K_r = -3$ dB:*

$$\overline{BER} = \frac{1 + K_r}{2(1 + K_r + \overline{\gamma}_B)} \exp\left(-\frac{K_r\overline{\gamma}_B}{1 + K_r + \overline{\gamma}_B}\right)$$
$$= 0.027 \tag{12.58}$$

Then, with $K_r = 0$ dB we get $\overline{BER} = 0.023$, and with $K_r = 10$ dB we get $\overline{BER} = 0.00056$. Note the strong decrease in BER for high Rice factors, even though we kept the SNR constant.

12.2.2 Computation of Average Error Probability – Alternative Method

In the late 1990s, a new method for computation of the average BER was proposed, and shown to be very efficient. It is based on an alternative representation of the Q-function, and allows easier averaging over different fading distributions ([Annamalai et al. 2000], [Simon and Alouini 2004]).

Alternative Representation of the Q-Function

When evaluating the classical definition of the Q-function

$$Q(x) = \frac{1}{\sqrt{2\pi}} \int_x^\infty \exp(-t^2/2)\, dt \qquad (12.59)$$

there is the problem that the argument of the Q-function is in the integration limit, not in its integrand. This makes evaluation of the integrals of the Q-function much more difficult, especially because we cannot use the beloved trick of exchanging the sequence of integration in multiple integrals. This problem is particularly relevant in BER computations. The problem can be solved by using an alternative formulation of the Q-function:

$$Q(x) = \frac{1}{\pi} \int_0^{\pi/2} \exp\left(-\frac{x^2}{2\sin^2\theta}\right) d\theta \quad \text{for } x > 0 \qquad (12.60)$$

This representation now has the argument in the integrand (in a Gaussian form, $\exp(-x^2)$), and also has finite integration limits. We will see below that this greatly simplifies evaluation of the error probabilities.

It turns out that Marcum's Q-function, as defined in Eq. (12.40), also has an alternative representation:

$$Q_M(a, b) = \begin{cases} \dfrac{1}{2\pi} \displaystyle\int_{-\pi}^{\pi} \dfrac{b^2 + ab\sin\theta}{b^2 + 2ab\sin\theta + a^2} \exp(-\tfrac{1}{2}(b^2 + 2ab\sin\theta + a^2))\, d\theta & \text{for } b > a \geq 0 \\[3mm] 1 + \dfrac{1}{2\pi} \displaystyle\int_{-\pi}^{\pi} \dfrac{b^2 + ab\sin\theta}{b^2 + 2ab\sin\theta + a^2} \exp(-\tfrac{1}{2}(b^2 + 2ab\sin\theta + a^2))\, d\theta & \text{for } a > b \geq 0 \end{cases}$$

$$(12.61)$$

Error Probability in Additive White Gaussian -Noise Channels

For computation of the BER in AWGN channels, we find that the BER for BPSK can be written as (compare also Eq. 12.30):

$$BER = Q(\sqrt{2\gamma_S}) \qquad (12.62)$$

$$= \frac{1}{\pi} \int_0^{\pi/2} \exp\left(-\frac{\gamma_S}{\sin^2\theta}\right) d\theta \qquad (12.63)$$

For QPSK, the SER can be computed as (compare Eq. 12.35):

$$SER = 2Q(\sqrt{\gamma_S}) - Q^2(\sqrt{\gamma_S}) \qquad (12.64)$$

$$= \frac{1}{\pi} \int_0^{3\pi/4} \exp\left(-\frac{\gamma_S}{\sin^2\theta} \sin^2(\pi/4)\right) d\theta \qquad (12.65)$$

and quite generally for M-ary PSK:

$$SER = \frac{1}{\pi} \int_0^{(M-1)\pi/M} \exp\left(-\frac{\gamma_S}{\sin^2\theta} \sin^2(\pi/M)\right) d\theta \qquad (12.66)$$

For binary orthogonal FSK, we finally find that

$$BER = Q(\sqrt{\gamma_S}) \qquad (12.67)$$

$$= \frac{1}{\pi} \int_0^{\pi/2} \exp\left(-\frac{\gamma_S}{2\sin^2\theta}\right) d\theta \qquad (12.68)$$

Example 12.5 *Compare the SER for 8-PSK as computed from Eq. (12.66) with the value obtained from the union bound for $\gamma_S = 3$ and $10\,dB$.*

First using Eq. (12.66) with $M = 8$ and $\gamma_S = 3\,$dB we have

$$SER = \frac{1}{\pi} \int_0^{(M-1)\pi/M} \exp\left(-\frac{\gamma_S}{\sin^2\theta}\sin^2(\pi/M)\right) d\theta$$

$$= 0.442 \tag{12.69}$$

and for $\gamma_S = 10\,$dB we get $SER = 0.087$, where the integral has been evaluated numerically.

The 8-PSK constellation has a minimum distance of $d_{\min} = 2\sqrt{E_S}\sin(\pi/8)$. The nearest neighbor union bound on the SER is then given by

$$SER_{\text{union-bound}} = 2 \cdot Q\left(\frac{d_{\min}}{\sqrt{2N_0}}\right)$$

$$= 2 \cdot Q\left(\frac{2\sqrt{E_S}\sin(\pi/8)}{\sqrt{2N_0}}\right)$$

$$= 2 \cdot Q\left(\sqrt{2\gamma_S}\sin(\pi/8)\right) \tag{12.70}$$

Thus, with $\gamma_S = 3\,$dB we get $SER_{\text{union-bound}} = 0.445$, and with $\gamma_S = 10\,$dB we get $SER_{\text{union-bound}} = 0.087$. In this example the union bound gives a very good approximation, even at the rather low SNR of 3 dB.

Error Probability in Fading Channels

For AWGN channels, the advantages of the alternative representation of the Q-function are rather limited. They allow a simpler formulation for higher order modulation formats, but do not exhibit significant advantages for the modulation formats that are mostly used in practice. The real advantage emerges when we apply this description method as the basis for computations of the BER in fading channels. We find that we have to average over the pdf of the SNR $pdf_\gamma(\gamma)$, as described in Eq. (12.50). We have now seen that the alternative representation of the Q-function allows us to write the SER (for a given SNR) in the generic form:

$$SER(\gamma) = \int_{\theta_1}^{\theta_2} f_1(\theta)\exp(-\gamma f_2(\theta))\,d\theta \tag{12.71}$$

Thus, the average SER becomes

$$\overline{SER} = \int_0^\infty pdf_\gamma(\gamma)SER(\gamma)\,d\gamma \tag{12.72}$$

$$= \int_0^\infty pdf_\gamma(\gamma)\int_{\theta_1}^{\theta_2} f_1(\theta)\exp(-\gamma f_2(\theta))\,d\theta\,d\gamma \tag{12.73}$$

$$= \int_{\theta_1}^{\theta_2} f_1(\theta)\int_0^\infty pdf_\gamma(\gamma)\exp(-\gamma f_2(\theta))\,d\gamma\,d\theta \tag{12.74}$$

Let us now have a closer look at the inner integral:

$$\int_0^\infty pdf_\gamma(\gamma)\exp(-\gamma f_2(\theta))\,d\gamma \tag{12.75}$$

We find that it is the moment-generating function of $pdf_\gamma(\gamma)$, evaluated at the point $-f_2(\theta)$. Remember (see also, e.g., Papoulis [1991]) that the moment-generating function is defined as the Laplace transform of the pdf of γ:

$$M_\gamma(s) = \int_0^\infty pdf_\gamma(\gamma) \exp(\gamma s)\, d\gamma \tag{12.76}$$

and the mean SNR is the first derivative, evaluated at $s = 0$:

$$\overline{\gamma} = \left. \frac{dM_\gamma(s)}{ds} \right|_{s=0} \tag{12.77}$$

Summarizing, the average SER can be computed as

$$\overline{SER} = \int_{\theta_1}^{\theta_2} f_1(\theta) M_\gamma(-f_2(\theta))\, d\theta \tag{12.78}$$

The next step is then finding the moment-generating function of the distribution of the SNR. Without going into the details of the derivations, we find that for a Rayleigh distribution of the signal amplitude, the moment-generating function of the SNR distribution is

$$M_\gamma(s) = \frac{1}{1 - s\overline{\gamma}} \tag{12.79}$$

for a Rice distribution it is

$$M_\gamma(s) = \frac{1 + K_r}{1 + K_r - s\overline{\gamma}} \exp\left[\frac{K_r s \overline{\gamma}}{1 + K_r - s\overline{\gamma}} \right] \tag{12.80}$$

and for a Nakagami distribution with parameter m:

$$M_\gamma(s) = \left(1 - \frac{s\overline{\gamma}}{m}\right)^{-m} \tag{12.81}$$

Having now the general form of the SER (Eq. 12.78) and the form of the moment-generating function, the computation of the error probabilities become straightforward (if sometimes a bit tedious).

Example 12.6 *BER of BPSK in Rayleigh fading.*

As one example, let us go through the computation of the average BER of BPSK in a Rayleigh-fading channel – a problem for which we already know the result from Eq. (12.52). Looking at Eq. (12.63), we find that $\theta_1 = 0$, $\theta_2 = \pi/2$:

$$f_1(\theta) = \frac{1}{\pi} \tag{12.82}$$

$$f_2(\theta) = \frac{1}{\sin^2(\theta)} \tag{12.83}$$

Since we consider Rayleigh fading:

$$M_\gamma(-f_2(\theta)) = \frac{1}{1 + \dfrac{\overline{\gamma}}{\sin^2(\theta)}} \tag{12.84}$$

so that the total BER is, according to Eq. (12.78):

$$\overline{SER} = \frac{1}{\pi} \int_0^{\pi/2} \frac{\sin^2(\theta)}{\sin^2(\theta) + \overline{\gamma}} \, d\theta \tag{12.85}$$

which can be shown to be identical to the first (exact) expression in Eq. (12.52) – namely:

$$\overline{BER} = \frac{1}{2} \left[1 - \sqrt{\frac{\overline{\gamma}}{1 + \overline{\gamma}}} \right] \tag{12.86}$$

Many modulation formats and fading distributions can be treated in a similar way. An extensive list of solutions, dealing with coherent detection, partially coherent detection, and noncoherent detection in different types of channels, is given in the book by Simon and Alouini [2004].

12.2.3   Outage Probability versus Average Error Probability

In the previous section, we computed the average bit error probability, where averaging was done over small-scale fading. Similarly, we could also average this distribution over large-scale fading – the mathematics are quite similar. But what is the physical meaning of these averaged values? In order to understand *what* we are computing here, and what averaging makes sense, we first have to have a look at the different operating scenarios that can occur in a wireless system.

As a first step, we have to compare the length of the "memory" of the system with the coherence time of the channel. The "memory" of the system in this context could, e.g., be caused by

1. perception of the human ear (error bursts on the order of a few microseconds or less tend to be "smoothed out" by the human ear); or
2. the size of typical data structures that are to be transmitted over the wireless connection;
3. further memory can also come from coding (in which case the block length of the block code or the constraint length of a convolutional code would determine memory duration), and interleaving, in which case the length of the interleaver would determine the memory (see also Chapter 14).

Let us first consider the case where the receiver or transmitter moves, so that the channel changes within a finite time, and system memory can thus extend over many channel realizations. Such system memories might typically see many-scale realizations of the channel, as well as large-scale variations of the channel. When we now want to see the average BER for a file transfer, we have to average the BER over the duration of that file transfer, and thus average it over the distribution of the SNRs seen during that time.[5] The BER is a single, deterministic value.

Next we consider the case where system memory is much shorter than coherence time. To name but one example: if we wish to transmit a file from a (stationary) laptop, and the objects in the environment are not moving, the channel seen by the wireless connection is static for the duration of the transmission. The SNR seen by the demodulator is thus constant, but random, depending on the position. It is thus meaningful to investigate the pdf of the SNR, in order to see what SNR is available in what percentage of locations. For each of the locations – i.e., for each realization of the SNR – we compute the BER simply as the BER of an AWGN channel with that specific SNR.

---

[5] For a coded system, where the code length is larger than channel coherence time, computations are a bit trickier (as discussed in Chapter 14).

In such a case, we will also get a *distribution* of the BER; note that this distribution is for a fixed transmit power.

From these considerations, we arrive at the concept of *outage probability*. For many applications, it is not important what the exact value of the BER is, as long as it stays below a certain threshold. For example, a file transfer is successful as long as the raw BER is small enough so that errors can be corrected by the error correction coding – in other words, a long as a certain threshold raw BER (typically on the order of a few percent) is not exceeded. It is then meaningful to determine the percentage of locations where successful file transfer will not be available. This percentage is known as outage probability.

In some situations, the memory is long enough to see multiple realizations of the small-scale fading, but only one realization of the large-scale fading. In this case, an outage occurs if the small-scale averaged BER lies below a certain threshold, and the probability for such an outage to occur is determined by the large-scale fading statistics only.

Computation of outage probability becomes much simpler if we define not a maximum BER but rather a minimum SNR γ_0, for the system to work properly. We can then find the outage probability as

$$\mathrm{Pr}_{\mathrm{out}} = P(\gamma < \gamma_0) = \int_0^{\gamma_0} pdf_\gamma(\gamma)d\gamma \tag{12.87}$$

Outage probability can also be seen as another way of establishing a fading margin: we need to find the mean SNR that guarantees a certain outage.

12.3   Error Probability in Delay- and Frequency-Dispersive Fading Channels

12.3.1   Physical Cause of Error Floors

In wireless propagation channels, transmission errors are caused not only by noise but also by signal distortions. These distortions are created on one hand by delay dispersion (i.e., echoes of the transmit signal arriving with different delays), and on the other hand by frequency dispersion (Doppler effect – i.e., signal components arriving with different Doppler shifts). For high data rates, delay dispersion is dominant; at low data rates, frequency dispersion is the main reason for signal distortion errors. In either of these cases, an increase in transmit power does not lead to a reduction of the BER; for that reason, these errors are often called *error floor* or *irreducible errors*. Of course, these errors can also be reduced or eliminated, but this has to be done by methods other than increasing power (e.g., equalization, diversity, etc.). In this section, we only treat the case when the receiver does not use any of these countermeasures, so that dispersion leads to increased error rates. Later chapters discuss in detail the fact that dispersion can actually be a benign effect if specific receiver structures are used.

Frequency Dispersion

We first consider errors due to frequency dispersion. For FSK, it is immediately obvious how frequency dispersion leads to errors: random Frequency Modulation (FM) (see Section 5.7.3) leads to a frequency shift of the received signal, and can push a bit over the decision boundary. Assume that a $+1$ was sent (i.e., the frequency $f_c + f_{\mathrm{mod}}$). Due to the random FM effect, the frequency $f_c + f_{\mathrm{mod}} f_{\mathrm{inst}}$ is received. If this is smaller than f_c, the receiver opts for a -1. Note that *instantaneous* frequency shifts can be significantly larger than the maximum Doppler frequency even though the statistics of the random FM are determined by the Doppler spectrum of the channel. Consider the following equation for the instantaneous frequency:

$$f_{\text{inst}}(t) = \frac{\text{Im}\left(r^*(t)\frac{dr(t)}{dt}\right)}{|r(t)|^2} \tag{12.88}$$

Obviously, this can become very large when the amplitude becomes very small. In other words, deep fading dips lead to large shifts in the instantaneous frequency, and thus higher error probability.

A somewhat different interpretation can be given for differential detection. As mentioned above, differential detection assumes that the channel does not change between two adjacent symbols. However, if there is a finite Doppler, then the channel *does* change – remember that the Doppler spectrum gives a statistical description of channel changes. Thus, a nonzero Doppler effect implies a wrong reference phase for differential detection. If this effect is strong, it can lead to erroneous decisions. Also in this case it is true that channel changes are strongest near fading dips.[6]

For MSK with differential detection, Hirade et al. [1979] determined the BER due to the Doppler effect:

$$\overline{BER}_{\text{Doppler}} = \frac{1}{2}(1 - \xi_s(T_B)) \tag{12.89}$$

where $\xi_s(t)$ is the normalized autocorrelation function of the channel (so that $\xi_s(0) = 1$) – i.e., the Fourier transform of the normalized Doppler spectrum. For small $v_{\max}T_B$ we then get a BER that is proportional to the squared magnitude of the product of Doppler shift and bit duration:

$$\overline{BER}_{\text{Doppler}} = \frac{1}{2}\pi^2(v_{\max}T_B)^2 \tag{12.90}$$

This basic functional relationship also holds for other Doppler spectra and modulation formats; it is only the proportionality constant that changes.

From this relationship, we find that errors due to frequency dispersion are mainly important for systems with a low data rate. For example, paging systems and sensor networks exhibit data rates on the order of 1 kbit/s, while Doppler frequencies can be up to a few hundred Hz. Error floors of 10^{-2} are thus easily possible. This has to be taken into account when designing the coding for such systems. For high-data-rate systems (which include almost all current cellular, cordless, and Wireless Local Area Network (WLAN) systems), errors due to frequency dispersion do not play a noticeable role.[7] Even for the Japanese JDC cellular system, which has a symbol duration of 50 μs the BER due to frequency dispersion is only on the order of 10^{-4}, which is negligible compared with errors due to noise.

Delay Dispersion

In contrast to frequency dispersion, delay dispersion has great importance for high-data-rate systems. This becomes obvious when we remember that the errors in unequalized systems are determined by the ratio of symbol duration that is disturbed by InterSymbol Interference (ISI) to that of the undisturbed part of the symbol. The maximum excess delay of a channel impulse response is determined by the environment, and independent of the system; let us assume in the following a maximum excess delay of 1 μs. In a system with a symbol duration of 20 μs, the ISI can disturb 5% of each symbol, while it can disturb 20% if the symbol duration is 5 μs.

---

[6] For general QAM, not only the reference phase but also the reference amplitude is relevant. However, in the following we will restrict our considerations to PSK and FSK, and thus ignore amplitude distortions.

[7] However, this does *not* mean that time variations in the channel are unimportant in such systems. Channel variations can also have an impact on coding, on the validity of channel estimation, etc.

Many theoretical and experimental investigations have shown that the error floor due to delay dispersion is given by the following equation:

$$\overline{BER} = K \left(\frac{S_\tau}{T_B}\right)^2 \tag{12.91}$$

where S_τ is the rms delay spread of the channel (see Chapter 6). Just as for frequency dispersion, errors mainly occur near fading dips. Section 12.3.2 gives an interpretation of this fact in terms of group delay, which reaches its largest values near fading dips (see also Chapter 5).

Equation (12.91) is only valid if the maximum excess delay of the channel is much smaller than the symbol duration, and the channel is Rayleigh fading. The proportionality constant K depends on the modulation method, filtering at transmitter and receiver, the form of the average impulse response, and choice of the sampling instant, as we will discuss in the sections below.

Choice of the Sampling Instant   In a flat-fading channel, the choice of sampling instant is obvious – sampling should always occur at those times where the SNR at the decision device is largest; this usually occurs either at the bit transitions or exactly in the middle between bit transitions; in either case we call this time henceforth $t_s = 0$.

For channels with delay dispersion, the choice of sampling time is no longer that obvious. Most theoretical derivations assume that either $t_s = 0$ (i.e., sampling occurs at the minimum excess delay),[8] or at the *average mean delay*. The latter actually is the optimum sampling time for some Power Delay Profiles (PDPs, see Chapter 6), as demonstrated in Figure 12.7.

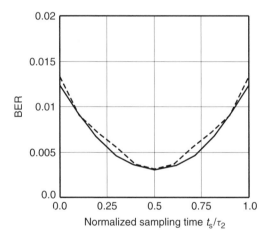

Figure 12.7   Dependence of delay-dispersion-induced error probability BER on choice of sampling instant in a two-spike channel.
Reproduced with permission from Molisch [2000] © Prentice Hall.

When the sampling instant is chosen adaptively, according to the instantaneous state of the channel, the error floor can be decreased considerably, and – in unfiltered systems – even completely eliminated.

Impact of the Shape of the Power Delay Profile   To a first approximation, only the *rms delay spread* determines the BER due to delay dispersion. A closer look reveals, however, that the actual

---

[8] Without restriction of generality, we assume that $\tau_0 = 0$.

Figure 12.8 Impact of the shape of the PDP on the error floor due to delay dispersion. The modulation method is MSK with differential detection.
Reproduced with permission from Molisch [2000] © Prentice Hall.

shape of the PDP also has an impact. The variation of the BER for different PDPs (for equal S_τ) is usually less than a factor of 2, and is thus often neglected. Figure 12.8 shows that a rectangular PDP leads to a slightly larger BER than a two-spike PDP; the difference is 75%.

An exponential PDP leads to an even larger error floor; however, for this shape, sampling at the average mean delay is not optimum.

Filtering   Filtering at the transmitter and/or receiver also leads to signal distortion, and thus makes the signal more susceptible to errors by the additional distortions caused by the channel. The narrower the filtering, the larger the error floor. Figure 12.9 shows the effect of filtering on QAM with Raised Cosine (RC) filters.

The question naturally arises as to the optimum filter bandwidth. Very narrow filters (bandwidth on the order of the inverse symbol duration or less) lead to strong ISIs by themselves. Even if all decisions can be made correctly in the absence of further disturbances, such a filter makes the system more susceptible to channel-induced ISI and noise. For wide filters, a lot of noise can pass through the filter, which also leads to high error rates. The optimum filter bandwidth depends on the ratio of delay dispersion to noise. Figure 12.10 shows an example of such a tradeoff.

Modulation Method   The modulation method also has an impact on the error floor: obviously, a modulation format is more sensitive to distortions by the channel the closer the signals are in the signal constellation diagram. QPSK shows a higher error floor than BPSK when the rms delay spread is normalized to the same *symbol* duration (see Figure 12.11). Higher order modulation formats fare better when equal *bit* durations are assumed.

12.3.2   Computation of the Error Floor Using the Group Delay Method

In this section, we present a very simple, approximate method for computing the BER due to delay dispersion. The method was introduced in Andersen [1991] for PSK and Crohn et al. [1993] for MSK. In the following, we present the method for differentially detected MSK.

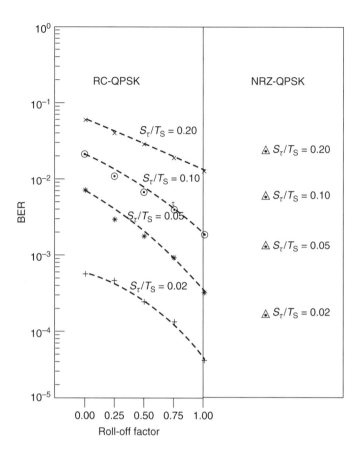

Figure 12.9 Error floor of quadrature-phase shift keying with coherent detection and RC filters as a function of the roll-off factor. For comparison purposes, we also show the BER of conventional QPSK. Note that there is no roll-off factor where RC QPSK reduces to conventional Non Return to Zero (NRZ) QPSK.

Reproduced with permission from Chuang [1987] © IEEE.

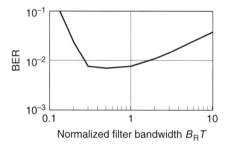

Figure 12.10 BER of filtered minimum shift keying with differential detection. The normalized rms delay spread is 0.1, and $SNR = 12$ dB.

Reproduced with permission from Molisch [2000] © Prentice Hall.

Figure 12.11   Error floor of different modulation formats as a function of normalized rms delay spread (normalized to symbol duration).

Reproduced with permission from Chuang [1987] © IEEE.

Group delay T_g is defined as:

$$T_g = - \left. \frac{\partial \Phi_c}{\partial \omega} \right|_{\omega=0} \tag{12.92}$$

where ω is angular frequency. Here, $\Phi_c(\omega)$ is the phase of the channel transfer function, which can be expanded into a Taylor series:

$$\Phi_c(\omega) = \Phi_c(0) + \omega \left. \frac{\partial \Phi_c}{\partial \omega} \right|_{\omega=0} + \frac{1}{2} \omega^2 \left. \frac{\partial^2 \Phi_c}{\partial \omega^2} \right|_{\omega=0} + \cdots \tag{12.93}$$

The first term of this series corresponds to the *average mean delay* and can be omitted if the sampling is done at the average mean delay. Terminating the Taylor series after the linear term, we obtain

$$\Phi_c(\omega) = -\omega T_g \tag{12.94}$$

Obviously, phase distortion at the sampling instant depends on the instantaneous frequency at these instants. If a $+1$ is transmitted, then the instantaneous frequency is $\pi/(2T_B)$; otherwise, it is $-\pi/(2T_B)$. When the same bits are transmitted (i.e., a $+1$ followed by a $+1$, or a -1 followed by a -1), then the difference between the phase distortions is $\Delta\Phi = 0$; when different bits are transmitted, then

$$\Delta\Phi = \pm \frac{\pi}{T_B} T_g \tag{12.95}$$

A decision error occurs when the size of channel-induced phase distortions (or rather; their difference at the sampling instants) is larger than $\pi/2$ – i.e., when:

$$|T_g| > T_B/2 \tag{12.96}$$

The statistics of group delay in a Rayleigh-fading channel is well known [Andersen et al. 1990] – namely, a *Student's t* distribution:

$$pdf_{T_g}(T_g) = \frac{1}{2S_\tau} \frac{1}{\left[1 + (T_g/S_\tau)^2\right]^{3/2}} \tag{12.97}$$

The probability for bit errors can thus be easily computed from Eq. (12.96). Furthermore, when averaging over the different possible bit combinations, we get:

$$BER = \frac{4}{9}\left(\frac{S_\tau}{T_B}\right)^2 \approx \frac{1}{2}\left(\frac{S_\tau}{T_B}\right)^2 \tag{12.98}$$

Example 12.7   *Consider a system using differentially detected MSK with $T_B = 35\,\mu s$, operating at 900 MHz, moving at 360 km/h (high-speed train), and an exponential power delay profile with $S_\tau = 10\,\mu s$. Compute the BER due to frequency dispersion and delay dispersion.*

The BER due to frequency dispersion can be computed by first calculating the maximum Doppler shift as

$$v_{max} = f_c \frac{v}{c}$$

$$= 9 \cdot 10^8 \frac{100}{3 \cdot 10^8}$$

$$= 300\,\text{Hz} \tag{12.99}$$

For MSK with differential detection, and assuming a classical Jake's Doppler spectrum, the BER due to frequency dispersion is given by Eq. (12.90), and hence:

$$\overline{BER}_{\text{Doppler}} = \frac{1}{2}\pi^2(v_{max}T_B)^2$$

$$= 5.4 \cdot 10^{-4} \tag{12.100}$$

The BER due to delay dispersion is given by Eq. (12.98), which means that

$$\overline{BER}_{\text{Delay}} = \frac{4}{9}\left(\frac{S_\tau}{T_B}\right)^2$$

$$= 3.6 \cdot 10^{-2} \tag{12.101}$$

12.3.3   General Fading Channels: The Quadratic Form Gaussian Variable Method

A general method for the computation of BERs in dispersive fading channels is the so-called Quadratic Form Gaussian Variable (QFGV) method.[9] The channel is Rayleigh or Rice fading, suffering from delay dispersion and frequency dispersion, and adds AWGN.

---

[9] This method is based on evaluation of certain quadratic forms of Gaussian variables. It is also often named after the groundbreaking papers of Bello and Nelin [1963] and Proakis [1968].

Mathematically speaking, the QFGV method determines the probability that a variable D:

$$D = A|X|^2 + B|Y|^2 + CXY^* + C^*X^*Y \tag{12.102}$$

is smaller than 0. Here, X and Y are complex Gaussian variables, A and B are real constants, and C is a complex constant. Defining the auxiliary variables:

$$\left.\begin{aligned} w &= \frac{AR_{xx} + BR_{yy} + CR_{xy}^* + C^*R_{xy}}{4(R_{xx}R_{yy} - |R_{xy}|^2)(|C|^2 - AB)} \\[2mm] v_{1,2} &= \sqrt{w^2 + \frac{1}{4(R_{xx}R_{yy} - |R_{xy}|^2)(|C|^2 - AB)}} \mp w \\[2mm] \alpha_1 &= 2(|C|^2 - AB)(|\overline{X}|^2 R_{yy} + |\overline{Y}|^2 R_{xx} - \overline{X}^*\overline{Y}R_{xy} - \overline{X}\overline{Y}^* R_{xy}^*) \\[2mm] \alpha_2 &= A|\overline{X}|^2 + B|\overline{Y}|^2 + C\overline{X}^*\overline{Y} + C^*\overline{X}\overline{Y}^* \\[2mm] p_1 &= \frac{\sqrt{2v_1^2 v_2(\alpha_1 v_2 - \alpha_2)}}{|v_1 + v_2|} \\[2mm] p_2 &= \frac{\sqrt{2v_1 v_2^2(\alpha_1 v_1 + \alpha_2)}}{|v_1 + v_2|} \end{aligned}\right\} \tag{12.103}$$

where R_{xy} is the second central moment $R_{xy} = \frac{1}{2}E\{(X - \overline{X})(Y - \overline{Y})^*\}$, the error probability becomes

$$P\{D < 0\} = Q_M(p_1, p_2) - \frac{v_2/v_1}{1 + v_2/v_1} I_0(p_1 p_2) \exp\left(-\frac{p_1^2 + p_2^2}{2}\right) \tag{12.104}$$

where Q_M is Marcum's Q-function (Eq. 12.40). If there is only Rayleigh fading, and no dispersion, then α_1 and α_2 are zero, so that $P(D < 0) = v_1/(v_1 + v_2)$. The biggest problem is often to formulate the error probability as $P\{D < 0\}$. This will be discussed in the following.

Canonical Receiver

It is often helpful to reduce different receiver structures to a "canonical" receiver [Suzuki 1982], whose structure is shown in Figure 12.12. The received signal (after bandpass filtering) is multiplied

Figure 12.12 Canonical receiver structure. *In this figure*: BPF, bandpass filter; LPF, low-pass filter.
Reproduced with permission from Suzuki [1982] © IEEE.

by reference signals: once by a "normal" reference signal, and once by the signal shifted by $\pi/2$ to get the quadrature component. The resulting signal is lowpass-filtered, and sent to the decision device. For coherent detection, the reference signal is obtained from a carrier recovery circuit; for differential detection, the reference signal is a delayed version of the incoming signal.

As an example, we consider the differential detection of MSK. It is fairly easy to describe a bit error in the form $D < 0$. An error is made if $a + 1$ was transmitted, but the phase difference of the signals at the sampling instants, $X = r(t_s)$ and $Y = r(t_s - T)$, lies between π and 2π.

The condition for an error is thus identical to

$$\text{Re}\{b_0 X Y^* \exp(-j\pi/2)\} < 0 \tag{12.105}$$

where b_0 is the transmitted bit. Since $\text{Re}\{Z\} = (Z + Z^*)/2$, Eq. (12.105) is a quadratic form D with $A = B = 0$.

12.3.4 Bit Error Probability

A first step in the application of the QFGV method to differential detection is computation of the correlation between the quantities $X = r(t_s)$ and $Y = r(t_s - T)$. For detection using a frequency discriminator, we define X as the sample value $r(t_s)$ and define Y as the derivative $dr(t)/dt$ at time $t = t_s$. Explicit equations for the resulting correlation coefficient are given in Adachi and Parsons [1989] and Molisch [2000].

After the correlation coefficient has been found, the mean BER can be computed from Eq. (12.104). For the case of differential detection of binary FSK in Rayleigh fading, these equations can be simplified to

$$\overline{BER} = \frac{1}{2} - \frac{1}{2} \frac{b_0 \, \text{Im}\{\rho_{XY}\}}{\sqrt{\text{Im}\{\rho_{XY}\}^2 + (1 - |\rho_{XY}|^2)}} \tag{12.106}$$

For $\pi/4$-DQPSK we obtain:

$$\overline{BER} = \frac{1}{2} - \frac{1}{4} \left\{ \frac{b_0 \text{Re}\{\rho_{XY}\}}{\sqrt{(\text{Re}\{\rho_{XY}\})^2 + (1 - |\rho_{XY}|^2)}} + \frac{b_0' \text{Im}\{\rho_{XY}\}}{\sqrt{(\text{Im}\{\rho_{XY}\}^2) + (1 - |\rho_{XY}|^2)}} \right\} \tag{12.107}$$

where b_0 and b_0' are the bits making up a symbol.

Summarizing, the BER can be computed in the following steps:

(i) reduce the actual receiver structure to canonical form;
(ii) formulate the condition for the occurrence of errors as $D < 0$;
(iii) compute the mean and the correlation coefficients of X and Y;
(iv) compute the BER according to the general Eqs. (12.104) and (12.103), or use the simplified Eqs. (12.107) or (12.106) for MSK and $\pi/4$-DQPSK in Rayleigh-fading channels.

Further Reading

The literature describing the bit error probability of digital modulation formats encompasses hundreds of papers. For the BER in AWGN, we again just refer to the textbooks on digital communications [Anderson 2005, Barry et al. 2003, Proakis 2005, and Sklar 2001]. An extremely rigorous mathematical description can be found in Gallagher [2008]. A discrete-time approach is used in Rice [2008]. Fundamental computation methods that are applicable both to nonfading and fading channels were proposed in Pawula et al. [1982], Proakis [1968], and Stein [1964]. The computation of the error probability in flat-fading channels is described in many papers: essentially, each modulation format, combined with the amplitude statistics of the channel, results in at least one paper. Some important examples in Rayleigh-fading channels include Chennakeshu and Saulnier [1993] for $\pi/4$-DQPSK, Varshney and Kumar [1991] and Yongacoglu et al. [1988] for Gaussian Minimum Shift Keying (GMSK), and Divsalar and Simon [1990] for differential detection of M-ary Phase Shift Keying MPSK.

An important alternative to the classical computation of the BER is computation via the moment-generating function described in Simon and Alouini [2004]. or via the characteristic function (see Annamalai et al. [2000]). Simon and Alouini's [2004] monograph gives a number of BER equations for different modulation formats, channel statistics, and receiver structures.

For delay-dispersive channels, a number of different computation methods have been proposed: we have already mentioned in the main part of the text the QFGV method ([Adachi and Parsons 1989], [Proakis 1968]), and the group delay method ([Andersen 1991], [Crohn et al. 1993]). Furthermore, a number of papers use the pdf of the angles between Gaussian vectors as derived by Pawula et al. [1982]. Chuang [1987] is a very readable paper comparing different modulation formats. A wealth of papers are also dedicated to modulation formats or detection methods that reduce the impact of delay dispersion in unequalized systems. A summary of these methods, and further literature, can be found in Molisch [2000].

For updates and errata for this chapter, see wides.usc.edu/teaching/textbook

13

Diversity

13.1 Introduction

13.1.1 Principle of Diversity

In the previous chapter, we treated conventional transceivers that transmit an uncoded bitstream over fading channels. For Additive White Gaussian Noise (AWGN) channels, such an approach can be quite reasonable: the Bit Error Rate (BER) decreases exponentially as the Signal-to-Noise Ratio (SNR) increases, and a 10-dB SNR leads to BERs on the order of 10^{-4}. However, in Rayleigh fading the BER decreases only linearly with the SNR. We thus would need an SNR on the order of 40 dB in order to achieve a 10^{-4} BER, which is clearly unpractical. The reason for this different performance is the fading of the channel: the BER is mostly determined by the probability of channel attenuation being large, and thus of the instantaneous SNR being low. A way to improve the BER is thus to change the effective channel statistics – i.e., to make sure that the SNR has a smaller probability of being low. Diversity is a way to achieve this.

The principle of diversity is to ensure that the same information reaches the receiver (RX) on statistically independent channels. Consider the simple case of an RX with two antennas. The antennas are assumed to be far enough from each other that small-scale fading is independent at the two antennas. The RX always chooses the antenna that has instantaneously larger receive power.[1] As the signals are statistically independent, the probability that both antennas are in a fading dip *simultaneously* is low – certainly lower than the probability that one antenna is in a fading dip. The diversity thus changes the SNR statistics at the detector input.

Example 13.1 *Diversity reception in a two-state fading channel.*

To quantify this effect, let us consider a simple numerical example: the noise power within the RX filter bandwidth is 50 pW, the average received signal power is 1 nW, the SNR is thus 13 dB. In an AWGN channel, the resulting BER is 10^{-9}, assuming that the modulation is differentially detected Frequency Shift Keying (FSK). Now consider a fading channel where during 90% of the time the received power is 1.11 nW, and the SNR is thus 13.5 dB, while for the remainder, it is zero. This means that during 90% of the time, the BER is 10^{-10}; the remainder of the time,

---

[1] We will see later on that this scheme is only one of many different possible diversity schemes.

Wireless Communications, Second Edition   Andreas F. Molisch
© 2011 John Wiley & Sons, Ltd

it is 0.5; the average BER is thus

$$0.9 \cdot 10^{-10} + 0.1 \cdot 0.5 = 0.05 \tag{13.1}$$

For the case of two-antenna diversity, the probability that the received signal power is 0 at both antennas simultaneously is $0.1 \cdot 0.1 = 0.01$. The probability that the received power is 1.11 nW at both antennas simultaneously is $0.9 \cdot 0.9 = 0.81$; the probability that it is 1.11 nW at one antenna and 0 at the other is 0.18. Assuming selection diversity, in both the latter cases, the SNR at the detector is 13.5 dB. The total BER is thus

$$0.01 \cdot 0.5 + 0.99 \cdot 10^{-10} = 0.005 \tag{13.2}$$

This is approximately the square of the BER for a single-antenna system. If we have three antennas, then the probability that the signal power is 0 at all three antennas simultaneously is 0.1^3; the total BER is then $0.5 \cdot 0.001 + 0.999 \cdot 10^{-10} = 0.0005$; this is approximately the third power of the BER for a single-antenna system.

Later sections will give exact equations for the BER with diversity in Rayleigh-fading channels. The general concepts, however, are the same as in the simple example described above. With N_r diversity antennas, we obtain a bit error probability $\propto BER_{oc}^{N_r}$, where BER_{oc} is the BER with just one receive channel.[2]

In the following, we first describe the characterization of correlation coefficients between different transmission channels. We then give an overview about how transmission over independent channels can be realized – spatial antenna diversity described above is one, but certainly not the only approach. Next, we describe how signals from different channels can best be combined, and what performance can be achieved with the different combining schemes.

13.1.2 Definition of the Correlation Coefficient

Diversity is most efficient when the different transmission channels (also called diversity branches) carry independently fading copies of the same signal. This means that the joint probability density function (pdf) of field strength (or power) $pdf_{r_1, r_2, \ldots}(r_1, r_2, \ldots)$ is equal to the product of the marginal pdfs for the channels, $pdf_{r_1}(r_1), pdf_{r_2}(r_2), \ldots$ Any correlation between the fading of the channels decreases the effectiveness of diversity.

The *correlation coefficient* characterizes the correlation between signals on different diversity branches. A number of different definitions are being used for this important quantity: complex correlation coefficients, correlation coefficient of the phase, etc. The most important one is the correlation coefficient of signal envelopes x and y:

$$\rho_{xy} = \frac{E\{x \cdot y\} - E\{x\} \cdot E\{y\}}{\sqrt{(E\{x^2\} - E\{x\}^2) \cdot (E\{y^2\} - E\{y\}^2)}} \tag{13.3}$$

For two statistically independent signals, the relationship $E\{xy\} = E\{x\}E\{y\}$ holds; therefore, the correlation coefficient becomes zero. Signals are often said to be "effectively" decorrelated if ρ is below a certain threshold (typically 0.5 or 0.7).

---

[2] Since in a Rayleigh-fading channel, $BER_{oc} \propto SNR^{-1}$, we find that for a diversity system with N_r independently fading channels, $BER \propto SNR^{-N_r}$. Quite generally in fading channels with diversity, $BER \propto SNR^{-d_{div}}$, where d_{div} is known as *diversity order*.

13.2 Microdiversity

As mentioned in the introduction, the basic principle of diversity is that the RX has multiple copies of the transmit signal, where each of the copies goes through a statistically independent channel. This section describes different ways of obtaining these statistically independent copies.

We concentrate on methods that can be used to combat small-scale fading, which are therefore called "microdiversity." The five most common methods are as follows:

1. *Spatial diversity*: several antenna elements separated in space.
2. *Temporal diversity*: transmission of the transmit signal at different times.
3. *Frequency diversity*: transmission of the signal on different frequencies.
4. *Angular diversity*: multiple antennas (with or without spatial separation) with different antenna patterns.
5. *Polarization diversity*: multiple antennas with different polarizations (e.g., vertical and horizontal).

When we speak of antenna diversity, we imply that there are multiple antennas at the *receiver*. Only in Section 13.6 (and later in Chapter 20) will we discuss how multiple *transmit* antennas can be exploited to improve performance.

The following important equation will come in handy: Consider the correlation coefficient of two signals that have a temporal separation τ and a frequency separation $f_1 - f_2$. As shown in Appendix 13.A (see *www.wiley.com/go/molisch*), the correlation coefficient is

$$\rho_{xy} = \frac{J_0^2(k_0 v \tau)}{1 + (2\pi)^2 S_\tau^2 (f_2 - f_1)^2} \qquad (13.4)$$

Note that for moving Mobile Stations (MSs), temporal separation can be easily converted into spatial separation, so that temporal and spatial diversity become mathematically equivalent. Equation (13.4) is thus quite general in the sense that it can be applied to spatial, temporal, and frequency diversity. However, a number of assumptions were made in the derivation of this equation: (i) validity of the Wide Sense Stationary Uncorrelated Scatterer (WSSUS) model, (ii) no existence of Line Of Sight (LOS), (iii) exponential shape of the Power Delay Profile (PDP), (iv) isotropic distribution of incident power, and (v) use of omnidirectional antennas.

Example 13.2 *Compute the correlation coefficient of two frequencies with separation (i) 30 kHz, (ii) 200 kHz, (iii) 5 MHz, in the "typical urban" environment, as defined in COST 207[3] channel models.*

For zero temporal separation, the Bessel function in Eq. (13.4) is unity, and the correlation coefficient is thus only dependent on rms delay spread and frequency separation. Rms delay spread is calculated according to Eq. (6.39), where the PDP of the COST 207 typical urban channel model is given in Section 7.6.3. We obtain

$$S_\tau = \sqrt{\frac{\int_0^{7 \cdot 10^{-6}} e^{-\tau/10^{-6}} \tau^2 \, d\tau}{\int_0^{7 \cdot 10^{-6}} e^{-\tau/10^{-6}} \, d\tau} - \left(\frac{\int_0^{7 \cdot 10^{-6}} e^{-\tau/10^{-6}} \tau \, d\tau}{\int_0^{7 \cdot 10^{-6}} e^{-\tau/10^{-6}} \, d\tau} \right)^2}$$

$$= 0.977 \, \mu s \qquad (13.5)$$

---

[3] European COoperation in the field of Scientific and Technical research.

The correlation coefficient thus becomes

$$\rho_{xy} = \frac{1}{1 + (2\pi)^2 (0.977 \cdot 10^{-6})^2 (f_1 - f_2)^2}$$

$$= \begin{cases} 0.97, & f_1 - f_2 = 30\,\text{kHz} \\ 0.4, & f_1 - f_2 = 200\,\text{kHz} \\ 1 \cdot 10^{-3}, & f_1 - f_2 = 5\,\text{MHz} \end{cases} \tag{13.6}$$

From this, we can see that the correlation between two neighboring 30-kHz bands – as used, e.g., in the old IS-136[4] Time Division Multiple Access (TDMA) cellular system – is very high; correlation over 200-kHz bands (two neighboring channels in Global System for Mobile communication (GSM) are 200 kHz apart, see Chapter 24) is also appreciable, but two neighboring channels in Wideband Code Division Multiple Access (WCDMA) (5 MHz apart) are uncorrelated in this environment. Furthermore, carriers that are, e.g., 45 MHz apart – used for Base Station (BS)–MS and MS–BS communication in GSM (see Chapter 24 and also Chapter 17) are completely uncorrelated.

13.2.1 Spatial Diversity

Spatial diversity is the oldest and simplest form of diversity. Despite (or because) of this, it is also the most widely used. The transmit signal is received at several antenna elements, and the signals from these antennas are then further processed according to the principles that will be described in Section 13.4. But, irrespective of the processing method, performance is influenced by correlation of the signals between the antenna elements. A large correlation between signals at antenna elements is undesirable, as it decreases the effectiveness of diversity. A first important step in designing diversity antennas is thus to establish a relationship between antenna spacing and the correlation coefficient. This relationship is different for BS antennas and MS antennas, and thus will be treated separately.

1. *MS in cellular and cordless systems*: it is a standard assumption that waves are incident from all directions at the MS (see also Chapter 5). Thus, points of constructive and destructive interference of Multi Path Components (MPCs) – i.e., points where we have high and low received power, respectively – are spaced approximately $\lambda/4$ apart. This is therefore the distance that is required for decorrelation of received signals. This intuitive insight agrees very well with the results from the exact mathematical derivation (Eq. (13.4), with $f_2 - f_1 = 0$), given in Figure 13.1: decorrelation, defined as $\rho = 0.5$, occurs at an antenna separation of $\lambda/4$. Compare also Example 5.3.

   The above considerations imply that the minimum distance for antenna elements in GSM (at 900 MHz) is about 8 cm, and for various cordless and cellular systems at the 1,800-MHz band it is about 4 cm. For Wireless Local Area Networks (WLANs) (at 2.4 and 5 GHz), the distances are even smaller. It is thus clearly possible to place two antennas on an MS of a cellular system.

2. *BS in cordless systems and WLANs*: in a first approximation, the angular distribution of incident radiation at indoor BSs is also uniform – i.e., radiation is incident with equal strength from all directions. Therefore, the same rules apply as for MSs.

3. *BSs in cellular systems*: for a cellular BS, the assumption of uniform directions of incidence is no longer valid. Interacting Objects (IOs) are typically concentrated around the MS (Figure 13.2, see also Chapter 7). Since all waves are incident essentially from one direction, the correlation

---

[4] IS-136 is a (now defunct) second-generation cellular system using TDMA.

coefficient (for a given distance between antenna elements d_a) is much higher. Expressed differently, the antenna spacing required to obtain sufficient decorrelation increases.

To get an intuitive insight, we start with the simple case when there are only two MPCs whose wave vectors are at an angle α with respect to each other (Figure 13.3, see also Chapter 5). It is obvious that the distance between the maxima and minima of the interference pattern is larger the smaller α is. For very small α, the connection line between antenna elements lies on a "ridge" of the interference pattern and antenna elements are completely correlated.

Numerical evaluations of the correlation coefficient as a function of antenna spacing are shown in Figure 13.4. The first column shows the results for rectangular angular power spectra; the results for Gaussian distributions are shown in the second column. We can see that antenna spacing has to be on the order of 2–20 wavelengths for angular spreads between $1°$ and $5°$ in order to achieve decorrelation. We also find that it is mostly rms angular spread that determines the required antenna spacing, while the shape of the angular power spectrum has only a minor influence.

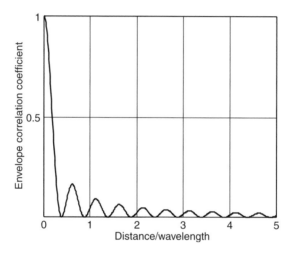

Figure 13.1 Envelope correlation coefficient as a function of antenna separation.

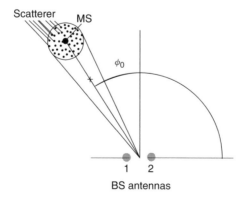

Figure 13.2 Scatterers concentrated around the mobile station.

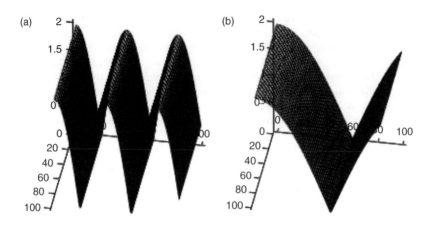

Figure 13.3 Interference pattern of two waves with 45° (a) and 15° (b) angular separation.

13.2.2 Temporal Diversity

As the wireless propagation channel is time variant, signals that are received at different times are uncorrelated. For "sufficient" decorrelation, the temporal distance must be at least $1/(2\nu_{max})$, where ν_{max} is the maximum Doppler frequency. In a static channel, where neither transmitter (TX), RX, nor the IOs are moving, the channel state is the same at all times. Such a situation can occur, e.g., for WLANs. In such a case, the correlation coefficient is $\rho = 1$ for all time intervals, and temporal diversity is useless.

Temporal diversity can be realized in different ways:

1. *Repetition coding*: this is the simplest form. The signal is repeated several times, where the repetition intervals are long enough to achieve decorrelation. This obviously achieves diversity, but is also highly bandwidth inefficient. Spectral efficiency decreases by a factor that is equal to the number of repetitions.
2. *Automatic Repeat reQuest (ARQ)*: here, the RX sends a message to the TX to indicate whether it received the data with sufficient quality (see Appendix 14.A at *www.wiley.com/go/molisch*). If this is not the case, then the transmission is repeated (after a wait period that achieves decorrelation). The spectral efficiency of ARQ is better than that of repetition coding, since it requires multiple transmissions only when the first transmission occurs in a bad fading state, while for repetition coding, retransmissions occur always. On the downside, ARQ requires a feedback channel.
3. *Combination of interleaving and coding*: a more advanced version of repetition coding is forward error correction coding with interleaving. The different symbols of a codeword are transmitted at different times, which increases the probability that at least some of them arrive with a good SNR. The transmitted codeword can then be reconstructed. For more details, see Chapter 14.

13.2.3 Frequency Diversity

In frequency diversity, the same signal is transmitted at two (or more) different frequencies. If these frequencies are spaced apart by more than the coherence bandwidth of the channel, then their fading is approximately independent, and the probability is low that the signal is in a deep fade at both frequencies simultaneously. For an exponential PDP, the correlation between two

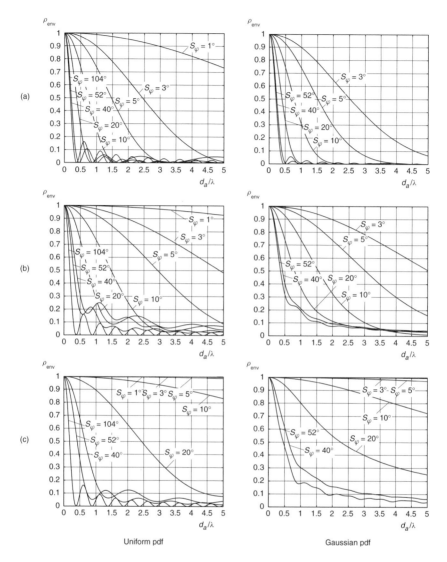

Figure 13.4 Envelope correlation coefficient at the BS for uniform and Gaussian probability density function (pdf) of the directions of arrival (a) $\phi_0 = 90°$, (b) $\phi_0 = 45°$, (c) $\phi_0 = 10°$, and different values of angular spread S_φ. Reproduced with permission from Fuhl et al. [1998] © IEE.

frequencies can be obtained from Eq. (13.4) by setting the numerator to unity as the signals at the two frequencies occur at the same time. Thus

$$\rho = \frac{1}{1 + (2\pi)^2 S_\tau^2 (f_2 - f_1)^2} \tag{13.7}$$

This again confirms that the two signals have to be at least one coherence bandwidth apart from each other. Figure 13.5 shows ρ as a function of the spacing between the two frequencies. For a more general discussion of frequency correlation, see Chapter 6.

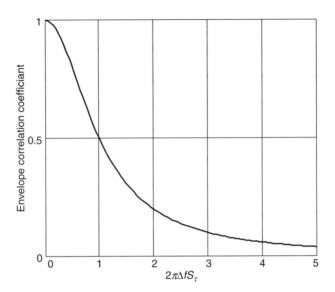

Figure 13.5 Correlation coefficient of the envelope as a function of normalized frequency spacing.

It is not common to actually repeat the same information at two different frequencies, as this would greatly decrease spectral efficiency. Rather, information is spread over a large bandwidth, so that small parts of the information are conveyed by different frequency components. The RX can then sum over the different frequencies to recover the original information.

This spreading can be done by different methods:

- *Compressing the information in time*: – i.e., sending short bursts that each occupy a large bandwidth – TDMA (see Chapter 17).
- *Code Division Multiple Access (CDMA)*: (Section 18.2).
- Multicarrier CDMA (Section 19.9) and coded orthogonal frequency division multiplexing (Section 19.4.3).
- *Frequency hopping in conjunction with coding*: different parts of a codeword are transmitted on different carrier frequencies (Section 18.1).

These methods allow the transmission of information without wasting bandwidth and will be described in greater detail in Chapters 17–19. For the moment, we just stress that the use of frequency diversity requires the channel to be frequency selective. In other words, frequency diversity (delay dispersion) can be exploited by the system to make it more robust, and decrease the effects of fading. This seems to be a contradiction to the results of Chapter 12, where we had shown that frequency selectivity leads to an *increase* of the BER, and even an error floor. The reason for this discrepancy is that Chapter 12 considered a very simple RX that takes no measures to combat (or exploit) the effects of frequency selectivity.

13.2.4 Angle Diversity

A fading dip is created when MPCs, which usually come from different directions, interfere destructively. If some of these waves are attenuated or eliminated, then the location of fading dips changes.

In other words, two colocated antennas with different patterns "see" differently weighted MPCs, so that the MPCs interfere differently for the two antennas. This is the principle of *angle diversity* (also known as *pattern diversity*).

Angular diversity is usually used in conjunction with spatial diversity; it enhances the decorrelation of signals at closely spaced antennas. Different antenna patterns can be achieved very easily. Of course, different types of antennas have different patterns. But even identical antennas can have different patterns when mounted close to each other (see Figure 13.6). This effect is due to *mutual coupling*: antenna B acts as a reflector for antenna A, whose pattern is therefore skewed to the left.[5] Analogously, the pattern of antenna B is skewed to the right due to reflections from antenna A. Thus, the two patterns are different.

The different patterns are even more pronounced when the antennas are located on different parts of the casing. While dipole antennas are usually restricted to the top of the casing, patch antennas and inverted-F antennas (see Chapter 9) can be placed on all parts of the casing (see Figure 13.7). In all of these cases, decorrelation is good even if the antennas are placed very closely to each other.

Figure 13.6   Angle diversity for closely spaced antennas.

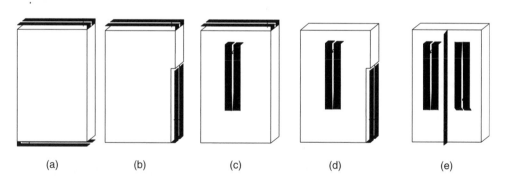

Figure 13.7   Configurations of diversity antennas at a mobile station.
Reproduced with permission from Erätuuli and Bonek [1997] © IEEE.

---

[5] This arrangement can also be considered as a Yagi antenna. It depends on the spacing of the two elements whether antenna B acts as director or reflector, and thus which direction the pattern is skewed into.

13.2.5   Polarization Diversity

Horizontally and vertically polarized MPCs propagate differently in a wireless channel,[6] as the reflection and diffraction processes depend on polarization (see Chapters 4 and 7). Even if the transmit antenna only sends signals with a single polarization, the propagation effects in the channel lead to depolarization so that both polarizations arrive at the RX. The fading of signals with different polarizations is statistically independent. Thus, receiving both polarizations using a dual-polarized antenna, and processing the signals separately, offers diversity. This diversity can be obtained without any requirement for a minimum distance between antenna elements.

Let us now consider more closely the situation where the transmit signal is vertically polarized, while the signal is received in both vertical and horizontal polarization. In that case, fading of the two received signals is independent, but the average received signal strength in the two diversity branches is *not* identical. Depending on the environment, the horizontal (i.e., cross-polarized) component is some 3–20 dB weaker than the vertical (co-polarized) component. As we will see later on, this has an important impact on the effectiveness of the diversity scheme. Various antenna arrangements have been proposed in order to mitigate this problem.

It has also been claimed that the diversity order that can be achieved with polarization diversity is up to 6: three possible components of the E-field and three components of the H-field can all be exploited [Andrews et al. 2001].[7] However, propagation characteristics as well as practical considerations prevent a full exploitation of that diversity order especially for outdoor situations. This is usually not a serious restriction for diversity systems, as we will see later on that going from diversity order 1 (i.e., no diversity) to diversity order 2 gives larger benefits than increasing the diversity order from 2 to higher values. However, it is an important issue for Multiple Input Multiple Output (MIMO) systems (see Section 20.2).

13.3   Macrodiversity and Simulcast

The previous section described diversity methods that combat small-scale fading – i.e., the fading created by interference of MPCs. However, not all of these diversity methods are suitable for combating large-scale fading, which is created by shadowing effects. Shadowing is almost independent of transmit frequency and polarization, so that frequency diversity or polarization diversity are not effective. Spatial diversity (or equivalently, temporal diversity with moving TX/RX) can be used, but we have to keep in mind that the correlation distances for large-scale fading are on the order of tens or hundreds of meters. In other words, if there is a hill between the TX and RX, adding antennas on either the BS or the MS does not help to eliminate the shadowing caused by this hill. Rather, we should use a separate base station (BS2) that is placed in such a way that the hill is not in the connection line between the MS and BS2. This in turn implies a large distance between BS1 and BS2, which gives rise to the word *macrodiversity*.

The simplest method for macrodiversity is the use of *on-frequency repeaters* that receive the signal and retransmit an amplified version of it. *Simulcast* is very similar to this approach; the same signal is transmitted simultaneously from different BSs. In cellular applications the two BSs should be synchronized, and transmit the signals intended for a specific user in such a way that the two waves arrive at the RX almost simultaneously (timing advance).[8] Note that synchronization

---

[6] For simplicity, we henceforth speak of horizontal and vertical polarization. However, the considerations are valid for any two orthogonal polarizations.

[7] Note that this result is surrounded by some controversy, and (at the time of this writing) there is no general agreement in the scientific community whether diversity order 6 can really be achieved from polarization only.

[8] If the RX cannot easily deal with delay dispersion, it is desirable that the two signals arrive exactly at the same time. For advanced RXs (Chapters 16–19) it can be desirable to have a small amount of delay dispersion; if the timing of the BSs and the runtimes of the signals to the RX are exactly known, this can also be achieved.

can only be obtained if the runtimes from the two BSs to the MS are known. Generally speaking, it is desirable that the synchronization error is no larger than the delay dispersion that the RX can handle. Especially critical are RXs in regions where the strengths of the signals from the two BSs are approximately equal.

Simulcast is also widely used for broadcast applications, especially digital TV. In this case, the exact synchronization of all possible RXs is not possible – each RX would require a different timing advance from the TXs.

A disadvantage of simulcast is the large amount of signaling information that has to be carried on landlines. Synchronization information as well as transmit data have to be transported on landlines (or microwave links) to the BSs. This used to be a serious problem in the early days of digital mobile telephony, but the current wide availability of fiber-optic links has made this less of an issue.

The use of on-frequency repeaters is simpler than that of simulcast, as no synchronization is required. On the other hand, delay dispersion is larger, because (i) the runtime from BS to repeater, and repeater to MS is larger (compared with the runtime from a second BS), and (ii) the repeater itself introduces additional delays due to the group delays of electronic components, filters, etc.

13.4   Combination of Signals

Now we turn our attention to the question of how to use diversity signals in a way that improves the total quality of the signal that is to be detected. To simplify the notation, we speak here only about the combination of signals from different antenna signals at the RX. However, the mathematical methods remain valid for other types of diversity signals as well. In general, we can distinguish two ways of exploiting signals from the multiple diversity branches:

1. *Selection diversity*, where the "best" signal copy is selected and processed (demodulated and decoded), while all other copies are discarded. There are different criteria for what constitutes the "best" signal.
2. *Combining diversity*, where all copies of the signal are combined (before or after the demodulator), and the combined signal is decoded. Again, there are different algorithms for combination of the signals.

Combining diversity leads to better performance, as all available information is exploited. On the downside, it requires a more complex RX than selection diversity. In most RXs, all processing is done in the baseband. Thus, an RX with combining diversity needs to downconvert all available signals, and combine them appropriately in the baseband. Thus, it requires N_r antenna elements as well as N_r complete Radio Frequency (RF) (downconversion) chains. An RX with selection diversity requires only *one* RF chain, as it processes only a single received signal at a time.

In the following, we give a more detailed description of selection (combination) criteria and algorithms. We assume that different signal copies undergo statistically independent fading – this greatly simplifies the discussion of both the intuitive explanations and the mathematics of the signal combination. Discussions on the impact of a finite correlation coefficient are relegated to Section 13.5.

In these considerations, we also have to keep in mind that the gain of multiple antennas is due to two effects: *diversity gain* and *beamforming gain*. Diversity gain reflects the fact that it is improbable that several antenna elements are in a fading dip simultaneously; the probability for very low signal levels is thus decreased by the use of multiple antenna elements. Beamforming gain reflects the fact that (for combining diversity) the combiner performs an averaging over the noise at different antennas. Thus, even if the signal levels at all antenna elements are identical, the combiner output SNR is larger than the SNR at a single-antenna element.

13.4.1   Selection Diversity

Received-Signal-Strength-Indication-Driven Diversity

In this method, the RX selects the signal with the largest instantaneous power (or *Received Signal Strength Indication – RSSI*), and processes it further. This method requires N_r antenna elements, N_r RSSI sensors, and a N_r-to-1 multiplexer (switch), but only one RF chain (see Figure 13.8). The method allows simple tracking of the selection criterion even in fast-fading channels. Thus, we can switch to a better antenna as soon as the RSSI becomes higher there.

1. If the BER is determined by noise, then RSSI-driven diversity is the best of all the selection diversity methods, as maximization of the RSSI also maximizes the SNR.
2. If the BER is determined by co-channel interference, then RSSI is no longer a good selection criterion. High receive power can be caused by a high level of interference, such that the RSSI criterion makes the system select branches with a low signal-to-interference ratio. This is especially critical when interference is caused mainly by one dominant interferer – a situation that is typical for Frequency Division Multiple Access (FDMA) or TDMA systems.
3. Similarly, RSSI-driven diversity is suboptimum if the errors are caused by the frequency selectivity of the channel. RSSI-driven diversity can still be a reasonable approximation, because

Figure 13.8   Selection diversity principle: (a) Received-signal-strength-indication-controlled diversity. (b) Bit-error-rate-controlled diversity.

we have shown in Chapter 12 that errors caused by signal distortion occur mainly in the fading dips of the channel. However, this is only an approximation, and it can be shown that (uncoded, unequalized) systems with RSSI-driven selection diversity have a BER that is higher by a constant factor compared with optimum (BER-driven) diversity.

For an exact performance assessment (Section 13.5), it is important to obtain the SNR distribution of the output of the selector. Assume that the instantaneous signal amplitude is Rayleigh distributed, such that the SNR of the nth diversity branch, γ_n, is (see Eq. 12.48)

$$\mathrm{pdf}_{\gamma_n}(\gamma_n) = \frac{1}{\overline{\gamma}} \exp\left(-\frac{\gamma_n}{\overline{\gamma}}\right) \tag{13.8}$$

where $\overline{\gamma}$ is the mean branch SNR (assumed to be identical for all diversity branches). The cumulative distribution function (cdf) is then

$$\mathrm{cdf}_{\gamma_n}(\gamma_n) = 1 - \exp\left(-\frac{\gamma_n}{\overline{\gamma}}\right) \tag{13.9}$$

The cdf is, by definition, the probability that the instantaneous SNR lies below a given level. As the RX selects the branch with the largest SNR, the probability that the chosen signal lies below the threshold is the product of the probabilities that the SNR at each branch is below the threshold. In other words, the cdf of the selected signal is the product of the cdfs of each branch:

$$\mathrm{cdf}_{\gamma}(\gamma) = \left[1 - \exp\left(-\frac{\gamma}{\overline{\gamma}}\right)\right]^{N_{\mathrm{r}}} \tag{13.10}$$

Example 13.3 *Compute the probability that the output power of a selection diversity system is 5 dB lower than the mean power of each branch, when using $N_{\mathrm{r}} = 1, 2, 4$ antennas.*

The threshold is $\gamma_{|\mathrm{dB}} = \overline{\gamma}_{|\mathrm{dB}} - 5\,\mathrm{dB}$, which in linear scale is $\gamma = \overline{\gamma} \cdot 10^{-0.5}$. Using Eq. (13.10), the probability that the output power is less than $\overline{\gamma} \cdot 10^{-0.5}$ becomes:

$$\mathrm{cdf}_{\gamma}(\overline{\gamma} \cdot 10^{-0.5}) = [1 - \exp(-10^{-0.5})]^{N_{\mathrm{r}}}$$

$$= \begin{cases} 0.27, & N_{\mathrm{r}} = 1 \\ 7.4 \cdot 10^{-2}, & N_{\mathrm{r}} = 2 \\ 5.4 \cdot 10^{-3}, & N_{\mathrm{r}} = 4 \end{cases} \tag{13.11}$$

Example 13.4 *Consider now the case that $N_{\mathrm{r}} = 2$, and that the mean powers in the branches are $1.5\overline{\gamma}$ and $0.5\overline{\gamma}$, respectively. How does the result change?*

In this case, the probability is

$$\mathrm{cdf}_{\gamma}(\overline{\gamma} \cdot 10^{-0.5}) = \left[1 - \exp\left(-\frac{1}{1.5} \cdot 10^{-0.5}\right)\right]\left[1 - \exp\left(-\frac{1}{0.5} \cdot 10^{-0.5}\right)\right] \tag{13.12}$$

$$= 8.9 \cdot 10^{-2} \tag{13.13}$$

This demonstrates that diversity is less efficient when the average branch powers are different.

Bit-Error-Rate-Driven Diversity

For BER-driven diversity, we first transmit a *training sequence* – i.e., a bit sequence that is known at the RX. The RX then demodulates the signal from each receive antenna element and compares it with the transmit signal. The antenna whose associated signal results in the smallest BER is judged to be the "best," and used for the subsequent reception of data signals. A similar approach is the use of the mean square error of the "soft-decision" demodulated signal, or the correlation between transmit and receive signal.

If the channel is time variant, the training sequence has to be repeated at regular intervals and selection of the best antenna has to be done anew. The necessary repetition rate depends on the coherence time of the channel.

BER-driven diversity has several drawbacks:

1. The RX needs either N_r RF chains and demodulators (which makes the RX more complex), or the training sequence has to be repeated N_r times (which decreases spectral efficiency), so that the signal at all antenna elements can be evaluated.
2. If the RX has only one demodulator, then it is not possible to continuously monitor the selection criterion (i.e., the BER) of all diversity branches. This is especially critical if the channel changes quickly.
3. Since the duration of the training sequence is finite, the selection criterion – i.e., bit error probability – cannot be determined exactly. The variance of the BER around its true mean decreases as the duration of the training sequence increases. There is thus a tradeoff between performance loss due to erroneous determination of the selection criterion, and spectral efficiency loss due to longer training sequences.

13.4.2 Switched Diversity

The main drawback of selection diversity is that the selection criteria (power, BER, etc.) of *all* diversity branches have to be monitored in order to know when to select a different antenna. As we have shown above, this leads to either increased hardware effort or reduced spectral efficiency. An alternative solution, which avoids these drawbacks, is *switched diversity*. In this method, the selection criterion of just the active diversity branch is monitored. If it falls below a certain threshold, then the RX switches to a different antenna.[9] Switching only depends on the quality of the active diversity branch; it does not matter whether the other branch actually provides a better signal quality or not.

Switched diversity runs into problems when both branches have signal quality below the threshold: in that case, the RX just switches back and forth between the branches. This problem can be avoided by introducing a hysteresis or hold time, so that the new diversity branch is used for a certain amount of time, independent of the actual signal quality. We thus have two free parameters: switching threshold and hysteresis time. These parameters have to be selected very carefully: if the threshold is chosen too low, then a diversity branch is used even when the other antenna might offer better quality; if it is chosen too high, then it becomes probable that the branch the RX switches to actually offers lower signal quality than the currently active one. If hysteresis time is chosen too long, then a "bad" diversity branch can be used for a long time; if it is chosen too short, then the RX spends all the time switching between two antennas.

Summarizing, the performance of switched diversity is worse than that of selection diversity; we will therefore not consider it further.

---

[9] The method is mostly applied when two diversity branches are available.

13.4.3   Combining Diversity

Basic Principle

Selection diversity wastes signal energy by discarding $(N_r - 1)$ copies of the received signal. This drawback is avoided by combining diversity, which exploits *all* available signal copies. Each signal copy is multiplied by a (complex) weight and then added up. Each complex weight w_n^* can be thought of as consisting of a phase correction,[10] plus a (real) weight for the amplitude:

- Phase correction causes the *signal amplitudes* to add up, while, on the other hand, noise is added incoherently, so that *noise powers* add up.
- For amplitude weighting, two methods are widely used: *Maximum Ratio Combining* (MRC) weighs all signal copies by their amplitude. It can be shown that (using some assumptions) this is an optimum combination strategy. An alternative is *Equal Gain Combining* (EGC), where all amplitude weights are the same (in other words, there is no weighting, but just a phase correction). The two methods are outlined in Figure 13.9.

Maximum Ratio Combining

MRC compensates for the phases, and weights the signals from the different antenna branches according to their SNR. This is the optimum way of combining different diversity branches – if several assumptions are fulfilled. Let us assume a propagation channel that is slow fading and flat fading. The only disturbance is AWGN. Under these assumptions, each channel realization can be written as a time-invariant filter with impulse response:

$$h_n(\tau) = \alpha_n \delta(\tau) \tag{13.14}$$

where α_n is the (instantaneous) gain of diversity branch n. This signals at the different branches are multiplied with weights w_n^* and added up, so that the SNR becomes

$$\frac{\left|\sum_{n=1}^{N} w_n^* \alpha_n\right|^2}{P_n \sum_{n=1}^{N} |w_n|^2} \tag{13.15}$$

where P_n is the noise power per branch (assumed to be the same in each branch). According to the Cauchy–Schwartz inequality, $\left|\sum_{n=1}^{N} w_n^* \alpha_n\right|^2 \leq \sum_{n=1}^{N} |w_n^*|^2 \sum_{n=1}^{N} |\alpha_n|^2$, where equality holds if and only if $w_n = \alpha_n$. Thus, the SNR is maximized by choosing the weights as

$$\text{w}_{\text{MRC}} = \alpha_n \tag{13.16}$$

i.e., the signals are phase-corrected (remember that the received signals are multiplied with w^*) and weighted by the amplitude. We can then easily see that in that case the output SNR of the diversity combiner is the *sum* of the branch SNRs:

$$\gamma_{\text{MRC}} = \sum_{n=1}^{N_r} \gamma_n \tag{13.17}$$

---

[10] Note that in our notation the signals are weighted with the *complex conjugate* of w; this simplifies notation later on.

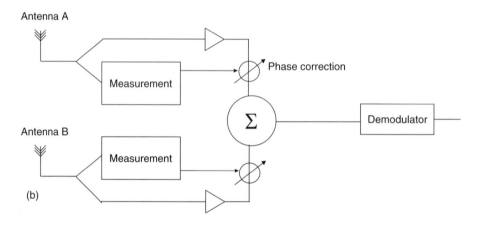

Figure 13.9 Combining diversity principle: (a) maximum ratio combining, (b) equal gain combining.

If the branches are statistically independent, then the moment-generating function of the total SNR can be computed as the product of the characteristic functions of the branch SNRs. If, furthermore, the SNR distribution in each branch is exponential (corresponding to Rayleigh fading), and all branches have the same mean SNR $\bar{\gamma}_n = \bar{\gamma}$, we find after some manipulations that

$$pdf_\gamma(\gamma) = \frac{1}{(N_r - 1)!} \frac{\gamma^{N_r - 1}}{\bar{\gamma}^{N_r}} \exp\left(-\frac{\gamma}{\bar{\gamma}}\right) \tag{13.18}$$

and the mean SNR of the combiner output is just the mean branch SNR, multiplied by the number of diversity branches:

$$\bar{\gamma}_{MRC} = N_r \bar{\gamma} \tag{13.19}$$

Figure 13.10 compares the statistics of the SNR for RSSI-driven selection diversity and MRC. Naturally, there is no difference between the diversity types for $N_r = 1$, since there is no diversity.

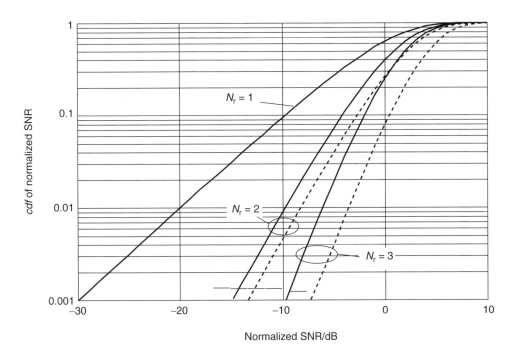

Figure 13.10 Cumulative distribution function of the normalized instantaneous signal-to-noise ratio $\gamma/\overline{\gamma}$ for received-signal-strength-indication-driven selection diversity (solid), and maximum ratio combining (dashed) for $N_r = 1, 2, 3$. Note that for $N_r = 1$, there is no difference between diversity types.

We furthermore see that the slope of the distribution is the same for MRC and selection diversity, but that the difference in the mean values increases with increasing N_r. This is intuitively clear, as selection diversity discards $N_r - 1$ signal copies – something that increases with N_r. For $N_r = 3$, the difference between the two types of diversity is only about 2 dB.

Equal Gain Combining

For EGC, we find that the SNR of the combiner output is

$$\gamma_{ECG} = \frac{\left(\sum_{n=1}^{N_r} \sqrt{\gamma_n}\right)^2}{N_r} \qquad (13.20)$$

where we have assumed that noise levels are the same on all diversity branches. The mean SNR of the combiner output can be found to be

$$\overline{\gamma}_{EGC} = \overline{\gamma}\left(1 + (N_r - 1)\frac{\pi}{4}\right) \qquad (13.21)$$

if all branches suffer from Rayleigh fading with the same mean SNR $\overline{\gamma}$. Remember that we only assume here that the *mean* SNR is the same in all branches, while instantaneous branch SNRs (representing different channel realizations) can be different. It is quite remarkable that EGC performs worse than MRC by only a factor $\pi/4$ (in terms of mean SNR). The performance difference between EGC and MRC becomes bigger when mean branch SNRs are also different.

Equations for the pdf of the SNR can be computed from Eq. (13.20), but become unwieldy very quickly (results can be found, e.g., in Lee [1982]).

Optimum Combining

One of the assumptions in the derivation of MRC was that only AWGN disturbs the signal. If it is interference that determines signal quality, then MRC is no longer the best solution. In order to maximize the Signal-to-Interference-and-Noise Ratio (SINR), the weights should then be determined according to a strategy called *optimum combining*, first derived in the groundbreaking paper of Winters [1984]. The first step is the determination of the correlation matrix of noise and interference at the different antenna elements:

$$\mathbf{R} = \sigma_n^2 \mathbf{I} + \sum_{k=1}^{K} E\{\mathbf{r}_k \mathbf{r}_k^\dagger\} \tag{13.22}$$

where expectation is over a time period when the channel remains constant, and $\mathbf{r}_k$ is the receive signal vector of the kth interferer. We furthermore need the complex transfer function (complex gain, since we assume flat-fading channels) of the N_r diversity branches; these are written into the vector $\mathbf{h}_d$. The vector containing the optimum receive weights is then

$$\mathbf{w}_{opt} = \mathbf{R}^{-1} \mathbf{h}_d \tag{13.23}$$

These weights have to be adjusted as the channel changes. It is easy to see that for a noise-limited system the correlation matrix becomes a (scaled) identity matrix, and optimum combining reduces to MRC.

A further interesting observation is that minimizing the Mean Square Error (MSE) is equivalent to maximizing the SINR, which can be written as

$$\text{SINR} = ((\mathbf{w}^\dagger \mathbf{h}_d \mathbf{h}_d^\dagger \mathbf{w})/(\mathbf{w}^\dagger \mathbf{R} \mathbf{w})) \tag{13.24}$$

This SINR is a generalized Rayleigh quotient, and is maximized by the generalized eigenvalue corresponding to the maximum generalized eigenvector of

$$\mathbf{h}_d \mathbf{h}_d^\dagger \mathbf{w} = \lambda \mathbf{R} \mathbf{w} \tag{13.25}$$

The weight vector obtained from this generalized eigenvalue problem is the same as $\mathbf{w}_{opt}$ in Eq. (13.22).

Optimum combining of signals from N_r diversity branches gives N_r degrees of freedom. This allows interference from $N_r - 1$ interferers to be eliminated. Alternatively, $N_s \leq N_r - 1$ interferers can be eliminated, while the remaining $N_r - N_s$ antennas behave like "normal" diversity antennas that can be used for noise reduction. This seemingly simple statement has great impact on wireless system design! As we discussed in Chapter 3, many wireless systems are interference limited. The possibility of eliminating at least some of the interferers by appropriate diversity combining opens up the possibility of drastically improving the link quality of such systems, or of increasing their capacity (more details can be found in Section 20.1).

Example 13.5 *Consider a desired signal with Binary Phase Shift Keying (BPSK) propagating through a frequency-flat channel with $\mathbf{h}_d = [1, 0.5 + 0.7j]^T$ and a synchronous, interfering BPSK signal with $\mathbf{h}_{int} = [0.3, -0.2 + 1.7j]^T$. The noise variance is $\sigma_n^2 = 0.01$. Show that weighting the received signal with the weights as computed from Eq. (13.23) leads to mitigation of interference.*

Let the desired transmitted signal be $s_d(t)$, the interfering transmitted signal be $s_{int}(t)$, and $n_1(t)$ and $n_2(t)$ be independent zero-mean noise processes. It is assumed that desired and interfering signals are uncorrelated, and that $E\{s_d s_d^*\} = 1$, and also $E\{s_{int} s_{int}^*\} = 1$.

The noise-plus-interference correlation matrix is computed according to Eq. (13.22):

$$\mathbf{R} = \sigma_n^2 \mathbf{I} + \sum_{k=1}^{K} E\{\mathbf{r}_k \mathbf{r}_k^\dagger\} \tag{13.26}$$

$$= 0.01 \begin{pmatrix} 1 & 0 \\ 0 & 1 \end{pmatrix} + \begin{pmatrix} 0.3 \\ -0.2 + 1.7j \end{pmatrix} (0.3 \quad -0.2 - 1.7j) \tag{13.27}$$

$$= \begin{pmatrix} 0.1 & -0.06 - 0.51j \\ -0.06 + 0.51j & 2.94 \end{pmatrix} \tag{13.28}$$

Using Eq. (13.23) and normalizing the weights, we obtain:

$$\mathbf{w} = \frac{\mathbf{R}^{-1} \mathbf{h}_d}{|\mathbf{R}^{-1} \mathbf{h}_d|} \tag{13.29}$$

$$= \begin{pmatrix} 0.979 + 0.111j \\ 0.041 - 0.165j \end{pmatrix} \tag{13.30}$$

Inserting this value for $\mathbf{w}$ into Eq. (13.24), we obtain an SINR of 78. This can be compared to the SNR = 100, which could be obtained if the receiver would not have to suppress an interferer, and the SINR of 0.6 that is available at a single antenna.

Hybrid Selection – Maximum Ratio Combining

A compromise between selection diversity and full signal combining is the so-called hybrid selection scheme, where the best L out of N_r antenna signals are chosen, downconverted, and processed. This reduces the number of required RF chains from N_r to L, and thus leads to significant savings. The savings come at the price of a (usually small) performance loss compared with the full-complexity system. The approach is called *Hybrid Selection/Maximum Ratio Combining* (H-S/MRC), or sometimes also *Generalized Selection Combining* (GSC).

It is well known that the output SNR of MRC is just the sum of the SNRs at the different receive antenna elements. For H-S/MRC, the instantaneous output SNR of H-S/MRC looks deceptively similar to MRC – namely:

$$\gamma_{H-S/MRC} = \sum_{n=1}^{L} \gamma_{(n)} \tag{13.31}$$

The major difference from MRC is that the $\gamma_{(n)}$ are *ordered* SNRs – i.e., $\gamma_{(1)} > \gamma_{(2)} > \cdots > \gamma_{(N_r)}$. This leads to different performance, and poses new mathematical challenges for performance analysis. Specifically, we have to introduce the concept of *order statistics*. Note that selection diversity (where just one of N_r antennas is selected) and MRC are limiting cases of H-S/MRC with $L = 1$ and $L = N_r$, respectively.

H-S/MRC schemes provide good diversity gain, as they select the best antenna branches for combining. Actually, it can be shown that the diversity *order* obtained with such schemes is proportional to N_r, not to L. However, they do not provide full beamforming gain. If the signals at all antenna elements are completely correlated, then the SNR gain of H-S/MRC is only L, compared with N_r for an MRC scheme.

The analysis of H-S/MRC based on a chosen ordering of the branches at first appears to be complicated, since the SNR statistics of the ordered branches are *not* independent. However, we can alleviate this problem by transforming ordered branch variables into a new set of random variables. It is possible to find a transformation that leads to *independently distributed* random variables (termed *virtual branch variables*). The fact that the combiner output SNR can be expressed in terms of independent identically distributed (iid) virtual branch variables enormously simplifies performance analysis of the system. For example, the derivation of Symbol Error Probability (SEP) for uncoded H-S/MRC systems, which normally would require evaluation of nested N-fold integrals, essentially reduces to evaluation of a *single* integral with finite limits.

The mean and variance of the output SNR for H-S/MRC is thus

$$\overline{\gamma}_{\text{H--S/MRC}} = L \left(1 + \sum_{n=L+1}^{N_r} \frac{1}{n} \right) \overline{\gamma} \tag{13.32}$$

and

$$\sigma_{\text{H--S/MRC}}^2 = L \left(1 + L \sum_{n=L+1}^{N_r} \frac{1}{n^2} \right) \overline{\gamma}^2 \tag{13.33}$$

13.5 Error Probability in Fading Channels with Diversity Reception

In this section we determine the Symbol Error Rate (SER) in fading channels when diversity is used at the RX. We start with the case of flat-fading channels, computing the statistics of the received power and the BER. We then proceed to dispersive channels, where we analyze how diversity can mitigate the detrimental effects of dispersive channels on simple RXs.

13.5.1 Error Probability in Flat-Fading Channels

Classical Computation Method

Analogous to Chapter 12, we can compute the error probability of diversity systems by averaging the conditional error probability (conditioned on a certain SNR) over the distribution of the SNR:

$$\overline{SER} = \int_0^\infty pdf_\gamma(\gamma) SER(\gamma) \, d\gamma \tag{13.34}$$

As an example, let us compute the performance of BPSK with N_r diversity branches with MRC. The SER of BPSK in AWGN is (see Chapter 12)

$$SER(\gamma) = Q(\sqrt{2\gamma}) \tag{13.35}$$

Let us apply this principle to the case of MRC. When inserting Eqs. (13.18) and (13.35) into Eq. (13.34), we obtain an equation that can be evaluated analytically:

$$\overline{SER} = \left(\frac{1-b}{2} \right)^{N_r} \sum_{n=0}^{N_r-1} \binom{N_r - 1 + n}{n} \left(\frac{1+b}{2} \right)^n \tag{13.36}$$

where b is defined as

$$b = \sqrt{\frac{\overline{\gamma}}{1 + \overline{\gamma}}} \tag{13.37}$$

For large values of $\overline{\gamma}$, this can be approximated as

$$\overline{SER} = \left(\frac{1}{4\overline{\gamma}}\right)^{N_r} \binom{2N_r - 1}{N_r} \tag{13.38}$$

From this, we can see that (with N_r diversity antennas) the BER decreases with the N_r-th power of the SNR.

Computation via the Moment-Generating Function

In the previous section, we averaged the BER over the distribution of SNRs, using the "classical" representation of the Q-function. As we have already seen (Chapter 12), there is an alternative definition of the Q-function, which can easily be combined with the moment-generating function $M_\gamma(s)$ of the SNR. Let us start by writing the SER conditioned on a given SNR in the form (see Chapter 12):

$$SER(\gamma) = \int_{\theta_1}^{\theta_2} f_1(\theta) \exp(-\gamma_{\text{MRC}} f_2(\theta)) \, d\theta \tag{13.39}$$

Since

$$\gamma_{\text{MRC}} = \sum_{n=1}^{N_r} \gamma_n \tag{13.40}$$

this can be rewritten as

$$SER(\gamma) = \int_{\theta_1}^{\theta_2} f_1(\theta) \prod_{n=1}^{N_r} \exp(-\gamma_n f_2(\theta)) \, d\theta \tag{13.41}$$

Averaging over the SNRs in different branches then becomes

$$\overline{SER} = \int d\gamma_1 pdf_{\gamma_1}(\gamma_1) \int d\gamma_2 pdf_{\gamma_2}(\gamma_2) \cdots \int d\gamma_{N_r} pdf_{\gamma_{N_r}}(\gamma_{N_r}) \int_{\theta_1}^{\theta_2} d\theta f_1(\theta) \prod_{n=1}^{N_r} \exp(-\gamma_n f_2(\theta))$$

$$\tag{13.42}$$

$$= \int_{\theta_1}^{\theta_2} d\theta f_1(\theta) \prod_{n=1}^{N_r} \int d\gamma_n pdf_{\gamma_n}(\gamma_n) \exp(-\gamma_n f_2(\theta)) \tag{13.43}$$

$$= \int_{\theta_1}^{\theta_2} d\theta f_1(\theta) \prod_{n=1}^{N_r} M_\gamma(-f_2(\theta)) \tag{13.44}$$

$$= \int_{\theta_1}^{\theta_2} d\theta f_1(\theta) [M_\gamma(-f_2(\theta))]^{N_r} \tag{13.45}$$

With that, we can write the error probability for BPSK in Rayleigh fading as

$$\overline{SER} = \frac{1}{\pi} \int_0^{\pi/2} \left[\frac{\sin^2(\theta)}{\sin^2(\theta) + \overline{\gamma}} \right]^{N_r} d\theta \tag{13.46}$$

Example 13.6 *Compare the symbol error probability in Rayleigh fading of 8-PSK (Phase Shift Keying) and four available antennas with 10-dB SNRs when using L = 1, 2, 4 receive chains.*

The SER for M-ary Phase Shift Keying (MPSK) with H-S/MRC can be shown to be

$$\overline{SER}_{e,\text{H-S/MRC}}^{\text{MPSK}} = \frac{1}{\pi} \int_0^{\pi(M-1)/M} \left[\frac{\sin^2 \theta}{\sin^2(\pi/M)\overline{\gamma} + \sin^2 \theta} \right]^L \prod_{n=L+1}^{N_r} \left[\frac{\sin^2 \theta}{\sin^2(\pi/M)\overline{\gamma}\frac{L}{n} + \sin^2 \theta} \right] d\theta$$

(13.47)

Evaluating Eq. (13.47) with $M = 8$, $\overline{\gamma} = 10\,\text{dB}$, $N_r = 4$, and $L = 1, 2, 4$ yields:

L	$\overline{SER}_{e,\text{H-S/MRC}}^{\text{MPSK}}$
1	0.0442
2	0.0168
4	0.0090

13.5.2 Symbol Error Rate in Frequency-Selective Fading Channels

We now determine the SER in channels that suffer from time dispersion and frequency dispersion. We assume here FSK with differential phase detection. The analysis uses the correlation coefficient ρ_{XY} between signals at two sampling times that was discussed in Chapter 12.

For binary FSK with selection diversity:

$$\overline{SER} = \frac{1}{2} - \frac{1}{2} \sum_{n=1}^{N_r} \binom{N_r}{n} (-1)^{n+1} \frac{b_0 \,\text{Im}\{\rho_{XY}\}}{\sqrt{(\text{Im}\{\rho_{XY}\})^2 + n(1 - |\rho_{XY}|^2)}}$$

(13.48)

where b_0 is the transmitted bit. This can be approximated as

$$\overline{SER} = \frac{(2N_r - 1)!!}{2} \left(\frac{1 - |\rho_{XY}|^2}{2(\text{Im}\{\rho_{XY}\})^2} \right)^{N_r}$$

(13.49)

where $(2N_r - 1)!! = 1 \cdot 3 \cdot 5 \ldots (2N_r - 1)$.

For binary FSK with MRC:

$$\overline{SER} = \frac{1}{2} - \frac{1}{2} \frac{b_0 \text{Im}\{\rho_{XY}\}}{\sqrt{1 - (\text{Re}\{\rho_{XY}\})^2}} \sum_{n=0}^{N_r-1} \frac{(2n-1)!!}{(2n)!!} \left(1 - \frac{(\text{Im}\{\rho_{XY}\})^2}{1 - (\text{Re}\{\rho_{XY}\})^2} \right)^n$$

(13.50)

which can be approximated as

$$\overline{SER} = \frac{(2N_r - 1)!!}{2(N_r!)} \left(\frac{1 - |\rho_{XY}|^2}{2(\text{Im}\{\rho_{XY}\})^2} \right)^{N_r}$$

(13.51)

This formulation shows that MRC improves the SER by a factor $N_r!$ compared with selection diversity. A further important consequence is that the errors due to delay dispersion and random Frequency Modulation (FM) are decreased in the same way as errors due to noise. This is shown by the expressions in parentheses that are taken to the N_r-the power. These terms subsume the errors due to all different effects. The SER with diversity is approximately the N_r-th power of the SER without diversity (see Figures 13.11–13.13).

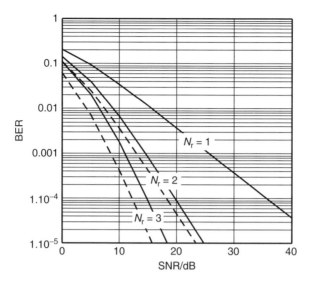

Figure 13.11   Bit error rate of minimum shift keying (MSK) with received-signal-strength-indication-driven selection diversity (solid) and maximum ratio combining (dashed) as a function of the signal-to-noise ratio with N_r diversity antennas.
Reproduced with permission from Molisch [2000] © Prentice Hall.

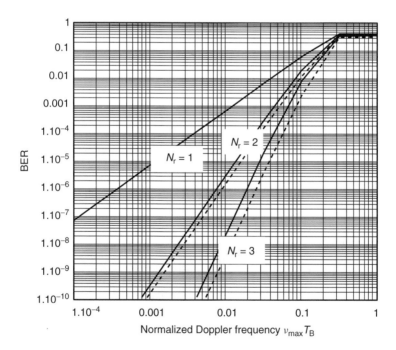

Figure 13.12   Bit error rate of MSK with received-signal-strength-indication-driven selection diversity (solid) and maximum ratio combining (dashed) as a function of the normalized Doppler frequency with N_r diversity antennas.
Reproduced with permission from Molisch [2000] © Prentice Hall.

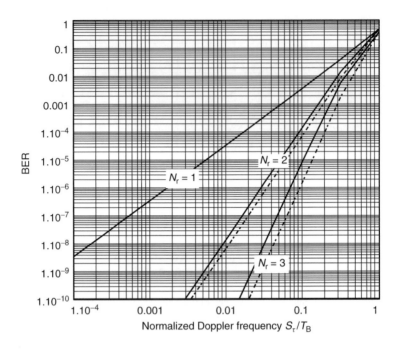

Figure 13.13 Bit error rate of MSK with received-signal-strength-indication-driven selection diversity (solid) and maximum ratio combining (dashed) as a function of the normalized rms delay spread with N_{r} diversity antennas. Reproduced with permission from Molisch [2000] © Prentice Hall.

For Differential Quadrature-Phase Shift Keying (DQPSK) with selection diversity, the average BER is

$$\overline{BER} = \frac{1}{2} - \frac{1}{4} \sum_{n=1}^{N_{\mathrm{r}}} \binom{N_{\mathrm{r}}}{n} (-1)^{n+1} \left[\frac{b_0 \mathrm{Re}\{\rho_{XY}\}}{\sqrt{(\mathrm{Re}\{\rho_{XY}\})^2 + n(1 - |\rho_{XY}|^2)}} \right.$$

$$\left. + \frac{b_0' \mathrm{Im}\{\rho_{XY}\}}{\sqrt{(\mathrm{Im}\{\rho_{XY}\})^2 + n(1 - |\rho_{XY}|^2)}} \right] \qquad (13.52)$$

where (b_0, b_0') is the transmitted symbol.

For DQPSK with MRC:

$$\overline{BER} = \frac{1}{2} - \frac{1}{4} \sum_{n=0}^{N_{\mathrm{r}}-1} \frac{(2n-1)!!}{(2n)!!} \left[\frac{b_0 \mathrm{Re}\{\rho_{XY}\}}{\sqrt{1 - (\mathrm{Im}\{\rho_{XY}\})^2}} \left(\frac{1 - |\rho_{XY}|^2}{1 - (\mathrm{Im}\{\rho_{XY}\})^2} \right)^n \right.$$

$$\left. + \frac{b_0' \mathrm{Im}\{\rho_{XY}\}}{\sqrt{1 - (\mathrm{Re}\{\rho_{XY}\})^2}} \left(\frac{1 - |\rho_{XY}|^2}{1 - (\mathrm{Re}\{\rho_{XY}\})^2} \right)^n \right] \qquad (13.53)$$

More general cases can be treated with the Quadratic Form Gaussian Variable (QFGV) method (see Section 12.3.3), where Eq. (12.103) is replaced by

$$P(D<0) = Q_M(p_1, p_2) - I_0(p_1 p_2) \exp\left[-\frac{1}{2}(p_1^2 + p_2^2)\right]$$

$$+ \frac{I_0(p_1 p_2) \exp\left[-\frac{1}{2}(p_1^2 + p_2^2)\right]}{(1 + v_2/v_1)^{2N_r - 1}} \sum_{n=0}^{N_r - 1} \binom{2N_r - 1}{n} \left(\frac{v_2}{v_1}\right)^n$$

$$+ \frac{\exp\left[-\frac{1}{2}(p_1^2 + p_2^2)\right]}{(1 + v_2/v_1)^{2N_r - 1}} \cdot \sum_{n=1}^{N_r - 1} I_n(p_1 p_2) \sum_{k=0}^{N_r - 1 - n} \binom{2N_r - 1}{k}$$

$$\times \left[\left(\frac{p_2}{p_1}\right)^n \left(\frac{v_2}{v_1}\right)^k - \left(\frac{p_1}{p_2}\right)^n \left(\frac{v_2}{v_1}\right)^{2N_r - 1 - k}\right] \tag{13.54}$$

RSSI-driven diversity is not the best selection strategy when errors are mostly caused by frequency selectivity and time selectivity. It puts emphasis on signals that have a large amplitude, and not on those with the smallest distortion.[11] In these cases, BER-driven selection diversity is preferable. For $N_r = 2$, the BER of Minimum Shift Keying (MSK) with differential detection becomes

$$\overline{BER} \approx \left(\frac{\pi}{4}\right)^4 \left(\frac{S_\tau}{T_B}\right)^4 \tag{13.55}$$

compared with the RSSI-driven result:

$$\overline{BER} \approx 3 \left(\frac{\pi}{4}\right)^4 \left(\frac{S_\tau}{T_B}\right)^4 \tag{13.56}$$

13.6 Transmit Diversity

For many situations, multiple antennas can be installed at just one link end (usually the BS). For the uplink transmission from the MS to BS, multiple antennas can act as receive diversity branches. For the downlink, any possible diversity originates at the *transmitter*. In this section, we will thus discuss ways of transmitting signals from several TX antennas and achieve a diversity effect with it. Note that this section specifically refers to *antenna* diversity (which encompasses spatial diversity, pattern diversity, and polarization diversity). Time diversity and frequency diversity inherently involve the TX, and thus need not be discussed again here.

The question of how transmit diversity can be combined with multiple antenna elements at the RX is discussed in Chapter 20.

13.6.1 Transmitter Diversity with Channel State Information

The first situation we analyze is the case where the TX knows the channel perfectly. This knowledge might be obtained from feedback from the RX, or from reciprocity principles; a more detailed discussion of this issue can be found in Section 20.1.6. In this case, we find that (at least for

---

[11] For a similar reason, MRC is not the best combining strategy.

the noise-limited case) there is a complete equivalence between transmit diversity and receive diversity. In other words, the optimum transmission scheme linearly weights signals transmitted from different antenna elements with the complex conjugates of the channel transfer functions from the transmit antenna elements to the single receive antenna. This approach is known as *maximum ratio transmission*.

13.6.2   Transmitter Diversity Without Channel State Information

In many cases, Channel State Information (CSI) is not available at the TX. We then cannot simply transmit weighted copies of the same signal from different transmit antennas, because we cannot know how they would add up at the RX. It is equally likely for the addition of different components to be constructive or destructive; in other words, we would just be adding up MPCs with random phases, which results in Rayleigh fading. We thus cannot gain any diversity (or beamforming).

In order to give benefits, transmission of the signals from different antenna elements has to be done is such a way that it allows the RX to distinguish different transmitted signal components. One way is *delay diversity*. In this scheme, signals transmitted from different antenna elements are delayed copies of the same signal. This makes sure that the effective impulse response is delay dispersive, even if the channel itself is flat fading. So, in a flat-fading channel, we transmit data streams with a delay of 1 symbol duration (relative to preceding antennas) from each of the transmit antennas. The effective impulse response of the channel then becomes

$$h(\tau) = \frac{1}{\sqrt{N_{\rm r}}} \sum_{n=1}^{N_{\rm t}} h_n \delta(\tau - nT_{\rm s}) \tag{13.57}$$

where the h_n are gains from the nth transmit antenna to the receive antenna, and the impulse response has been normalized so that total transmit power is independent of the number of antenna elements. The signals from different transmit antennas to the RX act effectively as delayed MPCs. If antenna elements are spaced sufficiently far apart, these coefficients fade independently. With an appropriate RX for delay-dispersive channels – e.g., an equalizer as described in Chapter 16, or a Rake RX as described in Chapter 18 – we get a diversity order that is equal to the number of antenna elements.

If the channel from a single transmit antenna to the RX is already delay dispersive, then the scheme still works, but care has to be taken in the choice of delays for different antenna elements. The delay between signals transmitted from different antenna elements should be at least as large as the maximum excess delay of the channel.

An alternative method is *phase-sweeping diversity*. In this method, which is especially useful if there are only two antenna elements, the same signal is transmitted from both antenna elements. However, one of the antenna signals undergoes a time-varying phase shift. This means that at the RX the received signals add up in a time-varying way; in other words, we are artificially introducing temporal variations into the channel. The reason for this is that – even if the TX, RX, and the IOs are stationary – the signal does not remain stuck in a fading dip. If this scheme is combined with appropriate coding and/or interleaving, it improves performance.

Yet another possibility for achieving transmit diversity is *space–time coding*. This method is discussed in Chapter 20.

13.7   Appendix

Please go to *www.wiley.com/go/molisch*

Further Reading

Spatial (antenna) diversity is the oldest form of diversity, and is discussed in textbooks on wireless propagation channels (see, e.g., Vaughan and Anderson [2003]). Evaluations of the antenna correlation coefficient for different angular spectra can be found in Fuhl et al. [1998] and Roy and Fortier [2004]. For antennas on a handset, results are given in Ogawa et al. [2001]. Taga [1993] also shows the effect of mutual coupling on pattern diversity and thus the correlation coefficient, as do a number of recent papers written in the context of MIMO systems [Waldschmidt et al. 2004 and Wallace and Jensen 2004]. Polarization diversity is discussed in more detail, e.g., in Narayanan et al. [2004] and Shafi et al. [2006]; joint spatial, polarization, and pattern (angle) diversity is discussed in Dietrich et al. [2001]. Different combination strategies for antenna signals, and the resulting channel statistics, are discussed in Proakis [2005], and in many papers in the primary literature. The situation is similar to that mentioned in Chapter 12: each type of fading statistics and each combination strategy merits at least one paper. Application of these fading statistics to computation of BERs gives rise to an even greater variety of papers. A nice, unified treatment can be found in Simon and Alouini [2004], who include different fading statistics and antenna combination strategies for a variety of modulation formats. For Rayleigh or Rice fading, the classical QFGV method of Proakis [1968] is applicable; Adachi and Ohno [1991] and Adachi and Parsons [1989] compute the performance for DQPSK and FSK, respectively. The virtual path method for H-S/MRC was introduced in Win and Winters [1999]; Simon and Alouini [2004] also investigate this case. General mathematical aspects of order statistics can be found in David and Nagaraja [2003]. Discussions on the impact of diversity on systems in frequency-selective channels can be found in Molisch [2000, ch. 13]. Maximum ratio transmission was first suggested in Lo [1999]. Delay diversity was suggested by Winters [1994] and Wittneben [1993]. An overview of various transmit diversity techniques can be found in Hottinen et al. [2003].

For updates and errata for this chapter, see wides.usc.edu/teaching/textbook

14

Channel Coding and Information Theory

14.1 Fundamentals of Coding and Information Theory

14.1.1 History and Motivation of Coding

Chapter 12 demonstrated that Bit Error Rates (BERs) on the order of 10^{-2} can occur for Signal-to-Noise Ratios (SNRs) typically encountered in wireless systems. Those high BERs are mostly due to the effect of multipath propagation. Advanced receiver structures can help to reduce those values: diversity combats fading dips, while equalizers and Rake receivers (see Chapters 16 and 18) improve the performance in frequency-selective channels. However, even those advanced receivers might not sufficiently reduce the BER. Data communications often require BERs on the order of $10^{-6} - 10^{-9}$. Such low values can only be achieved by employing coding of the data, i.e., introducing redundancy into the transmission. The use of error-correcting codes[1] leads to a reduction of the BER, or equivalently to a coding gain G_{code}, i.e., we have to use G_{code} decibel (dB) less transmit power to achieve the target BER than in an uncoded system.[2]

The history of coding starts with the seminal work of Claude Shannon [Shannon 1948] on "The mathematical theory of communication." He showed that it is possible to transmit data without errors as long as the bit rate is smaller than the *channel capacity*. The absence of errors is achieved by the use of "appropriate" codes. Shannon showed that (infinitely long) random codes achieve capacity. Unfortunately, such codes cannot be used in practice due to the enormous effort required for their decoding. For more than 50 years, the work of the coding theorists mainly consisted of finding practical codes that come close to the Shannon limit, i.e., allow communications with rates close to the channel capacity.

In the subsequent sections of this chapter, we will give a brief overview of error-correction coding. The basic coding theory holds for temporally constant and temporally varying channels; Sections 14.2–14.7 thus do not distinguish between those cases. Rather, they lay out the theoretical background of the most important classes of codes and their decoding: block codes, convolutional codes, Trellis Coded Modulation (TCM), turbo codes, and Low Density Parity Check (LDPC) codes. Sections 14.8–14.9 then deals specifically with the idiosyncrasies of fading channels and describes how the coding structures need to be adapted for this case.

---

[1] In the remainder of this chapter, we will say *coding* when we mean *error-correcting coding*. Note that this is different from Chapters 15 and 23, where coding will refer to source coding.

[2] Note that the coding gain can depend on the target error rate. Coding leads to a change of shape of the BER-over-SNR curve.

Wireless Communications, Second Edition   Andreas F. Molisch
© 2011 John Wiley & Sons, Ltd

14.1.2   Fundamental Concepts of Information Theory

Shannon's seminal work is the foundation of *information theory*, which explores the theoretical performance limits of optimum communication systems. Knowledge of information-theoretic limits is useful for system designers because it indicates how far a real system could be improved. Key concepts we will encounter are (i) a mathematically solid definition of *information*, and (ii) the *channel capacity*, which describes at what rate information can be transmitted, at best, over a given channel. We will find that suitable coding is required for communication to approach the channel capacity.

As a preliminary step, let us define the *Discrete Memoryless Channel* (DMC). We have an alphabet $\mathcal{X}$ of transmit symbols (the size of the alphabet is $|\mathcal{X}|$), where the probability for the transmission of each of the elements of the alphabet is known. We furthermore have a receive symbol alphabet $\mathcal{Y}$, with size $|\mathcal{Y}|$. Last, but not least, the DMC defines transition probabilities from each of the transmit symbols to each of the receive symbols. The most common form is the binary symmetric channel, which is characterized by $|\mathcal{X}| = |\mathcal{Y}| = 2$, and a transition probability p- e.g., if we transmit $X = +1$, then with $1 - p$ probability, the receive symbol is $Y = +1$, and with p probability, it is $Y = -1$; similarly if we transmit $X = -1$, probability for receiving -1 is $1 - p$, and for receiving $+1$ is p. A memoryless channel has furthermore the property that if we transmit a sequence of transmit symbols $\mathbf{x} = \{x_1, x_2, \ldots . x_N\}$ and observe a sequence of output symbols $\mathbf{y} = \{y_1, y_2, \ldots . y_N\}$, then

$$\Pr(\mathbf{y}|\mathbf{x}) = \prod_{n=1}^{N} \Pr(y_n|x_n) \qquad (14.1)$$

in other words, there is no InterSymbol Interference (ISI).

DMCs can describe, e.g., the concatenation of a Binary Phase Shift Keying (BPSK) modulator, an Additive White Gaussian Noise (AWGN) channel, and a demodulator with hard-decision output. Then, the variable X corresponds to the transmit symbols $+1/-1$. The variable Y corresponds to the output of the demodulator/decision device, which is also $+1/-1$. The amount of noise determines the symbol error probability (as computed in Chapter 12), which is in this case identical to the transition probability p. If we want to avoid the restrictions of hard-decision demodulators, a useful channel model is the discrete-time AWGN channel (as used in Chapters 11–13), with

$$y_n = x_n + n_n \qquad (14.2)$$

where the n_n are Gaussian-distributed random variables with variance σ^2, and the input variables are subject to an average power constraint, $E\{X^2\} \le P$.[3]

We now define the mutual information between two discrete random variables X and Y. If we observe a certain realization of Y, i.e., $Y = y$, then the mutual information is a measure for how much this observation tells us about the occurrence of an event $X = x$. The mutual information between the realizations x and y is defined as

$$I(x; y) = \log \frac{\Pr(x|y)}{\Pr(x)} \qquad (14.3)$$

where $\Pr(x|y)$ is the probability of x, conditioned on y. The logarithm in the above equation can either have base 2, in which case the mutual information is in units of *bits*, or it can have base e,

---

[3] Note that P is power, while p is the transition probability.

in which case the units of I are *nats*. The mutual information between the random variables X and Y is then the average of $I(x; y)$, namely

$$I(X; Y) = \sum_x \sum_y \Pr(X = x, Y = y) \log \frac{\Pr(x|y)}{\Pr(x)} \tag{14.4}$$

$$= \sum_x \sum_y \Pr(X = x, Y = y) \log \frac{\Pr(x, y)}{\Pr(x)\Pr(y)} \tag{14.5}$$

In the case that y and/or x are continuous variables, the summations are replaced by integrations.

An intuitive interpretation of the mutual information follows from the DMC: associate again X with the transmit symbols and Y with the received symbols. If X and Y are statistically independent, i.e., receiving a particular symbol gives us no information about which symbol was transmitted, then the mutual information is 0; this situation occurs in channels with very low SNR, in a binary symmetric channel this corresponds to, $p = 0.5$. A high mutual information occurs if receiving a certain symbol, say $+1$, tells us with high probability which symbol was transmitted. This situation occurs in channels with small transition probability p, or in other words, in channels with little noise. Note that the mutual information can never become larger than $\min[\log |\mathcal{X}|, \log |\mathcal{Y}|]$. If $X = Y$, then for discrete alphabets

$$I(X; Y) = -\sum_x \sum_y \Pr(X = x, Y = y) \log[\Pr(x)] \tag{14.6}$$

$$= -\sum_x \Pr(X = x) \log[\Pr(x)] \tag{14.7}$$

and is called the *entropy* $H(X)$. The entropy characterizes the information content of the random variable X. For discrete memoryless sources, the entropy is maximized if all symbols of the alphabet are equiprobable.

The general mutual information can also be written in terms of the entropy of the received symbols $H(Y)$ and the conditional entropy $H(Y|X) = -\sum_{x,y} \Pr(x, y) \log[\Pr(y|x)]$,

$$I(X; Y) = H(Y) - H(Y|X) = H(X) - H(X|Y) \tag{14.8}$$

14.1.3 Channel Capacity

For any communications over a DMC, the following theorem holds: reliable communications (i.e., communications with error probability tending to zero) is possible iff the communication rate $R < C$, where C is the channel capacity

$$C = \max_{\Pr(x)} I(X; Y) \tag{14.9}$$

i.e., the mutual information of X and Y maximized over all input distributions. If the rate is higher than the capacity, reliable communications is not possible.

The key to achieving, or at least approaching, the channel capacity is the use of coding with long codewords. As mentioned in the introduction, coding introduces redundancy, so that even if some bits in the codeword are received erroneously, it is still possible to recover the original codeword. Consider transmission of a codeword of length N over a binary symmetric channel: as $N \to \infty$, the probability of receiving $\approx Np$ errors tends to unity. There are

$$\binom{N}{Np} = \frac{N!}{(Np)!(N(1-p)!)} \tag{14.10}$$

$$\approx \frac{\sqrt{2\pi N} N^N e^{-N}}{\left[\sqrt{2\pi Np}(Np)^{Np} e^{-Np}\right]\left[\sqrt{2\pi N(1-p)}(N(1-p))^{N(1-p)} e^{-N(1-p)}\right]} \tag{14.11}$$

$$\approx \frac{1}{2^{N[p\log(p)]} 2^{N[(1-p)\log(1-p)]}} \tag{14.12}$$

$$= 2^{NH_b(p)} \tag{14.13}$$

different codewords with Np errors; the first approximation follows from Sterling's formula, and the second is a rearrangement of terms using $p = 2^{\log(p)}$. The last equality is simply the definition of the binary entropy function, $H_b(p) = -p\log(p) - [(1-p)\log(1-p)]$. The overall number of available sequences of length N is 2^N. Thus, the number of clearly distinguishable sequences is $M = 2^{N(1-H_b(p))}$. Therefore, using N symbols, we can transmit M different messages, and thus $\log(M)$ information bits. The possible data rate is thus

$$R = \frac{\log(M)}{N} = 1 - H_b(p) \tag{14.14}$$

For an AWGN channel, a similar argument can be made. For a long codeword, we know that with high probability

$$\frac{1}{N} \sum_n |n_n|^2 \rightarrow \sigma^2 \tag{14.15}$$

so that the received signal vector $\mathbf{y}$ lies near the surface of a sphere (called noise sphere) of radius $\sqrt{N\sigma^2}$ around the transmit signal vector $\mathbf{x}$. Reliable communications is possible as long as the spheres associated with the different codewords do not overlap, i.e., each received signal point can be uniquely associated with a particular transmit signal point. On the other hand, due to the average power constraint, we know that all received signal points must lie within a sphere of radius $\sqrt{N(P + \sigma^2)}$. We can thus conclude that the number of different received sequences that can be decoded reliably is equal to the number of noise spheres that fit into a sphere of radius $\sqrt{N(P + \sigma^2)}$. Since the volume of an N-dimensional sphere with radius ρ is proportional to ρ^N, the number of different messages that can be communicated with a codeword of length N is

$$M = \frac{[N(P + \sigma^2)]^{N/2}}{[N\sigma^2]^{N/2}} \tag{14.16}$$

so that the possible rate of communications is

$$R = \frac{\log(M)}{N} = \frac{1}{2} \log\left[1 + \frac{P}{\sigma^2}\right] \tag{14.17}$$

This equals the capacity of the AWGN channel. Without derivation, we state here also that this capacity is achieved when the transmit alphabet is Gaussian (and thus continuous).

It is also noteworthy that Eq. (14.17) is the capacity *per channel use* (i.e., per transmitted underlying symbol) for a *real* modulation alphabet and channel. When using complex modulation, the capacity per unit bandwidth becomes

$$C_{\text{AWGN}} = 2 \cdot \frac{1}{2} \log\left[1 + \frac{P}{N_0 B}\right] \text{ bits/s/Hz} \tag{14.18}$$

which can be expressed in terms of the SNR as

$$C_{\text{AWGN}} = \log\left[1 + \gamma\right] \text{ bits/s/Hz} \tag{14.19}$$

It is noteworthy that the capacity increases only logarithmically with the SNR. This implies that increasing transmit power has only limited value for improving the capacity of an AWGN channel. Remember also that this statement is different from our insights in Chapters 12 and 13: there we said that *for a given transmission rate*, increasing the transmit power is very effective for decreasing the symbol error rate of *uncoded* transmission over an AWGN channel. Here, we are stating that for an *ideally coded* system (which by definition has no symbol errors), increasing the transmit power is *not* an efficient way for *increasing the transmission rate*. Clearly, however, the admissible transmission rate increases linearly with the bandwidth that is available.

14.1.4 Power–Bandwidth Relationship

From the capacity equation, we also see that we can trade off spectral efficiency and energy efficiency of a transmission. We know that the transmission rate has to be smaller than the normalized bandwidth, multiplied by the invested bandwidth

$$R < B \log\left[1 + \frac{P}{N_0 B}\right] \tag{14.20}$$

Furthermore, the energy expended per transmitted information bit is

$$E_{\text{b}} = \frac{P T_{\text{seq}}}{\log(M)} = \frac{P}{\frac{\log(M)}{T_{\text{seq}}}} = \frac{P}{R} \tag{14.21}$$

where T_{seq} is the duration of the symbol sequence of length N. Denoting R/B as the spectral efficiency r, we can rewrite Eq. (14.20) as

$$r < \log\left(1 + r\frac{E_{\text{b}}}{N_0}\right) \tag{14.22}$$

so that

$$\frac{E_{\text{b}}}{N_0} > \frac{2^r - 1}{r} \tag{14.23}$$

This equation demonstrates the tradeoff between energy per bit, and the spectral efficiency. As the spectral efficiency decreases, so does the requirement for the E_{b}/N_0. In the limit of zero spectral efficiency, $E_{\text{b}}/N_0 \to \ln(2) = -1.59$ dB, as follows from L'Hopital's rule. *This −1.59 dB is the smallest value for the* E_{b}/N_0 *for which reliable communication is possible.* Note that orthogonal modulation (e.g., Frequency Shift Keying (FSK) or Pulse Position Modulation (PPM)) with a large number of dimensions approximates this case.

14.1.5 Relationship to Practical Systems

The above derivations of channel capacity were all using infinitely long codes. Furthermore, the results do not give a constructive way *how* good codes, with all the above properties, can be achieved. Shannon's derivations show that random codes achieve channel capacity; however, the only way of decoding truly random codes is a list decoding, and with N large, the lists quickly become unmanageable. In the sections below, we discuss practical coding methods that have worse performance than random codes, but have the advantage of being actually decodable with finite effort. This is particularly true for the block codes and convolutional codes, which have been used for many years but which – due to their relatively short codeword length – do not come close to the theoretical performance limits. The 1990s have finally seen codes that achieve practical decodability while having large effective length of codewords, and thus close-to-optimum performance: *turbo codes* and *LDPC codes* approach the Shannon limit within less than 1 dB; they are discussed in Sections 14.6–14.7.

14.1.6   Classification of Practical Codes

One way of classifying codes is to distinguish between *block codes*, where the redundance is added to blocks of data and *convolutional codes*, where redundancy is added continuously. Block codes are well suited for correcting burst errors – something that frequently occurs in wireless communications; however, error bursts can also be converted into random errors by interleaving techniques. Convolutional codes have the advantage that they are easily decoded by means of a Viterbi decoder. They also offer the possibility of joint decoding and equalization by means of the same algorithm. Turbo codes and LDPC codes easily fit into this categorization. As we will find in Section 14.6, turbo codes use two parallel, interleaved, convolutional codes in order to encode the information; however, the employed decoding structures are different because of the length of the memory of the encoder. A similar thing can be said about LDPC codes: as their name implies, they are block codes, but their large size necessitates different decoding structures; for that reason, they are treated in a separate section.

The research of the last years has actually led to a "smearing" of the boundary between block codes and convolutional codes. It has been shown that Viterbi decoders (the classical solution for convolutional codes) can be used to detect block codes [Wolf 1978], while belief propagation algorithms [Loeliger 2004] (commonly used for decoding block codes) can be generalized to decode convolutional codes. However, those are very advanced research topics and will not be treated further here. The interested reader is referred to the cited papers.

Another classification of codes can be based on whether the coder input and/or output are *hard* or *soft* information. Hard information just tells us the (binary) value that we detect for a bit. Soft information tells us also the confidence we have in our decision for this bit.

> **Example 14.1**   *Soft and hard information for repetition coding.*
>
> Let us consider a simple example: a repetition code that repeats a bit three times. Let the transmitted bit sequence be 1 1 1. Let the demodulated signal at the receiver be -0.05, -0.1, 1.0. One strategy is now to make a hard decision on each bit before decoding. After the slicer, the bits are then -1 -1 1. The decoder performs a majority decision and decides that -1 was transmitted. A soft decoder, on the other hand, might add up the demodulated signals,[4] giving 0.85, a hard decision on that results in a decision that $+1$ was transmitted.

For some applications (e.g., iterative and concatenated decoders), it is necessary that the decoder does not only use soft input information but also that it *puts out* soft information, and not just hard bits. In any case, the use of soft input information always leads to a performance improvement compared to the use of hard input information.

---

[4] This is purely for demonstration purposes. We will discuss more intelligent soft combining strategies later on.

14.2 Block Codes

14.2.1 Introduction

Block codes are codes that group the source data into blocks, and – from the values of the bits in that block – compute a longer codeword that is actually transmitted. The smaller the code rate – i.e., the ratio of the number of bits in the original datablock to that of the transmitted block – the higher the redundancy, and the higher the probability that errors can be corrected. The most simple codes are repetition codes with blocksize 1: for an input "block" x, the output block is xxx (for a repetition code with rate 1/3).

After this intuitive introduction, let us now give a more precise description. First, we define some important terms and notations:

- *Block coding*: for block coding, source data are parsed into blocks of K symbols. Each of these uncoded datablocks is then associated with a codeword of length N symbols.
- *Code rate:* The ratio K/N is called the *code rate* R_c (assuming the symbol alphabet of coded and uncoded data is the same).
- *Binary codes:* these occur when the symbol alphabet is binary, using only "0" and "1". Almost all practical block codes are binary, with the exception of *Reed–Solomon* (RS) *codes* (see below). If not stated otherwise, the remainder of the chapter always talks about binary codes. Therefore, in the following, "sum" means "modulo-2 sum", and "+" denotes a modulo-2 addition.
- *Hamming distance:* The Hamming distance $d_H(\mathbf{x}, \mathbf{y})$ between two codewords is the number of different bits:

$$d_H(\mathbf{x}, \mathbf{y}) = \sum_n |x_n - y_n| \qquad (14.24)$$

where $\mathbf{x}, \mathbf{y}$ are codevectors, $\mathbf{x} = [x_1 x_2 \cdots x_N]$, $x_n \in \{0, 1\}$. For example, the Hamming distance between the codewords 01001011 and 11101011 is 2. Note that it is common in coding theory to use row vectors instead of the column vectors commonly used in communication theory. In order to simplify cross-referencing to coding books, we follow this established notation in this chapter.

- *Euclidean distance:* The squared Euclidean distance between two codewords is the geometric distance between the codevectors $\mathbf{x}$ and $\mathbf{y}$ is:

$$d_E^2(\mathbf{x}, \mathbf{y}) = \sum_n |x_n - y_n|^2 \qquad (14.25)$$

- *Minimum distance:* the minimum distance d_{min} of a code is the minimum Hamming distance $\min(d_H)$, where the minimum is taken over all possible combinations of two codewords of the code. Note that this minimum distance is equal to the number of linearly independent columns in the parity check matrix (see below).
- *Weight:* the weight of a codeword is the distance from the origin – i.e., the number of 1's in the codeword. For example, the weight of codeword 01001011 is 4.
- *Systematic codes:* in a systematic code, the original, information-bearing bits occur explicitly in the output of the coder, at a fixed location. The parity check (redundant) bits, which are computed from the information-bearing bits, are at different (also fixed) locations. For transmission over an ideal (noise-free, nondistorting channel), the codeword could be determined without any information from the parity check bits. As an example, a systematic (7, 4) block code can be created in the following way:

k	k	k	k	m	m	m	m

where k represents information symbols and m represents parity check symbols.

- *Linear codes (group codes):* for these codes, the sum of any two codewords gives another valid codeword. Important properties can be derived from this basic fact:
  - o The all-zero word is a valid codeword.
  - o All codewords (except the all-zero word) have a weight equal to or larger than d_{min}.
  - o The distribution of distances – i.e., the Hamming distances between valid codewords – is equal to the weight distribution of the code.
  - o All codewords can be represented by a linear combination of basic codewords (*generator words*).
- *Cyclic codes*: cyclic codes are a special case of linear codes, with the property that any cyclic shift of a codeword results in another valid codeword. Cyclic codes can be interpreted either by codevectors, or by *polynomials* of degree $\leq N - 1$ where N is the length of the codeword. The nonzero coefficients correspond to the nonzero entries of the codevector; the variable x is a dummy variable.

  As an example, we show both representations of the codeword 011010:

  $$\mathbf{x} = [0\ 1\ 1\ 0\ 1\ 0] \tag{14.26}$$

  $$X(x) = 0 \cdot x^5 + 1 \cdot x^4 + 1 \cdot x^3 + 0 \cdot x^2 + 1 \cdot x^1 + 0 \cdot x^0 \tag{14.27}$$

- *Galois Fields*: A Galois field GF(p) is a finite field with p elements, where p is a prime integer. A field defines addition and multiplication for operating on elements, and it is closed under these operations (i.e., the sum of two elements is again a valid element, and similar for the product); it contains identity and inverse elements for the two operations; and the associative, commutative, and distributive laws apply. The most important example is GF(2). It consists of the elements 0 and 1 and is the smallest finite field. Its addition and multiplication tables are as follows:

+	0	1
0	0	1
1	1	0

×	0	1
0	0	0
1	0	1

Codes often use GF(2) because it is easily represented on a computer by a single bit. It is also possible to define extension fields GF(p^m), where again p is a prime integer, and m is an arbitrary integer.

- *Primitive polynomials:* we define as *irreducible* a polynomial of degree N that is not divisible by any polynomial of degree less than N and greater than 0. A primitive polynomial $g(x)$ of degree m is defined as primitive if it is an irreducible polynomial such that the smallest integer N for which $g(x)$ divides $(x^N + 1)$ is $N = 2^m - 1$.

14.2.2 Encoding

The most straightforward encoding is a mapping table: any K-valued information word is associated with an N-valued codeword; the table just checks the input, and reads out the associated codeword. However, this method is highly inefficient: it requires the storing of 2^K codewords.

For linear codes, any codeword can be created by a linear combination of other codewords, so that it is sufficient to store a subset of codewords. For example, for a K-valued information word, only K out of the 2^K codewords are linearly independent, and thus have to be stored. It is advantageous to select those codewords that have only a single 1 in the first K positions. This choice automatically leads to a systematic code. The encoding process can then best be described by a matrix multiplication:

$$\mathbf{x} = \mathbf{u}\mathbf{G} \tag{14.28}$$

Here, $\mathbf{x}$ denotes the N-dimensional codevector, $\mathbf{u}$ the K-dimensional information vector, and $\mathbf{G}$ the $K \times N$-dimensional generator matrix. For a systematic code, the leftmost K columns of the generator matrix are a $K \times K$ identity matrix, while the right $N - K$ columns denote the parity check bits. The first K bits of $\mathbf{x}$ are identical to $\mathbf{u}$. Note that – as discussed above – we use *row* vectors to represent codewords, and that a vector–matrix product is obtained by premultiplying the matrix with this row vector.

Example 14.2   *Encoding with a (7,4) Hamming code.*

To make things more concrete, let us now consider an example of a $(7, 4)$ code. We encode the source word [1011] with a generator matrix:

$$\mathbf{G} = \begin{bmatrix} 1 & 0 & 0 & 0 & 1 & 1 & 0 \\ 0 & 1 & 0 & 0 & 1 & 0 & 1 \\ 0 & 0 & 1 & 0 & 0 & 1 & 1 \\ 0 & 0 & 0 & 1 & 1 & 1 & 1 \end{bmatrix} \tag{14.29}$$

Computing $\mathbf{x} = \mathbf{uG}$:

$$\mathbf{x} = \begin{bmatrix} 1 & 0 & 1 & 1 \end{bmatrix} \begin{bmatrix} 1 & 0 & 0 & 0 & 1 & 1 & 0 \\ 0 & 1 & 0 & 0 & 1 & 0 & 1 \\ 0 & 0 & 1 & 0 & 0 & 1 & 1 \\ 0 & 0 & 0 & 1 & 1 & 1 & 1 \end{bmatrix} \tag{14.30}$$

the codeword becomes

$$\mathbf{x} = \begin{bmatrix} 1 & 0 & 1 & 1 & 0 & 1 & 0 \end{bmatrix} \tag{14.31}$$

14.2.3   Decoding

In order to decide whether the received codeword is a valid codeword, we multiply it by a *parity check matrix* $\mathbf{H}$. This results in a $N - K$ dimensional *syndrome vector* $\mathbf{s}_{\text{synd}}$. If this vector has all-zero entries, then the received codeword is valid.

Example 14.3   *Syndrome for (7,4) Hammung code.*

Let us demonstrate the computation of the syndrome with our above example. Let the received bit sequence (after hard decision) be

$$\hat{\mathbf{x}} = \begin{bmatrix} 1 & 0 & 0 & 0 & 1 & 0 & 1 \end{bmatrix} \tag{14.32}$$

A parity check matrix for the code of Example 14.2 is (we will describe later a constructive method for obtaining $\mathbf{H}$ from $\mathbf{G}$):

$$\mathbf{H} = \begin{bmatrix} 1 & 1 & 0 & 1 & 1 & 0 & 0 \\ 1 & 0 & 1 & 1 & 0 & 1 & 0 \\ 0 & 1 & 1 & 1 & 0 & 0 & 1 \end{bmatrix} \tag{14.33}$$

Let us now compute the expression $\mathbf{s}_{\text{synd}} = \hat{\mathbf{x}}\mathbf{H}^T$. Writing this expression component by component, we obtain three parity check equations, corresponding to the three parity check bits:

$$\mathbf{s}_{\text{synd}}^T = \hat{\mathbf{x}}\mathbf{H}^T = \begin{bmatrix} 0 & 1 & 1 \end{bmatrix} \tag{14.34}$$

The computation of the syndrome can thus be interpreted the following way:

- received information: 1000;
- computed parity check bits: 110 (parity bits computed from the received systematic bits);
- received parity check bits: 101;
- $\Rightarrow$ syndrome: 011.

Next, let us determine how to find an $\mathbf{H}$-matrix. The relationship $\mathbf{H} \cdot \mathbf{G}^T = \mathbf{0}$ has to hold, as each row of the generator matrix is a valid codeword, whose product with the parity check matrix has to be 0.[5] Representing $\mathbf{G}$ as

$$\mathbf{G} = (\mathbf{I} \; \mathbf{P}) \tag{14.35}$$

the relationship $\mathbf{H} \cdot \mathbf{G}^T = \mathbf{0}$ reduces to

$$\mathbf{H} \cdot \mathbf{G}^T = (\mathbf{H}_1 \quad \mathbf{H}_2) \begin{pmatrix} \mathbf{I} \\ \mathbf{P}^T \end{pmatrix} = (\mathbf{H}_1 + \mathbf{H}_2 \mathbf{P}^T) = \mathbf{0} \tag{14.36}$$

If we now select $\mathbf{H}_2 = \mathbf{I}$ and $\mathbf{H}_1 = -\mathbf{P}^T$ then the equation is certainly fulfilled: the subtraction of two identical matrices is the all-zero matrix. Note that different parity check matrices can exist for each generator matrix. The above constructive method gives just one possible solution.

Example 14.4 *Computation of parity check matrix for Hamming code.*

The procedure for computing the parity check matrix for the (7, 4) Hamming code (and actually for any systematic code) can also be described the following way: start with the $N - K$ rightmost columns of the generator matrix:

$$\begin{bmatrix} 1 & 1 & 0 \\ 1 & 0 & 1 \\ 0 & 1 & 1 \\ 1 & 1 & 1 \end{bmatrix} \tag{14.37}$$

and transpose them; note that in modulo-2 arithmetic, there is no difference between addition and subtraction. Appending now an $(N - K) \times (N - K)$ identity matrix gives $\mathbf{H}$:

$$\mathbf{H} = \begin{bmatrix} 1 & 1 & 0 & 1 & 1 & 0 & 0 \\ 1 & 0 & 1 & 1 & 0 & 1 & 0 \\ 0 & 1 & 1 & 1 & 0 & 0 & 1 \end{bmatrix} \tag{14.38}$$

A well-known class of linear codes are the Hamming codes. A Hamming code can be defined most easily through its parity check matrix. The columns of $\mathbf{H}$ contain all possible 2^{N-K} bit combinations of length K, with the exception of the all-zero word. Consequently, all columns of $\mathbf{H}$ are distinct. Note that the parity check matrix of Eq. (14.38) fulfills the condition, as can be easily verified by the reader. The size of a Hamming code is $(2^m - 1, 2^m - 1 - m)$, where m is a positive integer.

---

[5] In other words, $\mathbf{G}$ is the codespace, and $\mathbf{H}$ is the associated nullspace.

14.2.4 Recognition and Correction of Errors

Due to the linear properties of the code, each received word can be interpreted as the sum of a codeword $\mathbf{x}$ and an error word $\boldsymbol{\varepsilon}$. This implies that the syndrome depends only on the error word, not on the transmitted codeword. If a word $\hat{\mathbf{x}}$ is received, then the receiver computes the syndrome vector:

$$s_{\text{synd}} = \hat{\mathbf{x}}\mathbf{H}^T = (\mathbf{x} + \boldsymbol{\varepsilon})\mathbf{H}^T = \mathbf{x}\mathbf{H}^T + \boldsymbol{\varepsilon}\mathbf{H}^T = \mathbf{0} + \boldsymbol{\varepsilon}\mathbf{H}^T = \boldsymbol{\varepsilon}\mathbf{H}^T \tag{14.39}$$

An error ($\boldsymbol{\varepsilon} \neq 0$) is indicated by a nonzero syndrome. Decoding requires that we find the "correct" $\boldsymbol{\varepsilon}$ – by subtracting $\boldsymbol{\varepsilon}$ from $\hat{\mathbf{x}}$, we obtain the correct codeword. On the other hand, it is clear that a number of different error words $\boldsymbol{\varepsilon}$ can lead to the same syndrome; these error words are called a *coset*. The goal is thus to find the most probable $\boldsymbol{\varepsilon}$ corresponding to the observed syndrome. In a "reasonable" channel, the probability that a bit arrives correctly is higher than the probability that a bit arrives with an error. Therefore, the $\boldsymbol{\varepsilon}$ that has the minimum weight is the most probable $\boldsymbol{\varepsilon}$ in a given coset.

A different interpretation, which leads to an estimate of the number of detectable and correctable errors, starts with the representation in the *codespace* (see Figure 14.1): Each of the 2^N possible bit combinations of a binary (N, K) block code corresponds to a point in the codespace. 2^K of these bit combinations correspond to a valid codeword, each of which is separated at least by a distance $d_{\min}$. The other bit combinations can be interpreted as points located in "correction spheres" around the valid codewords. The minimum size of correction spheres t determines the number of corrigible errors.

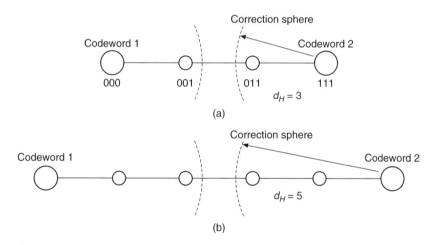

Figure 14.1 Cuts through a codespace (a) for a code with $d_H = 3$ and (b) a codespace for a code with $d_H = 5$ and a sketch of the correction spheres.

For codes of small size, decoding via lookup tables is possible – in other words, all possible syndromes and their corresponding minimum-weight error words are stored in a lookup table. For each received codeword, this table is then used to determine the transmitted codeword. Unfortunately, codes with small size usually show bad performance. For this reason, alternative decoding schemes that can be used for larger codes as well have been investigated. The most important one, the *belief propagation* algorithm, is described in more detail in Section 14.7. Furthermore, special cases of linear codes that allow a simplified decoding, like cyclic codes, might be preferable.

From these considerations, the following conclusions can be drawn: a code with minimum distance d_{min} always allows the detection of $d_{min} - 1$ errors in a codeword. Only errors that influence at least d_{min} bits can lead to another valid codeword. Alternatively, such a code can be capable of correcting $\lfloor (d_{min} - 1)/2 \rfloor$ errors – in other words, the received bit sequence has to lie in the correction sphere of the true codeword ($\lfloor x \rfloor$ denotes the largest integer smaller than x). If all bit combinations can be uniquely assigned to a correction sphere, the code is called *perfect*.

For a Hamming code, the location of a single error can be uniquely determined. Since all the columns in the parity check matrix are different and linearly independent, and a Hamming code can correct exactly one error, the syndrome tells us the location of the error.

14.2.5 Concatenated Codes

Error protection can be made stronger by using two codes: a so-called *inner code* protects the data in the usual way, as described above. Some errors can still remain (when the received word lies in the wrong correction sphere). We can thus interpret the combination of channel and inner code as a *superchannel* that exhibits a lower BER than the original channel. An *outer code* then provides protection against these errors (see Figure 14.2). The errors of the superchannel usually occur in bursts; if the inner code is an (N, K) block code, then the length of the burst is K bits. Thus, the outer code is usually a code that is especially well suited for combating burst errors. RS codes are efficient for these types of applications.[6]

Appropriate concatenation of codes is quite a difficult task. In particular, the relative capabilities of the two codes need to be carefully balanced. If the inner code is not strong enough, then it is useless, and might even increase the BER that the outer code has to deal with.

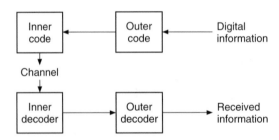

Figure 14.2   Concatenation of codes.

14.3   Convolutional Codes

14.3.1   Principle of Convolutional Codes

Convolutional codes do not divide (source) data streams into blocks, but rather add redundancy in a quasi-continuous manner. A convolutional encoder consists of a shift register with L memory cells and N (modulo-2) adders (see Figure 14.3). Let us assume that at the outset we have a clearly defined state in memory cells – i.e., they all contain 0. When the first data bit enters the encoder, it is put into the first memory cell of the shift register (the other zeros are shifted to the right, and the rightmost zero "falls out" of the register). Then a multiplexer reads out the output of all the adders $n = 1, 2, 3$. We thus get three output bits for one input bit. Then, the next source data bit is

---

[6] Code concatenation can also be applied when the inner code is a convolutional code, see Section 14.3.

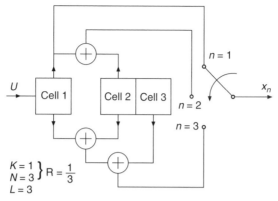

Figure showing convolutional encoder circuit with the following labels and text:

$$K = 1 \atop N = 3 \atop L = 3 \Big\} R = \frac{1}{3}$$

If shift-register initially is in all-zero state

$U_1 = 1 \rightarrow X_1 = 1, 1, 1$

$U_1 = 1 \quad U_2 = 1 \rightarrow X_2 = 1, 0, 0$

Number of states: $2^{L-1} = 4$

State	Cell 2	Cell 3
A	0	0
B	1	0
C	0	1
D	1	1

Figure 14.3 Example of a convolutional encoder.

Reproduced with permission from Oehrvik [1994] © Ericsson AB.

put into the register (and the contents of all memory cells are shifted to the right by one cell). The adders then have new outputs, which are again read by the multiplexer. The process is continued until the last source data bit is put into the register. Subsequently, zeros are used as register input, until the last source data bit has been pushed out of the register, and the memory cells are again in a clearly defined (all-zero) state.

A convolutional encoder is thus characterized by the number of shift registers and adders. Adders are characterized by their connections to memory cells. In the example of Figure 14.3, only element $l = 1$ is connected to the output $n = 1$, so that source data are directly mapped to the coder output. For the second output, the contents of memory cells $l = 1, 2$ are added. For the third output, the contents of elements $l = 1, 2, 3$ are combined.

This coder structure can be represented in different ways. One possibility is via generator sequences: we generate N vectors of length L each. The lth element of the nth vector has value 1 if the lth shift register element has a connection to the nth adder; otherwise it is 0. Generator sequences can immediately be interpreted for building an encoder.

For the decoder, the trellis diagram is a more useful description method. In this representation, the state of the encoder is characterized by the content of the memory cells. The trellis shows which input bits get the shift register into which state, and which output bits are created consequently. As an example, Figure 14.4 shows the trellis of the convolutional encoder of Figure 14.3. Only the states of the cells 2, ... L need to be described, since the content of cell 1 is identical to the input (information) bit. For that reason, the number of states that need to be distinguished is $2^{L-1} = 4$. Two lines originate from each state: the upper represents source data bit 0, and the lower source data bit 1. It is not possible to get from each state directly into each other state. For example, we can only get from state A to state A or B (but not C or D). This is the redundancy that can be

| Input (information) bit sequence | 1 | 0 | 1 | 0 | 0 |
| Output (coded) bit sequence | 111 | 011 | 110 | 011 | 001 |

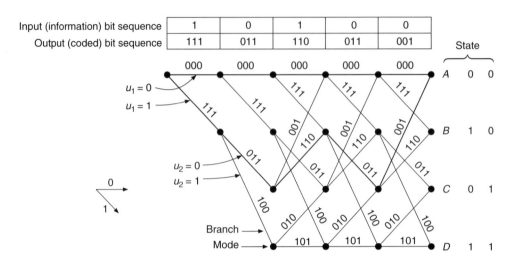

Figure 14.4   Trellis for the convolutional encoder of the previous figure.
Reproduced with permission from Oehrvik [1994] © Ericsson AB.

used for the reduction of error probability during decoding. We also find that the trellis is repeated periodically. It is thus unnecessary to plot an infinitely long trellis diagram, even if the input data sequence is infinitely long.

Decoding at the receiver would be very simple were the received data sequence identical to the transmit data sequence. As we would then know which coded sequence was received, we would just have to trace in the trellis the source data from which it was created. When there is additive noise, there are errors in the received bit sequence, and we have to discover from the perturbed receive data what sequence was most likely originally transmitted (*Maximum Likelihood Sequence Estimation*, MLSE). In practice, this estimation is usually done by the Viterbi algorithm that is discussed in the next section. Other implementations are the Fano algorithm, the stack algorithm, and the feedback algorithm (Lin and Costello [2004] and McEliece [2004]).

14.3.2   Viterbi Decoder – Classical Representation

The Viterbi algorithm [Viterbi 1967] is the most popular algorithm for MLSE. The goal of this algorithm is to find the sequence $\hat{\mathbf{s}}$ that was transmitted with the highest likelihood, if the sequence $\mathbf{r}$ was received:[7]

$$\hat{\mathbf{s}} = \max_{\mathbf{s}} \Pr(\mathbf{r}|\mathbf{s}) \qquad (14.40)$$

where maximization is done over all possible transmit sequences $\mathbf{s}$. If perturbations of the received symbols by noise are statistically independent,[8] then the probability for the sequence $\Pr(\mathbf{r}|\hat{\mathbf{s}})$ can be decomposed into the product of the probabilities of each symbol:

$$\hat{\mathbf{s}} = \max_{\mathbf{s}} \prod_{i} \Pr(r_i|s_i) \qquad (14.41)$$

---

[7] This is the definition for Maximum A Posteriori (MAP) detection. However, it is equivalent to MLSE for equiprobable sequences (see Chapter 12).
[8] If the noise is colored – i.e., correlated between the different samples – then the receiver has to use a so-called "whitening filter" (see Chapter 16).

Now, instead of maximizing the above product, maximizing its logarithm also finds the optimum sequence – this is true since the logarithm is a strictly monotonous function:

$$\hat{s} = \max_{s} \sum_i \log[\Pr(r_i|s_i)] \qquad (14.42)$$

The logarithmic transition functions $\log[\Pr(r_i|s_i)]$ are also known as *branch metrics*; in the case that the decoder puts out only "hard decision" (estimates which coded bits were sent), $r_i = \pm 1$, the branch metric is the Hamming distance between r_i and s_i.

The MLSE now determines the total metrics $\sum_i d_H(r_i, s_i)$ for all possible paths through the trellis – i.e., for all possible input sequences. In the end, the path with the smallest metric is selected. Such an optimum procedure requires large computational effort: the number of possible paths increases exponentially with the number of input bits. The key idea of the Viterbi algorithm is the following: instead of computing metrics for all possible paths (working from "top to bottom") in the trellis, we work our way "from the left to the right" through the trellis. More precisely, we start with a set of possible states of the shift register (A_i, B_i, C_i, D_i, where i denotes the time instant, or considered input bit). Let us now consider all paths that (from the left) lead into state A. We discard any possible path $s^{(1)}$ if it merges at state A_i with a path $s^{(2)}$ that has a smaller metric. As the paths run through the same state of the trellis, there is nothing that would distinguish the paths from the point of view of later states. We thus choose the one with the better properties. Similarly, we choose the best paths that run through the states B_i, C_i, and D_i. After having determined the *survivors* for state i, we next proceed to state $i+1$ (or rather, to the tuple of states A_{i+1}, B_{i+1}, C_{i+1}, D_{i+1}), and repeat the process. All paths in a trellis ultimately merge in a single, well-defined point, the all-zero state.[9] At this point, there is only a single survivor – the sequence that was transmitted with the highest probability.

Example 14.5   *Example for Viterbi decoding.*

Figure 14.5 shows an example of the algorithm. The basic structure of the trellis is depicted in Figure 14.5a; the bit sequence to be transmitted is shown in Figure 14.5b. The metrics shown are the Hamming distances of the received sequence, i.e., *after* the hard decision compared with the theoretically possible bit sequences in the trellis.

We assume that at the outset the shift register is in the all-zero state. Figure 14.5c shows the trellis for the first 3 bits. There are two possibilities for getting from state A_0 to state A_4: by transmission of the source data sequence 0, 0, 0 (which corresponds to the coded bit sequence 000, 000, 000, and thus via states A_2, A_3) or by transmission of the source data sequence 100 (coded bit sequence 111, 011, 001 – i.e., via states B_2, C_3). In the former case, the branch metric is 2; in the latter, it is 6. This allows us to immediately discard the second possibility. Similarly, we find that the transition from state A_0 to state B_4 could be created (with greatest likelihood) by the source data sequence 0, 0, 1, and not by 1, 1, 0. The following subfigures of Figure 14.5 show how the process is repeated for ensuing incoming bits.

The Viterbi algorithm greatly decreases storage requirements by elimination of nonsurviving paths, but they are still considerable. It is thus undesirable to wait for the decision as to which sequence was transmitted until the last bits of the source sequence. Rather, the algorithm makes decisions about bits that are "sufficiently" in the past. More precisely, during consideration of states A_i, B_i, C_i, D_i we decide about the symbols of the state tuple $A_{i-L_{Tr}}$, $B_{i-L_{Tr}}$, $C_{i-L_{Tr}}$, $D_{i-L_{Tr}}$,

---

[9] Actually, this only happens if the convolutional encoder appends enough zeros (tail bits) to the source data sequence to force the encoder into the defined state. If this is not done, then the decoder has to consider all possible finishing states, and compare the metrics of the path ending in each and all of them.

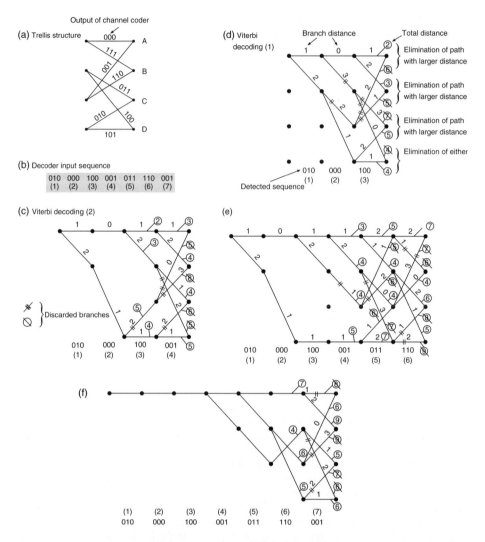

Figure 14.5 Example of Viterbi detection.

Reproduced with permission from Oehrvik [1994] © Ericsson AB.

where L_{Tr} is the *truncation depth*. This principle is shown in Figure 14.6. Data within the window of length L_{Tr} are stored. When moving to the next tuple in the trellis, the leftmost-state tuple moves out of the considered window, and we have to make a final decision about which bits were transmitted there. The decision is made in favor of the state that contains the path that has the smallest metric in the *currently* observed state – i.e., at the right side of the window. While this procedure is suboptimum, performance loss can be kept small by judiciously choosing the length of the window. In practice, a duration:

$$L_{Tr} = 6L \qquad (14.43)$$

has turned out to be a good compromise.

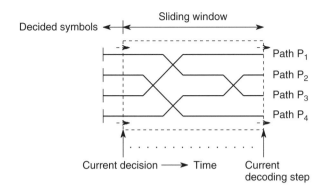

Figure 14.6 Principle of decision using a finite duration sliding window.
Reproduced with permission from Mayr [1996] © B. Mayer.

14.3.3 *Improvements of the Viterbi Algorithm*

Soft Decoding

The above example used the Hamming distance between received symbols and possible symbols as a metric. This means that received symbols first undergo a hard decision before being used in the decision process. This algorithm thus uses the redundancy of the code (not all combinations of bits are valid codewords), but does not use knowledge about the reliability of the received bits. In some cases, the bits might be close to the decision boundary, in which case they should have less impact on the finally chosen sequence than if they are far away from that boundary. "Soft information" can be taken into account by using a different metric in the Viterbi algorithm. It can be shown that in an AWGN channel as well as in a flat-fading channel the best decision metric is proportional to the squared Euclidean distance $d_i^2 = |r_i - s_i|^2$ in the signal space diagram. Proportionality constants are of no further importance, as they have no impact on finding the optimum metric:

$$\hat{\mathbf{s}} = \min_{\mathbf{s}} \sum_i |r_i - s_i|^2 \tag{14.44}$$

This equation assumes that the channel gain is the same for all received symbols and that the channel gain has been compensated by the receiver. If this is not the case, then we have to minimize the expression:

$$\min_{\mathbf{s}} \sum_i |r_i - \alpha_i s_i|^2 \tag{14.45}$$

where α_i is the channel gain of the ith symbol. A detailed example of a Viterbi algorithm with soft information is given in Chapter 16 (we will see there that MLSE equalization can also be done by means of this algorithm).

Tail Bits

As mentioned above, it is best if – at the end of the transmitted data sequence – the encoder ends up in a defined state, usually the all-zero state. In that case, it is clear which surviving path has to be chosen. Figure 14.7 shows that this approach significantly decreases the error probability for the

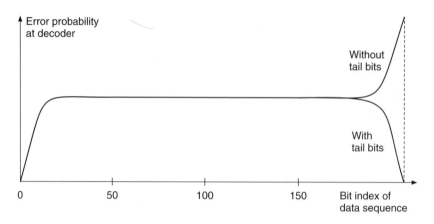

Figure 14.7 Error probability when tail bits are used.
Reproduced with permission from Oehrvik [1994] © Ercisson AB.

latest data bits in the sequence. The drawback is a loss in spectral efficiency, as non-information-bearing symbols (the zeros appended at the end) have to be transmitted. This can become a problem in systems with very short block durations.

If the datablocks that are to be transmitted are very long, the principle of tail bits can be used also within a block: at certain, predefined locations, a series of zeros is transmitted, and thus forces the shift register (and the trellis) into a defined state. Such *waists* in the trellis allow a reduction in error probability.

Puncturing

Normal convolutional codes can achieve only a certain subset of coding rates, like $R_c = 1/2, 1/3, 1/4$. However, for many applications, we need different code rates that are determined by the source rate and the available bandwidth. For example, the WiMedia ultrawideband standard uses a code rate of 11/32 for its default mode.

An easy way of adapting a code rate is puncturing of a code. It starts out with a code that has a rate that is lower than the desired. Then certain bits of the coded sequence are omitted from the transmission. Since the code contains considerable redundancy, this is not a problem – it is similar to the physical situation when codebits are erased due to fading. It is just required that the punctured bits are appropriately distributed throughout the codeword, so that not all information about a certain bit is eliminated.

14.4 Trellis Coded Modulation

14.4.1 Basic Principle

A main problem of coding is the reduction in spectral efficiency. Since we have to transmit more bits, the bandwidth requirement becomes larger as we add check bits. This problem can be avoided by the use of higher order modulation alphabets, which allow the transmission of more bits within the same bandwidth. In other words: using a rate 1/3 code, while at the same time changing the modulation alphabet from BPSK to 8-Phase Shift Keying (PSK), the number of *symbols* that are transmitted per unit time remains the same.

As we discussed in Chapter 12, increasing the symbol alphabet increases the probability of error; on the other hand, introducing coding decreases the error probability. A simplistic approach to solving the spectral efficiency problem would thus be to add parity check bits to the data bits, and map the resulting coded data to higher order modulation symbols. However, this usually does not give good results. In contrast, *Trellis coded modulation* adds to the redundancy of the code by increasing the dimension of the signal space, while disallowing some symbol sequences in this enlarged signal space. The important aspect here is that modulation and encoding are designed as a joint process. This allows the design of a modem-plus-codec that shows higher resilience to noise than uncoded systems with the same spectral efficiency.

Example 14.6 *Simple trellis-coded modulation.*

To understand the basic principle of TCM, consider first a simple example in an AWGN channel. We compare an uncoded system with a TCM system, both of which transmit two source data bits per symbol duration. In the uncoded case, Quadrature-Phase Shift Keying (QPSK) is used as the modulation format (see Figure 14.8). Every combination of two bits corresponds to one valid point in the signal constellation diagram. With the energy per bit being E_B, the distance between points in the signal constellation diagram is $2\sqrt{E_B}$. The error probability in an AWGN case can thus be represented (see Chapter 12) as

$$BER \approx Q(\sqrt{2\gamma_B}) \tag{14.46}$$

For TCM, we use a larger symbol alphabet, and thus a higher order modulation format (in our case, 8-PSK). As each symbol can transmit 3 bits, the code rate is $R_c = 2/3$. The allowed symbol sequences are determined by a shift register structure, like in a convolutional code. As with any other convolutional codes, the (coded) transmit symbol depends on the current source bit, as well as the current state of the memory. In our example, the memory is 2 bits long, so that there are four possible states: 00, 01, 10, 11. Depending on the state, as well as the source bit, different 8-PSK symbols are transmitted.

Figure 14.9 shows the possible transitions between states. If, e.g., memory is in state 11, and the information symbol 01 is to be transmitted, then the PSK symbol 3 is transmitted, and memory ends up in state 10. We also find from the diagram that so-called *parallel transitions* are possible. These are transitions where we can get from memory state A to memory state B by transmission of different 8-PSK symbols. However, not all combinations of states and 8-PSK symbols are possible: when in state 11, only transmission of symbols 1, 3, 5, 7 is allowed. Figure 14.10 shows the encoder structure and the trellis diagram for this system.

The trellis diagram also allows determination of the BER of TCM. As a first step, we determine the smallest squared Euclidean distance between symbol sequences that can lead to errors. Looking at a part of the trellis diagram, we find all allowed paths that lead from state 00 at time 0 to state 00 at time 3. The first possible pair are the parallel transitions using PSK symbols 0 and 4 (to get from state 00 at time 0 to state 00 at time 1; later symbols need not differ). The squared Euclidean distance between these two symbols is $d^2 = 8E_B$. Other transitions are not possible by moving just one step to the right in the trellis; rather, a whole sequence is required. For example, a transition from state A_0 (00) to state A_3 (00) is possible by either transmitting the PSK symbols $0 - 0 - 0$ or $2 - 1 - 6$. The squared Euclidean distance between symbols 0 and 2 is $d^2 = 4E_B$, between 0 and 1 it is $2E_B[2\sin(\pi/8)]^2$, and between 0 and 6 it is again $4E_B$. The sum of the squared magnitudes of the distances between the two paths is thus $(4 + 1.17 + 4) = 9.17$. Other symbol sequences that can lead to errors can be searched for in a similar fashion. It can be shown that the smallest squared Euclidean distance between sequences that begin and end in the same state is $d^2_{coded} = 8E_B$. This is twice the distance we had for the uncoded case.

Figure 14.8   Quadrature-phase shift keying signal constellation diagram.

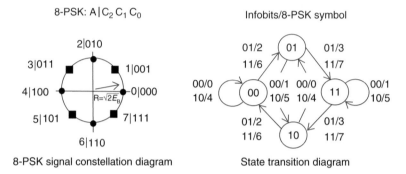

Figure 14.9   8-PSK signal constellation, and state transition diagram for a rate-2/3 8-PSK trellis-coded modulation.
Reproduced with permission from Mayr [1996] © B. Mayr.

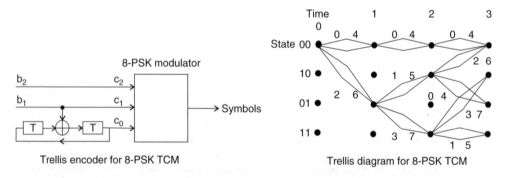

Figure 14.10   Encoder and trellis diagram for a rate-2/3 8-PSK trellis-coded modulation with four trellis states.
Reproduced with permission from Mayr [1996] © B. Mayr.

Generally, we can define the asymptotic coding gain – i.e., the coding gain at high SNRs – as

$$G_{\text{coded}} = 10 \log_{10} \left(\frac{d_{\text{coded}}^2}{d_{\text{uncoded}}^2} \right) \tag{14.47}$$

The above example achieves an asymptotic coding gain of 3 dB. In other words, for equal bandwidth requirement, equal source data rate, and equal target BER, the system needs 3 dB less transmit power. For finite target error rates, the coding gain is somewhat smaller; for a BER of 10^{-5}, the

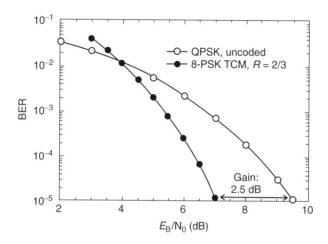

Figure 14.11 Simulated error probability of uncoded quadrature-phase shift keying rate-2/3 8-PSK trellis-coded modulation in an additive white Gaussian noise channel.
Reproduced with permission from Mayr [1996] © B. Mayr.

coding gain is only 2.5 dB (see Figure 14.11). It is, however, also noteworthy that at low SNRs, the performance is *worse* than for an uncoded system.

14.4.2 Set Partitioning

The previous subsection demonstrated the advantages of TCM, but the underlying code was ad hoc. For 8-PSK, such ad hoc construction is still feasible, and can lead to good results, but it becomes impossible for higher order modulation formats. Ungerboeck [1982] suggested a heuristic method for the construction of good codes, the so-called *set partitioning*. The basic principle of the method is as follows:

1. Double the modulation alphabet.
2. Select the allowable transitions so that the minimal squared Euclidean distance between sequences is maximized.

For maximization of the distance, the symbol alphabet is partitioned in several steps, and at each step, the minimum distance between symbols of the partitioned sets should increase maximally.

Example 14.7 *Set partitioning.*

This somewhat abstract principle can best be explained with an example. Consider the 8-PSK constellation of the previous section: the distance between two neighboring points in the signal constellation diagram is $d = \sqrt{E_B}[2\sqrt{2}\sin(\pi/8)] = 1.08\sqrt{E_B}$. As a first step, the existing symbols are partitioned into two sets (see Figure 14.12). In order to maximize the distance between elements, each set is a QPSK constellation that is rotated 45° with respect to the other constellation. This increases the Euclidean distance within a set to $2\sqrt{E_B}$; there is no constellation that would lead to a larger minimum distance. In the next step, the QPSK constellation is partitioned. This results in two BPSK constellations that are rotated 90° with respect to each other, and the Euclidean distance within the set is $2\sqrt{2E_B}$. This completes the set partitioning.

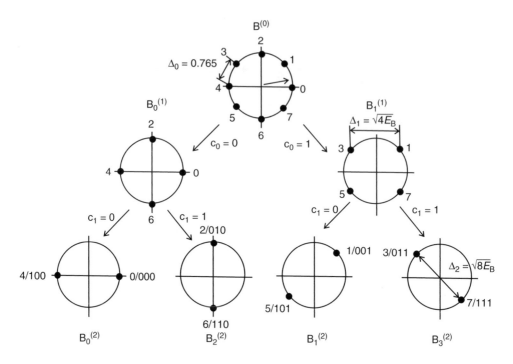

Figure 14.12   Partitioning of 8-PSK.

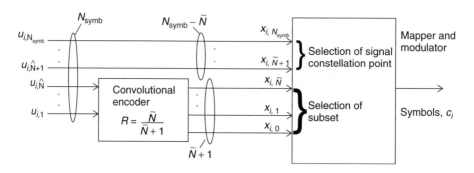

Figure 14.13   Structure of a trellis-coded modulation coder according to Ungerboeck.
Reproduced with permission from Mayr [1996] © B. Mayr.

Figure 14.13 shows the structure of a TCM encoder with such a set partitioning. We distinguish between N_{symb}, the number of information bits that can be transmitted per symbol, and $\tilde{N}$, the number of bits that are mapped into coded bits by a convolutional encoder of rate $R_{\text{c}} = \tilde{N}/(\tilde{N}+1)$. During the encoding process, the source bits $u_1 \cdots u_{\tilde{N}}$ are mapped to coded bits $x_0 \cdots x_{\tilde{N}}$. The remaining source bits $u_{\tilde{N}}+1 \cdots u_{N_{\text{symb}}}$ are directly mapped to "uncoded" bits $x_{\tilde{N}}+1 \cdots x_{N_{\text{symb}}}$. The uncoded bits are used to select a signal constellation point within a subset; such points have a large Euclidean distance, so that errors are improbable. The other bits select the subsets. The fewer bits are uncoded, the fewer parallel transitions are possible. Decoding is done by means of a Viterbi decoder (see Section 14.3.2).

14.5 Bit Interleaved Coded Modulation (BICM)

Another way of combining coding and higher order modulation is *Bit Interleaved Coded Modulation* (BICM). In this approach, the channel coder and the modulator are separated by an interleaver. As a consequence, encoded bits that are correlated through the coding process are mapped onto temporally separated symbols (which might increase diversity, see Section 14.8), and also provides greater robustness with respect to noise spikes.

Figure 14.14 shows a block diagram of a BICM transmission. The bits are encoded (by any binary encoder), and interleaved by the interleaver Π. The coded and interleaved bits are finally mapped onto the modulation symbols, which belong to an alphabet $\mathcal{X}$. The binary labeling scheme (discussed in more detail below) determines what bit combinations (as seen after the interleaver) are associated with which modulation symbol. The symbols are then transmitted over the channel, and the received signal is sent through a de-mapper, which computes reliability measures L (Log Likelihood Ratios, (LLRs) see below) for the estimated signals. The de-interleaver provides the "soft information" that forms the input of the decoder. Optionally, the output of the decoder can be re-interleaved and fed back to help the symbol de-mapper to refine its estimate of the LLRs. Such iterative decoding improves performance but increases complexity. It is similar in spirit to the turbo codes that will be discussed in Section 14.6, but will not be considered further in the current subsection.

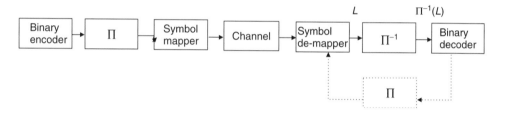

Figure 14.14 Block diagram of a bit-interleaved coded modulator. Feedback in the demodulator is optional.

The binary labeling is of critical importance for the performance of BICM. We assume in the following that higher order modulation (at least QPSK) is used. Numerical investigations as well as theoretical arguments show that for moderate-to-high SNR values, binary reflected Gray mapping is optimum. For low SNRs, set partitioning mapping is optimum..

The de-mapper computes a soft estimate for each particular (coded) bit; more precisely, the log-likelihood ratio (LLR)

$$L_i = \log\left[\frac{\Pr(b_i = +1|r)}{\Pr(b_i = -1|r)}\right] \tag{14.48}$$

$$= \log\left[\frac{\Pr(r|b_i = +1)}{\Pr(r|b_i = -1)}\right] + \log\left[\frac{\Pr(b_i = +1)}{\Pr(b_i = -1)}\right] \tag{14.49}$$

i.e., the logarithm of the ratios of the probabilities that bit b_i is $+1$ or -1, conditioned on the fact that the signal r was received. If we further assume equal a priori probabilities, and an AWGN model for the channel, the LLRs can be computed as

$$L_i = \log\left[\frac{\sum_{s \in X_{i,1}} \exp\left(-\gamma |\mathbf{r} - \alpha\mathbf{s}|^2\right)}{\sum_{s \in X_{i,0}} \exp\left(-\gamma |\mathbf{r} - \alpha\mathbf{s}|^2\right)}\right] \tag{14.50}$$

where α is the (complex) channel gain, and $\mathbf{s} \in X_{i,1}$ denotes modulation symbols whose associated binary representation (in the binary labeling) has the i-th bit equal to $+1$.

It has been shown that BICM, in particular when used in combination with iterative decoding, has excellent performance. It is thus widely used in practical wireless systems. The separation of the coder and the modulator allows a very flexible design (more flexible and easier to implement than TCM), and allows an easy implementation of adaptive modulation and coding (where, e.g., the modulation constellation is changed depending on the quality of the channel over which signaling is happening).

14.6 Turbo Codes

14.6.1 Introduction

Turbo codes are among the most important developments of coding theory since the field was founded. As we have already mentioned in Section 14.1, *very long codes* can approach the Shannon limit (channel capacity). However, the brute force decoding of such long codes is prohibitively complex. Turbo codes were the first practically used codes that came close to the Shannon limit using reasonable effort. Berrou et al. [1993] created *very long codes* by a combination of several parallel, simple codes. The codes are interleaved by a pseudorandom interleaver. The vital trick now lies in the decoder: because the code is a combination of several short codes, the decoder can also be broken up into several simple decoders that exchange soft information about the decoded bits and thus iteratively arrive at a solution.

The random interleaver in the code approximately realizes the idea of a random code, so that the total codeword has very little structure. This has the following advantages:

- Interleaving increases the effective codelength of the combined code. In other words, it is the interleaver (whose operation can easily be reversed), and not the constituent codes, that determines the length of the codes. This allows the construction of very long codes with simple encoder structures.
- The special structure of the total code – i.e., the composition of separate constituent codes – makes decoding possible with an effort that is essentially determined by the length of the constituent codes.

14.6.2 Encoder

Turbo codes use a combination of several codes. One method of combining codes, serial concatenation, was already discussed in Section 14.2. In this section, we concentrate on parallel codes. The principle is shown in Figure 14.15. The source data stream is put out directly, and also sent to several encoder branches. Each of the branches initially contains an interleaver that is different from branch to branch. After the interleaver, the data stream (in each parallel branch) is sent through a convolutional encoder, which maps the information vector $\mathbf{u}$ to the output. The number of these parallel encoders is – in principle – arbitrary. Due to concerns about the data rate, two encoders (resulting in code rate 1/3) are most common.

It is advantageous to use *Recursive Systematic Convolutional* (RSC) codes for the encoding of each branch. The structure of such a code is outlined in Figure 14.16. A shift register with feedback is used to compute the encoded bits. As these codes are systematic, there is one output that directly maps the source bits to the output. These bits are not actually transmitted, since the turbo code transmits the original data sequence anyway (see topmost branch in Figure 14.15); only encoded

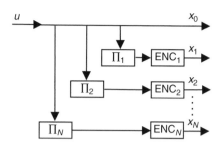

Figure 14.15 Structure of a turbo encoder. Π denotes interleavers.
Reproduced with permission from Valenti [1999] © M. Valenti.

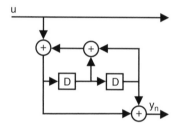

Figure 14.16 Structure of a recursive systematic convolutional encoder. D denotes delay elements.
Reproduced with permission from Valenti [1995] © M. Valenti.

bits are actually sent. Therefore, such a constituent encoder usually has code rate 1 – i.e., puts out one coded bit per source bit (see Figure 14.17).[10]

Summarizing, one of the encoders (the direct feedthrough), uses the original source data sequence as input, while in the other case, the sequence is first interleaved. The sequences (original sequences, plus the redundant bits from the other encoders) are then multiplexed. The resulting code is systematic, as it still contains the original data sequence. Since the systematic part is transmitted only once, the code rate of the total system is $1/(N+1)$.

Besides this basic structure, many other turbo encoders are possible. As research in this area is still very active, we refrain from giving a taxonomy of encoders here.

The code rate of the above encoder is $R_c = 1/3$. However, it can be increased by puncturing. For example, outputs x_1 and x_2 can be used as input to a multiplexer that alternatingly uses a bit from the first encoder x_1 and discards a bit x_2 from encoder 2; or that uses x_2 and discards x_1. This realizes an encoder with code rate $R_c = 1/2$.

14.6.3 Turbo Decoder

While encoding is always a comparatively simple process, decoding is the area where problems normally occur. Brute-force implementation of a decoder, where the combined code is viewed as a

---

[10] If we just used this encoder as a normal convolutional encoder, then the code rate would be 1/2, as the encoder would put out the source bit and the redundant bit for each incoming source bit.

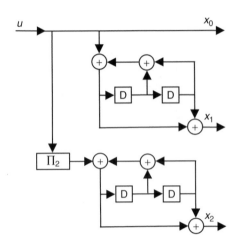

Figure 14.17   Structure of a turbo encoder based on recursive systematic convolutional encoders.
Reproduced with permission from Valenti [1999] © M. Valenti.

single, very complicated, code, would require prohibitive computational effort. It is here that turbo codes show their great advantage: it is possible to decode two constituent codes separately, and then combine the information from these two decoders.

Turbo codes are decoded iteratively, employing the exchange of soft information between constituent decoders. Figure 14.18 shows a block diagram of a turbo decoder. It consists of two Soft Input Soft Output (SISO) decoders for the constituent codes, an interleaver, and a de-interleaver. The SISO decoder puts out not only the various bits, but also the confidence it has in a specific decision. This confidence is described by the *LLR* which can be computed by a Viterbi-like algorithm, suggested by Bahl et al. [1974] and known as the BCJR algorithm. The LLR is (see Eq. (14.48))

$$\log \left[\frac{\Pr(b_i = +1|r)}{\Pr(b_i = -1|r)} \right] \tag{14.51}$$

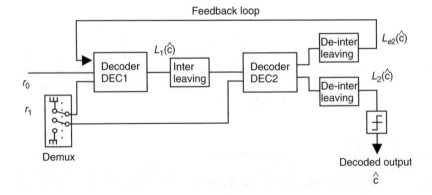

Figure 14.18   Structure of a turbo decoder.
Reproduced with permission from Sklar [1997] © IEEE.

i.e., the logarithm of the ratios of the probabilities that bit b_i is $+1$ or -1, conditioned on the fact that the value r was received. It can be shown that the LLR is the sum of three components:

- The LLR of the a priori probability for the bits (that LLR is usually 0, as all bits are equally likely).
- The information from the received raw data – i.e., the direct observation. This LLR also depends on the SNR.
- The extrinsic log likelihood, which contains the information from decoding. The extrinsic information from the second decoder helps the first decoder, and the first decoder helps the second decoder.

The first two components also exist for uncoded systems, the last one is the contribution from decoding of the other constituent code. It is important that the extrinsic information of a constituent decoder does not depend on the input of that decoder; otherwise, the information in the extrinsic information and in the direct observation would be strongly correlated, and the decoder would just confirm its own opinion.

The iteration now starts with the assumption that the extrinsic LLR is zero, and decoder 1 decodes the received signal like a "normal" convolutional decoder (though with soft output). From this soft output, the receiver then computes the extrinsic LLR at decoder 2. The extrinsic information computed from the output of decoder 2 is then used for the next iteration step at decoder 1. The extrinsic LLR from decoder 1 obtained in that next iteration is used for the subsequent iteration of decoder 2. This procedure is continued until convergence is achieved.

The BER improves as the number of iterations I increases. In an AWGN channel, $N_{it} = 5$ iterations are usually sufficient to achieve convergence, while a fading channel usually requires $N_{it} = 10$ iterations. The number of operations per information bit can be estimated as

$$N_{op} \leq N_{it}(62 + 8/R_c) \tag{14.52}$$

Typically 400–800 operations per information bit are required. This explains why the widespread adoption of turbo codes happened only after the turn of the century. Transmission rates of 400 kbit/s require up to 300 MIPS (Million Instructions Per Second) for decoding – this large amount of signal-processing power has only been available since then.

Turbo codes allow close approximation of channel capacity. As an example, Figure 14.19 shows the BER of a rate-1/2 turbo encoder in an AWGN channel. Six iterations are obviously sufficient to closely approach the converged value. After 18 iterations, the BER is 10^{-5} at a 0.7-dB SNR.

14.7 Low Density Parity Check Codes

When turbo codes were announced in 1993, they immediately drew a large amount of richly deserved attention. It seemed that for the first time the Shannon bound could be approached by practical codes. Yet it turns out that the problem had already been solved in the early 1960s! Gallagher [1961] in his thesis had designed linear block codes, called *Low Density Parity Check (LDPC) codes*, which allow the Shannon limit to be closely approached, and also proposed efficient iterative decoding mechanisms. However, this work was largely overlooked, because iterative decoding exceeded the available computer power at that time. However, several papers in the mid-1990s led to the rediscovery of these codes. And by that time the decoding complexity that once had seemed prohibitive looked quite reasonable. Since that time, a large number of papers have been published, and a considerable number of improvements have been proposed.

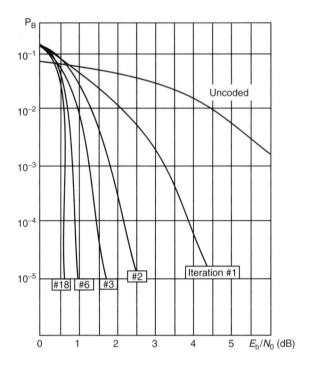

Figure 14.19   Bit error rate of a $R_c = 1/2$ turbo code with interleaver length 64,000 for different numbers of iterations in an additive white Gaussian noise channel.
Reproduced with permission from Sklar [1997] © IEEE.

14.7.1   Definition of Low Density Parity Check Codes

LDPC codes are linear block codes (as discussed in Section 14.2). One interesting aspect for them is that they are not defined via the generator matrix **G**, but rather via the parity check matrix **H**. This is a key trick, since it is normally decoding that causes the biggest problems, not encoding. It thus makes eminent sense to define a structure that allows for easy decoding! Blocksize, and thus the dimensions of the check matrix, are very large. However, the number of nonzero entries into that matrix is kept low. More precisely, the ratio of the number of nonzero elements to the total number of entries is small; this is the reason why the codes are called "low-density." Following Gallagher, let us define an (N, p, q) LDPC code as a code of length N, whose parity check matrix has p 1's in each column, and q 1's in each row. In order for the code to have good properties, it is necessary that $p \geq 3$. If all rows are linearly independent, then the resulting rate of the code is $(q - p)/q$.

Good parity check matrices can be constructed from a few simple rules. First, subdivide the matrix horizontally into p submatrices of equal size. Then put one "1" into each column of such a submatrix. Let the first submatrix be defined to be, e.g., q concatenated identity matrices, or a structure like:

$$\begin{bmatrix} 1 & 1 & 1 & 1 & 0 & 0 & 0 & 0 & 0 & 0 & 0 & 0 & 0 & 0 & 0 & 0 & 0 & 0 & 0 & 0 \\ 0 & 0 & 0 & 0 & 1 & 1 & 1 & 1 & 0 & 0 & 0 & 0 & 0 & 0 & 0 & 0 & 0 & 0 & 0 & 0 \\ 0 & 0 & 0 & 0 & 0 & 0 & 0 & 0 & 1 & 1 & 1 & 1 & 0 & 0 & 0 & 0 & 0 & 0 & 0 & 0 \\ 0 & 0 & 0 & 0 & 0 & 0 & 0 & 0 & 0 & 0 & 0 & 0 & 1 & 1 & 1 & 1 & 0 & 0 & 0 & 0 \\ 0 & 0 & 0 & 0 & 0 & 0 & 0 & 0 & 0 & 0 & 0 & 0 & 0 & 0 & 0 & 0 & 1 & 1 & 1 & 1 \end{bmatrix} \qquad (14.53)$$

Let, then, the other submatrices be random column permutations of this first submatrix. For example, we arrive at the following example of a (20, 3, 4) code [Davey 1999]:

$$
\mathbf{H} = \begin{bmatrix}
1 & 1 & 1 & 1 & 0 & 0 & 0 & 0 & 0 & 0 & 0 & 0 & 0 & 0 & 0 & 0 & 0 & 0 & 0 & 0 \\
0 & 0 & 0 & 0 & 1 & 1 & 1 & 1 & 0 & 0 & 0 & 0 & 0 & 0 & 0 & 0 & 0 & 0 & 0 & 0 \\
0 & 0 & 0 & 0 & 0 & 0 & 0 & 0 & 1 & 1 & 1 & 1 & 0 & 0 & 0 & 0 & 0 & 0 & 0 & 0 \\
0 & 0 & 0 & 0 & 0 & 0 & 0 & 0 & 0 & 0 & 0 & 0 & 1 & 1 & 1 & 1 & 0 & 0 & 0 & 0 \\
0 & 0 & 0 & 0 & 0 & 0 & 0 & 0 & 0 & 0 & 0 & 0 & 0 & 0 & 0 & 0 & 1 & 1 & 1 & 1 \\
1 & 0 & 0 & 0 & 1 & 0 & 0 & 0 & 1 & 0 & 0 & 0 & 1 & 0 & 0 & 0 & 0 & 0 & 0 & 0 \\
0 & 1 & 0 & 0 & 0 & 1 & 0 & 0 & 0 & 1 & 0 & 0 & 0 & 0 & 0 & 1 & 0 & 0 & 0 & 0 \\
0 & 0 & 1 & 0 & 0 & 0 & 1 & 0 & 0 & 0 & 0 & 0 & 1 & 0 & 0 & 0 & 1 & 0 & 0 & 0 \\
0 & 0 & 0 & 1 & 0 & 0 & 0 & 0 & 0 & 1 & 0 & 0 & 0 & 1 & 0 & 0 & 0 & 1 & 0 & 0 \\
0 & 0 & 0 & 0 & 0 & 0 & 0 & 1 & 0 & 0 & 0 & 1 & 0 & 0 & 0 & 1 & 0 & 0 & 0 & 1 \\
1 & 0 & 0 & 0 & 0 & 1 & 0 & 0 & 0 & 0 & 0 & 1 & 0 & 0 & 0 & 0 & 1 & 0 & 0 \\
0 & 1 & 0 & 0 & 0 & 0 & 1 & 0 & 0 & 0 & 1 & 0 & 0 & 0 & 0 & 1 & 0 & 0 & 0 & 0 \\
0 & 0 & 1 & 0 & 0 & 0 & 0 & 1 & 0 & 0 & 0 & 0 & 1 & 0 & 0 & 0 & 0 & 0 & 1 & 0 \\
0 & 0 & 0 & 1 & 0 & 0 & 0 & 0 & 1 & 0 & 0 & 0 & 0 & 1 & 0 & 0 & 1 & 0 & 0 & 0 \\
0 & 0 & 0 & 0 & 1 & 0 & 0 & 0 & 0 & 1 & 0 & 0 & 0 & 0 & 1 & 0 & 0 & 0 & 0 & 1 \\
\end{bmatrix}
\qquad (14.54)
$$

This structure obviously follows the rules of having p 1's in each column, and q 1's in each row; it also appears from inspection that the structure is reasonably random.

14.7.2 Encoding of Low Density Parity Check Codes

Since LDPC codes are defined via their parity check matrix, the encoding process is more complicated than for "normal" block codes. Remember that in normal block codes, we only have to multiply the codeword vector by the generator matrix. However, for LDPC codes, the generator matrix is not yet known. Fortunately, the computation is not very difficult: Using Gaussian elimination and reordering of columns, we can cast the parity check matrix in the form:

$$
\tilde{\mathbf{H}} = (-\mathbf{P}^T \quad \mathbf{I}) \qquad (14.55)
$$

The corresponding generator matrix is then

$$
\mathbf{G} = (\mathbf{I} \quad \mathbf{P}) \qquad (14.56)
$$

Note that, due to the process of Gaussian elimination, the generator matrix is typically *not* sparse. This also means that the encoding process requires more operations. Fortunately, all the required operations are simple because they are performed on discrete, known bits. Thus, encoding complexity is usually not an issue.

14.7.3 Decoding of Low Density Parity Check Codes

As we have mentioned above, the sparse structure of the parity check matrix is key to decoding that works with reasonable complexity. But it is still far from trivial! Performing an exact maximum likelihood decoding is an N-p hard problem (in other words, we have to check all possible codewords, and compare them with the received signal). It is therefore common to use an iterative algorithm called *belief propagation*. It is this algorithm that we describe in more detail in the following.[11]

---

[11] Note that the decoding algorithm can be described based on the syndrome vector (the method we will choose here) as well as the data vector.

Figure 14.20   Tanner graph for the parity check matrix $H = \begin{bmatrix} 1 & 0 & 1 & 1 \\ 0 & 1 & 1 & 1 \end{bmatrix}$.

Let the received signal vector be **r**. Let us next put the parity check equations into graphical form. In the so-called "Tanner graph" (see Figure 14.20),[12] we distinguish two kinds of nodes:

1. *Variable (bit) nodes*: each variable node corresponds to one bit, and we know that it can be either in state 0, or state 1. Variable nodes correspond to the *columns* of the parity check matrix. We denote these nodes by circles.
2. *Constraint nodes*: constraint nodes (checknodes) describe the parity check equations; we know that if there are no errors present, the inputs to constraint nodes have to add up to 0. This follows from the definition of the syndrome, which is 0 if no errors are present. Constraint nodes correspond to the *rows* of the parity check matrix. We denote these nodes by squares.

Because there are two different types of nodes, and no connections between nodes of the same type, such a graph is also called a "bipartite" graph. In addition to constraint nodes and variable nodes, there is also external evidence, obtained by observation of the received signal, which has to influence our decisions.

Constraint nodes are connected to variable nodes if the appropriate entries in the parity check matrix are 1 – i.e., constraint node i is connected to variable node j if $H_{ij} = 1$. "Soft" information from the observed signals – i.e., the external evidence – is connected to the variable nodes. We also need to know the probability density function (pdf) of the amplitude of the variables – i.e., the probability that a variable node has a certain state, given the value of the external evidence.

Decoding on such a graph is done by a procedure called *message passing* or *belief propagation*. Each node collects incoming information, makes computations according to a so-called *local rule*, and passes the result of the computation to other nodes. Essentially, the jth variable node tells the constraint nodes it is connected to what it thinks its – i.e., the variable node's – value is, given the external information r_j and the information from the other constraint nodes. This message is denoted λ_{ij}. In turn the ith constraint node tells the jth variable node what *it* thinks the variable node has to be, given the information that the constraint nodes have from all the *other* variable nodes; this message is called $\mu_{i,j}$. This is shown in Figure 14.21.

Let us formulate the decoding strategy mathematically, for an AWGN channel:

1. First, the data bits decide what value they *think* they are, given the external evidence **r** only. Knowing the statistics of the noise σ_n^2, the variable nodes can easily compute their probability to be 1 or a 0, respectively, and pass that information to the constraint nodes. Conversely, the constraint nodes cannot pass a meaningful message to the variable nodes yet. Therefore,

$$\mu_{i,j}^{(0)} = 0, \qquad \text{for all } i \tag{14.57}$$

$$\lambda_{i,j}^{(0)} = (2/\sigma_n^2)r_j, \qquad \text{for all } j \tag{14.58}$$

---

[12] There are two types of graphical representations: the "Tanner graph" used here, and the "Forney factor graph" [Loeliger 2004].

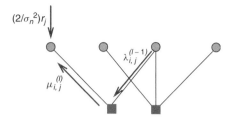

Figure 14.21 Message-passing in a factor graph.

2. Then, the constraint nodes pass a different message to each variable node. Elaborating on the principle mentioned above, let us look specifically at constraint node i: assume that a set of connections ends in node i, which originate from an ensemble $A(i)$ of variable nodes.

   Now, each of the checknodes has two important pieces of information: (i) it knows the values (or probabilities) of all data bits connected to this checknode; (ii) furthermore, it knows that all the bits coming into a checknode have to sum up to 0 mod 2 (that is the whole point of a parity check matrix). From these pieces of information, it can compute the probabilities for the value that it thinks data bit j has to have. Since we have an AWGN channel, with a continuous output, and not a binary channel, we have to use LLRs instead of simple probabilities that a bit is reversed, so that the message becomes

$$\mu_{i,j}^{(l)} = 2\tanh^{-1}\left(\prod_{k \in A(i)-j} \tanh\left(\frac{\lambda_{i,k}^{(l-1)}}{2}\right)\right) \tag{14.59}$$

where $A(i) - j$ denotes "all the members of ensemble $A(i)$ with the exception of j" – i.e., all constraint nodes that connect to the ith variable node, with the exception of the jth node. Superscript$^{(l-1)}$ denotes the $l-1$th iteration – i.e., we use the results from the previous iteration steps.

3. Next, we update our opinion of what the variable nodes are, based on the information passed by the constraint nodes, as well as the external evidence. This rule is very simple:

$$\lambda_{i,j}^{(l)} = (2/\sigma_n^2)r_j + \sum_{k \in B(j)-i} \mu_{k,j}^{(l)} \tag{14.60}$$

where $B(j) - i$ denotes all variable nodes that connect to the jth constraint node, with the exception of i.

4. From the above, we can compute the pseudoposterior probabilities that a bit is 1 or 0:

$$L_j = (2/\sigma_n^2)r_j + \sum_i \mu_{i,j}^{(l)} \tag{14.61}$$

based on which tentative decision we make about the codeword. If that codeword is consistent – i.e., its syndrome is 0 – then decoding stops.

Example 14.8 *Decoding of a low density parity check code.*

Let us now consider a very simple example for this algorithm. Let the parity check matrix be

$$\mathbf{H} = \begin{bmatrix} 0 & 0 & 1 & 1 & 1 & 1 \\ 1 & 1 & 1 & 1 & 0 & 0 \\ 1 & 1 & 0 & 0 & 1 & 1 \end{bmatrix} \tag{14.62}$$

Let the codeword:

$$\bar{y} = [0 \quad 1 \quad 1 \quad 0 \quad 1 \quad 0] \tag{14.63}$$

be sent through an AWGN channel with $\sigma_n^2 = 0.237$ corresponding to $\gamma = 6.25\,\mathrm{dB}$; the received word be

$$\bar{r} = [-0.71 \quad 0.71 \quad 0.99 \quad -1.03 \quad -0.61 \quad -0.93] \tag{14.64}$$

Then according to step 1 (mentioned above), likelihood values are computed from external evidence as

$$\overline{\lambda^{(0)}} = [-6.0 \quad 6.0 \quad 8.3 \quad -8.7 \quad -5.2 \quad -7.9] \tag{14.65}$$

Hard thresholding of the received likelihood values would result in codeword error with an error at bit position 5:

$$[0 \quad 1 \quad 1 \quad 0 \quad 0 \quad 0] \tag{14.66}$$

Figure 14.22 demonstrates how the message-passing algorithm iterates itself to the correct solution, following the recipe given above.

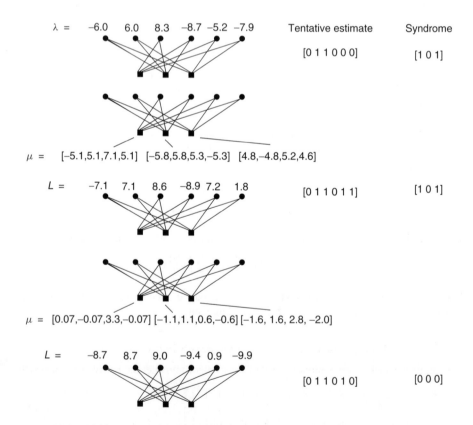

Figure 14.22   Example for the iterations of low density parity check message passing.

14.7.4 Performance Improvements

It can be shown that the belief propagation algorithm always converges to the maximum likelihood solution if the Tanner graph can be rolled up into a tree structure – i.e., each node is a "parent" or a "child" of another node, but not both at the same time. In other words, there should be no cycles in the Tanner graph. Short cycles, like the one shown in Figure 14.23, lead to problems with convergence: if nodes start out with a wrong belief, they tend to reinforce it among themselves, instead of being convinced by evidence from other variable nodes that they need to change their self-assessment. The construction of codes without shot cycles is one of the most important, and most challenging, tasks in the design of LDPC codes. Note, however, that codes that lead to a *pure* tree structure are usually not good codes, even though they can be decoded exactly by the belief propagation algorithm.

Figure 14.23 Tanner graph from Figure 14.20, showing a short cycle (with "girth" 4).

Another important area to improve convergence and performance is the use of irregular codes. This means that column and row weights are not fixed, as we have assumed up to now, but that we rather prescribe only the *mean* weights, and allow some of the columns to have more entries. These nodes associated with these "heavier" columns often converge faster, and can spread their "secure" knowledge to other nodes, which leads to improved convergence for the other nodes as well.

The performance that can be achieved with LDPC codes is very impressive, and (for large blocksizes) can approach the Shannon capacity within a fraction of a dB. There is also an interesting rule for computational complexity in the decoding of LDPC codes: when the used rate approaches $(1 - \delta)$ of Shannon capacity, then the complexity per bit goes like $(1/\delta)\log_2(1/\delta)$ [Richardson and Urbanke 2008]. For most other codes, the complexity grows with $\exp(1/\delta)$.

14.8 Coding for the Fading Channel

The structure of errors in a fading channel are different from that in an AWGN channel. The presence of error bursts is noticeable: when the channel shows (instantaneous) high attenuation due to the destructive interference of multipath components, then the error probability is much larger than in constructive interference. Correction of these error bursts can be achieved either by using codes that are especially suited for bursty errors. Alternatively, interleaving "breaks up" the bursty structure of errors.

14.8.1 Interleaving

Compared with a symbol duration, wireless channels change only slowly. Typically, a mobile station stays in a fading dip (which has spatial extension of about $\lambda/4$) for a duration of $10-100$ ms. As a consequence, a large number of bits are strongly attenuated (and thus more susceptible to errors by noise). A normal code usually cannot correct such a large number of errors. To understand the basic principle, consider a simple rate-1/3 repetition code.[13] The probability of a wrong decision in a coded bit is $P_{single} \sim Q(\sqrt{\gamma})$, where γ is the *instantaneous* SNR. Due to the majority rule, the probability of a wrong final decision is approximately $P_{single}^2 \sim Q^2(\sqrt{\gamma})$, where we have made use of the fact that the SNR of two subsequent bits is the same. Averaged over different channel states, the BER is then approximately:

$$BER \sim \int_0^\infty pdf_\gamma(\gamma)Q^2(\sqrt{\gamma})d\gamma \qquad (14.67)$$

When an interleaver is used, then the three bits associated with one source bit are transmitted at large intervals, so that the SNR for each of these transmissions is different (see Figure 14.24). Therefore, an error occurs only if two of these independent transmissions are in independent fading dips, which is very unlikely (see Figure 14.25). Mathematically, this can be written as

$$BER \sim \int_0^\infty \int_0^\infty pdf_{\gamma_1}(\gamma_1)Q(\sqrt{\gamma_1})pdf_{\gamma_2}(\gamma_2)Q(\sqrt{\gamma_2})\,d\gamma_1\,d\gamma_2$$

$$= \left(\int_0^\infty pdf_\gamma(\gamma)Q(\sqrt{\gamma})\,d\gamma \right)^2 \qquad (14.68)$$

We stress that an interleaver can reduce the mean BER only in combination with a codec. For an uncoded system, the interleaver still breaks up error bursts (which sometimes can be desirable), but it does not lead to a decrease in the mean BER. Another problem with interleavers lies in the fact that they increase latency of transmission. For speech communications, it is necessary to keep latency below 50 ms. In that case, it can happen that maximum latency is smaller than the duration of a fading dip, which greatly reduces the effectiveness of the interleaver.

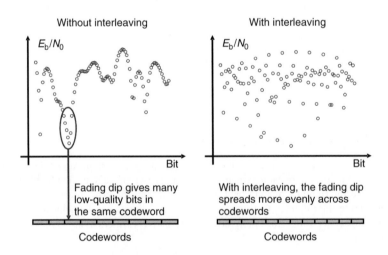

Figure 14.24   Effect of an interleaver.

---

[13] Each bit is transmitted three times. The receiver first performs a hard decision, and then makes a majority decision. If it decides two or three times that the coded bit was a $+1$, then it concludes that the source data bit was $+1$.

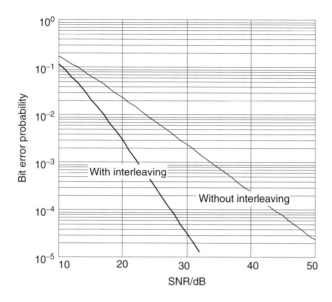

Figure 14.25   Bit error rate of a rate-1/3 repetition code with hard majority decision in a flat Rayleigh-fading channel, with and without interleaving.

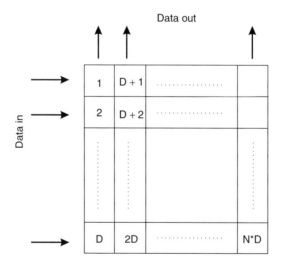

Figure 14.26   Structure of a block interleaver.

After these basic considerations, we now turn to the structure of the interleaver. Basically, two structures are possible: block interleaving and convolutional interleaving. For a block interleaver (Figure 14.26), a block of data of size $N_{\text{itnerleav}} D_{\text{interleav}}$ is interleaved at once. The structure of the interleaver is a matrix: bits are read in line-by-line, and read out column-by-column. The bits that are originally adjacent are now separated by $D_{\text{interleave}}$. Latency due to this interleaver is $N_{\text{interleave}} D_{\text{interleave}} T_B$ – it has to wait until the matrix has been filled up before it can read out the bits. Latency is thus considerably larger than the separation of bits that can be achieved.

For a convolutional interleaver, the data are interleaved in a continuous stream, similar to a convolutional encoder. This reduces latency compared with the block interleaver (for the same separation of bits). It is common to use block interleavers together with block codes, and convolutional interleavers together with convolutional codes.

14.8.2   Block Codes

We now apply the above-mentioned principles to block codes. Due to interleaving, each of the bits (symbols) in a codeword fades independently. The redundancy implicit in the code should thus make it possible to recover the information even when some of the bits are in fading dips and therefore have a very poor SNR. In the absence of interleaving, the effectiveness of the code would be much reduced.

As an example, let us consider a block code with hard decoding: let $K = 12, N = 23$, and the minimum distance be $d_{\min} = 7$, and the coding gain 2 dB. Due to the slowness of fading, all 23 codebits see the same channel if there is no interleaving. In other words, each and every codeword has a gain of 2 dB. An uncoded system needs a 26-dB SNR to achieve a BER of 10^{-3}, with coding, this reduces to 24 dB – which is still bad. In other words – even with an encoder, the BER decreases only proportionally to γ^{-1}, just the proportionality constant changes.

The use of interleaving dramatically changes the situation. Errors occur only when there are errors at least at four locations in the codeword (note that $\left\lceil \frac{d_{\min}-1}{2} \right\rceil = 3$ can be corrected). The BER is approximately proportional to [Wilson 1996]:

$$\sum_{i=4}^{23} K_i \left(\frac{1}{2 + 2\overline{\gamma}_B} \right)^i \left(1 - \frac{1}{2 + 2\overline{\gamma}_B} \right)^{23-i} \approx \frac{1}{\overline{\gamma}_B^4} \tag{14.69}$$

where the K_i are constants. Quite generally we find that a block code with minimum distance $d_{\min}$ and "hard" decoding achieves a diversity order $\left\lceil \frac{d_{\min}-1}{2} \right\rceil + 1$.

When interleaving together with soft decoding is used, the resulting diversity order is almost twice as large – namely, $d_{\min}$. To put this into more mathematical terms, we write the metric that we need to minimize as

$$\min_{\mathbf{s}} \sum_i |r_i - \alpha_i s_i|^2 \tag{14.70}$$

and the pairwise error probability becomes [Benedetto and Biglieri 1999]:

$$P(\mathbf{s} \to \mathbf{s}_E) = \prod_{i \in A} \frac{1}{1 + |s_i - s_{E,i}|^2/(4N_0)} \tag{14.71}$$

where A is the set of indices so that $s_i \neq s_{E,i}$. For a linear code the error probability becomes

$$P \leq \sum_{w \in W} \left(\frac{1}{1 + R_c \gamma} \right)^w \tag{14.72}$$

where R_c is the code rate, and W the set of nonzero Hamming weights of the code. This shows clearly that the minimum distance determines the diversity order (slope of the BER versus SNR curve).

14.8.3 Convolutional Codes

The performance measures and design rules for convolutional codes are quite similar to those of block codes. As a matter of fact, the mathematical description given above for block codes with soft decoding can directly be applied to convolutional codes. Thus, we find that the Euclidean distance of the code is no longer an important metric. Rather, we strive to find goals with good minimum (Hamming) distance properties. If the code sequences differ in many positions, then the probability is small that all of the distinguishing bits are in a fading dip simultaneously. Expressed in other words: we wish that two sequences that can be confused with each other (i.e., start and end in the same state of the trellis) are as long as possible. The minimum length of such sequences is often called the "effective length" of the code.

From these simple considerations, we can derive a few rules about code design for flat-fading channels:

- The minimal Euclidean distance does not occur in the equations for the BER. Thus, the design criteria for AWGN channels and fading channels are quite different. Since Rayleigh fading and AWGN channels are only limiting cases of the Rician channels that are practically important, it is important to find codes that work in both of these channel types.
- The most important parameter in the fading channel is effective length, which enters exponentially into the BER.

14.8.4 Concatenated Codes

An important class of codes for fading channels is the concatenation of convolutional codes with RS codes. Before 1995, this was actually the most popular code for cases where extremely high coding gains are required. Even today, this method is sometimes used for situations where high data rates or complexity restrictions prevent the use of turbo codes and LDPC codes.

RS codes are good at correcting bursts of bit errors – actually much better than at correcting distributed errors. It would seem that this makes them well suited for fading channels. However, for typical coherence times and bit rates, the length of the fades (in units of bits) in a wireless channel is typically much larger than the correction capability of an RS code. Thus, the way RS codes are applied is the following: first, the data are convolutionally encoded, and the resulting data stream is interleaved, to break up any possible error bursts (long interleavers are much easier to build than codes for long error bursts). While the convolutional code does not see long strings of severely impaired data (as it would without the interleaver), the decoding of a convolutional code results in error bursts, as this is a fundamental property of convolutional codes. These error bursts are then corrected by means of the outer code – namely, the RS code – which is well equipped to deal with these bursts.

14.8.5 Trellis Coded Modulation in Fading Channels

For TCM, the same design criteria are valid as for convolutional codes. Figure 14.27 shows that the effectiveness of the codes can be very different in different channels. We show a code by Ungerboeck [1982], designed for AWGN channels, and a code from Jamali and LeNgoc [1991] designed for fading channels. The superior performance of each code in its "natural" environment is obvious.

TCM is also very helpful in combating the errors due to delay dispersion that were discussed in Section 12.3. Chen and Chuang [1998] showed that these errors are determined essentially by the Euclidean distance between the codewords. Minimization of this metric gives codes that can combat these types of errors (see Figure 14.28). While the search for codes for flat-fading channels is relatively mature, code designs for frequency-selective channels are rather recent.

Figure 14.27 Bit error rate in a fading channel: uncoded quadrature-phase shift keying and 8-PSK trellis-coded modulation.

Reproduced with permission from Mayr [1996] © B. Mayr.

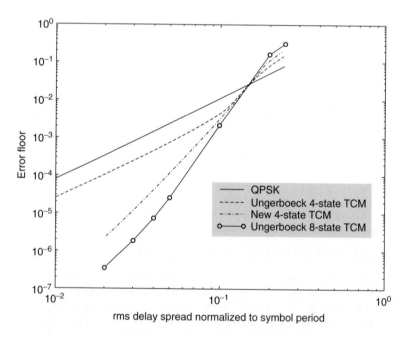

Figure 14.28 Error floor of trellis-coded modulation in frequency-selective channels.

Reproduced with permission from Chen und Chuang [1998] © IEEE.

14.9 Information-Theoretic Performance Limits of Fading Channels

After discussing the performance of practical codes in fading channels, we now analyze the information-theoretic limits in fading channels. We will consider for the most part frequency-flat fading channels, which can be described by a time-variant SNR γ. Only in the last subsection will we turn to frequency-selective channels.

14.9.1 Ergodic Capacity vs. Outage Capacity

We first reiterate two critical facts about AWGN channels: the normalized capacity of a (complex) channel is given by

$$C(\gamma) = \log_2 \left[1 + \gamma\right] \text{ bits/s/Hz} \tag{14.73}$$

and this capacity is derived under the assumption of codes with infinite codeword length.

Since the codewords are infinitely long, they extend over different realizations of the fading state. The *ergodic (Shannon) capacity* in a fading channel is therefore the expected value of the capacity, taken over all realizations of the channel. This quantity assumes an infinitely long code that extends over all the different channel realizations.

However, in practical situations we are often faced with the situation where the data are encoded with a near-Shannon-limit-achieving code that extends over a period that is much shorter than the channel coherence time. LDPC codes with reasonable block lengths (e.g., 10,000 bits) get close to the Shannon limit (see Section 14.7). For a data rate of 10 Mbit/s, such a block can be transmitted within 1 ms. This is much shorter than 10 ms, which is a typical coherence time of wireless channels (see Chapter 5). Thus, each channel realization can be associated with a capacity value. This capacity (sometimes called *instantaneous capacity*) is a random variable (r.v.) with an associated cumulative distribution function (cdf). We thus will look henceforth at this distribution function, or equivalently the capacity that can be guaranteed for $x\%$ of all channel realizations. The latter quantity is also known *outage capacity*.

Since the channel is time varying, it can also be meaningful to let the transmit power vary with time. Whether, and to what extent, this can be done depends on the amount of Channel State Information at the Transmitter (CSIT), i.e., what the transmitter knows about the SNR at the receiver.

14.9.2 Capacity for Channel State Information at the Receiver (CSIR) Only

If the transmitter has no channel state information, then the transmitter cannot adapt its transmit power to the channel state, so that the only thing it can do is transmit with a constant power. The Shannon capacity (per unit bandwidth) is then the AWGN capacity averaged over the SNR distribution, i.e.,

$$E\{C(\gamma)\} = \int_0^\infty \log_2 \left[1 + \gamma\right] pdf_\gamma(\gamma)d\gamma \text{ bits/s/Hz} \tag{14.74}$$

From Jensen's inequality, it follows immediately that $E\{C(\gamma)\} < C(E\{\gamma\})$; in other words, the Shannon capacity in a fading channel is smaller than the capacity in an AWGN channel with the same mean SNR.

We now turn to the computation of the outage capacity, or capacity cdf. The mapping $\gamma \to \log(1 + \gamma)$ constitutes a transformation between two random variables, so that we can derive the pdf of C using transformation of variables with the Jacobian, resulting in

$$pdf_C(C) = pdf_\gamma(2^C - 1)\ln(2)2^C. \tag{14.75}$$

For example, in a Rayleigh-fading channel,

$$pdf_C(C) = \frac{1}{\overline{\gamma}}\exp\left[-\frac{2^C - 1}{\overline{\gamma}}\right]\ln(2)2^C \tag{14.76}$$

and the resulting cdf is

$$cdf_C(C) = 1 - \exp\left[-\frac{2^C - 1}{\overline{\gamma}}\right]. \tag{14.77}$$

In the case that the admissible outage is given, we can compute the corresponding γ from $cdf_\gamma(\gamma)$, and insert this value into the capacity equation to obtain the outage capacity.

From this example, we see that there are channel constellations at which the capacity vanishes. Thus, it is not possible to guarantee a zero-outage transmission.

14.9.3 Capacity for CSIT and CSIR – Waterfilling

When the transmitter knows the channel state, it can adapt its power (subject to a long-term power constraint) to maximize the channel capacity. Thus, the ergodic capacity is given by

$$C = \max_{P(\gamma)} \int_0^\infty \log_2\left[1 + \gamma\frac{P(\gamma)}{\overline{P}}\right]pdf_\gamma(\gamma)d\gamma \tag{14.78}$$

with the constraint

$$\int_0^\infty P(\gamma)pdf_\gamma(\gamma)d\gamma \leq \overline{P} \tag{14.79}$$

This optimization problem can be solved by

$$\frac{P(\gamma)}{\overline{P}} = \begin{cases} \frac{1}{\gamma_0} - \frac{1}{\gamma} & \gamma > \gamma_0 \\ 0 & \text{otherwise} \end{cases} \tag{14.80}$$

where γ_0 is determined by inserting Eq. (14.80) the constraint Eq. (14.79). This power allocation strategy is known as "waterfilling"; we will discuss it in more detail in Section 19.8. Essentially, it means that the transmitter does not transmit if the channel takes on a very bad realization (SNR is below the threshold γ_0). If the SNR is above the threshold, then the transmitter will use the more power the better the SNR is.

When considering the outage capacity, one might try to improve the outage capacity by letting the transmit power compensate for the losses of the receiver, i.e., $P(\gamma)/\overline{P} = K_{inv}/\gamma$ where, K_{inv} is a proportionality constant that is determined from Eq. (14.79), so that $K_{inv} = 1/E\{1/\gamma\}$. This approach results in an instantaneous capacity $C_{inv} = \log_2(1 + K_{inv})$ that is independent of γ; therefore, there is no "outage," i.e., the capacity never falls below C_{inv}; The value C_{inv} is therefore also known as the "zero-outage capacity." Note that in Rayleigh fading, $E\{1/\gamma\} \to \infty$, so that $C_{inv} \to 0$.

When we allow outage in $x\%$ of all cases, then the best strategy is not to transmit at all for the $x\%$ worst channels, and for the remainder of channel realizations control the transmit power in such a way that the SNR is a constant.

Last, but not least, we find that in the case of diversity reception, the above equations remain true, with γ representing the SNR at the output of the diversity combiner (see Section 13.4).

14.10   Appendices

Please see companion website *www.wiley.com/go/molisch*

Further Reading

More details on channel coding can be found in the many excellent textbooks on this topic, e.g., Lin and Costello [2004], MacKay [2002], Moon [2005], McEliece [2004], Sweeney [2002], and Wilson [1996]. Steele and Hanzo [1999, ch. 4] give a detailed description of block codes, including the decoding algorithms for RS codes, while Sklar and Harris [2004] give an intuitive introduction; further details can also be found in Lin and Costello [2004]. Convolutional codes are detailed in Johannesson and Zigangirov [1999]. TCM was first invented in the early 1980s by Ungerboeck [1982]; excellent tutorial expositions can be found in Biglieri et al. [1991]. BICM is described in the tutorial booklet [Fabregas et al. 2008]. More details about turbo codes can be found in Schlegel and Perez [2003], and also in the excellent tutorial of Sklar [1997]. LDPC codes originally proposed by Gallagher [1961], were rediscovered by MacKay and Neal [1997], and described in a very understandable way in MacKay [2002]. The important improvement of irregular LDPC codes was proposed by Richardson et al. [2001]. Richardson and Urbanke [2008] give a detailed description of LDPC codes. Important theoretical foundations were laid by Bahl et al. in the 1970s [Bahl et al. 1974] and Hagenauer in the 1980s in their work of SISO decoding. Fundamentals of information theory are described, e.g., in Cover and Thomas [2006], Yeung [2006].

For updates and errata for this chapter, see wides.usc.edu/teaching/textbook

15

Speech Coding

Gernot Kubin

Signal Processing and Speech Communication Laboratory, Graz University of Technology, Graz, Austria

15.1 Introduction

15.1.1 Speech Telephony as Conversational Multimedia Service

When O. Nußbaumer succeeded in the first wireless transmission of speech and music in the experimental physics lab at Graz University of Technology in 1904, nobody would have predicted the tremendous growth in wireless multimedia communications 100 years after this historical achievement. Many new media types have emerged such as text, image, and video, and modern services range from essentially one-way media download, browsing, messaging, application sharing, broadcasting, and real-time streaming to two-way, interactive, real-time conversations by text (chat), speech, and videotelephony. Still, speech telephony is the backbone of all conversational services and continues as an indispensable functionality of any mobile communications terminal. It is for this reason that we will focus our discussion of source-coding methods on speech signals – i.e., on their efficient digital representation for transmission (or storage) applications.

The success story of digital speech coding started with the introduction of digital switching in the Public Switched Telephone Network (PSTN) using Pulse Code Modulated (PCM) speech at 64 kbit/s and continued with a cascade of advanced compression standards from 32 kbit/s in the early 1980s over 16 kbit/s to 8 kbit/s in the late 1990s, all with a focus on long-distance circuit multiplication while maintaining the traditional high-quality level of wireline telephony (*toll quality*). For wireless telephony, the requirements on digital coding were rather stringent with regard to bit rate and complexity from the very beginning in the mid-1980s whereas some compromises in speech quality seemed acceptable because users either had never made the experience of mobile telephony before or, if so, their expectations were biased from the relatively poor quality of analog mobile radio systems. This situation changed quickly, and the first standards introduced at the beginning of the 1990s were completely overturned within 5 years with significant quality enhancements through new coding algorithms, maintaining a bit rate of about 12 kbit/s where the accompanying complexity increase was mitigated by related advances in microelectronics and algorithm implementation.

Wireless Communications, Second Edition   Andreas F. Molisch
© 2011 John Wiley & Sons, Ltd

15.1.2   Source-Coding Basics

The foundations for source coding were laid by C. Shannon [1959], who developed not only channel-coding theory for imperfect transmission channels but also *rate–distortion theory* for signal compression. The latter theory is based on two components:

- a stochastic source model which allows us to characterize the *redundancy* in source information; and
- a distortion measure which characterizes the *relevance* of source information for a user.

For asymptotically infinite delay and complexity, and certain simple source models and distortion measures, it can be shown that there exists an achievable lower bound on the bit rate necessary to achieve a given distortion level and, vice versa, that their exists an achievable lower bound on the distortion to be tolerated for a given bit rate. While complexity is an ever-dwindling obstacle, delay is a substantial issue in telephony, as it degrades the interactive quality of conversations severely when it exceeds a few 100 ms. Therefore, the main insight from rate–distortion theory is the existence of a three-way tradeoff among the fundamental parameters *rate, distortion*, and *delay*. Traditional telephony networks operate in circuit-switched mode where transmission delay is essentially given by the electromagnetic propagation time and becomes only noticeable when dealing with satellite links. However, packet-switched networks are increasingly being used for telephony as well – as in Voice over Internet Protocol (VoIP) systems – where substantial delays can be accumulated in router queues, etc. In such systems, delay becomes the most essential parameter and will determine the achievable rate–distortion tradeoff.

Source coding with a small but tolerable level of distortion is also known as *lossy coding* whereas the limiting case of zero distortion is known as *lossless coding*. In most cases, a finite rate allows lossless coding only for discrete amplitude signals which we might consider for *transcoding* of PCM speech – i.e., the digital compression of speech signals which have already been digitized with a conventional PCM codec. However, for circuit-switched wireless speech telephony, such lossless coders have two drawbacks: first, they waste the most precious resource – i.e., the allocated radio spectrum – as they invest more bits than necessary to meet the quality expectations of a typical user; second, they often result in a bitstream with a *variable rate* – e.g., when using a Huffman coder – which cannot be matched efficiently to the *fixed rate* offered by circuit-switched transmission.

Variable-rate coding is, however, a highly relevant topic in packet-switched networks and in certain applications of joint source channel coding for circuit-switched networks (see Section 15.4.5). While Shannon's theory shows that, under idealized conditions, source coding and channel coding can be fully separated such that the two coding steps can be designed and optimized independently, this is not true under practical constraints such as finite delay or time-varying conditions where only a joint design of the source and channel coders is optimal. In this case, a fixed rate offered by the network can be advantageously split into a variable source rate and a variable channel code rate.

15.1.3   Speech Coder Designs

Source-coding theory teaches us how to use models of source redundancy and of user-defined relevance in the design of speech-coding systems. Perceptual relevance aspects will be discussed in later sections; however, the use of the source model gives rise to a generic classification of speech coder designs:

1. *Waveform coders* use source models only *implicitly* to design an adaptive dynamical system which maps the original speech waveform on a processed waveform that can be transmitted with

fewer bits over the given digital channel. The decoder essentially inverts encoder processing to restore a faithful approximation of the original waveform. All waveform coders share the property that an increase of the bit rate will asymptotically result in lossless transcoding of the original PCM waveform. For such systems, the definition of a coding error signal as the difference between the original and the decoded waveform makes sense (although it is no immediate measure of the perceptual relevance of the distortion introduced).

2. *Model-based coders* or *vocoders* rely on an *explicit* source model to represent the speech signal using a small set of parameters which the encoder estimates, quantizes, and transmits over the digital channel. The decoder uses the received parameters to control a real-time implementation of the source model that generates the decoded speech signal. An increase of the bit rate will result in saturation of the speech quality at a nonzero distortion level which is limited by systematic errors in the source model. Only recently, model-based coders have advanced to a level where these errors have little perceptual impact, allowing their use for very-low-rate applications (2.4 kbit/s and below) with slightly reduced quality constraints. Furthermore, due to the signal generation process in the decoder, the decoded waveform is not synchronized with the original waveform and, therefore, the definition of a waveform error is useless to characterize the distortion of model-based coders.

3. *Hybrid coders* aim at the optimal mix of the two previous designs. They start out with a model-based approach to extract speech signal parameters but still compute the modeling error explicitly on the waveform level. This model error or *residual waveform* is transmitted using a waveform coder whereas the model parameters are quantized and transmitted as *side information*. The two information streams are combined in the decoder to reconstruct a faithful approximation of the waveform such that hybrid coders share the asymptotically lossless coding property with waveform coders. Their advantage lies in the explicit parameterization of the speech model which allows us to exploit more advanced models than is the case with pure waveform coders which rely on a single invertible dynamical system for their design.

The model-based view of speech-coding design suggests that description of a speech-coding system should always start with the decoder that typically contains an implementation of the underlying speech model. The encoder is then obtained as the signal analysis system that extracts the relevant model parameters and residual waveform. Therefore, the encoder has a higher complexity than the decoder and is more difficult to understand and implement. In this sense, a speech-coding standard might specify only the decoder and the format for transmitted data streams while leaving the design of the best matching encoder to industrial competition.

Further Design Issues

The reliance on source models for speech coder design naturally results in a dependence of coder performance on the match between this model and the signal to be encoded. Any signal that is not clean speech produced from a single talker near the microphone may suffer from additional distortion which requires additional performance testing and, possibly, design modifications. Examples of these extra issues are the suitability of a coder for *music* (e.g, if put on hold while waiting for a specific party), for severe *acoustic background noise* (e.g., talking from the car, maybe with open windows), for *babble noise* from other speakers (e.g., talking from a cafeteria), for *reverberation* (e.g., talking in hands-free mode), etc.

Further system design aspects include the choice between *narrowband* speech as used in traditional wireline telephony (e.g., an analog bandwidth from 300 Hz to 3.4 kHz with 8-kHz sampling frequency) and *wideband* speech with quality similar to Frequency Modulation (FM) radio (e.g., an analog bandwidth from 50 Hz to 7 kHz with 16-kHz sampling frequency), which substantially enhances user experience with clearly noticeable speech quality improvements over and above the

Plain Old Telephone Service (POTS). Second, even with the use of sophisticated channel codes, the *robustness of channel errors* such as individual bit errors, bursts, or entire lost transmission frames or packets is also a source-coding design issue (as discussed below in Section 15.4.5). Finally, within the network path from the talker to the listener, there may be several *codec-tandeming* steps where transcoding from one coding standard to another occurs, each time with a potential further loss of speech quality.

15.2 The Sound of Speech

While the "sound of music" includes a wide range of signal generation mechanisms as provided by an orchestra of musical instruments, the instrument for generating speech is fairly unique and constitutes the physical basis for speech modeling, even at the acoustic or perception levels.

15.2.1 Speech Production

In a nutshell, *speech* communication consists of information exchange using a natural *language* as its code and the human *voice* as its carrier. Voice is generated by an intricate oscillator – the vocal folds – which is excited by sound pressure from the lungs. In the view of wireless engineering, this oscillator generates a nearly periodic, Discrete Multi Tone (DMT) signal with a fundamental frequency f_0 in the range of 100 to 150 Hz for males, 190 to 250 Hz for females, and 350 to 500 Hz for children. Its spectrum slowly falls off toward higher frequencies and spans a frequency range of several 1,000 Hz. From its relative bandwidth, it should be considered an *ultrawideband* signal,[1] which is one of the reasons why the signal is so robust and power-efficient (opera singers use no amplifiers) under many difficult natural environments.

The voice signal travels further down the *vocal tract* – the throat, and the oral and nasal cavities – which can be described as an acoustic waveguide shaping the spectral envelope of the signal by its resonance frequencies known as *formant frequencies*. Finally, the signal is radiated from a relatively small opening (mouth and/or nostrils) in a large sphere (our head) which gives rise to a high-pass radiation characteristic such that the far-field speech signal has a typical spectral rolloff of 20 dB per decade.

Sound Generation

Besides the oscillatory voice signal ("phonation"), additional sound sources may contribute to the signal carrier, such as turbulent noise generation at narrow flow constrictions (in "fricative" sounds like [s] in "lesson") or due to impulse-like pressure release after complete flow closures (in "plosive" sounds like [t] in "attempt"). If the vocal folds do not contribute at all to the sound generation mechanism, the speech signals are called *unvoiced*, otherwise they are *voiced*. Note that in the latter case, nearly periodic and noise-like excitation can still coexist (in "voiced fricative" sounds like [z] in "puzzle"), such sounds are often referred to as *mixed excitation* sounds.[2] Besides the three above-mentioned sound generation principles (phonation, frication, explosion) there is a fourth one which is often overlooked: *silence*. As an example, compare the words "Mets" and "mess." Their major difference lies in the closure period of the [t] which results in a silence interval between the vowel [ɛ] and the [s] in "Mets" which is absent in "mess." Otherwise, the two pronunciations are essentially the same, so the information about the [t] is carried entirely by the silence interval.

---

[1] This analogy becomes even more striking if we consider that the phase velocities of light and sound are related by a scale factor of approx. 10^6, showing that 1-kHz soundwaves have about the same wavelength as 1-GHz electromagnetic waves.

[2] Note that from our definitions, mixed excitation speech sounds are clearly a subclass of voiced speech.

Articulation

While the sound generation mechanisms provide the speech carrier, the *articulation* mechanism provides its modulation using a language-specific code. This code is both sequential and hierarchical – i.e., spoken language utterances consist of a sequence of phrases built of a sequence of words built of a sequence of syllables built of a sequence of speech sounds (called *phonemes* if discussed in terms of linguistic symbols). However, the articulation process is not organized in a purely sequential way – i.e., the shape of the vocal tract waveguide is not switched from one steady-state pose to another for each speech sound. Rather, the articulatory gestures (lip, tongue, velum, throat movements) are strongly overlapping and interwoven (typically spanning an entire syllable), mostly asynchronous, and continuously evolving patterns, resulting in a continuous modulation process rather than in a discrete shift-keying modulation process. This *co-articulation* phenomenon is the core problem in automatic speech recognition or synthesis where we attempt to map continuously evolving sound patterns on discrete symbol strings (and vice versa).

15.2.2   Speech Acoustics

Source Filter Model

The scientific study of speech acoustics dates back to 1791 when W. von Kempelen, a high-ranked official at Empress Maria Theresa's[3] court, published his description of the first mechanical speaking machine as a physical model of human speech production. The foundations of modern speech acoustics were laid by G. Fant in the 1950s [Fant 1970] and resulted in the *source filter* model for speech signals (see Figure 15.1).

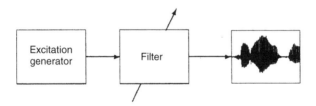

Figure 15.1   Source filter model of speech: The excitation signal generator provides the source which drives a slowly time-varying filter that shapes the spectral envelope according to formant frequencies to produce a speech waveform.

While this model is still inspired from our understanding of natural speech production, it deviates from its physical basis significantly. In particular, all the natural sound sources (which can be located at many positions along the vocal tract and which are often controlled by the local aerodynamic flow) are collapsed into a single source which drives the filter in an independent way. Furthermore, there is only a single output whereas the natural production system may switch between or even combine the oral and nasal branches of the vocal tract. Therefore, the true value of the model does not lie in its accuracy describing human physiology but in its flexibility in modeling speech acoustics. In particular, the typical properties of speech signals as evidenced in its temporal and spectral analyses are well represented by this structure.

Sound Spectrograms

An example of the typical time–frequency analysis for speech is shown in Figure 15.2.

---

[3] Empress Maria Theresa was Archduchess of Austria and Queen of Hungary in the 1700s.

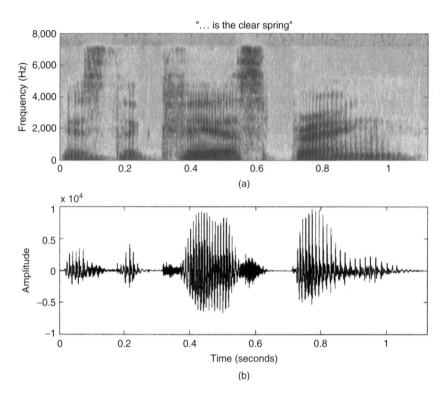

(a)

(b)

Figure 15.2 Spectrogram of the phrase *"is the clear spring"* spoken by a male speaker, analog bandwidth limited to 7 kHz, sampled at $f_s = 16$ kHz, horizontal axis = time, vertical axis = frequency, dark areas indicate high-energy density (a). Time domain waveform of the same signal, vertical axis = amplitude (b).

The lower graph shows the time domain waveform for the phrase *"is the clear spring,"* spoken by a male speaker and limited to an analog bandwidth of 7 kHz for 16-kHz sampling. This graph shows the marked alternation between the four excitation source mechanisms where we can note that the "nearly periodic" source of voiced speech can mean an interval of only three fundamental "periods" which show a highly irregular pattern in the case of the second voiced segment corresponding to *"the."* Furthermore, strong fluctuations of the envelope are visible, which often correspond to 20 to 30 dB.

The upper graph shows a *spectrogram* of the same signal which provides the time–frequency distribution of the signal energy where darker areas correspond to higher energy densities. This representation can be obtained either by means of a filterbank or short-time Fourier analysis. It illustrates global signal properties – like anti-aliasing low-pass filtering at 7 kHz – and the observation that a significant amount of speech energy is found above 3.4 kHz (in particular for fricative and plosive sounds) suggesting that conventional telephony is too narrow in its bandwidth and destroys natural speech quality in a significant way.[4] Local properties are found in:

- noise-like signal components visible as broadband irregular energy distributions;
- impulse-like components visible as broadband spikes;

---

[4] The impact of limited POTS bandwidth on intelligibility is evidenced from the need to use "alpha/bravo/charlie ..." spelling alphabets when communicating an unknown proper name over the phone.

- nearly periodic components visible as narrowly spaced (around 10 ms for a male voice) vertical striations, corresponding to the pulse trains seen in the time domain;
- formant frequencies visible as slowly drifting energy concentration due to vocal tract resonances which apparently are very context dependent as they differ even for the three occurrences of an [i] sound in this utterance.

The first three properties are related to the source or excitation signal of the source–filter model. The fourth property is related to the filter and its resonance frequencies or poles. The slow temporal variation of these poles models the temporal evolution of the formant frequencies.

15.2.3   Speech Perception

The ultimate recipient of human speech is the human hearing system, a remarkable receiver with two broadband, directional antennas shaped for spatiotemporal filtering (the outer ear) in terms of the individual, monaural *Head Related Transfer Functions*, HRTFs (functions of both azimuth angle and frequency) which along with interaural delay evaluation give rise to our spatial hearing ability. Second, a highly adaptive, mechanical impedance-matching network (the middle ear) covers a dynamic range of more than 100 dB and a 3,000-channel, phase-locking, threshold-based receiver (hair cells in the inner ear's cochlea) with very low self-noise (just above the level where we could hear our own blood-flow-induced noise) converts the signal to an extremely parallel, low-rate, synchronized representation useful for distributed processing with low-power and imprecise circuits (our nervous system).

Auditory Speech Modeling

Auditory models in the form of *psychoacoustic* – i.e., behavioral – models of perception are very popular in audiocoding (like those for the ubiquitous MP3 standard) because they allow us to separate relevant from irrelevant parts of the information. For instance, certain signal components may be masked by others to such an extent that they become totally inaudible. Naturally, for high-quality audiocoding, where little prior assumptions can be made about the nature of the signal source and its inherent redundancy, it is desirable to shape the distortion of lossy coders such that it gets masked by the relevant signal components.

In speech coding, the situation is reversed: we have a lot of prior knowledge about the signal source and can base the coder design on a source model and its redundancy, whereas the perceptual quality requirements are somewhat relaxed compared with audiocoding – i.e., in most speech coders some unmasked audible distortion is tolerated (note that we have long learnt to live with the distortions introduced by 3.4-kHz band limitation which would never work when listening to music). Therefore, perceptual models play a lesser role in speech coder design, although a simple *perceptual weighting filter*, originally proposed by Schroeder et al. [1979], has wended its way into most speech-coding standards and allows a perceptually favorable amount of noise shaping. More recent work on invertible auditory models for transparent coding of speech and audio is reviewed in Feldbauer et al. [2005].

Perceptual Quality Measures

The proof of a speech coder lies in listening. Till today, the best way of evaluating the quality of a speech coder is by controlled listening tests performed with sizable groups of listeners (a couple of dozens or more). The related experimental procedures have been standardized in International Telecommunications Union (ITU-T) Recommendation P.800 and include both *absolute category rating* and *comparative category rating* tests. An important example of the former is the so-called

Mean Opinion Score (MOS) test which asks listeners to rate the perceived quality on a scale from 1 = poor to 5 = excellent where traditional narrowband speech with logarithmic PCM coding at 64 kbit/s is typically rated with an MOS score in the vicinity of 4.0. With a properly designed experimental setup, high reproducibility and discrimination ability can be achieved. The test can be calibrated by using so-called anchor conditions generated with artificially controlled distortions using the ITU's Modulated Noise Reference Unit (MNRU).

Besides these one-way listening-only tests, two-way *conversational tests* are important whenever speech transmission suffers from delay (including source-coding delay!). Such tests have led to an overall planning tool for speech quality assessment, the so-called E-model standardized by ITU-T Recommendation G.107 which covers various effects from one-way coding and transmission losses to the impact of delay on conversations or the subjective user advantage obtained from mobile service access.

As a means of bypassing tedious listening and/or conversational tests, objective speech quality measures have gained importance. They are often based on perceptual models to evaluate the impact of coding distortions, as observed by comparing the original and the decoded speech signal. Such measures are also called *intrusive* because they require transmission of a specific speech signal over the equipment (or network connection) under test in order to make this original signal in its uncoded form available for evaluation at the receiver site. One standardized example is the *Perceptual Evaluation of Speech Quality*, PESQ (compare ITU-T Recommendation P.862).

Nonintrusive or single-ended speech quality measures are still under development. They promise to assess quality without having access to the original signal, just as we do as human listeners when we judge a telephone connection to have bad quality without direct access to the other end of the line. A first attempt (from May 2004) at standardization is the *Single Sided Speech Quality Measure* (3SQM) (ITU-T Recommendation P.563).

15.3 Stochastic Models for Speech

15.3.1 Short-Time Stationary Modeling

With all the physical insights obtained from speech production and perception, speech coding has only the actual acoustic waveform to work with and, as this is an information-carrying signal, a stochastic modeling framework is called for. At first, we need to decide how to incorporate the time-varying aspects of the physical signal generation mechanism which suggests utilization of a *nonstationary*[5] *stochastic process* model. There are two "external" sources for this nonstationarity, best described in terms of the time variations of the acoustic wave propagation channel:

- The movements of the articulators which shape the boundary conditions of the vocal tract at a rate of 10 to 20 speech sounds/second. As the vocal tract impulse response typically has a delay spread of less than 50 ms, a *short-time stationary representation* is adequate for this aspect of nonstationarity.
- The movements of the vocal folds at fundamental frequencies from 100 to 500 Hz give rise to a nearly periodic change in the boundary conditions of the vocal tract at its lower end – i.e., the vocal folds. In this case, comparison with delay spread suggests that the time variation is too fast for a short-time stationary model. Doppler-shift-induced modulation effects become important and only a *cyclostationary representation* can cope with this effect (compare Section 15.3.5).

---

[5] An alternative view would introduce a hypermodel that controls the evolution of time-varying speech production model parameters. A stationary hypermodel could reflect the stationary process of speaking randomly selected utterances such that the overall two-tiered speech signal model would turn out to be stationary.

Most classical speech models neglect cyclostationary aspects; so, let us begin by discussing the class of short-time stationary models. For these models, stochastic properties vary slowly enough to allow the use of a subsampled estimator – i.e., we need to reestimate the model parameters only once for each speech signal *frame* where in most systems a new frame is obtained every 20 ms (corresponding to $N = 160$ samples at 8 kHz). For some applications – like signal generation which uses the parameterized model in the decoder – the parameters need to be updated more frequently (e.g., for every 5-ms subframe) which can be achieved by interpolation from available frame-based estimates.

Wold's Decomposition

Within a frame, the signal is regarded as a sample function of a stationary stochastic process. This view allows us to apply *Wold's decomposition* [Papoulis 1985] which guarantees that *any stationary stochastic process* can be decomposed into the sum of two components: a *regular component* $x_n^{(r)}$ which can best be understood as *filtered noise* and which is not perfectly predictable using linear systems and a *singular component* $x_n^{(s)}$ which is essentially a *sum of sinusoids* that can be perfectly predicted with linear systems:

$$x_n = x_n^{(r)} + x_n^{(s)} \tag{15.1}$$

Note that the sinusoids need not be harmonically related and may have random phases and amplitudes. In the following three subsections, we will show that this result – already established in the theory of stochastic processes by 1938 – serves as the basis for a number of current speech models which only differ in the implementation details of this generic approach. These models are the Linear Predictive voCoder (LPC), sinusoidal modeling, and Harmonic + Noise Modeling (HNM).

15.3.2 Linear Predictive voCoder (LPC)

The first implementation of Wold's decomposition emphasizes its relationship to the Linear Prediction (LP) theory as first proposed by Itakura and Saito [1968]. The model derives its structure from the representation of the regular component as white noise run through a linear filter, which is causal, stable, and has a causal and stable inverse – i.e., a *minimum-phase filter*. The same filter is used for the generation of the singular component in voiced speech where it mostly consists of a number of harmonically related sinusoids which can be modeled by running a periodic pulse train through the linear filter. This allows flexible shaping of the spectral envelope of all harmonics but does not account for their original phase information because the phases are now determined by the phase response of the minimum-phase filter which is strictly coupled to its logarithmic amplitude response (via a Hilbert transform). Furthermore, as there is only one filter for two excitation types (white noise and pulse trains), the model simplifies their additive superposition to a *hard switch in the time domain*, requiring strict temporal segregation of noise-like and nearly periodic speech (compare the decoder block diagram shown in Figure 15.3). This simplification results in a systematic model mismatch for mixed excitation signals.

Linear Prediction Analysis

The LPC encoder has to estimate the model parameters for every given frame, and to quantize and code them for transmission. We will only discuss estimation of the LP filter parameters here; estimation of the fundamental period $T_0 = 1/f_0$ is treated later in this text (Section 15.4.3). As a

Figure 15.3   Linear predictive vocoder signal generator as used in decoder.

first step, specification of the minimum-phase filter is narrowed down to an *all-pole filter* – i.e., a filter where all transfer function zeros are concentrated in the origin of the z-plane and where only the poles are used to shape its frequency response. This is justified by two considerations:

- The vocal tract is mainly characterized by its resonance frequencies and our perception is more sensitive to spectral peaks than to valleys.
- For a minimum-phase filter, all the poles and zeros lie inside the unit circle. In this case, a zero at position z_0 with $|z_0|<1$ can be replaced by a geometric series of poles, which converges for $|z| = 1$:

$$(1 - z_0 z^{-1}) = \frac{1}{1 + z_0 z^{-1} + z_0^2 z^{-2} + \cdots} \tag{15.2}$$

This allows us to write the signal model in the z-transform domain as

$$X(z) = \frac{U(z)}{A(z)} \tag{15.3}$$

where the speech signal is represented by $X(z)$, the excitation signal by $U(z)$, and the filter transfer function by $1/A(z)$. In the time domain, this reads as

$$x_n = - \sum_{i=1}^{m} a_i x_{n-i} + u_n \tag{15.4}$$

where the filter or predictor order is chosen as $m = 10$ for speech sampled at 8 kHz and the parameters a_i are known as *predictor coefficients*, with normalization[6] $a_0 = 1$.

For a given speech frame, we estimate model parameters $\hat{a}_i$ such that model mismatch is minimized. This mismatch can be observed through the prediction error signal e_n:

$$e_n = x_n - \hat{x}_n = x_n - \left(- \sum_{i=1}^{m} \hat{a}_i x_{n-i} \right) = \sum_{i=1}^{m} (\hat{a}_i - a_i) x_{n-i} + u_n \tag{15.5}$$

For uncorrelated excitation u_n, the prediction error power achieves its minimum iff $\hat{a}_i = a_i$, for $i = 1, \ldots, m$. In this case, the prediction error signal e_n becomes identical to the excitation signal u_n. Note that we use the prediction framework only for model fitting, not for forecasting or extrapolation of future signal samples. To apply this estimator to short-time stationary speech data, for every frame

---

[6] The gain g can be included in the excitation signal amplitude.

with an update rate of N samples, a window of $L \geq N$ samples is chosen, where the greater sign means that the windows have some overlap or lookahead. Typically, a special window function w_n is applied to mitigate artifacts due to the data discontinuity introduced by the windowing mechanism. Asymmetric windows with their peak close to the most recent samples allow a better compromise between estimation accuracy and delay. The window can be applied to the data in two different ways, giving rise to two LP analysis methods:

1. The *autocorrelation method* defines the prediction error power estimate based on a windowed speech signal:

$$\hat{x}_n = \begin{cases} w_n \cdot x_n, & \text{for } n = 0, 1, \ldots, L-1 \\ 0, & \text{for } n \text{ outside the window} \end{cases} \tag{15.6}$$

as

$$\sum_{n=-\infty}^{\infty} e_n^2 = \sum_{n=-\infty}^{+\infty} \left(\tilde{x}_n + \sum_{i=1}^{m} \hat{a}_i \tilde{x}_{n-i} \right)^2 \tag{15.7}$$

where data windowing results in an implicit limitation of the infinite sum.

2. The *covariance method* defines the prediction error power estimate via a windowing of the error signal itself without explicit data windowing:

$$\sum_{n=m}^{L-1} (w_n \cdot e_n)^2 = \sum_{n=m}^{L-1} w_n \left(x_n + \sum_{i=1}^{m} \hat{a}_i x_{n-i} \right)^2 \tag{15.8}$$

where the summation bounds are carefully chosen to avoid the use of speech samples outside the window.

In both methods, minimization of the quadratic cost function leads to a system of linear equations for the unknown parameters $\hat{a}_i$. In the autocorrelation method, the system matrix turns out to be a proper correlation matrix with *Toeplitz* structure which allows a computationally efficient solution using the order-recursive *Levinson–Durbin algorithm*. This algorithm reduces the operation count of the estimator from $\mathcal{O}(m^3)$ to $\mathcal{O}(m^2)$ and guarantees that the roots of the polynomial $\hat{A}(z)$ lie inside the unit circle. In the covariance method, no such structure can be exploited such that higher complexity and additional mechanisms for stabilizing $\hat{A}(z)$ are required. Its advantage is the significantly increased accuracy of estimated coefficients because the absence of explicit data windowing avoids some systematic errors of the autocorrelation method. Finally, to enhance the numerical properties of LP analysis, additional (pre-)processing steps are routinely included such as high-pass prefiltering of input speech to remove unwanted low-frequency components, a bandwidth expansion applied to correlation function estimates, and a correction of the autocorrelation at lag 0 which corresponds to the addition of a weak white noise floor (at $-40\,$dB of the data).

15.3.3 Sinusoidal Modeling

The second implementation of Wold's decomposition emphasizes the idea of spectral modeling using sinusoids as proposed by MacAulay and Quatieri [1986]. In this model, both signal components are produced by the same *sum of sinusoids*. The noise-like regular component can well be approximated by such a sum (as suggested by spectral representation theory) if the relative phases of the sinusoids are randomized frequently, at least once for each frame. This scheme offers the advantage of retaining the original phase information in the singular component (if the available bit rate allows us to code it, of course) and it is more flexible in combining regular and singular components. Typically, even voiced speech contains a noise-like component in the higher frequency bands which can be modeled by using a fixed-phase model for the lower harmonics and a random-phase model for the higher harmonics. This *hard switch in the frequency domain* assumes the segregation of noise-like and nearly periodic signal components along the frequency axis. While this allows the modeling of mixed excitation speech signals, the transitions between excitation signal types are now a priori restricted to the frame boundaries (whereas the LPC allows in principle higher temporal resolution for voicing switch update).

A further development of sinusoidal modeling is the Multi Band Excitation (MBE) coder which allows separate voicing decisions in multiple frequency bands, thereby achieving a more accurate description of mixed excitation phenomena. This also establishes a relationship to the principles of subband coding or transform coding which do not rely on specific source signal models and which are most popular in coding of generic audio.

15.3.4   Harmonic + Noise Modeling

The third implementation of Wold's decomposition strives at a full realization of the additive superposition of regular and singular components *without enforcing a hard switch in either the time or frequency domains*. Thus, the HNM is maybe the simplest in its decoder structure which just follows Eq. (15.1), whereas the encoder is the most difficult as it has to solve the problem of simultaneous estimation of superimposed continuous spectra (the regular component) and discrete spectra (the singular component, assumed to be harmonically related). So far, use of this model has been confined to applications in speech and audio synthesis, whereas for speech coding it has been recognized that the level of sophistication achieved in the HNM-spectral representation needs to be complemented by a more detailed analysis of time domain variations, too, leading us beyond the conventional short-time stationary model.

15.3.5   Cyclostationary Modeling

The short-time stationary model of speech neglects the rapid time variations induced by vocal fold oscillations. For voiced speech, this nearly periodic oscillation not only serves as the main excitation of the vocal tract but also as a cyclic modulation of all signal statistics. Stochastic processes with periodic statistics are known as *cyclostationary processes*. For our purposes, where the fundamental period T_0 and the associated oscillation pattern may slowly evolve over time, we need to adapt this concept to short-time cyclostationary processes. For a given speech frame, the waveform will be decomposed in a cyclic mean signal (corresponding to the singular component of Wold's decomposition) and a zero-mean noise-like process with periodically time-varying correlation function. Most importantly, the variance of this noise-like component is a periodic function of time representing the periodic envelope modulation of the noise-like component of voiced speech. This effect is clearly audible for lower pitched male voices and constitutes one of the main improvements of cyclostationary speech models over HNMs. These models were originally introduced as (Prototype) Waveform Interpolation (PWI) coders by Kleijn and Granzow [1991] where the cyclic mean is interpreted as the "slowly evolving waveform" and the periodically time-varying noise-like component is termed the "rapidly evolving waveform". More recent developments work with filterbanks whose channels are adapted to multiples of f_0 by prewarping the signal in the time domain. A key aspect of cyclostationary signal modeling is the extraction of reliable "pitch marks" that allow explicit time domain *synchronization* of subsequent fundamental periods. As all speech properties evolve over time, this synchronization problem is a major challenge and leads to the characterization of speech signals in terms of an underlying self-oscillating nonlinear dynamical system with the added benefit of explaining certain nonadditive irregularities of speech oscillations such as jitter or period-doubling phenomena.

15.4 Quantization and Coding

Once the signal model and the parameter estimation algorithms have been selected, the encoder has to quantize the parameters and, in hybrid coders, samples of the modeling residual waveform as well.[7] To achieve source-coding efficiency, several techniques beyond the simple uniform quantization scheme found in most analog-to-digital converters have been developed.

15.4.1 Scalar Quantization

Scalar quantization refers to the quantization of a single random variable which can be either a waveform sample or a single model parameter. If there is more than one variable to be quantized then they are treated independently of each other. As a general example, the continued (or high-resolution discrete) value x of the random variable X is quantized to the ith quantization interval $[t_i, t_{i+1}]$ if $t_i \leq x < t_{i+1}$ and the index $i = Q(x)$ will be transmitted using a (binary) codeword. The receiver decodes the codeword back to the quantizer index and uses it to address a lookup table that stores a high-resolution reconstruction value $x^{(q)} = Q^{-1}(i)$. The latter operation is sometimes referred to as "inverse quantization" although the static nonlinearity of a quantizer is a many-to-one map and, as such, has no unique inverse function. A typical performance measure for scalar quantizers is the Signal-to-Quantization Noise Ratio (SQNR) based on the mean-square value of the quantization error $q = x^{(q)} - x$:

$$SQNR_{[\text{dB}]} = 10 \log_{10} \frac{E(x^2)}{E(q^2)} = 10 \log_{10} \frac{E(x^2)}{E((x^{(q)} - x)^2)} \tag{15.9}$$

where the operator E is the expectation operator performing an average weighted by the probability measure of the variable X. This definition includes both overflow distortions when the input value reaches the boundaries of the quantizer range and granular distortions related to the individual quantizer steps. In most designs, overflow will be avoided by using proper scaling such that granular distortions will be the quality-determining effect. Normalization of error power by input power makes the measure somewhat more relevant to human perception when dealing with waveform quantization.

In uniform quantization, all intervals have the same width $t_{i+1} - t_i = q = $ constant. The most prominent nonuniform scalar quantization law is 8-bit *logarithmic* quantization for PCM speech signals with two variants of the logarithmic characteristic, the μ-law in the U.S.A and Japan and the A-law in Europe. Logarithmic quantization has the remarkable property that the SQNR becomes nearly independent[8] of the single power $E(x^2)$ as long as overflows are avoided and the signal does

---

[7] Note that this is the case for *forward-adaptive* speech coders. For extremely low delay, *backward-adaptive* coders avoid the transmission of parameters as side information and rather estimate model parameters making exclusive use of speech samples which have already been decoded. This requires implementation of both encoder and decoder algorithms in the transmitter and works under the assumption of low channel error rates. The parameters estimated from decoded data only implicitly contain some delay relative to those parameters found from the current frame in forward-adaptive schemes. However, due to the short-time stationarity assumption, this delay results in an acceptable loss in modeling performance which is approximately compensated by the bit rate reduction obtained from suppression of the side information channel.

[8] Actually, for logarithmic quantization, the SQNR is essentially independent of the amplitude probability distribution of the input, thereby making it maximally *robust* to unknown or strongly varying signal properties.

not fall in the close vicinity of 0 (where the log function is approximated with a linear function, corresponding to the resolution of a 12-bit uniform quantizer). Thus, logarithmic quantizers are designed to cope with the strong temporal variations of the speech signal envelope which are both intrinsic (short-time stationary sequence of highly different speech sounds, soft versus loud voice) and extrinsic (variable distance to the microphone).

Adaptive quantization extends the dynamic range of a quantizer even further than logarithmic quantization such that input power fluctuations of 40 dB and more show no pronounced effect on SQNR performance. It is implemented with an adaptive gain control mechanism, operating on a sample-by-sample backward-adaptive basis in one of the most popular variants.

If the random variable X itself has a nonuniform amplitude probability distribution, the uniform quantizer does not achieve the minimal SQNR performance for a given number of quantization intervals I. The set of *optimal quantizer* thresholds $t_i, i = 1, \ldots, I$ and reconstruction values $x^{(q)}(i), i = 1, \ldots, I$ can be found using the iterative Lloyd–Max algorithm which alternates between the two following implicit conditions found from minimizing the mean-square quantization error for fixed input power:

1. The best quantization thresholds lie at equal distance between adjacent reconstruction values:

$$t_i = \frac{x^{(q)}(i) - x^{(q)}(i-1)}{2} \tag{15.10}$$

2. The best reconstruction values lie in the centers of mass of the quantization intervals, computed as a conditional expected value using the input amplitude probability density $f_X(x)$:

$$x^{(q)}(i) = E(x|t_i \leq x < t_{i+1}) = \int_{t_i}^{t_{i+1}} f_X(x) x \, dx \tag{15.11}$$

In practice, probability distributions are often not known such that the center-of-mass computations are done by evaluating the class-conditional means of a large set of training data samples collected in real-world experiments.

15.4.2 Vector Quantization

Vector Quantization (VQ) combines d scalar random variables X_k into a d-dimensional random vector $\mathbf{X}$. In the d-dimensional space, the scalar quantization intervals turn into multidimensional, convex *Voronoi cells* $V_i, i = 1, \ldots, I$. If an input vector $\mathbf{x} \in V_i$, then $Q(\mathbf{x}) = i$ and the value i is mapped to a binary codeword for transmission. Shannon [1949] showed that using VQ with asymptotically increasing dimension d can achieve the rate distortion bound for source coding – i.e., just like in channel coding the best coding performance is obtained when considering large blocks of signal samples simultaneously. VQ is known to have three different advantages over scalar quantization. For this comparison, we will study the number of bits per dimension $\log_2(I)/d$ needed to achieve a certain quantization distortion.

1. The *memory advantage* – i.e., the handling of statistical dependences: in general, a random vector exhibits statistical dependences among its components, which may go beyond linear correlations. Therefore, even if we use linear transforms (such as the Discrete Cosine Transform, DCT, or principal component analysis) to decorrelate the components of the random vector, the joint quantization of all components is still more efficient due to the redundancy contained in the remaining (nonlinear) dependences.
2. The *space-filling advantage*: in scalar quantization, there is not much choice in selecting the shape of quantization intervals, whereas, in multiple dimensions, Voronoi cells can be shaped

to achieve the best space-filling degree – e.g., according to a dense sphere-packing principle. This allows coverage of a certain volume in space with the same maximal distortion (given by a linear dimension of the Voronoi cell) but fewer cells – i.e., a lower number $\log_2(I)$ of codeword bits. This advantage approaches 0.25 bit/dimension for large vector dimension d. For low-rate speech coding at 4 kbit/s or less (which means less than 0.5 bit/sample) this becomes a very significant contribution.

3. The *shape advantage*: this is very similar to the gains achieved for optimal nonuniform quantizers of scalar variables where the shape (and size) of the quantizer cells are matched to the probability distribution of the random vector. This advantage is only relevant for resolution-constrained quantizers with a predetermined number of cells I. For entropy-constrained quantizers (where we only care about the average entropy of all codewords but we neither constrain the number of cells or codewords nor the length of the codewords), there is no shape advantage as this can equally be obtained by lossless entropy coding of the codewords themselves.

The design of optimal VQ proceeds along the same lines as the design of nonuniform quantizers for scalar variables. The generalization of the Lloyd–Max algorithm to multiple dimensions is known as the Linde–Buzo–Gray or LBG algorithm.

Suboptimum Vector Quantization

The larger the vector dimension d, the better the VQ performance. However, if we design a VQ for a certain *given number of bits per dimension*, this means that the complexity of the VQ grows exponentially with dimension d because I, the number of Voronoi cells, grows exponentially with d. As multidimensional Voronoi cells may have complicated shapes, we rather define them by their reconstruction vectors $\mathbf{x}_q(i)$ and perform quantization by selecting the index $i = Q(\mathbf{x})$ which minimizes the quantization distortion $D(\mathbf{x}_q(i), \mathbf{x})$ over all $i = 1, \ldots, I$. A typical distortion measure would be the quadratic distortion or Euclidean distance $||\mathbf{x}_q(i) - \mathbf{x}||^2$. Thus we need to memorize a table or *codebook* of I vectors of dimension d and we need to make a full search through this codebook to minimize distortion, which requires I distortion measure computations where each has a cost proportional to d in case of Euclidean distance. So even per dimension, the memory cost and the number of operations is directly proportional to I and, therefore, suffers from the curse of (= exponential growth with) dimensionality.

To address this issue, simplified schemes for VQ design and implementation have been developed. Surprisingly, it turns out that the suboptimal search for a good vector in a codebook filled with suboptimal entries can still outperform an optimal VQ system of lower dimension. While early attempts at VQ-based waveform quantization worked at 1 bit/dimension with a full search over $I = 2^{10} = 1,024$ entries of dimension $d = 10$, the new suboptimal systems work at 0.875 bit/dimension with a greedy search over $I = 2^{35} = 34.4 \cdot 10^9$ entries of dimension $d = 40$ and still achieve better performance.[9] Examples of such simplified, suboptimal designs are as follows:

- *Gain shape quantization* where the codevector is the product of a scalar gain and a unit vector (norm 1).
- *Multistage VQ* where a first codebook provides a coarse approximation and the subsequent $K - 1$ codebooks successively refine the quantization.
- *Split VQ* where the high-dimensional vector is split into K pieces of dimension d/K each, which are then quantized independently.

---

[9] Note that a full search through these 40-dimensional codebooks would result in an increase in complexity *per dimension* by a factor of $2^{23} = 8.4 \cdot 10^6$ when compared with the earlier 10-dimensional codebooks!

In the two latter cases, complexity is approximately reduced to that of a d/K-dimensional VQ, as nonexponential factors do not matter much for such comparison.

As one of the most successful simplifications – having made it into several international standards – we discuss *algebraic codebooks* where codevector amplitudes are restricted to a ternary alphabet $\{-1, 0, +1\}$ and where, furthermore, only a small percentage of nonzero amplitudes is allowed for (e.g., 10 in a vector of dimension 40). This design is inspired from the earlier techniques of "multipulse excitation" and "Regular Pulse Excitation" (RPE) where the source excitation of a source–filter model was pruned to a few nonzero amplitudes. As Euclidean distance computations consist in the evaluation of inner products, the algebraic structure reduces each evaluation from 40 multiply–add operations to 10 additions only. Furthermore, the selection of nonzero pulses follows a greedy scheme that does not try all possible combinations. Finally, they are encoded efficiently using several interleaved combs of subsampled pulse trains or polyphases (Interleaved Single Pulse Permutation, ISPP). Note that this codebook is built with a systematic structure in mind, which achieves a fairly uniform coverage of the entire vector space, so no training or optimization is done for codevector entries.

Line Spectrum Pairs (LSP)

When applying VQ to the parameters of a speech model (and not to waveform samples), special considerations regarding the distortions induced by parameter quantization errors are needed. A common application is the simultaneous quantization of the $m = 10$ LP coefficients a_i, which raises the following issues:

1. The quantized parameter vector should still correspond to a minimum-phase filter allowing stable operation of the filter and its inverse.
2. Quantization error is best measured in terms of induced *spectral distortion* which is the mean-square difference between the original spectral envelope and the one described by the quantized parameters, both expressed in decibels (dBs). From experience, this distortion should be less than 1 dB for most of the parameter vectors to achieve transparent coding quality.
3. Within the short-time stationary model, the transition from one frame to the next should allow simple interpolation mechanisms between two parameter vectors which still maintain stability at all times and avoid unexpected spectral excursions of the interpolated spectral envelopes.

All these issues are best resolved before VQ by applying a nonlinear transform from the predictor coefficients to a vector of line-spectral frequencies which occur in distinct pairs and, therefore, are also referred to as Line Spectrum Pairs (LSPs). They are derived by introducing two polynomials $F_1(z)$ and $F_2(z)$ for order $m + 1$ with even and odd symmetries in their coefficients, respectively:

$$F_1(z) = A(z) + z^{-1} \cdot z^{-m} A(z^{-1}) \tag{15.12}$$

$$F_2(z) = A(z) - z^{-1} \cdot z^{-m} A(z^{-1}) \tag{15.13}$$

where $z^{-m} A(z^{-1}) = \sum_{i=0}^{m} a_{m-i} z^{-i}$ is the *time-reversed predictor polynomial*. We recognize that this first step is invertible as we have $A(z) = (F_1(z) + F_2(z))/2$. However, for a minimum-phase filter $A(z)$, the mirror poynomial $F_1(z)$ and $F_2(z)$ have their roots not inside but exactly on the unit circle such that each of their locations can be fully specified by a single real-valued frequency. Furthermore, the root location sets of the two polynomials follow a joint sorting relationship such that, with increasing frequency, the two sets are strictly interleaved on the unit circle. These properties simplify the search for the polynomial roots significantly and also suggest the combination of pairs of adjacent line spectrum frequencies (one from each mirror polynomial). Furthermore, this sorting relationship helps in interpolation of estimated LSP vectors over time. Finally, the VQ for this vector of ten real numbers is often implemented as a split VQ with three parts sorted by increasing frequencies which contributes to the relative independence of vector parts and minimizes the effects of suboptimal quantization.

15.4.3 Noise Shaping in Predictive Coding

Hybrid speech coders do not rely entirely on the source signal model but observe and transmit the modeling error or residual waveform as well. If we use a predictive signal model, the *residual waveform* is obtained as the prediction error $e_n = x_n - \hat{x}_n$. If we quantize the prediction residual e_n for transmission, it results in $e_n^{(q)} = e_n + q_n$ from which the decoder reconstructs $x_n^{(d)} = \hat{x}_n + e_n^{(q)}$. The *decoder* computes the prediction $\hat{x}_n$ from the delayed reconstructed signal samples $x_{n-i}^{(d)}$:

$$\hat{x}_n = -\sum_{i=1}^{m} a_i x_{n-i}^{(d)} \tag{15.14}$$

which results in a recursive or *closed-loop* reconstruction filter:

$$x_n^{(d)} = -\sum_{i=1}^{m} a_i x_{n-i}^{(d)} + e_n^{(q)} \tag{15.15}$$

$$A(z)X^{(d)}(z) = E^{(q)}(z) = E(z) + Q(z) \tag{15.16}$$

However, when computing the predicted signal sample $\hat{x}_n$ in the *encoder*, we can do this in two different ways:

1. *Open-loop prediction* makes use of the delayed original signal samples as available in the transmitter only:

$$\hat{x}_n = -\sum_{i=1}^{m} a_i x_{n-i} \tag{15.17}$$

$$E(z) = A(z)X(z) \tag{15.18}$$

   from which we get the decoder output with Eq. 15.16:

$$X^{(d)}(z) = X(z) + \frac{Q(z)}{A(z)} \tag{15.19}$$

2. *Closed-loop prediction*, however, uses exactly the same reconstructed samples for prediction in the encoder as in the decoder (Eq. 15.14) to compute:

$$e_n = x_n - \hat{x}_n = x_n + \sum_{i=1}^{m} a_i x_{n-i}^{(d)} \tag{15.20}$$

   This results in the decoder output as

$$x_n^{(d)} = -\sum_{i=1}^{m} a_i x_{n-i}^{(d)} + e_n^{(q)} \tag{15.21}$$

$$= -\sum_{i=1}^{m} a_i x_{n-i}^{(d)} + x_n + \sum_{i=1}^{m} a_i x_{n-i}^{(d)} + q_n \tag{15.22}$$

$$X^{(d)}(z) = X(z) + Q(z) \tag{15.23}$$

In this setup, the encoder predictor duplicates the *closed-loop* structure of the decoder, hence its name.

Quantization noise $Q(z)$ is modeled with a white spectrum. According to Eq. (15.19), open-loop prediction results in an overall coding distortion $X^{(d)}(z) - X(z)$ with a spectrum shaped like the speech model spectrum $1/A(z)$ which reduces the audibility of this noise due to the masking properties of human hearing. However, filtering of the noise by $1/A(z)$ also results in an overall gain that makes the total noise power identical to that observed in direct quantization of the speech waveform without using prediction. Closed-loop prediction keeps a white noise spectrum without noise amplification according to Eq. (15.23). However, as quantization noise is not shaped like the speech spectrum, it may become audible in the spectral valleys of the latter. A compromise between these two extremal positions is found in the use of a *perceptual noise-shaping filter* [Schroeder et al. 1979]. To this end, quantization noise is observed as the difference between quantizer output and input, $q_n = e_n^{(q)} - e_n$, and a filtered version of this noise is fed back to the input of the quantizer $E(z) \rightarrow E(z) + (W(z) - 1)Q(z)$ (where the zero-delay gain is $w_0 = 1$). Thereby, the open-loop prediction system results in a decoder output equal to $X^{(d)}(z) = X(z) + \frac{W(z)}{A(z)} Q(z)$. If we choose the weighting filter $W(z) = A(z/\gamma)$, with $0 < \gamma \leq 1$, we can control the spectral shape of the decoder distortion to be anywhere between a white spectrum ($\gamma = 1$) and the speech model spectrum ($\gamma = 0$). Note that this *bandwidth expansion* for the perceptual weighting filter can be simply obtained by setting $w_i = a_i \gamma^i$.

Long-Term Prediction

The *Short Term Predictors* (STPs) discussed so far only address the statistical dependences among neighboring signal samples (short-term correlations) which can be related to a nonflat spectral envelope – i.e., the formant structure of speech. The harmonic structure of voiced speech results in an additional spectral fine structure which corresponds to long-term correlations in the speech signal. These correlations are visible in the repetitive nature of the waveform cycles for a nearly periodic signal. The simplest *Long Term Predictor* (LTP) $B(z)$ predicts the current sample x_n from a sample that lies one period T_0 back into the past:

$$\hat{x}_n = b \cdot x_{n-T_0} \tag{15.24}$$

As both the STP and LTP are stable linear systems, they can be cascaded in any sequence, but it often is more advantageous to first perform STP and then LTP. For LTP, closed-loop prediction is the preferred option. The best parameters b and T_0 can be obtained from analyzing the signal autocorrelation over a range of lags that spans from the shortest fundamental periods (high-pitched female voices) to the longest (low-pitched male voices). Even if the true pitch is not found in some cases (e.g., period doubling), the LTP will still contribute to coder performance by extracting any redundancy related to the autocorrelation maximum.

A specific problem in periodicity modeling with LTP occurs whenever the true fundamental period T_0 is not an integer multiple of the sampling interval T_s ("fractional pitch"). As a workaround, signal interpolation by a factor of 3 to 6 can be used to increase the effective sampling resolution. Another approach is to increase the order of the LTP filter $B(z)$ such that it simultaneously solves the problem of signal interpolation *and* prediction.

15.4.4   Analysis by Synthesis

The analysis-by-synthesis procedure has been introduced in the context of multipulse-excited linear predictive coding [Atal and Remde 1982] and, later on, it was combined by Atal and

Schroeder [1984] and Schroeder and Atal [1985] with VQ to result in *Code Excited Linear Prediction* (CELP). Essentially it consists in reformulation of the open-loop/closed-loop prediction principle with perceptual weighting, which replaces the LP analysis step of the encoder with a linear predictive synthesis step. Thereby, the basic structure of the underlying signal model becomes more directly visible and certain computational advantages can be realized when combined with VQ. Note, however, that this analysis-by-synthesis formulation is still exactly equivalent to the noise-shaping predictive coding ideas presented in the previous subsection.

The starting point is to generate or synthesize one (or more) decoded signal sample by selecting a quantized residual signal sample $e_n^{(q)}$ from an inventory of allowed values (codebook). To this end, all entries from the inventory are run through the LP synthesis filter $1/A(z)$ to produce a set of all decoded signal candidates $x_n^{(d)}$. The coding distortion can be computed from the difference $x_n^{(d)} - x_n$. From the above, it is evident that this distortion is proportional to quantization error q_n via the filters $W(z)/A(z)$. To select the best quantized residual signal, the coding distortion is filtered by the inverse weighting filters $A(z)/W(z)$ to obtain the quantization noise $q_n = e_n^{(q)} - e_n$. If we minimize the squared magnitude of this variable we effectively perform a nearest neighbor decision among all possible residual candidates $e_n^{(q)}$ from the codebook to find the entry closest to e_n.

The power of this analysis-by-synthesis interpretation is that it lends itself to any description of the residual waveform inventory as long as it can be enumerated. This is, in particular, the case for VQ of the residual waveform where, typically, instead of a single sample a number of $d = 40$ samples are collected into a subframe. The squared magnitude estimate of the perceptually weighted synthesis error is now computed by summing the squares of the d samples. The VQ application suggests the following two computational reorganizations that reduce numerical complexity without compromising the solution:

1. As we process blocks or subframes of d samples at a time, we have to make sure that all filters carry over their filter states from one subframe to the next. This means that the filter output for one subframe will be the sum of a *zero-input response* (the decaying filter states) and a *zero-state response* (the component driven by the current input). Only the latter component provides the information for selecting the best codevector from the VQ codebook. Therefore, it is convenient to first subtract the zero-input response from the speech signal x_n such that codevector selection is done on zero-state responses only.

2. For the zero-state response of each codevector, we have to run the vector through both the LP synthesis filter $1/A(z)$ and the (inverse) weighting filters $A(z)/W(z)$. This filter cascade can be simplified to the overall transfer function $1/W(z)$, resulting in the block diagram shown in Figure 15.4.

Adaptive Codebook

The analysis-by-synthesis principle can also be carried over to the LTP which tries to explain the current subframe with predictions from (scaled) samples delayed by approximately one fundamental period T_0. As the fundamental period can vary in a certain range, it is advantageous to store subframe-length vectors taken from this range of the residual signal itself in a buffer memory that is called the *adaptive codebook*. One residual subframe is then represented by the weighted sum of two vectors, one selected from this adaptive codebook and another one selected from the *fixed codebook*. In most systems, the two selections are made sequentially such that first the adaptive codebook entry is selected to model the periodicity of the signal and, based on this choice, the fixed codebook entry is selected to represent the remaining "stochastic" part of the residual (Figure 15.4). This approach is essentially identical to the use of an LTP, with minor differences for high-pitched voices where the fundamental period is shorter than a subframe length. Just as

Figure 15.4   Code-excited linear predictive coding of speech; simplified block diagram showing separate treatment of zero-input and zero-state response with adaptive and fixed codebook excitation; weighting filter $W(z) = A(z/\gamma)$.

for LTP, good quality requires interpolation of the signal to increase the temporal resolution of the adaptive codebook, at least in the vicinity of the long-term correlation maximum.

15.4.5   Joint Source Channel Coding

Unequal Error Protection

Once all the source parameters have been computed for an entire speech frame, their codewords need to be assembled for transmission and extra bits will be added for channel-coding purposes. Channel coding usually works with the assumption that the codeword sequence is independent identically distributed (iid) and that all codewords bear the same relevance for the receiver. The last condition is certainly not the case for hybrid speech coders where some codewords represent model parameters such as fundamental period, gains, or spectral envelopes (LSPs) whereas other codewords represent the residual waveform by entries in the fixed VQ codebook. A channel error in one of the more relevant parameters (where T_0 is maybe the most important one) will result in significant waveform errors in the decoded signal as, e.g., the reconstructed voice may sound much lower or higher than the original. A channel error in one of the less relevant codewords may hardly be noticed as a properly parameterized speech model will generate rather good speech even if the details of the residual waveform are wrong. In conclusion, *"all source bits are equal, but some are more equal than others."* This results in classification of the codewords in different importance or channel sensitivity classes which are handled by different levels of channel error protection, ranging from sophisticated convolutional codes to no error protection at all.

Adaptive Multi Rate Coding

Both channel and source statistics vary over time. Therefore, the optimum coding strategy has to adapt itself to the time-varying channel state and source state. For circuit-switched communication, fixed overall rates are desired on the channel, so the task of the rate adaptation mechanism is to determine the best tradeoff between the rate budget spent on source coding and on channel error protection such that the total rate remains constant (Adaptive Multi Rate AMR coding). Because source statistics change much faster (essentially they are different for each frame – i.e., every 20 ms) than channel-state information is updated, current systems will essentially select the best source versus channel rate split based on channel-state information only. Depending on this information, the source coder produces a different number of source bits per frame and the channel coder will use the remaining bits to perform adequate unequal error protection as described in the previous paragraph.

Source-Coding Optimization

There are two approaches to optimization of source coding in the light of joint source channel coding:

- *Encoder optimization*, where, for instance, the *index assignment* of the VQ is optimized such that bit errors in the codewords representing the VQ index will result in the minimum mean perceptual impact, very much like the Gray coding strategy for scalar index assignment. A further encoder technique that strives for equal relevance of all codewords is found in *multiple-description coding*.
- *Decoder optimization*, where residual redundancy among the received codewords is used to assist the channel decoder to increase the reliability of the overall decoding process.

15.5   From Speech Transmission to Acoustic Telepresence

The ultimate goal of speech transmission or telephony has always been to recreate the acoustic presence of a human talker at a geographically distant location. In that sense, telephony is just a special case of *telepresence* where we attempt to augment our local reality with the virtual presence of one or more remote persons, preferably recreating the remote environment to some extent. Our short discussion starts out from simple add-on functionalities for speech transmission and progresses to the most advanced services of three-dimensional virtual audio.

15.5.1   Voice Activity Detection

In a symmetric conversation between two persons, each of the participants is silent for about 50% of the time. This fact has long been observed and exploited for multiplexing telephone conversations over the same transmission channel in a technique known as Digital Speech Interpolation (DSI). For wireless communications, multiple access to the same shared radio spectrum requires the reduction of interference created among the users. This can be achieved by transmitting speech frames over the air interface only when the talker is actively speaking, a state which is detected by a Voice Activity Detector (VAD). For an inactive speaker, Discontinuous Transmission (DTX) results in:

1. less power consumption on the mobile terminal (saving of baseband-processing and radio transmitter power), thereby extending battery life;

2. less multiuser interference on the air interface, thereby enhancing mobile access network performance;
3. less network load if packet switching is used, thereby increasing backbone network capacity.

In a typical realization, speech transmission is cut after seven nonactive frames, and a "SIlence Descriptor " (SID, only 35 bits for 20 ms) frame is sent as a model for acoustic background noise which is used to regenerate Comfort Noise Generation (CNG) at the receiver end. This noise model is updated at least every 24 frames and can be seen as a first step to virtual reality rendering of the auditory scene, thereby maintaining coherence between the communication partners by letting the other know implicitly that the call was made while driving a car, etc.

While this DTX method is sometimes referred to as source-controlled rate adaptation, it operates at a very high source description level (i.e., only source on versus source off) and, e.g., is not able to rapidly vary the source rate according to the phonetic contents (e.g., voiced versus unvoiced speech). This is also not the same as AMR coding described above (with VAD, the source rate is either zero or the currently allowed maximum source rate, the latter still being adapted according to the channel state).

15.5.2   Receiver End Enhancements

If an entire speech frame is lost due to severe fading on the air interface, the resulting error cannot be corrected with channel codes targeted for individual or bursty bit errors. Rather, *error concealment* methods are required to interpolate such lost frames from received neighboring frames. This usually works very well for substitution of the first lost frame but may lead to unnatural sounding speech if continued over several frames lost in a row. In the latter case, later substituted frames will receive a gradual damping to achieve slow fadeout of the signal amplitude (typically over six frames = 120 ms).

A further receiver end signal enhancement is *adaptive postfiltering* which constitutes filters shaped according to the spectral envelope and LTP filters which will make noise introduced by lossy coding less audible to the listener. *Adaptive playout buffers* or *jitter buffers* help to conceal lost frames on packet-switched networks. If a terminal is equipped with wideband speech Input/Output (I/O) functionality (analog bandwidth from 50 Hz up to 7 kHz), speech received over a narrowband network may be augmented by artificially created frequency bands above 3.4 kHz and below 300 Hz using a synthesis mechanism known as *bandwidth extension*.

15.5.3   Acoustic Echo and Noise

The acoustic environment of a speaker may not only contain useful background information that adds to the presence of the auditory scene but also noise sources that are considered annoying, or reverberation and (acoustic) echo due to the fact that both speaking parties share a common acoustic space when using hands-free phones. In such situations, echo and noise need not only be controlled for the comfort of the human users but also for the speech-coding systems which, if strongly based on a speech signal model, will fail to handle background noise or echo appropriately. Solutions that perform *joint echo and noise control* will often achieve the best compromise in reducing the two impairments. An uncontrolled echo can have drastic consequences on the stability of the entire communication system; strict performance requirements have been set as standards by ITU-T in recommendations G.167 and G.168.

15.5.4   Service Augmentation for Telepresence

Speech-Enabled Services

In a telepresence framework, users will access a much wider range of services than in conventional telephony. Given the shrinking size of mobile terminals, service access benefits a lot from spoken language dialog with the machine offering the service. This can range from simple name dialling by voice (possibly activated by a spoken magic word) to Text To Speech (TTS) synthesis for email reading or *Distributed Speech Recognition* (DSR) where feature vectors for speech recognition are extracted on the mobile terminal, encoded with sufficient accuracy for pattern recognition (but not necessarily for signal regeneration at the central server end), and *transmitted as data* over the wireless link. This allows us to transmit recognition features based on higher quality signals than those used in telephony and it also allows us to use specific source- and channel-coding mechanisms that maximize benefits for the remote speech recognition server (rather than a human listener).

While DSR carries speech information over a data channel, the dual application of *Cellular Text Telephony* (CTT) allows the carriage of textual data over the speech channel so as to provide augmentative communication means for people with hearing or speaking impairments who still can exchange interactive text (not just short messages) in a chat-like style using a modem operating over the digital speech channel.

Personalization

For personalized services, *talker authentication* will be a must which should be distinguished from authentication of the mobile station or the infrastructure itself. The identity of a talker can be established via voice identification or verification techniques and, if needed, an additional personalized watermark might be inserted in the speech signal prior to encoding it.

For *talker privacy* the traditional phone booth might once be replaced by a virtual talker sphere outside of which active speech cancellation (using wearable loudspeaker arrays) would make the phone conversation hardly audible to bystanders.

Three-Dimensional Audio

The virtual talker sphere has already introduced the concept of advanced audioprocessing for speech telephony which has seen a tremendous boost from the use of microphone and loudspeaker arrays and virtual/augmented audio techniques that allow the spatial rendering (e.g., ambisonic) of three-dimensional sound fields, potentially converted to binaural headphone listening. The best effects are expected if personalized HRTFs can be used together with real-time tracking of head movements to place virtual sound sources in the context of the real environment, creating the immersive telepresence which blends the presence of virtual participants and local participants for successful teleconferencing.

Ultimately, this will require a significant shift beyond today's speech-coding standards to allow for high-quality, multichannel audio including metadata information. A first move in this direction has been made by standardization of the AMR wideband speech codec which has become the first speech-coding standard ever to be accepted almost simultaneously for wireline communication (ITU), wireless communication (European Telecommunications Standards Institute/Third Generation Partnership Project (ETSI/3GPP)), and Internet telephony (Internet Engineering Task Force, (IETF)).

Further Reading

The classical textbooks on speech coding are Jayant and Noll [1984] and Kleijn and Paliwal [1995], which, at the time of this writing, are unfortunately out of print. The excellent textbook by Vary et al. [1998] is written in German, but an improved edition in English is in preparation. Recently, a second edition of Kondoz [2004] has appeared. Another excellent overview of various aspects of speech processing is Rabiner [1994]. Source-coding theory is treated in Berger [1971], Gray [1989], and Kleijn [2005] which is maybe best suited for the needs of speech coding. For speech signal processing in general, we recommend the classic Rabiner and Schafer [1978] and the more recent textbooks of Deller et al. [2000], O'Shaughnessy [2000], and Quatieri [2002]. For extensions to nonlinear speech modeling, see Chollet et al. [2005] and Kubin [1995]. In Gersho and Gray [1992] an extensive discussion of VQ is provided. Hanzo et al. [2001] address joint solutions for source and channel coding for wireless speech transmission. The wider context of acoustic signal processing for telecommunications is treated in Gay and Benesty [2000], Haensler and Schmidt [2004], and Vaseghi [2000]. Bandwidth extension is the main topic of Larsen and Aarts [2004]. An excellent reference for spoken language processing is Huang et al. [2001] and Jurafsky and Martin [2000]. Gibson et al. [1998] treat the compression of general multimedia data.

For updates and errata for this chapter, see wides.usc.edu/teaching/textbook

16

Equalizers

16.1 Introduction

16.1.1 Equalization in the Time Domain and Frequency Domain

Wireless channels can exhibit delay dispersion – i.e., Multi Path Components (MPCs) can have different runtimes from the transmitter (TX) to the receiver (RX) (see Chapters 6 and 7). Delay dispersion leads to InterSymbol Interference (ISI), which can greatly disturb the transmission of digital signals. We already saw in Chapter 12 that even a delay spread that is smaller than the symbol duration can lead to a considerable Bit Error Rate (BER) degradation. If the delay spread becomes comparable with or larger than the symbol duration, as happens often in second- and third-generation cellular systems, then the BER becomes unacceptably large if no countermeasures are taken. Coding and diversity can reduce, but not completely eliminate, errors due to ISI (Chapters 13 and 14). On the other hand, delay dispersion can also be a positive effect. Since fading of the different MPCs is statistically independent, resolvable MPCs can be interpreted as diversity paths. Delay dispersion thus offers the possibility of *delay diversity*, if the RX can separate, and exploit, the resolvable MPCs. Using the fact that the transfer function is the Fourier transform of the impulse response with the Fourier transform pair $\tau \rightarrow f$, delay diversity can also be interpreted as frequency diversity (see Chapter 13).

Equalizers are RX structures that work both ways: they reduce or eliminate ISI, and at the same time exploit the delay diversity inherent in the channel. The operational principle of an equalizer can be visualized either in the time domain or the frequency domain.

For an interpretation in the frequency domain, remember that delay dispersion corresponds to frequency selectivity. In other words, ISI arises from the fact that the transfer function is not constant over the considered system bandwidth. The goal of an equalizer is thus to reverse distortions by the channel. In other words, the product of the transfer functions of channel and equalizer should be constant.[1] This can be expressed mathematically the following way: let the original signal be $s(t)$; it is sent through a (quasi-static) wireless channel with the impulse response $h(t)$, received, and sent through an equalizer with impulse response $e(t)$. We furthermore assume that the transmit signal uses Pulse Amplitude Modulation, PAM (see Chapter 11), so that

$$s(t) = \sum_i c_i g(t - iT) \tag{16.1}$$

---

[1] Actually, this is a special form of equalizer, the so-called "Zero-Forcing" (ZF) linear equalizer, that will be discussed below in more detail.

Wireless Communications, Second Edition   Andreas F. Molisch
© 2011 John Wiley & Sons, Ltd

where the c_i are the complex transmit symbols. We now require that

$$H(\omega)G(\omega)E(\omega) = \text{const} \tag{16.2}$$

where $E(\omega)$, $H(\omega)$, and $G(\omega)$ are the Fourier transforms of $e(t)$, $h(t)$, and $g(t)$, respectively.

An equivalent formulation in the *time domain* requires that the received signal be free of ISI at the sampling instants. Define $\eta(t)$ as the convolution of the channel impulse response $h(t)$ with the basis pulse $g(t)$:

$$\eta(t) = h(t) * g(t) \tag{16.3}$$

Then we require that

$$[\eta(t) * e(t)]_{t=iT_s} = \begin{cases} 1 & i = 0 \\ 0 & \text{otherwise} \end{cases} \tag{16.4}$$

If the channel were known and static, we could build (in hardware) a filter that performs the required equalizations of the transfer function. In wireless communications, however, the channel is (i) unknown and (ii) time variant. The former problem can be solved by transmission of a *training sequence* – i.e., a sequence of known bits. From the received signal $r(t)$ and knowledge of the shape of the basis pulses $g(t)$, the RX can estimate the channel impulse response $h(t)$. The problem of time variance is solved by repeating the transmission of the training sequence at "sufficiently short" time intervals, so that the equalizer can be adapted to the channel state at regular intervals. The concept is thus known as "adaptive equalization."

Over the years, many different types of equalizers have been developed. The simplest is the linear equalizer, which is usually a tapped-delay-line filter with coefficients that are adapted to the channel state (Section 16.2). Decision feedback filters make use of the fact that the ISI created by past symbols can be computed (and subtracted) from the received signal (Section 16.3). Finally, the optimum way of detection in a delay-dispersive channel is the Maximum Likelihood Sequence Estimation (MLSE) (Section 16.4). Blind equalizers, which do not need a training sequence, have been intensively investigated by researchers, but are not very popular for practical systems.

16.1.2 Modeling of Channel and Equalizer

The following sections, describing different equalizer structures, will require a discrete time model of the channel and equalizer. We now give such a model, together with the important concept of the *noise-whitening filter* that has great importance for optimum RXs.

The first stage of the RX consists of a filter that limits the amount of received noise power. This filter also should make sure that all information is contained in sample values at instances $t_s + iT_S$. This is achieved by a filter that is matched to $\eta(t)$ – i.e., the convolution of channel impulse response and basis pulse. The sequence of sample values at the output of the matched filter is then given by

$$\psi_i = c_i \zeta_0 + \sum_{n \neq i} c_n \zeta_{i-n} + \hat{n}_i \tag{16.5}$$

where ζ_i are the sample values of the AutoCorrelation Function (ACF) of $\eta(t)$, and $\hat{n}_i$ is a sequence of complex Gaussian random variables with ACF $N_0\zeta_i$ – i.e., the noise filtered by the matched filter.

The z-transform of the sampled ACF ζ_i can be factored as

$$\Xi(z) = F(z)F^*(z^{-1}) \tag{16.6}$$

This factorization is not unique; but it is advantageous to choose $F^*(z^{-1})$ in such a way that all roots are within the unit circle. In such a case, $1/F^*(z^{-1})$ is a stable, realizable filter.

Now, if the matched filter is concatenated with $1/F^*(z^{-1})$, then the noise at the output of this concatenation is white again, with an ACF given by $N_0\delta_k$. It is therefore also known as the *noise-whitening filter*. The sample values of the impulse response of concatenation of the basis pulse, wireless channel, matched filter, and noise-whitening filter is henceforth denoted as the *discrete time channel*. The impulse response of the discrete time channel is written as f_k and its z-transform as $F(z)$. Note that the impulse response of this channel is causal, so that the output signal can be written as

$$u_i = \sum_{n=0}^{L_c} f_n c_{i-n} + n_i \tag{16.7}$$

where L_c is the length of the impulse response of the discrete time channel. Therefore, the noise-whitening filter is also often known as the *precursor equalizer*.

We will use the following notation in this chapter:

c_i ith complex transmit symbol;
$\eta(t)$ convolution of basis pulse and channel impulse response;
e_i ith equalizer coefficients;
ζ_i ith sample value of the ACF of $\eta(t)$;
$\hat{n}_i$ ith sample of a sequence of complex Gaussian random variables with ACF $N_0\zeta_i$;
n_i ith sample of a sequence of uncorrelated complex Gaussian random variables;
f_i ith sample of the impulse response of the time-discrete channel;
u_i ith sample of the output signal of the time-discrete channel;
$\hat{c}_i$ estimate of transmit symbol c_i;
ε_i deviation between estimated and true transmit symbol $c_i - \hat{c}_i$.

16.1.3   Channel Estimation

A common strategy for data detection with an equalizer is to separate the estimation of **f** and **c**. In a first step, a training sequence (i.e., known **c**) is used to estimate **f**. During the subsequent transmission of the unknown payload data, we assume that the estimated impulse response is the true one, and solve the above equation for **c**.

In this subsection, we discuss estimation of the channel impulse response by means of a training sequence. Channel estimation shows strong similarities to the "channel-sounding" techniques described in Chapter 8. A very simple estimate can be obtained by means of *Pseudo Noise* (PN) sequences with period N_{per}. The ACF of the PN sequence approximates a Dirac delta function. More precisely, periodic continuation of the sequence $\{b_i\}$, convolved with a time-reversed version of itself,[2] gives a sum of Dirac pulses spaced N_{per} symbols apart:

$$\{b_{-i}\}*_{per}\{b_i\} \approx \sum_{n=-\infty}^{\infty} \delta_{i-nN_{per}} \tag{16.8}$$

where $*_{per}$ denotes periodic convolution. If this sequence is sent over a channel with impulse response **f**, the output of the correlator becomes

$$\{\hat{f}_i\} = (\{b_{-i}\}*_{per}\{b_i\}) * \{f_i\} \tag{16.9}$$

---

[2] Strictly speaking, we need to take the complex conjugate of the time-reversed sequence. However, as the transmit sequence is usually real $(+1/-1)$, bits and complex symbols can be considered to be equivalent.

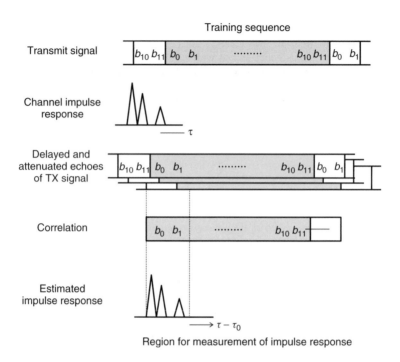

Figure 16.1   The principle of channel estimation by correlation.

Now, if the duration of the channel impulse response is shorter than N_{per}, then $\widehat{\mathbf{f}}$ is simply a periodic repetition of $\mathbf{f}$ (see Figure 16.1).

In practice, we do not transmit a periodic continuation of the PN sequence, but just a single realization. In order to avoid the possible influence of (unknown) payload bits on correlator output, known "buffer bits" have to be transmitted before and after the sequence.

Channel estimation by means of a training sequence technique has several drawbacks:

1. *A reduction in spectral efficiency*: the training sequence does not convey any payload information. For example, the Global System for Mobile communications (GSM) uses 26 bits in every 148-bit frame for the training sequence (see Chapter 24).
2. *Sensitivity to noise*: in order to keep spectral efficiency reasonable, the training sequence has to be short. However, this implies that the training sequence is sensitive to noise (longer training sequences can average out noise), and also to nonidealities in sounding sequences (remember that the peak-to-offpeak ratio of a PN-sequence increases with the length of such a sequence). If channel estimation is done by means of iterative algorithms, only algorithms with a fast convergence rate can be used; however, such algorithms lead to a high residual error rate.
3. *Outdated estimates*: if the channel changes after transmission of the training sequence, the RX cannot detect this variation. Use of an outdated channel estimate leads to decision errors.

Despite these problems, training-sequence-based channel estimation is used in practically every system. The reason for this is that the alternative (blind techniques, see Section 16.7) requires high computational effort and suffers from significant numerical problems.

16.2 Linear Equalizers

Linear equalizers are simple linear filter structures that try to invert the channel in the sense that the product of the transfer functions of channel and equalizer fulfills a certain criterion. This criterion can either be achieving a completely flat transfer function of the channel–filter concatenation, or minimizing the mean-squared error at the filter output.

The basic structure of a linear equalizer is sketched in Figure 16.2. Following the system model of Section 16.1.2, a transmit sequence $\{c_i\}$ is sent over a dispersive, noisy channel, so that the sequence $\{u_i\}$ is available at the equalizer input. We now need to find the coefficients of a Finite Impulse Response (FIR) filter (transversal filter, Figure 16.3) with $2K + 1$ taps. This filter should convert sequence $\{u_i\}$ into sequence $\{\hat{c}_i\}$:

$$\hat{c}_i = \sum_{n=-K}^{K} e_n u_{i-n} \tag{16.10}$$

that should be "as close as possible" to the sequence $\{c_i\}$. Defining the deviation ε_i as

$$\varepsilon_i = c_i - \hat{c}_i \tag{16.11}$$

we aim to find a filter so that

$$\varepsilon_i = 0 \quad \text{for } N_0 = 0 \tag{16.12}$$

which gives the *ZF equalizer*, or that

$$E\{|\varepsilon_i|^2\} \to \min \quad \text{for } N_0 \text{ having a finite value} \tag{16.13}$$

which gives the *Minimum Mean Square Error (MMSE) equalizer*.

Figure 16.2 Linear equalizer in the time domain (a) and time-discrete equivalent system in the z-transform domain (b).

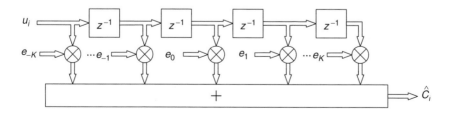

Figure 16.3 Structure of a linear transversal filter. Remember that z^{-1} represents a delay by one sample.

16.2.1 Zero-Forcing Equalizer

The ZF equalizer can be interpreted in the frequency domain as enforcing a completely flat (constant) transfer function of the combination of channel and equalizer by choosing the equalizer transfer function as $E(z) = 1/F(z)$. In the time domain, this can be interpreted as minimizing the maximum ISI (*peak distortion criterion*). Appendix 16.A (see *www.wiley.com/go/molisch*) shows that these two criteria are identical.

The ZF equalizer is optimum for elimination of ISI. However, channels also add noise, which is amplified by the equalizer. At frequencies where the transfer function of the channel attains small values, the equalizer has a strong amplification, and thus also amplifies the noise. As a consequence, the noise power at the detector input is larger than for the case without an equalizer (see Figure 16.4).

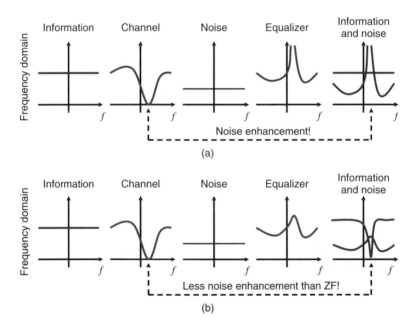

Figure 16.4 Illustration of noise enhancement in zero-forcing equalizer (a), which is mitigated in an MMSE linear equalizer (b).

The Fourier transform $\Xi(e^{j\omega T_s})$ of the sample ACF ζ_i is related to $\hat{\Xi}(e^{j\omega T})$, the Fourier transform of $\eta(t)$, as

$$\Xi(e^{j\omega T_s}) = \frac{1}{T_S} \sum_{n=-\infty}^{\infty} \left| \hat{\Xi}\left(\omega + \frac{2\pi n}{T_S}\right) \right|^2, \quad |\omega| \leq \frac{\pi}{T_S} \tag{16.14}$$

The noise power at the detector is

$$\sigma^2_{\text{n-LE-ZF}} = N_0 \frac{T_S}{2\pi} \int_{-\pi/T_S}^{\pi/T_S} \frac{1}{\Xi(e^{j\omega T_s})} \, d\omega \tag{16.15}$$

It is finite only if the spectral density Ξ has no (or only integrable) singularities.

16.2.2   The Mean Square Error Criterion

The ultimate goal of an equalizer is minimization, not of the ISI, but of the bit error probability. Noise enhancement makes the ZF equalizer ill-suited for this purpose. A better criterion is minimization of the *Mean Square Error* (MSE) between the transmit signal and the output of the equalizer.

We are thus searching for a filter that minimizes:

$$MSE = E\left\{|\varepsilon_i|^2\right\} = E\left\{\varepsilon_i \varepsilon_i^*\right\} \tag{16.16}$$

As shown in Appendix 16.B (see *www.wiley.com/go/molisch*), this can be achieved with a filter whose coefficients $\mathbf{e}_{opt}$ are given by

$$\mathbf{e}_{opt} = \mathbf{R}^{-1}\mathbf{p} \tag{16.17}$$

where $\mathbf{R} = E\{\mathbf{u}_i^*\mathbf{u}_i^T\}$ is the correlation matrix of the received signal, and $\mathbf{p} = E\{\mathbf{u}_i^* c_i\}$ the cross-correlation between the received signal and the transmit signal. Considering the frequency domain, concatenation of the noise-whitening filter with the equalizer $E(z)$ has the transfer function:

$$\tilde{E}(z) = \cfrac{1}{\Xi(z) + \frac{N_0}{\sigma_s^2}} \tag{16.18}$$

which is the transfer function of the Wiener filter. The MSE is then

$$\sigma_{\text{n-LE-MSE}}^2 = N_0 \frac{T_S}{2\pi} \int\limits_{-\pi/T_S}^{\pi/T_S} \cfrac{1}{\Xi(e^{j\omega T_S}) + \frac{N_0}{\sigma_s^2}} \, d\omega \tag{16.19}$$

Comparison with Eq. (16.15) shows that the noise power of an MMSE equalizer is smaller than that of a ZF equalizer (as illustrated in Figure 16.4).

Example 16.1   *Equalizer coefficients and noise enhancement for linear equalizers: consider a channel with impulse response* $h(\tau) = 0.4\delta(\tau) - 0.7\delta(\tau - T_S) + 0.6\delta(\tau - 2T_S)$, $N_0 = 0.3$, *and* $g(t) = g_R(t, T_S)$. *Compute the noise variance at the output of a ZF equalizer and an MMSE equalizer.*

First we note that $\eta(t) = h(t) * g(t) = 0.4g(t) - 0.7g(t - T_S) + 0.6g(t - 2T_S)$. For the given rectangular pulse, $\sigma_s^2 = 1$. The z-transform of the ACF of $\eta(t)$ is given by

$$\Xi(z) = 0.24z^2 - 0.7z + 1.01 - 0.7z^{-1} + 0.24z^{-2} = F(z)F^*(z^{-1})$$

Let us then choose $F(z) = 0.4 - 0.7z^{-1} + 0.6z^{-2}$. Thus, $F^*(z^{-1}) = 0.4 - 0.7z^1 + 0.6z^2$ has roots at $0.58 \pm 0.57j$, which are inside the unit circle.

The transfer function of the ZF equalizer is then

$$E_1(z) = \frac{1}{F(z)} = \frac{1}{0.4 - 0.7z^{-1} + 0.6z^{-2}} \tag{16.20}$$

The noise variance is given by Eq. (16.15):

$$\sigma^2_{\text{n-LE-ZF}} = N_0 \frac{T_S}{2\pi} \int\limits_{-\pi/T_S}^{\pi/T_S} \frac{1}{0.24e^{j2\omega T_s} - 0.7e^{j\omega T_s} + 1.01 - 0.7e^{-j\omega T_s} + 0.24e^{-j2\omega T_s}} \, d\omega$$

$$= N_0 \frac{1}{2\pi} \int\limits_{-\pi}^{\pi} \frac{1}{0.24e^{j2\omega} + 0.24e^{-j2\omega} - 0.7e^{j\omega} - 0.7e^{-j\omega} + 1.01} \, d\omega$$

$$= N_0 \frac{1}{2\pi} \int\limits_{-\pi}^{\pi} \frac{1}{0.48 \cos 2\omega - 1.4 \cos \omega + 1.01} \, d\omega$$

$$\approx 2.94. \tag{16.21}$$

The effective Signal-to-Noise Ratio (SNR) is thus $1/\sigma^2_{\text{n-LE-ZF}} = 0.34$.
   The MMSE equalizer is given by

$$E_2(z) = \frac{1}{F(z) + N_0} = \frac{1}{0.7 - 0.7z^{-1} + 0.6z^{-2}} \tag{16.22}$$

The noise variance is given by Eq. (16.19). We note that the only difference from the ZF case is the addition of N_o/σ^2_s in the denominator of the integrand:

$$\sigma^2_{\text{n-LE-MSE}} = N_0 \frac{1}{2\pi} \int\limits_{-\pi}^{\pi} \frac{1}{0.48 \cos 2\omega - 1.4 \cos \omega + 1.31} \, d\omega$$

$$\approx 0.46 \tag{16.23}$$

As expected, the noise variance is lower for the MMSE equalizer than for the ZF equalizer. The effective SNR is [Proakis 2005]

$$\gamma_\infty = \frac{1 - \sigma^2_{\text{n-LE-MSE}}}{\sigma^2_{\text{n-LE-MSE}}} = 1.17 \tag{16.24}$$

compared with $(1/N_0) = 3.33$ if there were no ISI (and thus, no equalizer). Thus, the necessity to equalize decreases the effective SNR (SNR at the output of the equalizer) by 4.5 dB and 10 dB for the MMSE and ZF equalizer, respectively.

16.2.3 Adaptation Algorithms for Mean Square Error Equalizers

In order to find the optimum equalizer weights, we can directly solve Eq. (16.17). However, this requires on the order of $(2K + 1)^3$ complex operations. To ease the computational burden, iterative algorithms have been developed. The quality of an iterative algorithm is described by the following criteria:

- *Convergence rate*: how many iterations are required to "closely approximate" the final result? It is usually assumed that the channel does not change during the iteration period. However, if an algorithm converges too slowly, it will never reach a stable state – the channel has changed before the algorithm has converged.

- *Misadjustment*: the size of deviation of the converged state of the iterative algorithm from the exact MSE solution.
- Computational effort per iteration.

In the following, we discuss two algorithms that are widely used – the Least Mean Square (LMS) and the Recursive Least Square (RLS).

Least Mean Square Algorithm

The LMS algorithm; also known as the *stochastic gradient method*, consists of the following steps:

1. Initialize the weights with values $\mathbf{e}_0$.
2. With this value, compute an approximation for the gradient of the MSE. The true gradient cannot be computed, because it is an expected value. Rather, we are using an estimate for $\mathbf{R}$ and $\mathbf{p}$ – namely, their instantaneous realizations:

$$\hat{\mathbf{R}}_n = \mathbf{u}_n^* \mathbf{u}_n^T \tag{16.25}$$

$$\hat{\mathbf{p}}_n = \mathbf{u}_n^* c_n \tag{16.26}$$

where subscript n indexes the iterations. The gradient is estimated as

$$\hat{\nabla}_n = -2\hat{\mathbf{p}}_n + 2\hat{\mathbf{R}}_n \mathbf{e}_n \tag{16.27}$$

3. We next compute an updated estimate of the weight vector $\mathbf{e}$ by adjusting weights in the direction of the negative gradient:

$$\mathbf{e}_{n+1} = \mathbf{e}_n - \mu \hat{\nabla}_n \tag{16.28}$$

where μ is a user-defined parameter that determines convergence and residual error.
4. If the stop criterion is fulfilled – e.g., the relative change in weight vector falls below a predefined threshold – the algorithm has converged. Otherwise, we return to step 2.

It can be shown that the LMS algorithm converges if

$$0 < \mu < \frac{2}{\lambda_{\max}} \tag{16.29}$$

Here $\lambda_{\max}$ is the largest eigenvalue of the correlation matrix $\mathbf{R}$. The problem is that we do not know this eigenvalue (computing it requires larger computational effort than inverting the correlation matrix). We thus have to guess values for μ. If μ is too large, we obtain faster convergence, but the algorithm might sometimes diverge. If we choose μ too small, then convergence is very probable, but slow. Generally, convergence speed depends on the condition number of the correlation matrix (i.e., the ratio of largest to smallest eigenvalue): the larger the condition number, the slower the convergence of the LMS algorithm.

The Recursive Least Squares Algorithm

In most cases, the LMS algorithm converges very slowly. Furthermore, the use of this algorithm is justified only when the statistical properties of the received signal fulfill certain conditions. The general Least Squares (LS) criterion, on the other hand, does not require such assumptions. It just analyzes the N subsequent errors ε_i, and chooses weights such that the sum of the squared errors is minimized. This general LS problem can be solved by a recursive algorithm as well – known as *RLS* – the details of which are given in Appendix 16.C (see *www.wiley.com/go/molisch*).

Comparison of Algorithms

There are two classes of algorithms for the determination of equalizer coefficients: direct implementation (Wiener filter, LS criterion) and iterative methods (LMS, RLS).

For the Wiener filter, we first have to determine the correlation matrix; this can be the major part of the numerical effort especially if the number of weights is small. The actual inversion of the matrix requires $(2K + 1)^3$ operations. Alternatively, we can use the data matrix directly, and invert it (LS algorithm). Construction of the data matrix requires less numerical effort than the correlation matrix; on the other hand the effort for inversion is much larger and depends on the number of used bits.

When comparing iterative algorithms, we find that the LMS algorithm usually converges too slowly. The RLS algorithm converges faster, but has a larger residual error. Figure 16.5 shows the typical example of the MSE for a Digital Enhanced Cordless Telecommunications (DECT) cordless telephone (see the companion website at *www.wiley.com/go/molisch*). We see that the RLS algorithm has converged after 10 bits, while the LMS algorithm requires almost 300 bits. Due to the temporal variance of the channel, as well as spectral efficiency considerations, fast convergence is more important than an extremely low residual error rate.

On the other hand, the LMS algorithm requires far fewer (complex) operations (see Figure 16.6; the terms "Decision Feedback Equalizer (DFE)" and "gradient lattice" will be explained below). One important conclusion from this figure is that for a small number of weights the complexity of the algorithms does not differ significantly. For up to 5–8 weights, the differences are less than 50% (with the exception of the LMS which is disqualified because of its slow convergence). In this regime, convergence, stability, and ease of implementation are the dominant criteria.

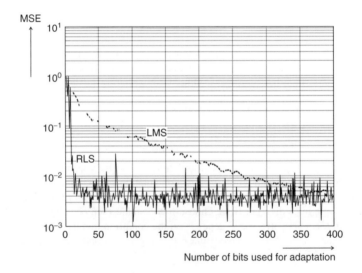

Figure 16.5 Mean-square error as a function of the number of iterations for a decision feedback equalizer (see below). For least mean square: $\mu = 0.03$; recursive least squares: $\lambda = 0.99, \delta = 10^{-9}$. From Fuhl [1994].

16.2.4   Further Linear Structures

Up to now, we have considered transversal FIR filters. However, linear filters can also be realized by means of other structures. One possibility is the use of recursive filters (*Infinite Impulse Response*

Figure 16.6 Number of operations per iteration step for different algorithms for a decision feedback equalizer (see below). *In this figure*: FTF, Fast Transversal Filter.

filters (IIR)). They have the advantage that fewer taps are required to achieve equalization. The major drawback is that these filters can show, not only zeros, but also poles in the transfer function, such that they can become unstable. For this reason, IIR filters are rarely used in practice.

A further possible structure is the *lattice filter*. The equations for the equalization algorithms are different from those of transversal filters (details can be found in Proakis [2005]).

16.3 Decision Feedback Equalizers

A *decision feedback equalizer (DFE)* has a simple underlying premise: once we have detected a bit correctly, we can use this knowledge in conjunction with knowledge of the channel impulse response to compute the ISI caused by this bit. In other words, we determine the effect this bit will have on subsequent samples of the receive signal. The ISI caused by each bit can then be subtracted from these later samples.

The block diagram of a DFE is shown in Figure 16.7. The DFE consists of a *forward filter* with transfer function $E(z)$, which is a conventional linear equalizer, as well as a *feedback filter* with transfer function $D(z)$. As soon as the RX has decided on a received symbol, its impact on all *future* samples (*postcursor ISI*) can be computed, and (via the feedback) subtracted from the received signal. A key point is the fact that the ISI is computed based on the signal *after* the hard decision; this eliminates additive noise from the feedback signal. Therefore, a DFE results in a smaller error probability than a linear equalizer.

One possible source of problems is *error propagation*. If the RX decides incorrectly for one bit, then the computed postcursor ISI is also erroneous, so that later signal samples arriving at the decision device are even more afflicted by ISI than the unequalized samples. This leads to a vicious cycle of wrong decisions and wrong subtraction of postcursors.

Error propagation does not usually play a role when the BER is small. Note, however, that small error rates are often achieved via coding. It may therefore be necessary to decode the bits, re-encode them (such that the signal becomes a noise-free version of the received signal), and use

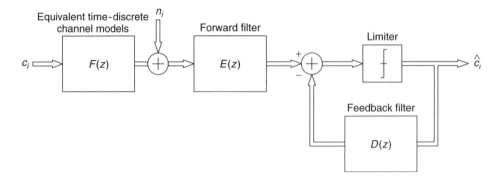

Figure 16.7   Structure of a decision feedback equalizer.

this new signal in the feedback from the DFE.[3] In the following, we will only consider uncoded systems without error propagation.

16.3.1 MMSE Decision Feedback Equalizer

The goal of the MMSE DFE is again minimization of the MSE, by striking a balance between noise enhancement and residual ISI. As noise enhancement is different in the DFE case from that of linear equalizers, the coefficients for the forward filter are different: as postcursor ISI does not contribute to noise enhancement, we now aim to minimize the sum of noise and (average) *precursor* ISI. Obviously, performance also differs.

The coefficients of the feedforward filter can be computed from the following equation:

$$\sum_{n=-K_{\rm ff}}^{0} e_n \left(\sum_{m=0}^{-l} f_m^* f_{m+l-n} + N_0 \delta_{nl} \right) = -f_{-l}^* \quad \text{for } l, n = -K_{\rm ff}, \dots, 0 \qquad (16.30)$$

where $K_{\rm ff}$ is the number of taps in the feedforward filter. The coefficients of the feedback filter are then

$$d_n = -\sum_{m=-K_{\rm ff}}^{0} e_m f_{n-m} \quad \text{for } n = 1, \dots, K_{\rm fb} \qquad (16.31)$$

where $K_{\rm fb}$ is the number of taps in the feedback filter.

Assuming some idealizations (the feedback filter must be at least as long as the postcursor ISI; it must have as many taps as required to fulfill Eq. (16.30); there is no error propagation), the MSE at the equalizer output is

$$\sigma_{\rm n}^2(DFE - MMSE) = N_0 \exp\left(\frac{T_{\rm S}}{2\pi} \int_{-\pi/T_{\rm s}}^{\pi/T_{\rm s}} \ln\left[\frac{1}{\Xi(e^{j\omega T}) + N_0} \right] d\omega \right) \qquad (16.32)$$

---

[3] As the decoder itself shows a delay, this can become a challenging task. Possible solutions to this include the design of joint equalization and decoding, with the exchange of soft information (see also Chapter 14).

16.3.2   Zero-Forcing Decision Feedback Equalizer

The ZF DFE is conceptually even simpler. As mentioned in Section 16.1.2, the noise-whitening filter eliminates all precursor ISI, such that the resulting effective channel is purely causal. Postcursor ISI is subtracted by the feedback branch. The effective noise power at the decision device is

$$
\sigma_n^2(DFE - ZF) = N_0 \exp\left(\frac{T_S}{2\pi} \int_{-\pi/T_s}^{\pi/T_s} \ln\left[\frac{1}{\Xi(e^{j\omega T})} \right] d\omega \right)
\tag{16.33}
$$

This equation demonstrates that noise power is larger than it is in the unequalized case, but smaller than that for the linear ZF equalizer.

Example 16.2 *Using the channel from Example 16.1, compute the noise enhancement for the MMSE DFE and ZF DFE*

From Example 16.1, remember that $\Xi(e^{j\omega T}) = 0.48 \cos 2\omega T_S - 1.4 \cos \omega T_S + 1.01$. Inserting this in Eq. (16.32), we obtain:

$$
\sigma_n^2(DFE - MMSE) = N_0 \exp\left(\frac{T_S}{2\pi} \int_{-\pi/T_s}^{\pi/T_s} \ln\left[\frac{1}{\Xi(e^{j\omega T}) + N_0} \right] d\omega \right)
\tag{16.34}
$$

$$
= N_0 \exp\left(\frac{1}{2\pi} \int_{-\pi}^{\pi} \ln\left[\frac{1}{0.48 \cos 2\omega - 1.4 \cos \omega + 1.31} \right] d\omega \right)
\tag{16.35}
$$

$$
\approx 0.33
\tag{16.36}
$$

so that the output SNR is 2. Thus, the SNR deteriorates by 2 dB compared with the Additive White Gaussian Noise (AWGN) case. For the ZF DFE, the noise variance at the output is 0.83, so that the SNR deteriorates by 4.4 dB.

16.4   Maximum Likelihood Sequence Estimation – Viterbi Detector

The equalizer structures considered up to now influence the decision about which *symbol* has been transmitted. For MLSE, on the other hand, we try to determine the *sequence of symbols* that has most likely been transmitted. This situation shows strong similarities to the decoding of convolutional codes. As a matter of fact, transmission through a delay-dispersive channel can be viewed as convolutional encoding with a code rate $R_c = 1/1$. MLSE estimators give the best performance of all equalizers.

Remember that the output signal of the time-discrete channel can be written as

$$
u_i = \sum_{n=0}^{L_c} f_n c_{i-n} + n_i
\tag{16.37}
$$

where n is Gaussian white noise with variance σ_n^2. For a sequence of N received values, the joint probability density function (pdf) of the vector of received signals **u** (conditioned on the data vector

c and impulse response vector **f**) is[4]

$$pdf(\mathbf{u}|\mathbf{c};\mathbf{f}) = \frac{1}{(2\pi\sigma_n^2)^{N/2}} \exp\left(-\frac{1}{2\sigma_n^2}\sum_{i=1}^{N}\left|u_i - \sum_{n=0}^{L_c} f_n c_{i-n}\right|^2\right) \tag{16.38}$$

The MLSE of **c** (for a given **f**) are the values of the vectors that maximize the joint pdf $pdf(\mathbf{u}|\mathbf{c},\mathbf{f})$. As the variables only occur in the exponent, it is sufficient to minimize:

$$\sum_{i=1}^{N}\left|u_i - \sum_{n=0}^{L_c} f_n c_{i-n}\right|^2 \tag{16.39}$$

As for convolutional decoding, various algorithms exist for determination of the optimum sequence. The RX first generates all possible sequences that can result from convolution of valid transmit sequences with the channel impulse response. We then try to find the sequence that has the smallest distance (best metric) from the received signal. The most straightforward (but also most computationally intensive) method is the *exhaustive search*. In practice, the Viterbi algorithm is used instead.

MLSE as described above is only optimum if the additive noise at MLSE input is white. Therefore, the sample values used at the detector have to be the output of a noise-whitening filter. This filter has to be adapted to the current channel state; and each channel realization requires spectral factorization. Due to these difficulties, the total input filter (matched filter and noise-whitening filter) is often replaced by a simple brickwall filter whose bandwidth is approximately the inverse symbol duration. Note, however, that in this case one sample per symbol no longer provides sufficient statistics.

Example 16.3 *Viterbi equalization.*

This example shows the working of the Viterbi detection of a symbol stream that went through a channel with a discrete time impulse response:

$$\mathbf{f} = \begin{pmatrix} 1 \\ -0.5 \\ 0.3 \end{pmatrix} \tag{16.40}$$

The channel can be viewed as a tapped delay line (shift register) with weights 1, -0.5, and 0.3; see the top part of Figure 16.8 (the left top part shows the tapped delay line model, the right part shows the "cell" model analogous to the convolutional codes discussed in Chapter 14). For ease of exposition, we chose a channel with a real impulse response, and Binary Phase Shift Keying (BPSK) as the modulation format. The lower part of Figure 16.8 shows possible transitions in the trellis diagram. We have to consider four states in the trellis, as $L_c = 2$ samples, and the number of possible states in a cell of the equivalent shift register is equal to the size of the modulation alphabet $M = 2$. We assume furthermore that we know the starting state of the trellis –i.e., -1 -1 (e.g., because known bits have been transmitted before the start of our decoding). The bottom part of Figure 16.9 shows the "unfolding" of the trellis diagram. The numbers next to the transitions are metrics of the considered sequence.

---

[4] Here and in the following, we assume that all transmit symbols are equally likely, such that MLSE and maximum-a-posteriori estimation are identical.

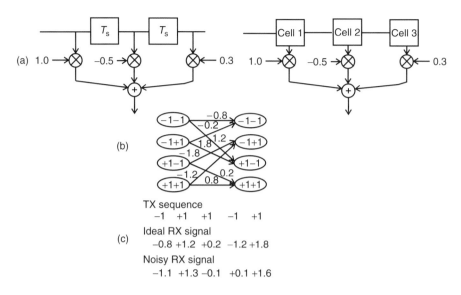

Figure 16.8 Representation of tapped delay line channel (a), transition probabilities (b), and transmitted and received signals (c).

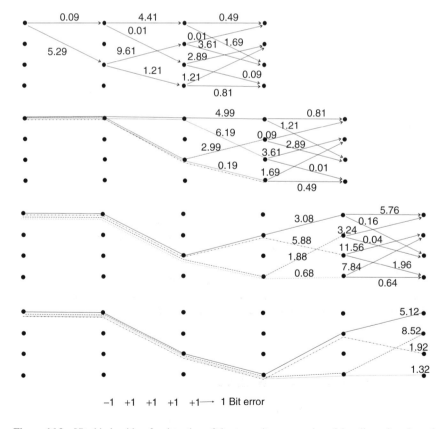

Figure 16.9 Viterbi algorithm for detection of the transmit sequence in a delay-dispersive channel.

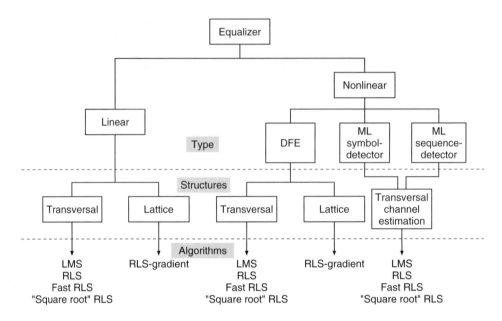

Figure 16.10   Taxonomy of equalizer structures.
Reproduced with permission from Proakis [1991] © IEEE.

16.5   Comparison of Equalizer Structures

Figure 16.10 shows a taxonomy of equalizer structures. When selecting an equalizer for a practical system, we have to consider the following criteria:

- *Minimization of the BER*: here MLSE is superior to all other structures. DFEs, though worse than MLSE estimators, are better than linear equalizers. The quantitative difference between the structures depends on the channel impulse response.
- *Can the channel deal with zeros in the channel transfer function?* ZF equalizers have problems, as they invert the transfer function and thus create poles in the equalizer transfer function. Neither MMSE nor MLSE equalizers have this problem.
- *Computational effort*: the effort for linear equalizers and DFEs is not significantly different. Depending on the adaptation algorithm, the number of operations increases linearly, quadratically, or cubically with equalizer length (number of weights). For MLSE, the computation effort increases exponentially with length of the impulse response of the channel. For short-impulse responses (e.g., impulse response is at most four symbol durations long, as in GSM), the computational complexity of MLSE is comparable with that of other equalizer structures.
- *Sensitivity to channel misestimation*: due to the error propagation effect, DFE equalizers are more sensitive to channel estimation errors than linear equalizers. Also, ZF equalizers are more sensitive than MMSE equalizers.
- *Power consumption and cost*: these can be deduced from the computational effort.

16.6   Fractionally Spaced Equalizers

In most cases, the RX samples and processes signals at the symbol frequency $(1/T_S)$. This is suboptimum if this sampling rate is lower than the Nyquist rate and the matched filter is matched only to the TX pulse (and not the received, distorted pulse). The spectrum of a signal that was filtered by a raised cosine filter (with roll-off factor α) extends up to a frequency $(1+\alpha)/2T_S$, such that the Nyquist rate is $(1+\alpha)/T_S$. A *fractionally spaced equalizer* is based on sampling at no less than the Nyquist rate. The taps of the equalizers are spaced at $T_S a/b$, where $a < b$ and a and b are integers.

Fractionally spaced equalizers can also be interpreted as performing equalization and matched filtering in one step. They are also less sensitive to errors in the sampling time. The drawback is that the number of required taps is larger than for a symbol-spaced equalizer, and thus the computational effort is higher.

16.7   Blind Equalizers

16.7.1   Introduction

"Normal" equalization works in two stages: a training phase and a detection phase. During the training phase, a known bit sequence is transmitted over the channel, and the distorted version at the RX is compared with a (locally generated) undistorted version; this in turn gives us information about the channel impulse response that is used for equalization (see Section 16.1.3 for a discussion of the pros and cons). In contrast, blind equalization exploits known statistical properties of the transmit signal to estimate both channel and data. Equalizer coefficients are adjusted in such a way that certain statistical properties of the equalizer output match known statistical properties of the transmit signal.

The advantages of blind equalization include the following:

- The whole timeslot is used for determination of the impulse response, not just a short training sequence.
- Spectral efficiency is improved, because no time is "wasted" on transmission of the training sequence.

The following signal properties can be used for blind equalization:

- *Constant envelope*: for many signals (Frequency Shift Keying (FSK), Minimum Shift Keying (MSK), Gaussian Minimum Shift Keying (GMSK)), the envelope (amplitude) is constant.
- *Statistical properties*: e.g., cyclostationarity.
- *Finite symbol alphabet*: only certain discrete values are valid points in the signal constellation diagram.
- *Spectral correlation*: signal spectra are correlated with shifted versions of themselves.
- A combination of these properties.

Well-studied blind algorithms include (i) *Constant Modulus Algorithm* (CMA), (ii) blind *MLSE*, and (iii) algorithms based on higher order statistics. In the following, we will discuss these classes of algorithms.

16.7.2   Constant Modulus Algorithm

CMAs are the oldest algorithms for blind equalization. In their simplest form, they use the LMS adaptation. The data-aided LMS algorithm was described in Section 16.2; it is based on minimization of the difference between a desired signal (known at the RX) and the output of an equalizer. In a blind LMS algorithm, the desired signal first has to be generated *from the output of the equalizer*.

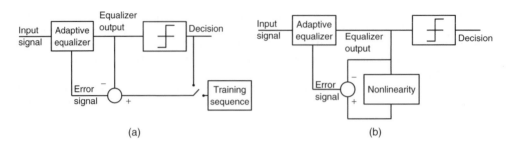

Figure 16.11 Structure of a conventional equalizer (a) and that of a constant-modulus-algorithm-based equalizer (b).

This is achieved by sending the equalizer output through a nonlinear function. The error signal is then the difference between the output of this nonlinear function and the equalizer output. The difference between a conventional equalizer and the CMA equalizer is shown in Figure 16.11.

The nonlinear function can either be memoryless, or contain mth order memory. Various types of algorithms (usually named for their inventors) are distinguished by different nonlinearities. The best known are the Sato algorithm (1975) and the Godard algorithm (1980).

The LMS algorithm used for blind adaptation has the same drawbacks as the conventional LMS: the convergence rate is slow (by increasing the stepwidth μ, convergence is improved and the error after convergence is worsened). It is also possible that the algorithm converges to a local minimum. These problems can be solved by the "analytic constant modulus algorithm" [van der Veen and Paulraj 1996]. This is a noniterative algorithm that provides exact solutions in the noise-free case, but is also robust with respect to noise.

16.7.3 Blind Maximum Likelihood Estimation

For conventional MLSE, a training sequence is used for determination of the channel impulse response **f**. This estimate is then used during the actual transmission of data, so that the Maximum Likelihood (ML) estimate only has to solve for **c**. For a blind estimate, **c** and **f** have to be estimated simultaneously. This can be achieved by either of the following methods:

1. The channel impulse response is estimated from $pdf(\mathbf{u}|\mathbf{f}, \mathbf{c})$ by averaging over all possible data sequences. This method has two drawbacks: it requires considerable computation time for the averaging, and it is suboptimum.
2. Alternatively, we can determine the ML estimate for **f** for all possible data sequences. Then we select the pair **f**, **c** that has the best overall metric. This method is even more computationally intensive than the averaging method; however, it is also more accurate. Furthermore, methods for reducing the computational effort have been proposed [Proakis 2005].

16.7.4 Algorithms Using Second- or Higher Order Statistics

In general, second-order statistics (ACFs) cannot provide information about the phase of the channel impulse response. An exception occurs when the ACF of the received signal is periodic. Cyclostationary properties are thus the basis for blind estimation methods using second-order statistics. Similar to our discussion in Chapter 6, we distinguish between strict-sense and wide-sense cyclostationarity. A process is strict-sense cyclostationary if *all* its statistical properties are invariant to shifts by integer multiples of the sampling period T_{per}. For wide-sense stationarity, only the mean and the ACF have to fulfill this condition:

$$E\{x(t + iT_{\text{per}})\} = E\{x(t)\} \tag{16.41}$$

$$E\{x^*(t_1 + iT_{\text{per}})x(t_2 + iT_{\text{per}})\} = E\{x^*(t_1)x(t_2)\} \tag{16.42}$$

If the signal is oversampled, then the resulting sequence of sample values is guaranteed to be cyclostationary. The actual channel estimate is then based on different correlation matrices of the received signal (for details, see Tong et al. [1994, 1995]). The use of higher order statistics does not require the applicability of cyclostationarity; however, its accuracy is usually much lower.

16.7.5   Assessment

Historically speaking, blind equalization was first developed in the 1970s for multiterminal computer networks (one central station linked to multiple terminals). During the 1980s and 1990s, a lot of theoretical work was devoted to this topic, as it offers some fascinating mathematical challenges. However, up to now, truly blind equalization has not been able to replace training-sequence-based equalization in practical systems. The main reason seems to be that the difference in computational effort and reliability is significant. In particular, blind equalizers require a long time to converge, and thus do not work well in quickly time-variant wireless channels. Another approach that has been developed in recent years uses a training sequence to obtain an initial estimate of the channel, and then refines these estimates by means of blind (decision-aided) equalization [Loncar et al. 2002].

16.8   Appendices

Please go to *www.wiley.com/go/molisch*

Further Reading

The first comprehensive description of adaptive equalizers, which is still worth reading today, was given in Lucky et al. [1968]. The description in this chapter – particularly, that of linear and DFE equalizers – was inspired by the excellent exposition in Proakis [2005], which also gives many more interesting details. Haykin [1991] describes adaptive filters, which is another way of interpreting equalizers. DFEs are also described in Belfiore and Park [1977]. Fractionally spaced equalizers were analyzed, e.g., in Gitlin and Weinstein [1981] and Ungerboeck [1976]. The Viterbi equalizer is an application of the Viterbi algorithm [Viterbi 1967]; the impact of channel estimation errors is discussed in Gorokhov [1998]. Vitetta et al. [2000] give a detailed description of a wide variety of equalizer structures, including the impact of channel-state information (perfect, estimated, or averaged). Spectral factorization techniques are surveyed in Sayed and Kailath [2001]. Blind algorithms for equalization are too numerous to list here; as examples, we just name Giannakis and Halford [1997], Liu et al. [1996], Sato [1975], and van der Veen and Paulraj [1996]. Iterative equalizers are described in [Wymeersch 2007].

For updates and errata for this chapter, see wides.usc.edu/teaching/textbook

Part IV

Multiple Access and Advanced Transceiver Schemes

In Part III, we described how a single transmitter can communicate with a single receiver. However, for most wireless systems, multiple devices should communicate simultaneously in the same area. It is therefore necessary to provide means for *multiple access*. For most first- and second-generation cellular systems, as well as for cordless phones and wireless Local Area Networks (LANs), different devices communicate either on different frequencies or at different times. The multiple-access schemes based on that principle (called frequency-division multiple access, time-division multiple access, and packet radio) are discussed in Chapter 17. A different type of multiple access is based on spreading the signal over a large bandwidth, and to make that spreading unique for each user. This allows multiple users to be on the air simultaneously, and the receiver can determine from the spreading which part of the "on-air" signal comes from which user. This *spread-spectrum* approach, described in Chapter 18, is used, e.g., for third-generation cellular systems in the form of *Code Division Multiple Access* (*CDMA*). The chapter also describes the Rake receiver, a device that enables CDMA systems to deal with the delay dispersion of the channel, and how multiuser detection can be exploited to significantly increase the performance in a multiple-access environment.

As the data rates of a wireless system increase to higher and higher values, both equalizers and Rake receivers become too complex for practical use. Orthogonal Frequency Division Multiplexing (OFDM) is one way of overcoming this problem. In this approach, the data stream is split up into a large number of substreams, each of which is modulated onto a different carrier – for this reason, OFDM is also often called a multicarrier modulation method. Chapter 19 describes its principles, and also some variants that combine CDMA with OFDM.

After exploiting the time domain, frequency domain, and code domain for signaling, "space is the final frontier." Multiple antenna elements allow to exploit the spatial domain to increase the data rate, allow more simultaneous users, and/or increase the transmission quality. One form for exploiting such multiple antennas, namely antenna diversity, was already treated in Chapter 13; it serves mainly to combat the impact of deep fades on the transmission quality. Now, Chapter 20 describes the theory of "smart antennas," which increase the number of users that can be served in a transmission system. Furthermore, this chapter also discusses the so-called Multiple Input Multiple Output (MIMO) systems, which have multiple antennas at transmitter and receiver, and can exploit those to transmit several data streams in parallel. This in turn allows to increase the data rate without requiring more spectrum.

It often occurs that a certain part of the spectrum is assigned to a particular service or operator, but not used in all locations. *Cognitive radio*, described in Chapter 21, aims to eliminate such inefficiencies, by enabling radios to monitor the spectral occupancy, and make use of (locally) unused parts of the spectrum.

Chapter 22 describes "user cooperation", where different devices do not compete with each other (and thus interfere), but rather help each other in forwarding messages to the destination. The simplest form of such a cooperation is a relay, whose sole purpose is to forward messages. In larger networks, relays can cooperate. Furthermore, in many networks nodes can change their role from information source to relay to destination, depending on the necessity.

As video is becoming a more and more important application in wireless communications systems, the properties of source coding for these applications influence the system performance. Finally, Chapter 23 gives an overview of video coding techniques.

17

Multiple Access and the Cellular Principle

17.1 Introduction

A wireless communications system uses a certain frequency band that is assigned to this specific service. Spectrum is thus a scarce resource, and one that cannot be easily extended. For this reason, a wireless system must make provisions to allow the simultaneous communication of as many users as possible within that band.

The problem of letting multiple users communicate simultaneously can be divided into two parts:

1. If there is only a *single Base Station* (BS), how can it communicate with many Mobile Stations (MSs) simultaneously?
2. If there are *multiple BSs*, how can we assign spectral resources to them in such a way that the total number of possible users is maximized? And how should these BSs be placed in a given geographical area?

As for the first question, there are different methods, called *Multiple Access* (MA) methods, that allow multiple users to talk to a BS simultaneously. In this chapter, we discuss the following three methods:

- *Frequency Division Multiple Access* (FDMA), where different frequencies are assigned to different users.
- *Time Division Multiple Access* (TDMA), where different timeslots are assigned to different users.
- *Packet Radio* can be viewed as a form of TDMA, where the assignment of timeslots to users is adaptive.

FDMA and TDMA are discussed in Sections 17.2 and 17.3. The fact that these sections are rather brief should not detract from the importance of TDMA and FDMA. However, these multiple access methods are conceptually easy to understand; furthermore, many of the concepts required for their treatment have been dealt with in Chapters 11, 12 and 16. A variant of TDMA, which is important for wireless data communications, is *packet radio*, which is discussed in Section 17.4. Besides the schemes mentioned above, *Code Division Multiple Access* (CDMA), where each user is assigned a different code, has gained increased popularity in recent years. We discuss this scheme, as well as other spread spectrum methods, in Chapter 18. Finally, *Space Division Multiple Access*

Wireless Communications, Second Edition   Andreas F. Molisch
© 2011 John Wiley & Sons, Ltd

(SDMA) is a multiple access format for systems with multiple antennas; it can be combined with all of the other multiple access methods. It is described in Section 20.1. The so-called "duplexing," which separates transmission and reception at a transceiver, is analyzed in Section 17.5.

The goal of all these methods is to maximize spectral efficiency – i.e., to maximize the number of users per unit bandwidth. As mentioned above, there is also a different (though related) question: how can we design a system so that the number of users per unit bandwidth *and unit area* is maximized? This goal obviously requires multiple BSs, and the assignment of spectral resources to them. All this leads us to the cellular principle, which requires reusing the same spectrum at different BSs; this is discussed in Section 17.6.

17.2 Frequency Division Multiple Access

17.2.1 Multiple Access via Frequency Division Multiple Access

FDMA is the oldest, and conceptually most simple, multiaccess method. Each user is assigned a frequency (sub)band – i.e., a (usually contiguous) part of the available spectrum (see Figure 17.1). The assignment of frequency bands is usually done during call setup, and retained during the whole call. FDMA is usually combined with the Frequency Domain Duplexing (FDD) (see Section 17.5), so that two frequency bands (with a fixed duplex distance) are assigned to each user: one for downlink (BS-to-MS) and one for uplink (MS-to-BS) communication.

Figure 17.1 Principle of frequency division multiple access.

Pure FDMA is conceptually very simple, and has some advantages for implementation:

- The transmitter (TX) and receiver (RX) require little *digital signal processing*. However, this is not so important in practice anymore, as the costs for digital processing are continuously decreasing.
- (*Temporal*) *synchronization is simple*. Once synchronization has been established during the call setup, it is easy to maintain it by means of a simple tracking algorithm, as transmission occurs continuously.

However, pure FDMA also has significant disadvantages, especially when used for speech communications. These problems arise from spectral efficiency considerations, as well as from sensitivity to multipath effects:

- *Frequency synchronization and stability are difficult*: for speech communications, each frequency subband is quite narrow (typically between 5 and 30 kHz). Local oscillators thus must be very accurate and stable; jitters in the carrier frequency result in adjacent channel interference. High spectral efficiency also requires the use of very steep filters to extract the desired signal. Both accurate oscillators and steep filters are expensive, and thus undesirable. If they are not admissible, guard bands can be used to mitigate filter requirements. This, however, reduces the spectral efficiency of the system.

- *Sensitivity to fading*: since each user is assigned a distinct frequency band, these bands are narrower than for other multiaccess methods (compare TDMA, CDMA) – i.e., 5–30 kHz. For such narrow subbands, fading is flat in practically all environments. This has the advantage that no equalization is required; the drawback is that there is no frequency diversity. Remember that frequency diversity is mainly provided by signal components that are more than one channel coherence bandwidth apart (see Chapters 13 and 16).
- *Sensitivity to random Frequency Modulation (FM)*: due to the narrow bandwidth, the system is sensitive to random FM: the Bit Error Rate (BER) due to random FM is proportional to $(v_{max}T_S)^2$ (see Chapter 12). Thus, it is inversely proportional to the square of the bandwidth. On the positive side, appropriate signal-processing schemes can not only mitigate these effects but even exploit them to obtain time diversity. Note that the situation here is dual to wideband systems, where delay dispersion can be a drawback, but equalizers can turn them into an asset by exploiting frequency diversity.
- *Intermodulation*: the BS needs to transmit multiple speech channels, each of which is active the whole time. Typically, a BS uses 20–100 frequency channels. If these signals are amplified by the same power amplifier, third-order modulation products can be created, which lie at undesirable frequencies – i.e., within the transmit band. We thus need either a separate amplifier for each speech channel, or a highly linear amplifier for the composite signal – each of these solutions makes a BS more expensive.

It is for these reasons that FDMA is mostly used for the following applications:

- *Analog communications systems*: here, FDMA is the only practicable multiple access method.
- *Combination of FDMA with other multiple access methods*: the spectrum allocated for a service (or a network operator) is divided into larger subbands, each of which is used for serving a *group* of users. Within this group, multiple access is done by means of another multiple access method – e.g., TDMA or CDMA. Most current wireless systems use FDMA in that way (see Chapters 24–28).
- *High-data-rate systems*: the disadvantages of FDMA are mostly relevant if each user requires only a small bandwidth – e.g., 20 kHz. The situation can be different for wireless Local Area Networks (LANs), where a single user requires a bandwidth on the order of 20 MHz, and only a few frequency channels are available.

17.2.2   Trunking Gain

We will now compute how many subscribers can be covered with an FDMA system by one BS. This seemingly simple question has become a separate branch of communications theory (often called *queuing theory*), with several textbooks dedicated to it (see, e.g., Gross and Harris [1998]). In this subsection, we describe in a very simplified way how to compute the required number of communications channels so that a given number of users can be served with "sufficient quality." We assume, for present purposes, a pure FDMA system that does not need any channels for signaling. Furthermore, we assume a system that is designed purely for speech communications. For the following, we define the offered traffic, as the product of the call arrival rate and the average call holding time (duration); the offered traffic is usually said to have the unit "Erlang," though this is really a dimensionless quantity (call arrival rate is in units of 1/s; call holding time has units of s).

There are two extreme cases in the planning of a cellular network:

1. *Worst case design*: it is assumed that all users want to call simultaneously. If the network operator wants to serve 700 users per cell, it has to provide 700 speech channels. Of course such a network should never be built in practice – this would be like designing a hospital that can treat all inhabitants of a city at the same time.
2. *Best case design*: we know that a typical user uses a phone only 20 min per day; if there are 700 potential users per cell, then 14,000 minutes of call time are actually used. A system with 10 speech channels per cell offers $24 * 60 * 10 = 14,400$ minutes of talk time, and could thus supply the required number of users. However, this computation assumes that all users call sequentially–i.e., new users dial in as soon as old ones have finished their calls – and they do so evenly distributed over a 24-hour period.

Obviously, neither of the two extreme cases is realistic. The art of network design is, to a considerable degree, to predict the call behavior of users, and derive the physical infrastructure (available number of speech channels) that guarantees an acceptable *grade of service*.

Several factors influence this planning process:

1. The number and duration of calls depend on the time of day. We therefore define a *busy hour* (usually around 10h00 and 16h00), which is defined as the hour when most calls are made. The traffic during that busy hour determines the required network capacity.
2. The spatial distribution of users is time variant. While business districts (city centers) usually see a lot of activity during the daytime, suburbs and entertainment districts experience more traffic during the nighttime.
3. Telephoning habits change over the years. While in the late 1980s, calls from cellular phones were usually limited to a few minutes, now hour-long calls have become quite common.
4. Changing user habits are also related to the offering of new services (e.g., data connections) and new pricing structures (e.g., free calls in the evening hours). The strategy of selling "minute accounts" that have to be used up each month also leads to longer talk times than a pricing strategy of charging per minute.

Based on statistical knowledge of user habits, we can now design a system that *with a certain probability* allows a given number of users per cell to make calls. If, through a statistical fluke, more users want to telephone simultaneously, some of the calls will be blocked. Note that the carried traffic is the offered traffic, multiplied by $1 - \text{Pr}_{\text{block}}$.

For the computation of the blocking probability of a simplified system, we make the following assumptions: (i) the times when the calls are placed are statistically independent, (ii) the duration of calls is an exponentially distributed random variable, (iii) if a user is rejected, his/her next call attempt is made statistically independent of the previous attempt (i.e., behaves like a new user).[1] Such a system is called an *Erlang-B* system; the *probability of call blocking* can be shown to be

$$\text{Pr}_{\text{block}} = \frac{T_{\text{tr}}^{N_c}/N_{\text{C}}!}{\sum_{k=0}^{N_{\text{C}}} T_{\text{tr}}^{k}/k!} \qquad (17.1)$$

where N_{C} is the number of speech channels (per cell), and T_{tr} is the average offered traffic. Figure 17.2 shows the relationship graphically. We see that the ratio of required channels to offered

---

[1] This obviously does not agree with reality. Typically, a blocked user retries immediately. After being blocked several times within a short time interval, she or he usually gives up and places the next call after a much longer wait-time.

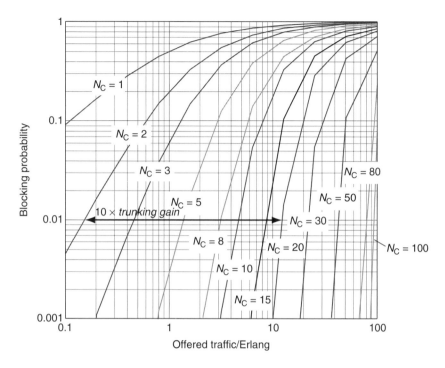

Figure 17.2 Blocking probability in an Erlang-B system. N_C is the number of available speech channels. As an example, the admissible traffic going from $N_C = 2$ to $N_C = 20$ increases from 0.15 to approx. 12 Erlang at 0.01 blockage probability; since the available number of speech channels increases only by a factor 10, the trunking gain is a factor 8.

traffic is very high if N_C is small, especially for low required blocking probabilities. For example, for a required blocking probability of 1%, the ratio of possible offered traffic to available channels is less than 0.1 if $N_C = 2$. If N_C is very large, then the ratio is only slightly less than unity, and becomes almost independent of the required blocking probability. Assuming again a required blocking probability of 1%, the ratio of admissible offered traffic to available channels is about 0.9 for $N_C = 50$.

An alternative model, called Erlang-C, assumes that any user that is not immediately assigned a channel is transferred to a waiting loop, and assigned a channel as soon as it becomes available. The *probability that a user is put on hold* is

$$\mathrm{Pr}_{\mathrm{wait}} = \frac{T_{\mathrm{tr}}^{N_C}}{T_{\mathrm{tr}}^{N_C} + N_C! \left(1 - \frac{T_{\mathrm{tr}}}{N_C}\right) \sum_{k=0}^{N_C-1} T_{\mathrm{tr}}^k / k!} \tag{17.2}$$

and the average wait-time is

$$t_{\mathrm{wait}} = \mathrm{Pr}_{\mathrm{wait}} \frac{T_{\mathrm{call}}}{N_C - T_{\mathrm{tr}}} \tag{17.3}$$

where T_{call} is the average duration of the call.

Example 17.1   *Consider an Erlang-C system where users are active 50% of the time, and the average call duration is 5 minutes. It is required that no more than 5% of all calls are put into a waiting loop. How many channels are required for $n_{user} = 1, 8, 30$ users? What is the average wait-time in each of these cases?*

Since $T_{tr} = 0.5 \cdot n_{user}$ is the average offered traffic, we need to find N_C that fulfills:

$$0.05 \geq \frac{(0.5 \cdot n_{user})^{N_C}}{(0.5 \cdot n_{user})^{N_C} + N_C! \left(1 - \frac{0.5 \cdot n_{user}}{N_C}\right) \sum_{k=0}^{N_C - 1} \frac{(0.5 \cdot n_{user})^k}{k!}} \tag{17.4}$$

This equation needs to be solved numerically; the results are given in Table 17.1 below. With $T_{call} = 5$ min, the average wait-time is

$$t_{wait} = Pr_{wait} \frac{5}{N_C - 0.5 \cdot n_{user}} \tag{17.5}$$

The required number of channels to fulfill inequality and the resulting average wait-time is

Table 17.1   Parameters of the considered Erlang-C system

n_{user}	1	8	30
N_C	3	9	23
Pr_{wait}	0.0152	0.0238	0.0380
t_{wait}	0.0304	0.0238	0.0238

The probability that a call can be placed, and is not blocked, is an important part of service quality: remember that quality of service is defined as 100% minus the percentage of blocked calls, minus ten times the percentage of lost calls (Section 1.3.7). In an FDMA system, a large system load can lead only to blocked calls, but not lost calls, as long as each user stays in the coverage area of his/her BS (note that in this section, we are neglecting call blocking or dropping due to too-high BER). However, calls can be dropped when a user with an ongoing call tries to move to a different cell whose BS is already fully occupied. And as we will see in Section 17.6, a fully loaded system also increases interference with neighboring cells, making the links in that cell more sensitive to fluctuations in signal strength, and possibly increasing the number of dropped calls due to insufficient Signal-to-Noise Ratio (SNR) and Signal-to-Interference Ratio (SIR).

Summarizing, we find that the number of users that can be accommodated with a given quality of service increases faster than linearly with the number of available speech channels. The difference between actual increase and linear increase is called the *trunking gain*. From a purely technical point of view, it is thus preferable to have a large pool of speech channels that serves all users. This situation could be fulfilled, e.g., were there only a single operator for cellular systems, owning the complete spectrum assigned to cellular services. The reasons for *not* choosing this approach are political (pricing, monopoly), not technical.[2]

Example 17.2   *In an Erlang-B system, 30 channels are available. A blocking probability of less than 2% is required. What is the traffic that can be served if there is one operator or three operators?*

---

[2] An approach envisioned in the U.S.A. in the early days of cellular telephony was to assign the available spectrum to exactly two providers, thus striking a compromise between political requirement (avoiding a monopoly) and technical expediency.

1. By inserting the required blocking probability $P_{block} = 0.02$ and the number of channels $N_C = 30$ into Eq. (17.1), we get:

$$0.02 = \frac{T_{tr}^{30}/30!}{\sum_{k=0}^{30} T_{tr}^{k}/k!} \qquad (17.6)$$

Solving this equation for T_{tr}, we get:

$$T_{tr} = 21.9 \qquad (17.7)$$

2. Similarly, sharing the 30 speech channels among the three operators, each having $N_C = 10$ speech channels, results in an average traffic T_{tr} of each operator of:

$$T_{tr} = 5.1 \qquad (17.8)$$

Hence, the total average traffic that can be handled by all three operators together is

$$T_{tr,tot} = 3 \cdot 5.1 = 15.3 \qquad (17.9)$$

17.3   Time Division Multiple Access

For TDMA, different users transmit not at different frequencies but rather at different times (see Figure 17.3). A time unit is subdivided into N timeslots of fixed duration, and each user is assigned one such timeslot. During the assigned timeslot, the user can transmit with a high data rate (as it can use the whole system bandwidth); subsequently, it remains silent for the next $N - 1$ timeslots, when other users take their turn. This process is then repeated periodically. At first glance, this approach has the same performance as FDMA: a user transmits only during $1/N$ of the available time, but then occupies N times the bandwidth. However, there are some important practical differences:

- Users occupy a larger bandwidth. This allows them to exploit the frequency diversity available within the bandwidth allocated to the system; furthermore, the sensitivity to random FM is

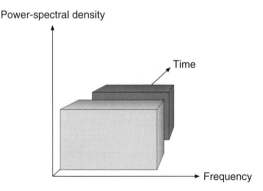

Figure 17.3   Principle behind time division multiple access.

reduced. On the flipside, equalizers are required to combat InterSymbol Interference (ISI) for most operating environments; this increases the effort needed for digital signal processing.

- Temporal guard intervals are required. A TX needs a finite amount of time to ramp up from 0-W output power to "full power" (typically between 100 mW and 100 W). Furthermore, there has to be sufficient guard time to compensate for the runtime of the signal between the MS and BS. It is possible that one MS is far away from the BS, while the one that transmits in the subsequent timeslot is very close to the BS and thus has negligible runtime. As the signals from the two users must not overlap at the BS, the second MS must not transmit during the time it takes the first signal to propagate to the BS.[3] Note, however, that there is no need for frequency guard bands, as each user completely fills up the assigned band.
- Each timeslot might require a new synchronization and channel estimation, as transmission is not continuous. Optimization of timeslot duration is a challenging task. If it is too short, then a large percentage of the time is used for synchronization and channel estimates (in GSM, 17% of a timeslot are used for this purpose). If the timeslot is too long, transmission delays become too long (which users find annoying especially for speech communications), and the channel starts to change during one timeslot. In that case, the equalizer has to track the channel during transmission of a timeslot, which increases implementation effort (this was required, e.g., in the – now defunct – Interim Standard (IS)-136 cellular standard). If the time between two timeslots assigned to one user is larger than coherence time, the channel has changed between these two timeslots, and a new channel estimate is required.
- For interference-limited systems, TDMA has a major advantage: during its period of inactivity, the MS can "listen" to transmission on other timeslots.[4] This is especially useful for the preparation of handovers from one BS to another, when the MS has to find out whether a neighboring BS would offer better quality, and has communications channels available.

TDMA is used in the worldwide cellular standard GSM (Chapter 24) as well as the cordless standard DECT (Digital Enhanced Cordless Telecommunications, see the companion website

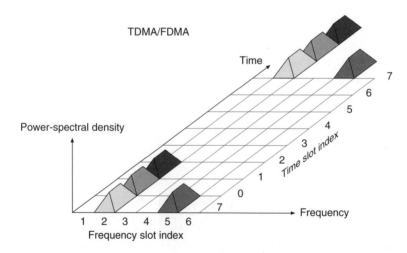

Figure 17.4 How the Global System for Mobile communications (GSM) combines time division multiple access with frequency division multiple access.

---

[3] The required guard time can be reduced by *timing advance* (as described in Chapter 24).

[4] In a mixed TDMA/FDMA system, the RX can also listen to activity on other frequencies.

at *www.wiley.com/go/molisch*), and – in a modified form – in the fourth-generation cellular standards 3GPP-LTE (Chapter 27) and WiMAX (Chapter 28). TDMA is often combined with FDMA, e.g., in GSM (see Figure 17.4). In contrast, pure FDMA is used mainly in analog cellular and cordless systems.

17.4 Packet Radio

Packet radio access schemes break data down into packets, and each of the packets is transmitted over the medium independently. In other words, each packet is like a new user that has to fight for its "own" resources. This allows the transport medium to be exploited much more efficiently when the data traffic from each user is bursty, as is the case for Web browsing, file downloads, and similar data applications.

Packet radio shows two main differences from TDMA and FDMA:

1. Each packet has to fight for its own resources, as described above. The most common methods for resource allocation are ALOHA systems, Carrier Sense Multiple Access (CSMA), and packet reservation (polling). These methods are described in Sections 17.4.1–17.4.4. In these sections, we consider the case where multiple users try to transmit packets to one BS.
2. Each packet can be routed to the RX in different ways – i.e., via different relay stations. This aspect does not play a major role in cellular systems, where connection can only be to the closest BS,[5] but it does play an important role in wireless ad hoc and sensor networks, where each wireless device can act as a relay for information originating from another wireless device. Appropriate routing is thus a very important aspect of sensor networks; this is described briefly in Chapter 22.

17.4.1 ALOHA

The first wireless packet radio system was the ALOHA system of the University of Hawaii; it was used to connect computer terminals in different parts of this archipelago to the central computer in Honululu. We consider in the following the case when multiple MSs try to transmit packets to a given BS; however, the principle also applies to ad hoc and sensor networks.

For an ALOHA system, each user sends packets to the BS whenever the data source makes them available. A TX does not take into consideration whether other users are already transmitting. Now, the situation can arise that several users want to transmit information simultaneously. When two TXs transmit packets at the same time, at least one of these packets suffers so much interference that it becomes unusable, and has to be retransmitted. Such collisions thus decrease the effective data rate of the system. Therefore, an ALOHA system becomes inefficient when the load is large, and thus the probability of collisions becomes large.

If the starting time of packet transmission is chosen completely at random by the TX, then the system is called a *pure* or *unslotted* ALOHA system. In the following, we determine the possible throughput of an unslotted ALOHA system. For that purpose, we first determine the possible collision time – i.e., the time during which collisions with packets of other users are possible; we assume that all packets have the same length T_p. Figure 17.5 shows that packet A (from TX 2) can suffer collisions with packets that are transmitted by TX 1 either before or after packet A. We assume here that even a short collision leads to such strong interference that the packet has to be retransmitted. In order to completely avoid collisions, a packet from TX 1 must start its transmission

---

[5] The routing of packets from one BS to another BS, or a fixed line RX, can be either packet based via a logical connection, or circuit based.

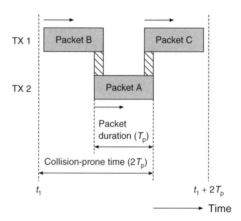

Figure 17.5   Collision-sensitive (-prone) time in an unslotted ALOHA system.
Reproduced with permission from Rappaport [1996] © IEEE.

at least T_p seconds before packet A, or has to start its transmission after packet A has finished – i.e., must start no sooner than T_p seconds after the start of packet A. The total collision-sensitive time is thus $2T_p$.[6]

For mathematical convenience, we assume now that all transmission times are completely random, and different transmitters are independent of each other. The average transmission rate of all TXs is denoted as λ_p packets per second; the offered rate $R = \lambda_p T_p$ is the normalized channel usage, which has to lie between 0 and 1. Under these assumptions, the probability that n packets are transmitted within time duration t is given by a Poisson distribution [Papoulis 1991]:

$$\Pr(n, t) = \frac{(\lambda_p t)^n \exp(-\lambda_p t)}{n!} \tag{17.10}$$

The probability that during time t zero packets are generated is thus

$$\Pr(0, t) = \exp(-\lambda_p t) \tag{17.11}$$

Effective throughput is the percentage of time during which the channel is used in a meaningful way – i.e., packets are offered, and transmitted successfully. A successful transmission occurs if no competing packets are generated during the "collision-sensitive" time of some arbitrary packet. As we have seen above, the possible collision-sensitive time is twice the packet length, so that the probability of not having a collision is $\exp(-2\lambda_p T_p)$. Effective channel throughput is thus the offered rate times the packet transmission success probability, i.e.,

$$\lambda_p T_p \exp(-2\lambda_p T_p) \tag{17.12}$$

It can easily be shown that the maximum effective throughput is $1/(2e)$, where e is Euler's number.

In a *slotted* ALOHA system, the BS prescribes a certain slot structure. Each TX has a synchronized clock that makes sure that the start of the transmission time coincides with the beginning of a slot. Thus, partial collisions cannot occur anymore: either two packets collide completely or they

---

[6] In the following, we also assume that T_p contains a guard period that accounts for the different physical runtimes of a packet in a cell.

do not collide at all. It is immediately obvious that the collision-sensitive time in such a system is T_p, so that effective throughput is

$$\lambda_p T_p \exp(-\lambda_p T_p) \tag{17.13}$$

and the maximum achievable throughput is $1/e$ – i.e., twice as large as in an unslotted ALOHA system.

17.4.2 Carrier Sense Multiple Access

Basic Principle

A TX can determine (*sense*) whether the channel is currently occupied by another user (*carrier*). This knowledge can be used to increase the efficiency of a packet-switched system: if one user is transmitting, no other user is allowed to send a signal. Such a method is called *CSMA*. It is more efficient than ALOHA, because a TX does not disturb other users that are already on the air.

The most important parameters of a CSMA system are detection delay and propagation delay. *Detection delay* is a relative measure for how long it takes a TX to determine whether the channel is currently occupied. It depends essentially on the hardware of the system, but also on the desired false alarm probability and the SNR. *Propagation delay* is the measure of how long a data packet takes to get from the MS to the BS. It can happen that at time t_1, TX 1 determines that the channel is free, and thus sends off a packet. At time t_2 another TX senses the channel. If $t_2 - t_1$ is shorter than the time it takes data packet A to get from TX 1 to TX 2, then TX 2 determines that the channel is free, and sends off data packet B. In such a case a collision occurs. This description makes it clear that detection delay and propagation delay should be much smaller than packet duration.

Implementation of Carrier Sense Multiple Access

There are different methods of implementing CSMA. The most popular are [Rappaport 1996] as follows:

- *Nonpersistent CSMA*: the TX senses the channel. If the channel is busy, the TX waits a random time duration until retransmission.
- *p-Persistent CSMA*: this method is applied in slotted channels. When a TX determines that a channel is available, it transmits with probability p in the subsequent frame; otherwise, it transmits one timeslot later.
- *1-Persistent CSMA*: the TX constantly senses the channel, until it realizes that the channel is free; then it immediately sends off the packet. This is obviously a special case of p-persistent transmission, with $p = 1$.
- *CSMA with collision detection*: in this method, a node observes whether two TXs start to transmit simultaneously. If that is the case, transmission is immediately terminated. This approach is not commonly used for wireless packet radio.
- *Data Sense Multiple Access* (DSMA): in this approach, the downlink includes a control channel, which transmits at periodic intervals a "busy/available" signal that indicates the state of the channel. If a user finds the channel to be free, it can immediately send off a data packet. Note that for peer-to-peer networks, implementation of the control channel is more difficult than in a scenario with a central node (BS).

17.4.3 Packet Reservation Multiple Access

In *Packet Reservation Multiple Access* (PRMA) each MS can send a request to transmit a data packet. A control mechanism (which can be centralized or noncentralized) answers by telling the MS when it is allowed to send off the packet. This eliminates the risk of collisions of data packets; however, the signals that carry the requests for time can collide. Furthermore, the system sacrifices some transmission capacity for the transmission of reservation requests. In order to maintain reasonable efficiency, the requests for time must be much shorter than the actual data packets. The method is also known as SRMA (*Split-channel Reservation Multiple Access*).

A variant of this method is that a user can keep the transmission medium until it has finished transmission of a datablock. Other methods hand out (time-varying) priorities, in order to avoid a situation where a single user hogs the transmission medium for a long time.

Another important method is *polling*. In this method, a BS asks (polls) one MS after another whether it wants to transmit a data packet. The shortest polling cycle occurs when no MS wants to transmit information; this is also the most inefficient case, as capacity is sacrificed for polling, and no payload is transmitted.

17.4.4 Comparison of the Methods

Figure 17.6 shows the efficiency of the various packet-switched methods. The abscissa shows channel usage, the ordinate the average packet delay. We see that ALOHA methods are only suitable if efficient channel use is of no importance. The maximum achievable capacity is only 36% in a slotted ALOHA system. CSMA and polling achieve better results.

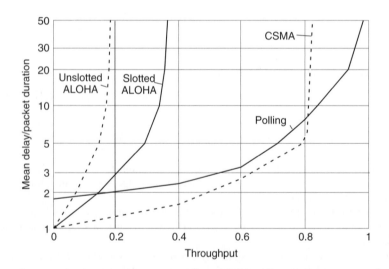

Figure 17.6 Channel usage and mean packet delay for different packet multiple-access methods.
Reproduced with permission from Oehrvik [1994] © Ericsson AB.

When comparing packet radio with TDMA and FDMA, we find that the latter schemes are very useful for the transmission of speech, since this usually requires low latency. Encoded speech data should arrive at their destination no later than about 100 ms after they have been spoken. This can easily be achieved by FDMA and TDMA systems with appropriate slot duration. Also, each TX can be certain of transmitting its data to the RX without significant blocking or delay on the line, since it has a channel (frequency or timeslot) exclusively reserved. On the other hand, TDMA and FDMA waste resources, especially for the transmission of data. A channel (timeslot) is *always* reserved for a single user even if this user does not have any data to transmit.

17.4.5 Routing for Packet Radio

We now describe the routing of packets in a packet radio system (see Figure 17.7); this principle holds both for wireless and wired systems. A packet-switched system builds up a *logical* connection between the TX and RX, but – in contrast to circuit-switched systems – not necessarily a constant *physical* connection. It is only important to get the packets from the TX to the RX in some way; the actual choice of route (the physical connection) can change with time. For this purpose, the message is broken into small pieces (packets), each of which can take a different route to the RX, depending on what transmission paths are currently available. Packets can thus take routes via different network nodes. Each of these nodes determines how it can pass the packet on. If no transmission path is currently available, then the packet is buffered until a path becomes available. This buffering can lead to considerable delays in transmission, and it can also happen that the sequence in which the data arrive is different from the sequence in which they were transmitted. For this reason, packet radio with routing over many nodes cannot easily be used for speech transmission. However, as the emergence of Voice over Internet Protocol (VoIP) telephony shows, it *is* possible to achieve packet-based voice telephony.

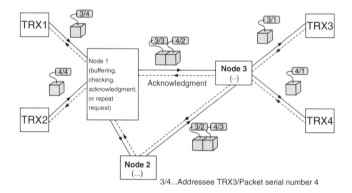

Figure 17.7   Principle behind a packet radio network. *In this figure*: TRX, transceiver.
Reproduced with permission from Oehvrik [1994] © Ericsson AB.

A data packet contains payload data as well as *routing information* (see Figure 17.8). The routing information spells out the origin and the destination of the message; furthermore, additional information about the buffering and the path that the packet has taken can be included.

Generally, routing methods can be classified in the following two categories:

- Source-driven routing: in this case, the header of the packet includes the complete route, and the nodes just follow the instructions for forwarding. The drawback is that the header can become quite long, especially for packets with little payload; this leads to a significant decrease in spectral efficiency.
- Table-driven routing: in this approach, each node stores in a table the nodes to which it should forward packets (depending on the destination address, and the node the packet came from). This

method has better spectral efficiency; the drawback is that the tables can become quite large, especially at nodes in the middle of a network.

A related topic is "route discovery" – i.e., determination of which route a packet should take from the transmitter to the receiver. Route discovery is typically done by means of special packets that are broadcast in the network, and record the quality of the links between different nodes. In order to achieve optimum performance, routing has to be changed whenever the channel between nodes changes significantly.

If a very low packet error rate is required, each node that acts as a relay stores the packets in a buffer and deletes them only after receiving an acknowledgment of successful transmission from the node it forwarded the packet to (see Figure 17.8).

A more detailed description of routing in ad-hoc networks is given in Section 22.4.

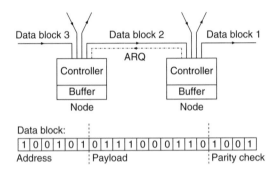

Figure 17.8 Structure of a datablock and the principle behind the buffering mechanism. ARQ, Automatic Repeat reQuest.

Reproduced with permission from Oehrvik [1994] © Ericsson AB.

17.5 Duplexing

In a cellular system, duplexing serves to separate the uplink (MS to BS) and downlink (BS to MS) transmission. We distinguish between *Time Domain Duplexing* (TDD) and *Frequency Domain Duplexing* (FDD). In TDD, uplink data are sent at times that are different from downlink transmission times (see Figure 17.9a). In FDD, uplink and downlink data are sent in different frequency bands (see Figure 17.9b).

TDD is often used in conjunction with TDMA. In such a case, the available time is divided not into N, but rather $2N$ timeslots, where each of the N users is assigned one timeslot for the

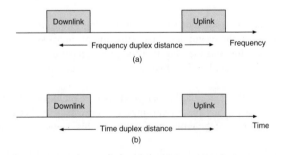

Figure 17.9 Frequency and time duplexing.

uplink, and one for the downlink. We also find that TDD can be used well in conjunction with packet radio schemes. On the other hand, the use of TDD in an FDMA system would counteract many of the advantages inherent in FDMA (continuous transmission and reception, which simplifies synchronization), while retaining the disadvantages.

FDD can be used in combination with any multiaccess method. In most cases, there is a fixed duplexing distance – i.e., a fixed frequency difference between the uplink and the downlink band.

For both TDD and FDD, it is interesting to consider the question as to whether the channel for uplink and downlink are identical. We find that for TDD, the requirement is that the duplexing time is much smaller than the coherence time, while for FDD systems, the requirement is that the frequency duplexing distance is much smaller than the coherence bandwidth (more details are given in Section 20.1.6). When considering practical system parameters, we find that the former condition can be fulfilled quite well, especially for wireless LANs that are operating in a quasi-static environment where neither BS nor MS are moving. The latter condition (for FDD) is practically never fulfilled. However, we find that there are considerable advantages when the channel for uplink and downlink is identical; we will find in Chapter 19 that it simplifies implementation of adaptive modulation, and in Chapter 20 that it is beneficial to implementation of various smart antenna systems.

On the other hand, TDD systems experience a *dead time* between transmission and reception. Imagine that a user at distance d (and thus runtime d/c_0) transmits data, and stops the transmission at time T. Then data will arrive at the BS until time $T + d/c_0$. Only now can the BS switch to transmission mode, and send out its own block of data. Since the data have to propagate back to the MS, they start arriving there at time $T + 2d/c_0$. Consequently, the MS has experienced a dead time of $2d/c_0$. This can result in a considerable loss of efficiency for the system when cell sizes are large.

A combination of FDD and TDD occurs in the case of semi-duplex systems, where transmission and reception at a single MS are distinguished by both time and frequency. While simultaneous transmission is possible when FDD is used, it requires high-quality duplex filters. By making sure that the transmission occurs *also* at a different time, the duplex filters do not have to be as good. An example of semi-duplex is GSM, where TX and RX operations use different frequency bands, and furthermore are offset by 2 timeslots from each other (see Chapter 24).

17.6 Principles of Cellular Networks

17.6.1 Reuse Distance

Let us now turn to the question of how a wireless system can cover a large area, and provide service to as many users as possible within this area. The first mobile radio systems were actually noise-limited systems with few users. Therefore, it was advantageous to put each BS on top of mountains or high towers, so that it could provide coverage for a large area. The next BS was so far away that interference was not an issue. However, this approach severely limited the number of users that could communicate simultaneously. The cellular principle, which we describe in this section, provides the solution to this problem. For this section, we use the example of FDMA as the multiaccess scheme within each cell; the same considerations hold for TDMA. CDMA systems will be discussed in Chapter 18.

In a cellular system, the coverage area is divided into many small areas called "cells." In each of these cells, there is one BS that provides coverage for this (and only this) cell area. Now, each frequency channel can be used in *multiple* cells. The question that naturally arises is: can we use each frequency channel in each cell? Typically, the answer is "no." Imagine the situation where user A is at the boundary of its assigned cell, so that distances from the "useful" BS and from a neighboring BS are the same. If the neighboring BS transmits in the same frequency channel (in order to communicate with user B in its own cell), then the SIR seen by user A is $C/I = 0\,\mathrm{dB}$. This is certainly not enough to sustain reliable communications, especially since $0\,\mathrm{dB}$ is the *median* SIR, and fading makes the situation worse 50% of the time.

The solution to this problem is to reuse a frequency channel not in every cell, but only in cells that have a certain minimum distance from each other. The normalized distance between two cells that can use the same frequency channels is called the *reuse distance, D/R*. This reuse distance can be computed from link budgets (as described in Section 3.2). We can also define a cluster of cells that all use different frequencies; therefore, there can be no co-channel interference within such a cluster. The number of cells in a cluster is called the *cluster size*. The total coverage area is divided into such clusters.

The cluster size also determines the *capacity* of the cellular system.[7] An operator that has licenses for 35 frequency channels, and uses cluster size 7, can support 5 simultaneous users in each cell. Maximization of capacity thus requires minimization of cluster size. Cluster size 1 (i.e., using each frequency in each cell) is the ultimate goal; however, we have seen above that this is not possible in an FDMA system due to the required link margins. Analog FDMA systems typically require an SIR of $18\,\mathrm{dB}$, which results in a cluster size of 21 (see below). Digital systems like GSM require less than $10\,\mathrm{dB}$, which decreases the reuse distance to 7 or less. This allows a dramatic increase in capacity, and was one of the most important reasons for the shift from analog to digital cellular systems in the early 1990s.

17.6.2   Cellshape

What shapes do cells normally take on? Let us first consider the idealized situation where path loss depends only on the distance from the BS, but not the direction. The most natural choice would be a disk (circle), as it provides constant power at the cell boundary. However, disks cannot fill a plane without either gaps or overlaps. Hexagons, on the other hand, have a shape similar to a circle, and they *can* fill up a plane, like in a beehive pattern. Thus, hexagons are usually considered the "basic" cell shape, especially for theoretical considerations.

We stress, however, that hexagonal structures are only possible under the following circumstances:

- The required traffic density is independent of the location. This condition is obviously violated whenever the population density changes.
- The terrain is completely flat, and there are no high edifices, so that path loss is influenced only by the distance from the BS. This is never fulfilled in Europe, and only rarely in the rest of the world – e.g., the American Midwest and Siberian tundra deserts. Figure 17.10 shows the terrain, and the power obtained from one BS, in a typical hilly terrain. According to the simplified considerations above, lines of equal power should be concentric circles around the BS. We see that the actual result is anything but – the power is not even necessarily a monotonic function of the distance in some directions. Thus, practical cell planning requires computer simulations or measurements of the received power. The modeling techniques described in Chapter 7 are thus a vital basis for realistic cell planning.

17.6.3   Cell Planning with Hexagonal Cells

For the case of hexagonal cells, some interesting conclusions can be drawn about the relationship between link margin and reuse distance. Consider the hexagon whose center is at the origin of

---

[7] Note that we use "capacity" in this chapter for the number of users or communication devices that can be supported simultaneously. We do *not* refer to information-theoretic capacity.

Figure 17.10   Example for received field strength in hilly terrain (western part of Vienna, Austria).
Reproduced with permission from Buehler [1994] © H. Buehler.

the coordinate system. Proceed now i hexagons in the y direction, turn $60°$ counterclockwise, and proceed k hexagons in that new direction (see Figure 17.11). This gets us to the cell whose center has the following distance from the origin:

$$D = \sqrt{3}\sqrt{(iR + \cos(60°)kR)^2 + (\sin(60°)kR)^2} \tag{17.14}$$

Note that the distance between the centers of two adjacent hexagons is $\sqrt{3}R$, where R is the distance from the center of a hexagon to its farthest corner. Also note that only integer values of i and k are possible.

The task of frequency planning is to find those values of i and k that make sure that the distance from Eq. (17.14) is larger than the required reuse distance. Of course there is an infinite manifold of such pairs – large values of i and k certainly satisfy the condition. What we want to find, however, is the pair of values that *minimizes* cluster size, and thus maximizes spectral efficiency, while still satisfying the minimum reuse distance.

Due to the hexagonal layout, the relationship between cluster size N and parameters i and k is

$$N = i^2 + ik + k^2, \quad i, k = 0, 1, 2, \ldots \tag{17.15}$$

This also establishes that not all integers are possible cluster sizes. Cluster size can only take on the numbers $N = 1, 3, 4, 7, 9, 12, 13, 16, 19, 21 \ldots$ The relationship between reuse distance D/R and cluster size results from Eqs. (17.15) and (17.14) as

$$D/R = \sqrt{3N} \tag{17.16}$$

see also Figure 17.12. Table 17.2 shows typical cluster sizes and reuse distances.

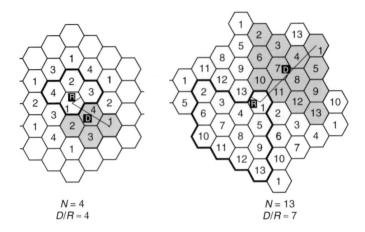

$N = 4$
$D/R \approx 4$

$N = 13$
$D/R \approx 7$

Figure 17.11 Interdependence between reuse distance D/R and cluster size N.
Reproduced with permission from Oehrvik [1994] © Ericsson AB.

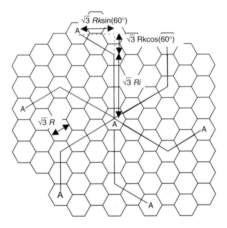

Figure 17.12 Minimum reuse distance.

Table 17.2 Typical cluster sizes and reuse distances, assuming omni-directional antennas

N	D/R	
1		CDMA
3	3	
4	3.46	
7	4.58	TDMA system (GSM)
9	5.2	
12		
13	6.24	
16	6.93	
19	7.55	
21	7.94	Analog system (NMT, AMPS)

NMT, Nordic Mobile Telephone; AMPS, Advanced Mobile Phone System.

Cell planning thus proceeds in the following steps: starting from the speci[
mum transmission quality, the link budget (Chapter 3) establishes the minim[
the desired BS and interferer. From this relationship, Eq. (17.16) provides the
it has to be the smallest integer number out of the set defined by Eq. (17.15).
obtaining nearest neighbors (move i cells into one direction, turn $60°$, move
frequencies for each cell can be determined.

Example 17.3 *Cell planning in an interference-limited system.*

The principle, as well as typical values, behind cell planning can be best understood from an example. In order to keep things simple (and work with a real system that uses pure FDMA), we consider the numbers of an analog Advanced Mobile Phone System (AMPS). Each frequency channel is 30 kHz wide, the SIR is 18 dB for satisfactory speech quality. The fading margin (for showing plus Rayleigh fading) is set to 15 dB. This implies that at the cell boundary the mean values of the signal power must be 33 dB $(2 \cdot 10^3)$ stronger than that of the interference power. The distance between the desired BS and the farthest corner of the hexagon is R, and the distance between the interfering BS and this corner is (approximately) $D - R$. Assuming further that power decreases with d^{-4}, we require that:

$$\frac{D - R}{R} = (2 \cdot 10^3)^{1/4} = 6.7 \tag{17.17}$$

so that the reuse distance must be $D/R = 7.7$. From Eq. (17.16), the reuse distance is $(3N)^{0.5}$, so that the cluster size is 19.8. The smallest $N \geq 19.8$ is $N = 21$. Now consider an operator with a licence for 5 MHz of spectrum. Having a total of $5,000/30 = 167$ possible frequency channels, only $167/21 \simeq 8$ can be used in each cell.

17.6.4 Methods for Increasing Capacity

System capacity is the most important measure for a cellular network. Methods for increasing capacity are thus an essential area of research. In the following, we give a brief overview, often referring to other chapters in this book:

1. *Increasing the amount of spectrum used*: this is the "brute force" method. It turns out to be very expensive, as spectrum is a scarce resource, and usually auctioned off by governments at very high prices. Furthermore, the total amount of spectrum assigned to wireless systems can change only very slowly; changes in spectrum assignments have to be approved by worldwide regulatory conferences, which often takes ten years or more.

2. *More efficient modulation formats and coding*: using modulation formats that require less bandwidth (higher order modulation) and/or are more resistant to interference. The former allows an increase in data rate for each user (or an increase in the number of users in a cell while keeping the data rate per user constant). However, the possible benefits of higher order modulation are limited: they are more sensitive to noise and interference (see Chapter 12), so that the reuse distance might have to be increased. The use of interference-resistant modulation allows a reduction in reuse distance. The introduction of near-capacity-achieving codes – turbo codes and low-density parity check codes (see Chapter 14) – is another way of achieving better immunity to interference, and thus increases system capacity.

3. *Better source coding*: depending on required speech quality, current speech coders need data rates between 32 kbit/s and 4 kbit/s. Better models for the properties of speech allow the data rate to be decreased without decreasing quality (see Chapter 15). Compression of data files and music/video compression also allows more users to be served.

4. *Discontinuous Voice Transmission* DTX: exploits the fact that during a phone conversation each participant talks only 50% of the time. A TDMA system can thus set up more calls than there are available timeslots. During the call, the users that are actively talking at the moment are multiplexed onto the available timeslots, while quiet users do not get assigned any radio resources.

5. *Multiuser detection*: this greatly reduces the effect of interference, and thus allows more users per cell for CDMA systems or smaller reuse distances for FDMA systems (for details see Chapter 18).

6. *Adaptive modulation and coding*: employs the knowledge at the TX of the transmission channel, and chooses the modulation format and coding rate that are "just right" for the current link situation. This approach makes better use of available power, and, among other effects, reduces interference (more details can be found in Chapter 19).

7. *Reduction of cell radius*: this is an effective, but very expensive, way of increasing capacity, as a new BS has to be built for each additional cell. For FDMA systems, it also means that the frequency planning for a large area has to be redone.[8] Furthermore, smaller cells also require more handovers for moving users, which is complicated, and reduces spectral efficiency due to the large amount of signaling information that has to be sent during a handover.

8. *Use of sector cells*: a hexagonal (or similarly shaped) cell can be divided into several (typically three) sectors. Each sector is served by one sector antenna. Thus, the number of cells has tripled, as has the number of BS *antennas*. However, the number of BS *locations* has remained the same, because the three antennas are at the same location (see Figure 17.13).

9. *Use of an overlay structure*: an overlay structure combines cells with different size and different traffic density. Therefore, some locations may be served by several BSs simultaneously. An *umbrella cell* provides basic coverage for a large area. Within that coverage area, multiple microcells are placed in areas of high traffic density. Within the coverage area of the microcells, most users are served by the microcell BS, but fast-moving users are assigned to the umbrella cell, in order to reduce the number of handovers between cells. Around 2010, so-called *femtocells* have been introduced, which are intended to be installed in apartments and offices. The BSs of those femtocells are connected to the cellular network via the internet connection (cable, Digital Subscriber Line, (DSL)) of the customer on whose premises the femtocell is located. The main purpose of such a femtocell is to provide good coverage/datarate for the apartment/office in which it is installed. The main challenge is the integration of femtocells into the overall network; since it can be anticipated that a large number of such cells will exist in the future, Self Organizing Networks (SONs) seem the best way for such integration.

(a) 120°-sectorization   (b) 60°-sectorization

Figure 17.13 Principle behind sector cells.

---

[8] For the case of hexagonal cells, algorithms are available that allow exact cell splitting without large-scale replanning of frequency assignments [Lee 1986].

10. *Multiple antennas*: these can be used to enhance capacity via different scenarios:

    (a) diversity (Chapter 13) increases the quality of the received signal, which can be exploited to increase capacity – e.g., by use of higher order modulation formats, or reduction of the reuse distance;

    (b) multiple-input multiple-output systems (Section 20.2) increase the capacity of each link;

    (c) space division multiple access (Section 20.1) allows several users in the same frequency channel in the same cell to be served.

11. *Fractional loading*: this system uses a small reuse distance, but uses only a small percentage of the available timeslots in each cell. This leads to approximately the same average capacity as the "conventional" scheme with large reuse distance and full loading of each cell. However, it has higher flexibility, as throughput can be made higher in some cells when throughput in other cells is low.

12. *Partial frequency reuse*: in this scheme, the available spectrum is divided into $N + 1$ subbands. One subband is used in *all* the cell centers, while the other subbands are used at the cell edges, employing a conventional frequency reuse (with cluster size N). The "cell edges" must be large enough so that interference from one cell center to another is sufficiently weak. We also note that the subbands need not all have the same bandwidth. Depending on the size of the cell center, the subband used in the center might be larger than the bands used at the edges.

17.7 Appendix

Please see companion website *www.wiley.com/go/molisch*

Further Reading

FDMA and TDMA systems are "classical" multiaccess schemes, and nowadays mostly analyzed in the context of specific systems (especially GSM, see Chapter 24); there is also a nice summary in Falconer et al. [1995]. An interesting performance analysis in interference is given in Xu et al. [2000]; message delays in TDMA and FDMA systems are analyzed in Rubin [1979]. The transition from TDMA/FDMA to CDMA is discussed in Sari et al. [2000]. The issue of queuing is discussed in the excellent textbook of Gross and Harris [1998]. Discussions of traffic distributions can be found, e.g., in Ashtiani et al. [2003].

The ALOHA system was first suggested in Abramson [1970], and analyzed for mobile radio applications in Namislo [1984]. A classical description of CSMA is given in Kleinrock and Tobagi [1975] and Tobagi [1980]. PRMA is described in Goodman et al. [1989].

The cellular principle and the basic concepts of cellular radio are described in detail in Lee [1995]. Though the description is by now somewhat outdated, and tries to cover both analog and digital systems, the principles have remained valid; Alouini and Goldsmith [1999] discuss more advanced aspects, including the effects of fading. Chan [1992] discusses the impact of sectorization. Frullone et al. [1996] give an overview of advanced cell-planning techniques; Woerner et al [1994] describe issues concerning the simulation of cellular systems and network planning. Frequency planning and scheduling for reducing interference (with a status of the year 2000) is described in Katzela and Naghshineh [2000]. More recent techniques, which are especially relevant for fourth-generation systems, are surveyed in Boudreau et al. [2009] and Necker [2008]. Femtocells are described, e.g., in Chandrasekhar et al. [2008], and several articles in the September 2009 issue of the IEEE Communications Magazine.

For updates and errata for this chapter, see wides.usc.edu/teaching/textbook

18

Spread Spectrum Systems

Spread spectrum techniques spread information over a very large bandwidth – specifically, a bandwidth that is much larger than the inverse of the data rate. In this chapter, we discuss various ways of providing multiple access by spreading the spectrum. We start out with the conceptually most simple approach, Frequency Hopping (FH). We then proceed to the most popular form of spread spectrum, Direct Sequence–Code Division Multiple Access (DS-CDMA). Finally, we elaborate on time-hopping impulse radio, a relatively new scheme that has gathered interest in recent years because of its application to ultrawideband systems.

We have stressed in previous chapters how important spectral efficiency is: we want to transmit as much information per available bandwidth as possible. Thus, it might seem like a strange idea to spread information over a large bandwidth in a commercial wireless system. After all, the term "spread spectrum" comes from the military area, where the main interest lies in keeping communications stealthy, safe from intercept, and safe from jamming efforts by hostile transmitters – issues that do not top the list of concerns of cellular operators.[1] It thus seems astonishing that spread spectrum approaches have attained such an important role in wireless communications.

This seeming paradox can be resolved when we recognize that different users can be spread across the spectrum in different ways. This allows multiple users to transmit in the same frequency band simultaneously; the receiver can determine which part of the total contribution comes from a specific user by looking only at signals with a specific spreading pattern. Thus, capacity (per unit bandwidth) is not necessarily decreased by using spread spectrum techniques, and can even be increased by exploiting its special features.

18.1 Frequency Hopping Multiple Access (FHMA)

18.1.1 Principle Behind Frequency Hopping

The basic thought underlying FH is to change the carrier frequency of a narrowband transmission system so that transmission is done in one frequency band only for a short while. The ratio between the bandwidth over which the carrier frequency is hopped and the narrowband transmission bandwidth is the spreading factor.

FH originated from military communications; it was invented by actress Hedi Lamar during the Second World War. It was inspired by the problem that emissions from radio transmitters could be used by the enemy to triangulate the position of transmitters, or that transmission could be jammed by the enemy with powerful (narrowband) transmitters. By changing the carrier frequency

---

[1] While security (insensitivity to intercept operations) is important, it can be achieved by cryptographic means, and does not require spectral spreading.

Wireless Communications, Second Edition   Andreas F. Molisch
© 2011 John Wiley & Sons, Ltd

frequently, the signal is in the vulnerable (observed or jammed) band only for a short while. The FH pattern has to be known to the desired receiver, but unpredictable for the enemy, making them unable to "follow" the FH. In addition to suppressing narrowband interferers, the FH also helps to mitigate the effect of deep fading dips through frequency diversity. Sometimes the system transmits on a "good" frequency – i.e., one with low attenuation between transmitter and receiver and low interference – and sometimes on a "bad" frequency – i.e., in a fading dip and/or with high interference.

There are two basic types of FH: "slow" and "fast." *Fast FH* changes the carrier frequency several times during transmission of one symbol; in other words, transmission of each separate symbol is spread over a large bandwidth. Consequently, the effects of fading or interference can be combated for each symbol separately. It follows from elementary Fourier considerations that transmission of each *part* of a symbol requires more bandwidth than that of a narrowband system. Furthermore, combining of the different contributions belonging to one symbol has to use processing that works faster than at the symbol rate. Fast FH is not in widespread use in commercial wireless systems; it has been mostly edged out by Code Division Multiple Access (CDMA).

Slow FH transmits one or several symbols on each frequency. This method is often used in conjunction with Time Division Multiple Access (TDMA): each timeslot is transmitted on a given carrier frequency; the next slot then changes to a different frequency. In such a case, the additional effort for synchronization is small, as the receiver has to synchronize for the next slot anyway. In order for the FH to be effective, interleaving and coding has to distribute information belonging to one source bit over several timeslots. Imagine simple repetition coding, where each bit is sent twice, in different timeslots (and thus on different carrier frequencies). If the first timeslot is transmitted in a deep fading dip, chances are that the second is at a frequency where channel attenuation is small; thus the information can be recovered. Slow FH is used, e.g., in the Global System for Mobile communications (GSM) (see Chapter 24).

Generally, FH leads to a *whitening* of received signal characteristics. There is implicit averaging over channel attenuation. Furthermore, at each carrier frequency, a different interferer is active, such that FH also leads to an averaging overall interferers. For many types of receivers this is advantageous, as it reduces the probability of "disastrous" scenarios, and thus decreases the required link margins. However, we will see in Sections 18.4 and 20.1 that there are some receiver structures that can actually exploit a specific (known) structure of interference in order to eliminate it. For example, the spatial structure of interferers can be exploited by smart antennas to form antenna patterns that have nulls in the direction the interference is coming from. FH can make such techniques more difficult to apply.

18.1.2   Frequency Hopping for Multiple Access (FHMA)

In the previous section, we looked at FH for the suppression of interference, and increasing frequency diversity. The price that has to be paid is that a larger bandwidth has to be used for transmission, which seems wasteful. In the following, we will show that FH can be used as a multiaccess method that is as spectrally efficient as TDMA and Frequency Division Multiple Access (FDMA). For these considerations, we distinguish between synchronized and unsynchronized systems.

Let us consider first the *synchronized* case – e.g., in the downlink of a cellular system – where the Base Station (BS) can always make sure that it emits to all Mobile Stations (MSs) at the same time. Figure 18.1 shows an example with three available bands (carrier frequencies). Clearly, during one time interval, the BS can transmit to three users simultaneously, but for ease of exposition, we assume just two active users (users A and B). During the first time interval (hopping period), the BS transmits to MS A in band 2. At the same time, band 3 is free, so the BS can transmit to MS B in that band. In the next timeframe, the BS now transmits to MS A in band 1, and MS B

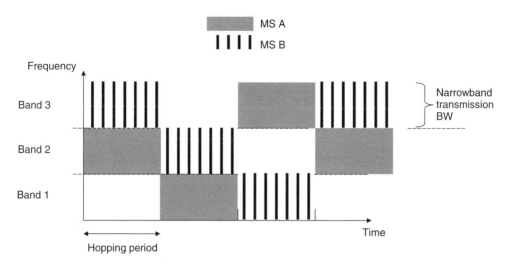

Figure 18.1 Principle behind frequency hopping for multiple access for synchronized users. *In this figure*: BW, bandwidth.

in band 2. In the third timeslot, MS A is serviced in band 3, and MS B in band 1. Then the whole sequence repeats. The signals for all the MSs use the same hopping sequence, and it can be made sure that there is never a collision between them. Thus, clearly, we have the same capacity as FDMA, with the added benefit of frequency diversity. In order to apply the same concept for the uplink, all MSs have to send their signals in such a way that they arrive at the BS synchronously, and thus recover the situation of Figure 18.1. This requires information about the runtime from each MS to the BS, which tells each MS exactly when to start transmission (timing advance, see also Chapter 24).

The situation is different when users are not synchronized – such a situation can either occur in simple networks where timing advance is not foreseen, for consideration of intercell interference,[2] or in ad hoc networks. For such a case, it is not a good idea to use the same hopping sequence for all users. Remember that, due to the lack of synchronization, any delay between the signals of different users is possible, including a zero-delay. If all users use the same hopping sequence, then such a zero-delay leads to *catastrophic collisions*, where different users interfere with each other all the time. In order to circumvent this problem, different hopping sequences are used for each user (see Figure 18.2). These sequences are designed in such a way that during each hopping cycle (i.e., one repetition of the hopping sequence; in our example, three times the hopping period), the duration of exactly one timeslot is disturbed, while the remainder of the time is guaranteed to be collision free.[3] Obviously, the performance of such a system is worse than that of a synchronized system (or an FDMA system). The design of hopping sequences that guarantee the low probability of collisions is also quite tricky, and still an active area of research.

18.2 Code Division Multiple Access

The origins of CDMA can also be traced to military communications research, especially the development of the *Direct Sequence–Spread Spectrum* (DS-SS). In Section 18.2.1, we discuss

---

[2] Users that design their uplink signals to arrive synchronously at one BS cannot be synchronous for another BS.

[3] By collision free we mean "free of collisions from user B" – other users might still interfere.

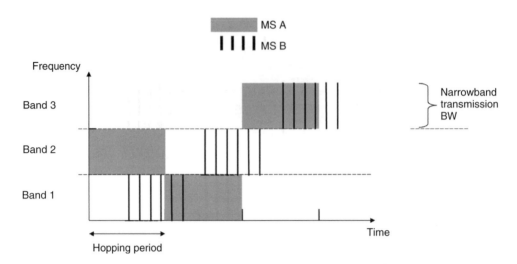

Figure 18.2 Principle behind frequency hopping for multiple access for unsynchronized users. *In this figure*: BW, bandwidth.

the basic spreading operation. Sections 18.2.2 and 18.2.3 then describe the principles and the mathematics of how a CDMA system uses DS-SS to enable multiple access. Section 18.2.4 analyzes behavior in frequency-selective channels. Sections 18.2.5 and 18.2.6 discuss synchronization and the selection of codes for CDMA. More details about the application in cellular networks can be found in Section 18.3.

18.2.1 Basic Principle Behind the Direct Sequence-Spread Spectrum

The DS-SS spreads the signal by multiplying the transmit signal by a second signal that has a very large bandwidth. The bandwidth of this total signal is approximately the same as the bandwidth of the wideband spreading signal. The ratio of the bandwidth of the new signal to that of the original signal is again known as the *spreading factor*. As the bandwidth of the spread signal is large, and the transmit power stays constant, the *power-spectral density* of the transmitted signal is very small – depending on the spreading factor and the BS–MS distance, it can lie below the noise power-spectral density. This is important in military applications, because unauthorized listeners cannot determine whether a signal is being transmitted. Authorized listeners, on the other hand, can invert the spreading operation and thus recover the narrowband signal (whose power-spectral density lies considerably *above* the noise power).

Figure 18.3 shows the block diagram of a DS-SS transmitter. The information sequence (possibly coded) is multiplied by a broadband signal that was created by modulating a sinusoidal carrier signal with a spreading sequence. This can be interpreted alternatively as multiplying each information symbol of duration T_S by a spreading sequence $p(t)$ before modulation.[4] We assume in the following that the spreading sequence is M_C *chips* long, where each chip has the duration $T_C = T_S/M_C$. As the bandwidth is the inverse of the chip duration, the bandwidth of the total signal is now also $W = 1/T_C = M_C/T_S$ – i.e., larger than the bandwidth of a narrowband-modulated signal by a factor M_C. As we assume that the spreading operation does not change the total transmit power, it also implies that the power-spectral density decreases by a factor M_C.

---

[4] This interpretation is valid if the narrowband signal and the wideband signal use the same modulation method.

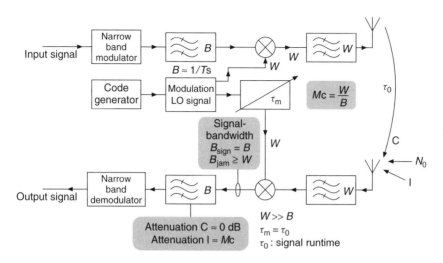

Figure 18.3 Block diagram of a direct sequence-spread-spectrum transmitter and receiver. *In this figure*: LO, Local Oscillator.

Reproduced with permission from Oehrvik [1994] © Ericsson AB.

In the receiver, we now have to invert the spreading operation. This can be achieved by correlating the received signal with the spreading sequence. This process reverses bandwidth spreading, so that after correlation, the desired signal again has a bandwidth of $1/T_S$. In addition to the desired signal, the received signal also contains noise, other wideband interferers, and possibly narrowband interferers. Note that the effective bandwidth of noise and wideband interferers is not significantly affected by the despreading operation, while narrowband interferers are actually spread over a bandwidth W. As part of despreading, the signal passes through a low-pass filter of bandwidth $B = 1/T_S$. This leaves the desired signal essentially unchanged, but reduces the power of noise, wideband interferers, and narrowband interferers by a factor M_C. At the symbol demodulator, DS-SS thus has the same Signal-to-Noise Ratio (SNR) as a narrowband system: for a narrowband system, the noise power at the demodulator is N_0/T_S. For a DS-SS system, the noise power at the receiver input is $N_0/T_C = N_0 M_C/T_S$, which is reduced by narrowband filtering (by a factor of M_C); thus, at the detector input, it is N_0/T_S. A similar effect occurs for wideband interference.

Let us next discuss the spreading signals for DS-SS systems. In order to perfectly reverse the spreading operation in the receiver by means of a correlation operation, we want the AutoCorrelation Function (ACF) of the spreading sequence to be a Dirac delta function. In such a case, the convolution of the original information sequence with the concatenation of spreader and despreader is the original sequence. We thus desire that $ACF(i)$ of $p(t)$ at times iT_C is

$$ACF(i) = \begin{cases} M_C & \text{for } i = 0 \\ 0 & \text{otherwise} \end{cases} \tag{18.1}$$

These ideal properties can only be approximated in practice. One group of suitable code sequences is a type of Pseudo Noise (PN) sequences called *maximum length sequence* (*m-sequence*). PN sequences have the following ACF:

$$ACF(i) = \begin{cases} M_C & \text{for } i = 0 \\ -1 & \text{otherwise} \end{cases} \tag{18.2}$$

see Figure 18.4.

Figure 18.4   Autocorrelation function of a m-sequence. *In this figure*: DAC, Digital to Analog Converter.
Reproduced with permission from Oehrvik [1994] © Ericsson AB.

18.2.2   Multiple Access

The Direct Sequence (DS) spreading operation itself – i.e., multiplication by the wideband signal – can be viewed as a modulation method for stealthy communications, and is as such mainly of military interest. CDMA is used on top of it, exploiting the spreading to achieve multiaccess capability. Each user is assigned a different spreading code, which determines the wideband signal that is multiplied by the information symbols. Thus, many users can transmit simultaneously in a wide band (see Figure 18.5).

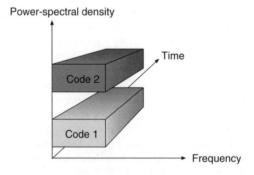

Figure 18.5   Principle behind code division multiple access.

At the receiver, the desired signal is obtained by correlating the received signal with the spreading signal of the desired user. Other users thus become wideband interferers; after passing through the despreader, the amount of interference power seen by the detector is equal to the *Cross Correlation Function* (CCF) between the spreading sequence of the interfering user and the spreading sequence of the desired user. Thus, we ideally wish for

$$CCF_{j,k}(t) = 0 \quad \text{for } j \neq k \tag{18.3}$$

for all users j and k. In other words, we require code sequences to be orthogonal. Perfect orthogonality can be achieved for at most M_C spreading sequences; this can be immediately seen by the fact that M_C orthogonal sequences span an M_C-dimensional space, and any other sequence of that duration can be represented as a linear combination.

If the spreading sequences are not orthogonal, the receiver achieves finite interference suppression – namely, suppression by a factor ACF/CCF. If the different spreading sequences are shifted m-sequences, then this suppression factor is M_C.

We can draw some important conclusions from the above description:

- The choice of spreading sequences is an essential factor for the quality of a CDMA system. Sequences have to have good ACFs (similar to a Dirac delta function) and a small cross correlation. One possible choice is m-sequences; with a shift register of length N_{reg}, it is possible to generate $2^{N_{reg}} - 1$ sequences that have an ACF given by Eq. (18.2) and $ACF/CCF = M_C$. Alternative sequences include Gold and Kasami sequences, which will be discussed in Section 18.2.6.
- CDMA requires accurate power control. If ACF/CCF has a finite value, the receiver cannot suppress interfering users perfectly. When the received interference power becomes much larger than the power from the desired transmitter, it exceeds the interference suppression capability of the despreading receiver. Thus, each MS has to adjust its power in such a way that the powers of all the signals arriving at the BS are approximately the same. Experience has shown that power control has to be accurate within about ± 1 dB in order to fully exploit the theoretical capacity of CDMA systems.

18.2.3   Mathematical Representation

In the following, we put the above qualitative description into a mathematical framework [Viterbi 1995]. We assume for this that Binary Phase Shift Keying (BPSK) modulation is used, and that perfect synchronization between the transmitter and receiver is available (discussed below in more detail). We then obtain four signal components at the receiver:

- *Desired (user) signal*: let $c_{i,k}$ represent the ith information symbol of the kth user, and $r_{i,k}$ the corresponding receive signal. Assuming that InterSymbol Interference (ISI), interchip interference (to be defined later) and noise are zero-mean processes, then the expected value of the received signal is proportional to the transmit symbol:

$$E\{r_{i,k}|c_{i,k}\} = \sqrt{(E_C)_k}c_{i,k} \int_{-\infty}^{\infty} |H_R(f)|^2 \, df \qquad (18.4)$$

where $(E_C)_k$ is the chip energy of the kth user, and $H_R(f)$ the transfer function of the receive filter normalized to $\int_{-\infty}^{\infty} |H_R(f)|^2 \, df = 1$.
- *Interchip interference*: the receive filter has a finite-duration impulse response, so that the convolution of a chip with this impulse response lasts longer than the chip itself. Thus, the signal after the receive filter exhibits interchip interference (we will see later on that delay dispersion of the channel also leads to interchip interference). If the spreading sequences are zero-mean, then the interchip interference increases the variance of the received signal by

$$(E_C)_k \sum_{i \neq 0} \left[\int_{-\infty}^{\infty} \cos(2\pi i f T_C)|H_R(f)|^2 \, df \right]^2 \qquad (18.5)$$

- *Noise* increases the variance by $N_0/2$.

- *Co Channel Interference (CCI)*: we assume that the mean of the received signal is not changed by the CCI, as the transmitted chips of interfering users are independent of the data symbols and chips of desired users. The CCI increases the variance by

$$\sum_{j \neq k} \frac{(E_C)_j}{2T_C} \int_{-\infty}^{\infty} |H_R(f)|^4 \, df \tag{18.6}$$

If we now assume that interchip interference and CCI are approximately Gaussian, then the problem of computing the error probability reduces to the standard problem of detecting a signal in Gaussian noise:

$$BER = Q\left(\sqrt{\frac{(E_C)_k M_C}{\text{Total variance}}}\right) \tag{18.7}$$

In a fading channel, the energy of the desired signal varies. For the uplink, the power control makes sure that these variations are compensated.

18.2.4 Effects of Multipath Propagation on Code Division Multiple Access

The above, strongly simplified, description of a CDMA system assumed a flat-fading channel. This assumption is violated under all practical circumstances. The basic nature of a CDMA system is to spread the signal over a large bandwidth; thus, it can be anticipated that the transfer function of the channel exhibits variations over this bandwidth.

The effect of frequency selectivity (delay dispersion) on a CDMA system can be understood by looking at the impulse response of the concatenation spreader–channel–despreader. If the channel is slowly time variant, the effective impulse response can be written as[5]

$$h_{\text{eff}}(t_i, \tau) = \tilde{p}(\tau) * h(t_i, \tau) \tag{18.8}$$

where the effective system impulse response $\tilde{p}(\tau)$ is the convolution of the transmit and receive spreading sequence:

$$\tilde{p}(\tau) = p_{TX}(\tau) * p_{RX}(\tau) = ACF(\tau) \tag{18.9}$$

In the following, we assume an ideal spreading sequence (Eq. 18.1). The despreader output then exhibits multiple peaks: more precisely, one for each Multi Path Component (MPC) that can be resolved by the receiver – i.e., spaced at least T_C apart. Each of the peaks contains information about the transmit signal. Thus, all peaks should be used in the detection process: just using the largest correlation peak would mean that we discard a lot of the arriving signal. A receiver that can use multiple correlation peaks is the so-called *Rake receiver*, which collects ("rakes up") the energy from different MPCs. As shown in Figure 18.6, a Rake receiver consists of a *bank of correlators*. Each correlator is sampled at a different time (with delay τ), and thus collects energy from the MPC with delay τ. The sample values from the correlators are then weighted and combined.

Alternatively, we can interpret the Rake receiver as a tapped delay line, whose outputs are weighted and added up. The tap delays, as well as the tap weights, are adjustable, and matched

---

[5] Note that this expression is identical to that in correlative channel sounders (see Chapter 8). Correlative channel sounders and CDMA systems have the same structure, it is just the goals that are different: the channel sounder tries to find the channel impulse response from the received signal and knowledge of the transmit data, while the CDMA receiver tries to find the transmit data from knowledge of the received signal and the channel impulse response.

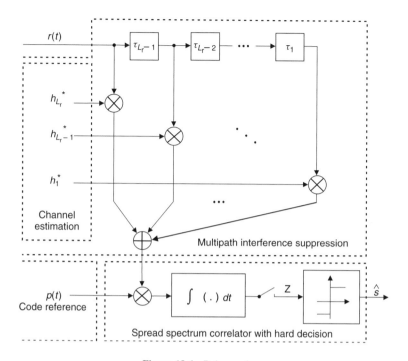

Figure 18.6   Rake receiver.
Reproduced with permission from Molisch [2000] © Prentice Hall.

to the channel. Note that the taps are usually spaced at least one chip duration apart, but there is no requirement for the taps to be spaced at regular intervals. The combination of the receiver filter and the Rake receiver constitutes a filter that is matched to the receive signal. The receive filter is matched to the transmit signal, while the Rake receiver is matched to the channel.

Independent of this interpretation, the receiver adds up the (weighted) signal from the different Rake fingers in a coherent way. As these signals correspond to different MPCs, their fading is (approximately) statistically independent – in other words, they provide delay diversity (frequency diversity). A Rake receiver is thus a diversity receiver, and all mathematical methods for the treatment of diversity remain valid. As for the performance of Rake receiver systems, we can now simply refer to the equations of Sections 13.4–13.5.

Example 18.1   *Performance of a Rake receiver: Compute the Bit Error Rate (BER) of BPSK in (i) a narrowband system and (ii) with a CDMA system that can resolve all multipaths, using a six-finger Rake receiver at a 15-dB SNR, in an International Telecommunications Union (ITU) Pedestrian-A channel.*

The tapped delay line model from an ITU Pedestrian-A channel is

$$|h(n)|_{\text{dB}} = \begin{bmatrix} 0 & -9.7 & -19.2 & -22.8 \end{bmatrix} \tag{18.10}$$

$$|h(n)| = \begin{bmatrix} 1 & 0.3273 & 0.1096 & 0.0724 \end{bmatrix} \tag{18.11}$$

The average channel gain of the flat-fading channel is

$$\Sigma|h(n)|^2 = 1 + 0.33^2 + 0.11^2 + 0.07^2 = 1.1 \tag{18.12}$$

and the transmit SNR has to be

$$\bar{\gamma}_{TX} = \frac{10^{1.5}}{1.1} = 28.75 \tag{18.13}$$

so that a receive SNR of 15 dB is achieved.

As can be found from Chapter 12, the BER for the flat-fading channel is

$$\overline{BER} = E[P_{BER}(\gamma_{Flat})] = \int_0^{\pi/2} \frac{1}{\pi} M_{\gamma_{Flat}}\left(-\frac{1}{\sin^2\theta}\right) d\theta \tag{18.14}$$

and

$$\overline{BER} = \int_0^{\pi/2} \frac{1}{\pi} \frac{\sin^2\theta}{\sin^2\theta + \bar{\gamma}_{TX}\Sigma|h(n)|^2} d\theta = 7.724 \times 10^{-3} \tag{18.15}$$

When combining the signals as done in the Rake receiver we have

$$\gamma_{Rake} = \gamma_1 + \cdots + \gamma_6 \tag{18.16}$$

Since only four MPCs carry energy, only four Rake fingers are effectively used. If the $\gamma_1, \ldots, \gamma_4$ are independent, the joint pdf of $f_{\gamma_1 \ldots \gamma_4}(\gamma_1, \ldots, \gamma_4) = f_{\gamma_1}(\gamma_1) \cdot \ldots \cdot f_{\gamma_4}(\gamma_4)$ (see also Eq. 13.39):

$$\overline{BER} = \int d\gamma_1 \, pdf_{\gamma_1}(\gamma_1) \int d\gamma_2 \, pdf_{\gamma_2}(\gamma_2) \cdots \int d\gamma_4 \, pdf_{\gamma_4}(\gamma_4) \int_0^{\pi/2} d\theta f_1(\theta) \prod_{k=1}^{N_r} \exp(-\gamma_k f_2(\theta))$$

$$= \int_0^{\pi/2} \frac{1}{\pi} \prod_{k=1}^4 \int_{\gamma_k} f_{\gamma_k}(\gamma_k) e^{\left(-\frac{\gamma_k}{\sin^2\theta}\right)} d\gamma_k \, d\theta$$

$$= \int_0^{\pi/2} \frac{1}{\pi} \prod_{k=1}^4 M_{\gamma_k}\left(-\frac{1}{\sin^2\theta}\right) d\theta \tag{18.17}$$

Thus,

$$\overline{SER} = \int_0^{\pi/2} \frac{1}{\pi} \prod_{k=1}^4 \left[\frac{\sin^2(\theta)}{\sin^2(\theta) + \bar{\gamma}_k}\right] d\theta \tag{18.18}$$

For the same transmit SNR $\bar{\gamma}_{TX} = 28.75$, we then get:

$$\overline{BER} = \int_0^{\pi/2} \frac{1}{\pi} \frac{\sin^2\theta}{\sin^2\theta + \bar{\gamma}_{TX}} \frac{\sin^2\theta}{\sin^2\theta + 0.33^2\bar{\gamma}_{TX}} \frac{\sin^2\theta}{\sin^2\theta + 0.1^2\bar{\gamma}_{TX}} \frac{\sin^2\theta}{\sin^2\theta + 0.07^2\bar{\gamma}_{TX}} d\theta$$

$$= 9.9 \times 10^{-4} \tag{18.19}$$

Another consequence of the delay diversity interpretation is the determination of the weights for the combination of Rake finger outputs. The optimum weights are the weights for maximum-ratio combining – i.e., the complex conjugates of the amplitudes of the MPC corresponding to each Rake finger. However, this is only possible if we can assign one Rake finger to each resolvable MPC (the term *all Rake* has been largely used in the literature for such a receiver). Up to $L_r = \tau_{max}/T_C$ taps, where τ_{max} is the maximum excess delay of the channel (see Chapter 6), are required in this case. Especially for outdoor environments, this number can easily exceed 20 taps. However, the number of taps that can be implemented in a practical Rake combiner is limited by power consumption, design complexity, and channel estimation. A Rake receiver that processes only a *subset* of the available L_r resolved MPCs achieves lower complexity, while still providing a performance that is better than that of a single-path receiver. The *Selective Rake* (*SRake*) receiver selects the L_b best paths (a *subset* of the L_r available resolved MPCs) and then combines the selected subset using maximum-ratio combining. This combining method is "hybrid selection: maximum ratio combining" (as discussed in Chapter 13); however, note that the average power in the different diversity branches is different. It is also noteworthy that the SRake still requires knowledge of the instantaneous values of *all* MPCs so that it can perform appropriate selection. Another possibility is the *Partial Rake* (*PRake*), which uses the first L_f MPCs. Although the performance it provides is not as good, it only needs to estimate L_f MPCs.

Another generally important problem for Rake receivers is interpath interference. Paths that have delay τ_i compared with the delay the Rake finger is tuned to are suppressed by a factor $ACF(\tau_i)/ACF(0)$, which is infinite only when the spreading sequence has ideal ACF properties. Rake receivers with nonideal spreading sequences thus suffer from interpath interference.

Finally, we note that in order for the Rake receiver to be optimal there must be no ISI – i.e., the maximum excess delay of the channel must be much smaller than T_S, though it can be larger than T_C. If there is ISI, then the receiver must have an equalizer (working on the Rake output – i.e., a signal sampled at intervals T_S) in addition to the Rake receiver. An alternative to this combination of Rake receiver and symbol-spaced equalizer is the chip-based equalizer, where an equalizer works directly on the output of the despreader sampled at the chip rate. This method is optimum, but very complex. As we showed in Chapter 16, the computational effort for equalizers increases quickly as the product of sampling frequency and channel delay spread increases.

18.2.5 Synchronization

Synchronization is one of the most important practical problems of a CDMA system. Mathematically speaking, synchronization is an estimation problem in which we determine the optimum sampling time out of an infinitely large ensemble of possible values – i.e., the continuous time. Implementation is facilitated by splitting the problem into two partial problems:

- *Acquisition*: a first step determines in which time interval (of duration T_C or $T_C/2$) the optimum sampling time lies. This is a hypothesis-testing problem: we test a finite number of hypotheses, each of which assumes that the sampling time is in a certain interval. The hypotheses can be tested in parallel or serially.
- *Tracking*: as soon as this interval has been determined, a control loop can be used to fine-tune the sampling time to its exact value.

For the acquisition phase, we use a special synchronization sequence that is shorter than the spreading sequence used during data transmission. This decreases the number of hypotheses that have to be tested, and thus decreases the time that has to be spent on synchronization. Furthermore, the synchronization sequence is designed to have especially good autocorrelation properties. For the tracking part, the normal spreading sequence used for data communications can be employed.

For many system design aspects, it is also important to know whether signals from different users are synchronized with respect to each other:

- *Synchronization within a cell*: the signals transmitted by a BS are always synchronous, as the BS has control over when to transmit them. For the uplink, synchronous arrival of the signals at the BS is usually not achieved. It would require that all MSs arrange their timing advance – i.e., when they start transmitting a code sequence – in such a way that all signals arrive simultaneously.

The timing advance would have to be accurate within one chip duration. This is too complicated for most applications, especially since movement of the MS leads to a change in the required timing advance.

- *Synchronization between BSs*: BSs can be synchronized with respect to each other. This is usually achieved by means of timing signals provided by GPS (*Global Positioning System*), so that each BS requires a GPS receiver and free line of sight to several GPS satellites. While the former is not a significant obstacle, the latter is difficult for microcells and picocells. The IS-95 system (see Chapter 25) uses such a synchronization.

18.2.6 Code Families

Selection Criteria

The selection of spreading codes has a vital influence on performance on a CDMA system. The m-sequences we used as examples up to now are among, but certainly not the only, possible choices. In general, the quality of spreading codes is determined by the following properties:

- *Autocorrelation*: ideally, $ACF(0)$ should be equal to the number of chips per symbol M_C, and zero at all other instances. For m-sequences $ACF(0) = M_C$, and $ACF(n) = -1$ for $n \neq 0$. Good properties of the ACF are also useful for synchronization. Furthermore, as discussed above, autocorrelation properties influence interchip interference in a Rake receiver: the output of the correlator is the sum of the ACF_S of the delayed echoes; a spurious peak in the ACF looks like an additional MPC.[6]
- *Cross-correlation*: ideally, all codes should be orthogonal to each other, so that interference from other users can be completely suppressed. In Section 18.2.6.3, we discuss a family of orthogonal codes. For unsynchronized systems, orthogonality must be fulfilled for arbitrary delays between the different users. Note that bandwidth spreading and separation of users can be done by different codes; in this case we distinguish between spreading codes and scrambling codes.
- *Number of codes*: a CDMA system should allow simultaneous communications of as many users as possible. This implies that a large number of codes have to be available. The number of orthogonal codes is limited by M_C. If more codes are required, worse cross-correlation properties have to be accepted. The situation is complicated further by the fact that codes in adjacent cells have to be different: after all, it is only the codes that distinguish different users. Now, if cell A has M_C users with all orthogonal codes, then the codes in cell B cannot be orthogonal to the codes in cell A; therefore, intercell interference cannot be completely suppressed. The codes in cell B can, however, be chosen in such a way that the interference to users in the original cell becomes noise-like; this approach then requires code planning instead of frequency planning. An alternative approach is the creation of a large number of codes with suboptimum CCFs (see the next subsection), and assigning them to wherever they are needed. In such a case, no code planning is required; however, system capacity is lower.

Pseudo Noise Sequences

The spreading sequences most frequently used for the uplink of CDMA systems are m-sequences, Gold sequences, and Kasami sequences. The autocorrelation properties of m-sequences are excellent; shifted versions of an m-sequence are again valid codewords with almost ideal cross-correlation

---

[6] As the ACF is known to the receiver, spurious peaks belonging to the first-detected MPC can be subtracted from the remaining signal, and thus their effect can be eliminated (this is essentially an interference cancellation technique, as discussed in Section 18.4).

properties. Gold sequences are created by the appropriate combination of m-sequences; the ACFs of Gold sequences can take on three possible values; both the offpeak values of the ACF and the CCF can be upper-bounded.

An even more general family of sequences is the Kasami sequences. In the following we distinguish between S (*small*), L (*large*), and VL (*very large*) Kasami sequences. The letters describe the number of codes within the family. S-Kasami sequences have the best CCFs, but only a rather small number of such sequences exists. VL-Kasami sequences have the worst CCF, but there is an almost unlimited number of such sequences. Table 18.1 shows the properties of the different code families.

Table 18.1 Properties of code division multiple access codes

Sequence	Number of codes	Maximum CCF/dB	Comment
m-Sequence	$2^{N_{reg}} - 1$		Good ACF
Gold	$2^{N_{reg}} + 1$	$\approx -3N_{reg}/2 + 1.5$	
S-Kasami	$2^{N_{reg}/2}$	$\approx -3N_{reg}/2$	Best CCF of all Kasami sequences
L-Kasami	$2^{N_{reg}/2}(2^{N_{reg}} + 1)$	$\approx -3N_{reg}/2 + 3$	
VL-Kasami	$2^{N_{reg}/2}(2^{N_{reg}} + 1)^2$	$\approx -3N_{reg}/2 + 6$	Almost unlimited number

N_{reg} is the size of the shift register used to create the sequences.

Walsh–Hadamard Codes

In the downlink, signals belonging to different users can be made completely synchronous as they are all emitted by the same transmitter (the BS). Under these circumstances, a family of codes that are all completely orthogonal to each other is given by Walsh–Hadamard matrices. Define the $n + 1$-order Hadamard matrix $\mathbf{H}_{had}^{(n+1)}$ in terms of the nth order matrix:

$$\mathbf{H}_{had}^{(n+1)} = \begin{pmatrix} \mathbf{H}_{had}^{(n)} & \mathbf{H}_{had}^{(n)} \\ \mathbf{H}_{had}^{(n)} & \overline{\mathbf{H}}_{had}^{(n)} \end{pmatrix} \tag{18.20}$$

where $\overline{\mathbf{H}}$ is the modulo-2 complement of $\mathbf{H}$. The recursive equation is initialized as

$$\mathbf{H}_{had}^{(1)} = \begin{pmatrix} 1 & 1 \\ 1 & -1 \end{pmatrix} \tag{18.21}$$

The columns of this matrix represent all possible Walsh–Hadamard codewords of length 2; it is immediately obvious that the columns are orthogonal to each other. From the recursion equation we find that $\mathbf{H}_{had}^{(2)}$ is

$$\mathbf{H}_{had}^{(2)} = \begin{pmatrix} 1 & 1 & 1 & 1 \\ 1 & -1 & 1 & -1 \\ 1 & 1 & -1 & -1 \\ 1 & -1 & -1 & 1 \end{pmatrix} \tag{18.22}$$

The columns of this matrix are all possible codewords of duration four; again it is easy to show that they are all orthogonal to each other. Further iterations give additional codewords, each of which is twice as long as that of the preceding matrix.

Orthogonal codes lead to perfect multiuser suppression at the receiver if the signal is transmitted over an Additive White Gaussian Noise (AWGN) channel. Delay dispersion destroys the orthogonality of the codes. The receiver can then either accept the additional interference (described by an *orthogonality factor*), or send the received signal through a chip-spaced equalizer that eliminates delay dispersion before correlation (and thus user separation) is performed.

Example 18.2 *Orthogonality of Orthogonal Variable Spreading Factor (OVSF) codes in frequency-selective channels: the codewords* $\begin{bmatrix} 1 & 1 & 1 & 1 \end{bmatrix}$ *and* $\begin{bmatrix} 1 & -1 & 1 & -1 \end{bmatrix}$ *are sent through a channel whose impulse response is* $(0.8, -0.6)$. *Assuming that* $\begin{bmatrix} 1 & 1 & 1 & 1 \end{bmatrix}$ *was sent, compute the correlation coefficient of the signal arriving at the receiver with the possible transmit signals.*

Assume that the BS communicates with two MSs and transmits spreading codes, $t_1(n) = \begin{bmatrix} 1 & 1 & 1 & 1 \end{bmatrix}$ and $t_2(n) = \begin{bmatrix} 1 & -1 & 1 & -1 \end{bmatrix}$. $t_1(n)$ and $t_2(n)$ are orthogonal to each other; e.g.,

$$t_1(n) \cdot t_2(n)^T = 0 \tag{18.23}$$

However, the time dispersive channel (multipath channel) $h(n) = \begin{bmatrix} 0.8 & -0.6 \end{bmatrix}$ will affect the orthogonality between the two codes. The received signal is the linear convolution of the impulse response and the transmitted signal (assuming that the channel is linear and time invariant); e.g.,

$$r(n) = t(n) * h(n) = \sum_{k=-\infty}^{\infty} t(k)h(n - k) \tag{18.24}$$

The received signal when transmitting $t_1(n)$ is then

$$r_1(0) = 1 \cdot 0.8 = 0.8 \tag{18.25}$$

$$r_1(1) = 1 \cdot (-0.6) + 1 \cdot 0.8 = 0.2 \tag{18.26}$$

$$r_1(2) = 1 \cdot (-0.6) + 1 \cdot 0.8 = 0.2 \tag{18.27}$$

$$r_1(3) = 1 \cdot (-0.6) + 1 \cdot 0.8 = 0.2 \tag{18.28}$$

$$r_1(4) = 1 \cdot (-0.6) = -0.6 \tag{18.29}$$

Thus, by correlating the received signals with $t_1(n)$ and $t_2(n)$ we got for $r_1(n)$:

$$\rho_{r_1 t_1} = \sum_{n=0}^{3} r_1(n)t_1(n) \tag{18.30}$$

$$= \begin{bmatrix} 0.8 & 0.2 & 0.2 & 0.2 \end{bmatrix} \begin{bmatrix} 1 & 1 & 1 & 1 \end{bmatrix}^T = 1.4 \tag{18.31}$$

$$\rho_{r_1 t_2} = \sum_{n=0}^{3} r_1(n)t_2(n) \tag{18.32}$$

$$= \begin{bmatrix} 0.8 & 0.2 & 0.2 & 0.2 \end{bmatrix} \begin{bmatrix} 1 & -1 & 1 & -1 \end{bmatrix}^T = 0.6 \neq 0 \tag{18.33}$$

In a similar manner, the signal that is received when $t_2(n)$ is transmitted is $r_2(n) = \begin{bmatrix} 0.8 & -1.4 & 1.4 & -1.4 & 0.6 \end{bmatrix}$. The correlation coefficient:

$$\rho_{r_2 t_1} = \sum_{n=0}^{3} r_2(n)t_1(n) = 0.6 \neq 0 \tag{18.34}$$

$$\rho_{r_2 t_2} = \sum_{n=0}^{3} r_2(n)t_2(n) = 5 \tag{18.35}$$

Correlation between the two received signals is

$$\rho_{r_1 r_2} = \sum_{n=0}^{3} r_1(n) r_2(n) \neq 0 \tag{18.36}$$

An additional challenge arises if different users require different data rates, so that codes of different length need to be used for the spreading. *Orthogonal Variable Spreading Factor (OVSF)* codes are a class of codes that fulfills these conditions; they are derived from Walsh–Hadamard codes.

Let us first define what we mean by orthogonality for codes of different duration. The chip duration is the same for all codes: it is given by the available system bandwidth, and independent of the data rate to be transmitted. Consider a code A that is two chips long $(1, 1)$, and a code B that is four chips long $(1, -1, -1, 1)$. The output of correlator A has to be zero if code B is at the input of the correlator. Thus, the correlation between code A and the first part of code B has to be zero, which is true: $1 \ 1 + 1 \cdot (-1) = 0$. Similarly, correlation of code A with the second part of code B has to be zero $1 \cdot (-1) + 1 \cdot 1 = 0$.

Let us now write all codewords of different Walsh–Hadamard matrices into a "code tree" (see Figure 18.7). All codes within one level of the tree (same duration of codes) are orthogonal to each other. Codes of different duration A, B are only orthogonal if they are in different branches of the tree. They are *not* orthogonal to each other if one code is a "mother code" of the second code – i.e., code A lies on the path from the "root" of the code tree to code B. Examples of such codes are $p_{2,2}$ and $p_{4,4}$ in Figure 18.7, whereas codes $p_{2,2}$ and $p_{4,1}$ are orthogonal to each other.

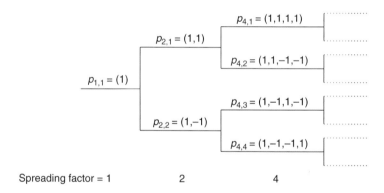

Figure 18.7 Code tree of orthogonal-variable-spreading-factor codes.

18.3 Cellular Code-Division-Multiple-Access Systems

18.3.1 *Principle Behind Code Division Multiple Access – Revisited*

When analyzing the multiaccess capability of a system, we are essentially asking the question "What prevents us from serving an infinite number of users at the same time?". In a TDMA/FDMA system, the answer is the limited number of available timeslots/frequencies. Users can occupy those slots, and not interfere with each other. But when all possible timeslots have been assigned to users, there are no longer free resources available, and no further users can be accepted into the system.

In a CDMA system, this mechanism is subtly different. We first analyze the uplink in the following, where spreading codes are imperfect, but there is a large number of them. Different users are distinguished by different spreading codes; however, as user separation is not perfect, each user in the cell contributes interference to all other users. Thus, as the number of users increases, the interference for each user increases as well. Consequently, transmission quality decreases gradually (*graceful degradation*), until users find the quality too bad to place (or continue) calls. Consequently, CDMA puts a soft limit on the number of users, not a hard limit like TDMA. Therefore, the number of users in a system depends critically on the Signal-to-Interference-and-Noise Ratio (SINR) required by the receiver. It also implies that any increase in SINR at the receiver, or reduction in the required SINR, can be immediately translated into higher capacity.

Most interference stems from within the same cell as the desired user, and is thus termed *intracell interference*. Total intracell interference is the sum of many independent contributions, and thus behaves approximately like Gaussian noise. Therefore, it causes effects that are similar to thermal noise. It is often described by *noise rise* – i.e., the increase in "effective" noise power (sum of noise and interference power) compared with the noise alone $(N_0 + I_0)/N_0$. Figure 18.8 shows an example of noise rise as a function of system load; here *system load* is defined as the number of active users, compared with the maximum possible number M_C. We see that noise rise becomes very strong as system load approaches 100%. A cell is thus often judged to be "full" if noise rise is 6 dB. However, as mentioned above, there is no hard limit to the number of active users.

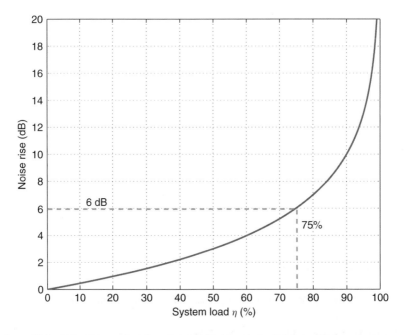

Figure 18.8   Noise rise as a function of system load in a code-division-multiple-access system.
Reproduced with permission from Neubauer et al. [2001] © T. Neubauer.

Some interference is from neighboring cells, and thus called *intercell interference*. A key property of a CDMA system is that it uses *universal frequency reuse* (also known as *reuse distance one*). In other words, the same frequency band is used in all cells; users in different cells are distinguished only by different codes. As discussed in Section 18.2.6.1, the amount of interference is mostly determined by the codes that are used in the different cells.

Many of the advantages of CDMA are related to the fact that interference behaves almost like noise, especially in the uplink. This noise-like behavior is due to several reasons:

- The number of users (and therefore, of interferers) in each cell is large.
- Power control makes sure that all intracell signals arriving at the BS have approximately the same strength (see also below).
- Interference from neighboring cells also comes from a large number of users. Spreading codes are designed in such a way that all signals in one cell have approximately the same cross-correlation with each signal in all neighboring cells. Note that this implies that we cannot simply reuse the same codeset in each cell; otherwise, there would be one user in the neighboring cell that would contribute much more interference (the user that uses the same code as the desired user in the desired cell).

Due to the above effects, total interference power shows very little fluctuations. At the same time, the power control makes sure that the signal strength from the desired user is always constant. The SINR is thus constant, and no fading margin has to be used in the link budget. However, note that making interference as Gaussian as possible is not always the best strategy for maximizing data throughput; multiuser detection (Section 18.4) actively exploits structure in interference and works best when there are only a few strong interferers.

In the downlink, spreading codes are orthogonal, so that (at least in theory) different users can be separated completely. However, in this case, the number of users in the cell is limited by the number of Walsh–Hadamard codes. The situation then becomes similar to a TDMA system: if the M_C available Walsh–Hadamard codes are used up, then no further users can be served. Furthermore, there is also interference from the Walsh–Hadamard codes of neighboring cells. The situation can be improved by multiplying the Walsh–Hadamard codes by a *scrambling code*. Walsh–Hadamard codes that are multiplied by the same scrambling code remain orthogonal; codes that are multiplied by different scrambling codes do not interfere catastrophically. Therefore, different cells use different scrambling codes (see also Chapter 26).

Downlink intercell interference does *not* come from a large number of independent sources. All the interference comes from the BSs in the vicinity of the considered MS – i.e., a few (at most six) BSs constitute the dominant source of interference. The fact that each of these BSs transmits signals to a large number of users within their cell does not alter this fact: the interfering signal still comes from a single geographical source that has a single propagation channel to the victim MS. Consequently, the downlink might require a fading margin in its link budget.

18.3.2   *Power Control*

As we mentioned above, power control is important to make sure that the desired user has a time-invariant signal strength, and that the interference from other users becomes noise-like. For further considerations, we have to distinguish between power control for the uplink and that for the downlink:

- *Power control in the uplink*: for the uplink, power control is vital for the proper operation of CDMA. Power control is done by a closed control loop: the MS first sends with a certain power, the BS then tells the MS whether the power was too high or too low, and the MS adjusts its power accordingly. The bandwidth of the control loop has to be chosen so that it can compensate for small-scale fading – i.e., has to be on the order of the Doppler frequency. Due to time variations of the channel and noise in the channel estimate, there is a remaining variance in the powers arriving at the BS; this variance is typically on the order of 1.5–2.5 dB, while the dynamic range that has to be compensated is 60 dB or more. This variance leads to a reduction in the capacity of a CDMA cellular system of up to 20% compared with the case when there is ideal power control.

Note that an open control loop (where the MS adjusts its transmit power based on its own channel estimate) cannot be used to compensate for small-scale fading in a Frequency Domain Duplexing (FDD) system: the channel seen by the MS (when it receives signals from the BS) is different from the channel it transmits to (see Section 17.5). However, an open loop can be used in conjunction with a closed loop. The open loop compensates for large-scale variations in the channel (path loss and shadowing), which are approximately the same at uplink and downlink frequencies. The closed loop is then used to compensate for small-scale variations.

- *Power control in the downlink*: for the downlink, power control is not necessary for CDMA to function: all signals from the BS arrive at one MS with the same power (the channel is the same for all signals). However, it can be advantageous to still use power control in order to keep the total transmit power low. Decreasing the transmit power for all users within a cell by the same amount leaves unchanged the ratio of desired signal power to intracell interference – i.e., interference from signals destined for other users in the cell. However, it does decrease the power of total interference to other cells. On the other hand, we cannot decrease signal power arbitrarily, as the SNR must not fall below a threshold. The goal of downlink power control is thus to minimize the total transmit power while keeping the BER or SINR level above a given threshold. The accuracy of downlink power control need not be as high as for the uplink; for many cases, open loop control is sufficient.

It is worth remembering that the power control of users in adjacent cells does not give constant power of the intercell interference. A user in an adjacent cell is power controlled by its own BSs – in other words, power is adjusted in such a way that the signal arriving at its desired BS is constant. However, it "sees" a completely different channel to the undesired BS, with temporal fluctuations that the desired BS neither knows nor cares about. Consequently, intercell interference is temporally variant.

Interference power from all users is the same only if all users employ the same data rate. Users with higher data rates contribute more interference power; high-data-rate users can thus be a dominant source of interference. This fact can be understood most easily when we increase the data rate of a user by assigning multiple spreading codes to him. In this case, it is obvious that the interference this user contributes increases linearly with the data rate. While this situation did not occur for second-generation cellular systems, which had only speech users, it is certainly relevant for third-generation cellular systems, which foresee high-data-rate services.

It should also be noted that power control is not an exclusive property of CDMA systems; it can also be used for FDMA or TDMA systems, where it decreases intercell interference and thus improves capacity. The major difference is that power control is *necessary* for CDMA, while it is *optional* for TMDA/FDMA.

Soft Handover

As all cells use the same frequencies, an MS can have contact with two BSs at the same time. If an MS is close to a cell boundary, it receives signals from two or more BSs (see Figure 18.9) and also transmits to all of these BSs. Signals coming from different MSs have different delays, but this can be compensated by the Rake receiver, and signals from different cells can be added coherently.[7]

---

[7] Note that different cells might use different codes. This is not a major problem; it just means that (for the downlink) different correlators in the fingers of the Rake receiver have to use different spreading sequences.

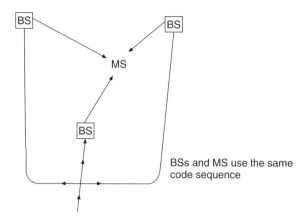

Figure 18.9 Principle behind soft handover.
Reproduced with permission from Oehrvik [1994] © Ericsson AB.

This is in contrast to the hard handover in an FDMA-based system, where an MS can have contact with only one BS at a time, because it can communicate only on one frequency at a time.

Consider now an MS that starts in cell A, but has already established a link to BS B as well. At the outset, the MS gets the strongest signal from BS A. As it starts to move toward cell B, the signal from BS A becomes weaker, and the signal from BS B becomes stronger, until the system decides to drop the link to BS A. Soft handover dramatically improves performance while the MS is near the borderline of the two cells, as it provides diversity (macrodiversity) that can combat large-scale as well as small-scale fading. On the downside, soft handover decreases the available capacity in the downlink: one MS requires resources (Walsh–Hadamard codes) in two cells at the same time, while the user talks – and pays – only once. Furthermore, soft handover increases the amount of signaling that is required between BSs.

18.3.3 Methods for Capacity Increases

- *Quiet periods during speech transmission*: for speech transmission, CDMA makes implicit use of the fact that a person does not talk continuously, but rather only about 50% of the time, the remainder of the time (s)he listens to the other participant. In addition, there are pauses between words and even syllables, so that the ratio of "talk time" to "total time of a call" is about 0.4. During quiet periods, no signal, or a signal with a very low data rate, has to be transmitted.[8] In a CDMA system, not transmitting information leads to a decrease in total transmitted power, and thus interference in the system. But we have already seen above that decreasing the interference power allows additional users to place calls. Of course, there can be a worst case scenario where all users in a cell are talking simultaneously, but, statistically speaking, this is highly improbable, especially when the number of users is large. Thus, pauses in the conversation can be used very efficiently by CDMA in order to improve capacity (compare also discontinuous transmission in TDMA systems).
- *Flexible data rate*: in an FDMA (TDMA) system, a user can occupy either one frequency (times-lot), or integer multiples thereof. In a CDMA system, arbitrary data rates can be transmitted by

---

[8] Actually, most systems transmit *comfort noise* during this time – i.e., some background noise. People speaking into a telephone feel uncomfortable (think the connection has been interrupted) if they cannot hear any sound while they talk.

an appropriate choice of spreading sequences. This is not important for speech communications, which operate at a fixed data rate. For data transmission, however, the flexible data rate allows for better exploitation of the available spectrum.

- *Soft capacity*: the capacity of a CDMA system can vary from cell to cell. If a given cell adds more users, it increases interference to other cells. It is thus possible to have some cells with high capacity, and some with lower; furthermore, this can change dynamically, as traffic changes. This concept is known as *breathing cells*.
- *Error correction coding*: the drawback of error correction coding is that the data rate that is to be transmitted is increased, which decreases spectral efficiency. On the other hand, CDMA consciously increases the amount of data to be transmitted. It is thus possible to include error correction coding without decreasing spectral efficiency; in other words, different users are distinguished by different error correction codes (coding by spreading). Note, however, that commercial systems (UMTS, Chapter 26) do not use this approach; they have separate error correction and spreading.

18.3.4   Combination with Other Multiaccess Methods

CDMA has advantages compared with TDMA and FDMA, especially with respect to flexibility, while it also has some drawbacks, like complexity. It is thus obvious to combine CDMA with other multiaccess methods in order to obtain the "best of both worlds." The most popular solution is a combination of CDMA with FDMA: the total available bandwidth is divided into multiple subbands, in each of which CDMA is used as the multiaccess method. Clearly, frequency diversity in such a system is lower than for the case where spreading is done over the whole bandwidth. On the positive side, the processing speed of the transmitters and receivers can be lower, as the chip rate is lower. The approach is used, e.g., in IS-95 (which uses 1.25-MHz-wide bands) and Universal Mobile Telecommunications System (UMTS), which uses 5-MHz-wide subbands.

Another combination is CDMA with TDMA. Each user can be assigned one timeslot (as in a TDMA system), while the users in different cells are distinguished by different spreading codes (instead of different frequencies). Another possibility is to combine several timeslots, and build up a narrowband CDMA system within them. This system works best when adding a CDMA component to an existing TDMA system (e.g., the TDD mode of UMTS).

18.4   Multiuser Detection

18.4.1   Introduction

Basic Idea Underlying Multiuser Detection

Multiuser detection is based on the idea of detecting interference, and exploiting the resulting knowledge to mitigate its effect on the desired signal. Up to 1986, it had been an established belief that interference from other users cannot be mitigated, and that in the best case, interference behaves like AWGN. Correct detection and demodulation in the presence of strong interference was thus considered impossible, just as it is impossible to correctly detect signals in strong noise. The work of Poor, Verdu and others – summarized in Verdu [1998] – demonstrated that it is actually possible to exploit the *structure* of multiuser interference to combat its effect. Using such a strategy, interference is *less* detrimental than Gaussian noise. Multiuser detection was intensely researched during the 1990s, and the first practical systems using this approach were introduced in the subsequent decade.

The conceptually simplest version of multiuser detection is serial *interference cancellation*. Consider a system where a (single) interfering signal is much stronger than the desired signal. The receiver first detects and demodulates this strongest signal. This signal has a good SINR, and can thus hopefully be detected without errors. Its effect is then subtracted (cancelled) from the total received signal. The receiver then detects the desired signal within the "cleaned-up" signal. As this cleaned-up signal consists only of the desired signal and the noise, the SINR is good, and detection can be done correctly. This example makes clear that detection at an SIR of less than 0 dB is possible with multiuser detection.

Other multiuser structures include Maximum Likelihood Sequence Estimation (MLSE) detectors, which try to perform optimum detection for the signals of *all* users simultaneously. These receivers show very good performance, but their complexity is usually prohibitive, as the effort increases exponentially with the number of users to be detected. The performance of MLSE can also be approximated by receivers that use the turbo principle (see Section 14.5).

A general classification of multiuser detection distinguishes between linear and nonlinear detectors. The former class includes the decorrelation receiver and the Minimum Mean Square Error (MMSE) receiver; the latter includes MLSE, interference cancellation, decision feedback receivers, and turbo receivers.

Assumptions

In our basic description here, we make a number of simplifying assumptions:

- The receiver has perfect knowledge of the channel from the interferer to the receiver. This is obviously a best case. Consider the situation in a serial interference cancellation receiver: only if the receiver has perfect channel knowledge of the interfering signal can it perfectly subtract it from the total signal. The stronger the interference, the larger is the impact of any channel estimation error on the subtraction process.
- All users employ CDMA as a multiaccess scheme. We stress, however, that multiuser detection is also possible for other multiaccess methods, like TDMA, and can be used, e.g., to decrease the reuse distance in a TDMA network. We will also see later that there is a strong similarity between multiuser detection and detection of spatial-multiplexing systems (Section 20.2).
- All users are synchronized.
- We only treat multiuser detection at the receiver. A related topic would be the design of transmission schemes so that interference at different receivers is mitigated. A surprising result of information theory is that if the transmitter knows the channel and the interference, then the capacity of the interfered channel is the same as the capacity of the noninterfered channel. This capacity can be realized by appropriate coding strategies, called "writing on dirty paper" coding [Peel 2003]. Such coding could be exploited to increase the capacity of the downlink in cellular systems.

18.4.2   Linear Multiuser Detectors

The block diagram of a linear multiuser system is sketched in Figure 18.10. It first estimates the signals from different users by despreading using the spreading sequences of the different users.[9] Note that this requires a number of parallel despreaders, each operating with a different spreading sequence. The outputs from these despreaders are then linearly combined. This combination step can be considered as filtering using a matrix filter, and used for elimination of interference. This approach shows a strong similarity to linear equalization for elimination of ISI. Therefore, concepts like zero-forcing, Wiener filtering, etc., which we discussed in Chapter 16, are also encountered in this context.

---

[9] This step might also include the combination of signals from multiple Rake fingers, and the elements of an antenna array.

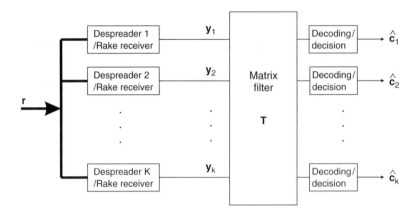

Figure 18.10   Linear multiuser detector.

Decorrelation Receiver

The decorrelation receiver is the simplest means of multiuser detection; it is the equivalent of a zero-forcing equalizer. We write the output of the receiver filter (after despreading) as

$$y = Rc + n \tag{18.37}$$

where the correlation matrix R can include possible antenna and/or delay diversity; n is the noise vector. Estimation of the symbols is now obtained simply by filtering with a matrix filter $T = R^{-1}$:

$$\hat{c} = R^{-1}y = c + R^{-1}n \tag{18.38}$$

The advantage of this approach is its simplicity, and the fact that it is not necessary to know the received amplitudes. Only the correlation matrix R needs to be determined. The drawback lies in noise enhancement (compare, again, the zero-forcing equalizer). The worse the conditioning of the correlation matrix, the more the noise is increased.

MMSE Receiver

Just like for the MMSE equalizer, the MMSE multiuser detector strikes a balance between interference suppression and noise enhancement. A measure for total disturbance is the mean quadratic error $E\{|c - \hat{c}|^2\}$. The matrix filter is thus

$$T = \left[R + \sigma_n^2 I\right]^{-1} \tag{18.39}$$

The MMSE does not lead to complete suppression of interference, but due to its smaller noise enhancement the signal distortions are still smaller than for the decorrelation receiver.

18.4.3   Nonlinear Multiuser Detectors

Linear multiuser detectors ignore part of the structure of transmit signals: they allow any continuous value for the estimate $\hat{c}$ of the transmit signal. Thus, they ignore the fact that the transmit signal can only contain members of a finite transmit alphabet. Nonlinear detectors also exploit this information.

Multiuser Maximum-Likelihood Sequence Estimation

The structure of multiuser MLSE is the same as that for conventional MLSE (usually decoded by a Viterbi detector). If we have K users, then M^K combinations of transmit symbols can be sent out by the users. The number of possible states in the trellis diagram thus also increases exponentially with the number of users. For this reason, multiuser MLSE is not used in practice. However, it gives important insights into the performance limits of multiuser detection (for more details, see Verdu [1998]).

Successive Interference Cancellation

Successive Interference Cancellation (SIC) detects users in the sequence of their signal strength. The signal of each user is subtracted from the total signal before the next user is detected (see Figure 18.11). The SIC is thus a special case of a decision feedback receiver. The receiver works the following way: the sum of all signals is received, and it is despread using the different spreading codes of each user. Then, the strongest signal is detected and decided on, so that we get the original bitstream, unaffected by noise or interference. This bitstream is then respread, and subtracted from the total signal. The "cleaned-up" signal is then sent through the despreaders again, the strongest user within this new signal is detected, respread, and subtracted. The process is repeated until the last user has been detected. Note that error propagation can seriously affect the performance of SIC: if the receiver decides wrongly about one bit, it subtracts the wrong contribution from the total signal, and the residual signal, which is further processed, suffers from more, not less, interference.

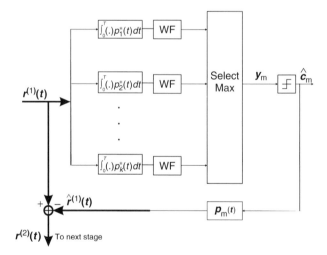

Figure 18.11 One stage of a successive interference cancellation receiver. Received signal **r** from K users is correlated with the spreading sequence of the kth user; the largest signal is selected and its impact – i.e., the respread signal – is subtracted from the received signal. WF denotes the Whitening Filter. The impact of delay dispersion is not shown.

There are two possibilities for subtracting interference: "hard" and "soft." For hard subtraction, interference is subtracted completely; for soft subtraction, only a scaled-down version of the signal is subtracted. This does not lead to complete cancellation of the signal, but makes error propagation less of an issue. Another method for reducing error propagation is to make sure that bits are not only demodulated but also decoded (i.e., undergo error correction decoding), then recoded, remodulated, and respread before subtraction. As the error probability after decoding is much lower than before, this obviously reduces error propagation. On the downside, the decoding process increases the latency of the detection process.

Parallel Interference Cancellation

Instead of subtracting interference in a serial (user-by-user) fashion, we can also cancel all users simultaneously. To achieve this, the first step makes a (hard or soft) decision for all users based on the total received signal. The signals are then respread, and contributions from all interferers to the total signal are subtracted. Note that a different group of interferers is active for each user: For user 1, users $2 \cdots K$ act as interferers; for user 2, users 1 and $3 \cdots K$ act as interferers, and so on. The next stage of the canceller then uses the "cleaned-up" signals as a basis for a decision, and again performs remodulation and subtraction. The process is repeated until decisions no longer change from iteration to iteration, or until a certain number of iterations are reached (see Figure 18.12).

Error propagation can be a significant problem for parallel interference subtraction as well. Note that only the first stage would be necessary if all decisions were correct. However, each signal in the first stage has a bad signal-to-interference ratio, so that there is a high probability of only the strongest signal being decided correctly. Moreover, wrong decisions distort the signal even more, so that later stages perform even worse than the first.

One approach to mitigating these problems is subdivision of the signals into power classes; only signals that are detected reliably are used for interference cancellation. Arriving signals are divided into several groups, depending on their SINR. For a decision in class m, we use feedback from classes $1 \cdots m - $ i.e., the more reliable decisions $-$ but not those of classes $m + 1$, $m + 2$, $\ldots$.

It is also possible to use partial cancellation. At each stage, signals are sent through a mapping $x \rightarrow \tanh(\lambda x)$ (instead of a hard-decision device), where the steepness of the mapping curve $-$ i.e., $\lambda -$ increases from stage to stage. Thus, only the last stages use de facto hard decisions. The *turbo-multiuser detector* further improves on this principle by feeding back the log-likelihood ratios for different bits. This greatly decreases the probability of error propagation: bits that are decided

Figure 18.12   Parallel interference cancellation.

wrongly usually have a considerable uncertainty about them (it is unlikely that noise is so large that we make a wrong decision with great confidence). The turbo detector does exploit this fact, by assigning a small log-likelihood ratio. It can actually be shown that a turbo-multiuser detector can approach the performance of a multiuser MLSE.

18.5   Time Hopping Impulse Radio

When the desired spreading bandwidth W is on the order of 500 MHz or higher, it becomes interesting to spread the spectrum by transmitting short pulses whose position or amplitude contains the desired information. This method, often called "impulse radio," allows the use of simple transmitters and receivers. It is mostly used for so-called "ultrawideband" communications systems [diBenedetto et al. 2005], see also Section 21.8.

18.5.1   Simple Impulse Radio

Let us start out with the most simple possible impulse radio: a transmitter sending out a single pulse to represent one symbol. For the moment, assume that the modulation method is orthogonal pulse position modulation (see Chapter 11). A pulse is either sent at time t, or at time $t + T_d$, where T_d is larger than the duration of a pulse T_C. The detection process in an AWGN channel is then exceedingly simple: we just need an energy detector that determines during which of the two possible time intervals

$$[t + \tau_{\text{run}}, t + \tau_{\text{run}} + T_C]. \tag{18.40}$$

or

$$[t + \tau_{\text{run}} + T_d, t + \tau_{\text{run}} + T_C + T_d] \tag{18.41}$$

we get more energy. Here, τ_{run} is the runtime between transmitter and receiver, which we determine via the synchronization process.

Obviously, a pulsed transmission achieves spreading, because the bandwidth of the transmit signal is given by the inverse of pulse duration $1/T_C$. And "despreading" is done in a very simple way: by only recording and using the arriving signal in the intervals given by Eqs. (18.40)–(18.41), we can make our decision about which bit was sent. The SNR occurring at the receiver is E_B/N_0: if the receiver is an energy detector, the observed peak power is $\bar{P}T_S/T_C$, and noise power is N_0/T_C, resulting in an SNR of $\bar{P}T_S/N_0 = E_B/N_0$. This can be interpreted as suppressing the (wideband) noise by a factor T_S/T_C.

However, there are two main drawbacks to such a simple scheme:

1. The peak-to-average ratio (crest factor) of the transmitted signal is very high, giving rise to problems in the design of transmit and receive circuitry.
2. The scheme is not robust to different types of interference. In particular, there are problems with multiaccess interference. In other words, what happens if the desired transmitter sends a 0 such that the signal would arrive at $t + \tau_{\text{run,desired}}$, while another user sends a 0 that arrives at $t + \tau_{\text{run,interfere}}$? It can happen that this interfering pulse arrives precisely at the time the receiver expects a pulse representing a 1 from the desired user. More exactly, this occurs when $t + \tau_{\text{run,interfere}} = t + \tau_{\text{run,desired}} + T_d$. Since it is very difficult to influence the runtimes from different users in a highly accurate way, such a situation occurs with finite probability. We term it "catastrophic collision," since the receiver has no way of making a good decision about the received symbol and error probability will be very high; compare also the catastrophic collisions in frequency-hopping systems (see Section 18.1).

In order to solve these problems we need to transmit multiple pulses for each symbol. The first idea that springs to mind is to send a regular pulse train, where the duration between pulses is T_f. This solves the peak-to-average problem. However, we can easily see that it does not decrease the probability of catastrophic collisions: if the first pulse of the train collides with a pulse from an interfering pulse train, then subsequent pulses collide as well. A solution to this problem is provided by the concept of time-hopping impulse radio, described in the following section.

18.5.2   Time Hopping

The basic idea of time-hopping impulse ratio (TH-IR) is to use a sequence of *irregularly spaced* pulses to represent a single symbol. More precisely, we divide the available symbol time into a number of so-called *frames* of duration T_f, and transmit one pulse within each frame (see Figure 18.13).[10] The key idea is now to vary the position of the pulses *within* the frames. For example, in the first frame, we send the pulse in the third chip of the frame; in the second frame, we send the pulse in the eighth chip of the frame, and so on. The positions of the pulses within a frame are determined by a pseudorandom sequence called a "time-hopping sequence." Mathematically speaking, the transmit signal is

$$s(t) = \frac{\sqrt{E_S}}{\sqrt{N_f}} \sum_i \sum_{j=1}^{N_f} g(t - jT_f - c_j T_C - b_i T_d - iT_S) = \frac{\sqrt{E_S}}{\sqrt{N_f}} \sum_i p(t - b_i T_d - iT_S) \quad (18.42)$$

where $g(t)$ is the transmitted unit energy pulse of duration T_C, N_f is the number of frames (and therefore also the number of pulses) representing one information symbol of length T_S, and b is the information symbol transmitted. The time-hopping sequence provides an additional timeshift of $c_j T_C$ seconds to the jth pulse of the signal, where c_j are the elements of a pseudorandom sequence, taking on integer values between 0 and $N_c - 1$ (see also Figure 18.13). In the receiver, we perform matched filtering (matched to the total transmitted waveform $p(t)$), and sample this output at times $t = T_s + \tau_{run}$, and $t = T_s + \tau_{run} + T_d$. A comparison of the sample values at these two times determines which sequence had been sent. As in the case of CDMA, the matched filter operation can also be interpreted as a correlation with the transmit sequence.

Figure 18.13   Transmit waveform of time-hopping impulse radio for one symbol, $p(t)$, indicating chip, frame, and symbol duration.

In this TH-IR, we have the same suppression of noise as for the "simple" impulse radio system – namely, T_S/T_C; however, now the gain comes from two different sources. One part of the gain stems from the fact that we observe the noise only over a short period of time – i.e., the same type of gain we have in the simple impulse radio. Its value is now T_f/T_C, and is thus

---

[10] Note that the notation "frame," while established in the impulse radio literature, can give rise to confusion. In impulse radio, one data symbol contains several frames. For TDMA or Packet Reservation Multiple Access (PRMA), a block of data is also often called a "frame"; when used in this context, a frame contains several symbols.

smaller than in simple impulse radio. The other type of gain stems from combining the pulses in the different frames: the desired signal components from the different frames add up coherently, while the noise components add up incoherently. Taken together, those two gains provide a total gain of T_S/T_C.

Now, what is the advantage of this approach, compared with a regular pulse train? We can ascribe *different* time-hopping sequences to different users. As we will show below, these sequences can be constructed in such a way that pulses collide only in a few ($N_{collide}$) frames, but not the others. And as the receiver correlates the incoming signal with the time-hopping sequence of the desired user, it sees interference only in those frames where there is actually a collision. This leads to a suppression of interference by a factor $N_{collide}/N_f$.

We have already mentioned that signals between desired user and interferer are not synchronized, and thus can have an arbitrary timeshift against each other. Thus, time-hopping sequences are constructed according to the following criterion: irrespective of the relative shift between the different sequences, the number of collisions between pulses must not exceed a threshold λ (usually, we choose $\lambda = 1$).[11] This way the system designer does not have to worry about runtime effects or synchronization between users; a good suppression of Multiple Access Interference (MAI) is always guaranteed. Designing such sequences is difficult, especially if we want a large number of sequences with small collisions; exhaustive computer searches are often the best method.

Example 18.3   *The time-hopping sequence of a TH-IR system with $N_f = 6$ is [1, 2, 4, 6, 3, 5]. Find another hopping sequence that has at most one collision for arbitrary shifts.*

By making a systematic search, we find that the sequence [1, 3, 5, 2, 6, 4] fulfills the requirements; this is shown in Figure 18.14.

Impulse radio can also be given a different, very useful, interpretation: it is a direct-sequence CDMA system (like the one in Section 18.2), where the spreading sequence has a large number of 0s, and a small number of 1s. Compare this with the conventional DS-CDMA systems, where

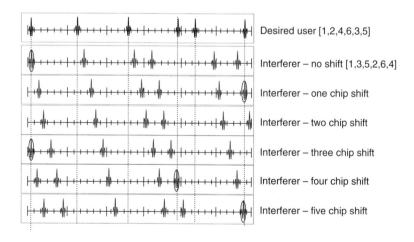

Figure 18.14   Interference between two users, for all possible (integer) shifts between the two users. Circles around a pulse indicate collisions.

[11] It is interesting that this problem of constructing good time-hopping sequences has strong similarities to constructing good frequency-hopping sequences for FH spread systems (see Section 18.1).

there is an almost equal number of $+1$s and -1s. The charm of this interpretation is that a lot more research has been performed for conventional DS-CDMA systems than for impulse radio. But, by interpreting TH-IR as a DS-CDMA system, many of these results can be adopted immediately.

Finally, we note that TH-IR can be used not only in conjunction with pulse position modulation, but also with pulse amplitude modulation. Furthermore, the polarity of transmitted pulses within a symbol can be randomized, so that the transmit signal then reads:

$$s(t) = \frac{\sqrt{E_S}}{\sqrt{N_f}} \sum_i b_i \sum_{j=-\infty}^{\infty} d_j g(t - jT_f - c_j T_C - iT_S) \tag{18.43}$$

where each pulse is multiplied by a pseudorandom variable d_j that can take on the values $+1$ or -1 with equal probability. Such a polarity randomization has advantages with respect to spectral shaping of the transmit signal. For such a system, the resemblance to DS-CDMA is even more striking: the spreading sequence is now a *ternary* sequence, with an (almost) equal number of $+1$'s and -1's, and a large number of 0's in between. The suppression of interference now occurs not just because of a small number of pulse collisions but also because the interference contributions from different frames (but the same symbol) might cancel out.

18.5.3 Impulse Radio in Delay-Dispersive Channels

Up to now, we have discussed impulse radio in AWGN channels. We have found that the transmitter as well as the receiver can be made very simple in these cases. However, TH-IR almost never works in AWGN channels. The purpose of such a system is the use of a very large bandwidth (typically 500 MHz or more), which in turn implies that the channel will certainly show variations over that bandwidth. It is then necessary to build a receiver that can work well in a dispersive channel.

Let us first consider coherent reception of the incoming signal. In this case, we need a Rake receiver, just like the one discussed in Section 18.2.3. Essentially, the Rake receiver consists of multiple correlators or fingers. In each of the fingers, correlation of the incoming signal is done with a delayed version of the time-hopping sequence, where the delay is equal to the delay of the MPC that we want to "rake up." The output of the Rake fingers is then weighted (according to the principles of maximum-ratio combining or optimum combining) and summed up. Because Impulse Radio (IR) systems usually use very large bandwidth, they always use structures whose number of fingers is smaller than the number of available MPCs (SRake, PRake, see Section 18.2.3). In order to get reasonable performance, it might be required to have 20 or more Rake fingers even in indoor environments. Differentially coherent (transmitted reference) or noncoherent schemes thus become an attractive alternative.

In order to understand the principle of *Transmitted Reference* (TR) schemes, remember that the ideal matched filter in a delay-dispersive channel is matched to the convolution of the time-hopping sequence with the impulse response of the channel. For coherent reception, we first have a filter matched to the time-hopping sequence, followed by the Rake receiver, which is matched to the impulse response of the channel. A so-called TR scheme creates this composite matched filter in a different way. A TR transmitter sends out two pulses in each frame: one unmodulated (reference) pulse, and, a fixed time T_{pd} later, the modulated (data) pulse. The sequence of reference pulses is convolved with the channel impulse response when it is transmitted through the wireless channel; this signal thus constitutes a noisy version of the system impulse response (convolution of transmit basis waveform and channel impulse response). In order to perform a matched filtering on the data-carrying part of the received signal, the receiver just multiplies the received signal by the received reference signal.

Let us now have a look at the mathematical expression of the transmit signal for one symbol:

$$p(t) = \sqrt{\frac{1}{2}} \sum_{j=0}^{N_f} d_j [g(t - jT_f - c_j T_C) + b \cdot g(t - jT_f - c_j T_C - T_{pd})] \qquad (18.44)$$

Inspection shows that correlation with the received reference signal can be done by just multiplying the total received signal by a delayed (by time T_{pd}) version of itself and integration. To be more precise: the first step at the receiver involves filtering by a receive filter $h_R(t)$; this filter should be wide enough not to introduce any signal distortions, and just limit the available noise. The filtered received signal $\hat{r}(t)$ is multiplied by a delayed version of itself, and integrated over a finite interval T_{int}, which should be long enough to collect most multipath energy, but short enough not to collect too much noise energy.

TR schemes have the advantage of being exceedingly simple (though implementation of the delay in the receiver may be nontrivial). On the downside, they show poorer performance than coherent schemes for two reasons: (i) reference pulses waste energy, in the sense that they do not carry any information (this results in a 3-dB penalty); (ii) the reference part of the signal is noisy, as is the data-carrying part of the signal. Multiplication of the two signals in the receiver gives rise to noise–noise cross-terms that worsen the SNR. Now remember that impulse radio is a spread spectrum system, so that the SNR of the received signal is negative. Noise–noise cross-terms can thus become large.

Finally, we note that noncoherent detection can be an attractive alternative as well. This is especially true if only a single user is to be served. However, noncoherent receivers have problems with multiaccess interference. Due to multipath propagation, energy is dispersed over several adjacent chips. Thus, energy detection will see much more interference (as exemplified in Figure 18.15).

Desired user – integration times

Energy from interferer
– three chips shift

Figure 18.15   Interference from a delay-dispersed interferer signal (lower row) to the desired signal, whose integration time (two chip durations per frame) is sketched in the upper row.

Further Reading

An overview of spread spectrum communications in general can be found in the classic monographs of Dixon [1994] and Simon et al. [1994], which provide good coverage of FH and DS systems. Books that are more tuned to cellular systems, especially CDMA, are Glisic and Vucetic [1997], Goiser [1998], Li and Miller [1998], Viterbi [1995], and Ziemer et al. [1995]; the overview paper [Kohno et al. 1995] gives a shorter description. Scholtz [1982] gives an overview of the history of the spread spectrum (though more from a military perspective, since cellular applications were not yet considered at the time of that article). Milstein [1988] discusses the interference rejection capabilities of various spread spectrum techniques.

Frequency-hopping codes are designed in Maric and Titlebaum [1992]. Estimates for the capacity of CDMA systems were first published in the widely cited paper of Gilhousen et al. [1991]. The effect of multipath on CDMA – namely, the use of Rake receivers – is reviewed in Goiser et al.

[2000], Swarts et al. [1998] particularly, the effect of a finite number of Rake fingers [Win and Chrisikos 2000]. Other important papers on Rake reception include Holtzman and Jalloul [1994] and Bottomley et al. [2000], among many others. For synchronization aspects, the papers by Polydoros and Weber [1984] are still interesting to read. Spreading codes are reviewed in Dinan and Jabbari [1998]. An authoritative description of sequence design (for radar as well as communications) is Golomb and Gong [2005]. While the Gaussian approximation is widely used for intercell interference, it can give significant deviations from reality when shadowing is present; a more refined model is presented in Singh et al. [2010]. A complete TH-IR radio system is described in Molisch et al. [2004].

For multiuser detection, the book by Verdu [1998] is a must-read. Also, the overview papers (Duel-Hallen et al. [1995], Moshavi [1996], and Poor [2001]) give useful insights. Advanced concepts like blind multiuser detection are described extensively in Wang and Poor [2003] and Honig [2009]; the interaction between multiuser detection and decoding is explored in Poor [2004].

Time-hopping impulse radio was pioneered by Win and Scholtz in the 1990s. For a description of the basic principles, see Win and Scholtz [1998], and more detailed results in Win and Scholtz [2000]. An overview of the different aspects of ultrawideband system design can be found in the monographs Shen et al. [2006], Ghavami et al. [2006], Reed [2005], diBenedetto et al. [2005], Roy et al. [2004]; TR receivers and other simplified receiver structures are discussed in the overview paper of Witrisal et al. [2009].

For updates and errata for this chapter, see wides.usc.edu/teaching/textbook

19

Orthogonal Frequency Division Multiplexing (OFDM)

19.1 Introduction

Orthogonal Frequency Division Multiplexing (OFDM) is a modulation scheme that is especially suited for high-data-rate transmission in delay-dispersive environments. It converts a high-rate data stream into a number of low-rate streams that are transmitted over parallel, narrowband channels that can be easily equalized.

Let us first analyze why traditional modulation methods become problematic at very high data rates. As the required data rate increases, the symbol duration T_s has to become very small in order to achieve the required data rate, and the system bandwidth becomes very large.[1] Now, delay dispersion of a wireless channel is given by nature; its values depend on the environment, but not on the transmission system. Thus, if the symbol duration becomes very small, then the impulse response (and thus the required length of the equalizer) becomes very long *in terms of symbol durations*. The computational effort for such a long equalizer is very large (see Chapter 16), and the probability of instabilities increases. For example, the Global System for Mobile communications (GSM) system (see Chapter 24) which is designed for peak data rates up to 200 kbit/s, uses 200 kHz bandwidth, while the IEEE 802.11 system (see Chapter 29), with data rates of up to 55 Mbit/s uses 20 MHz bandwidth. In a channel with 1 μs maximum excess delay, the former needs a two-tap equalizer, while the latter needs 20 taps. OFDM, on the other hand, increases the symbol duration on each of its carriers compared to a single-carrier system, and can thus have a very simple equalizer for each subcarrier.

OFDM dates back some 40 years; a patent was applied for in the mid-1960s [Chang 1966]. A few years later, an important improvement – the *Cyclic Prefix* (CP) – was introduced; it helps to eliminate residual delay dispersion. Cimini [1985] was the first to suggest OFDM for wireless communications. But it was only in the early 1990s that advances in hardware for digital signal processing made OFDM a realistic option for wireless systems. Furthermore, the high-data-rate applications for which OFDM is especially suitable emerged only in recent years. Currently, OFDM is used for *Digital Audio Broadcasting* (DAB), *Digital Video Broadcasting* (DVB), and *wireless Local Area Networks* (LANs) (IEEE 802.11a, IEEE 802.11g). It will also be used in fourth-generation cellular systems, including *Third Generation Partnership Project-Long-Term Evolution* (*3GPP-LTE*) and WiMAX.

---

[1] (This can be compounded by multiple access formals like Time Division Multiple Access (TDMA), which has a high peak data rate because it compresses data into bursts (see Chapter 17).

Wireless Communications, Second Edition   Andreas F. Molisch
© 2011 John Wiley & Sons, Ltd

19.2   Principle of Orthogonal Frequency Division Multiplexing

OFDM splits a high-rate data stream into N parallel streams, which are then transmitted by modulating N distinct carriers (henceforth called *subcarriers* or *tones*). Symbol duration on each subcarrier thus becomes larger by a factor of N. In order for the receiver to be able to separate signals carried by different subcarriers, they have to be orthogonal. Conventional Frequency Division Multiple Access (FDMA), as described in Section 17.1 and depicted again in Figure 19.1, can achieve this by having large (frequency) spacing between carriers. This, however, wastes precious spectrum. A much narrower spacing of subcarriers can be achieved. Specifically, let subcarriers be at the frequencies $f_n = nW/N$, where n is an integer, and W the total available bandwidth; in the most simple case, $W = N/T_s$. We furthermore assume for the moment that modulation on each of the subcarriers is Pulse Amplitude Modulation (PAM) with rectangular basis pulses. We can then easily see that subcarriers are mutually orthogonal, since the relationship

$$\int_{iT_s}^{(i+1)T_S} \exp(j2\pi f_k t)\exp(-j2\pi f_n t)\,dt = \delta_{nk} \tag{19.1}$$

holds.

Figure 19.1 shows this principle in the frequency domain. Due to the rectangular shape of pulses in the time domain, the spectrum of each modulated carrier has a $\sin(x)/x$ shape. The spectra of different modulated carriers overlap, but each carrier is in the spectral nulls of all other carriers. Therefore, as long as the receiver does the appropriate demodulation (multiplying by $\exp(-j2\pi f_n t)$ and integrating over symbol duration), the data streams of any two subcarriers will not interfere.

Figure 19.1   Principle behind orthogonal frequency division multiplexing: N carriers within a bandwidth of W.

19.3   Implementation of Transceivers

OFDM can be interpreted in two ways: one is an "analog" interpretation following from the picture of Figure 19.2a. As discussed in Section 19.2, we first split our original data stream into N parallel data streams, each of which has a lower data rate. We furthermore have a number of local oscillators (LOs) available, each of which oscillates at a frequency $f_n = nW/N$, where $n = 0, 1, \ldots, N-1$. Each of the parallel data streams then modulates one of the carriers. This picture allows an easy understanding of the principle, but is ill suited for actual implementation – the hardware effort of multiple local oscillators is too high.

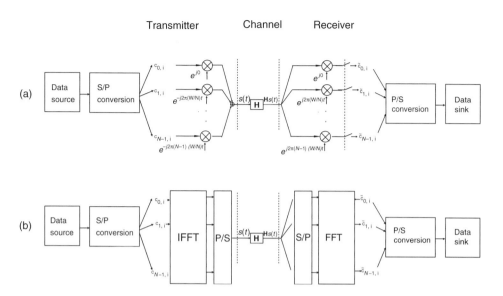

Figure 19.2 Transceiver structures for orthogonal frequency division multiplexing in purely analog technology (a), and using inverse fast Fourier transformation (b).

An alternative implementation is *digital*. It first divides the transmit data into blocks of N symbols. Each block of data is subjected to an *Inverse Fast Fourier Transformation* (IFFT), and then transmitted (see Figure 19.2b). This approach is much easier to implement with integrated circuits. In the following, we will show that the two approaches are equivalent.

Let us first consider the analog interpretation. Let the complex transmit symbol at time instant i on the nth carrier be $c_{n,i}$. The transmit signal is then:

$$s(t) = \sum_{i=-\infty}^{\infty} s_i(t) = \sum_{i=-\infty}^{\infty} \sum_{n=0}^{N-1} c_{n,i} g_n(t - iT_S) \tag{19.2}$$

where the basis pulse $g_n(t)$ is a normalized, frequency-shifted rectangular pulse:

$$g_n(t) = \begin{cases} \frac{1}{\sqrt{T_S}} \exp\left(j 2\pi n \frac{t}{T_S}\right) & \text{for } 0 < t < T_S \\ 0 & \text{otherwise} \end{cases} \tag{19.3}$$

Let us now – without restriction of generality – consider the signal only for $i = 0$, and sample it at instances $t_k = kT_S/N$:

$$s_k = s(t_k) = \frac{1}{\sqrt{T_S}} \sum_{n=0}^{N-1} c_{n,0} \exp\left(j 2\pi n \frac{k}{N}\right) \tag{19.4}$$

Now, this is nothing but the inverse Discrete Fourier Transform (DFT) of the transmit symbols. Therefore, the transmitter can be realized by performing an *Inverse Discrete Fourier Transform* (IDFT) on the block of transmit symbols (the blocksize must equal the number of subcarriers). In almost all practical cases, the number of samples N is chosen to be a power of 2, and the IDFT is realized as an IFFT. In the following, we will only speak of IFFTs and Fast Fourier Transforms (FFTs).

Note that the input to this IFFT is made up of N samples (the symbols for the different subcarriers), and therefore the output from the IFFT also consists of N values. These N values now have to be transmitted, one after the other, as temporal samples – this is the reason why we have a P/S (Parallel to Serial) conversion directly after the IFFT. At the receiver, we can reverse the process: sample the received signal, write a block of N samples into a vector – i.e., an S/P (Serial to Parallel) conversion – and perform an FFT on this vector. The result is an estimate $\tilde{c}_n$ of the original data c_n.

Analog implementation of OFDM would require multiple LOs, each of which has to operate with little phase noise and drift, in order to retain orthogonality between the different subcarriers. This is usually not a practical solution. The success of OFDM is based on the above-described digital implementation that allows an implementation of the transceivers that is much simpler and cheaper. In particular, highly efficient structures exist for the implementation of an FFT (so-called "butterfly structures"), and the computational effort (per bit) of performing an FFT increases only with log (N).

OFDM can also be interpreted in the time–frequency plane. Each index i corresponds to a (temporal) pulse; each index n to a carrier frequency. This ensemble of functions spans a grid in the time–frequency plane.

19.4   Frequency-Selective Channels

In the previous section, we explained how the OFDM transmitter and receiver work in an Additive White Gaussian Noise (AWGN) channel. We could take this scheme without any changes, and just let it operate in a frequency-selective channel. Intuitively, we would anticipate that delay dispersion will have only a small impact on the performance of OFDM we convert the system into a parallel system of narrowband channels, so that the symbol duration on each carrier is made much larger than the delay spread. But, as we saw in Chapter 12, delay dispersion can lead to appreciable errors even when $S_\tau / T_s < 1$. Furthermore, as we will elaborate below, delay dispersion also leads to a loss of orthogonality between the subcarriers, and thus to *Inter Carrier Interference* (ICI). Fortunately, both these negative effects can be eliminated by a special type of guard interval, called the *cyclic prefix (CP)*. In this section, we show how to construct this cyclic prefix, how it works, and what performance can be achieved in frequency-selective channels.

19.4.1   Cyclic Prefix

Let us first define a new base function for transmission:

$$g_n(t) = \exp\left[j2\pi n \frac{W}{N} t \right] \quad \text{for} - T_{cp} < t < \hat{T}_S \tag{19.5}$$

where again W/N is the carrier spacing, and $\hat{T}_S = N/W$. The symbol duration T_S is now $T_S = \hat{T}_S + T_{cp}$. This definition of the base function means that for duration $0 < t < \hat{T}_S$ the "normal" OFDM symbol is transmitted (Figure 19.3). It can be easily seen by substituting in Eq. (19.5) that, $g_n(t) = g_n(t + N/W)$. Therefore, during time $-T_{cp} < t < 0$, a copy of the last part of the symbol is transmitted. From linearity, it also follows that the *total* signal s(t) transmitted during time $-T_{cp} < t < 0$ is a copy of s(t) during the last part, $\hat{T}_S - T_{cp} < t < \hat{T}_S$. This prepended part of the signal is called the "cyclic prefix."

Now that we know what a cyclic prefix is, let us investigate why it is beneficial in delay-dispersive channels. When transmitting any data stream over a delay-dispersive channel, the arriving signal is the linear convolution of the transmitted signal with the channel impulse response. The cyclic prefix converts this *linear* convolution into a *cyclical* convolution. During

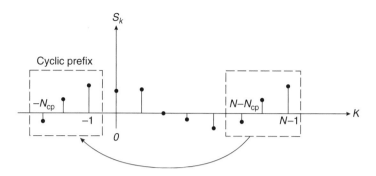

Figure 19.3 Principle of the cyclic prefix. $N_{cp} = NT_{cp}/(N/W)$ is the number of samples in the cyclic prefix.

the time $-T_{cp} < t < -T_{cp} + \tau_{max}$, where τ_{max} is the maximum excess delay of the channel, the received signal suffers from "real" InterSymbol Interference (ISI), as echoes of the last part of the preceding symbol interfere with the desired symbol.[2] This "regular" ISI is eliminated by discarding the received signal during this time interval. During the remainder of the symbol, we have *cyclical* ISI; especially, it is the last part of the current (not the preceding) symbol that interferes with the first part of the current symbol. In the following, we show how an extremely simple mathematical operation can eliminate the effect of such a cyclical convolution.

For the following mathematical derivation, we assume that the duration of the impulse response is exactly equal to the duration of the prefix; furthermore, in order to simplify the notation, we assume (without restriction of generality) $i = 0$. In the receiver, there is a bank of filters that are matched to the basis functions *without* the cyclic prefix:

$$\bar{g}_n(t) = \begin{cases} g_n^*(\hat{T}_S - t) & \text{for } 0 < t < \hat{T}_S \\ 0 & \text{otherwise} \end{cases} \tag{19.6}$$

This operation removes the first part of the received signal (of duration T_{cp}) from the detection process; as discussed above, the matched filtering of the remainder can be realized as an FFT operation. The signal at the output of the matched filter is thus convolution of the transmit signal with the channel impulse response and the receive filter:

$$r_{n,0} = \int_0^{\hat{T}_S} \left[\int_0^{T_{cp}} h(t, \tau) \left(\sum_{k=0}^{N-1} c_{k,0} g_k(t - \tau) \right) d\tau \right] g_n^*(t) \, dt + n_n \tag{19.7}$$

where n_n is the noise at the output of the matched filter. Note that the argument of g_k can attain values between $-T_{cp}$ and $\hat{T}_S$, which is the region of definition of Eq. (19.5). If the channel can be considered as constant during the time T_S, then $h(t, \tau) = h(\tau)$, and we obtain:

$$r_{n,0} = \sum_{k=0}^{N-1} c_{k,0} \int_0^{\hat{T}_S} \left[\int_0^{T_{cp}} h(\tau)(g_k(t - \tau)) \, d\tau \right] g_n^*(t) \, dt + n_n \tag{19.8}$$

The inner integral can be written as

$$\exp\left[j2\pi t k \frac{W}{N} \right] \int_0^{T_{cp}} h(\tau) \exp\left(-j2\pi \tau k \frac{W}{N} \right) d\tau = g_k(t) H\left(k \frac{W}{N} \right) \tag{19.9}$$

---

[2] In the following, we assume $\tau_{max} \leq T_{cp}$.

where $H\left(k\frac{W}{N}\right)$ is the channel transfer function at the frequency kW/N. Since, furthermore, the basis functions $g_n(t)$ are orthogonal during the time $0 < t < \hat{T}_S$:

$$\int_0^{\hat{T}_S} g_k(t)g_n^*(t)\,dt = \delta_{kn}(t) \tag{19.10}$$

the received signal samples r can be written as

$$r_{n,0} = H\left(n\frac{W}{N}\right)c_{n,0} + n_n \tag{19.11}$$

The OFDM system is thus represented by a number of parallel *nondispersive*, fading channels, each with its own complex attenuation $H\left(n\frac{W}{N}\right)$. Equalization of the system thus becomes exceedingly simple: it just required division by the transfer function at the subcarrier frequency, independently for each subcarrier. In other words, the cyclic prefix has recovered the orthogonality of the subcarriers.

Two caveats have to be noted: (i) we assumed in the derivation that the channel is static for the duration of the OFDM symbol. If this assumption is not fulfilled, interference between the subcarriers can still occur (see Section 19.7); (ii) discarding part of the received signal decreases the Signal-to-Noise Ratio (SNR), as well as spectral efficiency. For usual operating parameters (cyclic prefix about 10% of symbol duration), this loss is tolerable.

The block diagram of an OFDM system, including the cyclic prefix, is given in Figure 19.4. The original data stream is S/P converted. Each block of N data symbols is subjected to an IFFT, and then the last NT_{cp}/T_S samples are prepended. The resulting signal is modulated onto a (single) carrier and transmitted over a channel, which distorts the signal and adds noise. At the receiver, the signal is partitioned into blocks. For each block, the cyclic prefix is stripped off, and the remainder is subjected to an FFT. The resulting samples (which can be interpreted as the samples in the frequency domain) are "equalized" by means of one-tap equalization – i.e., division by the complex channel attenuation – on each carrier.

Figure 19.4   Structure of an orthogonal-frequency-division-multiplexing transmission chain with cyclic prefix and one-tap equalization.

19.4.2   Performance in Frequency-Selective Channels

The cyclic prefix converts a frequency-selective channel into a number of parallel flat-fading channels. This is positive in the sense that it gets rid of the ISI that plagues TDMA and CDMA systems. On the downside, an uncoded OFDM system does not show any frequency diversity at all. If a subcarrier is in a fading dip, then error probability on that subcarrier is very high, and dominates the Bit Error Rate (BER) of the total system for high SNRs.

Example 19.1   *Bit error rate of uncoded orthogonal frequency division multiplexing.*

Figure 19.5 shows the transfer function and the BER of a Binary-Phase Shift Keying (BPSK) OFDM system for specific realization of a frequency-selective channel. Obviously, the BER is highest in fading dips. Note that the results are plotted on a logarithmic scale – while the BER on "good" subcarriers can be as low as 10^{-4}, the BER on subcarriers that are in fading dips are up to 0.5. This also has a significant impact on average error probability; the error probability on bad subcarriers dominates the behavior. Figure 19.6 shows a simulation of the average BER (over many channel realization) for a frequency-selective channel. We find that the BER decreases only linearly as the SNR increases, closer inspection reveals that the result is the same as in Figure 12.6.

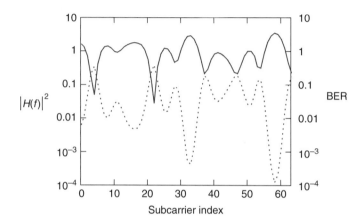

Figure 19.5 Normalized squared magnitude of the transfer function (solid), and bit error rate (dashed), for a channel with taps at [0, 0.89, 1.35, 2.41, 3.1] with amplitudes [1, −0.4, 0.3, 0.43, 0.2]. The average signal-to-noise ratio at the receiver is 3 dB; the modulation format is binary-phase shift keying. Subcarriers are at $f_k = 0.05k$, $k = 0....63$.

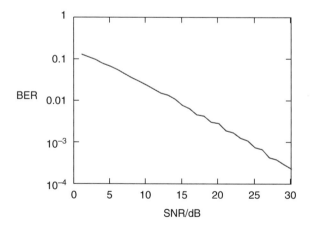

Figure 19.6 Bit error rate for a channel with taps at [0, 0.89, 1.35, 2.41, 3.1] with mean powers [1, 0.16, 0.09, 0.185, 0.04], each tap independently Rayleigh fading. The modulation format is binary-phase shift keying. Subcarriers are at $f_k = 0.05k$, $k = 0....63$.

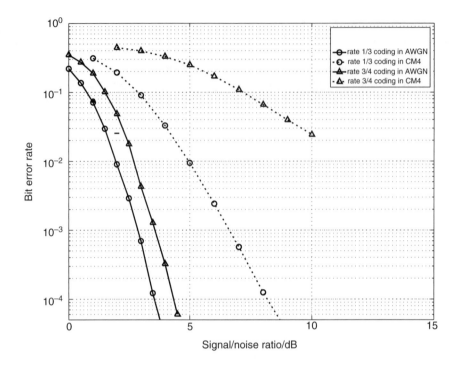

Figure 19.7   Bit error rate as a function of the signal-to-noise ratio for rate-1/3- and rate-3/4-coded orthogonal-frequency-division-multiplexing system. Channel is either additive white Gaussian noise, or channel model 4 of the IEEE 802.15.3a channels. The OFDM system follows the specifications of the WiMedia standard.
Reprinted with permission from Ramachandran et al. [2004] © JEEE.

More generally, we find that uncoded OFDM has the same average BER irrespective of the frequency selectivity of the channel. This can also be interpreted the following way: frequency selectivity gives us different channel realizations on different subcarriers; time variations give us different channel realizations at different times. Doubly selective channels have different realizations on different subcarriers as well as different times. But, for computation of the average BER, it does not matter how the different realizations are created, as long as the fading has the same statistics (e.g., Rayleigh), and the ensemble is large enough.[3]

From these examples, we see that the main problem lies in the fact that carriers with poor SNR dominate the performance of the system. Any of the following approaches circumvents this problem:

- *Coding across the different tones*: such coding helps to compensate for fading dips on one subcarrier by a good SNR in another subcarrier. This is described in more detail in Section 19.4.3.
- *Spreading the signal over all tones*: in this approach, each symbol is spread across all carriers, so that it sees an SNR that is the average of all tones over which it is spread. This method is discussed in more detail in Sections 19.10 and 19.11.

---

[3] Note that in a time-invariant, frequency-selective channel, the number of independent channel realizations depends on the ratio of system bandwidth to coherence bandwidth of the channel. If this value is small, there might not be a sufficiently ensemble to obtain good averaging.

- *Adaptive modulation*: if the transmitter knows the SNR on each of the subcarriers, it can choose its modulation alphabet and coding rate adaptively. Thus, on carriers with low SNR, the transmitter will send symbols using stronger encoding and a smaller modulation alphabet. Also, the power allocated to each subcarrier can be varied. This approach is described in more detail in Section 19.8.

19.4.3   Coded Orthogonal Frequency Division Multiplexing

Just as coding can be used to great effect in single-carrier systems to improve performance in fading channels, so can it be gainfully employed in OFDM systems. But now we have data that are transmitted at different frequencies as well as at different times. This gives rise to the question of how coding of the data should be applied.

To get an intuitive feeling for coding across different subcarriers, imagine again the simple case of repetition coding: each of the symbols that are to be transmitted is repeated on K different subcarriers. As long as fading is independent of the different subcarriers, K-fold diversity is achieved. In the most simple case, the receiver first makes a hard decision about symbols on each subcarrier, and then makes a majority decision among the K received symbols about which bit was sent. Of course, practical systems do not use repetition coding, but the principle remains the same.

We could now try and develop a whole theory for coding on OFDM systems. However, it is much easier to just consider the analogy between the time domain and the frequency domain. Remember the main lessons from Section 14.8: enough interleaving should be applied such that fading of coded bits is independent. In other words, we just need independent channel states over which to transmit our coded bits; this will automatically result in a high diversity order. It does not matter whether channel states are created by temporal variations of the channel, or as different transfer functions of subcarriers in frequency-selective channels. Thus, it is not really necessary to define new codes for OFDM, but it is more a question on how to design appropriate mappers and interleavers that assign the different coded bits in the time-frequency plane. This mapping, in turn, depends on the frequency selectivity as well as the time selectivity of the channel. If the channel is highly frequency selective, then it might be sufficient to code *only* across available frequencies, without any coding or interleaving along the time axis. This has two advantages: on one hand, this scheme also works in static channels, which occur quite often for wireless LANs and other high-rate data transmission scenarios; on the other, the absence of interleaving in the time domain results in lower latency of the transmission and decoding process.

Figure 19.7 shows a performance example. We see that for AWGN, both rate-1/3- and rate-3/4-coded systems exhibit good performance, with approximately a 1-dB difference. In fading channels, performance is dramatically different. While the rate-1/3 code has good diversity, and therefore the BER decreases fast as a function of the SNR, the rate-3/4 code has very little frequency diversity, and thus bad performance.

19.5   Channel Estimation

As for any other coherent wireless system, operation of OFDM systems requires an estimate of the channel transfer function, or, equivalently, the channel impulse response. Since OFDM is operated with a number of parallel narrowband subcarriers, it is intuitive to estimate the channel in the frequency domain. More precisely, we wish to obtain the N complex-valued channel gains on the subcarriers. Let us denote these channel attenuations as $h_{n,i}$, where n is the subchannel index and i is the time index. Assuming that we know the statistical properties of these channel attenuations, and some structure to the OFDM signal, we can derive good channel estimators.

In the following, we treat three approaches: (i) pilot symbols, which are mainly suitable for an initial estimate of the channel; (ii) scattered pilot tones, which help to track changes in channels over time; and (iii) eigenvalue-decomposition-based methods, which can be used to reduce the complexity of the first two methods.

19.5.1   Pilot-Symbol-Based Methods

The most straightforward channel estimation in OFDM is when we have a dedicated pilot *symbol* containing only known data – in other words, the data on each of the subcarriers is known. This approach is appropriate for initial acquisition of the channel, at the beginning of a transmission burst. The simplest channel estimate is then obtained by estimating the channel on each subcarrier separately. Denoting the known data on subcarrier n at time i as $c_{n,j}$, we can find a Least Squares (LS) channel estimate as

$$h_{n,i}^{\mathrm{LS}} = r_{n,i}/c_{n,i}$$

where $r_{n,i}$ is the received value on subchannel n.

We can improve the channel estimate by taking into account the correlation of the fading between different frequencies. Arranging the LS estimates in a vector $\mathbf{h}_i^{\mathrm{LS}} = \left(h_{1,i}^{\mathrm{LS}} \ h_{2,i}^{\mathrm{LS}} \ \cdots \ h_{n,i}^{\mathrm{LS}} \right)^T$, the corresponding vector of linear MMSE (LMMSE) estimate becomes

$$\mathbf{h}_i^{\mathrm{LMMSE}} = \mathbf{R}_{hh^{\mathrm{LS}}} \mathbf{R}_{h^{\mathrm{LS}}h^{\mathrm{LS}}}^{-1} \mathbf{h}_i^{\mathrm{LS}} \tag{19.12}$$

where $\mathbf{R}_{hh^{\mathrm{LS}}}$ is the covariance matrix between channel gains and the LS estimate of channel gains, $\mathbf{R}_{h^{\mathrm{LS}}h^{\mathrm{LS}}}$ is the autocovariance matrix of LS estimates. Given that we have AWGN with variance σ_n^2 on each subcarrier, $\mathbf{R}_{hh^{\mathrm{LS}}} = \mathbf{R}_{hh}$ and $\mathbf{R}_{h^{\mathrm{LS}}h^{\mathrm{LS}}} = (\mathbf{R}_{hh} + \sigma^2 \mathbf{I})$. Arranging channel attenuations in a vector $\mathbf{h}_i = \left(h_{1,i} \ h_{2,i} \ \cdots \ h_{n,i} \right)^T$, we can determine:

$$\mathbf{R}_{hh} = E\{\mathbf{h}_i \mathbf{h}_i^{\dagger}\} = E\{\mathbf{h}_i^* \mathbf{h}_i^T\}^* \tag{19.13}$$

which is independent of time i if the channel is wide-sense-stationary.

This estimation approach produces very good estimates, but computational complexity is high if the number of subcarriers is large: it requires N^2 multiplications – i.e. N multiplications per estimated channel gain (assuming that all correlation matrices and inversions are precalculated). This is quite a large complexity, even if this pilot-symbol-based estimation is usually done only at the beginning of a transmission burst. For this reason, there are several other suboptimal approaches available, where, e.g., smoothing Finite Impulse Response (FIR) filters of limited length (much less than N) are applied across LS-estimated attenuations to exploit the correlation between neighboring subchannels.

19.5.2   Methods Based on Scattered Pilots

After obtaining an initial estimate of the channel, we need to track changes in the channel as it evolves with time. In this case we would like to do two things: (i) reduce the number of known bits in an OFDM symbol (this improves spectral efficiency); and (ii) exploit the time correlation of the channel – i.e., the fact that the channel changes only slowly in time. An attractive way of tracking the channel is to use pilot symbols scattered in the OFDM time–frequency grid as illustrated in Figure 19.8, where pilots are spaced by N_f subcarriers and N_t OFDM symbols.[4]

When estimating the channel based on scattered pilots, we can start by performing LS estimation of the channel at pilot positions – i.e. $h_{n,i}^{\mathrm{LS}} = r_{n,i}/c_{n,i}$ is the received value and $c_{n,i}$ is the known

---

[4] We have used a rectangular pilot pattern in the illustration, but other pilot patterns can be used as well.

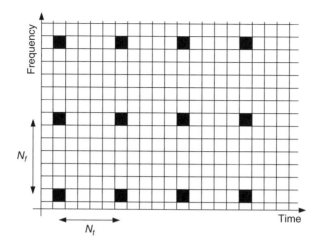

Figure 19.8   Scattered pilots in the orthogonal-frequency-division-multiplexing time–frequency grid. In this case the pattern is rectangular with pilot distances N_f subcarriers in frequency and N_t OFDM symbols in time.

pilot data in pilot position (n, i). From these initial estimates at pilot positions we then need to perform interpolation to obtain an estimate of the channel at all other positions. Interpreting the pilots as samples in a two-dimensional space, we can use standard sampling theory to put limits on the required density of our pilot pattern is [Nilsson et al. 1997]:

$$N_{\mathrm{f}} < \frac{N}{N_{\mathrm{cp}}}$$

$$N_{\mathrm{t}} < \frac{1}{2(1 + N_{\mathrm{cp}}/N)v_{\max}}$$

Since we need to reduce the effect of noise from the pilots and also help to reduce the complexity of estimation algorithms, it has been argued that a good tradeoff is to place twice as many pilots in each direction as required by the sampling theorem [Nilsson et al. 1997].

In principle, channel interpolation between these pilot positions can be done using the same estimation theory as for the all-pilot symbol case. When estimating a certain channel attenuation $h_{n,i}$ using a set of K pilot positions (n_j, i_j), $j − 1 \dots K$, we place the LS estimates in a pilot vector $\mathbf{p} = (h^{\mathrm{LS}}_{n_1, j_1} h^{\mathrm{LS}}_{n_2, j_2} \dots h^{\mathrm{LS}}_{n_k, j_k})^T$ and calculate the LMMSE estimate as

$$h^{\mathrm{LMMSE}}_{n,i} = \mathbf{r}_{hp} \mathbf{R}^{-1}_{pp} \mathbf{p}$$

where $\mathbf{r}_{hp}$ is the correlation (row) vector $E\{h_{n,i}\mathbf{p}^\dagger\}$ and $\mathbf{R}_{pp}$ is $E\{\mathbf{pp}^\dagger\}$. The complexity of this estimator grows with the number of pilot tones included in the estimation and requires K multiplications per estimated attenuation, again assuming that all correlation matrices and inversions are precalculated.

An alternative approach to the two-dimensional filtering above, where we use pilots in both the frequency direction and time direction at the same time, is to apply separable filters. This implies that we use two one-dimensional filters, one in the time direction and the other in the frequency direction. Many more pilots are thus influencing each estimated channel attenuation, for a given estimator complexity. The resulting increase in performance has been shown to dominate over loss in optimality when going from general two-dimensional filters to separable ones based on two-dimensional filters.

19.5.3   Methods Based in Eigen Decompositions

The structure of OFDM allows for efficient channel estimator structures. We know that the channel impulse response is short compared with the OFDM symbol length in any well-designed system. This fact can be used to reduce the dimensionality of the estimation problem. In essence, when using the LMMSE estimator in (Eq. 19.12), we would like to use the statistical properties of the channel to perform the matrix multiplication more efficiently. This can be done using the theory of optimal rank reduction from estimation theory, where an Eigen Value Decomposition (EVD) $\mathbf{R}_{hh} = \mathbf{U}\boldsymbol{\Lambda}\mathbf{U}^{\dagger}$ results in a new more computationally efficient version of (Eq. 19.12). The dimension of this space is approximately $N_{cp} + 1$ – i.e., one more than the number of samples in the cyclic prefix. We can therefore expect that, after the first $N_{cp} + 1$ diagonal elements in $\boldsymbol{\Lambda}$, the magnitude should decrease rapidly. Using the Singular Value Decomposition (SVD) to rewrite (Eq. 19.12) as

$$\mathbf{h}_i^{\mathrm{LMMSE}} = \mathbf{U}\boldsymbol{\Delta}\mathbf{U}^{\dagger}\mathbf{h}_i^{\mathrm{LS}}$$

where $\boldsymbol{\Delta}$ is a diagonal matrix containing the values $\delta_i = \lambda_i/(\lambda_i + 1/\gamma)$ on its diagonal. The diagonal elements δ_i will decrease rapidly after the first $N_{cp} + 1$ since the λ_i's do. By setting all but the p first λ_1's to zero – i.e. assigning $\delta_i = 0$ for $i > p$ – we get an optimal rank-p estimator for channel gains. The computational complexity of this estimator is $2Np$ multiplications, which is $2p$ per estimated attenuation. This should be compared with the N multiplications per estimated attenuation in the original estimator (Eq. 19.12). The estimator principle is illustrated in Figure 19.9.

In the case when the autocorrelation matrix $\mathbf{R}_{hh}$ is a circulant matrix, the resulting optimal transforms $\mathbf{U}^{\dagger}$ and $\mathbf{U}$ are the IDFT and DFT, respectively, and there are only N_{cp} nonzero singular values. The basic estimator structure stays the same, as shown in Figure 19.10, while the FFT processor already available in the OFDM receiver can be used to perform channel estimation as well.

Figure 19.9   The optimal rank-p channel estimator viewed as a transform ($\mathbf{U}^{\dagger}$) followed by p scalar multiplications and a second transform ($\mathbf{U}$).

Figure 19.10   Low-rank estimator for channels with circulant autocorrelation, implemented using fast Fourier transforms.

In many cases, when the channel correlation matrix is not circulant, the computational efficiency of DFT-based estimators may outweigh the suboptimality of their rank reduction. This general structure of estimators has also been used as one of the two one-dimensional estimators when performing two-dimensional estimation (see above). The gain here is that time direction smoothing can be done between two transforms, leading to a smaller number of filters that have to be applied in parallel. Instead of N filters (one per subcarrier), only p filters are needed in a rank-p estimator (as shown in Figure 19.11).

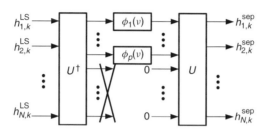

Figure 19.11   Two-dimensional (separable) channel estimation where time domain smoothing is done in the transform domain. This reduces the number of parallel filters needed, from N to p.

19.6   Peak-to-Average Power Ratio

19.6.1   *Origin of the Peak-to-Average Ratio Problem*

One of the major problems of OFDM is that the peak amplitude of the emitted signal can be considerably higher than the average amplitude. This *Peak-to-Average Ratio* (PAR) issue originates from the fact that an OFDM signal is the superposition of N sinusoidal signals on different subcarriers. On average the emitted power is linearly proportional to N. However, sometimes, the signals on the subcarriers add up constructively, so that the *amplitude* of the signal is proportional to N, and the power thus goes with N^2. We can thus anticipate the (worst case) power PAR to increase linearly with the number of subcarriers.

We can also look at this issue from a slightly different point of view: the contributions to the total signal from the different subcarriers can be viewed as random variables (they have quasi-random phases, depending on the sampling time as well as the value of the symbol with which they are modulated). If the number of subcarriers is large, we can invoke the central limit theorem to show that the distribution of the amplitudes of in-phase components is Gaussian, with a standard deviation $\sigma = 1/\sqrt{2}$ (and similarly for the quadrature components) such that mean power is unity. Since both in-phase and quadrature components are Gaussian, the absolute amplitude is Rayleigh distributed (see Chapter 5 for details of this derivation). Knowing the amplitude distribution, it is easy to compute the probability that the instantaneous amplitude will lie above a given threshold, and similarly for power. For example, there is a $\exp(-10^{6/10}) = 0.019$ probability that the peak power is 6 dB above the average power. Note that the Rayleigh distribution can only be an approximation for the amplitude distribution of OFDM signals: an actual OFDM signal has a bounded amplitude (N^* amplitude of signal on one subcarrier), while realizations of a Rayleigh distribution can take on arbitrarily large values.

There are three main methods to deal with the Peak-to-Average Power Ratio (PAPR):

1. Put a power amplifier into the transmitter that can amplify linearly up to the possible *peak* value of the transmit signal. This is usually not practical, as it requires expensive and power-consuming class-A amplifiers. The larger the number of subcarriers N, the more difficult this solution becomes.

2. Use a nonlinear amplifier, and accept the fact that amplifier characteristics will lead to distortions in the output signal. Those nonlinear distortions destroy orthogonality between subcarriers, and also lead to increased out-of-band emissions (*spectral regrowth* – similar to third-order inter-modulation products – such that the power emitted outside the nominal band is increased). The first effect increases the BER of the desired signal (see Figure 19.12), while the latter effect causes interference to other users and thus decreases the cellular capacity of an OFDM system (see Figure 19.13). This means that in order to have constant adjacent channel interference we can trade off power amplifier performance against spectral efficiency (note that increased carrier separation decreases spectral efficiency).

3. Use PAR reduction techniques. These will be described in the next subsection.

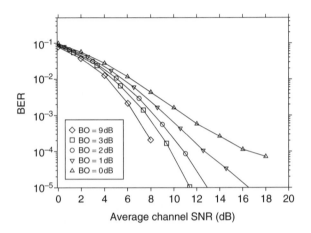

Figure 19.12   Bit error rate as a function of the signal-to-noise ratio, for different backoff levels of the transmit amplifier.

Reproduced with permission from Hanzo et al. [2003] © J. Wiley & Sons, Ltd

Figure 19.13   Interference power to adjacent bands (OFDM users), as a function of carrier separation, for different values of backoff of the transmit amplifier.

Reproduced with permission from Hanzo et al. [2003] © J. Wiley & Sons, Ltd.

19.6.2 Peak-to-Average Ratio Reduction Techniques

A wealth of methods for mitigating the PAR problem has been suggested in the literature. Some of the promising approaches are as follows:

1. *Coding for PAR reduction*: under normal circumstances, each OFDM symbol can represent one of 2^N codewords (assuming BPSK modulation). Now, of these codewords only a subset of size 2^K is acceptable in the sense that its PAR is lower than a given threshold. Both the transmitter and the receiver know the mapping between a bit combination of length K, and the codeword of length N that is chosen to represent it, and which has an admissible PAR. The transmission scheme is thus the following: (i) parse the incoming bitstream into blocks of length K; (ii) select the associated codeword of length N; (iii) transmit this codeword via the OFDM modulator. The coding scheme can guarantee a certain value for the PAR. It also has some coding gain, though this gain is smaller than for codes that are solely dedicated to error correction.

2. *Phase adjustments*: this scheme first defines an ensemble of phase adjustment vectors $\phi_l, l = 1, \ldots, L$, that are known to both the transmitter and receiver; each vector has N entries $\{\phi_n\}_l$. The transmitter then multiplies the OFDM symbol to be transmitted c_n by each of these phase vectors to get

$$\{\hat{c}_n\}_l = c_n \exp[j(\phi_n)_l] \tag{19.14}$$

and then selects

$$\hat{l} = \arg\min_l(PAR(\{\hat{c}_n\}_l)) \tag{19.15}$$

which gives the lowest PAR. The vector $\{\hat{c}_n\}_{\hat{l}}$ is then transmitted, together with the index $\hat{l}$. The receiver can then undo phase adjustment and demodulate the OFDM symbol. This method has the advantage that the overhead is rather small (at least as long as L stays within reasonable bounds); on the downside, it cannot guarantee to keep the PAR below a certain level.

3. *Correction by multiplicative function*: another approach is to multiply the OFDM signal by a time-dependent function whenever the peak value is very high. The simplest example for such an approach is the clipping we mentioned in the previous subsection: if the signal attains a level $s_k > A_0$, it is multiplied by a factor A_0/s_k. In other words, the transmit signal becomes

$$\hat{s}(t) = s(t)\left[1 - \sum_k \max\left(0, \frac{|s_k| - A_0}{|s_k|}\right)\right] \tag{19.16}$$

A less radical method is to multiply the signal by a Gaussian function centered at times when the level exceeds the threshold:

$$\hat{s}(t) = s(t)\left[1 - \sum_n \max\left(0, \frac{|s_k| - A_0}{|s_k|}\right)\exp\left(-\frac{t^2}{2\sigma_t^2}\right)\right] \tag{19.17}$$

Multiplication by a Gaussian function of variance σ_t^2 in the time domain implies convolution with a Gaussian function in the frequency domain with variance $\sigma_f^2 = 1/(2\pi\sigma_t^2)$. Thus, the amount of out-of-band interference can be influenced by the judicious choice of σ_t^2. On the downside, we find that the ICI (and thus BER) caused by this scheme is significant.

4. *Correction by additive function*: in a similar spirit, we can choose an additive, instead of a multiplicative, correction function. The correction function should be smooth enough not to introduce significant out-of-band interference. Furthermore, the correction function acts as additional pseudo noise, and thus increases the BER of the system.

When comparing the different approaches to PAR reduction, we find that there is no single "best" technique. The coding method can guarantee a maximum PAR value, but requires considerable overhead, and thus reduced throughput. The phase adjustment method has a smaller overhead (depending on the number of phase adjustment vectors), but cannot give a guaranteed performance. Neither of these two methods leads to an increase in either ICI or out-of-band emissions. The correction by multiplicative functions can guarantee performance – up to a point (subtracting the Gaussian functions centered at one point might lead to larger amplitudes at another point). Also, it can lead to considerable ICI, while out-of-band emissions are fairly well controlled.

19.7   Inter Carrier Interference

The cyclic prefix provides an excellent way of ensuring orthogonality of the carriers in a delay-dispersive (frequency-selective) environment – in other words, there is no ICI due to frequency selectivity of the channel. However, wireless propagation channels are also time varying, and thus time selective (= frequency-dispersive, due to the Doppler effect, see Chapter 5). Time selectivity has two important consequences for an OFDM system: (i) it leads to random Frequency Modulation (FM, see Chapter 5), which can cause errors especially on subcarriers that are in a fading dip; and (ii) it creates ICI. A Doppler shift of one subcarrier can cause ICI in many adjacent subcarriers (see Figure 19.14). The impact of time selectivity is mostly determined by the product of maximum Doppler frequency and symbol duration of the OFDM symbol. The spacing between the subcarriers is inversely proportional to symbol duration. Thus, if symbol duration is large, even a small Doppler shift can result in appreciable ICI.

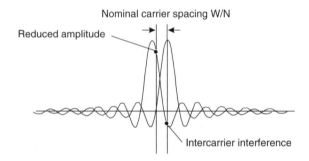

Figure 19.14   Intercarrier interference due to frequency offset.

Delay dispersion can be another source of ICI, namely if the cyclic prefix is shorter than the maximum excess delay of the channel. This situation can arise for various reasons. A system might consciously shorten or omit the cyclic prefix in order to improve spectral efficiency. In other cases, a system may originally be designed to operate in a certain class of environments (and thus a certain range of excess delays), and is later also deployed in other environments that have a larger excess delay. Finally, for many systems, the length of the cyclic prefix is a compromise between the desire to eliminate ICI, and the need to retain spectral efficiency – in other words, a cyclic prefix is not chosen to cope with the worst case channel situation.

In the following, we mathematically describe the received signal if ICI occurs either as a result of Doppler shift or insufficient cyclic prefix. Instead of Eq. (19.11), the relationship between data symbols c_n and receive samples after FFT is now given by

$$r_k = \sum_{n=0}^{N-1} c_n H_{k,n} + n_k \qquad (19.18)$$

where

$$H_{k,n} = \frac{1}{N} \sum_{q=0}^{N-1} \sum_{l=0}^{L-1} h[q,l] \exp\left[j\frac{2\pi}{N}(qn - nl - qk) \right] \mathcal{H}[q - l + N_{cp}] \tag{19.19}$$

where $h(n, l)$ is a sampled version of the time-variant channel impulse response $h(t, \tau)$, $\mathcal{H}[]$ denotes the Heaviside function, and L is the maximum excess delay in units of samples $L = \tau_{max} N / T_S$. Note also that Eq. (19.19) reduces to Eq. (19.11) for the case of a time-invariant channel and a sufficiently long guard interval.

Because ICI can be a limiting factor for OFDM systems, a large range of techniques for fighting it has been developed and can be classified as follows:

- *Optimum choice of carrier spacing and OFDM symbol length*: in this approach, we influence the OFDM symbol length in order to minimize its ICI. It follows from our statements above that short symbol duration is good for reduction of Doppler-induced ICI. On the other hand, spectral efficiency considerations enforce a minimum duration of T_S: the cyclic prefix (which is determined by the maximum excess delay of the channel) should not be shorter than approximately 10% of the symbol duration. The following equation gives a useful guideline on how to choose T_S. Let $R(k, l) = P_h(lT_c, kT_c)$ be the sampled delay cross power spectral density (see Chapter 6). Define furthermore a function:

$$w(q,r) = \frac{1}{N} \begin{cases} N - |r| & 0 \leq q \leq N_{cp} & 0 \leq |r| \leq N \\ N - q + N_{cp} - |r| & N_{cp} \leq q \leq N + N_{cp} & 0 \leq |r| \leq N - q + N_{cp} \\ N + q - |r| & -N \leq q \leq 0 & 0 \leq |r| \leq N + q \\ 0 & \text{elsewhere} \end{cases} \tag{19.20}$$

Then the desired signal power can be approximated as [Steendam and Moeneclaey 1999]:

$$P_{sig} = \frac{1}{N} \sum_l \sum_k w(k, l) R(k, l) \tag{19.21}$$

and the ICI and ISI powers as

$$P_{ICI} = \sum_k w(k, 0) R(k, 0) - P_{sig} \tag{19.22}$$

$$P_{ISI} = \sum_l [1 - w(k, 0)] R(k, 0) \tag{19.23}$$

and the SINR is

$$SINR = \frac{\frac{E_S}{N_0} P_{sig} \frac{N}{N_{cp} + N}}{\frac{E_S}{N_0} P_{sig} \frac{N}{N_{cp} + N} \frac{P_{ISI} + P_{ICI}}{P_{sig}} + 1} \tag{19.24}$$

The above equations allow an easy tradeoff between the ICI due to the Doppler effect, the ICI due to residual delay dispersion, and SNR loss due to the cyclic prefix.

- *Optimum choice of OFDM basis signal*: a related approach influences the OFDM basis pulse *shape* in order to minimize ICI. We know that a rectangular temporal signal has a very sharp cutoff in the temporal domain, but has a $\sin(x)/x$ shape in the frequency domain and thus decays slowly. And while in a perfect system each subcarrier is in the spectral nulls of all other subcarriers, the slope of the $\sin(x)/x$ is large near its zeros. Thus, even a small Doppler shift leads to large ICI. By choosing basis pulses whose spectrum decays faster and gentler, we decrease ICI due to the Doppler effect. On the downside, faster decay in the frequency domain is bought by slower decay in the time domain, which increases delay-spread-induced errors. Gaussian-shaped basis functions have been shown to be a useful compromise.

- *Self-interference cancellation techniques*: in this approach, information is modulated not just onto a single subcarrier but onto a group of them. This technique is very effective for mitigation of ICI, but leads to a reduction in spectral efficiency of the system.

- *Frequency domain equalizers*: if the channel and its variations are known, then its impact on the received signal, as described by Eq. (19.18), can be reversed. While this reversal can no longer be done by a single-tap equalizer, there is a variety of suitable techniques. For example, we can simply invert H, or use a minimum mean square error criterion. These inversions can be computationally expensive: as the channel is continuously changing, the inverse matrix has to be recomputed for every OFDM block. However, methods with reduced computational complexity exist. Another approach is to interpret different tones as different users, and then apply multiuser detection techniques (as described in Section 18.4) for detection of the tones. Figure 19.15 shows an example of the effect of different equalization techniques (Operator Perturbation Technique (OPT) denotes a linear inversion technique, while Parallel Interface Cancellation (PIC) and Successive Interface Cancellation (SIC) denote multiuser detection).

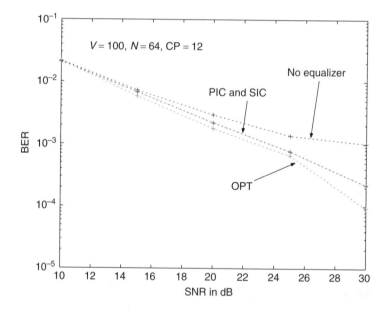

Figure 19.15 Bit error rate as a function of signal-to-noise ratio for an 802.11a-like orthogonal-frequency-division-multiplexing system with 64 carriers and 12 samples CP. Performance is analyzed in channel model F of the 802.11n channel models, with 100-m/s velocity.

In addition to time selectivity and delay dispersion, there is another effect that can destroy orthogonality between carriers: errors in the local oscillator (LO). Such errors can be produced by

- *Synchronization errors*: as we discussed in Section 19.4, synchronization is critical for retaining orthogonality between carriers. Any errors in the synchronization procedure will be reflected as deviation of the receiver's LO from the optimum frequency, and thus ICI.

- *Phase noise of the transmitter and receiver*: phase noise, which stems from inaccuracies in the oscillator, leads to deviation of the LO signal from its nominal, strictly sinusoidal shape. The distribution of phase noise is typically Gaussian, and is further characterized by its power-spectral density. Essentially, a narrow spectrum means that phase only changes very slowly, which can be more easily compensated by various receiver algorithms. The effect of phase noise is a spilling of the spectrum of subcarrier signals into adjacent subcarriers, and thus ICI.

Example 19.2 *Consider a system with a 5-MHz bandwidth, 128 tones, and a cyclic prefix that is 40 samples long. It operates in a channel with an exponential Power Delay Profile (PDP), $\tau_{rms} = 1\,\mu s$, $\nu_{rms} = 500\,Hz$, and an E_S/N_0 of 10 dB. What is the Signal-to-Interference-and-Noise Ratio (SINR) at the receiver? How do results change when the cyclic prefix is shortened to 12 samples?*

In a first step, we need to find the sampled delay cross power spectral density. For a bandwidth of 5 MHz, the sampling interval is 200 ns. Therefore, the rms delay spread is five samples, and the sampled PDP is described as $\exp(-k/5)$. The Doppler spectrum is assumed to have a Gaussian shape. Assuming furthermore that the Doppler spectrum is independent of delay, we obtain:

$$R(k, l) = \exp(-k/5) \cdot \exp\left(-\frac{l^2}{2 \cdot 10{,}000^2}\right) \qquad (19.25)$$

An accurate solution for interference power can then be found by inserting this sampled delay cross power spectral density into Eqs. (19.20)–(19.24). We obtain:

$$\frac{P_{ICI}}{P_{sig}} = 2.46 \cdot 10^{-5} \qquad (19.26)$$

$$\frac{P_{ISI}}{P_{sig}} = 1.18 \cdot 10^{-5} \qquad (19.27)$$

This shows that ISI and ICI are reasonably balanced, which overall leads to low interference power.

Furthermore, the cyclic prefix reduces the effective SNR by a factor $128/(128 + 40) = 0.762$. Thus, the total SINR becomes

$$\frac{7.62}{7.62(2.46 + 1.18) \cdot 10^{-5} + 1} = 7.6 \qquad (19.28)$$

This indicates that the major loss of SINR occurs due to the cyclic prefix. When we shorten it from 40 to 12 samples, the sum of ISI and ICI increases to

$$\frac{P_{ICI} + P_{ISI}}{P_{sig}} = 6.2 \cdot 10^{-3} \qquad (19.29)$$

On the other hand, the SNR becomes $10 \cdot 128/(128 + 12) = 9.14$. Thus the effective SNR becomes

$$\frac{9.14}{9.14 \cdot 6.2 \cdot 10^{-3} + 1} = 8.65 \qquad (19.30)$$

This shows that a long cyclic prefix is not always the best way to improve the SINR. Rather, it is important to correctly balance ISI, ICI, and duration of the cyclic prefix.

19.8 Adaptive Modulation and Capacity

Adaptive modulation changes the coding scheme and/or modulation method depending on channel-state information – choosing it in such a way that it always "pushes the limit" of what the channel can transmit. In OFDM, modulation and/or coding can be chosen differently for each subcarrier, and it can also change with time. We thus not only accept the fact that the channel (and thus the SNR) shows strong variations but even exploit this fact. On subcarriers with good SNR, transmission is done at a higher rate than on subcarriers with low SNR. In other words, such an adaptive modulation selects a modulation scheme and code rate according to the channel quality of a specific subcarrier.

Let us compare adaptive modulation for OFDM with coded OFDM and multicarrier CDMA. Both of the latter systems try to "smear" the data symbols over many subcarriers, so that each symbol sees approximately the same average SNR. It can be shown that – at least theoretically – systems using adaptive modulation perform better than systems whose modulation and coding are fixed once and for all.

19.8.1 Channel Quality Estimation

Adaptive modulation requires that the transmitter knows channel-state information. This requirement sounds trivial, but is quite difficult to realize in practice. It requires the channel to be reciprocal – e.g., the Base Station (BS) learns the channel state while it is in receiving mode; when it then transmits, it relies on the fact that the channel is still in the same state. This can only be fulfilled in systems with Time Domain Duplex (TDD) in slowly time-varying channels. Alternatively, feedback from the receiver to the transmitter can be used to inform the transmitter about the channel state (see also Sections 17.5 and 20.1.6). It is also noteworthy that the transmitter has to know the channel-state information *for the time instant when it will transmit* – in other words, it has to look into the future. Channel prediction is thus an important component for many adaptive modulation systems.

19.8.2 Parameter Adaptation

Once the channel state is known, the transmitter has to decide how to select the correct transmission parameter for each subcarrier – namely, coding rate and modulation alphabet. Furthermore, we also have to consider how much power should be assigned to each channel. In the following, we will assume that the channel is frequency selective, but time invariant. This will make the discussion easier, and the principles can be easily extended to doubly selective channels.

Let us first consider the question of how much power should be assigned to each channel. In order to find the answer, let us first reformulate the question in a more abstract way: "Given a number of parallel subchannels with different attenuations, what is the distribution of transmission power that maximizes capacity?" The answer to this latter question was given by Shannon in the 1940s, and is known as "waterfilling." Power allocation P_n of the nth subchannel is

$$P_n = \max\left(0, \varepsilon - \frac{\sigma_n^2}{|\alpha_n|^2}\right) \qquad (19.31)$$

where α_n is the gain (inverse attenuation) of the nth subchannel, σ_n^2 is noise variance, and the threshold ε is determined by the constraint of the total transmitted power P as

$$P = \sum_{n=1}^{N} P_n \qquad (19.32)$$

Waterfilling can be interpreted visually, according to Figure 19.16. Imagine a number of con- nected vessels. At the bottom of each vessel is a block of concrete with a height that is proportional to the inverse SNR of the subchannel that we are considering. Then take water, and pour it into the vessels; the amount of poured water is proportional to the total transmit power that is available. Because the vessels are connected, the surface level of the water is guaranteed to be the same in all vessels. The amount of power assigned to each subchannel is then the amount of water in the vessel corresponding to this subchannel. Obviously, subchannel 1, which has the highest SNR, has the most water in it. It can also happen that some subchannels that have a poor SNR (like channel 5), do not get any power assigned to them at all (the concrete block of that vessel is sticking out of the water surface). Essentially, waterfilling makes sure that energy is not wasted on subchannels that have poor SNR: in the OFDM context, this means not wasting power on subcarriers that are in a deep fade.

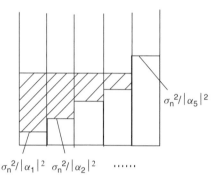

Figure 19.16   Principle behind waterfilling.

With waterfilling, power is allocated preferably to subchannels that have a good SNR ("give to the rich" principle). This is optimum from the point of view of theoretical capacity; however, it requires that the transmitter can actually make use of the large capacity on good subchannels.

In each subchannel (subcarrier), signaling as close to capacity as possible should be performed.[5] This means that the transmitter has to adapt the data rate according to the SNR that is available (note that waterfilling increases SNR differences between subcarriers). Consequently, the coding rate and the constellation size of the modulation alphabet have to be adjusted. For very high SNR, the constellation size, and thus the PAR, has to be very large. A Quadrature Amplitude Modulation (QAM) of alphabet size 64 currently seems to be the largest constellation size that can be used in practical systems. The capacity per subchannel is limited by $\log_2(N_a)$, where N_a is the size of the symbol alphabet. It is thus wasteful to assign more energy to one stream than can be actually exploited by the alphabet. If the available alphabet is small, a "giving to the poor" principle for power allocation is preferable – i.e., assigning power that cannot be exploited by good subchannels to bad subchannels.

Example 19.3   *Waterfilling: consider an OFDM system with three tones, with $\sigma_n^2 = 1$, $\alpha_n^2 = 1$, 0.4, 0.1, and total power $\sum P_n = 15$. Compute the power assigned to different tones according to (i) waterfilling, (ii) equal power allocation, (iii) predistortion (inverting the channel attention), and compute the resulting capacity.*

---

[5] We assume in the following that near-capacity-achieving codes are used. If this is not the case, we usually try to choose the data rate in such a way that a certain BER can be guaranteed.

From Eq. (19.31) we find that $\varepsilon = 9.25$ gives the correct solution: in that case, the power in the different subchannels are

$$P_1 = 8.25 \tag{19.33}$$

$$P_2 = 6.75 \tag{19.34}$$

$$P_3 = 0 \tag{19.35}$$

that is, no power is assigned to the channel that suffers from the strongest attenuation. The total capacity can be computed as

$$C_{\text{waterfill}} = \sum_{n=1}^{N} \log_2(1 + \alpha_n^2 P_n/\sigma_n^2) = 5.1 \text{ bit/s/Hz} \tag{19.36}$$

For equal power allocation:

$$P_1 = P_2 = P_3 = 5 \tag{19.37}$$

so that capacity becomes:

$$C_{\text{equal-power}} = 4.8 \text{ bit/s/Hz} \tag{19.38}$$

For the predistortion case, the powers become

$$P_1 = 1.1 \tag{19.39}$$

$$P_2 = 2.8 \tag{19.40}$$

$$P_3 = 11.1 \tag{19.41}$$

from which we obtain a capacity of:

$$C \text{ predistort} = 3.2 \text{ bit/s/Hz} \tag{19.42}$$

In this example, equal power allocation gives almost as high a capacity as (optimum) waterfilling, while predistortion leads to significant capacity loss.

19.8.3 Signaling of Chosen Parameters

Most information-theoretic investigations assume a continuum of modulation alphabets that can realize any arbitrary transmission rate. In practice, the transmitter has a finite and discrete set of modulation alphabets available (BPSK, Quadrature-Phase Shift Keying (QPSK), 16-QAM, and 64-QAM). There is also a finite set of possible code rates: the different codes are usually obtained form a "mother" code by different amounts of puncturing. Thus, the available data rates form a discrete set.

After the transmitter has decided which transmission mode – i.e., combination of signal constellation and encoder – to use on each tone, it has to communicate that decision to the receiver. There are three possibilities to achieve that task:

- *Explicit transmission*: the transmitter can send, in a predefined and robust format, the index of the transmission mode it intends to use. Transmission of this information itself should always be done in the same mode, and care should be taken that the message is well protected against errors during transmission.

- *Implicit transmission*: implicit transmission is possible when the transmitter gets its channel-state information from the receiver via feedback. In such a case, the receiver knows exactly what channel-state information is available to the transmitter, and thus the basis on which the decision for a transmission mode is being made. Thus, the receiver just needs to know the decision rule on which the transmitter bases its choice of transmission mode. If the receiver feeds back the mode that the transmitter should use, the situation is even simpler.

  The drawback to this method is that errors in channel-state feedback (from the receiver to the transmitter) not only lead to a wrong choice of transmission mode (which is bad, but usually not fatal) but also to detection and decoding using the wrong code, which leads to very high error rates.

- *Blind detection*: from the received signal, the receiver can try to determine the signal constellation. This can be achieved by considering different statistical properties of the received signal, including the PAR, autocorrelation functions, and higher order statistics of the signal.

19.9   Multiple Access – OFDMA

OFDM is a modulation format that allows the transmission of high data rates for a single user; it is usually not seen as a multiple-access format. But how can it be used for the purpose of multiple access, i.e., allowing simultaneous communication of several users? There are a number of different possibilities, most of which are just straightforward combinations with multiple-access technologies we encountered in Chapter 17:

- TDMA: each user occupies the whole system bandwidth, and different users are served at different times. At a minimum, one user transmits for one symbol duration, but it is also possible that one user transmits/receives multiple symbols before the system switches to the next user. This approach is used, e.g., in the OFDM mode of IEEE 802.16.
- Packet radio: in this mode, every user transmits complete packets, where the modulation format for each packet is OFDM. The access of the users to the channel is regulated by the packet access techniques discussed in Section 17.4, like ALOHA, and Carrier Sense Multiple Access (CSMA). This approach is used in IEEE 802.11a/g/n (see Chapter 29).

An alternative, known as Orthogonal Frequency Division Multiple Access (OFDMA) assigns different subcarriers to different users. There are essentially three ways of assigning the subcarriers to different users (see Figure 19.17):

1. Assigning adjacent subcarriers to one user, i.e., a set of $N^{(k)}$ carriers f_n, with $n = n_{\text{low},k}, n_{\text{low},k} + 1, \ldots n_{\text{low},k} + N^{(k)} - 1$, and $n_{\text{low},k+1} = n_{\text{low},k} + N^{(k)}$. The advantage of the method is that channel estimation is simplified: since adjacent subcarriers are correlated (they are close in frequency, and they are assigned to the same link. Furthermore, it is possible to perform "intelligent scheduling": each user is assigned the group of frequencies that provide the best channel quality for him/her. The fading of the channels for the different users is independent; thus if $h^{(k)}(f_{\tilde{n}})$ is very small (i.e., the channel for the k-th user is in a fading dip at frequency $\tilde{n}$), then it is unlikely that $h^{(j)}(f_{\tilde{n}})$ is also small for $j \neq k$. In other words, we obtain a high degree of selection diversity without suffering a loss of spectral efficiency – every subcarrier is assigned to somebody. This form of diversity is also known as "multiuser diversity" (compare also Section 20.1.9), because it requires the presence of multiple users to take effect. Note, however, that the BS has to first know the propagation channel (over the full system bandwidth) from each user; and as channels change, the channel assignments for the users have to change. This method is used, e.g., in 3GPP-LTE, see Chapter 27, and (optionally) in WiMAX, see Chapter 28.

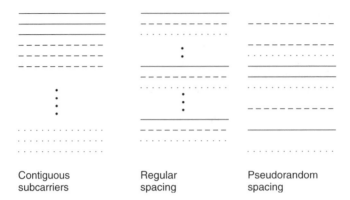

Contiguous subcarriers	Regular spacing	Pseudorandom spacing

Figure 19.17 Assignment of subcarriers to different users (indicated as solid, dashed, and dotted lines).

2. Assigning regularly spaced subcarriers to one user, i.e., the set f_n with $n = n_{\text{low},k}, n_{\text{low},k} + Q, \ldots n_{\text{low}}, k + Q(N^{(k)} - 1)$ and $n_{\text{low},k+1} = n_{\text{low},k} + 1$. This assignment has the advantage of a high degree of frequency diversity – actually, essentially the same as "standard" OFDM that has every single carrier assigned to one user, as long as $Q(f_{n+1} - f_n)$ is at most one coherence bandwidth. Compared to regular OFDM, a finer granularity of the data rates assigned to different users can be achieved, namely N/Q. Also the PAPR problem is reduced. Finally, the scheme does not require knowledge of all the channels at the BS for scheduling – essentially, each user can always be assigned to the same set of subcarriers. The drawback (compared to the adjacent subcarrier assignment) is that there is no multiuser diversity. Every assigned subcarrier has the probability of being in a fading dip; a user can exploit the frequency diversity only through appropriate coding, similar to the situation described in Section 19.4.
3. Assigning randomly spaced subcarriers to one user, so that the indices n of the subcarriers f_n are part of a random sequence $b(i)$, i.e., $n = b(i)$, with $i = i_{\text{low},k}, i_{\text{low},k} + 1, \ldots i_{\text{low},k} + N^{(k)} - 1$, and $i_{\text{low},k+1} = i_{\text{low},k} + N^{(k)}$. Largely, this assignment strategy has the same pros and cons as the regularly spaced subcarriers: (i) assignment of subcarriers to users can stay constant, (ii) no full channel knowledge is required for assignment strategy, (iii) full frequency diversity, and (iv) no multiuser diversity. The main advantage of this scheme lies in the behavior of adjacent-cell interference. If adjacent cells use different random sequences, then the adjacent-cell interference is randomized: interference on different subcarriers comes from different users in the neighboring cells. This method is used in WiMAX (see Chapter 28).

OFDMA is usually combined with TDMA, so that users are assigned different parts of the "time-frequency plane." It must be noted that the signaling required to let the users know which subcarriers (and which times) are intended for them can be a very significant overhead.

19.10 Multicarrier Code Division Multiple Access

Multi Carrier CDMA (MC-CDMA) spreads information from each data symbol over all tones of an OFDM symbol. At first glance, it is paradoxical to combine CDMA, which tries to spread a signal over a very large bandwidth, with multicarrier schemes, which try to signal over a very narrowband channel. But we will see in the following that the two methods can actually be combined very efficiently. We have mentioned repeatedly that uncoded OFDM has poor performance, because it is dominated by the high error rate of subcarriers that are in fading dips. Coding improves the situation, but in many cases a low coding rate – i.e., high redundancy – is not desirable. We thus need to find an alternative way of exploiting the frequency diversity of the channel. By spreading a modulation symbol over many tones, MC-CDMA becomes less sensitive to fading on one specific tone.

The basic idea of MC-CDMA is to transmit a data symbol on all available subcarriers simultaneously. In other words, a code symbol c is mapped to a vector $c\mathbf{p}$, where $\mathbf{p}$ is a predetermined vector. This can be interpreted as a repetition code (a symbol is repeated on each tone, but multiplied by a different known constant p_n), or as a spreading action, where each symbol is represented by a code sequence (but the sequence is now in the frequency domain, instead of the time domain). Irrespective of the interpretation, we obtain a bandwidth expansion by a factor of N.

Bandwidth expansion leads to a loss of spectral efficiency. However, we can eliminate this problem by transmitting N different symbols, and thus N different codevectors, *simultaneously*. The first symbol c_1 is multiplied by codevector $\mathbf{p}_1$, the second symbol c_2 by codevector $\mathbf{p}_2$, and so on, and those spread symbols are added up and transmitted. If all the vectors $\mathbf{p}_n$ are chosen to be orthogonal, then the receiver can recover the different transmitted symbols.

Let us now put this into a more compact mathematical form by writing the codevectors into a "spreading matrix" $\mathbf{P}$:

$$\mathbf{P} = [\mathbf{p}_1 \quad \mathbf{p}_2 \quad \cdots \quad \mathbf{p}_N] \tag{19.43}$$

where we will see later that it is advantageous if $\mathbf{P}$ is unitary. The *symbol spreader* performs a matrix multiplication:

$$\tilde{\mathbf{c}} = \mathbf{P}_\mathbf{c} \tag{19.44}$$

Now it is this modified signal that is OFDM-modulated – i.e., undergoes an IFFT and has the cyclic prefix prepended – and sent over the wireless channel.

In the receiver, we first perform the same operation as in "normal" OFDM: stripping off the cyclic prefix and performing an FFT. Thus, we have again transformed the symbols into the frequency domain. At this point, the received symbols are

$$\tilde{\mathbf{r}} = \mathbf{H}\tilde{\mathbf{c}} + \mathbf{n} \tag{19.45}$$

where $\mathbf{H}$ is a diagonal matrix with entries $H\left(n\frac{W}{N}\right)$ along the diagonal. The next step is "one-tap equalization." Let us assume for the moment that we use zero-forcing equalization (we will see below why this is more relevant in MC-CDMA than in conventional OFDM). Then employing the unitary properties of the spreading matrix, we can perform despreading by just multiplying the received signal by the Hermitian transpose of the spreading matrix to obtain:

$$\mathbf{P}^\dagger\mathbf{H}^{-1}\tilde{\mathbf{r}} = \mathbf{P}^\dagger\mathbf{H}^{-1}\mathbf{H}\mathbf{P}\mathbf{c} + \mathbf{P}^\dagger\mathbf{H}^{-1}\mathbf{n} \tag{19.46}$$

$$= \mathbf{c} + \tilde{\mathbf{n}} \tag{19.47}$$

We have thus recovered the transmit symbols. A summary of the transceiver structure can be seen in Figure 19.18.

Note that noise is no longer white, since the $\tilde{\mathbf{n}}$ includes noise enhancement from zero-forcing equalization. When the receiver uses MMSE equalization instead of zero-forcing, then noise enhancement is not as bad. However, MMSE equalization does not recover the orthogonality of different codewords the way that zero-forcing does. After equalization and despreading, we get

$$\mathbf{P}^\dagger \frac{\mathbf{H}^*}{|\mathbf{H}|^2 + \sigma_n^2}\tilde{\mathbf{r}} = \mathbf{P}^\dagger \frac{\mathbf{H}^*\mathbf{H}}{|\mathbf{H}|^2 + \sigma_n^2}\mathbf{P}\mathbf{c} + \mathbf{P}^\dagger \frac{\mathbf{H}^*}{|\mathbf{H}|^2 + \sigma_n^2}\mathbf{n} \tag{19.48}$$

The matrix $\mathbf{P}^\dagger\mathbf{H}^*\mathbf{H}/(|\mathbf{H}|^2 + \sigma_n^2)\mathbf{P}$ is not diagonal. This means that there is residual crosstalk from one codeword to the other.

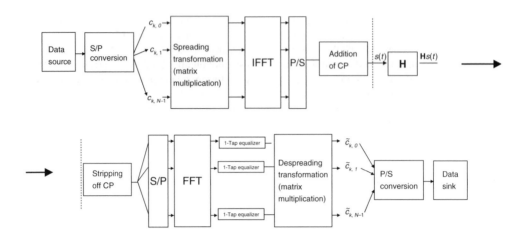

Figure 19.18   Block diagram of a multicarrier code-division-multiple-access transceiver.

What spreading matrices should be used for MC-CDMA systems? One obvious choice is the Walsh–Hadamard matrices that we discussed in Section 18.2. They are unitary and have an extremely simple structure. All the coefficients are ±1; furthermore, Walsh–Hadamard matrices of larger size can be computed via recursion from matrices of smaller size. This allows implementation of a Walsh–Hadamard transform with a "butterfly" structure, similar to implementation of an FFT.

How does spreading influence the PAR problem? In most cases, there is no significant impact. The explanation can again be found from the central limit theorem: the output (amplitudes of the I- and Q-components on different subcarriers) of the spreading matrix is approximately Gaussian distributed, as it is the sum of a large number of variables. The IFFT then just weights and sums those Gaussian variables. But the weighted sum of Gaussian variables is again a Gaussian variable. The amplitude distribution of the transmit signal of MC-CDMA is thus the same as the amplitude distribution of normal OFDM.

19.11   Single-Carrier Modulation with Frequency Domain Equalization

A special case of MC-CDMA occurs when the unitary transformation matrix **P** is chosen to be the FFT matrix. In such a case, multiplication by the spreading matrix and the IFFT inherent in the OFDM implementation cancel out. In other words, the transmit sequence that is transmitted over the channel is the original data sequence – plus a cyclic prefix that is just a prepending of a few data symbols at the beginning of each datablock.

This just seems like a rather contrived way of describing the single-carrier system that has already been discussed in Part III. However, the big difference here lies in the existence of the cyclic prefix, as well as in how the signal is processed at the receiver (see Figure 19.19). After stripping off the cyclic prefix, the signal is transformed by an FFT into the frequency domain. Due to the cyclic prefix, there are no residual effects of ISI or ICI. Then, the receiver performs equalization on each subcarrier (this can be zero-forcing or MMSE equalization), and finally transforms the signal back into the time domain via an IFFT (this is the despreading step of MC-CDMA). The receiver thus performs equalization in the frequency domain. Since FFTs or IFFTs can be implemented efficiently, the computational effort (per bit) for equalization goes only like $\log_2(N)$. This is a considerable

Figure 19.19   Block diagram of a single-carrier frequency domain equalization receiver.

advantage compared with the equalization techniques discussed in Chapter 16. On the downside, frequency domain equalization is a linear equalization scheme and therefore does not give optimum performance. Also, the extra overhead of a cyclic prefix has to be taken into account.

Further Reading

Several books describe multicarrier schemes, including OFDM and multicarrier CDMA, e.g., Hanzo et al. [2003] and Li and Stuber [2006]; another interesting description that also covers applications to various standardized systems is Bahai et al. [2004]. Review papers on that topic include Wang and Giannakis [2000], Jajszczyk and Wagrowski [2005], Hwang et al. [2009].

OFDM was invented by Chang [1966], and the cyclic prefix was proposed in Weinstein and Ebert [1971]. Optimization of a set of nonrectangular basis function that span the time-frequency plane according to different criteria (e.g., low sensitivity to time and frequency dispersion) can be found in Kozek and Molisch [1998]. Discussion of the cyclic prefix and its comparison to zero-padding can be found in Muquet et al. [2002]. The performance of a coded OFDM system in a multipath channel is analyzed, e.g., in Kim et al. [1999]. ICI is discussed in Cai and Giannakis [2003] and Choi et al. [2001]; the latter also discussed ICI mitigation techniques, as does a number of other papers (see, e.g., Schniter [2004] and Molisch et al. [2007a] and references therein.

Synchronization and channel estimation are very important topics. Some examples of time and frequency synchronization include Schmidl and Cox [1997], Speth et al. [1999], and van de Beek et al. [1999]. Channel estimation techniques are discussed in Edfors et al. [1998] and Li et al. [1998]. Approximate DFT estimators were introduced in van de Beek et al. [1995] and later analyzed in detail in Edfors et al. [2000]. Filtering after eigen transformation was introduced in Li et al. [1998] and extended to the transmit diversity case in Li et al. [1999]. Methods for PAR techniques are reviewed in May and Rohling [2000] and Jiang and Wu [2008]. Multi access considerations are described in Rohling [2005].

For adaptive modulation, the paper by Wong et al. [1999], and especially the overview paper by Keller and Hanzo [2000], gives a good account. Yang and Hanzo [2003] review MC-CDMA. Falconer et al. [2003] and Benvenuto et al. [2010] discuss single-carrier frequency equalization schemes; a combination of this scheme with diversity is detailed in Clark [1998]. A unified analysis of OFDM and single-carrier schemes with cyclic prefix is given in Akino et al. [2009].

For updates and errata for this chapter, see wides.usc.edu/teaching/textbook

20

Multiantenna Systems

Since the 1990s, there has been enormous interest in multiantenna systems. As spectrum became a more and more precious resource, researchers investigated ways of improving the capacity of wireless systems without actually increasing the required spectrum. Multiantenna systems offer such a possibility.

When discussing multiantenna systems, we distinguish between *smart antenna systems* (systems with multiantenna elements at one link end only), and *Multiple Input Multiple Output* (MIMO) systems, which have multiantenna elements at both link ends. Both of these systems are discussed in this chapter.

20.1 Smart Antennas

20.1.1 What are Smart Antennas?

Let us start out with a general definition: (receiver (RX)) smart antennas are "antennas with multiple elements, where signals from different elements are combined by an adaptive (intelligent) algorithm"; for the transmit case, the signals at the antenna elements are *created* by the algorithm.[1]

Intelligence (smartness) is not in the antenna, but rather in signal processing. In the simplest case, combination of antenna signals is a linear combination using a weight vector **w**. The ways of determining **w** essentially differentiates smart antenna systems. We see immediately that there is a strong relationship between multiantenna systems and diversity systems – as a matter of fact, an RX with antenna diversity *is* a smart antenna. This chapter thus reuses many results from Chapter 13, while concentrating more on system aspects and the impact smart antennas have on multiple access.

The definition of smart antennas as a combiner of different antenna signals stresses the fact that we exploit signals from different *spatial locations*. Alternatively, we can also say that smart antennas exploit the *directional properties of the channel*. Remember from Chapter 8 and 13 that an RX with multiantenna elements can distinguish Multi Path Components (MPCs) with different Directions Of Arrival (DOAs). Thus, one way to interpret smart antennas is as a spatial Rake[2] RX that distinguishes between MPCs with different DOAs and processes them separately. This allows the RX to coherently add up the different MPCs, and thus reduce fading; MPCs from interferers can also be suppressed.

---

[1] In most practical situations, smart antennas are located at the Base Station (BS), and this is also the case we assume in this section, unless otherwise specified.

[2] A device that enables Code Division Multiple Access (CDMA) systems to deal with a channel's MPCs (see Section 18.2).

Yet another interpretation is that the smart antenna adaptively forms an antenna array pattern that enhances (coherently combines) the desired MPCs, e.g., by pointing a beam with strong gain in the direction of desired MPCs. Furthermore, the antenna arrays can also form notches in their patterns. The latter is important for suppressing co-channel interference. If there are only few interferers, the ability to suppress interferers makes a bigger contribution to the improvement of the Signal-to-Interference Ratio (SIR) than the enhancement of the desired signal.

20.1.2   Purpose

Smart antennas can be used for various purposes:

1. *Increase of coverage*: assume that the smart antenna is at the RX. Now, if the spatial (angular) position of the transmitter (TX) is known, the RX can form an array pattern in the direction of the TX (beamforming). This results in higher receive power – e.g., an antenna array with eight elements can increase the Signal-to-Noise Ratio (SNR) by 9 dB compared with a single antenna. In a noise-limited cellular system, an improvement in the SNR increases the area that can be covered by one BS. Conversely, the coverage range can be kept constant while the transmit power is decreased.
2. *Increase of capacity*: smart antennas can increase the SIR, e.g., through the use of optimum combining (see Chapter 13); this in turn allows the number of users in the system to be increased. This is the practically most important advantage of smart antennas, and are discussed in more detail below.
3. *Improvement of link quality*: by increasing signal power and/or decreasing interference power, we can also increase the transmission quality on each single link.
4. *Decrease of delay dispersion*: by suppressing MPCs with large delays, delay dispersion can be reduced. This feature can be especially useful in systems with a very high data rate.
5. *Improvement of user position estimation*: knowledge of the DOAs, especially for the (quasi-) Line Of Sight (LOS) component, improves geolocation. This is of value both for *location-based services* and for the ability to locate users in case of emergency.

It is *not* possible to have all these advantages simultaneously to their fullest extent. For example, we can use the capability of smart antennas to reduce interference in order to *either* improve the quality of a single link *or* have more users in the system or to *slightly* increase both the quality per link and the number of users. For system design, the engineer has to decide which aspect is the most important.

20.1.3   Capacity Increase

As mentioned above, increasing the capacity is the most important application of smart antennas. Depending on whether the considered system uses Time Division Multiple Access (TDMA), Frequency Division Multiple Access (FDMA), or CDMA, there are different approaches for achieving such capacity gains (Figure 20.1):

1. *Spatial Filtering for Interference Reduction* (SFIR): this is used in TDMA/FDMA systems to reduce the reuse distance. A conventional TDMA/FDMA system cannot reuse the same frequency in each neighboring cell since the interference from the adjacent cells would be too strong (Section 17.6). Rather, there is a "cluster" of cells where each cell uses a different frequency, and then a neighboring cluster uses the same set of frequencies. Cluster size is a measure for how spectrally efficient a cellular network is: for Global System for Mobile communications

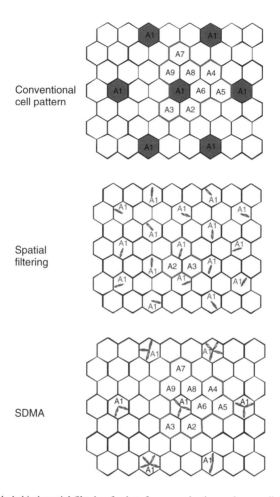

Figure 20.1   Principle behind spatial filtering for interference reduction and space division multiple access.

(GSM)-like systems, it is usually between 3 and 7. Smart antennas reduce interference. For this reason we can put cells with the same frequency closer together – in other words, reduce the cluster size. It is immediately obvious that this leads to an improvement in area spectral efficiency: when frequencies can be used in more cells, then the number of users per area is increased proportionately. Simulations have shown that eight antennas increase capacity by a factor of 3 [Kuchar et al. 1997], while field tests with prototypes [Dam et al. 1999] have shown a somewhat smaller gain of approximately 2.

2. *Space Division Multiple Access* (SDMA): this is an alternative way of increasing the capacity of TDMA/FDMA systems. In this method, cluster size (frequency reuse) remains unchanged, while the number of users within a given cell is increased. Multiple users can be served on the *same* time/frequency slot, because the BS distinguishes them by means of their different spatial signatures. To understand this mechanism, imagine a situation where there are multiple directional horn antennas at the BS, each of which can be pointed mechanically at a different user, independently of each other. An actual system realizes these directional antennas as the different beams of an antenna array (remember that an antenna array can form multiple beams

simultaneously).The capacity gain that can be achieved with SDMA is larger than with SFIR; however, the required modifications within a system, especially at the BS and the BS controller software, are considerably larger.

3. *Capacity increase in CDMA systems*: smart antennas can also be used to increase capacity in CDMA systems. The way this capacity increase is achieved is subtly different from TDMA/FDMA systems, in that it is based on the enhancement of the desired signal (and not the suppression of a few interferers). In the following, we develop an approximate mathematical model for the capacity enhancement. In a CDMA system, all Mobile Stations (MSs) use the same carrier frequency, and are distinguished only by the spreading codes they employ. The suppression of users with other codes is not perfect, and a cell is judged to be "full" when the residual interference from all the other K users becomes comparable to the admissible SIR (refer to Chapter 18 for more details and a discussion of the assumptions):

$$SIR_{\text{threshold}} = \frac{P_{\text{desired}}}{\sum P_{\text{interfere}}} \tag{20.1}$$

where, for the case of perfect power control, $\sum P_{interfere} = K \cdot P_{interfere}$. For typical (voice) CDMA systems, the number of users is on the order of 30 per cell. Thus, there are a large number of possible interferers, each of which contributes only a small part of the interference. It is thus *not* possible to completely suppress interference (remember that the maximum number of interferers that can be suppressed is smaller than the number of antenna elements). Rather, the goal of the smart antenna is to enhance the received signal power. An N_r– element antenna can increase the received signal power by a factor of N_r, so that approximately:

$$P_{\text{desired}} = M_{\text{c}} \cdot N_{\text{r}} \cdot P_{\text{interfere}} \tag{20.2}$$

where M_c is the spreading gain. In that case, the number of admissible users in the cell becomes

$$K = \frac{M_{\text{c}} \cdot N_{\text{r}}}{SIR_{\text{threshold}}} \tag{20.3}$$

and thus increases *linearly* with the number of antenna elements. In practice, such an enormous increase in capacity does not occur; typically, capacity can be doubled with an eight-element array. This is comparable with the gains that can be achieved in a TDMA system. Note that this derivation is based on numerous simplifying assumptions, especially with respect to the channel model. It furthermore assumes that there are a large number of users in a CDMA system, and that all of them contribute to interference in a similar way. This assumption holds for second-generation CDMA systems – e.g., IS-95[3] – which are mainly tuned to voice users.

4. *Capacity increase in third-generation CDMA systems*: in third-generation networks, high-rate data transmission is considered an important application. Such a high-rate user creates considerably more interference due to the lower spreading factor, such that suppression of such a user by placing a null in its direction becomes desirable. In other words, high-rate users lead to scenarios and results that are more similar to the SFIR approach than the traditional CDMA approach.

Example 20.1 *Consider a CDMA system with spreading factor 128 for voice users, and 4 for data users, and an SIR threshold of 9 dB. There is one data user. How many voice users can be accommodated without smart antennas? How many can be accommodated with a two-element smart antenna?*

---

[3] See Chapter 25.

The SIR threshold (on a linear scale) is 8. Without smart antennas, the capacity of the system (in units of voice users) can be obtained from

$$K = \frac{128}{8} = 16 \tag{20.4}$$

Since the spreading factor for data users is smaller by a factor of 32 compared with voice users, each data user contributes (approximately) 32 times the interference. The data user thus requires a capacity of $K_{\text{data}} = 32$, so that $16 - 32 = 0$ voice users can be accommodated.

Consider now the case with smart antennas. When smart antennas are used solely for enhancement of the desired signal power, the capacity of the cell (in units of voice users) becomes

$$K = \frac{128.2}{8} = 32 \tag{20.5}$$

so that now $32 - 32 = 0$ voice users can be serviced.

If, however, smart antennas are used to suppress data users, higher capacity can be achieved: as we saw in Chapter 13, we can suppress 1 interferer completely using two antennas, while retaining a diversity order of 1 for the desired signal. Thus, in our example, we can suppress the data user while retaining $N_r = 1$ for voice users. Thus, the capacity for voice users becomes

$$\frac{128}{8} = 16 \tag{20.6}$$

20.1.4 Receiver Structures

In this section, we describe RX structures that can be used for separation and processing of MPCs. Since our emphasis is on smart antennas at the BS, this implies that we consider the uplink case (see Section 20.1.6 for the downlink). We distinguish between *switched-beam antennas, adaptive spatial processing, space–time processing*, and *space–time detection*.

Switched-Beam Antennas

A switched-beam antenna is an antenna array that can form just a small set of patterns – i.e., beams pointing in certain discrete directions. A switch then selects one of the possible beams[4] for downconversion and further processing; it selects the beam that is *best* in the sense that it gives the highest SNR or the highest Signal-to-Interference-and-Noise Ratio (SINR). This approach greatly simplifies RX design.

There are many different ways of realizing switchable beams. For example, the antenna can contain multiple directional elements, oriented in different directions, and the switch selects from them. Another, more popular, variant is a linear array followed by a spatial Fourier transformation. The output of such a spatial Fourier transformation is a number of beams that are all orthogonal to each other, and point in different directions. The spatial Fourier transformation can be realized as a so-called *Butler matrix*. It has a structure that is essentially the butterfly structure that is well known from software implementations of the Fast Fourier Transform (FFT). The elements of each stage are simple phaseshifters that can be realized in the Radio Frequency (RF) domain.

The main advantage of the switched-beam approach is its simplicity: all processing (spatial FFT and selection) is done in the RF domain, so that only a single signal has to be downconverted to the baseband and processed there. Since downconversion circuits are among the most expensive

---

[4] More precisely, it selects the signal associated with one of the possible beams.

components in today's wireless systems, this is a significant advantage. The drawback of this scheme is its limited flexibility. The main beam can only be pointed in certain fixed directions, such that gain in the actual direction of the MS might not be the maximum achievable value. Even more importantly, nulls cannot be pointed in arbitrary directions, so that nulling of interferers is very ineffective. For these reasons, switched-beam antennas seem to be more suitable for CDMA applications, where signal enhancement is critical, and not for SDMA or SFIR applications, where interference suppression is essential.

Example 20.2 *For a switched-beam antenna, compute the maximum gain and the maximum relative loss due to mismatch of DOA and beam direction for a uniform linear array with Butler matrix and number of antenna elements (spaced $\lambda/2$ apart) equal to $N_r = 2, 4, 8$.*

Let us first write down the FFT matrix for N_r elements:

$$W_{\mathrm{FFT},k,n} = w_n(k) = \exp\left(-j\frac{2\pi}{N_r}\right)^{kn} \tag{20.7}$$

The array factor for a uniform linear array was given in Eq. (9.16):

$$M_k(\phi) = \sum_{n=0}^{N_r-1} w_n(k) \exp[-j\cos(\phi)2\pi d_a n/\lambda] \tag{20.8}$$

With $d_a = \lambda/2$, this becomes

$$M_k(\phi) = \sum_{n=0}^{N_r-1} \exp\left[-j2\pi n\left(\frac{\cos(\phi)}{2} + \frac{k}{N_r}\right)\right] \tag{20.9}$$

and the magnitude of the array factor is (compare Eq. 9.17)

$$|M_k(\phi)| = \left|\frac{\sin\left[\frac{N_r}{2}\left(\pi\cos\phi - 2\pi\frac{k}{N_r}\right)\right]}{\sin\left[\frac{1}{2}\left(\pi\cos\phi - 2\pi\frac{k}{N_r}\right)\right]}\right| \tag{20.10}$$

The crossover points between patterns at FFT output k and $k+1$ are at:

$$\left|\frac{\sin\left[\frac{N_r}{2}\left(\pi\cos\phi - 2\pi\frac{k}{N_r}\right)\right]}{\sin\left[\frac{1}{2}\left(\pi\cos\phi - 2\pi\frac{k}{N_r}\right)\right]}\right| = \left|\frac{\sin\left[\frac{N_r}{2}\left(\pi\cos\phi - 2\pi\frac{k+1}{N_r}\right)\right]}{\sin\left[\frac{1}{2}\left(\pi\cos\phi - 2\pi\frac{k+1}{N_r}\right)\right]}\right| \tag{20.11}$$

Figure 20.2 shows the array factor for different values of N_r. We see that in all cases the value of the array factor of the crossover point is approximately 0.7. The exact values can be found in Table 20.1:

Table 20.1

N_r	Max gain (dB)	Max relative loss due to direction-of-arrival mismatch (dB)
2	3.01	1.51
4	6.02	1.84
8	9.03	1.93

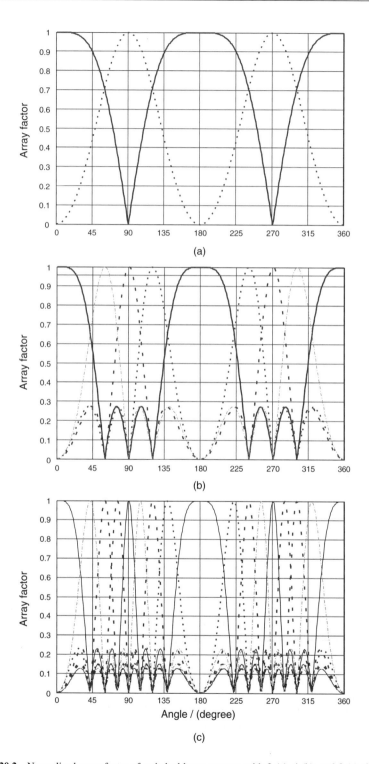

Figure 20.2 Normalized array factor of switched-beam antenna with 2 (a), 4 (b), and 8 (c) elements.

Adaptive Spatial Processing

The adaptive spatial processing approach uses a linear combination of signals (see Figure 20.3), where antenna weighting and summing is done in the baseband. For this reason, there are no restrictions with respect to weights, and they can also be adapted according to the current channel state. In other words, the RX can point maxima and nulls of the array pattern into (almost) arbitrary directions. On the downside, this approach requires in general N_r complete downconversion chains for N_r antenna elements, and is thus considerably more expensive, and consumes more power as well.

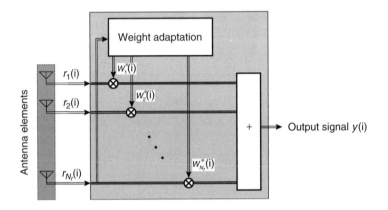

Figure 20.3 Linear combination of antenna signals.

If weights are adapted based on a training sequence, and change whenever the channel realization changes, then adaptive spatial processing is identical to diversity combining. It is thus possible to suppress $K = N_r - 1$ interferers, or alternatively to achieve high gain in the SNR (up to a factor of N_r) with such an approach (for more details, see Section 13.4.3).

Alternatively, antenna weights can be chosen based on angular spectrum (Angular Delay Power Spectrum (ADPS), see Section 6.7), but not adapted to the instantaneous amplitude of MPCs arriving from different directions. In this case, interference suppression capabilities are usually smaller. This aspect is discussed in more detail in Section 20.1.5.

Adaptive Space–Time Processing

The adaptive spatial processing described in the previous section first performs spatial processing and then puts out a single signal for further temporal processing. In other words, Rake reception or equalization is done only on the *single* signal that is put out from the spatial processor. This approach works well only if the temporal properties of the signal (delay dispersion) are the same at all antenna elements, or, conversely, if spatial properties are independent of delay. While this situation can occur in some cases (see also Chapter 7 for channel models that correspond to such an assumption), it is certainly not universally valid. If delay and angular properties cannot be decomposed multiplicatively, then spatial processing followed by temporal processing is no longer optimum.

In order to fully exploit the possibilities of different MPCs, adaptive *space–time processing* has to be used. The optimum linear RX is a two-dimensional Rake RX – i.e., a linear combiner that

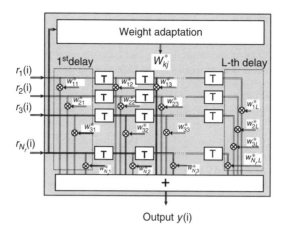

Figure 20.4 Space–time filter.

weights all *resolvable* (in the space–time domain) MPCs and adds them up. Figure 20.4 shows such a space–time processor. If the Rake RX (or equalizer) has L taps, then the total processor has $N_r L$ weights. The output from that processor is sent on to a decoder.

Space–Time Detection

For optimum reception, space–time processing and decoding/detection have to be done jointly. The special properties of the received signal, including finite-alphabet properties, can be taken into account for detection. The optimum detector is a generalized Maximum Likelihood Sequence Estimation (MLSE) RX. However, these structures are not always used in practice, because they are too complex, and the performance gain compared with linear processing does not always justify the additional effort.

20.1.5 Algorithms for Adaptation of Antenna Weights

Algorithms for adaptation of antenna weights can be broadly classified into Spatial Reference (SR) and temporal reference (TR) algorithms.[5] The distinction between these algorithms is subtle, but important. Let us first discuss their common features: both algorithms are based on *linear* weighting and addition of signals from different antenna elements. Both algorithms effectively form a beam pattern – the weights for any linear combining scheme can be associated with a "beam pattern" even when it sometimes looks somewhat strange, with many maxima and minima. The key difference is which information is used for the choice of the linear weights. In SR algorithms weights are chosen according to the spatial structure (DOAs) of the arriving signal combined with information about the array structure. TR algorithms optimize the SINR after the combiner with the help of a training sequence.

---

[5] In the literature, there is often a distinction between diversity and beamforming algorithms. However, this distinction is problematic as diversity (slope of the effective SNR distribution) and beamforming (change in mean SNR) are effects that, among others, depend on antenna arrangement and channel constellation.

Spatial Reference Algorithms

In the *spatial reference (SR)* case, the antenna tries to form a beam pattern that puts maxima in the direction of the main DOAs of the desired user, and nulls in the direction of MPCs coming from interferers. An SR algorithm thus proceeds in the following three steps (see Figure 20.5):

1. Determination of the DOAs, ϕ_n, of the MPCs: as we will see below, a main advantage of using these DOAs is that they show only small variations over time and frequency. The directly observable quantities are the signals at the antenna elements; it is thus necessary to extract the DOAs from these signals. This can be done by the same methods as those for channel sounding (see Chapter 8): spatial Fourier transformations (not recommended) or high-resolution algorithms (Minimum Variance Method MVM, Estimation of Signal Parameters by Rotational Invariance Technique ESPRIT, Space Alternating Generalized Expectation – maximization SAGE) can be applied. There are some important practical differences from the evaluation of channel sounder data: the SNR for the reception of data can be considerably worse than that used for channel sounding – this makes DOA determination more difficult. On the other hand, a lot of data are transmitted within a relatively short time. This allows the use of either averaging or tracking algorithms that observe the change in DOAs [Kuchar et al. 2002]. Note that some DOA estimation algorithms (e.g., ESPRIT) do not require a training sequence, since they are based exclusively on the correlation matrix of the received signal, and can thus estimate the DOA from user data that are unknown. Other algorithms, like SAGE, can eschew the training sequence only at the price of greatly complicating the algorithm.[6]
2. Association of DOAs with specific users: in contrast to channel sounding, where we have just a single possible source for the arriving signal, MPCs during communications can stem from different users – the desired user as well as interferers. It is thus necessary to separate desired and interfering users. For this identification problem, it is necessary to have a training sequence or some other property that is unique for each user.
3. Forming of the beam pattern that maximizes the SINR: this beam pattern is used for actual reception of data.

We note that many DOA estimation algorithms require the validity of a certain channel model (e.g., that the received signal can be written as a finite sum of plane waves). If this model is not valid, the performance of the algorithms, and of the ensuing beamforming, deteriorates.

Temporal Reference Algorithms

Temporal reference algorithms are based on the use of a training signal, which serves as a "temporal reference" to which we try to match the output from the smart antenna – thus the name. The antenna

Figure 20.5 Using smart antennas with spatial reference algorithms.

[6] However, when SAGE is used in conjunction with a training sequence, it usually outperforms ESPRIT.

weights **w** are adapted directly in such a way that the deviation of the combiner output from the (known) training sequence is minimized. As a criterion for the deviation we can use the SIR, the Minimum Mean Square Error (MMSE), the Bit Error Rate (BER) during the training sequence, or any other suitable criterion. These criteria and their advantages and disadvantages have already been discussed in Chapter 13 (we stress again that smart antennas with diversity combining at the RX is the "standard" diversity that we have discussed in Chapter 13). In particular, the SINR is maximized by optimum combining. The linear weights lead to the creation of an effective antenna array pattern, but this pattern need not have an intuitive interpretation. There are also no assumptions about the existence of discrete DOAs.

A major difficulty in TR algorithms is the fact that the RX must be synchronized to the incoming signal *before* the determination of the antenna weights can be done. This is required because the sampling instants for the determination of the training sequence have to be known before the weights can be adapted. However, the SINR for the (spatially unfiltered) receive signal can be very poor and can make synchronization to the desired training sequence difficult.

In summary, the TR algorithms proceed according to the following steps (see Figure 20.6):

1. During a training phase, the smart antenna receives the signal at all antenna elements, and adjusts the weights in such a way that the deviation from the known signal is minimized.
2. During the transmission of the user data, the antenna weights stay fixed and are used to weight the incoming signals before they are combined and decoded/demodulated.

Figure 20.6 Principle behind temporal reference algorithms.

We now look at more details of step 1. Assume that we have determined the channel coefficient, and we want to determine the optimum antenna weights. Paraphrasing the main insights from Section 13.4.3, those optimum antenna weights maximize the SINR. Define the power-weighted channel and interference correlation matrices as

$$\widetilde{\mathbf{R}}^{(k)} = P_k \mathbf{h}^{(k)} (\mathbf{h}^{(k)})^\dagger, \quad \widetilde{\mathbf{R}}_{\mathrm{ni},k} = \sigma_{\mathrm{n}}^2 \mathbf{I} + \sum_{l \neq k} P_l \mathbf{h}^{(l)} (\mathbf{h}^{(l)})^\dagger \tag{20.12}$$

where P_k is the power of the k-th user. Then the optimum antenna weights $w^{(k)}$ for the k-th user are obtained by maximization of the SINR, i.e.,

$$\mathbf{w}_{\mathrm{opt}}^{(k)} = \arg \max_{|\mathbf{w}|^2 = 1} \mathrm{SINR}_k = \arg \max_{|\mathbf{w}|^2 = 1} \frac{\mathbf{w}^\dagger \widetilde{\mathbf{R}}^{(k)} \mathbf{w}}{\mathbf{w}^\dagger \widetilde{\mathbf{R}}_{\mathrm{ni}}^{(k)} \mathbf{w}} \tag{20.13}$$

which is the solution of a generalized eigenvalue problem. An even simpler RX, called "decorrelating RX," can be achieved by setting $\sigma_n = 0$. This RX completely suppresses the interference, but considerably enhances the noise (compare also the discussion of zero-forcing and MMSE equalizers in Chapter 16).

The above derivation assumed that the interference for a specific user is fixed. We can get better results for the average throughput if we can influence the transmit powers of the different users – this requires feedback commands from the BS to the MS (to instruct the MSs about which power level to use), as well as power control capabilities of the MSs; most second- and third-generation cellular MSs have such capabilities. Finding the optimum MS transmission powers together with the optimum receive antenna weights is complicated because adjusting the power for one MS affects the SINR for all users. There is therefore no closed-form solution; rather, we iterate the following two steps until convergence:

1. Fix the transmit powers of the MSs and obtain the BS antenna weights for the reception of the different users by Eq. (20.13).
2. Then, leave the set of antenna weights $w^{(k)}$ unchanged and compute the optimum power allocations P_k. This computation itself does not have a closed-form solution but can be obtained through various numerical optimization methods that depend on the target of the optimization. For example, if we want to optimize the sum throughput, the RX iterates the following equation to convergence (where n is the iteration counter) [Chrisanthopoulou and Tsoukatos 2007]:

$$P_k^{(n+1)} = \min\left[P_{\max}, P_k^{(n)} + \alpha \left(\frac{1}{P_k^{(n)}} - \sum_{l \neq k} \frac{(\mathbf{w}^{(l)})^\dagger \widetilde{\mathbf{R}}^{(k)} \mathbf{w}^{(l)}}{\sigma_n^2 + \sum_{m \neq l} (\mathbf{w}^{(l)})^\dagger \widetilde{\mathbf{R}}^{(m)} \mathbf{w}^{(l)}} \right) \right] \tag{20.14}$$

where $P_{\max}$ is the maximum power an MS can transmit, and α is a parameter determining the convergence rate; if $P_k^{n+1} < 0$, we set it to 0.

Further performance improvements can be achieved when the RX is not restricted to linear combining of the antenna signals. In particular, the MMSE RX can be combined with a successive-interference canceller (see Section 18.4.3): the BS first decodes a particular user, and subtracts the contribution of this signal from the overall received signal. When now decoding the signal for the next user, a better SINR is available. This decoding strategy is theoretically optimum, but is (like all serial interference cancellation strategies) sensitive to error propagation. It is also noteworthy that the resulting RX has strong similarity to the Horizontal Bell labs LAyered Space Time (H-BLAST) RX that is discussed in Section 20.2.8.

Blind Algorithms

TR algorithms rely on the existence of a training sequence, while SR algorithms rely on the existence of a certain spatial structure of the arriving signal. Yet another group of algorithms, dubbed *blind* algorithms, makes neither of those assumptions. Rather, they exploit the statistical properties of the transmit signal.

Let us start out by describing the received signal by the equation

$$\mathbf{r}_i = \mathbf{h}_d s_i \tag{20.15}$$

where $\mathbf{r}_i$ is the receive signal vector created by the transmit signal s_i related to the ith bit, and $\mathbf{h}_d$ is the desired channel vector – we omit the noise contribution for convenience. A more general representation, which also includes possible intersymbol interference, is obtained by stacking signals corresponding to a large number of bits into matrices:

$$\mathbf{Y} = \mathbf{H}_{\text{stack}}\mathbf{X} \tag{20.16}$$

where the matrix H_{stack} is related to channel impulse responses at different antenna elements. The channel description is thus similar to the TR approach in the sense that there is no assumption about spatial structure, and the channel is characterized by impulse response (or transfer function) alone. However, a blind algorithm does *not* learn the values of H_{stack} from a training sequence. Rather, it tries to compute a factorization of the matrix $\mathbf{Y}$ according to Eq. (20.16), such that signal matrix $\mathbf{X}$ fulfills certain properties – namely, *known* properties of the transmit signal.

Example 20.3 *Consider a system with one transmit antenna and three receive antennas. Let the impulse response for each of the antenna elements last for two symbol durations, denoted as h_{il}, $l = 0, 1, i = 1, 2, 3$. For a five-symbol-long transmit signal, $x(n)$, $n = 1, \ldots, 5$, write the explicit form of $\mathbf{Y} = \mathbf{H}_{\text{stack}}\mathbf{X}$, assuming that impulse responses stay constant over the duration of the transmit signal.*

Writing the nth received symbol at the ith antenna element as $y_i(n)$, we can easily see that

$$y_i(n) = \sum_{l=0}^{L-1} h_{i,l} x(n-l) \tag{20.17}$$

where $h_{i,l}$ is the channel impulse response to the signal at the ith antenna element in the lth channel delay tap, L is the duration of the channel impulse response, and N is the number of transmitted symbols. Let us then define a matrix $\mathbf{X}$ as [Laurila 2000]

$$\mathbf{X} = \begin{bmatrix} x(1) & x(2) & \cdots & x(N) \\ x(0) & x(1) & \cdots & x(N-1) \\ \vdots & \vdots & \vdots & \vdots \\ x(-L+2) & x(-L+3) & \cdots & x(N-L+1) \end{bmatrix} \tag{20.18}$$

which in our case becomes

$$\mathbf{X} = \begin{bmatrix} x(1) & x(2) & x(3) & x(4) & x(5) \\ x(0) & x(1) & x(2) & x(3) & x(4) \end{bmatrix} \tag{20.19}$$

The channel matrix gets the block form:

$$\mathbf{H}_{\text{stack}} = \begin{bmatrix} h_{1,0} & h_{1,1} & \cdots & h_{1,L-1} \\ h_{2,0} & h_{2,1} & \cdots & h_{2,L-1} \\ \vdots & \vdots & \vdots & \vdots \\ h_{N_r,0} & h_{N_r,1} & \cdots & h_{N_r,L-1} \end{bmatrix} \tag{20.20}$$

which in our case becomes

$$\mathbf{H}_{\text{stack}} = \begin{bmatrix} h_{1,0} & h_{1,1} \\ h_{2,0} & h_{2,1} \\ h_{3,0} & h_{3,1} \end{bmatrix} \tag{20.21}$$

Inserting Eqs. (20.19) and (20.21) into Eq. (20.16), we get:

$$\mathbf{Y} = \begin{bmatrix} y_1(1) & y_1(2) & y_1(3) & y_1(4) & y_1(5) \\ y_2(1) & y_2(2) & y_2(3) & y_2(4) & y_2(5) \\ y_3(1) & y_3(2) & y_3(3) & y_3(4) & y_3(5) \end{bmatrix} \tag{20.22}$$

$$= \begin{bmatrix} h_{1,0}x(1) + h_{1,1}x(0) & \cdots & h_{1,0}x(5) + h_{1,1}x(4) \\ h_{2,0}x(1) + h_{2,1}x(0) & \cdots & h_{2,0}x(5) + h_{2,1}x(4) \\ h_{3,0}x(1) + h_{3,1}x(0) & \cdots & h_{3,0}x(5) + h_{3,1}x(4) \end{bmatrix} \tag{20.23}$$

which can easily be seen to be in agreement with Eq. (20.17).

Depending on which signal properties are exploited, we get different algorithms for the determination of $\mathbf{X}$. If there is oversampling in the temporal and/or spatial domain, then cyclostationarity of the sampled signal can be used. Another group of algorithms exploits higher order statistics. Also the finite-alphabet property can be exploited: e.g., we know that for a Binary Phase Shift Keying (BPSK) signal, the elements of $\mathbf{X}$ have to be ± 1.

Blind algorithms have a number of advantages:

- No assumptions about the channel are required. The method works even for channels that do not exhibit discrete DOAs or separable MPCs.
- Blind algorithms do not require any calibration of the antenna array, and make no assumptions about the specific structure of the array.
- The detection process can be done in one step, yielding directly the desired user data.
- Blind algorithms eliminate the need for a training sequence, thus increasing the spectral efficiency of the system.

However, these advantages are bought at a price:

- Most of the blind algorithms assume that the channel impulse response does not change over the time it takes to establish the statistics of $\mathbf{Y}$. For a good estimate of these statistics, we have to collect samples over a long time, and the assumption of a time-invariant impulse response might be violated. The number of samples that forms the basis of the statistics is thus a compromise between the need to have a large-enough basis for obtaining good statistics and the need to use only samples that correspond to the same channel state.
- Initialization of the estimate for the channel is critical. For this reason, so-called *semi-blind* algorithms have been proposed, which use a very short training sequence at the beginning of transmission. This sequence is chosen to be short enough not to significantly influence the spectral efficiency of the system, but helps to avoid possible convergence problems of the blind algorithms.

These drawbacks are essentially the same as for blind equalization (see Chapter 16), and have prevented widespread use of this method.

20.1.6 Uplink versus Downlink

Up to now, we have considered smart antennas at the RX; since we consider cellular systems in which smart antennas are at the BS, this means that we have been looking at the uplink. The next question is how a smart antenna system behaves in the downlink. The answer to this question depends critically on whether we are considering a Time Domain Duplex (TDD) or a Frequency Domain Duplexing (FDD) system.

Time Domain Duplexing

In a TDD system in a static environment – i.e., no movement of either the TX or the RX – the channel impulse response is the same for uplink and downlink. Let us make this statement more precise. The transfer function from the MS antenna to the mth BS antenna, $(\mathbf{h}_d)_m$, is the same as the transfer function from the mth BS antenna to the MS antenna. Thus, if we have chosen antenna weights in such a way that they constructively add signals from the antenna elements during the uplink, the *same* antenna weights, when used in the downlink, will ensure that the signals from different BS antenna elements add up constructively at the MS antenna. Furthermore, if BS antenna weights are chosen to supress interference from other MSs during the uplink, the BS will also cause little or no interference to these other MSs in the downlink transmission phase.

However, interference is not completely reciprocal for the uplink and downlink: the interference that the MS sees stems from other *BSs*, not from other MSs (see Figure 20.7). However, antenna weights **w** that are determined during the uplink cannot take this into account – the desired BS does not even "see" these other BSs. Thus, interference suppression for the downlink cannot be as effective as suppression in the uplink.[7]

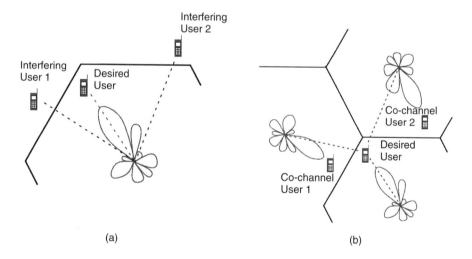

(a)   (b)

Figure 20.7 Interference situation for the uplink (a) and downlink (b) for a base station with smart antennas.

All we have stated up to now is that the *channel* is reciprocal. However, since antenna weights are determined in the baseband, we also have to require that RF elements and frequency conversion chains – i.e., everything from antenna connectors to the baseband – are reciprocal. This is not self-evident and, in fact, will not be the case for many transceivers. The problem can be eliminated or, at least, mitigated by a calibration procedure: during the setup of the call, the BS estimates the channel

---

[7] An exception would be if an MS found itself in a position to notify all the BSs in its surroundings about the details of the interference it sees from them, and the BSs were in a position to cooperate in order to provide as little interference as possible to all users. Such schemes have been proposed, but are rather complex.

impulse response both from the reciprocity principle and by means of feedback from the MS. From the difference of these two measurement methods, the impact of RF circuits can be determined.

Finally, we have to take into account that the channel need not be completely static in the time between uplink and downlink transmission (duplexing time). If the duplexing time is too long (becomes comparable with the coherence time of the channel), then the impulse response of the channel changes due to small-scale fading. The impact of such a decorrelation between the uplink and downlink channel is discussed in the next subsection.

Frequency Domain Duplexing

In an FDD system, the uplink and downlink happen at different frequencies; the frequency separation (duplexing distance) for cellular systems is typically on the order of 100 MHz. This is much larger than the coherence bandwidth of the channel (typically on the order of 1 MHz); as a consequence, small-scale fading of the channel impulse response in the uplink and in the downlink are completely decorrelated. In other words, small-scale fading is created by superposition of MPCs with different phases. The relative phase shifts depend, among other things, on the carrier frequency. For a large-enough duplexing distance, phase shifts are sufficiently different in uplink and downlink that superposition of MPCs occurs in different ways.

For this reason, we find that small-scale fading and, thus, the instantaneous impulse response of the channel, are different for uplink and downlink (a similar situation occurs in a TDD system when the duplexing time is much larger than the coherence time of the channel). What remains constant in uplink and downlink is the *average* channel state: in other words, the correlation matrix (averaged over small-scale fading), as well as the DOAs, delays, and mean powers of the MPCs.

It is still possible to obtain performance enhancements in the downlink for FDD. Antenna weights are now chosen based on the average channel state. For example, the beam pattern is chosen so that it points in the direction of the MPC that is strongest *on average*. Now, it is possible that this MPC is not the strongest *instantaneously*, and that it would therefore be better to choose a different beam pattern. However, since we do not know the instantaneous channel, optimizing on average is the best that we can do with the available information.

These considerations also show the usefulness of SR algorithms. Since these algorithms extract average *Channel State Information* (CSI), they can be easily used in FDD systems.

Feedback

Up to now, we have relied on reciprocity (either for the channel impulse response or for DOAs) to provide information about the channel. An alternative approach is to use feedback. Consider an FDD system. Let the BS transmit a training sequence from the first antenna element, then from the second antenna element, and so on. This enables the MS to learn all the separate impulse responses $(\mathbf{h}_d)_m$. It then feeds the information back to the BS on a separate feedback channel. In that case, the BS knows exactly how to adjust weights such that signals from antenna elements add up in the right way at the MS antenna.

This approach is very effective, but has two drawbacks: first, we have to make sure that feedback occurs within a time that is less than the coherence time of the channel. Second, the feedback channel decreases the spectral efficiency of the system. The information carried on that channel is not user data. System designers thus have to consider carefully how often to feed CSI back, and with what accuracy. Naturally, the performance of the smart antenna gets better the more accurate the feedback information is, but the performance penalty because of wasted resources can become prohibitive. A more detailed discussion can be found in Section 20.2.11.

20.1.7 Algorithms for the Adaptation of the Antenna Weights in the Downlink

Determination of the optimum transmission strategy for the downlink can be considerably more complicated than for the uplink, because changing the transmit strategy (e.g., the antenna weights) for one user influences the SINR for every other user.[8]

We will in the following only consider linear (beamforming) transmission schemes. The BS has a constraint on the *total* (summed over the signals for all users) transmit power $\sum_k P_k \leq P_{\max}$, which is reasonable, because the total power is limited both by the characteristics of the power amplifier, and by the intercell interference that we are willing to allow. We assume that the BS has complete CSI, i.e., it knows $\mathbf{h}^{(k)}$ for all the k users. For the following mathematical treatment, it is useful to create an *overall channel matrix* $\mathbf{H} = [(\mathbf{h}^{(1)})^T, (\mathbf{h}^{(2)})^T, \dots (\mathbf{h}^{(K)})^T]^T$. We also define $\widetilde{\mathbf{s}}$ as the signal vector transmitted to the BS antenna array, which in the case of beamforming (linear precoding) is

$$\widetilde{\mathbf{s}} = \mathbf{Ts}. \tag{20.24}$$

Let us first establish how many users can communicate simultaneously with the BS. We know that in the uplink, the number of users is limited by the number of antenna elements in the BS, N_{BS}. This follows from the principle of "optimum combining" that every additional antenna elements can suppress one interfering signal; thus, one set of BS antenna weights can deal with 1 desired user and $N_{\mathrm{BS}} - 1$ interferers simultaneously. The key insight is now that this admissible number of users is the same for the downlink, i.e., a BS with N_{BS} antenna elements can transmit to N_{BS} different MSs at the same time. Intuitively, this result follows from reciprocity: if – by adapting its weights – the BS in the uplink forms an antenna pattern that nulls out an interferer, then the same BS will not create any interference at this user when the BS transmits.[9]

Let us now formulate these insights mathematically. If the k-th MS should see no interference intended for other users, the transmit signal for the k-th MS must lie in the nullspace of the channel vectors for all other users. The precoding vector for the k-th MS is then the k-th column of the pseudoinverse of $\mathbf{H}$, so that the transmit signal is

$$\widetilde{\mathbf{s}} = \mathbf{H}^{\dagger} \left(\mathbf{H}\mathbf{H}^{\dagger} \right)^{-1} \Xi \mathbf{s} \tag{20.25}$$

where Ξ is a diagonal matrix that contains the power allocations for the different users. Thus, the beamforming can be combined with a power allocation strategy that either maximizes the sum throughput (for the given shape of the antenna patterns), optimize some "sum MSE," or tries to guarantee a certain minimum quality of service (SINR) for each user.

Clearly, zero interference can only be achieved if no more than N_t users are served at one given time, so we assume henceforth $K \leq N_t$. The key disadvantage of zero-forcing TXs is that in the case of ill-conditioned channel matrices, the transmit power along a particular dimension can become very large, and since the total transmit power is limited, this means that transmit power is used very inefficiently. In other words, if two users are "effectively" very close together (in the sense that

---

[8] There are some useful dualities between the uplink and downlink case, which can be exploited to derive good transmission schemes. For example, it can be shown that the same SINRs can be achieved for all users with the same overall transmit power (though the powers associated with each user in the uplink and downlink are different). However, most algorithms developed in the literature for the downlink antenna weights do not make explicit use of this duality.

[9] Note, however, that the optimum power assignments are different for uplink and downlink, because noise and interference levels are not necessarily reciprocal.

their channel vectors are almost the same – they do not have to be close together geographically), then it is very difficult to transmit to one user while simultaneously canceling out transmissions to the other. This situation is dual to the effect of "noise enhancement" by a decorrelation RX.

The performance can be improved by so-called "regularization" that trades off noise enhancement and interference. The transmit signal becomes

$$\widetilde{\mathbf{s}} = \mathbf{H}^\dagger \left(\mathbf{H}\mathbf{H}^\dagger + \zeta \mathbf{I} \right)^{-1} \Xi \mathbf{s} \tag{20.26}$$

where ζ is the so-called "regularization parameter." This is obviously a dual of an MMSE (optimum combining) RX.

A joint beamforming/power allocation strategy that leads to a maximization of the sum throughput is given – for the case $N_t = K$ in [Stojnic et al. 2006]. It is noteworthy that if we want to maximize the sum rate, fairness is no criterion – in other words, the data streams for some MSs have very high data rates, while other MSs are not serviced at all. Alternative beamforming/power allocation criteria thus try to guarantee a minimum rate for each user. Furthermore, the random beamforming discussed later in this section can be used to provide greater fairness.

There are many similar, but subtly different, optimization problems. For example, the "power control problem" aims to minimize the total transmitted power while guaranteeing a certain minimum SINR (and thus a certain rate) for every user. A number of such resource allocation problems can be tackled by "convex optimization" techniques [Boyd and Vandenberghe 2004].

20.1.8   Network Aspects

In order to really increase capacity by means of smart antennas, it is usually not sufficient to just insert an applique – i.e., replace the single-antenna RF front end and baseband demodulator with their multiantenna counterparts. Rather, the network has to provide additional functionalities. Consider a TDMA/FDMA system that uses SDMA. It can put two different users on the same time/frequency slot and separate them by their DOAs. However, when the two users get too close, such a separation is no longer possible. In such a case, the network has to swap one of these two users with one that is on a different time/frequency slot and sufficiently far away in the angular domain. In other words, the network has to provide intracell handover capabilities. Thus, the control software of the BS has to be designed specifically for the use of SDMA.

Another major problem is the call setup phase. In this phase, the position and thus DOAs of the various users are not known yet, and the call setup information has to be transmitted in all possible directions simultaneously. Compared with actual data transmission, network planning, link budgets, etc., can thus be different for this phase.

20.1.9   Multiuser Diversity and Random Beamforming

Throughout this book, we have assumed that the BS and the MS transmit at predetermined times, irrespective of the channel state. However, it has recently been established that the performance of cellular systems can be improved significantly by scheduling transmission depending on the channel state. It can also be shown that the use of multiple antennas can enhance this idea even further. In the following, we will explain this idea in more detail.

Scheduling and Multiuser Diversity

Let us first consider the downlink in a system where the BS as well as the MSs have only a single antenna, and multiple MSs communicate with one BS. All links undergo flat Rayleigh fading;

we assume for the moment that the mean power is the same for all users. Let us further assume a conventional TDMA system, so that each user is assigned a fixed timeslot and communicates during that timeslot. The probability for user k to "see" a certain SNR γ is then given by the exponential distribution:

$$\Pr(r_k \leq \gamma) = 1 - \exp\left(-\frac{\gamma}{\bar{\gamma}}\right) \qquad (20.27)$$

This probability is independent of user k, and is also independent of the *number* of users K. The scheme, also called *round-robin*, is the best we can do if the TX (BS) does not know anything about the channel. Obviously, it treats users fairly in the sense that each user gets the same amount of time to communicate.

However, we can do better if the BS knows the CSI for the downlink of all users. At any given point in time, chances are that there is a very good link to some users, while there are bad links to other users. The optimum strategy for the BS is now to communicate always with the user who has the best link – i.e., the highest instantaneous path gain – and thus the highest instantaneous SNR γ. As the channel changes, different users k become "instantaneously best." The cumulative distribution function for the active user – i.e., the instantaneously best user – is the same as for selection diversity (see Eq. 13.10). Multiple users thus act like diversity branches – however, we do not need any additional antennas to realize that diversity. This scheme is thus called *multiuser diversity*; it is optimum in the sense that capacity is maximized.

In a cellular system, users that are very close to the BS have a much better SNR than users that are far away. In that case, the danger of selecting the user with the best SNR is that the MS closest to the BS gets to use the channel most often, while MSs that are far away are almost never "instantaneously best," and thus are chosen much less frequently. As a consequence, the data rate for users at the cell boundary is much lower than for users near the BS. This problem can be solved by *proportionally fair* scheduling. The BS communicates not with the absolutely best user but rather with the user whose ratio of instantaneous to average SNR $\gamma/\bar{\gamma}$ is the highest. In such a case, each user is assigned the same amount of time if the *normalized* fading statistics of the different users are identical – e.g., all users are Rayleigh fading.

A key requirement for scheduling is that the BS has knowledge of the instantaneous downlink channel. As we have discussed in Section 20.1.6, in an FDD system this requires that the MSs feed back their instantaneous SNRs to the BS. The rate of this feedback depends on the coherence time of the channel. Furthermore, it also involves a system design tradeoff. If feedback is done too rarely, then the BS picks a suboptimum user. If feedback is done too often, then the overhead for this feedback becomes prohibitive. It is also noteworthy that *all* MSs have to feed back their SNRs, even though only one will ultimately be picked for communication.[10]

All the above considerations were for the downlink; however, the scheme works for the uplink as well. In this case, the BS determines the quality of the uplink channel, and then broadcasts which MS is allowed to transmit.

Delay Considerations and Random Beamforming

Another key problem of multiuser diversity is the inherent delay. If the BS always uses proportionally fair scheduling, it retains communication with the chosen user for approximately one coherence time of the channel. If user mobility is low, the coherence time can be quite large. Remember that

---

[10] Some tricks can be used to alleviate this problem: e.g., an MS that sees an SNR below a certain threshold does not need to feed its information back since the chance that it will be picked is very low. Again, the choice of this threshold involves an engineering tradeoff: if the threshold is too low, it does not lead to significant savings; if it is too high, then there are times when no MS provides feedback, and the BS thus cannot pick the best user.

each user has to wait until its $\gamma/\overline{\gamma}$ becomes instantaneously largest and it can communicate; this waiting time (latency) is roughly KT_{coh}. In a system with many users and large coherence times, this latency can become prohibitively large.

An ingenious solution to this problem, called *random beamforming*, uses multiple antennas at the BS [Viswanath et al. 2002]. Signals transmitted from the different antenna elements of the BS are multiplied by time-varying complex coefficients. This can be interpreted in two (equivalent) ways:

1. Each vector of transmit weight coefficients can be associated with a beam pattern. No matter what beam is formed, it will enhance the channel to *some* MS. Scheduling (as discussed above) then makes sure that the enhanced MS is chosen for communication. Every time the coefficients are changed, the beam pattern changes, and a different MS gets enhanced.
2. The combination of multiple antennas (with weighting coefficients) and physical channel can be viewed as an "equivalent" channel. By varying the coefficients for antenna elements, we are enforcing time variations on the channel, and thus reducing the coherence time T_{coh}. The effective channel exhibits Rayleigh fading with a coherence time that is approximately the smaller of the coherence time of the physical channel and the time over which antenna weights change.

The main difference from transmit diversity or beamforming (as discussed at the beginning of this section) is the following:

- For *conventional transmit diversity*, the BS needs to know the amplitude and phase of the transfer function from all MSs. It then chooses weights for the signals from different antenna elements in such a way that the different transmit signals add up in an optimum way at the destined MS. Coefficients are changed only if the transfer function between the BS and the chosen MS changes, or if the BS decides to communicate with another MS.
- For *random beamforming*, the BS chooses the coefficients completely at random and changes them according to system parameters (related to latency), but independently of the channel.

The time over which weighting coefficients are kept constant constitutes a compromise between the resulting latency and system overhead for feeding back channel quality.

Impact on System Design

Multiuser diversity is mainly useful for data communications. For voice, stringent delay requirements (total delay less than about 100 ms) preclude the use of "conventional" multiuser diversity, especially when users are stationary or slowly moving (pedestrian speeds), though random beamforming can alleviate this problem somewhat. On the other hand, the scheme seems very well suited to data communications, where larger delays are acceptable.

From a scientific point of view, it is also noteworthy that multiuser diversity leads to a paradigm change in physical-layer design. Many of the transceiver structures and signal-processing methods we have encountered in Chapters 10–20 aim at combating fading by *reducing* variations in SNR for a specific link. For multiuser diversity, however, we exploit and possibly enhance the variations in SNR.

20.2   Multiple Input Multiple Output Systems

20.2.1   Introduction

MIMO systems are systems with *Multiple Element Antennas* (MEAs) at *both* link ends. Originally suggested in Winters [1987], they attracted great attention through theoretical investigations in the

1990s ([Foschini and Gans 1998] and [Telatar 1999]). Since that time, research in these systems has exploded and practical systems based on MIMO have been developed.

The MEAs of a MIMO system can be used for four different purposes: (i) beamforming, (ii) diversity, (iii) interference suppression, and (iv) spatial multiplexing (transmission of several data streams in parallel). The first three concepts are the same as for smart antennas. Having multiple antennas at both link ends leads to some interesting new technical possibilities, but does not change the fundamental effects of this approach. Spatial multiplexing, on the other hand, is a new concept, and has thus drawn the greatest attention. It allows direct improvement of capacity by simultaneous transmission of multiple data streams. We will show below that the (information-theoretic) capacity *for a single link* increases linearly with the number of antenna elements.

In the early years of MIMO research, the main emphasis was on information-theoretic limits, and this section will also mostly concentrate on these aspects. After 2000, emphasis shifted more to the question of how the theoretical gains of MIMO can be realized in practice. Advances in practical implementation of MIMO systems have also greatly helped in their adoption by international standards organizations. MIMO was included in fourth-generation cellular systems (see Chapters 27 and 28) as well as high-throughput wireless Local Area Networks (LANs) (IEEE 802.11n, see Chapter 29).

20.2.2   How Does Spatial Multiplexing Work?

Spatial multiplexing uses MEAs at the TX for transmission of parallel data streams (see Figure 20.8). An original high-rate data stream is multiplexed into several parallel streams, each of which is sent from one transmit antenna element. The channel "mixes up" these data streams, so that each of the receive antenna elements sees a combination of them. If the channel is well behaved, the received signals represent *linearly independent* combinations. In this case, appropriate signal processing at the RX can separate the data streams. A basic condition is that the number of receive antenna elements is at least as large as the number of transmit data streams. It is clear that this approach allows the data rate to be drastically increased – namely, by a factor of $\min(N_t, N_r)$.

Figure 20.8   Principle behind spatial multiplexing.

For the case when the TX knows the channel, we can also develop another intuition (see Figure 20.9). With N_t transmit antennas, we can form N_t different beams. We point all these beams at different Interacting Objects (IOs), and transmit different data streams over them. At the RX, we can use N_r antenna elements to form N_r beams, and also point them at different IOs. If all the beams can be kept orthogonal to each other, there is no interference between the data streams; in other words, we have established parallel channels. The IOs (in combination with the beams pointing in their direction) play the same role as wires in the transmission of multiple data streams on multiple wires.

Figure 20.9   Transmission of different data streams via different interacting objects.

From this description, we can also immediately derive some important principles: the number of possible data streams is limited by $\min(N_t, N_r, N_s)$, where N_s is the number of (significant) IOs. We have already seen above that the number of data streams cannot be larger than the number of transmit antenna elements, and that we need a sufficient number of receive antenna elements (at least as many as data streams) to form the receive beams and, thus, be able to separate the data streams. But it is also very important to notice that the number of IOs poses an upper limit: if two data streams are transmitted to the same IO, then the RX has no possibility of sorting them out by forming different beams.

The above intuitive pictures are somewhat simplified. A more exact mathematical treatment follows in subsequent sections.

20.2.3   System Model

Before going into further details, let us first establish the generic system that will be considered for capacity computations. Figure 20.10 exhibits a block diagram. At the TX, the data stream enters an encoder, whose outputs are forwarded to N_t transmit antennas. From the antennas, the signal is sent through the wireless propagation channel, which is assumed to be quasi-static and frequency-flat if not stated otherwise. By quasi-static we mean that the coherence time of the channel is so long that "a large number" of bits can be transmitted within this time.

We denote the $N_r \times N_t$ matrix of the channel as

$$\mathbf{H} = \begin{pmatrix} h_{11} & h_{12} & \cdots & h_{1N_t} \\ h_{21} & h_{22} & \cdots & h_{2N_t} \\ \vdots & \vdots & \cdots & \vdots \\ h_{N_r1} & h_{N_r2} & \cdots & h_{N_rN_t} \end{pmatrix} \tag{20.28}$$

whose entries h_{ij} are complex channel gains (transfer functions) from the jth transmit to the ith receive antenna.

The received signal vector

$$\mathbf{r} = \mathbf{Hs} + \mathbf{n} = \mathbf{x} + \mathbf{n} \tag{20.29}$$

contains the signals received by N_r antenna elements, where $\mathbf{s}$ is the transmit signal vector and $\mathbf{n}$ is the noise vector.

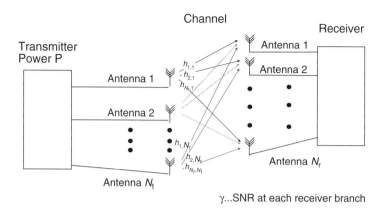

Figure 20.10 Block diagram of a multiple-input multiple-output system.

20.2.4 Channel State Information

Algorithms for MIMO transmission can be categorized by the amount of CSI that they require. We distinguish the following cases:

1. *Full CSI at the TX (CSIT) and full CSI at the RX (CSIR)*: in this ideal case, both the TX and the RX have full and perfect knowledge of the channel. This case obviously results in the highest possible capacity. However, it is difficult to obtain the full CSIT (as we have discussed in Section 20.1.6).
2. *Average CSIT and full CSIR*: in this case, the RX has full information of the instantaneous channel state, but the TX knows only the average CSI – e.g., the correlation matrix of $\mathbf{H}$ or the angular power spectrum. As we have discussed in Section 20.1.6, this is easier to achieve and does not require reciprocity or fast feedback; however, it does require calibration (to eliminate the nonreciprocity of transmit and receive chains) or slow feedback.
3. *No CSIT and full CSIR*: this is the case that can be achieved most easily, without any feedback or calibration. The TX simply does not use any CSI, while the RX learns the instantaneous channel state from a training sequence or using blind estimation.
4. *Noisy CSI*: when we assume "full CSI" at the RX, this implies that the RX has learned the channel state perfectly. However, any received training sequence will be affected by additive noise as well as quantization noise. It is thus more realistic to assume a "mismatched RX," where the RX processes the signal based on the *observed* channel $\mathbf{H}_{obs}$, while in reality the signals pass through channel $\mathbf{H}_{true}$

$$\mathbf{H}_{true} = \mathbf{H}_{obs} + \boldsymbol{\Delta} \tag{20.30}$$

Some papers have taken this into account by ad hoc modification of noise variance (replacing σ_n^2 by $\sigma_n^2 + \sigma_e^2$, where σ_e^2 is the variance of the entries of $\boldsymbol{\Delta}$).
5. *No CSIT and no CSIR*: it is remarkable that channel capacity is also high when neither the TX nor the RX have CSI. We can, e.g., use a generalization of differential modulation. For high SNR, capacity no longer increases linearly with $m = \min(N_t, N_r)$, but rather increases as $\tilde{m}(1 - \tilde{m}/T_{coh})$, where $\tilde{m} = \min(N_t, N_r\lfloor T_{coh}/2\rfloor)$, and T_{coh} is the coherence time of the channel in units of symbol duration.

20.2.5   Capacity in Nonfading Channels

The first key step in understanding MIMO systems is the derivation of the capacity equation for MIMO systems in nonfading channels, often known as "Foschini's equation" [Foschini and Gans 1998]. Let us start with the capacity equation for "normal" (single-antenna) Additive White Gaussian Noise (AWGN) channels. As Shannon showed, the information-theoretic (ergodic) capacity of such a channel is (see also Chapter 14)

$$C_{\text{shannon}} = \log_2\left(1 + \gamma \cdot |H|^2\right) \tag{20.31}$$

where γ is the SNR at the RX, and H is the normalized transfer function from the TX to the RX (as we are for now dealing with the frequency-flat case, the transfer function is just a scalar number). The key statement of this equation is that capacity increases only logarithmically with the SNR, so that boosting the transmit power is a highly ineffective way of increasing capacity.

Consider now the MIMO case, where the channel is represented by matrix Eq. (20.28). Let us then consider a *singular value decomposition*[11] of the channel:

$$\mathbf{H} = \mathbf{W}\mathbf{\Sigma}\mathbf{U}^\dagger \tag{20.32}$$

where $\mathbf{\Sigma}$ is a diagonal matrix containing singular values, and $\mathbf{W}$ and $\mathbf{U}^\dagger$ are unitary matrices composed of the left and right singular vectors, respectively. The received signal is then

$$\mathbf{r} = \mathbf{H}\mathbf{s} + \mathbf{n} \tag{20.33}$$

$$= \mathbf{W}\mathbf{\Sigma}\mathbf{U}^\dagger\mathbf{s} + \mathbf{n} \tag{20.34}$$

Then, multiplication of the transmit data vector by matrix $\mathbf{U}$ and the received signal vector by $\mathbf{W}^\dagger$ diagonalizes the channel:

$$\mathbf{W}^\dagger\mathbf{r} = \mathbf{W}^\dagger\mathbf{W}\mathbf{\Sigma}\mathbf{U}^\dagger\mathbf{U}\widetilde{\mathbf{s}} + \mathbf{W}^\dagger\mathbf{n}$$

$$\widetilde{\mathbf{r}} = \mathbf{\Sigma}\widetilde{\mathbf{s}} + \widetilde{\mathbf{n}} \tag{20.35}$$

Note that – because $\mathbf{U}$ and $\mathbf{W}$ are unitary matrices – $\widetilde{\mathbf{n}}$ has the same statistical properties as $\mathbf{n}$ – i.e., it is independent identically distributed (iid) white Gaussian noise. The capacity of the system Eq. (20.35) is the same as that of system Eq. (20.29). Now, computation of the capacity of Eq. (20.35) is rather straightforward. The matrix $\mathbf{\Sigma}$ is a diagonal matrix with R_{H} nonzero entries σ_k, where R_{H} is the rank of $\mathbf{H}$ (and thus defined as the number of nonzero singular values), and σ_k is the kth singular value of $\mathbf{H}$. We have therefore R_{H} *parallel* channels (eigenmodes of the channel), and it is clear that the capacity of parallel channels just adds up.

The capacity of channel $\mathbf{H}$ is thus given by the sum of the capacities of the eigenmodes of the channel:

$$C = \sum_{k=1}^{R_{\text{H}}} \log_2\left[1 + \frac{P_k}{\sigma_{\text{n}}^2}\sigma_k^2\right] \tag{20.36}$$

where σ_{n}^2 is noise variance, and P_k is the power allocated to the kth eigenmode; we assume that $\sum P_k = P$ is independent of the number of antennas. This capacity expression can be shown to

---

[11] Singular value decomposition is similar to eigenvalue decomposition, but exists also for rectangular matrices (more rows than columns, or vice versa). It decomposes any matrix into a product of three matrices: a unitary matrix corresponding to the row space, a diagonal matrix describing the strength of different eigenmodes, and a unitary matrix describing the column space.

be equivalent to

$$C = \log_2\left[\det\left(\mathbf{I}_{N_r} + \frac{\overline{\gamma}}{N_t}\mathbf{H}\mathbf{R}_{ss}\mathbf{H}^\dagger\right)\right] \qquad (20.37)$$

where $\mathbf{I}_{N_r}$ is the $N_r \times N_r$ identity matrix, $\overline{\gamma}$ is the mean SNR per RX branch, and $\mathbf{R}_{ss}$ is the correlation matrix of the transmit data (if data at the different antenna elements are uncorrelated, it is a diagonal matrix with entries that describe the power distribution among antennas).[12] The distribution of power among the different eigenmodes (or antennas) depends on the amount of CSIT; we also assume for the moment that the RX has perfect CSI. The equations above confirm our intuitive picture that capacity increases linearly with $\min(N_t, N_r, N_s)$, as the number of nonzero singular values R_H is upper-limited by $\min(N_t, N_r, N_s)$.

No Channel State Information at the Transmitter and Full CSI at the Receiver

When the RX knows the channel perfectly, but no CSI is available at the TX, it is optimum to assign equal transmit power to all TX antennas, $P_k = P/N_t$, and use uncorrelated data streams. Capacity thus takes on the form:

$$C = \log_2\left[\det\left(\mathbf{I}_{N_r} + \frac{\overline{\gamma}}{N_t}\mathbf{H}\mathbf{H}^\dagger\right)\right] \qquad (20.38)$$

It is worth noting that (for sufficiently large N_s) the capacity of a MIMO system increases linearly with $\min(N_t, N_r)$, irrespective of whether the channel is known at the TX or not.

Let us now look at a few special cases. To make the discussion easier, we assume that $N_t = N_r = N$:

1. All transfer functions are identical – i.e., $h_{1,1} = h_{1,2} = \ldots = h_{N,N}$. This case occurs when all antenna elements are spaced very closely together, and all waves are coming from similar directions. In such a case, the rank of the channel matrix is unity. Then, capacity is

$$C_{MIMO} = \log_2(1 + N\overline{\gamma}) \qquad (20.39)$$

We see that in this case the SNR is increased by a factor of N compared with the single antenna case, due to beamforming gain at the RX. However, this only leads to a logarithmic increase in capacity with the number of antennas.

2. All transfer functions are different such that the channel matrix is full rank, and has N eigenvalues of equal magnitude. This case can occur when the antenna elements are spaced far apart and are arranged in a special way. In this case, capacity is

$$C_{MIMO} = N\log_2(1 + \overline{\gamma}) \qquad (20.40)$$

and, thus, increases linearly with the number of antenna elements.

3. Parallel transmission channels – e.g., parallel cables. In this case, capacity also increases linearly with the number of antenna elements. However, the SNR per channel decreases with N, so that total capacity is

$$C_{MIMO} = N\log_2\left(1 + \frac{\overline{\gamma}}{N}\right) \qquad (20.41)$$

Figure 20.11 shows capacity as a function of N for different values of SNR.

---

[12] Note that the $\mathbf{H}$ and $\mathbf{R}$ must be normalized to ensure that $\overline{\gamma}$ is the mean SNR.

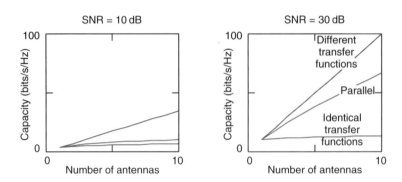

Figure 20.11   Capacity of multiple-input-multiple-output systems in additive white Gaussian noise channels.

Full Channel State Information at the Transmitter and Full CSI at the Receiver

Let us next consider the case where both the RX and TX know the channel perfectly. In such a case, it can be more advantageous to distribute power not uniformly between the different transmit antennas (or eigenmodes) but rather assign it based on the channel state. In other words, we are faced with the problem of optimally allocating power to several parallel channels, each of which has a different SNR. This is the same problem that we considered in Section 19.7.2, and therefore the answer is the same: *waterfilling*. This is another nice example of how the same mathematics can be applied to different communications problems: replace the word "subchannel" or "subcarrier in an Orthogonal Frequency Division Multiplexing (OFDM) system" by "eigenmode in a MIMO system," and we can apply the whole discussion of Section 19.7.2 to MIMO systems with CSIT.

20.2.6   Capacity in Flat-Fading Channels

General Concepts

In the previous section, we considered capacity for one given channel realization – i.e., for one channel matrix **H**. In wireless systems, we are, however, faced with channel fading. In this case, entries in channel matrix Eq. (20.28) are random variables. If the channel is Rayleigh fading, and fading is independent at different antenna elements, the h_{ij} are iid zero-mean, circularly symmetric complex Gaussian random variables with unit variance – i.e., the real and imaginary part each has variance 1/2. This is the case we will consider for now, unless stated otherwise. Consequently, the power carried by each h_{ij} is chi-square-distributed with 2 degrees of freedom. This is the simplest possible channel model; it requires the existence of "heavy multipath" – i.e., many MPCs of approximately equal strength (see Chapter 5) as well as a sufficient distance between the antenna elements. Since fading is independent, there is a high probability that the channel matrix is full rank and the eigenvalues are fairly similar to each other; consequently, capacity increases linearly with the number of antenna elements. Thus, the existence of heavy multipath, which is usually considered a drawback, becomes a major advantage in MIMO systems.

Because the entries of the channel matrix are random variables, we also have to rethink the concept of information-theoretic capacity. As a matter of fact, two different definitions of capacity exist for MIMO systems:

- *Ergodic (Shannon) capacity*: this is the expected value of the capacity, taken over all realizations of the channel. This quantity assumes an infinitely long code that extends over all the different channel realizations.
- *Outage capacity*: this is the minimum transmission rate that is achieved over a certain fraction of the time – e.g., 90% or 95%. We assume that data are encoded with a near-Shannon-limit-achieving code that extends over a period that is much shorter than the channel coherence time. Thus, each channel realization can be associated with a (Shannon) capacity value. Capacity thus becomes a random variable (rv) with an associated cumulative distribution function (cdf); see also the discussion in Section 14.9.1.

No Channel State Information at the Transmitter and Perfect CSI at the Receiver

Now, what is the capacity that we can achieve in a fading channel without CSI? Figure 20.12 shows the result for some interesting systems at a 21-dB SNR. The (1, 1) curve describes a Single Input Single Output (SISO) system. We find that the median capacity is on the order of 6 bit/s/Hz, but the 5% outage capacity is considerably lower (on the order of 3 bit/s/Hz). When using a (1, 8) system – i.e., 1 transmit antenna and 8 receive antennas – the mean capacity does not increase that significantly – from 6 to 10 bit/s/Hz. However, the 5% outage capacity increases significantly from 3 to 9 bit/s/Hz. The reason for this is the much higher resistance to fading that such a diversity system has. However, when going to a (8, 8) system – i.e., a system with 8 transmit and 8 receive antennas – both capacities increase dramatically: the mean capacity is on the order of 46 bit/s/Hz, and the 5% outage probability is more than 40 bit/s/Hz.

The exact expression for the *ergodic capacity* was derived in [Telatar 1999] as

$$E\{C\} = \int_0^\infty \log_2\left[1 + \frac{\overline{\gamma}}{N_t}\lambda\right] \sum_{k=0}^{m-1} \frac{k!}{(k+n-m)!}\left[L_k^{n-m}(\lambda)\right]^2 \lambda^{n-m}\exp(-\lambda)\,d\lambda \qquad (20.42)$$

Figure 20.12 Cumulative distribution function of capacity for 1×1, 1×8, and the 8×8 optimum scheme.

where $m = \min(N_t, N_r)$ and $n = \max(N_t, N_r)$ and $L_k^{n-m}(\lambda)$ are associated Laguerre polynomials. Exact analytical expressions for the cdfs of capacity are rather complicated; therefore, two approximations are in widespread use:

- Capacity can be well approximated by a Gaussian distribution, such that only the mean – i.e., the ergodic capacity given above – and variance need to be computed.
- From physical considerations, the following upper and lower bounds for capacity distribution have been derived in [Foschini and Gans 1998] for the case $N_t \geq N_r$:

$$\sum_{k=N_t-N_r+1}^{N_t} \log_2\left[1 + \frac{\overline{\gamma}}{N_t}\chi_{2k}^2\right] < C < \sum_{k=1}^{N_t} \log_2\left[1 + \frac{\overline{\gamma}}{N_t}\chi_{2N_r}^2\right] \qquad (20.43)$$

where χ_{2k}^2 is a chi-square-distributed random variable with $2k$ degrees of freedom.[13] These two bounds have very clear physical interpretations. The lower bound corresponds to capacity that can be achieved with a Bell labs LAyered Space Time (BLAST) system; this system and its operating principle are described below. The upper bound corresponds to an idealized situation where there is a separate array of receive antennas for each transmit antenna; it receives the signal in such a way that there is no interference from other transmit streams. As can be seen from Figure 20.13, the lower bound is fairly tight, while the upper bound can become rather loose especially for large number of antennas.

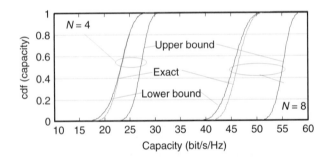

Figure 20.13   Exact capacity, upper bound, and lower bound of a multiple-input-multiple-output system in an independent identically distributed channel at a 21-dB signal-to-noise ratio, with $N_r = N_t = N$ equal to 4 and 8. Reproduced with permission from Molisch and Tufvesson [2005] © Hindawi.

Perfect Channel State Information at the Transmitter and Receiver

The capacity gain by waterfilling (compared with equal-power distribution) is rather small when the number of transmit and receive antennas is identical. This is especially true in the limit of large SNRs: when there is a lot of water available, the height of "concrete blocks" in the vessel has little influence on the total amount that ends up in the vessels. When N_t is larger than N_r, the benefits of waterfilling become more pronounced (see Figure 20.14). We can interpret this the following way: if the TX has no channel knowledge, then there is little point in having more transmit than receive antennas – the number of data streams is limited by the number of receive antennas. Of course, we can transmit the same data stream from multiple transmit antennas, but this does not

---

[13] Equation 20.43 is a slight abuse of notation, indicating as it does that the cdf of the capacity is bounded by the cdfs of the random variables given by the equations on the left and right sides.

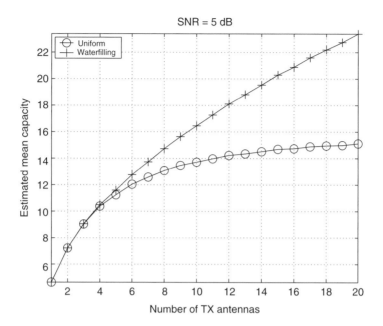

Figure 20.14 Capacity with and without channel state information at the transmitter with $N_t = 8$ antennas and a signal-to-noise ratio of 5 dB.

Reproduced with permission from Molisch and Tufvesson [2005] © Hindawi.

increase the SNR for that stream at the RX; without channel knowledge at the TX, the streams add up incoherently at the RX.

On the other hand, if the TX has full channel knowledge, it can perform beamforming, and direct the energy better toward the receive array. Thus, increasing the number of TX antennas improves the SNR, and (logarithmically) capacity. Thus, having a larger TX array improves capacity. However, this also increases the demand for channel estimation.

20.2.7 Impact of the Channel

Up to now, we have discussed the capacity of a channel with flat-fading and iid zero-mean complex Gaussian coefficients. Channels occurring in practice are more complicated, and deviations of the channel from idealized assumptions can have a significant impact on capacity. In the following, we describe some of the more important effects.

Channel Correlation

Correlation of the signals at different antenna elements can significantly reduce the capacity of a MIMO system. This can be shown the following way: the capacity is determined by the distribution of the singular values of the channel matrix. For a given SNR, maximum capacity is achieved when the channel transfer matrix has full rank and the singular values of **H** are equally strong. If the coefficients of the channel matrix are iid Rayleigh fading, then this situation is *approximately* fulfilled (though the ordered eigenvalues still have different values). But if the fading of the channel coefficients is correlated, then the singular value spread – i.e., the difference between the largest and the smallest singular values – becomes much bigger. This, in turn, leads to a reduction in system capacity because some of the "parallel channels" of Eq. (20.36) have extremely low SNRs.

Correlation is influenced by the angular spectrum of the channel as well as the arrangement and spacing of antenna elements (see Chapter 13). For antennas that are spaced half a wavelength apart, a uniform angular power spectrum leads approximately to a decorrelation of incident signals. A smaller angular spread of the channel leads to an increase in correlation. Since we are now looking at a MIMO system, we have to consider correlation both at the TX and at the RX. A popular model (the so-called Kronecker model) assumes that correlation at the TX is independent of correlation at the RX (see Section 7.4.7).[14] Realization of the channel matrix can then be obtained as

$$\mathbf{H}_{\text{kron}} = \frac{1}{E\{\text{tr}(\mathbf{H}\mathbf{H}^{\dagger})\}} \mathbf{R}_{\text{RX}}^{1/2} \mathbf{G}_{\text{G}} \mathbf{R}_{\text{TX}}^{1/2} \qquad (20.44)$$

where $\mathbf{G}_{\text{G}}$ is a matrix with iid complex Gaussian entries with unit variance.

Analytical computation of capacity is much more complicated in the case of correlated channels. It is easier to obtain simulation results by generating realizations of the channel matrix from Eq. (20.44), and then inserting them into Eq. (20.38) or Eq. (20.37).

Figure 20.15 shows the ergodic capacity of a 4×4 MIMO system with uniform linear arrays at the TX and RX as a function of angular spread at one link end.[15] We see that a small angular spread leads to a drastic reduction in capacity.

Frequency-Selective Channels

The previous sections assumed frequency-flat channels. Fortunately, generalization to the frequency-selective case is straightforward. As Shannon has shown, the use of an OFDM-like scheme is optimum for dealing with the frequency-selective channel, converting it into a number of parallel flat-fading channels. Thus, capacity per unit bandwidth is a straightforward generalization of Eq. (20.37)

$$C = \frac{1}{B} \int_B \log_2 \left[\det \left(\mathbf{I}_{N_r} + \frac{\overline{\gamma}}{N_t} \mathbf{H}(f) \mathbf{R}_{ss}(f) \mathbf{H}(f)^{\dagger} \right) \right] df \qquad (20.45)$$

where B is the bandwidth of the considered system.

This equation also implies that frequency selectivity offers additional diversity that increases the slope of the capacity cdf. If one of the frequency subchannels shows poor capacity, there is a good chance that another subchannel has high quality. Figure 20.16 shows an example of a measured capacity cdf in a microcellular environment. We see that the capacity cdf becomes steeper as the bandwidth increases.

Line-of-Sight Versus Non-Line-of-Sight

In some situations, there is an LOS connection between the TX and RX, resulting in different fading statistics. As we saw in Chapter 5, the fading statistics of an SISO link becomes Ricean instead of Rayleigh. For a MIMO system, the channel matrix can be written as

$$\mathbf{H} = \sqrt{\frac{K_{\text{LOS}}}{K_{\text{LOS}} + 1}} \hat{\mathbf{H}}_{\text{LOS}} + \sqrt{\frac{1}{K_{\text{LOS}} + 1}} \hat{\mathbf{H}}_{\text{res}} \qquad (20.46)$$

---

[14] For a discussion of this assumption, and a more general model, see Weichselberger et al. [2006]

[15] Note that this figure is based on the idealized assumption that there is no mutual coupling between antenna elements; investigations have shown that mutual coupling influences capacity by introducing pattern diversity as well as changing the average power received in different antenna elements.

Figure 20.15 Average capacity of a 4×4 multiple-input-multiple-output system with a $10°$ root-mean-square angular spread as seen from the base station as a function of antenna spacing at the BS.
Reproduced with permission from Molisch and Tufvesson [2005] © Hindawi.

Figure 20.16 Capacity in a measured microcellular channel, for different system bandwidths: 4×4 multiple-input-multiple-output system.
Reproduced with permission from Molisch et al. [2002] © IEEE.

where K_{LOS} is the ratio of powers in the LOS to those in residual components, $\hat{\mathbf{H}}_{LOS}$ is a purely deterministic matrix, and $\hat{\mathbf{H}}_{res}$ has (uncorrelated or correlated) zero-mean Gaussian entries.[16] If the distance between the TX and RX is large – i.e., larger than the Rayleigh distance (see Chapter 4) – the LOS gives rise to a matrix $\hat{\mathbf{H}}_{LOS}$ that has rank 1 (a single wave can only be associated with one singular value of the channel matrix!). This in turn implies that the singular value spread of matrix Eq. (20.46) is much larger than for a Non Line Of Sight (NLOS) matrix. Consequently, the capacity of a LOS channel is lower than for an NLOS channel *when assuming equal SNR*. One should however note that the SNR is often better in the LOS case compared with the NLOS case. For a power-limited scenario with realistic channels, the LOS case often gives the highest capacity, despite the imbalance between singular values.

It is also noteworthy that a strong LOS component leads to a larger spread of eigenvalues if the LOS component is a *plane* wave. A *spherical* wave leads to a transfer function matrix that can have full rank if antenna elements are spaced appropriately. The curvature of waves is noticeable up to one Rayleigh distance – i.e., typically a few meters.

Example 20.4 *Capacity in a channel with LOS. The transmit and receive arrays are uniform linear arrays with element spacing λ between the elements and $N_r = N_t = 8$. The arrays are perpendicular to the LOS connection. The DOA of the NLOS components is uniformly distributed between 0 and 2π. Estimate the mean capacity if the TX does not have CSI for an SNR of 20 dB, and $K_{LOS} = 0$ and 20 dB.*

As a first step, we have to determine the channel matrix. Since the transmit and receive arrays are linear arrays that are oriented perpendicular to the LOS, we find that $\hat{\mathbf{H}}_{LOS}$ is the all-1's matrix if transmit and receive arrays are sufficiently far apart from each other. Furthermore, $\hat{\mathbf{H}}_{res}$ has unit energy, iid complex Gaussian entries because the angular spectrum is uniformly distributed and the antenna elements are more than $\lambda/2$ apart from each other. From Eq. (20.38), capacity is then:

$$C = \log_2\left[\det\left(\mathbf{I}_{N_r} + \frac{\overline{\gamma}}{N_t}\mathbf{H}\mathbf{H}^\dagger\right)\right] \tag{20.47}$$

$$= \log_2\left[\det\left(\mathbf{I}_{N_r} + \frac{\overline{\gamma}}{N_t}\left[\frac{K_{LOS}}{K_{LOS}+1}\hat{\mathbf{H}}_{LOS}\hat{\mathbf{H}}_{LOS}^\dagger\right.\right.\right.$$
$$\left.\left.\left. + \frac{\sqrt{K_{LOS}}}{K_{LOS}+1}\left(\hat{\mathbf{H}}_{res}\hat{\mathbf{H}}_{LOS}^\dagger + \hat{\mathbf{H}}_{LOS}\hat{\mathbf{H}}_{res}^\dagger\right) + \frac{1}{K_{LOS}+1}\hat{\mathbf{H}}_{res}\hat{\mathbf{H}}_{res}^\dagger\right]\right)\right] \tag{20.48}$$

Using Jensen's inequality, the expected value of the capacity can be approximated as

$$E\{C\} \simeq \log_2\left[\det\left(\mathbf{I}_{N_r} + \frac{\overline{\gamma}}{N_t}\left[\frac{K_{LOS}}{K_{LOS}+1}E\{\hat{\mathbf{H}}_{LOS}\hat{\mathbf{H}}_{LOS}^\dagger\}\right.\right.\right. \tag{20.49}$$

$$\left.\left.\left. + \frac{\sqrt{K_{LOS}}}{K_{LOS}+1}E\{\hat{\mathbf{H}}_{res}\hat{\mathbf{H}}_{LOS}^\dagger + \hat{\mathbf{H}}_{LOS}\hat{\mathbf{H}}_{res}^\dagger\} + \frac{1}{K_{LOS}+1}E\{\hat{\mathbf{H}}_{res}\hat{\mathbf{H}}_{res}^\dagger\}\right]\right)\right] \tag{20.50}$$

$$= \log_2\left[\det\left(\mathbf{I}_{N_r} + \frac{\overline{\gamma}}{N_t}\left[\frac{K_{LOS}N_t}{K_{LOS}+1}\mathbf{1} + \frac{1}{K_{LOS}+1}E\{\hat{\mathbf{H}}_{res}\hat{\mathbf{H}}_{res}^\dagger\}\right]\right)\right] \tag{20.51}$$

$$= \log_2\left[\det\left(\left[1 + \frac{\overline{\gamma}}{K_{LOS}+1}\right]\mathbf{I}_{N_r} + \left[\overline{\gamma}\frac{K_{LOS}}{K_{LOS}+1}\mathbf{1}\right]\right)\right] \tag{20.52}$$

---

[16] Both $\hat{\mathbf{H}}_{LOS}$ and $\hat{\mathbf{H}}_{res}$ are normalized to $E\{|\hat{\mathbf{H}}|_F^2\} = N_tN_r$, where $|\hat{\mathbf{H}}|_F$ is the Frobenius norm of $\hat{\mathbf{H}}$.

where $\mathbf{1}$ is an $N_r \times N_r$ matrix where each entry is 1; such a matrix has only a single nonzero eigenvalue, whose magnitude is N_r. Equation (20.52) can be further approximated as

$$E\{C\} \simeq N_r \log_2\left[1 + \frac{\overline{\gamma}}{K_{LOS} + 1}\right] + \log_2\left[1 + \frac{\overline{\gamma}K_{LOS}N_r}{K_{LOS} + 1}\right] \qquad (20.53)$$

Using Eq. (20.53), we find that capacity for $K_{LOS} = 0$ and 20 dB is 54 and 17 bit/s/Hz, respectively. Note that the above is a rather crude approximation, but it gives correct trends. Monte Carlo simulations give 40 and 15 bit/s/Hz, respectively.[17]

Limited Number of Interacting Objects

In the intuitive picture of Section 20.2.1, we have already seen that a certain number of IOs are required to act as relays for data streams. Relating this to the mathematics of the previous section, we note that a certain number of IOs are needed in order to guarantee that the channel coefficients in matrix $\mathbf{H}$ are independent. While the number of existing IOs is always large, the number of *significant* IOs might be limited in practice. After all, IOs that are too weak to provide appreciable SNR (and thus capacity) are not useful in carrying data streams.

Figure 20.17 shows some measurement results for capacity as a function of the number of antenna elements. Measurements were taken in a microcellular scenario where the number of IOs was rather small. We find that capacity, especially for the LOS scenario, does not increase linearly with the number of antenna elements N when N exceeds 4. This is a clear sign that the number of IOs limits the achievable capacity.

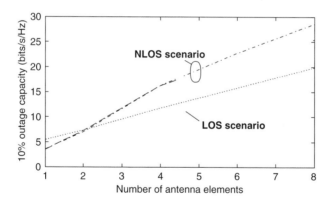

Figure 20.17   Ten percent outage capacity as a function of the number of antenna elements at the transmitter and RX in measured line-of-sight and non-line-of-sight scenarios.
Reproduced with permission from Molisch et al. [2002] © IEEE.

Keyhole Channels

There are some special cases where capacity is low even though the signals at antenna elements are uncorrelated. These cases are often referred to as *keyholes* or *pinholes*. An example of a keyhole scenario is a rich scattering environment at both the TX and RX ends; between them there is only

---

[17] Note that the step from Eq. (20.52) to (20.53) involves approximations that are not detailed further here. It leads to a result that is somewhat similar to the result of [Ayadi et al. 2002]

one propagation path with just one degree of freedom. Such a scenario can occur when the TX and RX are surrounded by IOs, and the two IO groups are separated by a long stretch of empty space (greenfield). Another scenario is the case when IO areas are connected by a single-mode waveguide or by a diffraction edge. In all these cases, the total transfer function is proportional to

$$\mathbf{H} = \mathbf{R}_{RX}^{1/2} \mathbf{G}_{G,1} \mathbf{R}_{RX-TX}^{1/2} \mathbf{G}_{G,2} \mathbf{R}_{TX}^{1/2} \tag{20.54}$$

where $\mathbf{G}_{G,1}$ and $\mathbf{G}_{G,2}$ are both iid complex Gaussian matrices, and where $\mathbf{R}_{RX-TX}$ describes the correlation of the channel matrix between the TX and RX environments; in a keyhole channel, this is a low-rank matrix. It can also be seen from this description that the statistics of the entries are no longer Gaussian – this explains why it can be possible to have low correlation and low capacity at the same time. It should, however, be noted that keyhole channels occur very seldom in practice. While Almers et al. [2003] measured one in a controlled environment, it seems to be a rare occurrence in "normal" environments.

20.2.8   Layered Space–Time Structure

Up to now, we have only discussed the information-theoretic limits of MIMO systems. The big question is how to realize these capacities in practice. One possibility is joint encoding of the data streams that are to be transmitted from different antenna elements, combined with maximum-likelihood detection. When this technique is combined with (almost) capacity-achieving codes, it can closely approximate the capacity of a MIMO system. For a small number of antenna elements and for a small modulation alphabet (BPSK or Quadrature-Phase Shift Keying (QPSK)), such a scheme can actually be gainfully employed. However, for most practical cases, the complexity of a joint MLSE is rather high. For this reason, so-called layered space–time structures have been proposed, which allow us to break up the demodulation process into several separate pieces, each of which has lower complexity. These structures are also widely known under the name of *BLAST architectures*.

Horizontal BLAST

H-BLAST is the simplest possible layered space–time structure.[18] The TX first demultiplexes the data stream into N_t parallel streams, each of which is encoded *separately*. Each encoded data stream is then transmitted from a different transmit antenna. The channel mixes up the different data streams; the RX separates them out by successive nulling and interference subtraction. In other words, the RX proceeds in the following steps (Figure 20.18):

- It considers the first data stream as the useful one, and regards the other data streams as interference. It can then use *optimum combining* for suppression of interfering streams (see Chapter 13). The RX has $N_r \geq N_t$ antenna elements available. If $N_r = N_t$, it can suppress all $N_t - 1$ interfering data streams and receive the desired data stream with diversity order 1. If the RX has more antennas, it can receive the first data stream with better quality. But in any case, interference from the other streams can be eliminated.
- The desired stream can now be demodulated and decoded. Outputs from that process are firm decisions on the bits of stream 1. Note that since we have separate encoding of different data streams, we only need knowledge of the first stream to complete the decoding process.

---

[18] This scheme was originally called *V-BLAST* (for Vertical BLAST) but the name was changed later by Foschini et al. [2003].

Figure 20.18   Block diagram of a horizontal BLAST transceiver.

- The bits that have thus been decoded are now re-encoded, and remodulated. Multiplying this symbol stream by the transfer function of the channel, we obtain the contribution that stream 1 has made to the total received signal at the different antenna elements.
- We subtract these contributions from the signals at the different antenna elements.
- Now we consider the "cleaned-up" signal and try to detect the second data stream. We again have N_r received signals, but only $N_t - 2$ interferers. Using optimum combining again, we can now receive the desired data stream with diversity order 2.
- The next step is again decoding, recoding, and remodulating the considered data stream (stream 2 now), and subtraction of the associated signal from the total signal at the receive antenna elements obtained in the previous step. This cleans up the received signal even more.
- The process is repeated until the last data stream is decoded.

This scheme is actually very similar to multiuser detection (Section 18.4): if different transmit streams were to come from different users, then H-BLAST would be normal serial interference cancellation. Note also that the encoding scheme does not require "cooperation" between different antenna elements (or users). Similar to serial interference cancellation, H-BLAST also faces the problem of error propagation, especially since the first decoded data stream has the worst quality. In other words: if data stream 1 is decoded incorrectly, then we subtract the "wrong" signal from the remaining signals at the antenna elements. Thus, instead of "cleaning up" the receive signal, we introduce even more interference. This in turn increases the likelihood that the second data stream is decoded incorrectly, and so on. In order to mitigate this problem, stream ordering should be used: the RX should first decode the stream that has the best SINR, then the one with the next best, and so on.

Diagonal BLAST

The main problem with H-BLAST is that it does not provide diversity. The first stream, which has diversity order 1, dominates the performance at high SNRs. A better performance can be achieved with the so-called D-BLAST scheme. In this approach, streams are cycled through the different transmit antennas, such that each stream sees all possible antenna elements. In other words, each single transmit stream is subdivided into a number of subblocks. The first subblock of stream 1 is transmitted from antenna 1, the next subblock from antenna 2, and so on (compare Figure 20.19).

Decoding can be done stream by stream; again, each decoded block can be subtracted from signals at the other antenna elements and, thus, enhances the quality of the residual signal.

The difference from H-BLAST is that each stream is sometimes in a "good" position in the sense that the other streams have already been subtracted, and thus the SINR is very high, while sometimes

Figure 20.19   Assignment of bitstreams to different antennas for horizontal BLAST and diversity BLAST.

it is in a bad position, in the sense that it suffers full interference. Thus, each stream experiences full diversity. More precisely, each data stream alternatingly sees a channel with diversity order 1 (whose SNR has a pdf that is a chi-square distribution with 2 degrees of freedom χ_2^2), diversity order 2 (chi-square with 4 degrees of freedom, χ_4^2), and so on. Therefore, total capacity in the case when $N_t = N_r$:

$$C_{\text{D-BLAST}} = \sum_{k=1}^{N_t} \log_2 \left[1 + \frac{\bar{\gamma}}{N_t} \chi_{2k}^2 \right] < C \tag{20.55}$$

This actually achieves the lower capacity bound of Eq. (20.43) [Ariyavisitakul 2000].

Structures for Channel State Information at the Transmitter

When full CSI is available at the TX, transceiver schemes become much simpler, at least conceptually. By multiplying transmitted and received signal vectors by the right and left singular vectors of the channel matrix, respectively, diagonalization of the channel is achieved. Therefore, the different data streams do not interfere with each other; rather, we just have a number of parallel channels, each of which can be separately encoded and decoded. The difficulties lie instead in the practical aspects of obtaining and using CSIT (as discussed in Section 20.1.6).

20.2.9   Diversity

Multiple antennas can also be used to provide pure diversity. Again, we have to distinguish between systems that have CSIT and systems that do not. The former can achieve beamforming gain in addition to diversity gain, while the latter is restricted to achieving better resistance to fading.

Diversity with Channel State Information at the Transmitter

Having the CSIT leads to a conceptually simple transceiver structure in the diversity case, just as it did for spatial multiplexing. The transmit vector consists of weighted replica of a single data symbol, $\mathbf{s} = \mathbf{u}c$. Consider again the singular value decomposition of the channel matrix, Eq. (20.34). We then find that the received signal is

$$\mathbf{r} = \mathbf{W}\boldsymbol{\Sigma}\mathbf{U}^\dagger \mathbf{u}c + \mathbf{n} \tag{20.56}$$

The RX performs a linear combination (summation) of the signals at the different antenna elements, such that the output of the combiner is $\tilde{c} = \mathbf{wr}$. Choosing $\mathbf{w} = (\mathbf{W})_1$, and $\mathbf{u} = (\mathbf{U})_1$ maximizes the SNR at the RX, where $(\mathbf{W})_1$ is the left singular vector corresponding to the largest singular value (and similarly for $\mathbf{U}$). In other words, we choose the transmit and receive weights according to the singular value decomposition of the channel. The SNR that can be achieved this way is

$$\gamma = \frac{P}{\sigma_n^2} \tilde{\sigma}_{max}^2 \tag{20.57}$$

where $\tilde{\sigma}_{max}$ is the largest singular value of matrix **H**. For the case when we have only one transmit antenna, the scheme reduces to maximum-ratio combining; for the case of a single receive antenna, it becomes maximum-ratio transmission. Figure 20.20 shows the cdf of the SNR that can be achieved this way. We can see that the slope of the largest eigenvalue is identical to the slope of the sum of the eigenvalue. In a 4×4 MIMO system, the diversity order is 16, and SNR distribution is almost a step function – in other words, the required fading margin is very small. We also find that there is an enhancement of the average SNR, but it is smaller than $N_r N_t = 16$ (12 dB) (see also Section 20.2.10).

Figure 20.20 Cumulative distribution function of the signal-to-noise ratio in 2×2 (left) and 4×4 (right) multiple-input-multiple-output diversity with independent identically distributed Rayleigh fading at all antenna elements.

Example 20.5 *Consider a 2×2 MIMO system with a 10-dB average SNR at each of the antenna elements. What is the probability that the SNR is smaller than 7 dB?*

As shown in Andersen [2000], the pdf for the largest eigenvalue of matrix $\mathbf{HH}^\dagger$ in a 2×2 system is

$$pdf_{\lambda_{\max}}(\lambda) = \exp(-\lambda)[\lambda^2 - 2\lambda + 2] - 2\exp(-2\lambda) \qquad (20.58)$$

and the cdf is

$$cdf_{\lambda_{\max}}(x) = 1 - \exp(-x)(x^2 + 2) + \exp(-2x) \qquad (20.59)$$

Here λ describes the normalized SNR. We now want to identify the probability that the largest eigenvalue is 3 dB below the mean value of $\mathbf{HH}^\dagger$; i.e.,

$$cdf_{\lambda_{\max}}(0.5) = 3 \cdot 10^{-3} \qquad (20.60)$$

Of course, we can also ascribe capacity to a diversity system. This capacity follows immediately from SNR statistics:

$$C = \log_2 \left[1 + \frac{P}{\sigma_n^2} \max_k (\tilde{\sigma}_k^2) \right] \qquad (20.61)$$

Diversity Without Channel State Information at the Transmitter – Space–Time Coding

If the channel is unknown at the TX, then we try to transmit different versions of the data stream from the different transmit antennas. In Chapter 13, we have already encountered some methods that achieve this – especially, delay diversity. Another approach that has gained enormous attention is space–time coding. In this approach, redundancy is introduced by sending from each transmit antenna a differently encoded (and fully redundant) version of the same signal. There are multiple ways in which encoding can be done. In the following, we will first describe the *Alamouti* code, the most popular form of *space–time block codes*. Subsequently, we mention the basic principles of *space–time trellis codes*. The codes we consider here work independently of the number of receive antennas, and can thus be seen as a form of transmit diversity (see also Chapter 13).

Orthogonal Space–Time Block Codes

The idea behind Space Time Block Codes (STBCs) is to transmit data in a way that guarantees high diversity, while allowing a simple decoding process. The most popular STBC is the Alamouti code [Alamouti 1998]. Its idea is simple yet ingenious. Consider a flat-fading channel, where the complex channel gain from TX antenna 1 is given by h_1 and the gain from TX antenna 2 to the RX by h_2. Now, transmit the two symbols, c_1 and c_2, from the two transmit antennas at time instant 1:

$$\mathbf{s}_1 = \frac{1}{\sqrt{2}} \begin{pmatrix} c_1 \\ c_2 \end{pmatrix} \qquad (20.62)$$

At the second time instant, transmit the vector:

$$\mathbf{s}_2 = \frac{1}{\sqrt{2}} \begin{pmatrix} -c_2^* \\ c_1^* \end{pmatrix} \qquad (20.63)$$

The factor $1/\sqrt{2}$ stems from the necessity to keep the energy constant, as we are now using two transmit antennas. Then the received signal can be written as

$$\mathbf{r} = \begin{pmatrix} r_1 \\ r_2^* \end{pmatrix} = \frac{1}{\sqrt{2}} \begin{pmatrix} h_1 & h_2 \\ h_2^* & -h_1^* \end{pmatrix} \begin{pmatrix} c_1 \\ c_2 \end{pmatrix} + \mathbf{n} = \mathbf{Hc} + \mathbf{n} \qquad (20.64)$$

We have thus created a "virtual" MIMO system. It is important to note that the columns of the "virtual" channel matrix $\mathbf{H}$ are orthogonal. Therefore, $\mathbf{H}^\dagger \mathbf{H}$ becomes a scaled identity matrix $\alpha \mathbf{I}$. For decoding, we thus first multiply the received signal vector by $\mathbf{H}^\dagger$. Then we get

$$\tilde{\mathbf{r}} = \mathbf{H}^\dagger \mathbf{r} = \mathbf{H}^\dagger \mathbf{Hc} + \mathbf{H}^\dagger \mathbf{n} = \alpha \mathbf{c} + \tilde{\mathbf{n}} \qquad (20.65)$$

Since the columns of $\mathbf{H}$ are orthogonal, the components of $\tilde{\mathbf{n}}$ are still uncorrelated zero-mean Gaussian, and have variance $\alpha \sigma_n^2$. Therefore, decoding of data c_1 and c_2 becomes decoupled; which greatly reduces the computational effort in the RX.

As far as performance is concerned, we find that α describes the "effective" channel power gain, which is

$$\alpha = \frac{|h_1|^2 + |h_2|^2}{2} \qquad (20.66)$$

Consequently, the diversity order is 2: both h_1 and h_2 would have to be in a fading dip for the effective channel gain to be low. We also see that the scheme can only increase diversity, but not beamforming gain, since it does not have CSIT.

There have been many attempts to generalize the Alamouti code to more than two transmit antennas. Unfortunately, it can be shown that orthogonal STBCs for more than two antennas have a rate smaller than 1 – in other words, we cannot even achieve the rate that we could have with a single-antenna system [Tarokh et al. 1999].

Space–Time Trellis Codes STBCs provide full diversity order, but no coding gain. Such coding gain can be obtained from Space Time Trellis Codes (STTCs). Given N_t transmit antennas, the STTC maps each symbol from the information source to a *vector* of N_1 transmit symbols that are sent from the different antenna elements. Decoding requires a vector Viterbi decoder. The error probability – i.e., the probability of confusing one codeword $\mathbf{C} = (\mathbf{c}_1, \mathbf{c}_2, \cdots, \mathbf{c}_L)$ of length L_c with another codeword $\tilde{\mathbf{C}}$ – is upper-bounded by

$$P(\mathbf{C} \to \tilde{\mathbf{C}}) \leq \left(\prod_{i=1}^{R_e} \lambda_i \right)^{-N_r} \left(\frac{E_S}{4N_0} \right)^{-R_e N_r} \qquad (20.67)$$

where R_c and λ_i are the rank and eigenvalues, respectively, of the error matrix:

$$\sum_{i=1}^{L_c} (\mathbf{c}_i - \tilde{\mathbf{c}}_i)(\mathbf{c}_i - \tilde{\mathbf{c}}_i)^\dagger \qquad (20.68)$$

The term

$$\left(\prod_{i=1}^{R_e} \lambda_i \right)^{-N_r} \qquad (20.69)$$

represents the coding gain, while the term

$$\left(\frac{E_S}{4N_0} \right)^{-R_e N_r} \qquad (20.70)$$

describes diversity gain. In order to achieve full diversity, the rank of the error matrix should thus be as high as possible (*rank criterion*); in order to achieve high coding gain, the determinant of the error matrix should be maximized (*determinant criterion*). The rank criterion and the determinant criterion provide important guidelines on how to design STTCs.

The main drawback with STTCs is their complexity. The requirement for a vector Viterbi RX has proved to be a major obstacle in their application in practical systems.

20.2.10 Tradeoffs between Diversity, Beamforming Gain, and Spatial Multiplexing

MIMO systems can be used to achieve spatial multiplexing, diversity, and/or beamforming. However, it is not possible to attain all of those goals simultaneously at their full extent. First, we find that there is a tradeoff between beamforming and diversity gain; this tradeoff also depends on the environment in which we are operating. Consider first an LOS scenario. In this case, it is obvious that the achievable beamforming gain is $N_t N_r$: we form beams at the TX (with gain N_t) and at the RX (with gain N_r), and point them at each other. The gains thus multiply. On the other hand, there is obviously no diversity gain, since there is no fading in an LOS scenario – in other words, the slope of the SNR distribution curve does not change.

In a heavily scattering environment, the slope of SNR distribution changes drastically due to the use of multiantenna elements. If the channel is known at the TX, we can choose our antenna weights in such a way that it is unlikely that the receive signals are in a fading dip; furthermore, even if this happens, it is unlikely that *all* the receive signals are in a fading dip. In other words, it is easy to make sure that *at least one* receive signal has good quality. As discussed in Chapter 13, the diversity order is the slope of the BER versus SNR curve for very high SNRs:

$$d_{\text{div}} = - \lim_{\bar{\gamma} \to \infty} \frac{\log[BER(\bar{\gamma})]}{\log(\bar{\gamma})} \tag{20.71}$$

it can also be related to the slope of the SNR distribution at very low values of the SNR, and is thus a measure for the likelihood that all signals are in a bad fading dip simultaneously. The maximum diversity order in a heavily scattering environment can be shown to be $N_t N_r$. On the other hand, it also turns out that the maximum beamforming gain in such a heavily scattering environment is upper limited by $(\sqrt{N_t} + \sqrt{N_r})^2$. The reason is that it is not possible to form the transmit beam pattern in such a way that MPCs overlap constructively *at all receive antenna elements simultaneously*. Note the key tradeoff here between achieving full diversity order and beamforming gain.

There is also a fundamental tradeoff between spatial multiplexing and diversity. Zheng and Tse [2003] showed that the optimum tradeoff curve between diversity order and rate r is piecewise linear, connecting the points:

$$d_{\text{div}}(r) = (N_t - r)(N_t - r), \quad r = 0, \ldots, \min(N_t, N_r) \tag{20.72}$$

This implies the maximum diversity order, $N_t N_r$, and maximum rate, $\min(N_t, N_r)$, cannot be achieved simultaneously.

20.2.11 Feedback for MIMO

As we have discussed previously, performance of MIMO systems can be improved by CSIT. Such channel knowledge might be obtainable from the principle of reciprocity in TDD systems, but requires significant amount of feedback (and thus overhead) in FDD systems, see also Section 20.1.6.

A naive approach to CSI feedback requires quantizing the entries of the transfer function matrix, resulting in a large number of bits in the feedback message.

The problem is compounded by the fact that quantization of the channel coefficients has to be done with a fine resolution, because the eigenstructure of a channel matrix is fairly sensitive to changes in the coefficients. In other words, if some of the entries of the channel matrix change even by a small amount (due to the quantization), then the eigenvectors computed from this perturbed matrix can be quite different from the eigenvectors of the original matrix. Since the optimal beam-forming vectors are usually along the eigenvectors of the matrix, we thus find a strong sensitivity of beamforming vectors (and resulting SNR) to the quantization.

Example 20.6 *We consider a typical example for a cellular MIMO system: let the BS have 8 antenna elements, and each MS have 2 antenna elements. The system has 5 MHz bandwidth centered at 2 GHz carrier frequency, and operates in a channel with 250 kHz coherence bandwidth. The coherence time is 5 ms, corresponding to typical vehicular speeds. With 30 users in the cell, what is the total overhead data rate for the feedback?*

Assume that real and imaginary part are quantized with 6 bits each, and a rate 2/3 code is used to protect the feedback information. The total number of feedback bits per second is

$$2 \cdot 6 \cdot \frac{1}{2/3} \cdot K \cdot N_t \cdot N_r \cdot \frac{B}{B_{coh}} \cdot \frac{1}{T_{coh}} \tag{20.73}$$

$$= 2 \cdot 6 \cdot \frac{1}{2/3} \cdot 30 \cdot 8 \cdot 2 \cdot \frac{5000}{250} \cdot \frac{1}{0.005} \tag{20.74}$$

$$= 34.6 \cdot 10^6 \tag{20.75}$$

Thus, the total feedback load is tens of Mbit/s, which is clearly unacceptable. Note that we have assumed here that the CSI from all the MSs has to be fed back, even though at a given point in time the number of data streams is limited to 8 by the dimensionality constraint mentioned previously; the feedback of all MSs' CSI might be required for the purpose of scheduling as discussed below. But even when considering only 4 active users, the overall feedback data rate is 4.6 Mbit/s – about as much as the typical cell capacity for a 5 MHz system. This example shows clearly that feedback reduction techniques are extremely important.

If the TX uses linear precoding (beamforming), an effective way for reducing feedback is the use of limited *feedback codebooks*. In this approach the RX does not feed back the channel transfer function matrix itself, but rather the index of a specific setting of the precoder matrix that it wants the TX to use. Codebook sizes (i.e., number of possible precoder settings) are usually very small (8 or 16 entries, requiring 3 or 4 bits to communicate the index of the entry), and thus the savings can be significant: instead of feeding back $12 \cdot N_t \cdot N_r$ bits in each coherence time and each coherence bandwidth (192 bits in the example above), only 3 or 4 bits need to be transmitted – a saving of almost two orders of magnitude.

The advantages of using codebooks are compounded if the number of spatial streams is smaller than the number of antenna elements. The channel coefficients are the entries into a matrix, whose size is proportional to $N_t N_r$. On the other hand, the number of codebook indices that have to be fed back is L, i.e., the BS has to know for each spatial stream what codeword (precoder setting) to use.

Good performance can, however, only be achieved with a good design of the entries of the codebook. The basic idea of codebook design is to exploit the intrinsic geometry of the underlying vector spaces. Note that the correlation structure of the channel (which in turn depends on the antenna array and the propagation channel) influences the characteristics of the vector space.

In the following, we consider only feedback for a single user, and assume a flat-fading channel. A generalization to frequency-selective channels is given at the end of this section.

Feedback for Single-Data-Stream Transmission

We first discuss the situation where the BS transmits a single data stream to an MS with multiple antenna elements. Assume we have a codebook $\mathcal{C}$ (set of quantized precoding vectors $\mathbf{t}$ with unit norm) of size Q, so that feedback of one codebook entry takes $\log_2 Q$ bits. For a given $\mathbf{t}^{(q)}$, the SNR available at the MS is

$$SNR_q = \gamma \|\mathbf{H}\mathbf{t}^{(q)}\|_2^2 \tag{20.76}$$

where γ is again the SNR obtained in a SISO AWGN channel of unit gain, and $\mathbf{H}$ is the $N_r \times N_t$ transfer function matrix. The RX measures the transfer function matrix, and defines the best possible precoding vector as

$$\widetilde{\mathbf{t}} = \arg \max_{\mathbf{t}^{(q)} \in \mathcal{C}} SNR_q \tag{20.77}$$

where $\widetilde{\mathbf{t}}$ depends on $\mathbf{H}$, even though we do not explicitly write this dependence. The MS can determine the desired $\widetilde{\mathbf{t}}$ for a given (measured) $\mathbf{H}$, e.g., by a brute-force search over the whole codebook.

But before the codebook can be used, we first have to define what its entries are (clearly, the codebook has to be known a priori to the TX and the RX). It is desirable to have a codebook that minimizes – on average – the loss of SNR at the RX

$$\zeta = E_{\mathbf{H}} \left\{ \gamma \|\mathbf{H}\mathbf{t}^{(\mathrm{opt})}\|_2^2 - \gamma \|\mathbf{H}\widetilde{\mathbf{t}}\|_2^2 \right\} \tag{20.78}$$

Now it can be shown that this loss can be upper-bounded as

$$\zeta \leq \widehat{\zeta} \leq E_{\mathbf{H}} \left\{ \gamma \|\mathbf{H}\|_2^2 \right\} E_{\mathbf{H}} \left\{ 1 - \max_{\mathbf{t}^{(q)} \in \mathcal{C}} |\mathbf{v}^* \mathbf{t}^{(q)}|^2 \right\} \tag{20.79}$$

where $\mathbf{v}$ is the right singular vector of $\mathbf{H}$ corresponding to the largest singular value. Furthermore $\widehat{\zeta}$ can be upper-limited as

$$\widehat{\zeta} \leq E_{\mathbf{H}} \left\{ \gamma \|\mathbf{H}\|_2^2 \right\} \left(\left(\frac{\delta}{2}\right)^2 \left(\frac{\delta}{2}\right)^{2(N_t-1)} Q + \left[1 - \left(\frac{\delta}{2}\right)^{2(N_t-1)} Q \right] \right) \tag{20.80}$$

where

$$\delta = \min_{p \neq q} \sqrt{1 - |[\mathbf{t}^{(q)}]^\dagger \mathbf{t}^{(p)}|^2} \tag{20.81}$$

i.e., the minimum over the subspace distance between $\mathbf{t}^{(p)}$ and $\mathbf{t}^{(q)}$. Finding the optimum codebook thus means that we have to maximize the minimum of the subspace distances – a problem that is related to so-called *Grassmannian line packing*. An example codebook for $N_t = 2$ and $Q = 4$ is given in Table 20.2.

Note that the codebook design is not unique – a multiplication with a constant phase (for all entries) does not change the SNR (and that multiplication is undone at the RX anyway).

The above derivation derives the codebook for iid complex Gaussian fading channels. Let us now consider the case where the channel is correlated at TX and RX, and let us consider the Kronecker model of Section 7.4.7.

$$\mathbf{H} = \mathbf{R}_{\mathrm{RX}}^{1/2} \mathbf{G}_{\mathrm{G}} \mathbf{R}_{\mathrm{TX}}^{1/2}, \tag{20.82}$$

Table 20.2 Codebook entries for limited feedback with $N_t = 2$ and $Q = 4$. [Love et al. 2003]

$q = \backslash$	$t_1^{(q)}$	$t_2^{(q)}$
1	$-0.1612 - 0.7348j$	$-0.5135 - 0.4128j$
2	$-0.0787 - 0.3192j$	$-0.2506 + 0.9106j$
3	$-0.2399 + 0.5985j$	$-0.7641 - 0.0212j$
4	-0.9541	0.2996

where $\mathbf{G}_G$ is a matrix with independent identically distributed (iid) complex Gaussian entries. It is intuitive that the transmit correlation should have a strong impact on the codebook: there is a relationship between the correlation matrix and the angles in which the radiation is transmitted. Clearly, there is no sense in quantizing finely in directions which are never used to transmit energy. This statement can be put into a more quantitative form by the following precoder design recipe: the codebook vectors $\hat{\mathbf{t}}^{(q)}$ are the codebook vectors from the uncorrelated case, scaled by the correlation matrix

$$\widehat{\mathbf{t}}^{(q)} = \frac{\mathbf{R}_{TX}^{\dagger/2}\mathbf{t}^{(q)}}{||\mathbf{R}_{TX}^{\dagger/2}\mathbf{t}^{(q)}||_2}. \tag{20.83}$$

Note, however, that this approach requires the TX and the RX to know the TX correlation matrix.

An even simpler codebook consists simply of the coefficients of a Discrete Fourier Transform (DFT) matrix. In that case, the entries of the codebook are just beams pointing to different directions. Clearly such a codebook works well if the angular spread of the signal is small, as often happens at macrocellular BSs.

Feedback for Space–Time Block Codes and Spatial Multiplexing

The design of the precoding codebook becomes more complicated when the TX employs orthogonal space–time block coding. In that case, the precoding vector $\mathbf{t}$ has to be replaced by a precoding matrix $\mathbf{T}$ of size $N_t \times \tilde{L}$, where $\tilde{L}$ is a dimension of the space–time code (e.g., 2 for an Alamouti code). The optimization goal is still the maximization of the overall SNR at the RX (note that there is only a single data stream, even though it is space–time coded)

$$SNR_q = \gamma ||\mathbf{HT}^{(q)}||_2^2 \tag{20.84}$$

A bound to the SNR loss due to the quantization can be derived as

$$\zeta = E_{\mathbf{H}}\left\{\gamma||\mathbf{HT}^{(\text{opt})}||_2^2 - \gamma||\mathbf{H}\widetilde{\mathbf{T}}||_2^2\right\} \leq E_{\mathbf{H}}\left\{\gamma\lambda_1^2(\mathbf{H})\right\}\left\{\tilde{L} + \left(\frac{\delta}{2\sqrt{L}}\right)^{2N_t\tilde{L}+\mathcal{O}(N_t)}Q\left[\left(\frac{\delta}{2}\right)^2 - \tilde{L}\right]\right\} \tag{20.85}$$

where the minimum subspace distance is now defined as the minimum chordal distance

$$\delta = \min_{p\neq q}\frac{1}{\sqrt{2}}||\mathbf{T}^{(p)}(\mathbf{T}^{(p)})^{\dagger} - \mathbf{T}^{(q)}(\mathbf{T}^{(q)})^{\dagger}||_F \tag{20.86}$$

The codebook entries are not unique – clearly a multiplication with a unitary matrix does not change the characteristics of $\mathbf{T}$. This is similar to the case of a single data stream, where multiplication with a constant phase does not change the optimality.

One might now ask the question why to consider space–time block codes, in particular Alamouti codes, in a setting where feedback is available – after all, Alamouti codes perform worse than maximum-ratio transmission (which is approximated by the beamforming based on the feedback). However, there are situations where CSI becomes uncertain (due to noise) or quickly outdated. In this case, it is advantageous to supplement the partial-CSI-based beamforming with extra diversity from the space–time coding. If average CSI is available, then the precoder should be used to appropriately "color" the transmission signal (i.e., align with the correlation matrix information) before transmission.

When spatial multiplexing is used in the transmission, finding the optimum precoding codebook becomes even more difficult, because – depending on the RX type – different criteria can be used for optimization.

Finally, we note that antenna selection at the transmitter can be interpreted as precoding with very simple precoder settings (entries of $\mathbf{T}$ are 1 or 0. Consequently, feedback of the precoder settings is also very simple.

Feedback in Frequency-Selective Channels

Consider now an OFDM system operating in a frequency-selective channel. The brute-force approach would be to feed back the codebook entry index for each subcarrier separately. However, this would be highly wasteful, since channels on adjacent subcarriers are correlated, and therefore also the optimum beamformers are correlated. As an intuitive improvement is thus easy to group the subcarriers, and just feed back one index for each group (e.g., the index for the subcarrier at the center of the group). More sophisticated approaches use interpolation (each group might have to be multiplied with a different phase factor; this phase factor also has to be fed back).

20.3 Multiuser MIMO

We now turn our attention to the question of how MIMO systems work in a cellular scenario where the BS communicates with multiple users at the same time. As we see in the following, this situation requires some new paradigms for the usage of the degrees of freedom provided by the multiple antenna elements. As a matter of fact, many of the issues are more related to SDMA (see Section 20.1.5 and 20.1.7) than to the single-user MIMO discussed previously in Section 20.2.

The system model we consider in the following is outlined in Figure 20.21. A single BS with N_{BS} antenna elements communicates with K MSs with $N_{\mathrm{MS}}^{(k)}$ antenna elements each. We assume that $N_{\mathrm{BS}} > N_{\mathrm{MS}}^{(k)}$, but that $N_{\mathrm{BS}} < \sum_k N_{\mathrm{MS}}^{(k)}$. This is the case normally occurring in a cellular network, and also the one with the most interesting effects for multiuser MIMO. A typical fourth-generation

Figure 20.21 Multiuser MIMO system setup.

cell would have $N_{BS} = 8$, $N_{MS}^{(k)} = 2$, and $K = 20$. There is no requirement that all MSs are active all the time; rather the BS can schedule the communication with specific users according to their channel state. However, certain fairness and delay criteria must be fulfilled for the different users.

It is tempting to think of a multiuser MIMO system simply as a single-user system where the antennas of the RX are distributed over different locations. This point of view is helpful in many aspects but must be taken with a grain of salt, since it can lead to some erroneous conclusions especially for the downlink case. The key differences between single-user MIMO and multiuser MIMO are as follows:

- Cross-layer design (scheduling) is critical, and can greatly reduce the probability of a breakdown of the overall capacity due to peculiarities of the channel state. If there are more MSs than there are BS antennas (a typical case in a cellular setting), then a judicious scheduling algorithm can greatly reduce the possibility that the channels to two users are (almost) linearly dependent (which would make spatial multiplexing to them inefficient).
- CSIT has a dramatic impact on the performance of a multiuser MIMO system in the downlink. One might wonder why the CSIT is so critical for the performance of this situation – after all, in the single-user case, the capacity difference between systems with and without CSIT is marginal. If the different MSs could collaborate in the reception of the signals, then indeed the capacity with and without CSIT would hardly differ; since by assumption $N_{BS} < \sum_k N_{MS}^{(k)}$, the RX could decode all the data streams that the BS sends out. However, in reality, the MSs can*not* collaborate, and each MS can decode at most $N_{MS}^{(k)}$ data streams. Thus, without CSIT, the following situations could occur: (i) the BS transmits only $\min_k(N_{MS}^{(k)})$ streams, which drastically limits the capacity of the downlink; (ii) the BS transmits N_{BS} data streams, but since the MSs cannot decode that many data streams, the SINR at each MS is extremely bad, and thus the capacity is low. With CSIT, however, the BS can "null out" undesired streams in the direction of the different MSs, and thus transmit more data streams while still enabling a good SINR at each MS.

20.3.1 Performance Limits

In a first step, we want to explore the fundamental limits of multiuser MIMO, i.e., how many users can be supported, and at what rate. We will not discuss in detail concepts of information theory, but mostly restrict ourselves to intuitive results that we can draw from the analogy with single-user MIMO and smart antennas.

Uplink

First consider the uplink: We have K MSs with $N_{MS}^{(k)}$ antenna elements each, and a BS with N_{BS} antenna elements. Each of the MSs is capable of transmitting multiple data streams, or of transmitting one data stream from multiple antenna elements. It is clear that such an arrangement is very similar to a single-user MIMO system with $\sum_k N_{MS}^{(k)}$ transmit antenna elements and N_{BS} receive antenna elements. The only difference lies in the fact that the different MSs cannot cooperate in transmitting their information; however, this hardly influences the capacity: consider the extreme case that each MS has only a single antenna: then the system is identical to an H-BLAST system; it is known that as long as the transmit rates from the different antenna elements are allowed to be different, an H-BLAST system can achieve capacity. Thus the capacity of the uplink system grows with N_{BS}, the number of BS antenna elements (remember that we assume $N_{BS} < \sum_k N_{MS}^{(k)}$). The RX at the BS can be, e.g., an H-BLAST RX that combines MMSE reception with serial interference cancellation.

The optimum power assignments for the different users follows from the principle of "iterative waterfilling" [Yu et al. 2004], in which we iterate the following algorithm to convergence:

- for $k = 1$ to K
  - for a given user k, compute the covariance matrix of the noise and the interference from all other users;
  - obtain the optimum signal correlation matrix from the waterfilling principle as described in Section 20.2.5.

Downlink

Let us now turn to the downlink. The received signal $\mathbf{r}^{(k)}$ at the k-th MS is

$$\mathbf{r}^{(k)} = \mathbf{H}^{(k)}\widetilde{\mathbf{s}} + \mathbf{n} \tag{20.87}$$

Here $\widetilde{\mathbf{s}}$ is the signal transmitted to the users; it consists of a superposition of the data intended for all the different users $\widetilde{\mathbf{s}} = \sum_k \widetilde{\mathbf{s}}^{(k)}$. The $\widetilde{\mathbf{s}}^{(k)}$ are linearly or nonlinearly encoded versions of the signal streams intended for the different users. The RXs, i.e., the MSs, cannot cooperate for the *decoding* of the datastreams. This has much more significant consequences for the practical implementation than the inability to cooperate for the encoding: access of the decoder to all the data streams is critical for interference suppression. Consequently, the only way to avoid interference is via a *TX precoding* that eliminates interference at the RX (and which requires CSIT).

In the 1980s, a remarkable result was discovered: if interference is known at the TX, then the desired signal can be precoded in such a way that the RX does not "see" any interference. This result is also known as "writing on dirty paper," and codes that achieve or approximate such interference suppression are called "dirty paper codes" or "Costas codes." The name derives from the principle that if one knows the position of dirt on a paper, one can write letters in such a way that the dirt does not hinder the reading of the letters [Peel 2003]. There are numerous ways on how such dirty paper codes can be approximated, e.g., through Tomlinson-Harashima precoding.

When performing the encoding of multiple data streams, the order in which the streams are encoded is very important. The encoder for the k-th MS treats the interference from data streams for MS $1, \ldots k - 1$ as known (which can therefore be completely suppressed), while the interference for users $k + 1, \ldots K$ has the same effect as noise. Consequently, the first encoded stream has a poor SINR, while later encoded streams have good SINR. This situation is dual to the one in the uplink (with a Successive Interface Cancellation (SIC) RX), where the first-decoded data stream has a poor SINR, while the later-decoded streams (which are derived from the "cleaned-up" signals) have a good SINR.

20.3.2 Scheduling

If the number of data streams for users is larger than N_{BS}, then the BS has to determine which users it wants to serve at a given time. The criterion for the serving might be the maximization of the overall throughput (without regard to fairness), or it might want to ensure that each user is served with a certain minimum data rate. Different ways of selecting users (and possibly, grouping users together in certain timeslots) lead to different overall capacity and quality of service. Optimizing such criteria in principle requires an exhaustive search over all scheduling assignments which becomes infeasible if the number of users is large. It is thus popular to perform a "greedy" search, where the first user to be selected is the one that gives, e.g., the highest overall capacity. The next user that is selected from the remaining ones is the user that increases the capacity by the most significant amount, and so on. Simulations indicate that such a greedy algorithm gives performance that is quite close to the optimum scheduling.

20.3.3 Linear Precoding – Uplink

The simplest practical implementation of multiuser MIMO processing is based on linear processing at TX and RX. The RX (the BS) has all signals available, and can thus perform optimum (linear) processing for interference suppression. The optimum transmit and receive weights depend on each other (changing the TX weights at the MS changes the amount of useful power and interference for all the users at the BS), therefore an iterative determination of the optimum weights is necessary. The details depend on the optimization criterion; in the following we consider the minimization of the overall Mean Square Error (MSE).

The MSE can be written as

$$\text{MSE} = tr \left\{ \sum_{k=1}^{K} \left\{ \sum_{j=1}^{K} (\mathbf{T}^{(j)})^\dagger (\mathbf{H}^{(j)})^\dagger \mathbf{W}^{(k)} (\mathbf{W}^{(k)})^\dagger \mathbf{H}^{(j)} \mathbf{T}^{(j)} - (\mathbf{T}^{(k)})^\dagger (\mathbf{H}^{(k)})^\dagger \mathbf{W}^{(k)} \right. \right. \tag{20.88}$$

$$\left. \left. - (\mathbf{W}^{(k)})^\dagger \mathbf{H}^{(k)} \mathbf{T}^{(k)} + \mathbf{I} + \sigma_n^2 (\mathbf{W}^{(k)})^\dagger \mathbf{W}^{(k)} \right\} \right\}$$

where $(\mathbf{W}^{(k)})^\dagger$ is the receive matrix for the k-th user. The goal is then to minimize this MSE under the power constraints

$$tr \left\{ (\mathbf{T}^{(k)})^\dagger \mathbf{T}^{(k)} \right\} \leq P_k^{\max} \tag{20.89}$$

The TX weights are functions of the receive weights of *all* users, while the optimum RX weights depend on the transmit weights of *all* users. Thus, the RX weights and transmit weights can be computed with the following iteration:

1. Update for all users ($k = 1, \ldots K$)

$$(\mathbf{W}^{(k)})^\dagger = (\mathbf{T}^{(k)})^\dagger (\mathbf{H}^{(k)})^\dagger \left[\sigma_n^2 \mathbf{I} + \sum_{j=1}^{K} \mathbf{H}^{(j)} \mathbf{T}^{(j)} (\mathbf{T}^{(j)})^\dagger (\mathbf{H}^{(j)})^\dagger \right]^{-1} \tag{20.90}$$

2. Update for all users ($k = 1, \ldots K$)

$$\mathbf{X}^{(k)}(\mu_k') = \left[\mu_k' \mathbf{I} + \sum_{j=1}^{K} (\mathbf{H}^{(k)})^\dagger \mathbf{W}^{(j)} (\mathbf{W}^{(j)})^\dagger \mathbf{H}^{(k)} \right]^{-1} (\mathbf{H}^{(k)})^\dagger \mathbf{W}^{(k)} \tag{20.91}$$

$$\mu_k = \max \left[\arg_{\mu_k'} \left(tr \left\{ \mathbf{X}^{(k)}(\mu_k') \left(\mathbf{X}^{(k)}(\mu_k') \right)^\dagger \right\} = P_k^{\max} \right), 0 \right] \tag{20.92}$$

$$\mathbf{T}^{(k)} = \left[\mu_k \mathbf{I} + \sum_{j=1}^{K} (\mathbf{H}^{(k)})^\dagger \mathbf{W}^{(j)} (\mathbf{W}^{(j)})^\dagger \mathbf{H}^{(k)} \right]^{-1} (\mathbf{H}^{(k)})^\dagger \mathbf{W}^{(k)} \tag{20.93}$$

Typically, 10 to 20 iterations are sufficient for convergence.

20.3.4   Linear Precoding – Downlink

Also for the downlink, the linear precoding is the method that can be implemented most easily. In the case of beamforming (linear precoding) at the TX, the total transmit signal intended for the k-th user is the source signal, multiplied with a beamforming matrix $\mathbf{T}^{(k)}$

$$\widetilde{\mathbf{s}}^{(k)} = \mathbf{T}^{(k)} \mathbf{s}^{(k)} \tag{20.94}$$

where the $\mathbf{T}^{(k)}$ is the precoding matrix (beamformer) for the k-th user. The correlation matrix of the k-th user's signal is $\mathbf{R}_{\widetilde{\mathbf{ss}}}^{(k)} = E\{\widetilde{\mathbf{s}}^{(k)}\widetilde{\mathbf{s}}^{(k)\dagger}\}$, and the power allocated to the k-th user is $P_k = tr\{\mathbf{R}_{\widetilde{\mathbf{ss}}}^{(k)}\} = Trace\{\mathbf{T}^{(k)}\mathbf{T}^{(k)\dagger}\}$, since we assume that the modulation symbols have unit energy $\mathbf{R}_{ss}^{(k)} = \mathbf{I}$. Usually the BS has a constraint on the total transmit power $\sum_k P_k \leq P_{max}$.

Block Diagonalization

Consider the case where each MS has multiple antenna elements ($N_r^{(k)} \geq 1$) and the BS transmits $N_r^{(k)}$ data streams to each user. A simple solution can be achieved by imposing the dimensionality constraint $\sum_k N_r^{(k)} = N_t$. This constraint is very much along the lines of interpreting the multiple MSs as a "distributed array," and the dimensionality constraint ensures that the number of data streams does not exceed the number of RX chains that can demodulate the signal. To couch it in the language of Section 20.1: each antenna is treated as a separate (single-antenna) user, and the BS preprocessing makes sure that data streams arriving at the antennas are well separated. This eases the design of the MS, which does not have to do any additional processing.

However, the constraint $\sum_k N_r^{(k)} = N_t$ is actually too strict. According to this constraint, adding receive antennas at the MS decreases the number of users that can be transmitted to at a given point in time; this is obviously not a reasonable constraint, as additional receive antennas serve to *improve* the reception quality. Rather, it is the *number of data streams* that intended for the MSs that is limited, and data streams intended for different MSs have to be kept apart by appropriate beamforming at the BS side.

Keeping the data streams of different MSs apart is achieved by a *block diagonalization* technique [Spencer et al. 2004b]. Define for each user an interference channel matrix $\widetilde{\mathbf{H}}^{(k)}$

$$\widetilde{\mathbf{H}}^{(k)} = [(\mathbf{H}^{(1)})^T, \ldots (\mathbf{H}^{(k-1)})^T (\mathbf{H}^{(k+1)})^T, \ldots (\mathbf{H}^{(K)})^T]^T \tag{20.95}$$

Any precoding matrix $\mathbf{T}^{(k)}$ that lies in the nullspace of $\widetilde{\mathbf{H}}^{(k)}$ leads to a situation where the data streams intended for the other MSs do not influence the k-th MS. Consequently, there is now a new dimensionality constraint. Let J_k be the rank of $\widetilde{\mathbf{H}}^{(k)}$. Then the block diagonalization can be achieved if

$$N_t > \max_k (J_1, J_2, \ldots J_K) \tag{20.96}$$

If the propagation channels are all full rank, this means that the number of transmit antennas must be no smaller than the number of data streams. Define then the SVD of $\widetilde{\mathbf{H}}^{(k)}$

$$\widetilde{\mathbf{H}}^{(k)} = \widetilde{\mathbf{U}}^{(k)} \widetilde{\mathbf{\Sigma}}^{(k)} [\widetilde{\mathbf{V}}_{fc}^{(k)} \quad \widetilde{\mathbf{V}}_{lc}^{(k)}]^{\dagger} \tag{20.97}$$

where $\widetilde{\mathbf{V}}_{fc}^{(k)}$ contains the first J_k and $\widetilde{\mathbf{V}}_{lc}^k$ the last $N_t - J_k$ right singular vectors; $\widetilde{\mathbf{V}}_{lc}^k$ thus forms an orthonormal basis of the nullspace of $\widetilde{\mathbf{H}}^{(k)}$. We define a new "overall effective channel matrix"

$$\widehat{\mathbf{H}} = \begin{bmatrix} \mathbf{H}^{(1)}\widetilde{\mathbf{V}}_{lc}^{(1)} & \mathbf{0} & \mathbf{0} \\ \mathbf{0} & .. & \mathbf{0} \\ \mathbf{0} & \mathbf{0} & \mathbf{H}^{(K)}\widetilde{\mathbf{V}}_{lc}^{(K)} \end{bmatrix} \tag{20.98}$$

We next find the SVD of this overall effective matrix; due to the block-diagonal structure of $\widehat{\mathbf{H}}$, this can be done in an efficient (block-by-block) manner. For the k-th block,

$$\widehat{\mathbf{H}}^{(k)} = \widehat{\mathbf{U}}^{(\mathbf{k})} \begin{bmatrix} \widehat{\Sigma}^{(k)} & \mathbf{0} \\ \mathbf{0} & \mathbf{0} \end{bmatrix} [\widehat{\mathbf{V}}_{\text{fc}}^{(k)} \quad \widehat{\mathbf{V}}_{\text{lc}}^{(k)}]^{\dagger} \tag{20.99}$$

The total beamforming/power allocation matrix $\mathbf{T}$ is then

$$\mathbf{T} = \begin{bmatrix} \widetilde{\mathbf{V}}_{\text{lc}}^{(1)}\widehat{\mathbf{V}}_{\text{fc}}^{(1)} & \widetilde{\mathbf{V}}_{\text{lc}}^{(2)}\widehat{\mathbf{V}}_{\text{fc}}^{(2)} & \cdots\cdots & \widetilde{\mathbf{V}}_{\text{lc}}^{(K)}\widehat{\mathbf{V}}_{\text{fc}}^{(K)} \end{bmatrix} \Lambda^{1/2} \tag{20.100}$$

where Λ is a diagonal matrix that performs waterfilling on the elements of

$$\begin{bmatrix} \widehat{\Sigma}^{(1)} & & & \mathbf{0} \\ & \widehat{\Sigma}^{(2)} & & \\ & & \cdots & \\ \mathbf{0} & & & \widehat{\Sigma}^{(K)} \end{bmatrix} \tag{20.101}$$

Coordinated Beamforming

Note that if the RX has a linear filter, then $\mathbf{H}^{(j)}$ should be replaced by the concatenation of filter and channel matrix. Consequently, the rank of $\widetilde{\mathbf{H}}^{(\mathbf{k})}$ is influenced by all the RX filters. Let signals at the k-th MS be multiplied with a matrix $\mathbf{W}^{(k)\dagger}$. The overall received signal is then

$$\mathbf{r}^{(k)} = \mathbf{W}^{(k)\dagger}\mathbf{H}^{(k)}\mathbf{T}^{(k)}\mathbf{s}^{(k)} + \mathbf{W}^{(k)\dagger}\mathbf{H}^{(k)}\sum_{l\neq k}\mathbf{T}^{(l)}\mathbf{s}^{(l)} + \mathbf{W}^{(k)\dagger}\mathbf{n} \tag{20.102}$$

The second term on the r.h.s. now describes the interference. Essentially, the RX processing means the creation of an "effective" channel $\mathbf{W}^{(k)\dagger}\mathbf{H}^{(k)}$ for each user. Changing this effective channel indirectly requires changes in the transmit beamforming matrix as well. The computation of the weights $\mathbf{T}$ and $\mathbf{W}$ thus has to be done in an iterative fashion: start out with a set $\mathbf{W}^{(k)\dagger}$ (e.g., the weights for maximum-ratio combining at the k-th MS), and compute the optimum transmit weights according to one of the strategies described above. With the new $\mathbf{T}$ compute the signal statistics at the MSs, and recompute the optimum linear weights according to a certain criterion (e.g., MMSE). These new RX weights create a new set of effective channel, which then forms the basis for a recomputation of $\mathbf{T}$, and so on. The iteration is stopped when the weights (or the performance measures) do not change appreciably anymore.

Figure 20.22 shows the capacity for various types of precoding. We see that with a pure channel inversion, the capacity first increases as the number of users increases, but then starts to decrease again as K starts to approach the number of antenna elements at the BS. This is due to the increased probability of an ill-conditioned channel matrix. When the RX has more antennas, the difference between channel inversion and regularized channel inversion (see below) becomes much smaller (see Figure 20.23).

Joint Wiener Filtering

Block diagonalization can be seen as a generalization of zero-forcing. Similarly, channel regularization (which aims to minimize the MSE at the RX) can be generalized to *joint Wiener filtering*. The MSE (MSE) metric is, as usual, defined as

$$\text{MSE}_k = \mathbb{E}\left[\|\mathbf{r}_k - \mathbf{z}_k\|^2\right] \tag{20.103}$$

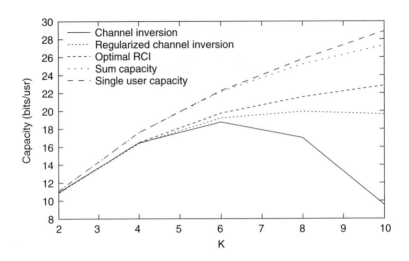

Figure 20.22 Multiuser MIMO capacity as function of the number of users K when each MS has only one antennas. $N_t = 10$, $\gamma = 10\,\text{dB}$. Different precoding methods are shown.
Reproduced from Tsoulos [2006] © CRC Press.

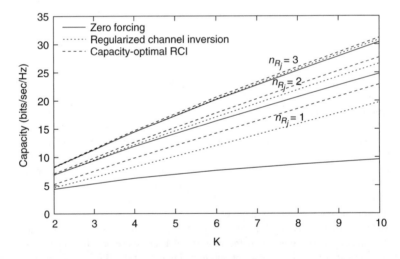

Figure 20.23 Multiuser MIMO capacity as function of the number of users K when each MS has multiple antenna; $N_t = 10$, $\gamma = 10\,\text{dB}$. Different precoding methods are shown.
Reproduced from Tsoulos [2006] © CRC Press.

where the expectation is over the random data vectors, $\{\mathbf{s}_k\}_{k=1}^{K}$, and the noise, $\{\mathbf{n}_k\}_{k=1}^{K}$. The optimization problem is then

$$\left\{\mathbf{T}_k^{\text{opt}}\right\}_{k=1}^{K} = \arg\min_{\{\mathbf{T}_k\}_{k=1}^{K}} \sum_k \text{MSE}_k \quad \text{for } k = 1, \dots, K. \tag{20.104}$$

$$\text{s. t.} \quad \text{Tr}\left\{\sum_k \mathbf{T}_k^{\dagger}\mathbf{T}_k\right\} \le P_{\max}$$

To solve this optimization problem, take an eigendecomposition $\mathbf{V}\Lambda\mathbf{V}^\dagger$ of the following matrix

$$\mathbf{H}_k^\dagger \left[\sigma_n^2 \mathbf{I} + \sum_{j\neq k} \mathbf{H}_i \mathbf{T}_j \mathbf{T}_j^\dagger \mathbf{H}_j^\dagger \right] \mathbf{H}_k = [\mathbf{V}_k \quad \overline{\mathbf{V}}_k] \begin{bmatrix} \Lambda_k & \\ & \overline{\Lambda}_k \end{bmatrix} [\mathbf{V}_k \quad \overline{\mathbf{V}}_k]^\dagger \tag{20.105}$$

where $\mathbf{V}_k$ is a square matrix of dimension $rank(\mathbf{H}_k)$ Then the optimum precoding matrix is

$$\mathbf{T}_k = \mathbf{V}_k \begin{bmatrix} 0 & 0 & .. & 0 & \xi_{1,k} & 0 & 0 \\ .. & .. & .. & .. & 0 & .. & 0 \\ 0 & .. & .. & 0 & 0 & 0 & \xi_{L,k} \end{bmatrix} \tag{20.106}$$

where L is the number of spatial streams and

$$\xi_{i,k} = \max \left[\frac{1}{\mu^{1/2}[\Lambda_k]_{i,i}^{1/2}} - \frac{1}{[\Lambda_k]_{i,i}}, 0 \right] \tag{20.107}$$

and μ is chosen to satisfy the total power constraint.

For a given transmit precoder, the optimum RX filter matrix is the Wiener filter

$$\mathbf{W}_k = \left[\mathbf{H}_k \mathbf{T}_k \mathbf{T}_k^\dagger \mathbf{H}_k^\dagger + \left[\sigma_n^2 \mathbf{I} + \sum_{j\neq k} \mathbf{H}_i \mathbf{T}_j \mathbf{T}_j^\dagger \mathbf{H}_j^\dagger \right] \right]^{-1} \mathbf{H}_k \mathbf{T}_k \tag{20.108}$$

Joint Leakage Suppression

In joint leakage suppression, the precoding matrices are designed to maximize the ratio of the power of the desired signal received by the k-th MS and the sum of the noise and the total interference power (leakage) due to the transmit signal intended for the k-th user at all the other MSs (note that this is different from the SINR at the k-th MS). The key motivation for this criterion is that it allows an easier optimization (often in closed form) than the SINR at the RX. However, performance of systems that maximize Signal-to-Leakage and Noise Ratio (SLNR) is slightly worse than systems that maximize SINR. The SLNR of the k-th user is

$$SLNR_k = \frac{E\{\mathbf{s}_k \mathbf{T}_k^\dagger \mathbf{H}_k^\dagger \mathbf{H}_k \mathbf{T}_k \mathbf{s}_k\}}{N_r \sigma_n^2 + E\{\sum_{i\neq k} \sum_{j\neq k} \mathbf{s}_k \mathbf{T}_k^\dagger \mathbf{H}_i^\dagger \mathbf{H}_j \mathbf{T}_k \mathbf{s}_k\}} \tag{20.109}$$

which can be simplified to

$$SLNR_k = \frac{Tr\{\mathbf{T}_k^\dagger \mathbf{H}_k^\dagger \mathbf{H}_k \mathbf{T}_k\}}{Tr\{\mathbf{T}_k^\dagger \left[N_r \sigma_n^2 \mathbf{I} + \widetilde{\mathbf{H}}_k^\dagger \widetilde{\mathbf{H}}_k \right] \mathbf{T}_k\}} \tag{20.110}$$

The BS thus needs to find precoding matrices $\mathbf{T}_k$ that maximize the $SLNR_k$ subject to a transmit power constraint $Tr\{\mathbf{T}_k^\dagger \mathbf{T}_k\} = P_{max}$. Note that we assume here an absence of power control, so that the powers for each user are fixed to a certain value – this decouples the optimization of the different users. We then define an auxiliary matrix $\mathbf{A}$ as a matrix that fulfills

$$\mathbf{A}_k^\dagger \mathbf{H}_k^\dagger \mathbf{H}_k \mathbf{A}_k = \Lambda_k$$

$$\mathbf{A}_k^\dagger \left[N_r \sigma_n^2 \mathbf{I} + \widetilde{\mathbf{H}}_k^\dagger \widetilde{\mathbf{H}}_k \right] \mathbf{A}_k = \mathbf{I} \tag{20.111}$$

where Λ_k is a diagonal matrix with (arbitrary) nonnegative entries, sorted in descending order. The precoding matrix $\mathbf{T}$ is then the first L_k columns of $\mathbf{A}$.

For the special case that each user has only one data stream, $L_k = 1$, the beamforming matrix becomes a vector, and can be obtained in a simple way: perform an eigendecomposition of the matrix

$$\left[N_r \sigma_n^2 \mathbf{I} + \tilde{\mathbf{H}}_k^\dagger \tilde{\mathbf{H}}_k \right]^{-1} \mathbf{H}_k^\dagger \mathbf{H}_k \tag{20.112}$$

The optimum precoding vector is the eigenvector associated with the largest eigenvalue of this matrix.

Iterative Waterfilling

If we want to maximize the sum rate over all users, the optimization problem is

$$\left\{ \mathbf{T}_k^{\text{opt}} \right\}_{k=1}^K = \arg \max_{\{\mathbf{T}_k\}_{k=1}^K} \sum_{k=1}^K R_k, \tag{20.113}$$

$$\text{s. t.} \quad \text{Tr} \left\{ \sum_{k=1}^K \mathbf{T}_k^\dagger \mathbf{T}_k \right\} \le P_{\max}, \quad \text{for } k = 1, \dots, K.$$

The bandwidth-normalized information rate, R_k, of MS k is given as

$$R_k = \log \left| \mathbf{I}_{N_R} + \left[\sigma_n^2 \mathbf{I} + \sum_{j \neq k} \mathbf{H}_i \mathbf{T}_j \mathbf{T}_j^\dagger \mathbf{H}_j^\dagger \right]^{-1} \mathbf{H}_k \mathbf{T}_k \mathbf{T}_k^\dagger \mathbf{H}_k^\dagger \right|. \tag{20.114}$$

There is no closed-form solution for this problem, but it can be tackled by iterative solution techniques.

20.3.5   Closed-Loop Systems and Quantized Feedback

As discussed in Section 20.2.11, limited-feedback codebooks provide a good way of reducing the feedback overhead for transmit beamforming. However, for multiuser MIMO, the technique is much more difficult. First of all, the quantization of the transmit precoding vector has to be much finer than in the single-user case. This can be explained the following way:[19] in a single-user case the BS forms a beam pattern that shows a maximum in the direction of the targeted MS. Slight deviations from the optimum beam pattern, due to the quantization effects, do not lead to a significant loss of SNR. In the multiuser case, in order to achieve good SINR at the MSs, we need to guarantee low interference by signals from other users; in other words, each transmit beam pattern needs to place nulls toward the users it should *not* cover. Now we know already from Section 20.1 that nulls are much more sensitive to perturbations of the antenna weights than "main beams." Thus, quantization in multiuser settings must be finer.

Another complication arises from the fact that the optimum settings of the precoder depend on the channels of *all* users, and the MS therefore cannot easily compute the precoding matrix settings that it needs to optimize its performance. A feasible way out of this dilemma are *projection techniques*. Here the BS first sends out a number of training signals on $Q = K$ different beams $\mathbf{g}^{(q)}$. The k-th MS then forms the quantities

$$\gamma_k = \max_q \frac{|\mathbf{h}^{(k)} (\mathbf{g}^{(q)})^T|^2}{\sigma_n^2 + \sum_{p \neq q} |\mathbf{h}^{(k)} (\mathbf{g}^{(p)})^T|^2} \tag{20.115}$$

---

[19] Compare also the description of smart-antenna beamforming in Section 20.1

which is simply the maximum SINR it can achieve if the BS transmits to this MS on the beam that is optimum for it, and to all the other MSs on different beams, without any power control. The k-th MS then just feeds back the index q with which it achieves this SINR. Of course, problems can occur when two MSs claim the same beam for themselves. This problem can be tackled, e.g., by having each MS transmit not just the index for the best beam but also for the second best, so that the BS is aware of a "fallback solution" that gives a certain MS an acceptable (though not optimum) SINR.

In the spirit of the "random beamforming" approach described in Section 20.1.9, it can be advantageous to change the beams onto which the projection takes place. In other words, the $\mathbf{g}^{(q)}$ can be generated randomly. Then, the probability is more evenly distributed for each MS to "see" a channel that maximizes its SINR and thus increase its throughput.

If the BS knows the long-term CSI (correlation matrix), then it can also adjust the vectors $\mathbf{g}^{(q)}$ to reduce the probability that a beam is not aligned with any user. This is especially important if there are few users, and they are not spatially uniformly distributed.

20.3.6 Base Station Cooperation

In a cellular system, the existence of multiple users – and thus interferers – influences the capacity, and decreases the data rate that is possible for a single user. A first investigation into this topic by [Catreux et al. 2001] looked at a cellular TDMA system with MMSE detection. Under these assumptions, it was shown that the *cellular* capacity of a MIMO system is hardly larger than that of a system with multiple antennas at the BS only (as described in Section 20.1). The reason for this somewhat astonishing result is that in a cellular system with multiple antennas at the BS only, those antennas can be used to suppress adjacent cell interference, and thus decrease the reuse distance. For a cellular MIMO system with $N_{\mathrm{r}} = N_{\mathrm{t}}$, the degrees of freedom created by the multiple BS antennas are all used for the separation of the multiple data streams from a single user, and none for the suppression of interfering users. Thus, we can use neither SDMA nor SFIR for the increase of the cellular capacity (remember that those can easily double or triple the capacity).

However it seems possible to combine MIMO with other techniques for multiple-access interference. For example, multiuser interference can be eliminated by BS cooperation. For the uplink, cooperating BSs can be viewed as a giant MIMO system with $N_{\mathrm{r}}N_{\mathrm{BS}}$ antenna elements at one link end, where N_{BS} is the number of cooperating BSs. The capacity for such a system can be approximated by inserting the generalized channel matrix into Eq. (20.36).

For the downlink, the effect of interfering BSs can be minimized by appropriate preprocessing if the BSs cooperate and each BS has the CSI for all MSs. Those methods are mathematically similar to the ones applied in multiuser MIMO. One fundamental distinction of multi-BS systems arises from the fact that the interference arriving at the undesired MSs is fundamentally asynchronous. Assuming perfect timing synchronization among cooperative BSs, the timing-advance mechanisms can ensure that the *desired* signals for an MS that are transmitted from multiple BSs reach the MS at the same time. However, the undesired signals do not arrive synchronously anymore; this leads to a change in the interference statistics, and requires modifications of how to compute the precoding coefficients. Another key difference to multiuser MIMO is that we have multiple power constraints – one for each BS; this complicates the precoder design.

Base Station Cooperation is also known in the literature as "network MIMO" or "Cooperative Multi-Point" (CoMP). Since it can greatly enhance performance in particular at the cell edge, it is envisioned for use in fourth-generation cellular systems.

Further Reading

An overview of smart antennas is given in the paper by Godara [1997], and the paper collections [Rappaport 1998] and [Tsoulos 2001]. The topic of space–time processing, which forms a basis for smart antenna systems, involves many topics we have already discussed in previous chapters (diversity in Chapter 13, Rake RXs in Chapter 18; an overview can be found in Paulraj and Papadias [1997]. Smart antennas for CDMA systems are discussed in Liberti and Rappaport [1999].

For the topic of MIMO systems, the books by Paulraj et al. [2003] and Tse and Visvanath [2005] give a good introduction, especially into the spatial-multiplexing aspects. A number of recent books cover a wide range of MIMO-related topics, from fundamental considerations to space–time codes and RX designs. In particular the monographs [Oestges and Clerckx 2007], [Biglieri et al. 2007] and the edited books [Tsoulos 2006] and [Boelcskei et al. 2008] and an upcoming book by Heath [2011]. Also the review papers by Gesbert et al. [2003] and Diggavi et al. [2004] are "must-reads" for an introduction to the topic. For understanding the basic concept of capacity in MIMO systems and the derivations of the capacity distributions and bounds, the original papers by Foschini and Gans [1998] and Telatar [1999] are still worth reading. Also, Andersen [2000] gives important intuitive insights both into capacity and diversity aspects. For the case of channels that are known at the TX, Raleigh and Cioffi [1998] is the first paper, and still very interesting to read. The possibility for a signaling scheme that allows MIMO communications without CSI at the RX was first pointed out by Marzetta and Hochwald [1999]. Goldsmith et al. [2003] review the information-theoretic capacity for different assumptions about the CSI. The impact of channel correlation on the capacity is described in Shiu et al. [2000] and Chuah et al. [2002]. In frequency-selective channels, MIMO is most often combined with OFDM, see Stueber et al. [2004] and Jiang and Hanzo [2007]. The book chapter [Lozano et al. 2008] debunks several common myths about MIMO capacity.

Foschini et al. [2003] is devoted to reviewing different layered space–time structures. MIMO systems with antenna selection are reviewed in Molisch and Win [2004]. For space–time coding, Diggavi et al. [2004] gives an extensive introduction to space–time coding, while Tarokh et al. [1998] and Tarokh et al. [1999] give more mathematical details for space–time trellis codes and space–time block codes, respectively. The books of Jafarkhani [2005] and Larsson and Stoica [2008] are mostly dedicated to space–time codes. Limited-feedback considerations for MIMO were initiated by the seminal paper of Love et al. [2003]; a comprehensive review of limited-feedback can be found in Love et al. [2008]. Overviews of the multiuser downlink can be found in Spencer et al. [2004a], 2006] and Gesbert et al. [2007]. Leakage-based precoding was introduced by Sayed and coworkers, see, e.g., Sadek et al. [2007]. Iterative waterfilling was first suggested in the context of Asymmetric Digital Subscriber Line (ADSL) in Yu et al. [2002]; see also the discussion in Kobayashi and Caire [2006] for MMSE-based multiuser downlinks, see Zhang and Lu [2006] as well as Shi et al. [2007] and references therein. An early (though nonpublic) suggestion of BS cooperation is Molisch [2001]. A number of algorithms for BS cooperation are reviewed in Zhang and Dai [2004].

For updates and errata for this chapter, see wides.usc.edu/teaching/textbook

Part V

Standardized Wireless Systems

One of the main reasons for the success of wireless systems has been the development of widely accepted standards, especially for cellular communications. Those standards make sure that the same type of equipment can be used all over the world, and also for different operators of wireless networks within a country – be those operators nationwide cellular operators or a private person who installed an access point for a Wireless Local Area Network (WLAN) in his/her house, and lets friends access it with their laptops. In this part of the book, we will describe the most important standards for cellular systems, cordless phones, and WLANs.

Two things are noteworthy whenever studying wireless standards: the first is that the standard documents themselves are unreadable for anybody but the standards experts. They are not written in a style that any scientist would use – they do not care about logical derivations, understandability, etc. Rather, they are semilegal documents that only aim for an unambiguous description of what should be done in order to ensure compliance with the standard. The reason for choosing a certain modulation format, coding strategy, etc., is only known to those who heard the discussions in a standards meeting, and is more likely to be political than technical anyway. The second point that will occur to even the casual student of a standards document is the huge number of acronyms – another reason why such documents are almost unreadable. The situation is compounded by the fact that each standard uses different acronyms. For this reason, there is a list of acronyms at the end of each chapter in this part.

Global System for Mobile communications (GSM), the most popular second- generation (2G) cellular standard, is the most successful wireless standard, with more than 3 billion users worldwide. While little research will be done on this topic in the future, GSM's ubiquity still makes it mandatory for most wireless engineers to have at least a basic understanding of its working. For this reason, Chapter 24 gives an introduction to GSM, including both the physical-layer side and some aspects of the networking operations. A rival 2G system, *Interim Standard 95*, *IS-95* (often erroneously referred to as Code Division Multiple Access (CDMA) in the newspapers)[1], achieved considerable popularity in the U.S.A. and Korea, and is described in Chapter 25. In the late 1990s, the third-generation (3G) cellular system Universal Mobile Telecommunications System (UMTS) (also known as Wideband Code Division Multiple Access (WCDMA); Third Generation Partnership Project-Frequency Domain Duplexing (3GPP-FDD) mode or simple 3GPP) was standardized, and

---

[1] Of course IS-95 is based on CDMA, but not every CDMA-based wireless system is an IS-95 system.

has been built up since the early 2000s. At the time of this writing, it had about 250 million users. Chapter 26 describes both physical-layer aspects and Medium Access Control (MAC) layer and networking considerations, since those are closely intertwined. Around 2008, interest in broadband wireless internet access sharply increased, leading to the development of fourth-generation cellular systems: 3GPP Long-Term Evolution (LTE) (Chapter 27) and Wimax (Chapter 28). These systems are not widely deployed yet, but are anticipated to gain great popularity in the future and ultimately replace the 2G and 3G systems. Finally, Chapter 29 describes the *Institute of Electrical and Electronics Engineers, IEEE 802.11* standard (also known as Wireless Fidelity (WiFi)), which defines devices for wireless communications between computers, and from computers to access points.

This part describes only the most important standards that have been designed for wireless standards. Myriads of other standards exist: on one hand, there are other standards for cellular, cordless, and WLAN applications; on the other hand, there are standards for many other wireless services that we do not mention here. For example:

- For *2G cellular systems, IS-136* Time Division Multiple Access (TDMA) [Coursey 1999], and *PDC* (Pacific Digital Cellular) standards exist, and were used by an appreciable number of users. However, they have never achieved GSM's popularity, as they are used only in specific countries (IS-136 in the U.S.A., PDC in Japan), and started to be phased out in the early 2000s.
- For *3G cellular systems*, the 3GPP standard foresees 5 different "modes." Actually, each mode is de facto a different standard. In Chapters 25 and 26, we only describe the most important of those modes. In particular, the Time Division Synchronous Code Division Multiple Access TD-SCDMA standard is anticipated to gain popularity in China; but at the time of this writing (2009), it has less than 1 million users.
- For *cordless systems*, the *Personal Handyphone System, PHS* standard is in widespread use in Japan, while the *Personal Access Communications System, PACS* standard was used in the U.S.A. [Yu et al. 1997], [Noerpel et al. 1996]; furthermore, CDMA cordless phones operating in the 2.45 GHz range are now in use in the U.S.A. The most widely used standard is the Digital Enhanced Cordless Telecommunications (DECT) standard; a brief introduction to this standard can be found on the companion website *www.wiley.com/go/molisch*
- For *WLANs*, there had been a battle between IEEE 802.11 and the European *HIgh PERformance Local Area Network, HIPERLAN* standard [Khun-Jush et al. 2002], [Doufexi et al. 2002]. However, this battle has been decided, with 802.11 being the clear winner.
- For *fixed wireless access and mobile broadband*, the *IEEE 802.20* standard (mobile broadband wireless access) is, in principle, a rival to LTE and Wimax. However, it does not seem to obtain widespread usage.
- For *personal area networks*, which allow wireless communications in a range up to about 10 m, the IEEE 802.15 standards have been developed [Callaway et al. 2002]. The *IEEE 802.15.1* standard, also known as Bluetooth, is now used for wireless links between headsets and cellphones, and similar applications [Chatschik 2001]. For higher data rates, the *Multi Band OFDM Alliance, MBOA* standard allows data rates between 100 and 500 Mbit/s, and is used as the physical-layer basis for *wireless Universal Serial Bus, USB* and *WiMedia*, two higher-layer standards for linking computer components and home entertainment systems, respectively. However, problems and delays in the development of suitable chips have drastically reduced its market opportunities.
- For *sensor networks*, the IEEE 802.15.4 standard, and the associated networking protocol Zigbee, are starting to get widespread usage. An alternative physical-layer standard IEEE 802.15.4a, based on ultrawideband signaling, was approved in 2007, but is not yet widely used.
- For *trunking radio* systems, the *TErrestrial Trunked RAdio, TETRA* standard was widely used in Europe, in addition to a large number of proprietary systems [Dunlop et al. 1999].

21

Cognitive Radio

21.1 Problem Description

The efficient use of available spectrum is a key requirement for wireless system design, since spectrum is a finite resource. Due to the propagation characteristics of electromagnetic waves (see Part II), it is mostly the frequencies between 10 MHz and 6 GHz that are of interest for wireless communications purposes. While this might sound like a lot of spectrum, we have to keep in mind that it is used for a large variety of wireless services, and that furthermore the lower frequency ranges (up to 1 GHz), which are most suitable for large-area services, offer little absolute bandwidth (BW).

The regulatory approach for spectrum usage up to now has been to assign a spectral band to a particular service, and even a particular operator; furthermore, most users employ the same transmission technique (PHYsical layer (PHY)). Such a scheme allows detailed network planning and good quality of service, but it is a waste of resources. An alternative approach is *cognitive radio*, in which users adapt to the environment, including existing spectrum usage. For the further discussion, it is useful to distinguish between two different definitions for cognitive radio:

- A fully cognitive radio (also known as "Mitola radio," in recognition of the engineer who first proposed it), adapts *all* transmission parameters to the environment, i.e., modulation format, multiple-access method, coding, as well as center frequency, bandwidth, transmission times, and so on. While a fully cognitive radio is interesting from a scientific point of view, it currently seems too complicated for practical purposes.
- A "spectrum-sensing cognitive radio," only adapts the transmission frequency, bandwidth, and time according to the environment. Such cognitive radio is also often called *Dynamic Spectrum Access* (DSA).

A key motivation for DSA comes from the fact that with the current, fixed, assignments, spectrum is not exploited to its full extent at all times. If, e.g., no active users are present in a cell of a cellular system, the spectrum is unused in this particular area. Similarly, a lot of the spectrum assigned to TV transmission is not used. In general, this problem of *white-space* spectrum is quite pervasive. For example, investigations in the U.S.A. estimate that in most locations, only 15% of spectrum is used. Figure 21.1a shows the maximum, minimum, and average spectrum usage in an outdoor environment, demonstrating enormous variations of interference power. Figure 21.1b shows that in an indoor environment, the spectrum usage is even smaller, and even on average, mostly thermal noise is present.

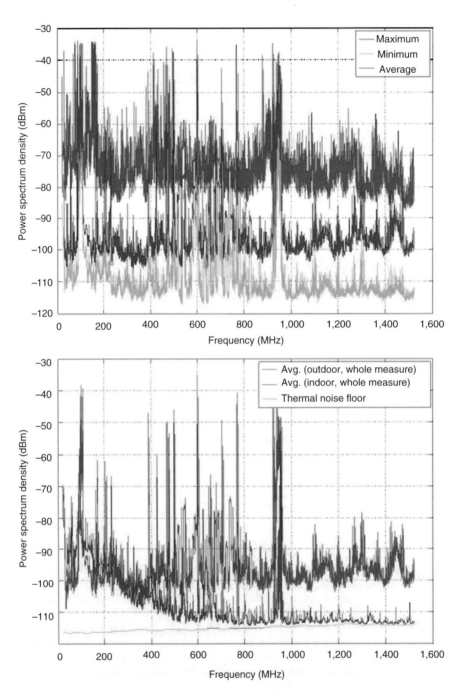

Figure 21.1 Maximum, minimum, and average received power spectral density in the frequency band 20–1, 520 MHz with a 200-kHz resolution bandwidth of the receiver. Outdoor location: on top of 10-storey building in Aachen, Germany. Indoor location: inside the office building in Aachen.

Models for DSA can be classified as follows [Zhao and Sadler 2007]:

- *Dynamic exclusive model*: here, a frequency band is still reserved for the exclusive use of a particular *service*, but different providers can share the spectrum. As we have seen in Section 17.2.2, due to trunking gain, joint usage of a single large frequency band (either by a single cellular operator or by a consortium of operators) results in higher spectral efficiency than the use of N smaller bands by N separate operators. The sharing of the spectrum can be done by a trading (buying/selling or auctioning approach). Another possibility is that a regulator – depending on the usage statistics in a given location – assigns spectrum to a particular user or service on an exclusive but time-varying basis. For example, a cellphone provider might have the right to use 50 MHz of spectrum in the morning and only 20 MHz at noon.[1]
- *Open sharing model*: here, all users can access the spectrum equally, subject to certain constraints on the characteristics of the transmit signal. Such an approach is used today, e.g., in the Industrial, Scientific, and Medical (ISM) bands. The proliferation of Wireless Fidelity (WiFi) devices in this band shows both the advantages and disadvantages of this approach. The free access and ease of type approval contributed to the popularity of WiFi when it was first introduced. However, due to the large number of WiFi devices, it has become almost impossible for other services (especially medical and industrial services) to operate in this frequency band with a reasonable quality of service.
- *Hierarchical access model*: this model assigns different priorities to different users. *Primary users* should be served in such a way that they experience the same service quality as if the spectrum were reserved exclusively for their usage. *Secondary users* are allowed to transmit, but only in such a way that they do not (or only "insignificantly") affect the performance or service quality of the primary users. The secondary user *adaptively* decides whether they might use parts of a spectrum that is assigned by default to primary users. In other words, a cognitive radio exploits spectrum that is assigned to a primary user, but is temporarily available. Therefore it has to sense the current channel usage first and then determine a transmission strategy that does not disturb the current primary users (spectrum management).

One might now ask why primary users would agree to allow secondary users to employ "their" spectrum. Possible reasons are as follows:

- *Profit*: spectrum owners might be able to charge secondary users. Auction systems have been proposed in the literature where secondary users could buy – in real time – the right to specific parts of spectrum for a short time. This is promising for some applications, but might not always be practical as the costs for monitoring and billing could become higher than the revenue from the auctioning of the spectrum.
- *Regulatory requirements*: the frequency regulator can mandate that a certain spectrum range can be used by cognitive devices as long as they do not interfere with primary users. Such an approach is likely for parts of the spectrum that primary users never paid for – especially TV. In the U.S.A., as well as in many other countries, TV stations did not buy the spectrum they use, but rather got it for free because they are deemed to perform a public service. This makes it easy for the frequency regulator to demand that TV stations coexist with other services "in public interest."
- *Emergency services*: another form of cognitive radio occurs in times of emergency, when services that normally count as "primary users" have to give up spectrum for emergency services.

---

[1] Note that this approach is somewhat similar to the partial frequency reuse described in Section 17.6.4. In partial frequency reuse, the spectrum assigned to a service and a provider is constant, but the assignment to specific cells changes depending on the traffic conditions.

The key principle of hierarchical cognitive radio is that the secondary users do not disturb the primary users. Such nondisturbance can be achieved by three fundamental approaches: *interweaving*, *overlay*, and *underlay*. In the interweaving approach, the radio first identifies those parts of the spectrum that are not being used at a certain time, and transmits in those; thus, such a radio is a spectrum-sensing radio. In an overlay approach, the cognitive radio detects the actually transmitted signal of the primary user, and adjusts its own signal in such a way that it does not disturb the primary receiver (RX) even though it transmits in the same band.[2] In the underlay approach, the secondary radio actually does not adapt to the current environment, but always keeps its transmit Power Spectral Density (PSD) so low that its interference to primary users is insignificant. Those concepts are discussed in more detail in the remainder of this chapter.

21.2 Cognitive Transceiver Architecture

We next consider the basic structure of a cognitive transceiver. Figure 21.2 divides the usual transceiver structure (compare Chapter 10) into three parts: Radio Frequency (RF), Analog to Digital Converter (ADC), and baseband processing. Depending on the particular type of cognitive radio, one or more of those components is made adaptive. In a spectrum-sensing cognitive radio, only the RF front end is different from a conventional RX. Figure 21.3 shows a typical RF RX front end, part of the RX shown in Figure 10.3. Its components, including antennas, Low Noise Amplifier (LNA), local oscillator, and automatic gain control, all have to be wideband enough so that they can operate at all possible frequencies at which the cognitive radio might want to operate. The front end also must be able to select the channel which the cognitive radio is to use, by having an adjustable local oscillator (e.g., a Voltage Controlled Oscillator, VCO).

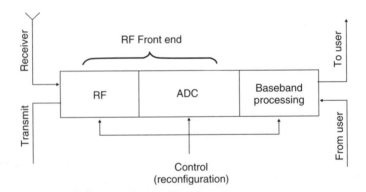

Figure 21.2 Basic structure of a cognitive transceiver. *In this figure:* ADC, Analog to Digital Converter; RF, Radio Frequency.
Reproduced from Akyildiz et al. [2006] © Elsevier.

Since the channel selection occurs only after the downconverter (mixer), a major challenge in cognitive radio lies in the possibility of RX saturation through strong out-of-band signals. RF components like the LNA can be saturated by a signal that is not in the band that the RX wants to demodulate at a specific time, but still within the overall receive band of the cognitive radio. One possibility of reducing such strong interference are tunable RF notch filters that can be placed before the LNA. However, the hardware cost of such filters is high, and the tunability is rather limited.

---

[2] A number of papers actually use the expression "overlay" for what is "interwaving" in the notation of this book.

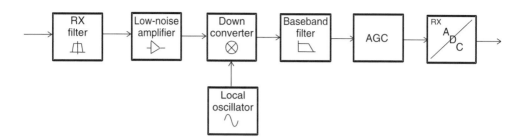

Figure 21.3 RF receiver front end. *In this figure:* AGC, Automatic Gain Control.

For a fully cognitive radio, the baseband processing also has to be adaptive. This can be most easily achieved by implementing the baseband processing as software on a Digital Signal Processor (DSP). Consequently, fully cognitive radio has also been called *software radio* by many of its proponents.

21.3   Principles of Interweaving

In an overlay system, a secondary user tries to identify spectral white space, and transmit at the times/locations/frequencies where primary users are not active; see Figure 21.4. This approach can be seen as "filling the holes" in the time–frequency plane. Clearly, the strategy involves three steps:

1. *Spectrum sensing*: the cognitive radio has to identify which parts of the time–frequency plane are not used by primary users. The sensing has to be done in the presence of noise, so that the sensing cannot be completely reliable. Furthermore, it is important to determine the signal level of the detected radiation. Spectrum sensing is discussed in more detail in Section 21.4.
2. *Spectrum management*: here, the secondary system decides when, and in which part of the spectrum, to transmit. This decision is difficult for several reasons: (i) the secondary system has only causal knowledge of the spectrum occupation, i.e., it knows only which parts of the spectrum were free in the past and the present, but it has to make *assumptions* (that are not necessarily correct) about how the primary system will work in the future, i.e., at the time when the secondary system will actually transmit, (ii) the sensing information on which the decisions of the secondary system is based are not perfect. Spectrum management is discussed in Section 21.5.
3. *Spectrum sharing*: a decision on how to divide up the free spectrum that the secondary system uses. It must be noted that spectrum sharing and spectrum management are strongly related – spectrum might be usable for a particular secondary transmitter (TX), but not for another.

21.4   Spectrum Sensing

As mentioned above, a key requirement for cognitive radio is spectrum sensing, i.e., detecting the presence of other users. This is particularly relevant in a hierarchical cognitive radio, where the secondary devices are obligated to avoid interference to primary users. Furthermore, spectrum sensing gives important information for spectrum sharing. Interference at a specific RX is often characterized by the interference temperature, which describes the interfering PSD divided by Boltzmann's constant k_B.

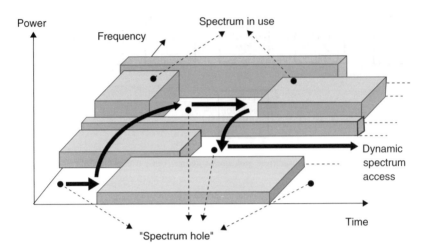

Power

Frequency

Spectrum in use

Dynamic
spectrum
access

Time

"Spectrum hole"

Figure 21.4 Concept of spectrum holes.
Reproduced from Akyildiz et al. [2006] © Elsevier.

21.4.1 Spectrum Sensing in a Hierarchical System

One of the key problems of spectrum sensing in a hierarchical cognitive radio system is that any device can only sense *transmitted* radiation – a completely passive RX is invisible to any sensor. This important dilemma has not been solved completely, though a number of possible solutions have been proposed:

1. Assume that any primary device acts as both TX and RX. Then the presence (and possibly the location) of the primary device can be deduced from the radiation transmitted by this device. Note that a secondary system needs knowledge about the duplexing policies of the primary device. For example, if a frequency-division duplexing primary transmits in a particular band, the secondary system would know in which band such a primary system device would receive.
2. Observe spurious emissions of the primary RX. For example, the local oscillators of TV RXs are not perfectly isolated from the antenna, and thus their emissions can be observed by the secondary system. This approach has the drawback that the level of the observed signal cannot easily be mapped to the location of the RX, and that it actually punishes good RX design (i.e., designs that do not have significant spurious emissions).
3. Assume that primary RXs are in all possible locations where primary transmissions can be heard at a level sufficient for demodulation. Such a policy is extremely conservative: RXs that need to be protected are assumed to be everywhere in the coverage region of the primary system.
4. Let the secondary device start to transmit in any band, and monitor whether the radiation of the primary system changes. For example, an increase in the transmit power of a primary Code Division Multiple Access (CDMA) TX would indicate that a secondary device is interfering with the primary RX (the primary system increases transmit power in order to keep the Signal-to-Interference-and-Noise Ratio (SINR) constant). This approach has, however, two major drawbacks: (i) it interferes, at least initially, with a primary user, (ii) it is difficult to separate changes in the primary TX characteristics due to secondary interference from changes that are caused by other environmental effects, (iii) it does not work in systems that do not change TX characteristics according to interference (and just provide bad transmission quality).

5. A common control channel that allows dissemination of information about spectrum usage. This scheme allows highly efficient utilization of the spectrum, but has the drawback that existing devices need to be modified to allow signaling on this channel.
6. A database that makes geographic locations of primary RXs available to secondary systems.

Like most of the literature, the remainder of this section will ignore the problem of sensing RXs, and just assume that detecting the primary TXs is sufficient.

21.4.2 Types of Detectors

Three types of detectors are commonly used for sensing, depending on the amount of a priori knowledge that a sensor has about the transmitted waveform. Energy detection is used when the sensor has no information about the signal structure. Matched filters provide better performance, but can only be used when the transmit signal waveform is known. Detection of cyclostationary properties is a compromise solution that does not require knowledge of the waveform, and gives a performance that is somewhat better than that of energy detection. It must be noted that a fully cognitive radio requires a matched-filter detection.

Energy Detection

If the sensor knows nothing about the waveform transmitted by other users in a frequency band of interest, it can only measure the energy present in that band. If no other user is present, the sensor measures only thermal noise energy; otherwise, the sensor measures signal-plus-noise energy. The decision of whether a signal is present or not is thus a classical detection problem that can be formulated as hypothesis testing. For simplicity, we only consider a single narrowband channel, where the received signal is

$$r_n = n_n \qquad \mathcal{H}_0 : \text{noise-only received} \tag{21.1}$$

$$r_n = h s_n + n_n : \qquad \mathcal{H}_1 : \text{noise-plus-signal received} \tag{21.2}$$

where r_n and s_n is the received and transmitted signal at time n, and $E\{|s_n|^2\} = 1$. Furthermore h is the channel gain (assumed time independent), and n_n is the observed noise. $\mathcal{H}_0$ and $\mathcal{H}_1$ are the two hypotheses between which the sensor needs to decide. The received signal samples can be averaged over a number of time instances N, so that the decision variables

$$y = \sum_{n=1}^{N} |r_n|^2 \tag{21.3}$$

are used. If the number of samples is large, y are Gaussian random variables, with expected value and variance[3]

$$E\{y\} = \begin{cases} N\sigma_n^2 & \mathcal{H}_0 \\ N\left[|h|^2 + \sigma_n^2\right] & \mathcal{H}_1 \end{cases} \tag{21.4}$$

$$\sigma_y^2 = \begin{cases} 2N\sigma_n^4 & \mathcal{H}_0 \\ 2N\sigma_n^2\left[2|h|^2 + \sigma_n^2\right] & \mathcal{H}_1 \end{cases} \tag{21.5}$$

---

[3] Keep in mind, though, that the y can never take on negative values, as they are sums of energy.

The decision rule is then

$$y \underset{\mathcal{H}_0}{\overset{\mathcal{H}_1}{\gtrless}} \theta \qquad (21.6)$$

where θ is a threshold. Because the noise is random, it is not possible to set the threshold in such a way that the presence of a signal is detected perfectly. There can be *false alarms*, i.e., even though there is only noise, the decision variable exceeds the threshold, and consequently the sensor thinks that a signal is present. The probability for such an event is

$$P_f(\theta) = \Pr(Y > \theta | \mathcal{H}_0) = Q\left(\frac{\theta - N\sigma_n^2}{\sigma_n^2 \sqrt{2N}}\right) \qquad (21.7)$$

where $Q(x)$ is the Q-function defined in Eq. (12.59). Similarly, there can be missed detections, which happen when the decision variable remains below the threshold even if a signal is present. The probability for this is

$$P_{md}(\theta) = \Pr(Y < \theta | \mathcal{H}_1) = 1 - Q\left(\frac{\theta - N\left[|h|^2 + \sigma_n^2\right]}{\sigma_n \sqrt{2N\left[2|h|^2 + \sigma_n^2\right]}}\right) \qquad (21.8)$$

By adjusting the threshold, we can then trade off false-alarm probability and missed-detection probability. In a hierarchical cognitive system, the missed-detection probability will usually be prescribed by the frequency regulator, because a missed detection of the spectrum sensor implies that the secondary system will transmit even though a primary user is active in the considered band.

Another interesting question is how many sensing values N should be obtained before a decision is made. If the number is too small, then the decision variable has a large variance, and a high P_f (or P_{md}) must be accepted. However, if the number is too large, the decision process takes too much time, reducing the spectrum usage and increasing the chance that the primary system starts to transmit even though the channel was free in the beginning. This problem can be treated by means of "optimum stopping theory."

Matched Filter

The performance of the spectrum sensing can be improved considerably if the sensor knows the transmitted waveform. In that case, matched filtering is the best way of improving the Signal-to-Noise Ratio (SNR) of the detection process, similar to the arguments made in Chapter 12. For a spectrum-sensing cognitive radio, the output of the matched filter is used as a test statistics to detect whether signal energy is present in the considered band or not. A fully cognitive radio might have to demodulate/detect the symbols transmitted by the primary user.

In some circumstances, a sensor might want to detect/decode only pilots or beacons, since those have usually better SNR than the actual payload data, and might also contain information about spectrum usage (e.g., the duration of a packet) that can be exploited by a secondary system.

Cyclostationarities

Modulated signals usually have cyclostationary statistics, i.e., certain statistical properties are periodic; see also Section 16.7.4. This fact can be employed to enhance the performance of an energy detector. In other words, the incoming signal is correlated with itself before being sent to an energy detector.

Wavelet Detection

If the PSD of the signal that is to be detected is smooth within a frequency band but shows steep decay at an edge, a wavelet transform can expose the edges of the PSD. The drawbacks of this technique are the high computational costs and the fact that the method does not work well for spread-spectrum signals [Latief and Zhang 2007].

21.4.3   Multinode Detection

In many cases, the secondary system consists of multiple nodes, which then have the possibility of helping each other reach a better sensing decision. In the simplest case, each of the secondary nodes listens and makes an independent decision concerning whether it can sense a certain channel to be occupied. The nodes then exchange this binary information, and come to a joint ("fused") decision as to whether the spectrum is available or not. In order to best protect the primary system, the secondary system should decide that a frequency band is occupied if at least one node senses occupation. Clearly, the overall false-alarm probability increases, becoming [Latief and Zhang 2007].

$$P_{\text{f, network}} = 1 - \prod_{k=1}^{K} \left[1 - P_{\text{f},k}\right] \tag{21.9}$$

while the probability of missed detection will decrease

$$P_{\text{md, network}} = \prod_{k=1}^{K} P_{\text{md},k}. \tag{21.10}$$

A more sophisticated joint decision can be made if the nodes communicate not just a binary decision about spectrum occupancy (yes/no) but perform a linear combination of the observed averaged samples y (Eq. 21.3). Introducing linear weights w_k for the signal from the k-th node, we form a total decision variable

$$z = \sum w_k y_k = \mathbf{w}^T \mathbf{y} \tag{21.11}$$

Because the decision variable is a weighted sum of Gaussians (see above for a discussion why y is Gaussian), it is itself Gaussian, and its mean and variance is

$$E\{z\} = \begin{cases} N\sigma_n^2 \mathbf{w}^T \mathbf{1} & \mathcal{H}_0 \\ N\mathbf{w}^T \left[\mathbf{g} + \sigma_n^2 \mathbf{1}\right] & \mathcal{H}_1 \end{cases} \tag{21.12}$$

$$\sigma_z^2 = \begin{cases} 2N\sigma_n^4 \mathbf{w}^T \mathbf{w} & \mathcal{H}_0 \\ 2N\sigma_n^2 \mathbf{w}^T \left[2\text{diag}(\mathbf{g}) + \sigma_n^2 \mathbf{I}\right] \mathbf{w} & \mathcal{H}_1 \end{cases} \tag{21.13}$$

where $\mathbf{g} = [|h_1|^2, |h_2|^2, ...|h_K|^2]^T$, and diag $(\mathbf{g})$ is a diagonal matrix whose entries are the $|h_k|^2$. The probabilities for false alarm and missed detection are then

$$P_{\text{f}}(\theta, \mathbf{w}) = Q\left(\frac{\theta - N\sigma_n^2 \mathbf{w}^T \mathbf{1}}{\sigma_n^2 \sqrt{2N\mathbf{w}^T \mathbf{w}}}\right) \tag{21.14}$$

$$P_{\text{md}}(\theta, \mathbf{w}) = 1 - Q\left(\frac{\theta - N\mathbf{w}^T \left[\mathbf{g} + \sigma_n^2 \mathbf{1}\right]}{\sigma_n \sqrt{2N\sigma_n^2 \mathbf{w}^T \left[2\text{diag}(\mathbf{g}) + \sigma_n^2 \mathbf{I}\right] \mathbf{w}}}\right) \tag{21.15}$$

Now we have to specify the optimum weights and thresholds for the detection. What is "optimum" under the circumstances depends on the system goals. We might wish, e.g., to maximize the throughput of the secondary system $(1 - P_f(\theta, \mathbf{w}))$, under the constraint that the probability of interference to the primary system $P_{md}(\theta, \mathbf{w})$ stays below a certain threshold. Standard optimization techniques can then be used to find out the optimum parameters $\theta, \mathbf{w}$.

Clearly, communicating the full decision variables from each sensing node requires more overhead than just communicating a binary decision. On the other hand, it must be noted that it is rarely possible to really just send a single "bit" over a wireless channel. Every transmitted packet needs headers, synchronization sequences, etc. Compared to this necessary overhead, it does not make much difference whether we want to transmit, say, 1 bit or 8 bits of information. Communicating the full decision variable from each node thus might not be a significantly higher effort than a binary decision.

21.4.4 Cognitive Pilots

The sensing of primary (and other secondary) users would be greatly facilitated by the introduction of Cognitive Pilot Channels (CPCs) [Zhang et al. 2008]. The procedure for using the CPC can include the following three phases:

- The wireless network/terminal first listens to the CPC at the initialization.
- The wireless network/terminal gets the information and selects the most suitable one to setup its communications.
- The CPC is broadcasted to a wide area (e.g., Personal Area Network (PAN)).

However, a CPC has to be agreed upon by a wide range of standardized devices, and thus constitutes a formidable logistics/standardization problem.

21.5 Spectrum Management

21.5.1 Spectrum Opportunity Tracking

A cognitive system has to make decisions about which frequency band and bandwidth to use, and how long to transmit in this band. However, spectrum sensing only gives information about the spectrum occupancy at a given point in time. It is thus important for the cognitive system to keep a record of the history of spectrum usage, and to have detailed models for the traffic statistics. Ideally, a secondary system observes the environment at all possible frequencies and at all times in order to get the best possible statistics; however, energy and hardware constraints might prevent such an approach.

A simple, analytically tractable model is a Markov model, which just assumes that each subchannel is in one of two states (occupied or unoccupied), and has a certain transition probability (which is obtained from observations) from one state to another. A more detailed model might include variations of the spectrum occupancy with time-of-day, or take typical durations of packets or voice calls into account.

If the cognitive radio recognizes that the current band in which it operates has to be released (either because of worsening propagation conditions or because the band is needed by other users), it either moves to a different band or (if no other band is available) ceases transmission. In the former case, the *spectrum handover* should be done in such a way that the performance of the link is affected as little as possible.

In a hierarchical system, the secondary system has to cease transmission when a primary user reappears and wants to use the spectrum. The duration for which a secondary system transmits in a certain band before doing another round of sensing is thus an important parameter of the system, which depends on the previously measured statistics of the primary channel.

21.6   Spectrum Sharing

21.6.1   Introduction

A key question is now how to distribute the detected spectral resources to the different users of a secondary system. A large variety of methods exist, depending on the degree of coordination between the users. At the one extreme, a centrally controlled scheme, where a master assigns the spectral resources to the different users, provides the best performance of the system at the price of significant overhead. On the other hand, uncoordinated competition between the different users does not require any message exchange, but has low efficiency.

A key mathematical tool in the analysis of spectrum sharing is *game theory*. A game consists of a number of players, a strategy for each player, and a payoff (utility function); each user adjusts its strategy in such a way that the payoff is maximized. It is noteworthy that the optimum strategies depend on the amount of coordination, i.e., how much each user knows about the other users, and whether there is a central authority that can enforce certain rules of the game.

In many cases, the ideal outcome of a game is to achieve a *Pareto optimum* (also known as *social optimum*); in that case, there is by definition no other outcome that has a better payoff for at least one player, and at the same time does not reduce the payoff for any other player.

21.6.2   Non-Cooperative Games

A noncooperative game is defined as "one in which any player is unable to make enforceable contracts outside of those specifically modeled in the game" [Han et al. 2007]. Consequently, any cooperation must be self-enforcing; there is no outside party that can enforce a particular policy. An important quantity in the context of such games is the *Nash equilibrium*, which is defined as an operating point where no user has a motivation to unilaterally change its current strategy.

Noncooperative games can lead to Nash equilibria that are very inefficient. A typical example is a heavily loaded ALOHA system (compare Chapter 17). If every user sends out packets at a high rate, there is a high number of collisions, and the system will become congested. However, a user that tries to behave in a "socially responsible" manner and sends out fewer packets will achieve an even lower throughput as long as the other users keep up their high channel access rate. Since each of the users faces the same dilemma (similar to the famous "prisoners' dilemma"), none will reduce their channel access rate, and the congestion will persist. The users are then in a Nash equilibrium, but have low efficiency.

21.6.3   Games with Partial Coordination

In cooperative games, good behavior of the users can be enforced through various means. This enforcement can be done, e.g., though a central authority, or through "punishment" by other players. In the following, we will outline some of the most widespread methods [Han et al. 2007].

Referee-Based Solution

The inefficiency of noncooperative games can be mitigated if a *referee* enforces certain types of behavior and alleviates the collisions and high-interference situations that plague standard competitive behavior. The amount of intervention depends on the amount of information that the referee has, and the amount of control information that is to be sent to the users. In the extreme case that the referee "micromanages" the resource allocation to all users, the referee-based solution becomes equivalent to the centralized solution described below. An example for a low-overhead referee algorithm is given in Figure 21.5.

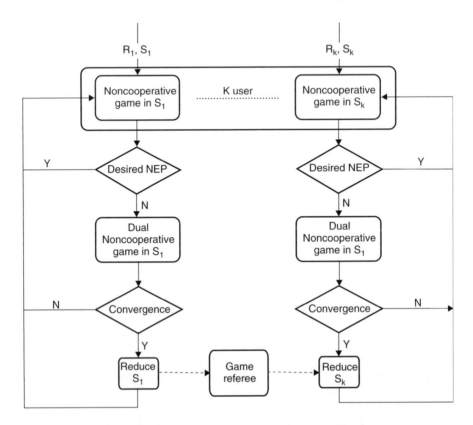

Figure 21.5   Flow diagram of a noncooperative game with referee.
Reproduced from Han et al. [2007] © IEEE.

Threat and Punishment in Repeated Games

The Nash equilibrium normally assumes a "static game," i.e., players play only once. Better results can be achieved in repeated games, where the players have the possibility to learn from the past, i.e., make their decisions conditioned on the results of the outcomes of previous games. Consequently, a user will not sacrifice its long-term benefits if it is going to be severely punished by the other users. The total payoff is the weighted average over time

$$V = \sum_{t=1}^{T} \beta^{t-1} u_t \tag{21.16}$$

where β is the "discount factor" and u_t is the payoff at time t; T is the total time that the game is played. It can be shown that as the discount factor approaches 1, any individually rational (and achievable) payoff can be enforced by an equilibrium.

An implementation of a punishment scheme is the "trigger punishment strategy" that proceeds in the following way:

- All players start in a cooperative mode.
- At the end of the first phase $t = 1$ of the game, the result of the game (e.g., the total payoff for all users) is made public. More cooperation will result in a higher total payoff; on the other hand, if one user behaves selfishly, the overall system behavior will become bad.
- Thus, if the total payoff is below a threshold, users switch to a noncooperative mode, and use as their strategy the Nash equilibrium strategy for a time T. If the total payoff is above the threshold, users stay with the cooperative strategy.

Problems can arise if the users are mobile. In that case, a nonsocial user can get high payoff in a certain area by behaving badly, and then escape the punishment by simply moving to a different area.

Spectrum Auction

A spectrum auction in a cognitive radio system works, in principle, like any auction in real life. The players (users) are bidding a certain price for the available spectrum; clearly, a user that pays more will get a larger assignment of resources. It must be noted that the "price" need not be actual money, and that it can depend on the quality of the user channel. For example, it has been proposed that each user is charged according to SINR or received power.

Bargaining Solutions

Another assignment algorithm that draws its inspiration from economic theory is bargaining. In this approach, different users can negotiate with each other (not with a central authority) about who gets which spectral resources assigned. In one-to-one bargaining, two neighboring users can exchange channels. Since the channel qualities are not necessarily the same for different users, an exchange of channels might be beneficial for both users. When more users are involved, a bargaining with one buyer and multiple sellers can be applied.

21.6.4 Centralized Solutions

The conceptually simplest case is one where a central authority has all the information about the bandwidth requirements of the different users as well as the available bandwidth, and assigns the resources in such a way as to optimize the spectral efficiency of the (secondary) system; it is assumed that the secondary users obey commands from the central authority. Such a solution involves a high overhead for sending all the required information to the central authority and for the channel control information.

Mathematically, the spectrum allocation can be written as a (constrained) optimization problem

$$\min_{\mathbf{x} \in \Omega} f(\mathbf{x}) \qquad (21.17)$$

$$g_i(\mathbf{x}) \leq 0 \quad \text{for } i = 1, ...I$$

$$h_j(\mathbf{x}) = 0 \quad \text{for } j = 1, ...J$$

where $\mathbf{x}$ is the parameter vector for the optimization of $f()$, and has to lie in the space Ω. The $g_i(\mathbf{x})$ and $h_j(\mathbf{x})$ are inequality constraints and equality constraints for the parameter vector. Depending on the application and the optimization goal, these quantities can mean different things. To give one example, **x** can be a vector containing the power and bandwidth assignment for the different users. The inequality constraints limit the power of the different users, and possibly the bandwidth assigned to the different users. Different mathematical techniques exist for solving constrained optimization problems. If, e.g., the functions f, g, and h are linear functions of **x**, then the optimization problem is a *linear program* that can be readily solved by standard mathematical packages. Similarly, if the function $f(\mathbf{x})$ is convex, the problem can be solved by numerical optimizations that rapidly converge to global optima. The situation is somewhat more difficult when the optimization function is nonconvex – in this case, numerical techniques usually can converge to local optima. If the admissible parameter space Ω is discrete, the optimization problem is usually cannot be solved in polynomial time (essentially, all possible points in the parameter space need to be tried out).

We now turn to one concrete example [Acharya and Yates 2007]. We have a situation with multiple secondary systems (BSs) i, where the i-th system provides bandwidth b_{ij} to the j-th user. A user then transmits its data to the i-th BS using that bandwidth. The channel between BS and user has a channel gain $|h_{ij}|^2$. Communication is assumed to happen at Shannon capacity, so that the data rate of a link is

$$R_{ij} = b_{ij} \log_2 \left[1 + \frac{|h_{ij}|^2 P_{ij}}{b_{ij}} \right] \tag{21.18}$$

where P_{ij} is the power on each link. The goal is the maximization of the overall throughput. The optimization problem can then be written as

$$\max_{b_{ij}, P_{ij}, X_i} \sum_j \sum_i R_{ij} \tag{21.19}$$

$$\sum_j b_{ij} \leq B_i \tag{21.20}$$

$$\sum_i P_{ij} \leq P_j \tag{21.21}$$

$$\sum_i B_i \leq B_{\text{tot}} \tag{21.22}$$

$$b_{ij} \geq 0, P_{ij} \geq 0, B_i \geq 0 \tag{21.23}$$

where Eq. (21.19) says that we want to optimize the total throughput in the system; Eq. (21.20) states that the total bandwidth allocated to the different users within a secondary system is limited to the total bandwidth B_i assigned to that secondary system by a central authority, and Eq. (21.22) limits the sum of the bandwidths assigned to the secondary systems to the total available bandwidth. Eq. (21.21) limits the sum power that each user j can employ to a certain value P_j. Finally, all bandwidth assignments and powers must be positive. The solution of the above-described problem can be obtained via standard Lagrangian optimization techniques.

21.7 Overlay

A very different approach to cognitive radio is the *overlay* principle. Here, we do not try to avoid transmission at the same time and frequency by primary and secondary user, but rather we exploit it. By using appropriate types of cooperation, the secondary user can help the primary user, and still at the same time transmit information of its own.

Let us start out with a very simple cognitive system: a primary TX/RX pair, as well as a secondary TX/RX pair tries to communicate over an Additive White Gaussian Noise (AWGN) channel. We furthermore assume that the secondary TX has full knowledge of the message of the primary user.

This latter assumption is clearly unrealistic in almost any wireless setting, but it can be seen as a reasonable approximation in the following two cases:

1. The primary user uses a special type of code ("rateless code"), which allows an RX to decode the message as soon as it has gathered enough mutual information. In other words, the RX does not have to wait with the decoding until the TX has finished sending out a packet of data. Rather, the better the channel between primary TX and secondary RX, the sooner can the RX decode. Thus, if secondary TX is very close to the primary TX, it can have almost instantaneous knowledge of the primary message.
2. The primary user employs Automatic Repeat reQuest (ARQ), i.e., it retransmits the same message multiple times, until the primary RX finally gets it (it takes repeated attempts because primary RX has poor SINR). The secondary TX can get the primary message at the first attempt (again, assuming that it has a very good channel to the primary TX) and then has noncausal knowledge of the primary message for the retransmission.

A secondary TX can now pursue two possible strategies:

1. *Selfish approach*: the secondary TX uses all its power to transmit the secondary message. It uses its knowledge of the primary message to make sure that there is no effective interference to the secondary RX. Such elimination of interference at the TX side can be achieved by "dirty paper coding" [Peel 2003]. It is noteworthy that in such a system there is still interference from the secondary TX to the primary RX, which violates the basic tenet of hierarchical cognitive radio.
2. *Selfless approach*: the secondary TX uses part of its power to help convey the primary message to the primary RX; the remainder of the power is used to send the secondary message to the secondary RX. Depending on the fraction of power used for sending the primary message, the secondary system can actually boost the rate of the primary message. An especially interesting case occurs when the secondary system makes sure that the primary rate remains unchanged (compared to the case when there is no secondary system) [Devroye et al. 2007]. In that case, the primary system is completely oblivious to the fact that a secondary system is present at all, and its capacity in an AWGN channel is [Devroye et al. 2007, Jovicic and Viswanath 2009]

$$R_1 = \log_2 \left[1 + \frac{P_1}{N_0} \right] \tag{21.24}$$

and still the secondary system can transmit some of its data, which can be done at a rate

$$R_2 = \log_2 \left[1 + (1 - \alpha') \frac{P_2}{N_0} \right] \tag{21.25}$$

where

$$\alpha' = \left[\frac{\sqrt{P_1} \left(\sqrt{1 + |h_{21}|^2 \frac{P_2}{N_0} (1 + \frac{P_1}{N_0})} - 1 \right)}{|h_{21}| \sqrt{P_2} (1 + \frac{P_1}{N_0})} \right]^2 \tag{21.26}$$

where we have assumed $|h_{11}| = |h_{22}| = 1$, and $|h_{21}| < 1$. In other words, as long as the crosstalk channel from the secondary TX to the primary RX is weaker than the primary channel, it is possible for the secondary system to transmit its own information "for free," i.e., without reducing the rate of the primary system.

Overlay systems are fascinating from a theoretical point of view; however, at least at the time of this writing, they seem quite far away from practical implementation: (i) the question of how the secondary system obtains the noncausal knowledge of the primary system is at least partially unanswered (though progress has been made on cognitive systems with causal message knowledge); (ii) even more importantly, complete channel state information for *all* relevant channels (including the one from primary TX to primary RX) has to be available at the secondary TX; such knowledge seems to be difficult to obtain without explicit cooperation from the primary system.

21.8   Underlay Hierarchical Access – Ultra Wide Bandwidth System Communications

In an underlay system, the secondary users have severe restrictions on the transmit PSD, so that the effect on a primary RX is a "minor" increase of the noise floor that the RX sees. Such low PSD can be achieved by keeping the transmit power low (which is feasible only if the secondary users communicate only over short distances) and/or by spreading the signal over a very large bandwidth. Only the part of the secondary signal that falls within the RX bandwidth of a primary user acts as interference. Underlay radios are actually not "cognitive" in the sense that they adapt their transmission parameters to the environment, but because of their use as secondary radios they are still often mentioned in the cognitive category.

Frequency Regulations and Transmit Power Constraints of UWB Signals

The underlay principle is realized in Ultra Wide Bandwidth system (UWB) communications, where signals have extremely large bandwidth. Such a large bandwidth offers the possibility of very large spreading factors: in other words, the ratio of the signal bandwidth to the symbol rate is very large. For a typical sensor network application with 5 ksymbol/s throughput, a spreading factor of 10^5 to 10^6 is achieved for transmission bandwidths of 500 MHz and 5 GHz, respectively. Spreading over such a large bandwidth means that the PSD of the radiation, i.e., the power per unit bandwidth, can be made very low while still maintaining good SNR at the secondary RX. A primary (narrowband) RX will only see the secondary-signal power within its own system bandwidth, i.e., a small part of the total secondary transmit power; see Figure 21.6. This implies that the interference to primary (narrowband) systems is small.

Figure 21.6   Interference between a UWB system and a narrowband (IEEE 802.11a) local area network.

Frequency regulators have defined UWB signals as signals with a minimum of 500 MHz absolute bandwidth.[4] They stipulate that UWB can be used as *unlicensed* underlay systems as

---

[4] Signals are also defined as UWB if they have more than 20% relative bandwidth. However, for the subsequent discussion, we assume large absolute bandwidth.

long as the PSD is limited to $-41.3\,$dBm/MHz EIRP (Equivalent Isotropically Radiated Power; see Chapter 3). This PSD is so low that it does not significantly disturb primary RXs that are more than approximately 10 m away from a UWB TX. Take, e.g., a signal with 6 GHz carrier frequency. It suffers a free-space attenuation of 68 dB at a 10 m distance, resulting in a received PSD of approximately $-109\,$dBm/MHz, which is comparable to the PSD of white noise. Consequently, even a primary RX operating at the thermal-noise sensitivity limit would hardly see an impact on its performance. RXs with nonideal noise figures, and/or impacted by co-channel interference, e.g., from neighboring cells, would not be affected even if the UWB TXs were somewhat closer to the primary RX. It is also noteworthy that transmission at $-41\,$dBm/MHz is allowed only in certain frequency ranges, while others (e.g., the Global Positioning System, GPS band, and also most cellular bands) are more strongly protected.

In the U.S.A., the Federal Communications Commission (FCC) allows emission between 3.1 and 10.6 GHz. Limits for indoor and outdoor communication systems differ as shown in Figure 21.7a. For outdoor systems, UWB devices are required to operate without a fixed infrastructure. In Europe, the Radio Spectrum Committee (RSC) of the European Commission (EC) imposes a spectral mask shown in Figure 21.7b. Emission between 6 and 8.5 GHz with EIRP of $-41.3\,$dBm/MHz is allowed for devices without some type of additional interference mitigation techniques (techniques similar to those of interweaving systems). The same limit is valid for the shaded frequency region (4.2–4.8 GHz) until the end of 2010. UWB systems *with* interference mitigation techniques or low duty cycle operation are allowed to transmit at $-41.3\,$dBm/MHz in the 3.4–4.8 GHz band. In Japan, operation between 3.4 and 4.8 GHz is admissible as shown in Figure 21.7c, if the UWB TX uses interference mitigation. However, for 4.2 GHz through 4.8 GHz, interference mitigation techniques were not required until the end of December 2008. Operation between 7.25 and 10.25 GHz is admissible also without special techniques.

The high spreading factor of UWB systems helps not only in mitigating interference to primary users but also enables a UWB RX to suppress narrowband (primary-user) interference by a factor that is approximately equal to the spreading factor. These principles are well understood from the general theory of spread-spectrum systems, see Chapter 18. The distinctive feature of UWB is that it goes to extremes in terms of the spreading factor. It must be kept in mind that the spreading factor is a function of both the transmission bandwidth and the data rate. Consequently, UWB systems with high data rates ($>100\,$Mbit/s) exhibit a rather small spreading factor at and can thus be used to communicate over very short distances.

Methods of UWB Signal Generation

There are a number of different ways to spread signals to large bandwidths.

1. *Frequency Hopping (FH)*: FH uses different carrier frequencies at different times. In slow FH, one or more symbols are transmitted on a given frequency; in fast FH, the frequency changes several times per symbol; see also Section 18.1. The bandwidth of the resulting signal is determined by the range of the oscillator, not the bandwidth of the original signal that is to be transmitted. Implementation of an FH TX is fairly simple: it is just a conventional narrowband modulator followed by a mixer with the output of a frequency-agile oscillator. An FH RX can be constructed in a similar way; such a simple RX is efficient as long as the delay spread of the channel is shorter than the hopping time (otherwise, multipath energy is still arriving on one subcarrier while the RX has already hopped to a different frequency). Consequently, a hopping rate of approximately 1 MHz or less would be desirable. However, such slow FH can lead to significant interference to primary RXs, since – at a given time – a victim RX "sees" the full power of the UWB signal. For this reason, FH for UWB has been explicitly prohibited by several frequency regulators.
2. *Orthogonal Frequency Division Multiplexing (OFDM)*: in OFDM, the information is modulated onto a number of parallel subcarriers (in contrast to FH, where the carriers are used one after

Figure 21.7   Spectral masks for UWB transmission in different continents.

the other); see Chapter 19. For this reason, OFDM has no innate spectral spreading. Rather, spreading can be achieved by low-rate coding, e.g., by a spreading code similar to CDMA, or by a low-rate convolutional code. The bandwidth of the resulting signal is determined by the employed code rate and the data rate of the original (source) signal. In modern implementations, the subcarriers are produced by a fast Fourier transformation; see Section 19.3. However, this implies that signal generation at the TX, as well as sampling and signal processing at the RX, has to be done at a rate that is equal to the employed bandwidth, i.e., at least 500 MHz.

3. *Direct Sequence–Spread Spectrum (DS–SS)*: also known as CDMA, multiplies each bit of the transmit signal with a spreading sequence. The bandwidth of the overall signal is determined by the product of the bandwidth of the original signal and the spreading factor. At the RX, despreading is achieved by correlating the received signal with the spreading sequence; see Section 18.2. The key implementation challenge lies in the speed at which the RX has to sample and process (despread) the signal.

4. *Time Hopping Impulse Radio (TH-IR)*: TH-IR represents each data symbol by a sequence of pulses with pseudorandom delays. The duration of the pulses essentially determines the width of the transmit spectrum; see also Section 18.5. The key implementation challenge lies in building coherent RXs that keep complexity low while still maintaining adequate performance.

Summarizing, we find that there is a strong duality between FH and TH-IR. FH sequentially hops in the frequency domain, while TH-IR hops in the time domain. Similarly, OFDM and DS–SS are dual, in that they perform low-rate coding operations in the frequency and time domains, respectively.

Further Advantages of UWB Transmission

In addition to the low interference to primary users, UWB signals also offer a number of other advantages.

1. A UWB RX can suppress narrowband interference by a factor that is approximately equal to the spreading factor.

2. A large absolute bandwidth can result in a high resilience to fading. First of all, a large absolute bandwidth allows to resolve a large number of (independently fading) Multi Path Components (MPCs), and thus a high degree of frequency diversity, i.e., the fading at sufficiently separated frequencies is independent. We can also give an alternative interpretation that is especially useful for TH-IR and CDMA systems. An RX with a large absolute bandwidth has a fine delay resolution, and can thus resolve many MPCs. The number of resolvable, and independently fading, MPCs can be up to τ_{max}/B, where τ_{max} is the maximum excess delay of the channel and B is the system bandwidth. By separately processing the different MPCs, the RX can make sure that all those components add up in an optimum way, giving rise to a smaller probability of deep fades. As an additional effect, we have seen in Section 6.6 that in UWB, the number of *actual* MPCs that constitute a *resolvable* MPC is rather small; for this reason, the fading statistics of each resolvable MPC does not have a complex Gaussian distribution anymore, but shows a lower probability of deep fades.

3. A large absolute bandwidth also leads to high accuracy of ranging and geolocation. Most ranging systems try to determine the flight time of the radiation between TX and RX. It follows from elementary Fourier considerations that the accuracy of the ranging improves the bandwidth of the ranging signal. Thus, even without sophisticated high-resolution algorithms for the determination of the time-of-arrival of the first path, a UWB system can achieve centimeter accuracy for ranging.

4. The large spreading factor and low PSD also provide increased protection against eavesdropping.

UWB Dynamic Spectrum Access

Despite the low interference created by a UWB underlay system, the residual interference to nearby victim RXs can still be too large. It is thus often necessary to combine UWB with a detect-and-avoid scheme. Such a strategy is potentially capable of enabling coexistence, compatibility, interference avoidance, and compliance with regulation – in particular European and Japanese frequency regulations that mandate Detect and Avoid (DAA) schemes for UWB TXs in some frequency ranges. The good ranging (and thus geolocation) capabilities of UWB nodes can help to determine whether nodes are close to potential victim nodes, in particular if the location of such victim nodes is kept in a database that can be accessed by the UWB nodes.

Further Reading

The field of cognitive radio is still very much in flux, and new research as well as new books are appearing on an almost monthly basis. At the time of this writing, the following papers can be especially recommended: The edited monographs [Hossain and Barghava 2008], [Xiao and Hu 2008] give a broad overview of all the aspects of cognitive radio, including spectrum sensing, spectrum allocation and management, OFDM-based implementations of cognitive systems, and UWB-based cognitive systems as well as protocol- and Medium Access Control (MAC) design. Good concise overviews of these topics are also in Akyildiz et al. [2006], Haykin [2005], Zhao and Sadler [2007]. Spectral sensing is discussed in Quan et al. [2008]; spectrum opportunity tracking in Zhao et al. [2007]. Application of game theory to cognitive radio system is reviewed in Han et al. [2007] as well as in Ji and Liu [2007]. Overlay systems are described in Jovocic and Visvanath [2007] as well as in Devroye and Tarokh [2007]. UWB communications and its many applications are described, e.g., in diBenedetto et al. [2006]; "detect-and-avoid" methods for UWB are described in Zhang et al. [2008].

For updates and errata for this chapter, see wides.usc.edu/teaching/textbook

22

Relaying, Multi-Hop, and Cooperative Communications

22.1 Introduction and Motivation

22.1.1 Principle of Relaying

Traditional wireless communications are based on point-to-point communication, i.e., only two nodes are involved in the communication of data. These two nodes are, e.g., the Base Station (BS) and Mobile Station (MS) in a cellular setting, access point and laptop in wireless Local Area Networks (LANs), or two MSs in peer-to-peer communications. Other wireless transmitters (TXs) and receivers (RXs) that are in the surroundings *compete* for the same (spectral) resources, giving rise to interference.

In contrast, this chapter deals with the situation that some nodes consciously *help* other nodes to get the information from the message source to the intended destination. This help can be provided either by

- *dedicated relays,* i.e., relays that never act as source or destination of the information, but whose sole purpose is to facilitate the information exchange of other nodes;
- *peer nodes acting as relays*. Such peer nodes, e.g., mobile handsets or sensor nodes, can change their roles depending on the situation at hand – sometimes they help to forward information and sometimes they act as a source or destination.

The introduction of relay nodes creates more degrees of freedom in the system design, which can help to improve the performance, but also complicates the design process. We now show how cooperative communications arises as a logical final result of a network structure that strives to get better and better efficiency and coverage. As a starting point, consider the three-node network in Figure 22.1a, where, for simplicity, we assume that only free-space attenuation (and no fading) occurs on each link. Imagine a situation where node A does not have enough transmit power to send a data packet directly to node C. However, it can, in a first step, transmit the data packet to the intermediate node B. Node B retransmits the packet (e.g., by completely demodulating/decoding it, and then re-encoding and retransmitting), and this retransmission can then be received by node C. This simple two-hop approach doubles the range of the network. Extending this approach to

Wireless Communications, Second Edition   Andreas F. Molisch
© 2011 John Wiley & Sons, Ltd

larger networks, the range that the network covers can be increased even more by multi-hopping, i.e., sending the message to a first relay, from there to a second relay, and so on, until it finally arrives at the destination. When having a large network like in Figure 22.1b, a key question is which nodes should be used in forwarding the information – a topic that is discussed in greater detail in Section 22.4.

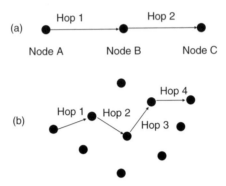

Figure 22.1 Two-hop network (a) and multi-hop ad hoc network (b).

A key feature of the wireless propagation channel is the *broadcast effect*: when one node transmits a signal, it can be received by any node in the vicinity – a fact that *can* be positively exploited in multinode networks.[1] While the multi-hopping strategy described above does *not* make use of the broadcast effect, the more advanced cooperative communications approach *does* use it. Consider Figure 22.2a, which just slightly redraws Figure 22.1a. When node *A* transmits, the signal reaches not only node *B* but also (in weaker form) node *C*. This weak signal might not be enough by itself for node *C* to decode, but it can be used to *augment* the signal received in a subsequent transmission from node *B* to node *C*. The broadcast effect has even more significant impact in larger networks, e.g., the situation depicted in Figure 22.2b: if the first node transmits, the signal reaches nodes *B* and *D* at about equal strength. Reaching those two nodes in the network thus does not "cost" anything more (i.e., does not require more transmit power) than reaching a single node. The two nodes *B* and *D* can now cooperate to forward the information to node *C*, and – as we will show later on – such a cooperative transmission can be more efficient than if only a single node transmits. The same principle holds in even larger networks, like the one depicted in Figure 22.2c.

In the subsequent sections, more detailed descriptions of the scenarios of Figure 22.2 are given. We start out with the three-node relaying network, which is the most fundamental cooperative system, while Section 22.3 considers multiple parallel relays. The subsequent two subsections deal with larger networks, where a message is not transmitted in just two "hops" (transmission from source to relay(s), and a second transmission from relay(s) to destination), but where multiple transmissions from one relay (or set of relays) to another is done. For such transmission, via a sequence of relays, the *routing* problem arises, i.e., which nodes should be used for relaying, and in what sequence. Finally, Section 22.6 describes the applications of relays in cellular networks and ad hoc networks.

---

[1] The flip side of this coin is that the transmission of a message from a node creates interference to other nodes that want to receive a different message. This negative effect is always present, regardless of whether the positive aspect of the broadcast effect is exploited or not.

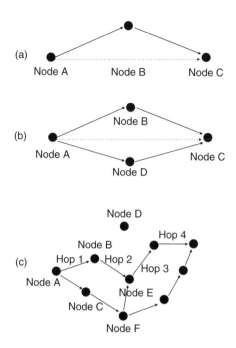

Figure 22.2   Demonstrating the broadcast effect in relay networks.

22.2   Fundamentals of Relaying

22.2.1   Fundamental Protocols

Consider the three-node network shown in Figure 22.3. A source is connected to a relay and a destination, with the channels between the nodes given by h_{sr}, h_{sd}, and h_{rd}, respectively. The relay can now help in various ways in forwarding the information.

Figure 22.3   The fundamental relay channel.

In *Amplify-and-Forward* (AF), the relay amplifies the received signal by a certain factor, and retransmits it. In *Decode-and-Forward* (DF), the relay decodes the packet and then subsequently re-encodes and retransmits it. In *Compress-and-Forward* (CF), the relay creates a quantized (compressed) version of the signal it obtains from the source and forwards that to the destination; the destination combines this compressed signal with the directly transmitted signal from the source. We assume in the following that all relaying nodes operate in a *half-duplex mode*, i.e., they cannot transmit and receive in the same frequency band at the same time. This is reasonable because the transmit and receive levels of wireless signals are so different that the transmitted signal would "swamp" the RX and make it impossible to detect the receive signal.[2]

These relay processing methods can now be combined with various transmission protocols that prescribe when what information blocks are transmitted from which nodes. We list them here in order of increasing performance (and at the same time of increasing complexity).

- *Multi-hop xF*[3] *(MxF)*: in the first timeslot, the source transmits, and *only* the relay is listening. In the second timeslot, only the relay is transmitting and the destination is listening.
- *Split-Combine xF (SCxF)*: in the first timeslot, the source transmits and only the relay is listening (just as in MxF). In the second timeslot, both source and relay transmit, and the destination is listening.
- *Diversity xF (DxF)*: in the first timeslot, the source transmits and both relay and destination are listening. In the second timeslot, only the relay is transmitting and the destination is listening. Thus, the destination gets two copies of the original signal.
- *Nonorthogonal Diversity xF (NDxF)*: in this scheme the source sends *differently encoded* information in the second timeslot. Take, e.g., the case where the source encodes the information with a rate 1/3 convolutional code that is punctured to rate 2/3 by omitting the first, third, fifth, and so on, bits. The relay recovers the original information, but then encodes with the same rate of 1/3 convolutional code, but now punctures out the second, fourth, sixth, and so on, bits. The RX thus sees, from source and relay, differently encoded versions of the same information.
- *Intersymbol Interference xF (IxF)*: this is a scheme that works only if the relay has full-duplex capability (contrary to our assumption above). In timeslot i, the source sends an information block to the relay, and in timeslot $i + 1$, the relay forwards this block to the destination, while at the same time, the source sends the next information block to the relay. The destination is continuously listening, and in each timeslot hears the superposition of the "current" information block directly from the source, and a "previous" information block from the relay.

The above-mentioned protocols are by no means the only ones possible, and a number of variations have been proposed that are either aimed to reduce the penalty of half-duplex operation, and/or are intended to allow simplified encoding and decoding.

	Phase 1			Phase 2		
	S	R	D	S	R	D
MxF	TX	RX	–	–	TX	RX
SCxF	TX	RX	–	TX	TX	RX
DxF	TX	RX	RX	–	TX	RX
NDxF	TX	RX	RX	TX	TX	RX
IxF	TX	RX+TX	RX	TX	TX+RX	RX

---

[2] There are some special cases when a relay (repeater) can transmit and receive at the same time: namely when TX and RX are using different antennas on the relay, and those antennas point to different directions. Such an arrangement can achieve sufficient isolation in the relay that – together with sophisticated interference cancellation – a simultaneous transmission and reception is made possible.

[3] xF stands for either AF, DF, or CF.

A further subclassification of the fundamental protocols can be done according to the following criteria:

- *Fixed vs adaptive power*: is the power expended by the source and by the relay fixed, or is it considered to be adaptive to the channel state? If it is adaptive, the power can be adapted
  o over the nodes, i.e., some node is assigned more power, while another is assigned less;
  o over time, i.e., according to the temporal changes of the channel state, a node can transmit with more power at some time, and less power at another;
  o over frequency: in a wideband system using, e.g., Orthogonal Frequency Division Multiplexing (OFDM) (Chapter 19), a node can transmit more power at some frequencies, and less on others;
  o any combination of the above.
  Constraints are then usually imposed on the sum (or average) transmit power, where the summation can be over nodes, time, and/or frequency. For example, constraining the power summed over all the nodes (and imposing that constraint for any arbitrary time) imposes a restriction on the instantaneous power of the network; this is important because that instantaneous power is a measure for the interference *to other networks*. A summation over nodes and time (and frequency, if applicable) is a constraint on the energy efficiency of the communication. A summation over time (and frequency) only is a measure for the energy consumption of a particular node, which determines its battery lifetime.
- *Fixed vs. adaptive allocation of time and spectral resources*: the most common form of half-duplex protocols assumes that the network spends half the time sending a packet from the source to the relay, and the other half of the time forwarding the relay to the destination. This might not, however, be the most effective use of the available time. If the relay power is fixed, then transmission should occur at a rate close to the capacity of the relay-destination channel. If the relay power can be varied (as discussed above), then power and time (or bandwidth) should be optimized simultaneously.

22.2.2 Decode-and-Forward

The most important of all relaying schemes is DF. The relay receives a packet and decodes it, thus eliminating the effects of noise, before re-encoding and retransmitting the packet. In the following, we will analyze the capacity of several of the implementations.

The simplest scheme to analyze is Multi-hop Decode-and-Forward (MDF). If we assume fixed transmit power, and equal splitting of the available time between the two phases, the overall data rate per unit bandwidth is

$$R = \frac{1}{2} \min \left[\log \left(1 + \frac{P_s |h_{sr}|^2}{P_n} \right), \log \left(1 + \frac{P_r |h_{rd}|^2}{P_n} \right) \right] \qquad (22.1)$$

where P_s and P_r are the powers used by source and relay, and P_n is the noise power. In other words, the link with the smallest Signal-to-Noise Ratio SNR becomes the "bottleneck" that determines the overall capacity; the factor 1/2 arises from the half-duplex constraint. The terms inside the min operation are the capacities of the source-relay and the relay-destination link. In order for a transmission to be successful, a data packet has to pass through *both* links; the link with the smaller capacity is therefore the bottleneck that determines the achievable transmission rate.

The values of P_s and P_r can be fixed or can be optimized given the power constraints and the values of the channel coefficients h_{sr} and h_{rd}. In the latter case the powers should be adjusted in such a way that the capacity of the source-relay link is the same as the one for the relay-destination link, i.e.,

$$P_s = P_0 \frac{|h_{rd}|^2}{|h_{sr}|^2 + |h_{rd}|^2}, \quad P_r = P_0 \frac{|h_{sr}|^2}{|h_{sr}|^2 + |h_{rd}|^2} \qquad (22.2)$$

Further optimizations for this scheme (as well as the ones we discuss below) can be achieved by dividing an available timeslot for data transmission into unequal parts, and optimizing the duration of those two parts [Stankovic et al. 2006].

In Diversity-Decode-and-Forward (DDF),[4] the destination listens during both the phases, and thus can add up the signals it received from the source (in phase 1) and the relay (in phase 2). We then have to distinguish two important cases:

1. *Transmission from the relay using repetition coding*: in this case, the relay uses the same encoder as the source. Consequently, the source can add up the received signals before the decoding, resulting in an improved SNR. Let us assume now the further restriction transmission is successful only if the destination correctly receives the message from both source and relay. In that case, the maximum achievable rate is thus

$$R_{DDF} = \frac{1}{2} \min \left[\log \left(1 + \frac{P_s |h_{sr}|^2}{P_n} \right), \log \left(1 + \frac{P_r |h_{rd}|^2}{P_n} + \frac{P_s |h_{sd}|^2}{P_n} \right) \right] \qquad (22.3)$$

and its optimum power allocation is

$$P_s = P_0 \frac{|h_{rd}|^2}{|h_{sr}|^2 + |h_{rd}|^2 - |h_{sd}|^2} \qquad (22.4)$$

$$P_r = P_0 \frac{|h_{sr}|^2 - |h_{sd}|^2}{|h_{sr}|^2 + |h_{rd}|^2 - |h_{sd}|^2}$$

A more intelligent protocol would employ the relay only if it can actually help, or otherwise keep it idle. Such a protocol is called adaptive DDF. An even better performance can be achieved by "incremental relaying," where the relay does not transmit if the destination can already decode the packet after the first transmission phase.

2. *Transmission from the relay using incremental-redundancy encoding*: in this case, the relay decodes the packet, and re-encodes it with a different coder. Intuitively, the RX can add up the mutual information from the two transmission phases – in other words, it sees a low-rate code, where some of the information bits and parity-check bits arrive in the first phase, and some in the second phase. The capacity of such a protocol is

$$R_{DDF,IR} = \frac{1}{2} \min \left[\log \left(1 + \frac{P_s |h_{sr}|^2}{P_n} \right), \log \left(1 + \frac{P_r |h_{rd}|^2}{P_n} \right) + \log \left(1 + \frac{P_s |h_{sd}|^2}{P_n} \right) \right] \qquad (22.5)$$

In the following, when we speak about DDF, we mean the "regular" DDF as described in Eq. (22.3).

Using the concept of outage capacity (see Section 14.9), we define the whole system to be in outage if the achievable rate falls below the desired threshold R_{th}. For nonadaptive DDF, the

---

[4] According to our notation, DDF is "Diversity-Decode-and-Forward." However, sometimes the acronym DDF is used in the literature for "dynamic decode and forward," which is a different protocol.

probability for such an outage is

$$
\Pr(R_{DDF} < R_{th}) = \Pr\left[|h_{sr}|^2 < \left(2^{2R_{th}} - 1\right)\frac{P_n}{P_s}\right]
$$

$$
+ \Pr\left[|h_{sr}|^2 > \left(2^{2R_{th}} - 1\right)\frac{P_n}{P_s}\right]\Pr\left[|h_{sd}|^2 P_s + |h_{rd}|^2 P_r < \left(2^{2R_{th}} - 1\right) P_n\right]
$$

$$(22.6)$$

where the first term on the right-hand side corresponds to the case where the source-relay connection is too weak (since according to the protocol the relay transmits in the second phase anyway, it creates an outage), and the second term corresponds to the case when the source-relay connection is strong enough, but the relay-destination and source-destination connections are too weak to sustain sufficient information flow to the destination. Since the various links in the system are independent, those probabilities add up. If all the links are Rayleigh fading, then in the limit of high SNRs, the first term dominates because it provides only diversity order 1; the second term vanishes if *either* the source-relay *or* the relay-destination link provides sufficient quality. The overall diversity order 1 is linked to the fact that the protocol absolutely requires the source-relay link to have sufficient strength. For adaptive DDF, on the other hand, diversity order 2 can be achieved: two independent paths (source-destination, or source-relay-destination) are available, and transmission is successful if *either* of those paths provides sufficient quality.

22.2.3 Amplify-and-Forward

The basic principle of AF is that the relay takes the (noisy) signal y_r that it receives and amplifies it with a gain β; there are no other manipulations of y_r (like decoding, demodulating, etc.). It is assumed that the AF processing at the relay leads to a delay of half a timeslot.[5] Thus, in the first phase, the signal received at the destination is simply the (attenuated) signal from the source, plus noise (additional terms occur for Intersymbol interference Amplify-and-Forward (IAF), which we do not consider further). In the second phase, the signal is the sum of the direct signal from the source, and the signal from the relay, which is the amplified source word of the *previous* phase of that slot:

$$
y_d^{(2)} = h_{sd}x_s^{(2)} + h_{rd}x_r^{(2)} + n_d^{(2)}
$$

$$
= h_{sd}x_s^{(2)} + \beta h_{sr}h_{rd}x_s^{(1)} + \beta h_{rd}n_r^{(1)} + n_d^{(2)}
$$

$$(22.7)$$

where x_s and x_r are the transmit signals from source and relay, respectively, superscript $^{(1)}$ and $^{(2)}$ denote the first and second phase of the transmission, respectively, and n_r and n_d are noise at relay and destination, respectively ($x_s^{(2)}$ can be zero, depending on the protocols outlined above).

The amplification factor is limited by power constraints. In the case of an instantaneous power constraint, we require that

$$
|\beta|^2 \leq \frac{P_r}{P_n + P_s|h_{sr}|^2}
$$

$$(22.8)$$

while in the case of an instantaneous power constraint, the constraint has to be averaged over the fading realizations.

---

[5] In the practical case of a purely analog repeater, the delay is actually much smaller.

Despite the seeming simplicity of the scheme, there are a number of possible protocol realizations, as indicated in the classification above. Let us first analyze the performance of Multi-hop Amplify-and-Forward (MAF) with $P_r = P_s = P$. The signal arriving at the RX during the first phase is simply the source signal multiplied with the channel coefficient h_{sd}, and the noise has variance P_n. In the second phase, the signal arriving at the destination is the source signal multiplied with $\beta h_{sr} h_{rd}$, and the noise has variance

$$P'_n = \left(|\beta|^2 |h_{rd}|^2 + 1 \right) P_n \tag{22.9}$$

The SNR is thus simply given by $\gamma = P|\beta h_{sr} h_{rd}|^2 / P'_n$. Note, however, that the optimum power allocation in MAF is given by

$$\frac{P_s}{P_r} = \sqrt{\frac{|h_{rd}|^2 P_0 + P_n}{|h_{sr}|^2 P_0 + P_n}} \tag{22.10}$$

In Diversity-Amplify-and-Forward (DAF), we wish to combine the signals from the two phases such as to maximize the SNR, i.e., we want to perform maximum-ratio combining. Therefore, the signal from the first phase has to be multiplied with

$$\sqrt{\frac{P_s}{P_n}} h^*_{sd} \tag{22.11}$$

i.e., phase adjusted, and multiplied with the square root of the receive SNR during that phase. Using a similar motivation, the signal from the second phase has to be multiplied by

$$\sqrt{\frac{P_s \beta^2}{P'_n}} (h_{sr} h_{rd})^* \tag{22.12}$$

Assuming $P_s = P_r = P$, the corresponding SNRs for the two phases are

$$\gamma_1 = \frac{P|h_{sd}|^2}{P_n} \tag{22.13}$$

and

$$\gamma_2 = \frac{(\frac{P}{P_n})^2 |h_{sr} h_{rd}|^2}{\frac{P}{P_n}|h_{sr}|^2 + \frac{P}{P_n}|h_{rd}|^2 + 1} \tag{22.14}$$

and the overall SNR is, as always with Maximum Ratio Combining (MRC), $\gamma = \gamma_1 + \gamma_2$. The resulting capacity is $\frac{1}{2} \log_2(1 + \gamma)$, where the factor 1/2 stems from the half-duplex constraint of the relay. We also note that the scheme has diversity order 2, because the signal can reach the destination on two independent paths: directly, or via the relay. However, note that for coverage-extension relays with practical SNRs (as opposed to the infinite SNRs used for diversity order computations), the direct path is not really helpful – if we could achieve good reception with the direct path, there would be no need for the relay in the first place.

22.2.4 Compress-and-Forward

CF is similar to AF in the sense that the relay does not decode a message, but forwards whatever it receives (including noise). The main *difference* compared to AF is that the forwarded signal is a *quantized, compressed* version of the signal received at the relay. The process of quantization and compression can be seen as a source encoding problem, i.e., the received signal is an (analog) source, and its information should be encoded into a digital signal with as few distortions as possible – but on the other hand, the rate at which that signal can be transmitted is limited (and depends on the relay-destination channel). At the destination, this signal is used together with the signal received directly from the source node to reconstruct the original signal.

CF has been shown to provide higher capacity than DF or AF for some specific channel configurations. However, it is considerably more complicated than either of those formats, and thus will not be considered further in this book. Further details can be found in the references cited at the end of the chapter.

22.3   Relaying with Multiple, Parallel Relays

In many situations, more than one relay is available for forwarding the information. In that case, a cooperation between the different relays can be used to greatly enhance the performance of the relaying scheme, in particular in fading channels. Essentially, the multiple relays provide diversity paths that bring better robustness with respect to fading and interference. At the same time, the cooperation between the relays necessitates the exchange of Channel State Information (CSI) and control information. There are thus a number of different schemes, which trade off the overhead with the system performance in different ways.

Also for the arrangement with multiple relays, various transmission schemes do exist – AF, DF, CF, with the various protocols discussed in the previous section. However, to keep the discussion focused, we restrict the discussion to DF (in most cases, restricted to MDF) with semi-duplex relays. In the current section, we will only deal with two-hop networks, since in this case the relaying problem can be viewed as a physical-layer problem. Networks with more hops require, in addition, routing, and are treated in subsequent sections.

Figure 22.4 shows the fundamental setup that we consider in this section. Transmission occurs in two phases: in phase 1, the source broadcasts the information. This phase takes advantage of the broadcast effect, i.e., the signal arrives at several relays (possibly with different strengths), even though only a single node (i.e., the source) transmits. In the second phase, one or more of the relays forward the information to the destination. We see that this second phase strongly resembles a smart-antenna system, specifically the transmission from a multiantenna TX to a single-antenna destination – the main difference being that the antennas are distributed over a larger area in space. This analogy to multiantenna systems will be helpful in the discussion below.

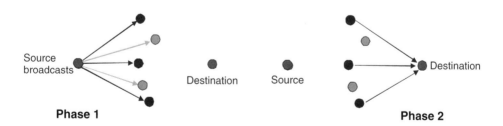

Figure 22.4   Two-phase transmission with parallel relays.

For phase 1, we always assume that the k-th relay knows the channel from the source to it (i.e., CSI at the Receiver (CSIR), is available). CSI at the Transmitter (CSIT) is only useful if the source can either adapt its power or its transmission time, in accordance to the channel states. For phase 2, we again assume CSIR (i.e., the destination knows the channels between the k-th relay and the destination). However, we distinguish different cases with respect to knowledge of CSIT, i.e., CSI that is available at the relays. Depending on the type of CSIT, different transmission schemes can be used:

- *Full CSIT available*: relays know both amplitude and phase of the channel to the destination. In that case, "virtual beamforming," similar to maximum-ratio transmission in a multiple-antenna system, can be used. This method ensures the maximum SNR at the RX for a given sum power expenditure at the relays. This case will be discussed in Section 22.3.2.
- *Amplitude CSIT available*: in that case, the relays know the amplitude (strength) of the channel to the destination, but not the phase. In this case, the best strategy (for a sum-power constraint) is to select a single relay that provides the best transmission quality (see Section 22.3.1).
- *No CSIT available*: in this case, the relays can transmit Space–Time (ST) encoded versions of the data packet – they act like antennas in a transmit-diversity system without CSIT (see Section 22.3.3). Alternatively, the relays can send out incremental-redundancy encoded bits of the same codeword (see Section 22.3.4). Note that for a sum-power constraint, the SNR for space–time codes is worse than for relay selection: transmit diversity provides an effective channel whose SNR is the *average* of the individual relay-destination channels, while relay selection provides a channel with an SNR that is the *maximum* of that of the individual channels.
- *Average CSIT available*: in this case, only the mean channel gain but not the instantaneous realization is available. This case is interesting because average CSIT can be acquired much more easily than instantaneous CSIT, particularly in fast-varying channels. Modifications of the no-CSIT schemes can be used.

22.3.1   Relay Selection

In relay selection, we simply pick the "best" of all the available relays, and then perform relaying the same way as described in Section 22.2. This approach sounds deceptively simple; the challenges are (i) defining what we mean by "best relay," (ii) actually finding the relay for a given set of channel states.

Let us first turn to the question of defining a criterion for the "best" relay. We have to distinguish two cases: if the source has fixed transmit power and data rate, (i.e., modulation format and coding of the packet are fixed), then we cannot influence which relay will receive the packet correctly during transmission phase 1; rather, we simply consider the set of relays that *do* get the packet, and select the one for forwarding that has the strongest channel to the destination.

If the source can adapt to the channels, then we can make sure that a specific relay (which is selected a priori as the one that will do the forwarding) gets the message in phase 1. Choosing this relay requires a balancing between the source-relay and relay-destination channel strengths: in MDF, where the rate is given by Eq. (22.1), we should aim to avoid bottlenecks, and thus pick the relay that provides the best

$$\eta_k = \min\left[|h_{s,k}|^2, |h_{k,d}|^2\right] \tag{22.15}$$

An alternative criterion considers a "smoothed-out" version of this criterion

$$\eta_k = \frac{2}{\frac{1}{|h_{s,k}|^2} + \frac{1}{|h_{k,d}|^2}} \tag{22.16}$$

It is often assumed that a central control node knows all the $|h_{s,k}|^2$ and $|h_{k,d}|^2$. In practice, this would require considerable signaling overhead, and is therefore undesirable. It is therefore preferable to use algorithms similar to the ones that used to control multiple access for packet radio systems (see Chapter 17), in a distributed manner:

1. In a first step, the destination sends out a brief broadcast signal that allows the relays to determine the $|h_{k,d}|^2$ (assuming channel reciprocity, see Section 20.1).
2. Next, the source sends the data packet, as well as a "Clear To Send" (CTS) message after it is finished. Each relay tries to receive the packet, and also determines its $|h_{s,k}|^2$, and from that, the η_i, according to one of the criteria Eq. (22.15) or (22.16).
3. Each relay starts a timer with an initial value K_{timer}/η_k (where K_{timer} is a suitably chosen constant) and counts down, while listening for possible on-air signals from the other relays. When the timer reaches 0, the relay starts to transmit – unless another relay has already started to transmit (and thus occupies the channel).

Clearly, the relay with the "best" channel (highest η_k) is the first – and therefore only – relay node to transmit. In practical networks, the performance is not ideal, since a second node might start to transmit between the time that the first node transmits and the time that signal actually arrives at the second node (again, compare Chapter 17), but such collisions can be resolved by repeating step 3 with different K_{timer}.

Relay selection performs remarkably well, and provides the same diversity order as other, more complicated relaying schemes discussed below. This is analogous to antenna selection (see Chapter 13): antenna selection provides the same slope of the Bit Error Rate (BER) vs. SNR curve (i.e., diversity order) as the (optimum) MRC. For K relays, the outage probability can be computed for the case of Rayleigh fading on all links as

$$Pr[I < R_{\text{th}}] = \prod_{k=1}^{K} \left[1 - \exp\left[-\frac{2^{2R_{\text{th}}} - 1}{P/P_{\text{n}}} \left(\frac{1}{\overline{\gamma}_{s,k}^2} + \frac{1}{\overline{\gamma}_{k,d}^2} \right) \right] \right] \tag{22.17}$$

where $\overline{\gamma}_{s,k}^2$ and $\overline{\gamma}_{k,d}^2$ are the mean channel gains of the source-relay and relay-destination channels.

22.3.2 *Distributed Beamforming*

This protocol consists of two phases: in phase 1 the source broadcasts the information and a set $\mathcal{D}$ (with size $|\mathcal{D}|$) receives the packet in good order. For an MDF protocol, the optimal transmission coefficient at each selected relay k in phase 2 can then be shown to be proportional to

$$\frac{h_{k,d}^*}{(\sum_{k \in \mathcal{D}} |h_{k,d}|^2)^{1/2}} \tag{22.18}$$

once the CSIT at the relays is available. The $|\mathcal{D}|$ nodes cooperate, i.e., transmit coherently, to send data to the destination. This is similar to beamforming or maximum ratio transmission in transmit diversity systems.

In the case that the relays are using AF, the optimum gain applied at relay k is

$$w_k = K^{\text{AF}} \frac{|h_{sk}||h_{kd}|}{1 + P_s|h_{sk}|^2 + P_k|h_{kd}|^2} \frac{h_{sk}^*}{|h_{sk}|} \frac{h_{kd}^*}{|h_{kd}|} \tag{22.19}$$

where the constant K^{AF} is chosen such that the total power constraint $\sum_k |w_k|^2(1 + P_s|h_{sk}|^2) = P_r$ is met. The power on the k-th relay is

$$P_k \propto \frac{|h_{sk}|^2|h_{kd}|^2 \left[P_s|h_{sk}|^2 + 1 \right]}{\left[1 + P_s|h_{sk}|^2 + P_k|h_{kd}|^2 \right]^2} \tag{22.20}$$

Obtaining the CSIT at the relays is nontrivial: not only does each relay need to know its channel to the destination but it also needs to know the sum of the channel gains from all the relays that will be active in the forwarding of the data (i.e., the denominator in Eq. 22.18). This can be implemented through consecutive transmission of training sequences (pilots) from the relays, followed by a feedback from the destination.

The situation is, again, more complicated if the source can adjust its power, because then it determines the set of possible active relays, $\mathcal{D}$. This leads to a tradeoff between the two phases in the relaying: if the source expends little energy on the broadcast, then $|\mathcal{D}|$ is too small, and the diversity order available in phase 2 is low – in other words, there is a risk that all the relays that received the packet have a bad channel to the destination, and thus have to expend a large amount of power to get the packet to the destination. On the other hand, spending too much energy on the broadcast is wasteful. An exact optimization of the best power allocation is somewhat complicated, but as a rule of thumb, $|\mathcal{D}|$ should be 3.

The above discussion assumes that the various relay nodes can co-phase their transmit signals in such a way that they superpose constructively at the intended destination. This is a very difficult endeavor in practice, since the relay nodes are not colocated, yet still have to be phase synchronous (in addition to being frequency and time synchronous). Typically, one node in the network would work as a master that periodically sends out synchronization signals and forces all other nodes to adapt their frequency and phase to this synchronization signal. Adjusting for the phase shift created by the runtime between nodes has to be done on a link-by-link case.

An alternative way of dealing with the problem of phase adjustment is the use of random beamforming (compare Section 20.1). If no special measures are taken (i.e., no specific phase adjustment), the beams created by the relays point into random directions, and by changing the relative phases of the nodes, the main direction of the beams changes. In the spirit of opportunistic beamforming, a destination node that finds itself in the main lobe of the beam sends a feedback signal, and asks for the relays to send payload data intended for this node.

22.3.3 *Transmission on Orthogonal Channels*

When CSI is not available at the TX, one possible solution is to have each relay transmit on an orthogonal channel. This clearly eliminates the interference between the different relay channels; however, it also leads to a drastic reduction of the spectral efficiency. In particular, we consider a DDF scheme where every relay has a reserved channel – whether it can decode the message or not. Its capacity (or more precisely, the mutual information using Gaussian codebooks) is

$$I = \frac{1}{K+1} \log \left[1 + \gamma_{s,d} + \sum_{k \in \mathcal{D}} \gamma_{k,d} \right] \tag{22.21}$$

where $\mathcal{D}$ is the set of the relays that can decode the message from a particular source. When all links are Rayleigh fading, the outage probability conditioned on a particular decoding set for this scheme is for high SNR

$$\Pr[I < R_{\text{th}} | \mathcal{D}] \sim \left[2^{(K+1)R_{\text{th}}} - 1 \right]^{|\mathcal{D}(s)|+1} \frac{1}{\gamma_{s,d}} \prod_{k \in \mathcal{D}} \frac{1}{\gamma_{k,d}} \frac{1}{[|\mathcal{D}|+1]!} \tag{22.22}$$

The probability of obtaining a particular decoding set is given by

$$\Pr[\mathcal{D}] \sim \left[2^{(K+1)R_{\text{th}}} - 1 \right]^{K-|\mathcal{D}(s)|} \prod_{k \notin \mathcal{D}} \frac{1}{\gamma_{s,k}} \tag{22.23}$$

The overall outage probability is then Eq. (22.22) unconditioned by Eq. (22.23); this expression can be bounded by

$$\left[\frac{2^{(K+1)R_{th}} - 1}{\overline{\gamma}^{lb}} \right]^{K+1} \sum_{k \in \mathcal{D}} \frac{1}{[|\mathcal{D}| + 1]!} \lesssim \Pr[I < R_{th}] \lesssim \left[\frac{2^{(K+1)R_{th}} - 1}{\overline{\gamma}^{ub}} \right]^{K+1} \sum_{k \in \mathcal{D}} \frac{1}{[|\mathcal{D}| + 1]!}$$

(22.24)

where

$$1/\overline{\gamma}_k^{lb} = \min\{1/\overline{\gamma}_{s,k}, 1/\overline{\gamma}_{k,d}\} \quad 1/\overline{\gamma}_k^{ub} = \max\{1/\overline{\gamma}_{s,k}, 1/\overline{\gamma}_{k,d}\} \quad \overline{\gamma}_s^{lb} = \overline{\gamma}_s^{ub} = \overline{\gamma}_{s,d} \quad (22.25)$$

and $\overline{\gamma}^{lb}$ is the geometric mean of the $\overline{\gamma}_k^{lb}$, $k = 1 \dots K + 1$, and similarly for $\overline{\gamma}^{ub}$. To again draw an analogy with multiple-antenna systems, transmission on orthogonal channels can be compared to *antenna cycling*, where only one antenna element is used (for one particular message) at each time.

We now turn to the situation where there are multiple nodes that all act as sources, as well as relays and which can transmit at different frequencies as well as times. Consider the situation depicted in Figure 22.5. Also in that case, each node is transmitting information from a particular source only for 1/(K+1) of the available time. In other words, the spectral efficiency is not improved compared to the situation discussed above.

Figure 22.5 Multiplexing of multiple signals on multiple relays.
Reproduced from Laneman and Wornell [2003] © IEEE.

22.3.4 Distributed Space–Time Coding

An alternative approach for the no-CSIT case is to have the relays use space–time codes during the transmission. Consider the following situation: in phase 1, the source sends the information to the relays. In phase 2, the relays now perform a space–time coded transmission to the destination. In other words, each relay node acts as a "virtual antenna," and sends out the signal that – in a Multiple Input Multiple Output MIMO setting – would be sent out by one of antenna elements of the transmit antenna array. For example, if two relays are used, then the used space–time code could be an Alamouti code. That means that two symbols, c_1 and c_2, are transmitted from the two relays at time instant 1:

$$\mathbf{s}_1 = \frac{1}{\sqrt{2}} \begin{pmatrix} c_1 \\ c_2 \end{pmatrix}.$$

(22.26)

where $\mathbf{s}$ is the vector containing the symbols sent from the relays. At the second time instant, the signal vector

$$\mathbf{s}_2 = \frac{1}{\sqrt{2}} \begin{pmatrix} -c_2^* \\ c_1^* \end{pmatrix}$$

(22.27)

is transmitted (compare Section 20.2). Of course, the communication protocol must have a means to assign to each relay which "antenna" it is, and therefore, which sequence of data $(c_1 \ -c_2^* \ ....)$ or $(c_2 \ c_1^* \ ....)$ it should send out.

Since the Alamouti code is a rate-1 code, the spectral efficiency of the transmission is better than for the relaying on orthogonal channels, where the rate (during the second phase of the relaying) is only 1/2 (for the case of two relays). When using more relays, the spectral efficiency of relaying with orthogonal space–time codes decreases somewhat: for $K > 2$, no rate-1 orthogonal space–time codes exist. For $K = 3$ or 4, the achievable rate decreases to 3/4. Still, this is much better than orthogonal relaying, where the rate decreases as $1/K$.

A further practical problem arises from the fact that the number of participating relays changes, depending on how many relays are able to decode the message from the source. Fortunately, this does not impact the operation of distributed space–time codes significantly: if a relay does not receive a message from the source, it simply does not transmit (which for the RX looks like that particular "antenna" is in a deep fade). The decoding operation of the RX is therefore not impacted.

22.3.5   Coded Cooperation

In coded cooperation, relaying and error correction coding are integrated, leading to enhanced diversity. A data packet from a source is encoded with a Forward Error Correction FEC code (see Chapter 14), and different parts of the codewords are sent via two (or more) different paths in the network.

To give more details, let us consider the example of Figure 22.6. At node 1, a source data block is encoded with an FEC, and the resulting codeword is split up into two parts, with N_1 and N_2 bits, respectively. It is important that it is possible to reconstruct the source data from the first N_1 bits alone. For example, the FEC can be a rate 1/3 convolutional code, which is then punctured to result in a rate 2/3 code that is transmitted in the first N_1 bits; the bits that are punctured out are transmitted in the second N_2 bits. A similar encoding and splitting is done for a different block of source data at node 2.

Now the transmission interval available for one block of source data is divided into two parts during the first interval, node 1 broadcasts the first N_1 bits. They are received by the destination, as well as by node 2. At the same time, node 2 transmits (on an orthogonal channel, e.g., a different frequency channel) its first N_1 bits. If node 1 can successfully decode the source word of node 2 (this is checked with a Cyclic Redundancy Check CRC), then it computes the second N_2 bits associated with that source data of node 2 and transmits it in the second time interval. If it cannot decode successfully, then it sends the N_2 bits associated with its own codeword. Node 2 behaves in a completely analogous way. Since there is no feedback between nodes 1 and 2, the four situations depicted in Figure 22.7 can arise. Summarizing, each node always sends $N_1 + N_2$ bits; if the channel between the two nodes is good, then part of the transmitted bits are helping a partner node; this is the case that we will consider in the following (the other case, where each node just transmits $N_1 + N_2$ of its own data, is a regular coded Frequency Division Multiple Access (FDMA) transmission of two users to an RX, see Chapter 17).

Since the different parts of the codeword are transmitted from different locations, the transmission has a diversity order of 2 (if the two nodes 1 and 2 are sufficiently widely separated, they might provide macrodiversity as well as microdiversity (see Chapter 13)). The diversity order is reflected in the asymptotic expressions for the outage probability. Assuming that all channels are Rayleigh fading, and the channels from node 1 to node 2 and node 2 to node 1 are independent (as is usually the case in a frequency duplexing), the outage probability is approximated in the high-SNR regime by

$$\Pr[I < R_{\text{th}}] = \frac{(2^{2R_{\text{th}}} - 1)^2}{\gamma_{\text{A,d}}\gamma_{\text{A,B}}} + \frac{R_{\text{th}} \ln(2) 2^{2R_{\text{th}}+1} - 2^{2R_{\text{th}}} + 1}{\gamma_{\text{A,d}}\gamma_{\text{B,d}}}. \tag{22.28}$$

For the case of reciprocal inter-user channels (i.e., if transmission from node A to node B and node B to node A is done on the same frequency channel, e.g., using Time Division Multiple

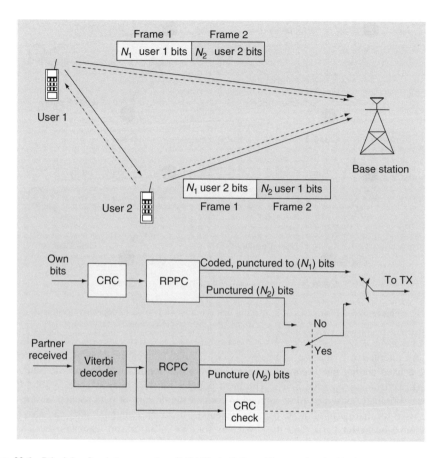

Figure 22.6 Principle of coded cooperation. Solid (dashed) lines: bits associated with the payload user 1 (2) has generated.

Reproduced from Nosratinia et al. [2004] © IEEE.

Access (TDMA)), the outage probability is

$$\Pr[I < R_{th}] = \frac{(2^{R_{th}} - 1)(2^{2R_{th}} - 1)}{\gamma_{A,d}\gamma_{A,B}} + \frac{R_{th}\ln(2)2^{2R_{th}+1} - 2^{2R_{th}} + 1}{\gamma_{A,d}\gamma_{B,d}}. \tag{22.29}$$

The transmission of the data packet for one particular user can be considered as DF relaying with incremental redundancy; this is more efficient than conventional DF, where the relay repeats the originally transmitted bits, as discussed in Section 22.2 above. This is also visible in Figure 22.8, which compares the block error rate of coded cooperation with rate 1/4 encoding to AF and DF with rate 1/2 encoding (i.e., all schemes have the same spectral efficiency).

22.3.6 Fountain Codes

The virtual-MIMO techniques described in the previous section suffer from a number of drawbacks, including the necessity to coordinate simultaneous transmissions to achieve cooperative gain, and comparatively low efficiency of the collaboration (RXs accumulate energy from the cooperating

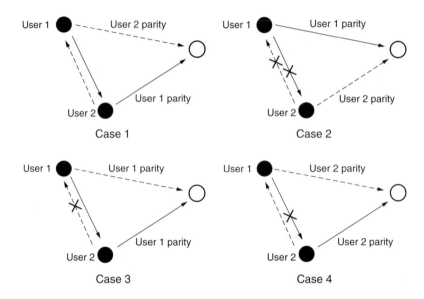

Figure 22.7   Four cases of messaging that can arise in two-user cooperative coding.
Reproduced from Hunter and Nosratinia [2006] © IEEE.

nodes). An alternative is the use of rateless codes. Unlike conventional codes, rateless codes do not have a fixed coding rate (hence, the name) and are not optimized for a specific SNR. Rather, they work well for all possible SNRs. The most popular form of rateless codes, called Fountain codes, was originally designed to be applied to combat the erasure of data packets on the Internet (and are used for that purpose in 3G cellular systems). However, Fountain codes can also be made to work on a bit-by-bit basis, and then work as follows in an erasure channel: the TX creates an (infinitely long) bitstream from a finite-length block of source data. Receiving nodes observe the bitstream, and accumulate the bits that were not erased in the channel. RXs can recover the original information from the observed, unordered subset of the codestream, just as long as the total received number of bits is larger than the number of bits in the original source word. It turns out that bit-by-bit Fountain codes also work well in Additive White Gaussian Noise (AWGN) channels, and fading channels. In that case, successful reception of a source word is achieved if the received mutual information equals the source entropy.

Since Fountain codes are "universal" codes that work at *any* SNR rate, the same code design can be used for broadcasting from one source to multiple RXs whose links to the source have different attenuations. At the same time, Fountain codes are useful in relay networks because they allow each node in the network to accumulate mutual information (and not just energy) from multiple transmitting nodes (source and other relays that received the same information at earlier times). Intuitively, the difference between energy accumulation and mutual-information accumulation can be most easily understood from the following simple example that uses binary signaling over erasure channels with erasure probability p_e. If the RX accumulates energy, then each bit will be erased with probability p_e^2, so $1 - p_e^2$ bits on average are received per transmission. On the other hand, if the RX can use mutual-information accumulation, it obtains on average $2(1 - p_e)$ bits (which exceeds $1 - p_e^2$ bits) per transmission.

Summarizing, cooperative communications with Fountain codes is different from "regular" cooperative communications in the following way:

1. The transmission duration of a packet is a random variable, depending on the channel state.

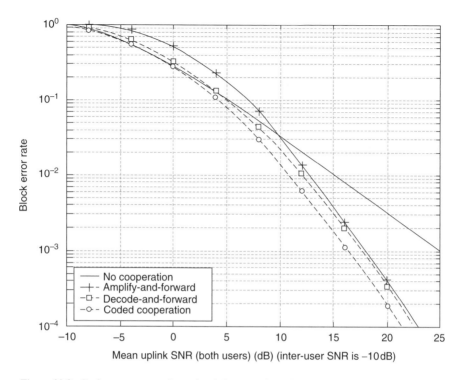

Figure 22.8 Performance comparison of coded cooperation with AF, DF, and direct transmission.
Reproduced from Nosratinia et al. [2004] © IEEE.

2. Successful "one-shot" transmission can be guaranteed, without the need to know the SINR (Signal-to-Interference-and-Noise Ratio) at the TX. The TX just keeps transmitting until it gets a 1-bit feedback from the RX that the message was successfully decoded.
3. An RX can accumulate mutual information (not just energy) from multiple relays.
4. Different parallel relays can be active for different amounts of time. This is in contrast to, e.g., distributed space–time coding, where the parallel nodes transmit for the same amount of time.

22.4 Routing and Resource Allocation in Multi-Hop Networks

We now turn our attention to larger networks, where a single relay, or a number of *parallel* relays, are not sufficient to get the information from the source to the destination. Rather, we have to use multiple relays sequentially. In particular, this section considers multi-hop systems, where a packet of data is transferred via multiple relays using MDF, in the manner of a bucket brigade: the packet is transmitted from the source to the first relay node, where it is decoded, re-encoded, and sent to the next relay, which also does decode/re-encode, then transmits it to the next relay, and so on, until finally the last relay transmits to the destination. Each of the hops is operated as a point-to-point link, i.e., the broadcast effect is not exploited.

The first task is to determinate which nodes should act as relays (i.e., forward the message) and in what sequence; in other words, what is the *route* of the packet to its destination? Related to this issue is the question as to what resources (power, bandwidth) should be allocated to those nodes. The combined *routing and resource allocation* is a cross-layer design problem, i.e., involving

PHYsical layer (PHY), Medium Access Control (MAC) layer, and networking. However, many of the studies in the literature treat only subaspects: routing algorithms tend to assume a fixed physical layer, for which an optimal route is constructed. Conversely, other papers assume a given route, and try to optimize the PHY for this route.

The subsequent derivations are done for unicast situations, i.e., each packet has exactly one source and one destination. In many cases, the algorithms can be generalized to cover multicast or broadcast (i.e., multiple destinations), but for the sake of simplicity, we will not deal with this.

22.4.1 Mathematical Preliminaries

We start out with networks where each link between two nodes can be treated as essentially "binary": it can either support error-free transmission of a packet (in which case there is a valid connection between the nodes), or it cannot. The network is then represented as a *graph*: each node corresponds to a vertex, and each (working) link corresponds to an edge. We furthermore assume that the links are reciprocal, i.e., that the transfer function from node A to node B is the same as from node B to node A; in that case the graph is undirected. Each edge has a *weight*: for the simple case of fixed transmit power, and assuming a link either works or doesn't work, the edge weights of all working links are unity, so that the cost of transmission is the "hop count."[6] In the case that different nodes can use different power (to ensure that certain links work), the edge weight could represent the power expenditure for sending a packet over a certain link. In either case, the routing problem becomes a shortest-path problem – more exactly, the problem of finding the path with the smallest "distance" (sum of the edge weights) between two vertices in a graph. These shortest-path problems have been considered in computer science literature, and several algorithms have been developed to solve them. The *Dijkstra* algorithm provides a fast solution if all edge weights are positive (as they usually are); in those cases where negative edge weights need to be considered, the *Bellman–Ford* algorithm should be used.

Dijkstra Algorithm

The Dijkstra algorithm finds the shortest-path weights from a source node to every node in the network, based on the principle of "greedy relaxation." It proceeds in the following steps:

1. Assign the source node the distance $d_s = 0$, and all other nodes the distance $d_i = \infty$ (note that d is the "distance from the source node," and as such is a property of the nodes, not of the edges). Furthermore, mark all nodes as "unvisited" and declare the source node the "current" node.
2. Loop until all nodes are visited.
   (a) Consider all unvisited neighbor nodes i of the current node c, i.e., nodes with (direct) links to the current node. For each neighbor node i, compute the $\widehat{d}_i = d_c + w_{c,i}$ where $w_{c,i}$ is the edge cost between nodes c and i. If $\widehat{d}_i < d_i$, replace d_i by $\widehat{d}_i$. Let the node store a pointer to the previous node on the route that led to the lowest cost.
   (b) Mark the current node as visited.
   (c) Pick the unvisited node with the smallest distance as current node.

The algorithm is very efficient; depending on the particular implementation, the runtime is proportional to $\mathcal{O}(|V|^2 + |E|)$ or $\mathcal{O}(|V|\log(|V|) + |E|)$, where $|V|$ is the number of vertices, and $|E|$ is the number of edges.

---

[6] A somewhat better metric is the Expected Number of Transmissions (ETX), which takes into account the fact that not all packet transmissions over a link are successful, and therefore require a retransmission.

Bellman–Ford Algorithm

The Bellman–Ford is an alternative algorithm for finding the shortest path. It proceeds in the following steps:

1. We again start by assigning the source node the distance $d_0 = 0$, and all other nodes the distance $d_i = \infty$.
2. Repeat $|V| - 1$ times.
   (a) Do for all edges in the graph.
       (i) Call the starting point of the edge j, and the end point i; then compute the $\widehat{d_i} = d_j + w_{j,i}$; if $\widehat{d_i} < d_i$, replace d_i by $\widehat{d_i}$. Let the node store a pointer to the previous node on the route that led to the lowest cost.
3. Check whether further reductions in the distances are still possible. If yes, this indicates that the graph contains "negative cycles," and the weights will not converge. Otherwise, the algorithm is finished. This step can be skipped if we know that all edge weights are positive.

> **Example 22.1** *Figure 22.9 shows a network with positive edge weights. The left side of the figure shows the various stages of the Dijkstra algorithm. The right side shows the working of Bellman–Ford algorithm, where each graph corresponds to one iteration of the outer loop; the tables indicate the evolutions of the weights as the inner loop (stepping through all the edges in the graph). Note that for the Bellman–Ford algorithm, the sequence of the edges in the inner loops does not play a role; in other words, the numbering of the edges as (1,2,3 ...) is arbitrary.*

22.4.2   Goals and Classifications of Routing Protocols

Routing protocols have been investigated extensively in the context of wired packet networks, in particular the Internet. There, the main goal is to transmit the information to the destination in the shortest possible time. In wireless ad hoc networks, additional constraints have to be considered. We can thus have a combination of the following goals (i) the energy consumption for the forwarding of the information should be kept as small as possible, (ii) the lifetime of the network, i.e., the time until the first node runs out of battery power, should be maximized, (iii) the protocol should be distributed, i.e., not require centralized control, (iv) the protocol should be able to react quickly to changes in the topology or link states, (v) the protocol should be bandwidth efficient, i.e., achieve high throughput in the bandwidth allocated to the network, (vi) the end-to-end transmission time (delay) should be minimized.

Taking the changing topology and link states into account can be done in one of the following two ways: (i) *proactive*: in this case, the network keeps track of the optimum routes from all possible sources to all possible destinations at all times. Thus, the actual transmission of packets can be done very quickly, as the optimum route is immediately available. On the downside, the overhead required to keep track of all the routes can be significant. (ii) *reactive*: in this approach, a route to a destination is only determined when there is actually a packet to be sent to that particular destination, i.e., *on demand*. This approach is more efficient, but clearly leads to a slower delivery of packets.

22.4.3   Source Routing

In source routing, each information-generating node can determine, e.g., from a lookup table, the sequence of nodes that this packet should take through the network. The sequence of nodes is added to the data packet, so that each node on the route can learn to which node the packet should be

Dijkstra algorithm   **Bellman-Ford algorithm**

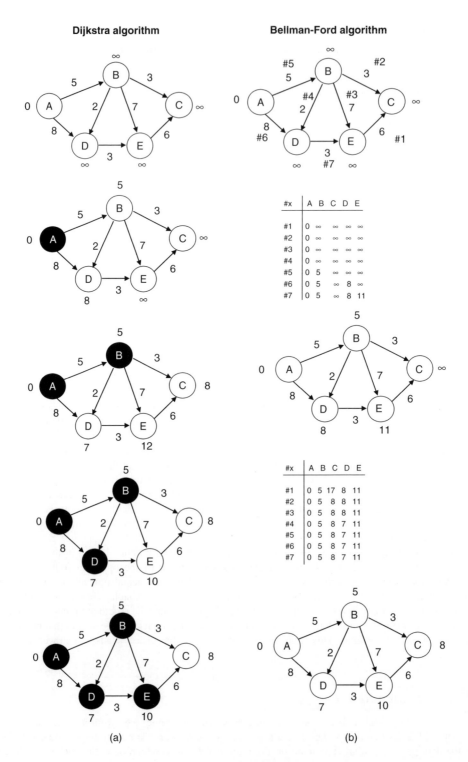

(a)   (b)

Figure 22.9   Dijkstra (a) and Bellman–Ford (b) algorithm.

transmitted next. A key advantage of source routing is that it is loop-free, i.e., there is no danger that a packet returns to an intermediate node that it had already visited. This absence of loops can be guaranteed trivially, since the source determines the route through the network, and therefore can make sure during the route setup that there are no loops. This point may sound trivial, but we will see below that other routing methods can suffer from loops, which increase energy consumption and latency, and can cause instabilities.

In *proactive source routing*, link-state advertisements are sent out periodically by each node to neighboring nodes. Those nodes compare the received link state to the one they have stored in their local tables, if it is fresher (i.e., has a higher sequence number), then they update their table, and forward the information to their neighbors, and so on. Thus, refreshed link-state information is propagated throughout the network. It is obvious that the amount of information that has to be distributed grows enormously as the number of nodes in the network increases. For this reason, proactive source routing algorithms are not well suited for large networks, and in particular not for networks whose states change frequently.

Dynamic Source Routing (DSR) can greatly reduce the overhead by performing on-demand routing. The routing procedure consists of two steps: an initial *route discovery*, followed by *route maintenance* that reacts to changes in the link states in the network. During the route discovery, the network is flooded with so-called *route request packets*. The route request contains the Identification (ID) of the intended information destination, a unique packet ID, as well as a list of nodes that have already been visited by the message. When a node receives a route request packet, it checks whether it is either the intended destination or has a path to the destination stored in its own routing table. If that is *not* the case, the node rebroadcasts the route request, adding its own address to the list of visited nodes in the message. If the node *is* the destination (or has a path to the destination), then the node answers with a *route reply packet*, which tracks back along the identified path to the source, and finally informs the source about the sequence of nodes that have to be taken from source to destination.[7] Note that this method of route request – route response requires reciprocal links (see the discussion in Section 20.1 when this is fulfilled). The route request packets can record the quality of the links, which then can be used for the route determination through the "edge weights" as described above for the Dijkstra algorithm.

During route maintenance, the protocol observes whether links in the established route are "broken" (which might include that the throughput of a particular link has gone below a certain threshold). It then either uses an alternative stored route for the destination, or initiates another full route discovery process.

A problem in DSR is the so-called reply storm, which occurs when a lot of neighbors know the route to the target, and try to simultaneously send the information. This wastes network resources. Route reply storms are prevented by making each node that wants to transmit wait by an amount of time that is inversely proportional to the quality of the route it can offer. In other words, a node that wants to suggest a good route is allowed to transmit sooner than a node that suggests an alternative, worse, route. Furthermore, nodes listen to route replies transmitted by other nodes, and do not transmit if another node has already transmitted a better route.

22.4.4 Link-State Based Routing

In link-state routing, each node gathers information about the states of the links in the whole network. Based on this information, a node can then construct the most efficient routes through the network to all other nodes, e.g., by means of the Dijkstra algorithm.

---

[7] Intermediate nodes on that route might also store the route from them to the destination nodes; this can accelerate future route searches.

Link-state based routing requires acquisition and distribution of the link states throughout the network. A node first has to learn its link state, e.g., from training sequences transmitted by other nodes. The distribution of the link-state information to other nodes is then achieved by means of short messages called *link-state advertisements*. Those messages contain the following pieces of information: (i) the ID of the node that is creating the advertisement, (ii) the nodes to which the advertising node is connected, as well as the link quality (edge weights) of that link, (iii) the sequence number, which indicates how "fresh" the information is (the sequence number is incremented every time the node sends out a new advertisement).

A popular algorithm for implementing link-state based routing for wireless networks is the *Optimized Link State Routing protocol* OLSR. It is a proactive protocol, where messages between nodes are exchanged on a regular basis, to create routing tables at every node. While classical link-state protocols flood the link-state information throughout the network, OLSR limits this information. This fact makes it particularly suitable for wireless networks, since the amount of link-state information in wireless ad hoc networks can be particularly large. OLSR therefore uses the concept of *Multi-Point Relays* (MPRs). The set of MPRs of a particular node is defined as a subset of the one-hop neighbors, where the subset must be chosen in such a way that the node can reach all two-hop neighbors via an MPR. In other words, we do not need to reach each two-hop neighbor by all routes that would be possible, but reaching it via one route is deemed sufficient. In order to sense the environment of a node, OLSR periodically lets the node transmit "Hello" messages. Those messages are sensed by the surrounding nodes, but not rebroadcast. Since a "Hello" message contains information about all known links and neighbors of a particular node, every node can find *two-hop neighbors* by passively listening to the Hello messages.

As mentioned above, each node selects a subset of its neighbors as MPRs. The set of MPRs is regularly broadcast by means of the "Topology Control" messages, to other MPRs, which can then forward that information. Thus, any node in the network can be accessed through a series of MPRs. By reducing the ways that a node can be reached, the overall amount of link-state information that has to be sent through the network is reduced.

22.4.5   Distance Vector Routing

In distance vector routing, each node maintains a list of all destinations that *only* contains the cost of getting to that destination, and the *next* node to send the message to. Thus, the source node only knows to which node to hand the packet, which in turn knows the next node, and so on. This approach has the advantage of vastly reduced storage costs compared to link-state algorithms (remember that in link-state routing, every node has to store, for each destination, the complete sequence of nodes via which to send the message). It follows that distance vector algorithms are easier to implement and require less storage space. The actual determination of the route is based on the Bellman–Ford algorithm, which was discussed in Section. 22.4.1.

However, distance vector routing also has a number of drawbacks:

1. Slow convergence: compared to the Dijkstra algorithm, Bellman–Ford requires multiple passes of the cost information. In a network with quickly changing topology, this can lead to situations where the link states have changed before an optimum route has been set up.
2. "Counting to infinity": in the extreme case that part of the network becomes separated, the network can form loops, so that information comes back to a node that it had already passed (remember that such a situation cannot occur in source routing). The essence of the problem is that even if node B tells node A that it has a route to the destination, node A does not know if that route contains node A (which would make it a loop). Under "normal," converged circumstances this is not a problem because a route containing a loop has a higher total cost

than a route that "cuts out" that loop. However, in case that a node "goes down" (either because of node failure or because of fading of links), a loop can be created. Consider a linear network, connected as A-B-C-D-E-F, and let the edge cost be unity for each hop. Now if node A goes down, node B learns during the update process of Bellman–Ford that its link to node A is down, because it does not receive an update. However, at the same time, node B gets an update from node C, which tells node B that node A is only two jumps from node C (since node C does not know about the outage yet). This misinformation thus propagates through the network, until the nodes have increased their costs to infinity.

A solution of the counting-to-infinity problem can be obtained by means of *destination sequences*, resulting in the DSDV (*Destination-Sequenced Distance Vector*) algorithm. For this case, the nodes store not only the cost of getting to the destination and the next node on the route but also a sequence number. Each node then periodically advertises routes to other destinations; the destination increases the sequence number and propagates the route to other nodes in the network. The route that is selected is the one with the largest sequence number in the network. If a link is broken, the sequence number is increased and the node cost is increased to infinity. This change is immediately propagated to other nodes.

The *Ad hoc On-demand Distance Vector* (*AODV*) routing algorithm is the reactive version of DSDV. Routes are built up using route requests and route replies (similar to DSR): when a source needs to send a packet to a destination for which it does not already have a route, it broadcasts a route request. Nodes that hear this request update their information for the source node and set up backward pointers to the source node in the route tables. If the node has a route to the destination, it replies to the source; otherwise, it broadcasts the information to further nodes.

22.4.6   Geography-Based Routing

In a number of cases, the nodes in the network know their geographic position. This can be achieved either by Global Positioning System GPS localization (if the nodes have built-in GPS RXs), or by other localization mechanisms (e.g., based on field strength maps, time-of-flight ranging, etc.). Routes can be designed based on this geographic information – though it is noteworthy that geographic distance between two nodes forming a link and SNR of that link is not monotonic if there is fading – a fact that is sometimes ignored in the networking literature. In the (over-) simplified picture of a path gain that is determined by distance only, each transmitting node is surrounded by a "coverage disk" of radius R, such that every node in the disk can hear the transmission, and all nodes outside the disk cannot hear it.

Geographical routing schemes are based on the concept of "progress" toward the destination. "Greedy" schemes, which are the most popular geographical schemes, pick the node within the coverage disk that has the smallest distance to the destination. Alternatively, they can maximize the projection of the link connecting the transmitting node to a particular receiving node, onto the line connecting the transmitting node to the destination; this approach is called "most forward within radius." An example of how those two algorithms behave is given in Figure 22.10; however, it must be emphasized that in most cases the two algorithms find the same path.

When greedy approaches fail to advance the message to the destination, networks can go into "recovery mode." This occurs when the message arrives at so-called "concave" nodes, which have no neighbor that is closer to the destination than themselves. An easy way out is to momentarily use flooding, i.e., concave nodes flood their neighbors with the message, and subsequently reject any further copies of the message (to avoid that the neighbor sends the message back to the concave node, which after all is close to the destination). After this flooding step, the routing continues using a greedy mode.

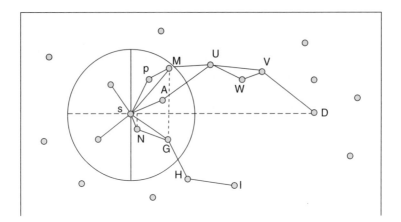

Figure 22.10   Criteria for choosing nodes in geographic routing: a greedy algorithm picks the path SGHI (which fails to deliver); the "most forward within radius" algorithm picks SMUVD.
Reproduced from Stojmenovic [2002] © IEEE.

22.4.7   Hierarchical Routing

Up to now, we have assumed that all nodes are equal with respect to forwarding messages. However, this need not be the case. In hierarchical routing, the nodes of the network are subdivided into clusters, i.e., group of nodes. One node in the cluster is the so-called *clusterhead*; all other nodes in the cluster can only communicate with this particular node. When a message is to be passed between clusters, this can only be done by communications via the clusterheads. A node joins the cluster whose clusterhead it can talk to with the least energy expenditure. If all nodes are battery powered, then the role of the clusterhead is randomly rotated between the nodes in a cluster; more specifically the probability of becoming a clusterhead is chosen to be proportional to the remaining battery power of the node. If there are some nodes that are connected to power mains (instead of batteries), so that energy consumption is not a significant concern, then those nodes are always used as clusterheads. One popular hierarchical algorithm (mainly used for data-driven routing, see below) is the LEACH protocol (see Figure 22.11).

22.4.8   Impact of Node Mobility

In most ad hoc networks, the nodes are fairly static during operation. However, there are also some networks where nodes show a very high degree of mobility – e.g., in vehicular ad hoc networks. This mobility has both advantages and drawbacks:

- The key advantage is that data packets can "hitchhike" on nodes that are moving. Imagine that a data packet has to be sent from node A to node B, both of which are static, and widely separated. Node A then transmits the packet to a highly mobile node C when it is passing by node A. Node C stores the message, and when it comes into the vicinity of node B, transmits it. Thus, the large distance between nodes A and B can be bridged without either high transmit power for direct transmission, or multiple transmissions. Note, however, that the latency of the packet transmission is significant.
- The key drawback of high node mobility is that the network can become temporarily disconnected, especially in sparse networks, where there are only few possible routes between a source and a destination. If a few nodes are moving, it can easily happen that there is no valid path

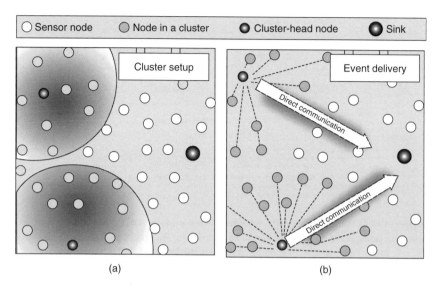

Figure 22.11 Clustering of nodes in hierarchical routing.
Reproduced from Martirosyan et al. [2008] © IEEE.

for a multi-hop connection between source and destination anymore. On the other hand, there are also sparse networks that are never connected in a static sense (i.e., at no point in time is there a route between TX and RX); still, exploiting node mobility by the "hitchhiking" process described above can allow packet delivery to the destination.

A routing algorithm that takes account of the node mobility is "epidemic routing." Whenever two nodes come into range of each other, they exchange all the packets that they do not have in common. In this way, every packet is ultimately distributed to every node in the network. This approach is quite wasteful of resources, in particular if it is mostly unicast messages that are to be distributed. A more efficient algorithm is "spray and wait," where L copies of the message are transmitted, and then we just wait until one of the nodes that received a copy comes into range for direct transmission to the destination.

22.4.9 Data-Driven Routing

The ultimate task of sensor networks is different from those in ad hoc networks. While ad hoc networks have to transmit data files from one node to another, the purpose of sensor networks is to get sensing data (typically data about the environment in which the sensor is located), to a data sink. In this process, it is not important from which particular node the data originated; all that matters is that the underlying environmental information arrives correctly. As a concrete example, take temperature measurements in a factory hall. Say that the hall has 100 temperature sensors in it, even though we know that the temperature is the same everywhere in the room. Then all that matters is that we get a temperature reading to a central monitoring station. The data from all the temperature sensors are correlated, so it might be overkill to ensure that each single sensor can communicate with the monitoring station. Thus, routing in sensor networks is *data driven* (or application driven), not connection driven.

The most popular data-driven routing approach is *directed diffusion* (Figure 22.12). The monitoring station requests data by sending out an *interest* message, which contains specifics of the type

(a) Interest propagation   (b) Initial gradients set up   (c) Data delivery along reinforced path

Figure 22.12   Directed diffusion.
Reproduced from Intanagonwiwat et al. [2003] © IEEE.

of information that it wants, intervals at which the data are to be collected, and geographical area. The message is propagated throughout the network; during this propagation, the nodes also set up gradients, i.e., reply links toward the nodes from which the interest statement was received. Note that multiple paths can be set up back to the monitoring station. Particular paths can be reinforced during the actual transmission of the data. During the data transmission, each node can cache and process data, and can aggregate data from different sources.

Further refinements of the protocol can include "meta-data negotiations," which essentially make sure that nodes only transmit new data to the clusterhead. The transmission of data occurs in three steps: advertising, request, and data transmission. In the advertising step, a node uses meta-data to advertise the existence of new data; in the request stage, the recipient requests the data if they are useful (e.g., not already collected from another node that has similar/correlated data), and in the data transmission stage, the actual data are sent.

22.4.10   Power Allocation Strategies

In the description of the above algorithms, we have assumed that the transmit power of the nodes is fixed – an assumption that is often fulfilled in ad hoc networks, where the nodes should be as simple as possible. However, there are cases where the nodes can adapt their power and/or their transmission rate. In that case, we should try to optimize the transmission power as much as possible. We can distinguish the following cases:

1. *Routes fixed, transmission rate fixed*: in this case, all we can achieve with power control is a reduction of the expended power. The transmit power at each node should be lowered as much as possible with the limitation that the SNR at the receiving node has to be high enough to guarantee decoding. As a side benefit of the power control, the interference to the rest of the network is reduced (this becomes important when multiple messages are to be sent through the network at the same time).
2. *Routes fixed, but transmission rate variable*: in this case, reducing the transmit power increases the necessary transmission time. If minimization of the energy expenditure is the ultimate goal of the optimization process, then transmit power should be lowered as much as possible. If the message has to be delivered to the destination within a certain deadline, then an optimization of the power allocation for a given delivery time can be done.
3. *Routes and transmission rates variable*: by changing the power and/or the rate, we can adapt the edge weights of the graph representing the network. Thus, a route that is optimal for a certain set of transmit powers might not be optimum for a different set. Thus, routing and power allocation has to be done in one joint step. Actually, there is an even more general and harder problem,

which occurs when there are *multiple* messages for which routing, transmission rate control, and power control need to be done. This general problem is discussed in Section 22.4.11.

22.4.11   *Routing for Multiple Messages – Stochastic Network Optimization*

All the routing algorithms described above are essentially based on the assumption that a single message in the network is to be transmitted to one or more destination nodes. The situation becomes much more complicated when multiple messages are to be transmitted, since each transmission creates interference on other active links. Thus, a link that provides high capacity when only a single message is routed over it might become a bottleneck when multiple messages are trying to be transmitted over this same link. In other words, the routes that are found to be optimum for the different messages separately (e.g., by the Dijkstra algorithm) are not necessarily the best (or even good) routes when multiple messages are being transmitted in the network simultaneously.

A logical next step is then to determine a *joint* routing and power allocation, which essentially makes sure that messages along a route do not interfere with each other significantly. By keeping routes "sufficiently" separated (where the transmit power influences how big a separation is sufficient), the throughput can be improved significantly. Still, such an algorithm misses another vital component of network design, namely *scheduling*. Different messages *can* be sent over the same node; we just have to ensure that this happens at different times – similar to road traffic at an intersection: the scheduling algorithm corresponds to a traffic light that tells the traffic from one road to stop while traffic on an intersecting road is using the intersection, so that no collisions occur. The general problem of finding routes, scheduling, and power/rate control for multiple messages is thus a very complicated, but also a practically very important problem. A number of different algorithms have been designed to solve this. As one example, the following will describe a stochastic optimization approach called the *backpressure algorithm*.

The backpressure algorithm is a stochastic network optimization algorithm that – despite its simplicity – turns out to be optimal under certain assumptions. Let us describe the network by its states, and by control actions that determine what and how much data are transmitted with what power. The formulation as a control problem allows to bring to bear useful techniques from control theory like Lyapunov functions. The essence of the approach is the following: each node has a buffer in which it stores arriving data, and it tries to transmit data to empty the buffer. The algorithm then computes link weights that take into account (i) the difference of the queue size between the transmitting and receiving node of a link and (ii) the data rate that can be achieved over this link.

Consider the following setup: a network contains N nodes, connected by L links. The transmission of the messages occurs in a slotted manner, where t is the slot index. The link between two nodes a and b is characterized by a transmission rate $\mu_{ab}(t)$; the rates are summarized in the transmission matrix $\mu(t) = \mathbf{C}(\mathbf{I}(t), S(t))$, where C is the transmission rate function that depends on the network topology state $S(t)$, which describes all the effects that the network cannot influence (e.g., fading) and the link control action $I(t)$, which includes all the actions of the network that can be influenced, like power control, etc. The most important example of a transmission rate function is capacity-achieving transmission, so that on the l-th link

$$
C_l(\mathbf{P}(t), S(t)) = \log_2 \left[1 + \frac{P_l(t)\alpha_{ll}(S(t))}{P_{\mathrm{n}} + \sum_{k \neq l} P_k(t)\alpha_{kl}(S(t))} \right]
\tag{22.30}
$$

where $\alpha_{kl}(S(t)) = |h_{kl}|^2$ is the power gain (inverse attenuation) of the signal transmitted by the intended TX of link k to the intended RX of link l, when the network topology is in the state $S(t)$, and $P_l(t)$ is the power used for transmission along the l-th link.

During each timeslot, an amount of data R_q^{out} is being transmitted by a node. At the same time, data are arriving from an external source (e.g., sensors, or computers that want to transmit data); and the node is receiving data via the wireless links from other sources; the total amount of data arriving during one timeslot is R_q^{in}.[8] Each node has an infinitely large buffer, which can store the arriving messages before they are sent out over the wireless links. The amount of data in the buffer (also called queue backlog) is written as $Q_q(t)$, where q is the index of the considered queue; the backlogs are all written into a vector $\mathbf{Q}(t)$. During one timeslot, the backlog of a queue at a particular node changes as

$$Q_q(t+1) \leq \max\left[Q_q(t) - R_q^{\text{out}}(I(t), S(t)), 0\right] + R_q^{\text{in}}(I(t), S(t)) \tag{22.31}$$

where the $\max[., 0]$ operator is used to ensure that the length of the queue cannot become negative. We can then define a general "penalty function" $\bar{\xi}$, which is defined as the time-average of an instantaneous penalty function – e.g., the overall power consumption of the network. An additional set of constraints $\bar{x}_i$ can be defined, e.g., the (time-averaged) power consumption per node. We then aim to solve the following optimization problem:

$$\text{Minimize } \bar{\xi} \tag{22.32}$$

$$\text{subject to } \bar{x}_i \leq x_{\text{av}} \text{ for all } i \tag{22.33}$$

$$\text{and network stability}$$

where x_{av} is, e.g., the maximum admissible mean power.

Here, network stability means that the size of the queues remains bounded – in other words, that on average not more data flow into the queue than can be "shoveled out." Note that it is important to ultimately deliver the data to their final destination, since such data delivery is the only way of making the data "vanish" from the network. As long as data are sent on their way through a multi-hop network, they are part of some queue, and therefore detrimental to achieving network stability. We also note that a possible optimization goal is simply the achievement of network stability without any further constraints (this is formally achieved by setting $\bar{\xi} = 1$).

The first step in solving the problem above is to convert the additional constraints $\bar{x}_i \leq x_{\text{av}}$ into "virtual queues" (these are not true data queues, but simply update equations):

$$Z_i(t+1) = \max\left[Z_i(t) - x_{\text{av}}, 0\right] + x_i(I(t), S(t)) \tag{22.34}$$

The optimization problem is then converted into the problem of minimizing $\bar{\xi}$ while stabilizing both the actual queues $Q(t)$ and the virtual queues $Z(t)$, combined into vector $\Theta(t)$. Defining now the Lyapunov drift

$$\Delta_Z(\Theta(t)) = E\left\{L_Z((\Theta(t+1)) - L_Z((\Theta(t))|\Theta(t)\right\} \tag{22.35}$$

$$\Delta_Q(\Theta(t)) = E\left\{L_Q((\Theta(t+1)) - L_Q((\Theta(t))|\Theta(t)\right\}$$

$$\Delta(\Theta(t)) = \Delta_Z(\Theta(t)) + \Delta_Q(\Theta(t))$$

where

$$L_Z(\Theta(t)) = \frac{1}{2}\sum_i [Z_i(t)]^2, \quad L_Q(\Theta(t)) = \frac{1}{2}\sum_q [Q_q(t)]^2 \tag{22.36}$$

---

[8] The half-duplex constraint can be included, e.g., by assuming that two orthogonal channels exist for each node, one on which data can be received and one on which only transmission can happen.

Upper bounds for the Lyapunov drift are given by

$$\overline{\Delta}_Z(\Theta(t)) = B_Z - \sum_i Z_i(t) E\left\{x_{\mathrm{av}} - x_i(I(t), S(t)) | \Theta(t)\right\} \tag{22.37}$$

$$\overline{\Delta}_Q(\Theta(t)) = B_Q - \sum_q Q_q(t) E\left\{R_q^{\mathrm{out}}(I(t), S(t)) - R_q^{\mathrm{in}}(I(t), S(t)) | \Theta(t)\right\} \tag{22.38}$$

where B_Z and B_Q are finite constants.

The "generalized max-weight policy," i.e., the determination for the optimum control policy for the backpressure algorithm, in each step greedily minimizes

$$\overline{\Delta}_Z(\Theta(t)) + \overline{\Delta}_Q(\Theta(t)) + V E\left\{\xi(I(t), S(t)) | \Theta(t)\right\} \tag{22.39}$$

where V is a control parameter. This can be written as minimizing

$$V\widehat{\xi}(I(t), S(t)) + \sum_i Z_i(t)\widehat{x}_i(I(t), S(t)) - \sum_q Q_q(t) \left[\widehat{R}_q^{\mathrm{out}}(I(t), S(t)) - \widehat{R}_q^{\mathrm{in}}(I(t), S(t))\right] \tag{22.40}$$

where $\widehat{\xi}(I(t), S(t)) = E\left\{\xi(I(t), S(t)) | I(t), S(t)\right\}$ and similarly for the other "hat" functions in the equation above. These functions of $I(t)$ and $S(t)$ are assumed to be known. For a given slot t, the network state $S(t)$ and queue backlogs $Q(t)$ and $Z(t)$ are known, though the pdf of S does not need to be known.

The "network stabilization" only guarantees that – as a time average – the length of the queues does not grow to infinity. There is, however, no guarantee about the delay of a particular data packet. Even the average delay of the data packets can be quite large. The control parameter V allows a tradeoff between this average delay and how close the solution of Eq. (22.40) comes to the theoretical minimum of the penalty function.

After this rather general description, let us turn to simpler special cases: we wish to minimize the average network power (without additional constraints on the per-node power). In that case, there are no "virtual" queues. We thus define now for each node Ω_n as the set of links for which node n acts as TX; we furthermore realize that at each node there can be multiple queues, each for one particular message. We thus aim to choose the power vector $\mathbf{P}(t)$ that maximizes the expression

$$\sum_n \left[\sum_{l \in \Omega_n} C_l(\mathbf{P}(t), S(t)) W_l^*(t) - V \sum_{l \in \Omega_n} P_l\right] \tag{22.41}$$

where P_l is the temporal average over $P_l(t)$, and the weights W_l^* are

$$W_l^*(t) = \max[Q_l^{\mathrm{t}} - Q_l^{\mathrm{r}}, 0] \tag{22.42}$$

and Q_l^{t} and Q_l^{r} are the backlogs of the queue at the transmitting node and receiving node, respectively, of link l for the packet stream (commodity) that has the biggest backlog differential at this particular link.

In the even simpler case of network stabilization only, the backpressure algorithm serves the queue whose product of "queue length difference" at the two link ends, and the transmission rate over that link, is maximum.[9]

It is also noteworthy that the backpressure algorithm provides an inherent routing; data packets will ultimately end up in their intended sinks, since this is the only way to get "out of the network."

---

[9] For one-hop problems we have $Q_l^{\mathrm{r}} = 0$, and the algorithm reduces to maximizing a weighted sum of transmission rates over each link, minus a power cost.

However, there is no guarantee that the routes taken by the packets follow a shortest path; especially in lightly loaded networks, data packets can take very circuitous routes. This is especially pronounced if we try to simply stabilize the network (without energy minimization). This problem can be alleviated by introducing a "bias" for shorter routes in the penalty function. Figure 22.13 shows the average backlog as a function of the generation rate of packets at the nodes, when no power minimization is attempted. We see that the shortest-path algorithm becomes unstable (backlog becomes infinite) at much lower packet generation rates. The "enhanced" backpressure algorithm (one that includes a bias term) performs much better than the regular algorithm at low packet generation rates.

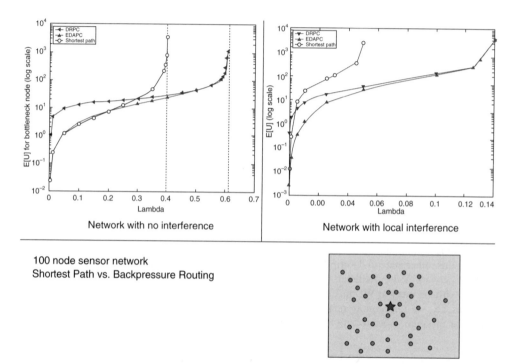

Figure 22.13 Performance of different routing algorithms in an ad hoc network: DRPC (backpressure algorithm), EDRPC (enhanced backpressure), and shortest-path.

Reproduced from Georgiadis et al. [2006] © NOW publishing.

22.4.12   Scaling Laws

As the number of nodes that want to communicate with each other in a network increases, the interference between the messages becomes worse and worse. A very active area of information-theoretic research is to investigate scaling laws, i.e., the functional dependence of the overall throughput of the network as a function of the number of nodes. Such a scaling law does not tell us the absolute throughput that can be achieved; it only tells us how much the throughput can be increased by, e.g., doubling of the number of nodes.

For ad hoc networks, the scaling laws depend very much on the assumed underlying protocol. A simple model uses the idea of "guard zones": a transmission from node A to node B (which are separated by a distance d) is successful if no other RX is active in a disk centered around node B, with radius $d(1 + \Delta)$. For multi-hop transmission, where each hop covers only a small distance, the guard zone can be smaller (which allows more nodes to be on simultaneously). On the other hand, a message needs to be sent over several hops, thus creating interference in the network for a longer time. If we now furthermore assume that all nodes are equal, generating traffic that needs to be forwarded to destination nodes, the overall transport capacity of a network covering an area A scales like $\sqrt{AN}$, which also implies that the capacity per node *decreases* like $1/\sqrt{N}$, and thus vanishes as N becomes large. Note that the definition of transport capacity assumes that the location of the nodes can be optimized.

Another interesting case is when the nodes are located randomly (uniformly, independently distributed) in a unit area. The "feasible throughput" is then a set of data generation rates at each node such that the resulting messages can be sent through the network and reach their destinations without driving the buffers into complete congestion. The admissible data generation rate per source scales then as $1/\sqrt{N \log(N)}$.

22.5 Routing and Resource Allocation in Collaborative Networks

When collaborative communications is used for the forwarding of messages, the routing problem even for a single message becomes much more complex. In multi-hop routing, all we needed to determine was the sequence of nodes that forward the message, and (possibly) the transmit power used by each node. The question of when, and for how long, a node should transmit was answered very simply: it should start when it had received and decoded the message from the previous node on the route, and it should stop when the subsequent node had decoded the message. When cooperative communications are used, start- and stop-time of transmission from a relay are (almost)[10] arbitrary parameters that need to be optimized.

In contrast to the multi-hop case an optimum solution can only be found by trying out all possible routes. For this reason, a number of heuristic algorithms have been proposed that can often come close to ideal performance. Furthermore, it must be noted that the field of "collaborative routing" is far from mature; there is in particular a dearth of solutions for the case when multiple messages are to be transmitted through the network simultaneously.

22.5.1 *Edge-Disjoint Routing and Anypath Routing*

A simple form of cooperative routing is obtained if the goal of the cooperation is increased robustness. In that case, it is advantageous to route the message on several parallel routes through the network; these routes should have as few links and nodes in common as possible, so that the failure of a particular link cannot lead to a blockage of all routes to a destination. *Edge-disjoint shortest-path routing* is a way of identifying routes that do not share any links; a suitable algorithm is a minor modification of the Bellman–Ford algorithm. Note, however, that this approach does not make significant use of the broadcast effect.

Anypath routing exploits the broadcast effect to achieve diversity: each node broadcasts the data packet to a group of neighbors, called the *forwarding set*. As long as at least one of the nodes in the forwarding set receives the message, the next step on the route can be made; one of the successful nodes then acts as the next relay on the route. Note, however, that only a single node from a forwarding set retransmits. In other words, the broadcast effect is used to obtain selection diversity. Thus, even if a particular link on the "nominal" shortest path goes down, the data packet can reach its destination, without the necessity of finding a new route. Anypath routing (see Figure 22.14) is thus especially useful for links that frequently change in quality.

Instead of a specific route, anypath routing results in an ensemble of possible routes; depending on the outcomes of the transmissions from the various nodes, a packet can take different routes

---

[10] Of course, a node can only start to transmit after it has decoded the message that needs to be relayed.

Figure 22.14 An anypath route (gray), and one possible trajectory taken by a packet (bold).
Reproduced from Dubois-Ferriere [2006] © EPFL Switzerland.

through the network. Formally speaking, an anypath route is the union of all possible trajectories along which a packet can travel from the source to the destination. Finding the best anypath route involves a tradeoff: increasing the candidate set gives a better robustness (and might therefore help to decrease the required link margin, and thus transmit power, while retaining the same outage probability for message delivery). On the other hand, a larger forwarding set increases the danger that the packet is routed farther away from the true shortest path (which we would take if we had perfect, completely current knowledge of all the network states). Additional complications arise when the data rate is also allowed to vary – changing the data rate changes the set of nodes that are, in principle, able to correctly receive the packet from a certain preceding node. In any case, however, variations of the Bellman–Ford algorithm can find the optimum anypath route (and associated rates) in polynomial time.

22.5.2 Routing with Energy Accumulation

Another way of exploiting diversity is energy accumulation at the relay nodes. This occurs when a node stores a received signal of a packet that is too weak for decoding and combines it with another signal of the same packet that arrives later. When using energy accumulation at the nodes instead of simple multi-hopping, the optimum route changes. A simple example is given in the following.

Example 22.2 *Consider a linear network with three nodes, where direct transmission from node A to node C has a path gain of 0.1, while transmission from node A to node B has a path gain of 0.19, and similarly from node 2 to node 3.*

Let the threshold of the received signal energy for decodability of the signal be 1 J. Then in a classical multi-hop scenario, the optimum route is direct transmission, with transmit energy of node A equal to 10; multi-hop with route $A \rightarrow B \rightarrow C$ would require $2(1/0.19)=10.53$ J. However, when the destination node can perform energy accumulation, the route $A \rightarrow B \rightarrow C$ becomes preferable: the source node 1 uses $1/0.19=5.26$ J for its transmission (so that node B can decode). During that transmission, node C receives $0.1 \cdot 5.26 = 0.53$ J energy from "overhearing" the packet. Thus, it requires only 0.47 J during the second transmission, which it obtains if node B transmits with $0.47/0.19=2.47$ J Thus, total transmission energy is 7.63 J; less than the 10 J required for direct transmission.

The problem of finding the optimum route involves the issues of finding (i) what nodes should participate, and (ii) in what sequence, and with what power. Unfortunately, the former problem is NP-hard, i.e., it can be solved exactly only by trying out all possible node combinations, and picking the best one.[11] A number of heuristic algorithms exist for finding the best route. Some of those algorithms start out with the optimum multi-hop route (which is determined by the Dijkstra or Bellman–Ford algorithm), and then add on nodes that reduce the overall energy consumption. Another type of algorithm builds up the route from scratch, starting from the source node. When it adds the next relay on the path, it reduces the signal energy still required at all the other nodes; this energy reduction depends on how much energy those other nodes can "overhear" when the new node transmits. In either case, the energy savings from the energy accumulation (and associated routing) increase as the density of nodes increases: the closer the nodes, the more energy a node can "overhear," (see Figure 22.15).

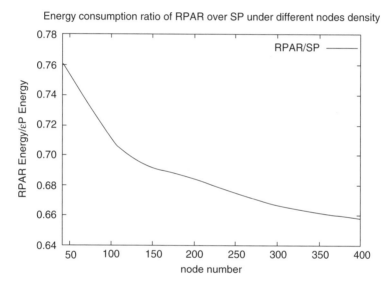

Figure 22.15 Energy savings of routing taking into account energy accumulating compared to shortest-path algorithm when network density increases. *In this figure:* RPAR, Relay PAth Routing; SP, Shortest Path. Reproduced from Chen et al. [2005] © IEEE.

A somewhat different case of energy accumulation occurs when multiple nodes transmit, synchronously, in parallel, to effect higher receive power:

1. Parallel nodes use orthogonal transmission (Section 22.3.3) or distributed space–time coding (Section 22.3.4) for transmission.
2. Parallel nodes use distributed beamforming (Section 22.3).

Also in this case, finding the optimum route is NP-hard. A heuristic method for finding a route is to subsume the nodes acting in parallel into a "super-node," and then try to find the best route of supernodes.

---

[11] For the case of cooperative broadcasting, the first problem is easy (since all nodes participate in the transmission), but the determination of the correct order of node participation is NP-hard.

22.5.3   Fountain Codes

When Fountain codes are used, relay nodes can exploit "overhearing" the signals intended for other relay nodes in an even more efficient manner: they accumulate mutual information, instead of energy, as discussed in Section 22.3.6. Still, routing with mutual-information accumulation shares two important properties with energy accumulation: (i) finding an optimum route is NP-hard, and (ii) for heuristic algorithms, it is useful to break down the problem into two subproblems: determination of the physical route or order of nodes through which packets propagate, and the allocation of resources (time, power) among the nodes. Under the assumption that each node has a fixed transmission power, the determination of the optimum resource allocation (time) can be done by a Linear Program (LP) for a specific routing order. A simple algorithm then revises the routing order based on the results of the LP. Iterating between the two subproblems (resource allocation and routing order) yields a very efficient approach to good route finding even in very large networks.

The LP can be set up the following way: by the end of the k time interval, defined as the time at which the kth node decodes the transmitted packet, the total information flow to the k-th node from the $k - 1$ nodes ahead of it in the route must exceed the packet payload of B bits. Formally,

$$\sum_{i=0}^{k-1} \sum_{n=0}^{k} A_{i,n} C_{i,k} \geq B \tag{22.43}$$

where $A_{i,n}$ is the resource (time or bandwidth) allocated to TX i in the nth time interval,[12] and $C_{i,k}$ is the data rate (a function of channel quality) from node i to node k. These constraints, together with the goal "minimization of total energy" (or other goals) constitute an LP that can be solved by standard software packages. The solution of the LP is subsequently used to update the route: if the start time of the $k + 1$ time interval becomes identical to that of the k-th time interval, the sequence of the k-th and $k + 1$-th node on the route are swapped; if a relay node swaps its place in the decoding sequence with the destination, it is not used at all (it would only become active after the destination has already decoded the message).

22.5.4   Other Collaborative Routing Problems

Different Optimization Criteria

In the previous parts of this section, we always used "overall energy consumption" as a criterion for optimization of a route. However, other criteria can be used in practice. To give but a few examples:

- *Network lifetime maximization*: network lifetime is usually defined as the time during which all nodes have sufficient energy to operate properly. Nodes in the center of a network are in particular danger of running out of energy, since they have the highest likelihood of acting as relays.
- *Message delay*: while the use of many hops through the network can decrease the energy consumption, it also increases the latency; in particular when the communication rate in the network is fixed[13].
- *Network throughput*: when multiple messages are being transmitted, then the resulting interference decreases the overall throughput of the network. The amount of interference depends on the route as well as on the particular collaboration scheme.

---

[12] Note that the duration of a time interval is *not* fixed, but variable, and actually an output of the LP. It just denotes a time during which the transmission parameters are constant.

[13] With adaptive modulation and coding, a short link allows the use of a higher communication rate, so that the overall time for a message to reach the destination might actually decrease when many short (instead of one long) link is used.

Routing with Selfish Nodes

Up to now, we have considered routing with the goal of decreasing the *overall* energy consumption of the network. For either centralized or distributed routing algorithms, it is assumed that the nodes will obey an algorithm that maximizes the social benefits, not the individual benefits of the nodes. This assumption works well in ad hoc networks in industrial or military/security settings, where all nodes are under the control of a single operator. However, in ad hoc networks made up of, e.g., laptops of individual users, the situation is different: every user asks "what is the benefit for me," or, in other words "why should I exhaust my battery in order to forward messages for somebody else?" There have to be proper incentives for users (typically, that their messages will also be forwarded by somebody else). Designing proper rules that maximize benefits for each user, while at the same time discouraging "rule breaking," is thus an interesting problem of networking.

It is easy to see from the above description that *game theory* can be applied to this problem. One class of game-theoretic approaches uses "virtual payments and credits": whenever a node acts as relay for somebody else's message, it receives a "virtual credit"; it can then spend it as payment to other nodes for forwarding its own message when the need arises. A second class of game-theoretic algorithms uses "enforcement" of good behavior either by a watchdog (centralized), or by "reputation-based" algorithms: a node that does not forward messages gets a bad reputation, which negatively affects its own ability to ask other nodes to forward messages for it. In either case, antisocial behavior is discouraged by appropriate punishment through the other nodes.

Clustering and Partitioning

A cluster of nodes can use cooperative communications to bridge larger distances (via collaborative beamforming) than a single node can achieve. Thus, there is a better connectivity of a network, i.e., the probability that a node is isolated (cannot be reached by any route) is smaller for a network allowing collaborative beamforming than for a noncollaborative (multi-hop) network. An example for this is shown in Figure 22.16.

Figure 22.16 Improved connectivity through use of cooperative clusters: collaborative beamforming of the cluster in the center increases the possible range. A denotes the required range of radio coverage to achieve connectivity with high probability; α is the path loss exponent.

Another case where clustering of nodes comes in handy is in networks where the number of hops is restricted (e.g., to two hops), but there are still a large number of nodes. In that case, it is required to find suitable cooperation nodes; in other words, which nodes should be "paired up" for the forwarding of a message. Choosing such node pairs can be considered a special case of so-called "matching problems on graphs," for which there is a rich literature in computer science and operations research. Particular examples include (i) minimal weighted matching, (ii) greedy matching, and (iii) random matching [Scaglione et al. 2006].

22.5.5 Scaling Laws

Cooperation between nodes leads to a change in the scaling of the network throughput as the node density increases. In Section 22.4.11, we saw that for multi-hop transmission, the throughput per node tends to zero as the node density increases. For cooperative communications, the feasible throughput per node is at least constant; in other words, the aggregate network throughput increases linearly. More precisely, it has been shown that the network throughput cannot increase faster than $N \log(N)$; and furthermore there is a known, constructive scheme that achieves a network throughput that scales as N.

This well-scaling scheme is hierarchical cooperation, which as its basic building block contains a three-phase cooperation scheme based on clustering:

1. In the first phase, the source node transmits the information to surrounding nodes. To be more precise, we divide the area containing nodes into cells, and the source node sends the information to the nodes located in the cell. Note that by application of the cellular principle, and an appropriate reuse distance, transmission from source node to surrounding nodes can happen in many cells in parallel. The source then divides the information into M blocks, and sends one such block to a particular node in the cluster (this does not exploit the broadcast effect).[14]
2. In the second phase, the cluster of nodes performs MIMO transmission to the cell (cluster) in which the destination node is located. Each node independently encodes the information block it received in the first phase, thus providing (distributed) spatial multiplexing. The nodes in the receiving cluster quantize the received signal.
3. In the third phase, the nodes in the receiving cluster send this quantized information within the cluster; by appropriate decoding of the spatial-multiplexing signal, a node can recover the original information.

This three-phase cooperation scheme can now be applied in a recursive manner, if distances need to be covered that are larger than what can be achieved (under given power constraints) with a single application of the three-phase scheme. The recursion starts with a small area, and is applied over consecutively larger areas until it can encompass the whole network area (see Figure 22.17).

22.6 Applications

Relaying and multi-hopping can be applied either in an infrastructure-based (cellular) setting, facilitating the communication between a BS and an MS, or they are an integral part of ad hoc networks.

22.6.1 Dedicated Relays

Dedicated Relay Stations (RSs) are mostly used in the context of cellular networks. The introduction of the RS might be beneficial in one or more of the following respects:

---

[14] Some special cases arise when the source node and the destination node are in adjacent cells; for details see [Ozgur et al. 2007].

- *Increase of coverage area*: since the relay improves the SNR, MSs that are far away from the BS can still receive a decodable signal. This helps to extend the effective radius of a cell, or eliminate "coverage holes," i.e., provide coverage in areas within a cell that due to the peculiarities of the topography are not covered by the BS. Dedicated RSs are usually placed at locations where they have good connection to the BS, e.g., on rooftops. This allows the RS to receive the signal with good quality, and forward it (after amplifying and/or "cleaning up" by decoding and re-encoding) to the destination; in either case, the signal arriving at the MS has improved SNR.
- *Improvement of indoor coverage*: similarly, relays can help to improve the indoor coverage. While coverage on the streets of urban/metropolitan areas is usually very good, indoor coverage on the same streets is often spotty or nonexistent due to the additional path loss suffered by signals when they penetrate into the building. Thus, relays (which are usually placed close to the building perimeter) often are necessary to fully cover the inside of office or residential buildings. In a related application, relays can be used to improve connections to the inside of trains, buses, and other vehicles: a single relay can connect to the BS, using complex signal processing to compensate for the high Doppler shifts encountered, e.g., in high-speed trains. The connection from the relay to the MSs inside the train, on the other hand, is fairly simple.
- *Increase of reliability*: by improving the SNR, relays help to improve the reliability. Furthermore, relays can add diversity (see Section 22.2), and thus increase reliability.
- *Increase of throughput*: if both the BS-RS and RS-MS links have good SNR, then higher throughput can be achieved, e.g., by using a higher order modulation alphabet and higher coding rate. However, we have to keep in mind that most relays lose rate because of the half-duplexing constraint, as discussed in Section 22.2. The tradeoff between the duplexing loss and the gains in per-link data rate determine whether a relay can help to increase the capacity and throughput of a cellular system or not. A different way of enhancing throughput is the use of relays to redirect traffic from overloaded BSs to less-congested BSs. Imagine a situation where a cell is temporarily overloaded (e.g., due to a large number of people congregating in that cell because of a special event). Then calls from some of the MSs can be connected via relays to BSs in neighboring cells that are momentarily underutilized.

Many of the practical implementation issues of dedicated RSs are related to the MAC-layer protocol and the control information that is being transmitted. In the downlink of *transparent relays*, the RS receives the signal from the BS and repeats it (with exactly the same control information); the uplink operates analogously. Care must be taken that the retransmission is consistent with the

Figure 22.17 Principle of hierarchical cooperation.
Reproduced from Ozgur et al. [2007] © IEEE.

timing and frequency information contained in the associated control information. In *nontransparent relays*, the RS adds control information of its own, and thus appears to the MS like a distinct BS.

While, in principle, multiple hops through a network are possible in infrastructure-based systems, this is rarely used in practice. Two-hop relaying (i.e., use of a single relay) is by far the most common situation. Of course, one physical relay can serve multiple MSs.

22.6.2   Relaying and User Cooperation in Ad hoc Networks

A wireless ad hoc network is a decentralized wireless network that does not use fixed infrastructure (BSs, etc.). Instead, each node can function as source, relay, and destination, depending on the requirements of particular data packets. In other words, "all nodes are created equal." The discussion of routing and resource allocation, as expounded in Section 22.4 and 22.5, is tuned to the needs of ad hoc networks (as mentioned above, RSs in cellular networks usually use two-hop relaying, so that routing does not occur, and optimum resource allocation can be done according to the equations in Section 22.2). The use of multiple nodes increases the reliability (there is no single point of failure) and reduces the costs.

Ad hoc networks also form the basis of sensor networks. Sensor networks, in general, are networks of nodes that have the ability to sense physical data (temperature, pressure, but also images), and communicate them to a data sink; this data sink often is a computer–human interface, or an automated control/monitoring system. Since most sensing nodes cannot reach the data sink in a single hop, data packets have to be relayed by other nodes in the network, just like in ad hoc networks. However, there are also some crucial differences:

- The data sink is (usually) the same, even though the source nodes differ.
- The information obtained from the different sensing nodes is correlated. For example, temperature-sensing nodes in one room will record very similar temperatures. For this reason, data aggregation and compression can be performed during the data-forwarding action.

22.7   Network Coding

The basic principle of network coding is that nodes in the network form combinations of messages that they receive, and forward those combinations; the destination then recovers the original messages from different combinations it receives. Thus, network coding can be seen as the epitome of collaboration: not only the resources (power, airtime) are shared between the nodes but also the messages.

Network coding turns out to be useful mainly for multicast situations, i.e., where multiple sources transmit messages, and *all* nodes want to learn those messages. For this case, the key result of network coding information theory says that if a network – without network coding – can provide a certain rate to each RX in isolation, then with an appropriate choice of network code, it can support all RXs at that rate simultaneously.

22.7.1   Two-Way Relaying

The simplest example of a network code is the bidirectional relay. Consider the situation sketched in Figure 22.18. There are two messages *a* and *b*, which are to be exchanged between nodes *A* and *B* via relay node *R*. The conventional way of relaying uses, e.g., TDMA to separate the messages, and thus requires four packet durations (time slots) for the exchange: (i) in slot 1, node *A* sends message *a* to the relay, (ii) in slot 2, node *B* sends message *b* to the relay, (iii) in slot 3, the relay sends message *a* to node *B* (though, strictly speaking, this transmission is a broadcast – in a

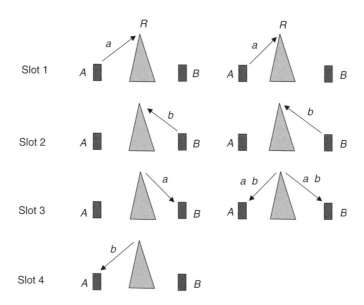

Figure 22.18   Bidirectional relaying: conventional method (left side) and network coding (right side).

wireless setting, a relay cannot help but send messages to multiple nodes), (iv) in slot 4, the relay sends message b to node A.[15]

A more efficient approach is the following: slots (i) and (ii) are as above, i.e., nodes A and B separately send their messages to the relay. However, in the third slot, the relay broadcasts the *sum* of the two messages, $s = a + b$. Since node A already knows message a, it can easily determine message b from the sum signal, $b = s - a$. Similarly, node B can determine message a from the sum signal s. With this approach, we have improved the spectral efficiency of the transmission: we now need only three timeslots instead of four to relay two messages. Note that the sum signal s really is a symbol-by-symbol summation of the messages, not a concatenation of the two messages, and thus has the same length as, e.g., the individual message a.

In the above (oversimplified) example, we implicitly assumed that the relay sums the complex modulation symbols of the two messages. This approach leads to a higher power value of the sum signal, and is thus undesirable. A better approach is for the relay to demodulate the messages a and b, and then perform a modulo-2 addition (XOR) of the information bits. It is this approach that is generally used for network coding. Other approaches include the use of novel modulation constellations onto which received complex symbol combinations are mapped.

22.7.2   Basics of Network Coding

Network coding in more complicated networks is similar in spirit to the above two-way relaying. Essentially, each node creates a linear combination of the incoming signals (i.e., analogously to the addition of signals in two-way relaying), and forwards them to other nodes. The linear combination is actually done on a finite field (i.e., using modulo-additions), similar to the above example where additions were interpreted as XOR (i.e., additions in Galois-Field 2). In order to reconstruct the signal intended for it, a destination node has to receive a sufficient number of *linearly independent combinations* of packets.

---

[15] Of course, slot 2 and 3, or slot 3 and 4, can be interchanged.

Another classical example of network coding is shown in Figure 22.19. Two messages, a and b, are to be transmitted to two receiving nodes; all links are assumed to be interference-free (i.e., no broadcast advantage). Normally, the link in the middle would constitute a bottleneck: it lies on the only path for message a to get from source A to the destination on the right, and similarly that link lies on the only path for message b from source B to the destination on the left. One thus might think that this link supports only half the data rate of the other links for each message, and thus limits the overall data rate. However, in network coding, we only transmit the signal $a \oplus b$ over this link, which requires the same rate as the rate on all other links. The source node on the left can then gets message a directly from its link to node A, and recovers message b from $a \oplus b$ and message a. Similarly, the destination node on the right can recover both messages b and a.

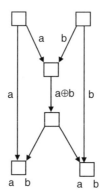

Figure 22.19   Multicast in a butterfly network with network coding.

After these simple examples, we turn to a more detailed mathematical description. Consider a set of N data packets $\mathbf{Z}^{(n)}$, $n = 1, ...N$, generated by one or several sources. A node generates a new message from multiple received packets as

$$\mathbf{X} = \sum_n g_n \mathbf{Z}^{(n)}. \tag{22.44}$$

Remember that the addition is done on a symbol-by-symbol basis, so that for the k-th symbol of the message, $X_k = \sum g_n Z_k^{(n)}$ The coefficients g_n depend on the path over which the packet reaches the considered node, and are summarized in the local *encoding vector* $\mathbf{g}$.

For the decoding of the original messages, we need $M \geq N$ independent linear packet combinations $\mathbf{X}^{(m)}$, and knowledge of the (global) encoding vectors. From those, we can obtain, e.g., by Gaussian elimination, the original $\mathbf{Z}^{(n)}$. If not all combinations $\mathbf{X}^{(m)}$ are linearly independent, we need more observations of packet combinations. Whether the $\mathbf{X}^{(m)}$ are independent is determined by the (global) encoding matrix $\mathbf{G}$. Fortunately, it can be shown that if the encoding vectors are chosen randomly, then with high probability the $\mathbf{X}$ are linearly independent for large field sizes. Thus, there is no need for a planning of the codevectors in the network, and the codevectors can be chosen in a decentralized manner.

Commonly, a forwarded data packet contains the encoding vector together with the information vector $\mathbf{X}$. Alternatively, the set of encoding vectors can be designed and made known to the whole network. The former approach has the advantage that no global information has to be distributed through the network, and temporal changes in the network structure do not constitute a problem. Also note that the encoding process can be either done on the original packets $\mathbf{Z}^{(n)}$, or on the already-encoded packets $\mathbf{X}$.

Network coding provides considerable robustness for the transmission of packets through a network. As long as the destination nodes correctly receive a sufficient number of linearly independent combinations of packets, they can do a decoding – irrespective of which packets get lost or irreparably damaged in transit.

22.7.3 Applications to Wireless Systems

Despite the theoretical promise of network coding, its application to practical wireless systems is not straightforward. Firstly, the largest benefits are achieved in multicasting scenarios (remember that the main theorem above showed the optimality of network coding for the case that all nodes want to learn all messages). However, most wireless traffic today is unicast (single-source to single RX), for which network coding is not necessarily optimal. Secondly, most of the current network coding theory ignores not only noise but also interference. In other words, in an ad hoc network, the zero-interference assumption is fulfilled approximately if only one node that is adjacent to an intended receiving node can transmit at one time; the other adjacent nodes have to be silent (otherwise, the Signal-to-Interference Ratio (SIR) at the receiving node would be too bad for proper packet reception). But such an approach reduces the spectral efficiency of the network and requires additional overhead for the coordination of the transmission times.

An ingenious solution to this latter problem was recently proposed as *Compute and Forward* (*CAF*). Instead of having the node receive two packets separately and do a linear combination in hardware, CAF exploits the fact that a wireless channel *inherently* performs a linear combination of on-air signals, i.e., when two source nodes transmit simultaneously, the received signal is the sum of the two signals weighted with their respective complex channel gains. Combining this fact with appropriate lattice codes, a more effective system can be designed.

22.7.4 Interference Alignment

Another revolutionary idea that emerged in 2008 is *interference alignment*. With this approach, it becomes possible to provide each (unicast) user with *half* the capacity it would obtain in an interference-free environment, irrespective of the number of users. In other words, the sum capacity of a network increases linearly with the number of users, while with a conventional multiple-access approach, the capacity stays constant (i.e., each of the N users is allowed a rate $1/N$). A number of different approaches have been proposed for implementing such interference alignment. The simplest, and intuitively clearest, exploits time variations of the wireless propagation channel: characterize the channels between the N TXs and the N RXs with a $N \times N$ matrix (transfer function matrix, analogous to a MIMO system):

$$\begin{pmatrix} h_{11} & h_{12} & h_{13}. & . & . & h_{1N} \\ h_{21} & h_{22} & h_{23} & . & . & h_{2N} \\ . & . & . & . & . & . \\ h_{N1} & h_{N2} & & & & h_{NN} \end{pmatrix} \tag{22.45}$$

The signals transmitted when the channel has one particular realization are repeated when the channel is in its *complementary* state:

$$\begin{pmatrix} h_{11} & -h_{12} & -h_{13} & . & . & -h_{1N} \\ -h_{21} & h_{22} & -h_{23} & . & . & -h_{2N} \\ . & . & . & . & . & . \\ -h_{N1} & -h_{N2} & & & & h_{NN} \end{pmatrix} \tag{22.46}$$

The RXs then just have to add up the signals from those two transmissions, to get signals that are effectively transmitted over a diagonal (i.e., interference-free) channel

$$
\begin{pmatrix}
h_{11} & 0 & 0 & . & . & 0 \\
0 & h_{22} & 0 & . & . & 0 \\
 & . & . & . & . & . \\
0 & 0 & & & & h_{NN}
\end{pmatrix}.
\tag{22.47}
$$

The signal for each user is thus interference-free (allowing transmission with capacity of the noise-only channel). However, due to the required signal repetition, the capacity is decreased by a factor 2.

As mentioned above, the concept can be implemented with very little hardware effort. However, it can lead to large delays in the signal transmission, because the TX has to wait until the channel takes on the complementary state. The slower the temporal channel variations, the longer that delay can be. There are other forms of interference alignment, which can achieve lower latency at the price of higher complexity.

Further Reading

The preceding chapter covers a wide variety of topics, most of which are – at the time of this writing (2009) – very much in flux, and among the hottest research topics in wireless communications. The literature list below is thus not only necessarily incomplete but also likely to be soon outdated.

General overviews of relaying and cooperative communications are found in the monograph [Liu et al. 2009] and the review papers [Hong et al. 2007], [Stankovic et al. 2006].

Repeaters, which are a primitive form of relays, have been used for almost 100 years. Theoretically sound studies started, however, with the work of van der Meulen [1971] that introduced the relay channel and Cover and El Gamal [1979] which explored its information-theoretic limits in greater detail. A number of relaying protocols were introduced, and their performance analyzed in Kramer et al. [2005], Laneman and Wornell [2003], and Laneman et al. [2004]. A summary of all these facts can be found in the monograph [Kramer et al. 2007]. SCxF and NDxF have more complicated forms for the capacity equations; they are discussed in Nabar et al. [2004] (called protocols III and I, respectively). Optimum power allocations for the two-hop case with multiple relays and average CSIT only is derived for various relaying methods in Annavajjala et al. [2007].

For relaying with multiple parallel relays, relay selection was analyzed in Bletsas et al. [2006, 2007]. Practical aspects of distributed beamforming, including the critical synchronization issues, are discussed in Mudumbai et al. [2009]. The impact of training overhead on beamforming is analyzed in Madan et al. [2009]. Transmission on orthogonal channels and distributed space–time coding are analyzed in detail in Laneman and Wornell [2003]. Various forms of user cooperation were introduced in Sendonaris et al. [2003], Hunter et al. [2006], and Nosratinia et al. [2004]. Fountain codes for relaying are discussed in Molisch et al. [2007].

There is a rich literature on routing in multi-hop networks. Many of the basics, which relate to computer networks in general, are explained in textbooks of computer science and operations research literature, e.g., Peterson and Davie [2003], though the specifics of wireless routing are scattered in various research papers. A taxonomy of various protocols is given in Boukerche et al. [2009]. Source routing is discussed in Johnson et al. [2001]. AODV routing was introduced in Perkins and Royer [1999]. Geography-based routing is discussed in Stojmenovic [2002]. The spray-and-wait algorithm was proposed in Spyropoulos et al. [2005] and directed diffusion in Intanagonwiwat et al. [2003]. Joint routing and resource allocation with the backpressure algorithm are described in tutorial form in Georgiadis et al. [2006] and backpressure with average-power optimization is treated in Neely [2006]. A practical implementation of the backpressure algorithm is described in [Moeller et al. 2010]. A large number of convex-optimization-based papers exist for this problem as well (see, e.g., Cruz and Santhanam [2003], Chiang [2005]).

A Bellman–Ford approach to anypath routing is developed in Lott and Teneketzis [2006]. Anypath routing is described in Laufer et al. [2009], and backpressure with anypath is in Neely and Urgaonkar [2009]. Routing on cooperative networks for unicast is discussed in Khandani et al. [2003], Chen et al. [2005]. Cooperative routing for multicast with energy accumulation is analyzed in Maric and Yates [2004]. Routing in networks with information accumulation is derived in Draper et al. [2008]. Routing in networks with selfish nodes is described in Han and Poor [2009].

The scaling laws for (noncooperative) multi-hop networks were derived in the landmark paper of Gupta and Kumar [2000]. The constructive method for achieving higher throughput was proposed in Ozgur et al. [2007]. For network coding, the primer of Fragouli et al. [2005] gives an excellent introduction. CAF was proposed in Nazer and Gastpar [2008]. Interference alignment was proposed by Cadambe and Jafar [2008]; the most easily implementable form, described in this book, is in Nazer et al. [2009].

For updates and errata for this chapter, see wides.usc.edu/teaching/textbook

23

Video Coding

Anthony Vetro

Mitsubishi Electric Research Labs, Cambridge, MA, USA

23.1 Introduction

Digital video is being communicated over wireless networks at an increasing rate. A major source of digital video is from broadcast services, which are now being offered in a number of countries around the world. This includes a wide range of services from high-definition video for home entertainment systems to mobile video for cellular phones and smartphone devices. Video is also streamed over wireless networks within the home and office, e.g., from a television receiver to a flat-panel display and from file servers to laptops. These different use case scenarios will generally be supported by a specific wireless infrastructure and associated communication techniques, each providing their own quality of service. An important task for service and network providers is to match their offerings with the capabilities of their network. When video is involved, they must consider the overall system design including the characteristics of the digital video signal, the compression scheme, and its capabilities, as well as the robustness to transmission errors.

In contrast to other digital signals such as speech, digital video requires a relatively large bandwidth. For instance, a typical high-definition video signal has a raw data rate in the range of 600–800 Mbit/s depending on the specific spatial resolution and frame rate. To transmit video in a practical manner over existing networks, compression of the digital video signal is necessary. This section introduces the digital video representation and describes the fundamental compression architecture. The major components for video compression are elaborated on in subsequent sections. A summary of major video coding standards is then provided. The chapter concludes with sections devoted to robust video transmission and video streaming.

23.1.1 Digital Video Representation and Formats

Video is a multidimensional signal that represents emitted and reflected light intensity from objects in a scene over time. The samples of a digital video signal are typically recorded by an imaging system, such as a camera. The capture system will typically record two-dimensional (2D) sample values in a particular color space, which correspond to a specific set of wavelengths, at discrete instants of time. The number of samples at each time instant is referred to as the spatial resolution, while the rate at which samples are taken is referred to as the frame rate.

Wireless Communications, Second Edition   Andreas F. Molisch
© 2011 John Wiley & Sons, Ltd

It is well established that most colors can be produced with three properly chosen color primaries. The RGB primary, which includes red, green, and blue colors, is perhaps the most popular set for both capture and display. However, video coding and transmission systems utilize a color coordinate system that is based on luminance (Y) and chrominance components (Cb and Cr). One reason is historical: in older analog systems, the luminance signal was backward compatible with the monochrome black-and-white television signals. The other advantage of this color space for transmission is that some of the information in the chrominance samples could be discarded since the human visual system is less sensitive to the color components relative to the luminance. There is a well-defined linear relation between the RGB and YCbCr color spaces.

The color sampling formats specified by the ITU-R recommendation BT.601 [ITU 1998] are illustrated in Figure 23.1. These formats are utilized by all international video coding standards. In the 4:4:4 color sampling format, the luminance and chrominance samples have the same resolution. In the 4:2:2 and 4:2:0 color sampling formats, the chrominance samples have a reduced resolution relative to the luminance samples. The 4:4:4 and 4:2:2 color sampling formats are primarily used in studio and production environments, where it is important to maintain the fidelity of the color components. For distribution to the consumer, current systems utilize the 4:2:0 sampling format. As display technology and distribution capabilities increase, it is expected that video with higher color fidelity will be transmitted to the consumer, e.g., in a 4:4:4 color sampling format.

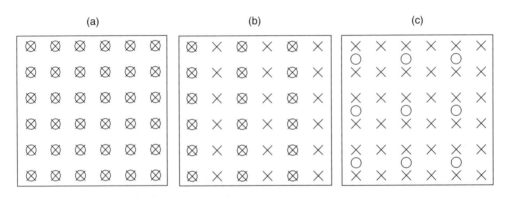

Figure 23.1 Color sampling structures with the location of luminance samples represented by a ∘ and the location of chrominance represented by a ×. 4:4:4 color sampling in which chrominance samples have the same resolution as luminance samples (a). 4:2:2 color sampling in which chrominance samples have half the horizontal resolution as luminance samples (b). 4:2:0 color sampling in which chrominance samples have half the horizontal and vertical resolution as luminance samples (c).

23.1.2 Video Coding Architecture

There is a great deal of redundancy in a video signal, in both spatial and temporal dimensions. Video compression schemes exploit that redundancy and determine a compact binary representation. The most popular and effective scheme for video compression is known as a block-based hybrid video coder, which uses a combination of both transform coding and temporal prediction to code the video signal. Since this scheme has been adopted by all international video coding standards to date, this chapter will focus exclusively on this scheme. A common architecture for the block-based hybrid video coder is shown in Figure 23.2.

There are several types of pictures that are commonly referred to in the video coding literature. Intracoded pictures, or I-pictures, refer to pictures that are coded without reference to any other picture in the video. Pixels within the I-picture may still be predicted from other neighboring pixels

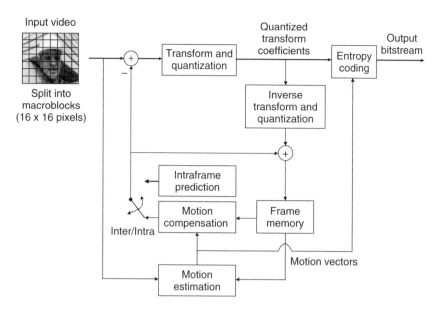

Figure 23.2   Block diagram of a typical video encoder.

to exploit spatial redundancy. Due to their independence, they are often inserted periodically in the bitstream to facilitate random access since the decoding can start from these pictures. Another type of picture is referred to as a P-picture, which utilizes a unidirectional temporal prediction. Since pixels in neighboring pictures are correlated with the current picture, temporal prediction is a very effective means to reduce the energy of the signal to be coded. Biprediction, or the prediction of the current picture from two references, is another effective means of temporal prediction. Pictures that use biprediction are referred to as B-pictures. An example prediction structure is shown in Figure 23.3, where P-pictures are predicted from the previous I/P-picture and the B-pictures are predicted from neighboring I/P-pictures.

Figure 23.3   Sample video prediction structure.

The first step in the video coding process is to divide each picture into fixed size blocks of pixels, or macroblocks (MBs). The most common size of an MB is 16×16. Then, on a block basis, a prediction is performed. For I-pictures, an intraframe prediction is utilized, where neighboring pixels are used to predict pixels in the current block. For P- and B-pictures, a motion-compensated temporal prediction is performed. The result of the prediction is a residual signal, which is then subject to a transform and quantization. The quantized transform coefficients are then entropy coded to produce the resulting video bitstream. Since the motion-compensated prediction is based on previously coded pictures that must also be available at the decoder, the inverse operations to reconstruct those reference pictures are also performed. These reference pictures are stored in a frame memory.

The remainder of this chapter describes the major components of a video coder in more detail, and discusses the particular set of coding tools that are supported by different video coding standards. Extensions to accommodate scalable and Multiview Video Coding (MVC) are also briefly described. Finally, a brief review of error-resilient video transmission and video streaming is given.

23.2   Transform and Quantization

In the previous section, a basic video coding architecture was introduced. In the following, further discussion on the transform and quantization modules are discussed. We specifically focus on block-based transforms and the Discrete Cosine Transform (DCT) in particular, as well as scalar quantization, since these methods have proven to be especially effective in modern image and video codecs.

23.2.1   Discrete Cosine Transform

The use of a transform in an image or video codec attempts to modify the input signal in two ways: (i) compact the energy of the signal and (ii) decorrelate the signal. The energy compaction property generally leads to a set of transform coefficients with few coefficients having a large magnitude and many coefficients becoming small or zero, which has obvious benefits for compression. The decorrelation property is advantageous since each coefficient could be quantized independently in an optimal manner, e.g., to either minimize Mean Square Error (MSE) or according to a perceptual sensitivity.

The transform of an image or video block represents the signal as a linear combination of basis functions. The coefficient that corresponds to a particular basis function denotes the weighting or contribution of that function in the overall signal representation.

Let $x_n, n = 0, 1, \ldots, N - 1$, be an input block of length N. Further, assume that x is a Markov-1 process with correlation coefficient, ρ. The covariance matrix of this random process is then given by

$$R_x(i, j) = \rho^{|i-j|} \tag{23.1}$$

The input signal can be transformed by matrix A to yield an output signal in the transform-domain, $y = Ax$, which has a covariance matrix given by

$$R_y = A R_x A^T \tag{23.2}$$

The gain of the transform, which is a typical measure of transform performance relative to Pulse Code Modulation (PCM), is defined as

$$G_T = \frac{\frac{1}{N} \sum_{i=0}^{N-1} \sigma_{y_i}^2}{(\prod_{i=0}^{N-1} \sigma_{y_i}^2) 1/N} \tag{23.3}$$

This metric is the ratio of the arithmetic mean of the transform coefficient variances to the geometric mean of these variances, and essentially measures the reduction in MSE that the transform provides relative to PCM.

A transform that maximizes the transform coding gain is the Karhunen Loève Transform (KLT). Its basis functions are the eigenvectors of the covariance matrix of the input signal, R_x. The transformed signal not only achieves optimum energy concentration, but the transform coefficients are also fully decorrelated, i.e., the covariance matrix R_y is diagonal. Despite these excellent properties of the KLT, there are notable disadvantages that have made it unusable for image and video coding in practice. Most importantly, it is dependent on the signal statistics and can only be computed for stationary sources with known covariance matrix. Also, it is not a separable transform, so additional complexity would be incurred when applying it to 2D image blocks.

Fortunately, the DCT has been shown to perform very close to the KLT for natural image and video data and has a basis function defined as follows [Rao and Yip 1990]:

$$a_{i,j} = \alpha_i \cos \frac{(2j+1)\pi i}{2N} \tag{23.4}$$

for $i, j = 0, 1, \ldots, N - 1$, with $\alpha_0 = \sqrt{1/N}$ and $\alpha_i = \sqrt{2/N}$ for $i \neq 0$. An illustration of the 2D basis function of the DCT is given in Figure 23.4. The DCT achieves similar transform gain to that of the KLT, while not being dependent on signal statistics.

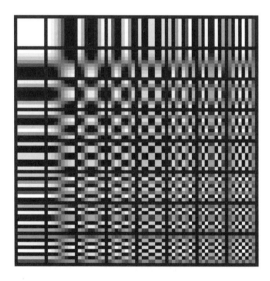

Figure 23.4 Basis functions of the 2D 8×8 DCT.

The 8-point DCT has been specified as the transform in a number of popular image and video coding standards including H.261, Joint Photographic Experts Group (JPEG), Moving Pictures Expert Group-1 (MPEG-1), and MPEG-2. There exist various factorizations and fixed-point implementations of the DCT [Arai et al. 1988], [Reznik et al. 2007]. The more recent MPEG-4/H.264 Advanced Video Coding (AVC) standard employs a 4-point transform as well as an 8-point transform, which are essentially scaled integer approximation of the DCT [Malvar et al. 2003].

23.2.2 Scalar Quantization

In contrast to the transformation process, which is generally invertible and does not incur loss, quantization is inherently a lossy process and introduces distortion into the reconstructed signal. The basics of scalar quantization have already been introduced in the chapter on speech coding. A very brief review is provided here for completeness and to introduce some specifics on quantization related to video coding.

In most video coding schemes, quantization is applied to the transform coefficients. The dynamic range of the transform coefficients will vary with the bit depth of the input video and the length of the transform[1]. Assuming transform coefficients in the range $[t_{min}, t_{max}]$ with dynamic range B, the quantization process will assign a quantization index $i = Q(x)$ to an input value x subject to a quantization function $Q(\cdot)$. A sample quantization function is shown in Figure 23.5, which is defined by the number of quantization indices, i.e., reconstruction values, and the boundary of each interval.

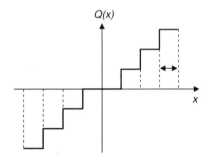

Figure 23.5   Illustration of scalar quantizer with uniform quantization step size Δ.

Consider a uniform quantizer with L quantization indices over the range of possible values and equal distance between adjacent boundary values, and let the distance of each interval be denoted by $\Delta = B/L$. With a fixed-length binary representation, each quantization index requires $R = \lceil \log_2 L \rceil$ bits. For a uniformly distributed source, it can be shown that the variance of quantization errors is equal to $\sigma_q^2 = \Delta^2/12 = \sigma_x^2 2^{-2R}$. The Signal-to-Noise Ratio (SNR) of the quantizer can then be computed as

$$SNR = 10 \log_{10} \frac{\sigma_x^2}{\sigma_q^2} = 20 \log_{10} 2R = 6.02R \qquad (23.5)$$

which reveals a classic result in coding theory that every additional bit in a uniform quantizer yields a 6.02-dB gain in SNR for a uniform source.

The quantization scheme utilized in most state-of-the-art video codecs are essentially uniform quantizers with some minor adjustments. One such adjustment adds a *dead-zone* which expands the interval of the quantization index that corresponds to a zero-level reconstruction. This has the effect of forcing more transform coefficients to zero. Second, a weighting matrix is often used to account for the fact that the human visual system is less sensitive to higher frequencies than lower frequencies. In this way, a courser quantization step size is applied to higher frequencies. Most standards allow the quantization matrix used for coding a picture to be signaled as part of the bitstream. Finally, a quantization offset could be utilized to shift the location of the reconstruction value within the quantization interval, i.e., different than the center of that interval. The reconstruction offset has shown to provide benefits when quantizing sources with a Laplacian distribution, which are typical of transform coefficients in video coding.

---

[1] For 8-bit image samples and 8-point DCT, the transform coefficients have a dynamic range of 11 bits with coefficient values between $[-1024, 1023]$. When the input to the transform is a prediction signal that may be in the range $[-255, 255]$, the dynamic range of the transform coefficients increases to 12 bits.

23.3   Prediction

Prediction is perhaps one of the most powerful and effective means of reducing redundancy in
a video signal. This section briefly introduces two commonly used approaches. Intraprediction
refers to the prediction of data from data within the same picture, while interprediction or motion-
compensated prediction refers to temporal predictions that utilize data from neighboring pictures.

23.3.1   Intraframe Prediction

Intracoded pictures are important for random access, which requires that they be decoded independently of any other pictures, i.e., there are no temporal dependencies. To code such pictures more
effectively, intraprediction techniques that make use of neighboring data within the current picture
are typically utilized to reduce the signal energy, thereby reducing the number of bits required to
represent that portion of the signal. There exist both transform-domain prediction schemes as well
as spatial domain prediction schemes. A few commonly used block-based intraprediction schemes
are shown in Figure 23.6.

Figure 23.6   Various types of block-based intraprediction schemes.

For natural images and video, the average value of pixels within a block are highly correlated
within a given picture. Therefore, an intuitive and effective way to exploit this correlation is in the
transform domain. When a block-based transform, such as the DCT discussed in Section 23.2.1,
is applied to a block in the picture, a DC (zero frequency) or average component of the block
is computed. As shown in Figure 23.6, the DC component can be predicted from the preceding
block in the same row. This prediction results in a set of residual DC components, which follow a
Laplacian-like distribution around zero. Without prediction, the distribution of DC components for
natural images is closer to a uniform distribution. The DC prediction has the effect of reducing the
entropy, thereby reducing the required bits to represent that signal component.

Rather than always predicting from the preceding block, one can also adaptively select either
the DC value of the left block or above block as a reference. A method to adaptively select the
direction would then be needed. For example, the DC prediction direction could be selected based
on a comparison of the horizontal and vertical DC gradients around the block to be coded.

For many natural images, the low-frequency components of neighboring transformed blocks are also correlated. As shown in Figure 23.6, the first rows and first columns of a transformed block, which represent low-frequency AC (non-zero frequency) coefficients, could also be predicted from neighboring blocks. When combined with adaptive DC prediction, i.e., prediction from either the left or above block, the prediction of AC coefficients would typically use the same direction. In this way, AC prediction is applied only to the rows or the columns of any block.

Another common method of intraprediction is to apply prediction in the spatial domain. Optimal linear predictors could be derived based on the correlation of the source, but in practice a fixed set of predictors is typically used. The latest video coding standards employ directional predictors, where multiple prediction directions could be selected adaptively on a block basis. Two sample prediction modes for spatial prediction are shown in Figure 23.6. In these examples, the neighboring pixels from nearby blocks are used to predict the pixels in the current block according to the chosen direction. The are many variations and different ways to form such predictors. When presented with such options, a method to determine the optimal mode is needed and the selected mode must then be signaled as part of the bitstream. This overhead is usually what limits one from devising too many different prediction modes to select from.

23.3.2 Interframe Prediction

The previous subsection discussed methods to predict data within a frame of the video. However, video also has a very high correlation in the temporal direction. In contrast to intraframe prediction in which both transform-domain and spatial domain predictors have shown to be effective, most interframe prediction techniques tend to focus on spatial domain methods.

The simplest type of temporal or interframe predictor is to use the data from a colocated block in the previous frame to predict data in the current frame. For stationary scenes or parts of the scene, this is actually quite effective. It is so effective and commonly used in practice that special signaling is used to indicate that a block to be coded in the current frame can simply be copied from a colocated position in a reference frame. The corresponding coding mode is referred to as a *skip* mode.

Most natural video scenes do have motion, including global motion that would result from camera pans and zooms as well as local motion that results from objects moving within the scene. Under such conditions, motion-compensated temporal prediction becomes an effective means to predict frames in the video from neighboring frames. The basic concept of motion-compensated prediction is to determine the displacement of pixels in one frame, which we will refer to as the current frame, relative to the set of pixels in a neighboring frame, which we refer to as reference frame. Note that there may be multiple reference frames used for the prediction of the current frame.

Determining the way that pixel intensities move from one time instant to another has been a rich area of study among the video coding community. One can imagine trying to estimate the movement of each pixel, but this is an ill-defined problem since there could be many pixels that match the current pixel intensity in neighboring frames, especially considering regions of a scene with constant intensity. Two common approaches to get around this problem are (i) use regularization techniques to enforce smoothness constraints on the motion field, and (ii) assume that pixels within a small neighborhood have the same motion. One drawback of the first option is that there would be significant overhead in signaling the motion for each pixel. Due to the block-based nature of most video coding standards, the second option has thus become a more popular and practical approach for compression. Specifically, for any block of a frame, a translational motion model is assumed and a motion vector that indicates the 2D displacement is determined.

Motion vectors are determined through a block matching process, which is illustrated in Figure 23.7. There is a search region associated with each block in the current frame, from which the best match between the current block and candidate blocks in the search region of the

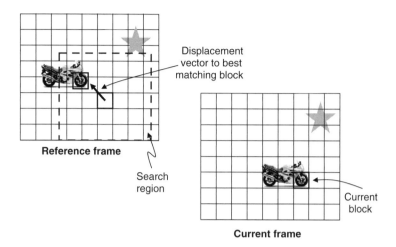

Figure 23.7   Illustration of block-based matching for interframe prediction.

reference frame are found. There is typically not a perfect match between the current block and the reference block, so the residual signal that represents the difference after prediction must also be coded to reconstruct the video. Additionally, the motion vector must be coded and signaled as part of the bitstream so that the prediction can be formed at the decoder based on previously decoded frames, which serve as the reference frame.

23.4   Entropy Coding

The entropy of a given source is a limit on the data compression rate that could be achieved. If all the symbols of the source data have equal probability, then a fixed-length binary representation for each symbol would achieve the lower limit on the rate. However, it is typical in video coding and other applications that the symbols have a nonuniform probability distribution. Therefore, a lower bit rate could be achieved by assigning shorter codewords to higher probability symbols and longer codewords to lower probability symbols. This is the basic principle of entropy coding, which is also referred to as Variable Length Coding (VLC).

VLC has two important properties: (i) the code should be uniquely decodable, and (ii) the code should be instantaneously decodable. The first property requires that there is only one possible set of source symbols that the codeword represents, while the second implies that no codeword is the prefix of any other codeword. Codes that satisfy this second property are referred to as prefix codes.

In the case of video coding, the symbols to be encoded include the quantized transform coefficients and any side information that is required to reconstruct the video signal, e.g., motion vectors, block coding modes, etc. It is noted that in contrast to quantization, which introduces distortion to the signal, entropy coding is a lossless process. Huffman coding and arithmetic coding are two popular entropy coding schemes that are widely used for video coding. The basic principles of each are described in the following subsections.

23.4.1   Huffman Coding

Huffman [1952] published an algorithm to construct an optimal prefix code for a given distribution of symbols [Huffman 1952]. Consider a finite alphabet source $\chi = x_1, x_2, \ldots, x_N$ with probabilities

$p(x_i)$. The optimal binary code should assign longer codewords to higher probability symbols and shorter codewords to those that appear more frequently. The procedure is outline as follows:

1. Let each symbol be a leaf node of a tree with an assigned probability, $p(x_i)$. Arrange the symbols so that their occurrence probabilities are in a decreasing order.
2. While there is more than one node,
   (a) identify the two nodes with smallest probabilities and arbitrarily assign a binary 1 and a binary 0 to these nodes;
   (b) merge the two nodes to create a new node with probability equal to the sum of those nodes.
3. The last remaining node is the root node. The codeword for each symbol can be traced from the root node to each leaf node.

An example of a Huffman code construction is shown in Figure 23.8. This example illustrates the successive assignment of 0 and 1 to each of the leaf nodes with lowest probability and merging of those nodes until the root node is reached. This example demonstrates that none of the resulting codewords are the prefix of any other codeword. The code shown in this example has an average length of 2.4 bits.

Figure 23.8   Example of Huffman coding based on probabilities of symbols to be coded.

A proof of the optimality of the Huffman code as well as other properties and examples of Huffman code constructions can be found in Cover and Thomas [2006]. It can be shown that the average length of the Huffman code, L_{avg}, satisfies the following condition, where $H(X)$ is the entropy of the source.

$$H(X) \le L_{avg} < H(X) + 1 \qquad (23.6)$$

One well-known drawback of Huffman codes is that when coding individual symbols, each symbol always requires a minimum of 1 bit. One way to overcome this is to consider the coding for blocks of symbols, i.e., a vector. In this way, each vector would be assigned a probability, which is a joint probability of each symbol in the block. The procedure to construct a code follows the same as that outlined above, only that the symbols represent vectors. Another way to code sets of symbols more efficiently is to condition them on a context. A simple example would be to condition on the previous sample, but more complex contexts are considered in practical video codecs.

23.4.2   Arithmetic Coding

It was noted in the previous section that Huffman coding could suffer from some coding loss relative to the optimal representation since each symbol must be coded with at least 1 bit. In an extreme case, consider an alphabet with two symbols, one with probability that is close to 1 and the other with probability close to 0. Since there is little uncertainty, the entropy of this source would be very small and close to 0. However, Huffman coding would still require 1 bit for each symbol, which shows that there is still further room for improvement. While blocks of symbols could be used to improve the coding efficiency, the complexity grows exponentially with the block length.

Rather than representing a symbol or block of symbols by a sequence of bits, arithmetic coding represents a sequence of symbols by a subinterval in the interval between 0 and 1, where the size of each subinterval is proportional to the probability of each symbol. Starting with an initial division, the first subinterval is selected based on the first symbol to be coded. This subinterval is then recursively divided based on each new symbol to be coded. A subinterval is represented by its lower and upper boundaries and a binary representation of those boundaries is used to code the sequence of symbols at each stage. In particular, when the most significant bit of the lower and upper boundaries are the same, that bit is written to the output. In this way, higher probability symbols will be associated with larger intervals, which require fewer bits to represent.

An example of the arithmetic coding process is shown in Figure 23.9. The source data includes three symbols, "A," "B," and "C," each with their respective probabilities given as $p(A) = 1/2$, $p(B) = 1/3$ and $p(C) = 1/6$. To begin the encoding process, the initial interval [0, 1] is first divided according to these probabilities. Assume we would like to code the sequence of symbols "ABAC." Since the first symbol is "A," the subinterval associated with this symbol [0, 1/2] is carried to the next stage and again subdivided according to the symbol probabilities. The next symbol to be coded is "B" so we take the second subinterval in the range [1/4, 5/12] and further subdivide it according to the probabilities. Following this process for each symbol in the sequence, we can arrive at the final subinterval that represents the sequence of symbols and its binary representation.

At first glance, it seems that arithmetic encoding would require extremely high precision to encode a long sequence of symbols. However, in practice, a fixed-point precision based on integer

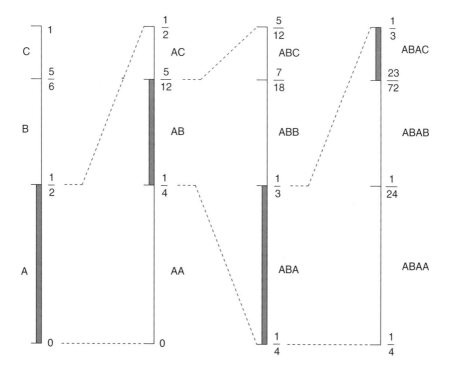

Figure 23.9 Example of arithmetic coding of the source sequence "ABAC" according to the probabilities of each symbol.

arithmetic can be maintained by renormalization of the upper and lower boundaries when bits are written out.

One advantage of arithmetic coding relative to Huffman coding is that it is incremental, i.e., additional symbols can be coded based on the code for the previous sequence of symbols. This property allows it to closely approach the entropy rate in practice since longer sequences can be effectively coded without maintaining large codebooks. Another merit is that it can easily adapt to changes in the statistics of the source data. It is only required that the encoder and decoder update their probabilities tables in a synchronized manner. In practical video coding schemes, context-based arithmetic coding is also used to improve the coding efficiency. In this case, a set of probability tables are maintained for each context. Due to the higher coding efficiency and capability to adapt better to the signal statistics, arithmetic coding is generally favored over Huffman coding for platforms that could tolerate the computational requirements.

23.5 Video Coding Standards

Since the early 1990s, a number of video coding standards have been developed to satisfy industry needs. These standards have been developed by two major international standardization organizations: the MPEG of International Standards Organization/International Electrotechnical Commission (ISO/IEC), and the Video Coding Expert Group (VCEG) of ITU-T, the Telecommunications Standardization Sector of the International Telecommunications Union ITU.

The video coding standards developed by ISO/IEC include MPEG-1, MPEG-2, and MPEG-4. The standards developed by ITU-T fall under the H.26x series of recommendations and include H.261, H.262, H.263, and H.264. It should be noted that Recommendation H.262 is the same as MPEG-2; this standard was jointly developed by both organizations. The most recent AVC standard, known as H.264 and MPEG-4 Part 10, was also developed through the Joint Video Team (JVT) including experts from both standard bodies. These standards have found many successful applications such as digital television broadcast, optical disc storage including CD, DVD and Blu-ray Disc, digital telephony, video streaming, and mobile video. In the following, we provide a short review of the main coding tools used by each standard.

The first version of H.261 was completed in 1990 with later amendments published in 1993 [ITU 1993] and was mainly designed for low-delay video conferencing over Integrated Services Digital Network (ISDN) lines with bit rates between 64 and 320 Kbps. It was the first standard to define the basic video coding architecture as shown in Figure 23.2. This standard utilized 16×16 MBs, 8×8 DCT with uniform quantization, and supported unidirectional forward motion-compensated prediction with integer-pixel precision. A VLC scheme known as run-level coding was used to code quantized transform coefficients. With run-level coding, the 2D transform coefficients are first scanned into a one-dimensional (1D) vector. Huffman coding is then applied to symbols that comprise a pair of numbers indicating the *run* of zeros followed by the *level* of the next nonzero coefficient. A special symbol is used to indicate the last nonzero coefficient in a block. H.261 also defined an optional loop filter that applied a low-pass filtering on the motion-compensated prediction to decrease prediction error and blocking artifacts at high compression ratios.

The MPEG-1 standard was completed in 1991 [MPEG 1]. The target application of MPEG-1 was digital storage media on CD-ROM at bit rates between 1 and 2 Mbps. MPEG-1 specified motion compensation with half-pixel accuracy as well as the use of B-frames and bidirectional prediction. The DC coefficients in MPEG-1 are predicted from the left neighbor.

MPEG-2 was completed in 1994, with later amendments in 2000 [MPEG 2]. The standard was developed jointly with ITU-T and is also known as H.262. MPEG-2 is an extension of MPEG-1 and allows for greater input format flexibility and higher data rates that include support for standard-definition and high-definition resolutions. Target bit rates are in the range of 4–30 Mbps. This standard is widely used for television broadcast and DVD applications. MPEG-2 adds coding tools that specifically support the efficient encoding of interlaced material. It also defined various modes of scalability, which are outlined briefly in the next section.

The first version of H.263 was completed in 1996 [ITU 1996] and it is based on the H.261 framework. It defined more computationally intensive and efficient algorithms to increase the coding performance in telecommunication applications. New technical features included advanced prediction, which supported overlapped block motion compensation and optional use of four motion vectors per MB, motion vector prediction to code motion vector data with less rate, and improved entropy coding that integrates the end of block symbol into the run-level coding. H.263 also allowed for arithmetic coding in place of Huffman coding.

A second version of H.263 referred to as H.263+ was approved in 1998. Several new optional features were added to improve coding efficiency. Most notably, advanced intraprediction, in which spatial prediction from neighboring blocks was employed, and an in-loop deblocking filter that is applied to block boundaries of 8×8 blocks of the reconstructed images that is used for reference. Additionally, significant improvements in error resilience were realized through a variety of tools including flexible resynchronization marker insertion, reference picture selection, data partitioning, reversible VLC, and header repetition.

A first version of MPEG-4 Part 2 was completed in 2000, with later editions in 2004 [MPEG 4]. It was the first object-based video coding standard and is designed to address the highly interactive multimedia applications. Specific profiles of MPEG-4 Part 2, targeting lower bit rate video, have been used for some mobile and Internet streaming applications. Among the new technical features are adaptive DC/AC prediction for intrablocks and quarter-pixel motion compensation. It also adds support for improved error resilience and generally has coding performance similar to that of H.263.

H.264 is the current state-of-the-art in video coding. The standard was developed jointly between MPEG and ITU-T and is also known as MPEG-4 Part 10 [ITU 2009]. Another name for this standard is Advanced Video Coding, or simply AVC. H.264 has significantly improved the coding performance over both MPEG-2 and H.263. The standard improves intraprediction with directionally adaptive spatial prediction, it defines a 4×4 and 8×8 integer transform that could be selected adaptively, there are more powerful motion-compensated prediction capabilities including support for various block partitioning schemes and multiple reference pictures, as well as a context-adaptive binary arithmetic coding scheme. H.264 is capable of achieving the same quality as prior standards such as MPEG with approximately half the bit rate. It is being deployed widely for television broadcast, Blu-ray Disc and mobile applications. For further details on the current state-of-the-art standard in video compression, interested readers are referred to the overview paper by Wiegand et al. [2003].

23.6   Layered Video Coding

23.6.1   Scalable Video Coding

The traditional dimensions of scalability include quality scalability, temporal scalability, and spatial scalability. A key objective of a scalable representation is to encode the video source signal once, then decode many times according to specific delivery, and receiver capabilities. Such functionality is highly desirable in any dynamic and heterogeneous communication environment, and especially for mobile video delivery. Scalable video coding typically incurs some loss relative to single nonlayered video coding. One challenge is to minimize this loss, another is to keep the complexity to a minimum.

Scalable video coding was first introduced in the MPEG-2 standard, and revisited in MPEG-4 Part 2. These scalable extensions were not very successful since the coding efficiency loss relative to nonlayered video was relatively high, and there was a notable increase in complexity to support such modes. The scalable video coding extension of the H.264/AVC standard has overcome these drawbacks, so we will focus on fundamental aspects of scalable video coding with an emphasis toward features introduced in the H.264 standard. A more detailed overview of this scalable extension of the H.264/AVC standard could be found in Schwarz et al. [2007].

Temporal scalability is very easily supported in the context of current standards with a hierarchical prediction structure. In older standards such as MPEG-2, B-frames were not used as reference and were at the bottom of a simple hierarchy between I- and P-frames, so they could easily be dropped without any impact on the decoding of other frames. In H.264/AVC, the prediction dependency is more flexible so deeper hierarchies and hence more temporal layers could be supported. An example of a hierarchical prediction structure is shown in Figure 23.10. Interestingly, it has been found that such hierarchical prediction structures actually improve coding efficiency provided that the quantizers for each level are selected appropriately, i.e., finer quantizers should be used at lowest temporal layers and coarser quantizers at the highest layers.

Group of pictures (N)

Figure 23.10   Hierarchical coding structure supporting temporal scalability.

To support spatial scalability, one performs a multilayered coding of each spatial scale, where each spatial scale support conventional motion-compensated prediction as well as an interlayer prediction. There are a number of ways to perform the interlayer prediction in a multiscale framework. One method is to simply up-sample the reference data in the lower layer and use the up-sampled data for prediction. Another method is to infer block-level data such as the motion vectors from lower reference layers. Finally, the residual from the lower reference layer could also be used to predict the residual that is derived at the higher spatial layer. All three forms of interlayer prediction are supported by the H.264 standard. Another major innovation in this standard to overcome the complexity issues that have plagued past standards was to constrain the interlayer prediction so that single-loop decoding could be enabled rather than having a decoding loops for each scale. Finally, combined spatial/temporal scalability is also possible since lower layer pictures need not be present at every time instant. A figure illustrating this is provided in Figure 23.11.

The main purpose of quality scalability is to refine the SNR of the video with increasing quality layers. This form of scalability is often referred to as SNR scalability as well. To achieve a coarse-grained scalability, a similar multiscale architecture as used for spatial scalability could be used without the up-sampling operation. In this way, each layer would use a different quantizer to achieve the desired level of quality at each layer. The base layer would be coded with a coarse quantization and finer levels of quantization would then be used for higher layers. A finer level of quality control could also be imposed within each layer by coding fragments of transform coefficients. This enables successive refinement of quality within a layer.

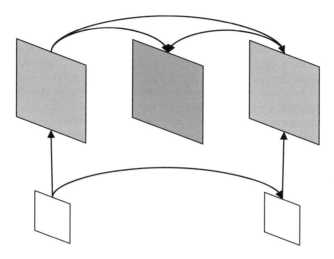

Figure 23.11 Combined spatial and temporal scalability including interlayer prediction and motion-compensated prediction.

23.6.2   Multiview Video Coding

Multiview video is used to support three-dimensional (3D) video applications. A special case of multiview video is stereo video in which only two views are present, a left view and right view corresponding to each eye. For most stereoscopic displays, glasses are needed to view the 3D scene. Auto-stereoscopic displays are capable of rendering multiple views of a 3D scene simultaneously and so do not require glasses. Three-dimensional services are becoming more popular for home entertainment systems and in mobile environments.

Performing efficient compression relies on having good predictors. While the correlation between temporally neighboring pictures is often very strong, including spatially neighboring pictures offers some advantages. For example, spatially neighboring pictures are useful predictors in uncovered regions of the scene, during fast object motion, or when objects appear in one view that are already present in neighboring views at the same time instant. Interview prediction is employed in all related works on efficient MVC, and aims to exploit both spatial and temporal redundancy for compression. The prediction is adaptive, so the best predictor among temporal and interview references is selected on a block basis. It is also noteworthy that there exists a base layer that can be independently decoded and possibly used as a 2D representation of the 3D scene.

It has been shown that coding multiview video with inter-view prediction does give significantly better results compared to independent coding of each view. Specifically, improvements of more than 2 dB have been reported relative to independent encoding of views using the multiview extensions of H.264/AVC. Furthermore, subjective testing has indicated that the same quality could be achieved with approximately half the bit rate.

23.7   Error Control

The previous sections focused mainly on the techniques to efficiently compress video and discussed related standards. These compression standards rely heavily on predictive coding and VLC techniques, both of which are not necessarily favorable characteristics when transmitting the compression video bitstream over error-prone channels. Prediction creates a dependency in the bitstream, whereby errors in one segment propagate to other segments. Errors in the VLC could also lead to synchronization issues and ultimately decoding failure.

This section covers three basic levels of error control that could be utilized to overcome errors during transmission: mechanisms that are available at the transport layer to protect the video, error-resilient features within the video layer, and techniques to conceal errors in a reconstructed video. We consider random bit errors that may result from characteristics of the physical channel, as well as the loss of packets that typically impact a greater portion of the bitstream. A more thorough treatment of this subject could also be found in Wang and Zhu [1998].

23.7.1   Transport Layer Mechanisms

A well-known method for error detection and correction is Forward Error Correction (FEC). FEC could be applied directly to the compressed bits to protect against bit errors or across data packets to recover from erasures. When applied to compressed bits, the FEC code is typically capable of correcting a single bit error within a frame comprising several hundred bits. When applied across data packets, a typical approach is to combine Reed–Solomon (RS) coding with block interleaving, where the RS code is first applied to blocks of data and then those blocks of data are interleaved into packets. In this way, a loss of one packet is essentially dispersed over multiple blocks of data and could be recovered. Since FEC increases the data transmission rate, which in turn reduces the available rate for the coded video, an appropriate balance between rate for source and channel coding must be considered in the overall design.

Unequal error protection is another effective means for increasing the robustness of the video transmission at the transport layer. It must first be noted that not all bits of the video bitstream are equally important. For instance, certain header information and other side information are much more critical to the final picture quality than some of the other block data. Also, in layered coding schemes described in Section 23.6, the base layer is much more critical than the enhancement layers since the enhancement layers are of no use without the base layer. Therefore, when error correction is used, important parts of the video bitstream should be coded with greater levels of protection. For networks that allow prioritization of data, higher prioritization can also be assigned accordingly so that such aspects including congestion control, retransmission, and power control can be optimized based on the priority of the data.

23.7.2   Error-Resilient Encoding of Video

While coding efficiency is the most important aspect in the design of any video codec, the transmission of compressed video through noisy channels must be considered. There are many error-resilience tools available in today's video coding standards. A brief review of some of the most relevant tools is provided in the following.

Localization

The basic principle of localization is to remove the spatial and temporal dependency between segments of the video to reduce error propagation. Such techniques essentially break the predictive coding loop so that if an error does occur, then it is not likely to affect other parts of the video. Obviously, a high degree of localization will lead to lower compression efficiency. There are two methods for localization of errors in a coded video: spatial localization and temporal localization; these methods are illustrated in Figure 23.12.

Spatial localization considers the fact that most video coding schemes make heavy use of VLC to reach high coding performance. In this case, even if 1 bit is lost or damaged, the entire bitstream

Error propagation
within a frame

Error propagation over time

Spatial
localization

Temporal localization

Figure 23.12 Illustration of spatial and temporal localization to minimize error propagation within a frame and over time, respectively. Spatial localization is achieved by means of resynchronization markers, while temporal localization is achieved by means of intrablock coding.

may become undecodable due to the loss of synchronization between the decoder and the bitstream. To regain synchronization after a transmission error has been detected, resynchronization markers are added periodically into the bitstream at the boundary of particular MBs in a frame. This marker would then be followed by essential header information that is necessary to restart the decoding process. When an error occurs, the data between the synchronization point prior to the error and the first point where synchronization is reestablished are typically discarded. For portions of the image that have been discarded, concealment techniques could be used to recover the pixel data, e.g., based on neighboring blocks that have been successfully decoded. For resynchronization markers to be effective in reducing error propagation, all predictions must be contained within the bounds of the marker bits. The restriction on predictions results in lowered compression efficiency. In addition, the inserted resynchronization markers and header information are redundant and they lower the coding efficiency. Spatial localization is supported in a number of standards through a *slice* structure, which is essentially a group of independently decodable MBs.

While resynchronization marker insertion is suitable to provide a spatial localization of errors, the insertion of intracoded MBs is used to provide a temporal localization of errors by decreasing the temporal dependency in the coded video sequence. While this is not a specific tool for error resilience, the technique is widely adopted and recognized as being useful for this purpose. The higher percentage of intrablocks used for coding the video will reduce the coding efficiency, but reduce the impact of error propagation on successively coded frames. In the most extreme case, all blocks in every frame are coded as intrablocks. In this case, there will be no temporal propagation of errors, but a significant increase in bit rate would be expected. The selection of intracoded blocks may be cyclic, in which the intracoded blocks are selected according to a predetermined pattern; the intracoded blocks may also be randomly chosen or adaptively chosen according to content characteristics.

Another form of temporal localization is reference picture selection, which was introduced in the H.263 and MPEG-4 standards for improved error resilience. Assuming a feedback-based system, the encoder receives information about corrupt areas of the picture from the decoder, e.g., at the slice level, and then alters its operation by choosing a noncorrupted reference for prediction or applying intracoding to the current data. In a similar spirit, the support for multiple reference pictures in H.264/AVC could be used to achieve temporal localization as well.

Data Partitioning

The objective of data partitioning is to group coded data according to relative importance to allow for unequal error protection or transport prioritization as discussed in the previous subsection. Data partition techniques have been developed to group together coded bits according to their importance to the decoding such that different groups may be more effectively protected or handled. For example, during the bitstream transmission over a single channel system, the more important partitions can be better protected with stronger channel codes than the less important partitions. Alternatively, with a multichannel system, the more important partitions could be transmitted over the more reliable channel.

In MPEG-2, data partitioning divides the coded bitstream into two parts: a high-priority partition and low-priority partition. The high-priority partition includes picture type, quantization scale, and motion vector ranges, without which the rest of the bitstream is not decodable. It may also include some MB header fields and DCT coefficients. The low-priority partition contains everything else. In MPEG-4, the data partitioning is achieved by separating the motion and MB header information away from the texture information. This approach requires that a second resynchronization between motion and texture information, which may further help localize the error. For instance, if the texture information is lost, the motion information may still be used to conceal these errors.

Redundant Coding

With this approach, segments of the video signal or syntactic elements of the bitstream are coded with added redundancy to enable robust decoding. The redundancy may be added explicitly, such as with the Redundant Slices tool, or implicitly in the coding scheme, as with Reversible Variable Length Codes (RVLC) and Multiple Description Coding (MDC).

RVLC has been developed for the purpose of data recovery at the receiver. Using this tool, the VLC are designed so that they can be read both in the forward and reverse directions. This allows the bitstream to be decoded backward from the next synchronization marker until the point of error. Examples of 3-bit codewords that satisfy this requirement include 111, 101, 010. It is obvious that this approach will reduce the coding efficiency compared with using normal VLC due to the constraints imposed in constructing the RVLC tables, which is the primary reason we classify the RVLC approach as a redundant coding technique. It also shares the benefit with other tools that robust decoding could be performed. However, since this tool is designed to recover from bit errors, it is not helpful for packet-erasure channels. RVLC has been adopted to both H.263 and MPEG-4 Part 2 standards.

MDC encodes a source with multiple bitstreams such that a basic-quality reconstruction is achieved if any one of them is correctly received, while enhanced-quality reconstructions are achieved if more than one of them is correctly received. With MDC, the redundancy may be controlled by the amount of correlation between descriptions. Generally, MDC video streams are suitable for delivery over multiple independent channels in which the probability of failure over one or more channels is likely. Some limited forms of MDC can be achieved with H.264/AVC.

Redundant slice is a new tool adopted into the H.264/AVC standard that allows for different representations of the same source data to be coded using different encoding parameters. For instance, the primary slice may be coded with a fine quantization, while the redundant slice may be coded with a coarse quantization. If the primary slice is received, the redundant slice is discarded, but if the primary slice is lost, the redundant slice would be used to provide a lower level of reconstructed quality. In contrast to MDC, the two slices together do not provide an improved reconstruction.

23.7.3 Error Concealment at the Decoder

In any video transmission system, there will inevitably be errors in the received bitstream. If they are not corrected by transport layer mechanisms or suppressed during the decoding of the video bit-stream, there could be severe damage to the reconstructed video. If the errors are appropriately local-ized, it is possible to conceal the effects of the transmission loss in the video signal. In most cases, it is assumed that the locations of errors have been detected and erroneous data have been discarded.

Perhaps the most widely studied error concealment approach is the recovery of texture infor-mation from neighboring data. By exploiting the inherent spatial and temporal redundancy in the video signal, it is often possible to recover missing blocks or even larger slices of the picture from neighboring data. A straightforward approach recovers missing data from one picture by copying colocated data from a neighboring picture, e.g., the previously decoded picture. The method could be applied when motion data is also lost and provides reasonable results for static parts of the video, but does not provide satisfactory quality when there is significant motion in the scene. When the motion vector data is available, a much better recovery of the missing texture could be performed from reference pictures that are used for prediction of the current picture.

Another means to recover texture information is by utilizing the valid data in the same picture. Spatial interpolation methods recover missing texture in damaged blocks based on pixel values in neighboring blocks. There are also more sophisticated methods that optimize the recovery by imposing smoothness constraints and accounting for any nonerroneously received DCT coefficients.

Since the motion vector data could also be viewed as smooth field for natural scenes, there also exist methods that attempt to recover motion information for damaged blocks so that the temporal recovery schemes discussed above could be applied. For example, the motion vector for a damaged block could be estimated based on an average or median of motion vectors in neighboring blocks, or the motion vector from the corresponding MB in the above row of MBs could be copied.

23.8 Video Streaming

Video streaming refers to the real-time transport of video to a receiving device. The video itself may be live or stored on a server. In either case, the delivery of video over wireless and wired networks poses a number of unique challenges to ensure that the best Quality of Service (QoS) could be achieved. There are of course unique challenges to transmitting and receiving video over a wireless network including multipath propagation, interference, energy, and power management as well as user mobility. Further discussion on these characteristics of a wireless communication system is provided in Chapter 2.

Generally speaking, QoS requirements are typically specified in terms of bandwidth, delay, and error rates. These parameters are likely to be time varying depending on the channel characteristics. The bandwidth is essential to ensure that the video at a particular resolution could be represented with high enough quality given the rate constraints. Since video must be played out continuously, there are also strict timing constraints imposed on the delivery and decoding processes. As discussed in the previous section, error loss also has a notable impact on the final reconstructed quality of the video.

Since the bandwidth and loss characteristics over a given channel are often fluctuating, the rate of the compressed video that is being streamed should ideally change based on such dynamics. There are a variety of techniques that could be used to regulate the rate of the video bitstreams depending on the bitstream characteristics, system level constraints, and application requirements.

For instance, given a nonlayered MPEG-2 video bitstream, one may apply transcoding operations so that the source rate matches the channel bandwidth. This could be achieved by requantizing the transform coefficients, which requires a partial decoding of the bitstream, or by simply dropping frames. If the video is encoded using a scalable format, adjustments to the rate could be made with simpler operations as described in Section 23.6. In multicast networking, receiver-based strategies could also be used to regulate the stream data to be processed locally.

Synchronization is another important aspect of video streaming. In most applications, the video is accompanied by an associated audio stream and in some cases there are other graphics and text that are delivered as separate streams that correspond to the video as well. It is very critical that the temporal relationship between the different streams should be maintained, and if lost, for them to be resynchronized at a later point in time. For instance, in broadcasting applications, even minor differences between the audio and video stream can lead to what is known as lip synchronization problems, which are annoying to the viewer.

There is a very rich set of networking protocols that supports the streaming of video. A brief discussion of relevant transport and session control protocols is given in the following.

- Transmission Control Protocol (TCP) is the dominant protocol for Internet Protocol (IP)-based data transfer and handles such functions as multiplexing, error control, and flow control. While TCP could be used for video streaming, there are several aspects that prevent it from providing reliable and good quality video streaming. For one, it utilizes retransmission for packet loss so end-to-end delay may be relatively large. Also, TCP does not handle variability in the data rate well. These problems can be addressed by buffering the data. An optimal buffer size could be determined based on a target delay, smoothness of playback, and data loss. In general, a small buffer size implies smaller delay. On the other hand, a larger buffer will provide for smoother playback since larger variations in the bit rate and transmission time could be tolerated.
- User Datagram Protocol (UDP) has become a preferred network protocol for video streaming. In contrast to TCP, UDP allows damaged or lost packets to be dropped. While this feature allows for reduced delay, it does not guarantee packet delivery; therefore, packet loss should be expected and error concealment techniques described in Section 23.7.3 would be needed to recover those losses.
- Real-time Transport Protocol (RTP) is a protocol for the transport of real-time data, including audio and video. RTP consists of two parts, a data part and a control part which is called RTCP. The data part of RTP supports real-time transmission for continuous media such as video and audio. It provides timing reconstruction, loss detection, security, and content identification. The control part provides source identification and support for gateways like audio and video bridges as well as multicast-to-unicast translators. While it offers QoS feedback from receivers to the multicast group as well as support for the synchronization of different media streams, it does not provide QoS guarantees. RTP/RTCP is commonly built on the top of UDP.
- Real Time Streaming Protocol (RTSP) is a session control protocol for media streaming. It has similar functions as the Hyper Text Transfer Protocol (HTTP) for text and graphics. This protocol is designed to initiate a session and direct the delivery of the video stream. One of its main functions is to select the delivery channel and mechanism. It also controls the playback of the video stream with support of so-called trick-play operations such as pause, fast-forward, and reverse play.

The above protocols are used in a variety of mobile delivery standards including Third Generation Partnership Project (3GPP), 1-seg in Japan, Digital Video Broadcasting – Handheld (DVB – H) in Europe and Advanced Television Systems Committee – Mobile/Handheld (ATSC – M/H) in North America. An overview of some important mobile video application standards, including Multimedia Messaging Service (MMS), streaming, video telephony, and multicast and broadcast is provided by Wang et al. [2007].

Further Reading

Further information about image and video processing can be found in the classic textbooks by Jain [1989], Gonzalez and Woods [2008], and Tekalp [1995]. For a more comprehensive overview of the principles of video compression, as well as corresponding algorithms and video compression standards, we recommend the textbooks by Wang et al. [2002] and Shi and Sun [2000]. The edited book by Sun and Reibman [2001] provides an excellent collection of works on networking and transport of compressed video.

For updates and errata for this chapter, see wides.usc.edu/teaching/textbook

24

GSM – Global System for Mobile Communications

24.1 Historical Overview

The *Global System for Mobile communications* (GSM) is by far the most successful mobile communication system worldwide. Its development started in 1982. The European Conference of Postal and Telecommunications Administrations (CEPT), predecessor of the *European Telecommunication Standards Institute* (ETSI), founded the *Groupe Speciale Mobile*, with the mandate to develop proposals for a pan-European digital mobile communication system. Two goals were supposed to be achieved:

- First, a better and more efficient technical solution for wireless communications – it had become evident at that time that digital systems would be superior in respect to user capacity, ease of use, and number of possible additional services compared with the then-prevalent analog systems.
- Second, a single standard was to be realized all over Europe, enabling roaming across borders. This was not possible before, as *incompatible* analog systems were employed in different countries.

In the following years, several companies developed proposals for such a system. These proposals covered almost all possible technical approaches in different technical areas. For *multiple access*, Time Division Multiple Access (TDMA), Frequency Division Multiple Access (FDMA), and Code Division Multiple Access (CDMA) were suggested. The proposed *modulation techniques* were Gaussian Minimum Shift Keying (GMSK), 4-Frequency Shift Keying (4-FSK), Quadrature Amplitude Modulation (QAM), and Adaptive Differential Pulse Modulation (ADPM). *Transmission rates* varied from 20 kbit/s to 8 Mbit/s. All of the proposed systems were tested both in field tests and with a channel simulator (in Paris in 1986). Apart from technical considerations, marketing and political arguments influenced the decision-making process. FDMA could not be employed, as it would have required antenna diversity at the Mobile Station (MS). Even though the technical feasibility of this diversity had been proven by the Japanese digital system, increased antenna sizes did not make it a desired option. CDMA was ultimately excluded, because the necessary signal processing seemed to be too expensive and unreliable at that time. Therefore, only a TDMA system could survive the selection process. However, the final TDMA system was not a proposal from a single company, but rather a compromise system was developed. The reasons for this were of a political and not technical nature: selecting the proposal of one company as the standard would

Wireless Communications, Second Edition   Andreas F. Molisch
© 2011 John Wiley & Sons, Ltd

have given this specific company a large competitive advantage. Specific details of the compromise system were developed by a – now permanent – committee over the following two years and served as the basis for systems implemented in Europe after 1992.

In the early 1990s, it was realized that GSM should have functionalities that had not been included in the original standard. Therefore, the so-called phase-2 specifications, which included these functions, were developed until 1995. Further enhancements, which include packet radio (General Packet Radio Service (GPRS), see Appendix 24.C on the companion website: *www.wiley.com/go/molisch*) and the more efficient modulation of Enhanced Data rates for GSM Evolution (EDGE), have been introduced since then. Because of these extensions GSM is often referred to as the *2.5th generation system*, as its functionalities are beyond those of a second-generation system, but do not enable all third-generation functionalities (Universal Mobile Telecommunications System (UMTS) (compare with Chapter 26)).

The success of GSM exceeded all expectations. Though it was originally developed as a European system, it has spread all over the world in the meantime. Australia was the first non-European country that signed the basic agreement (*Memorandum of Understanding* (MoU)). Since then, GSM has become *the* worldwide mobile communication standard,[1] with a number of subscribers that approached 3.5 billon in 2009. A few exceptions remain in Japan and Korea, where GSM was never implemented. In the U.S.A., GSM was competing with the CDMA-based Interim Standard-95 (IS-95) system. In contrast to most countries where spectral licenses were provided on condition that the network operator would use GSM, the licenses in the U.S.A. were sold without requiring companies to implement a specific system. In 2009, there were two major operators offering GSM-based services, while another two were using rival technologies (see Chapter 25).

There are three versions of GSM, each using different carrier frequencies. The original GSM system uses carrier frequencies around 900 MHz. GSM1800, which is also called Digital Cellular System at the 1800-MHz band (DCS1800), was added later to support the increasing numbers of subscribers. Its carrier frequencies are around 1,800 MHz, the total available bandwidth is roughly three times larger than the one around 900 MHz, and the maximal transmission power of MSs is reduced. Apart from this, GSM1800 is identical to the original GSM. Thus, signal processing, switching technology, etc. can be reused without changes. The higher carrier frequency, which implies a smaller path gain, and reduced transmission power reduce the sizes of the cells significantly. This fact, combined with the bigger available bandwidth, leads to a considerable increase in network capacity. A third system, known as GSM1900 or PCS-1900 (*Personal Communication System*) operates on the 1,900-MHz carrier frequency, and is mainly used in the U.S.A.

GSM is an open standard. This means that only the interfaces are specified, not the implementation. As an example, we consider the modulation of GSM, which is GMSK. The GSM standard specifies upper bounds for out-of-band emission, phase jitter, intermodulation products, etc. *How* the required linearity is achieved (e.g., by feedforward linearization, by using a class-A amplifier – which is unlikely because of the small efficiency – or by any other method) is up to the equipment manufacturer. Thus, this open standard ensures that all products from different manufacturers are compatible, though they can still differ in quality and price. Compatibility is especially important for service providers. When using proprietary systems, a provider is able to choose the equipment supplier only once – at the beginning of network implementation. For GSM (and other open standards), a provider can first purchase Base Stations (BSs) from one manufacturer but later on buy BSs to extend the capacity of his network from a different manufacturer, which might offer a better price. A provider may also buy some components from one company and other components from another company.

---

[1] Hence the reinterpretation of GSM from "Groupe Speciale Mobile" to "Global System for Mobile communications."

24.2 System Overview

A GSM system consists essentially of three parts – namely, the *Base Station Subsystem* (BSS), the *Network and Switching Subsystem* (NSS), and the *Operation Support System* (OSS).

24.2.1 Base Station Subsystem

The BSS consists of *Base Transceiver Stations* (BTSs) and the *Base Station Controllers* (BSCs) (see Figure 24.1). The BTS establishes and maintains the connection to the MSs within its cell. The interface between the MS and the BTS is the air interface, called the *Um-interface* in the GSM context. The BTS hosts, at a minimum, the antennas and the Radio Frequency (RF) hardware of a BS, as well as the software for multiple access. Several – or, rarely, one – BTSs are connected to one BSC; they are either colocated, or connected via landline, directional microwave radio links, or similar connections. The BSC has a control functionality. It is, among other things, responsible for HandOver HO) between two BTSs that are connected to the same BSC. The interface between BTS and BSC is called the *Abis-interface*. In contrast to the other interfaces, this interface is not completely specified in the standard.[2] Distribution of the functionalities between BTS and BSC may differ depending on the manufacturer. In most cases, one BSC is connected to several BTSs. Therefore, it is possible to increase the efficiency of implementation by shifting as much functionality as possible to the BSC. However, this implies increased signaling traffic on the link between the BTS and the BSC, which might be undesirable (remember that these links are often rented landline connections). In general, the BSS covers a large set of functionalities. It is responsible for channel assignment, maintenance of link quality and handover, power control, coding, and encryption.

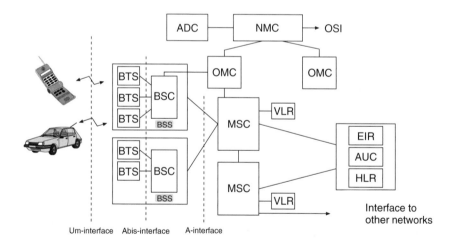

Figure 24.1 Block diagram of a Global System for Mobile communication system. *In this figure:* ADC, Administration Center; NMC, Network Management Center; OSI, Operator System Interface.
Adapted with permission from HP [1994] © Hewlett Packard.

---

[2] Therefore, a set of BTSs always *has* to be combined with a BSC of the same manufacturer.

24.2.2   Network and Switching Subsystem

The main component of the NSS is the *Mobile-services Switching Center* (MSC), which controls the traffic between different BSCs (see Figure 24.1). One function of the MSC is *mobility management*, which comprises all the functions that are necessary to enable true mobility for subscribers. To give but one example, one function of the MSC is the management of HOs that occur when an MS is leaving the area of one BSC and moving into the area covered by another BSC. Other functions are the so-called *paging* and *location update*. All interactions with other networks – especially the landline *Public Switched Telephone Network* (PSTN) – are also performed by the MSC.

The NSS includes some databases, too. The *Home Location Register* (HLR) contains all the numbers of the mobile subscribers associated with one MSC and information about the location of each of these subscribers. In the event of an incoming call, the location of the desired subscriber is looked up in the HLR and the call is forwarded to this location.[3] Therefore, we can conclude that from time to time a traveling MS has to send updates of its location to its HLR. The *Visitor Location Register* (VLR) of one MSC contains all the information about mobile subscribers from other HLRs that are in the area of this MSC and are allowed to roam in the network of this MSC. Furthermore, a temporary number will be assigned to the MS to enable the "host" MSC to establish a connection to the visiting MS.

The *AUthentication Center* (AUC) verifies the identity of each MS requesting a connection. The *Equipment Identity Register* (EIR) contains centralized information about stolen or misused devices.

24.2.3   Operating Support System

The OSS is responsible for organization of the network and operational maintenance. More specifically, the OSS mainly covers the following functions:

1. *Accounting*: how much does a specific call cost for a certain subscriber? There are also plenty of different services and features, from which each subscriber may choose an individual selection included in a specific plan. While this rich choice of services and prices is vital in the marketplace, the administrative support of this individualism is rather complicated. Examples are discussed in Section 24.10.
2. *Maintenance*: the full functionality of each component of the GSM network has to be maintained all the time. Malfunctions may either occur in the hardware or in the software components of the system. Hardware malfunctions are more costly, as they require a technician to drive to the location of the malfunction. In contrast, software is nowadays administrated from a central location. For example, new versions of switching software can be installed in the complete BSS from a central location, and activated all over the network at a specific time. Revision and maintenance software often constitutes a considerable part of the overall complexity of GSM control software.
3. *MS management*: even though all MSs have to pass a type approval, it may happen that "bad apple" devices, which cause systemwide interference, are operating in the network. These devices have to be identified and their further activities have to be blocked.
4. *Data collection*: the OSS collects data about the amount of traffic, as well as the quality of the links.

---

[3] Actually the call is only forwarded to the BSC in whose area the subscriber is. Routing to and selection of one BTS is the responsibility of the BSC.

24.3   The Air Interface

GSM employs a combined FDMA/TDMA approach which further combines with Frequency Domain Duplexing (FDD) (see Chapter 17). Let us elaborate on these acronyms.

FDD

In the first GSM version, frequencies from 890 to 915 MHz and from 935 to 960 MHz were available. The lower band is used for the uplink (connection from the MS to the BS). The upper band is used for the downlink. The frequency spacing between the uplink and downlink for any given connection is 45 MHz. Therefore, relatively cheap duplex filters are sufficient for achieving very good separation between the uplink and downlink.

For GSM1800, the frequency ranges are 1,710–1,785 MHz for the uplink, and 1,805–1,880 MHz for the downlink. In North America, 1,850–1,910 MHz are used for the uplink and 1,930–1,990 MHz for the downlink. Other bands are added as they become available, see also Chapter 27.

FDMA

Both uplink and downlink frequency bands are partitioned into a 200-kHz grid. The outer 100 kHz of each 25-MHz band are not used,[4] as they are *guard bands* to limit interference in the adjoined spectrum, which is used by other systems. The remaining 124 200-kHz subbands are numbered consecutively by the so-called *Absolute Radio Frequency Channel Numbers* (ARFCNs).

TDMA

Due to the very-bandwidth-efficient modulation technique (GMSK, see below), each 200-kHz subband supports a data rate of 271 kbit/s. Each subband is shared by eight users. The time axis is partitioned into timeslots, which are periodically available to each of the possible eight users (Figure 24.2). Each timeslot is 576.92 μs long, which is equivalent to 156.25 bits. A set of eight

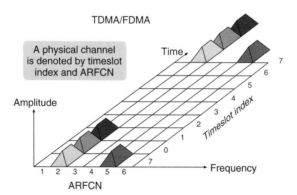

Figure 24.2   Time Division Multiple Access/Frequency Division Multiple Access system.
Adapted with permission from HP [1994] © Hewlett Packard.

---

[4] This applies to a GSM900 system, and analogously to other frequency bands.

timeslots is called a *frame*; it has a duration of 4.615 ms. Within each frame, the timeslots are numbered from 0 to 7. Each subscriber periodically accesses one specific timeslot in every frame on one frequency subband. The combination of timeslot number and frequency band is called the *physical channel*. The kind of data that are transmitted over one such physical channel depends on the *logical channel* (see also Section 24.4).

The important features of the air interface are now described in a step-by-step manner.

The Assignment of Timeslots in the Uplink and Downlink

A subscriber utilizes the timeslots with the same number (index) in the uplink and downlink. However, numbering in the uplink is shifted by three slots relative to the numbering in the downlink. This facilitates the design of the MS transmitter/receiver, because reception and transmission do not occur at the same time (compare Figure 24.3).

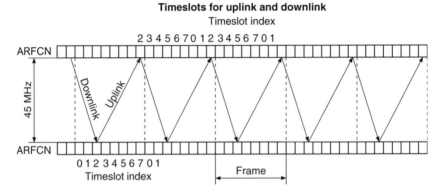

Figure 24.3 The alignment of timeslots in the uplinks and downlink.

Adapted with permission from HP [1994] © Hewlett Packard.

The Modulation Technique

GSM uses GMSK as a modulation format. GMSK is a variant of Minimum Shift Keying (MSK); the difference is that the data sequence is passed through a filter with a Gaussian impulse response (time bandwidth product $B_G T = 0.3$) (see Chapter 11).

This filtering is rather hard. Therefore, the spectrum is rather narrow, but there is a significant amount of Inter Symbol Interference (ISI). On the other hand, the ISI due to delay dispersion of the wireless channel is usually much more severe. Thus, some kind of equalization has to be used anyway. Figure 24.4 illustrates a typical example of a phase trellis of this kind of GMSK and of pure MSK for comparison. The detection method is not specified by the standard. Differential detection, coherent detection, or limiter–discriminator detection might be employed.

Power Ramping

Were a transmitter to start data transmission right at the beginning of each timeslot, it would have to be able to switch on its signal within a very short time (much shorter than a symbol period). Similarly, at the end of a timeslot, it would have to stop transmitting abruptly, so as not to create

Figure 24.4 Phase diagram for the bit sequence 1011011000 for Gaussian minimum shift keying with $B_G T = 0.3$ (solid line) and pure minimum shift keying (dashed line).

interference with the next timeslot. This is difficult to implement in hardware and – even if it could be realized – the sharp transition in the time domain would lead to broadening of the emission spectrum. Therefore, GSM defines a time during which signals are smoothly turned off and on (see Figure 24.5). Nevertheless, the hardware requirements are still tremendous. In the case when a transmitter is emitting the maximal signal power it has to ramp up from 2×10^{-7} W to 2 W within 28 µs. During the actual transmission of data, on the other hand, signal power may only deviate by 25% (1 dB) from its nominal value.

Signal Power and Power Control

GSM provides power control for transmission power. While power control is usually associated with CDMA systems, it also has major benefits in GSM (and other TDMA/FDMA systems):

1. It increases the possible operating time of the batteries. The transmit power amplifier is a major factor in the power consumption of an MS. Therefore, the possible operating time without recharging the batteries depends critically on the emitted signal level. Thus, emitting more power than necessary for maintaining good quality of the received signal at the other link end is a waste of energy.
2. Transmitting at too high a power level increases the interference level in adjacent cells. Because of the cellular concept, every transmitter is a possible interferer for users in other cells that use the same time/frequency slot. However, in contrast to CDMA systems, power control is not *essentially necessary* for operation of the system.

GSM specifies different types of MSs, with different maximum transmission powers though 2-W stations (peak power) are most common. Power control can reduce emitted signal power by about 30 dB; adjustment is done in 2-dB steps. Control is adaptive: the BS periodically informs the MS about the received signal level and the MS uses this information to increase or reduce its transmit power. Maximum power levels of the BSs can vary between 2 and more than 300 W. Also, BSs have a similar power control loop that can decrease output power by some 30 dB.

Figure 24.5   Power ramping during a timeslot.
Adapted with permission from HP [1994] © Hewlett Packard.

Out-of-Band Emission and Intermodulation Products

The limits for out-of-band emissions are not as severe as, e.g., for analog systems. The maximum permitted out-of-band signal power at both BS and MS is roughly $-30\,$dBm, which is a very high value for wireless communications. However, in the band from 890 to 915 MHz (the uplink band), the power emitted by the BS must not exceed $-93\,$dBm. This is necessary because the BS has to receive signals from MSs, with signal levels as low as $-102\,$dBm, in this band. Furthermore, transmit antennas are located close to receive antennas (or even colocated) at the BS, and therefore any out-of-band emission in this band causes severe interference.[5] Similar limits apply for the intermodulation products.[6]

Structure of a Timeslot

Figure 24.6 illustrates the data contained in a timeslot with a length of 148 bits. However, not all of these bits are payload data. Payload data are transmitted over two blocks of 57 bits. Between these blocks is the so-called *midamble*. This is a known sequence of 26 bits and provides the training for equalization, which will be covered in Section 24.7. Furthermore, the midamble serves as an identifier of the BS. There is an extra control bit between the midamble and each of the two data-containing blocks; the purpose of these control bits are explained in Section 24.4. Finally, the transmission *burst* starts and ends with three *tail bits*. These bits are known, and enable termination of Maximum Likelihood Sequence Estimation (MLSE) in defined states at the beginning and end

---

[5] Similarly, emissions in the bands used by other systems such as UMTS have strict limits.
[6] Note that intermodulation products can only be found at the BS, as only the BS transmits on several frequencies simultaneously.

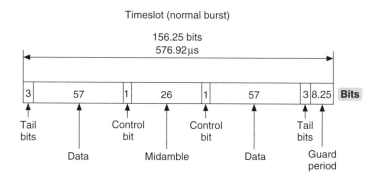

Figure 24.6   Functions of the bits of a normal transmission burst.

of the detection of burst data. This reduces the complexity and increases the performance of decoding (see also Chapter 14). The timeslots end with a *guard period* of 8.25 bits. Apart from "normal" transmission bursts, there are other kinds of bursts. MSs transmit *access bursts* to establish initial contact with the BS. *Frequency correction bursts* enable frequency correction of the MSs. *Synchronization bursts* allow MSs to synchronize to the frame timing of BSs. These bursts will be explained in more detail in Section 24.4.2.

24.4   Logical and Physical Channels

In addition to the actual payload data, GSM also needs to transmit a large amount of signaling information. These different types of data are transmitted via several *logical channels*. The name stems from the fact that each of the data types is transmitted on specific timeslots that are parts of *physical channels*. The first part of this section discusses the kind of data that is transmitted via logical channels. The second part describes the *mapping* of logical channels to physical channels.

24.4.1   Logical Channels

Traffic CHannels (TCHs)

Payload data are transmitted via the TCHs. The payload might consist of encoded voice data or "pure" data. There is a certain flexibility regarding the data rate: *Full-rate Traffic CHannels* (TCH/F) and *Half-rate Traffic CHannels* (TCH/H). Two half-rate channels are mapped to the same timeslot, but in alternating frames.

Full-Rate Traffic CHannels

- Full-rate voice channels: the output data rate of the voice encoder is 13 kbit/s. Channel coding increases the effective transmission rate to 22.8 kbit/s.
- Full-rate data channels: the payload data with data rates of 9.6, 4.8, or 2.4 kbit/s are encoded with Forward Error Correction (FEC) codes and transmitted with an effective data rate of 22.8 kbit/s.

Half-Rate Traffic CHannels

- Half-rate voice channels: voice encoding with a data rate as low as 6.5 kbit/s is feasible. Channel coding increases the transmitted data rate to 11.4 kbit/s.

- Half-rate data channels: payload data with rates of 4.8 or 2.4 kbits/s can be encoded with an FEC code, which leads to an effective transmission rate of 11.4 kbit/s.

Broadcast CHannels (BCHs)

BCHs are only found in the downlink. They serve as *beacon* signals. They provide the MS with the initial information that is necessary to start the establishment of any kind of connection. The MS uses signals from these channels to establish a synchronization in both time and frequency. Furthermore, these channels contain data regarding, e.g., cell identity. As the BSs are not synchronized with respect to each other, the MS has to track these channels not only before a connection is established, but all the time, in order to provide information about possible HOs.

Frequency Correction CHannels (FCCHs)   The carrier frequencies of the BSs are usually very precise and do not vary in time, as they are based on rubidium clocks. However, dimension considerations and price considerations make it impossible to implement such good frequency generators in MSs. Therefore, the BS provides the MS with a frequency reference (an unmodulated carrier with a fixed offset from the nominal carrier frequency) via the FCCH. The MS tunes its carrier frequency to this reference; this ensures that both the MS and the BS use the same carrier frequency.

Synchronization CHannel (SCH)   In order to transmit and receive bursts appropriately, an MS not only has to be aware of the carrier frequencies used by the BS but also of its frame timing on the selected carrier. This is achieved with the SCH, which informs the MS about the frame number and the *Base Station Identity Code* (BSIC). Decoding of the BSIC ensures that the MS only joins admissible GSM cells and does not attempt to synchronize to signals emitted by other systems in the same band.

Broadcast Control CHannel (BCCH)   Cell-specific information is transmitted via the BCCH. This includes, e.g., *Location Area Identity* (LAI),[7] maximum permitted signal power of the MS, actual available TCH, frequencies of the BCCH of neighboring BSs that are permanently observed by the MS to prepare for a handover, etc.

Common Control CHannels (CCCHs)

Before a BS can establish a connection to a certain MS, it has to send some signaling information to all MSs in an area, even though only one MS is the desired receiver. This is necessary because in the initial setup stage, there is no *dedicated* channel established between the BS and a MS. CCCHs are intended for transmission of information to all MSs.

Paging CHannel (PCH)   When a request – e.g., from a landline – arrives at the BS to establish a connection to a specific MS, the BSs within a location area send a signal to all MSs within their range. This signal contains either the permanent *International Mobile Subscriber Identity* (IMSI) or the *Temporary Mobile Subscriber Identity* (TMSI) of the desired MS. The desired MS continues the process of establishing the connection by requesting (via a Random Access CHannel (RACH)) a TCH, as discussed below. The PCH may also be used to broadcast local messages like street traffic information or commercials to all subscribers within a cell. Evidently, the PCH is only found in the downlink.

---

[7] A Location Area (LA) is a set of cells, within which the MS can roam without updating any location information in its HLR.

Random Access CHannel (RACH)   A mobile subscriber requests a connection. This might have two reasons. Either the subscriber wants to initiate a connection, or the MS was informed about an incoming connection request via the PCH. The RACH can only be found in the uplink.

Access Grant CHannel (AGCH)   Upon the arrival of a connection request via the RACH, the first thing that is established is a Dedicated Control CHannel (DCCH) for this connection. This channel is called the *Standalone Dedicated Control CHannel* (SDCCH), which is discussed below. This channel is assigned to the MS via the AGCH, which can only be found in the downlink.

Dedicated Control CHannels (DCCHs)

Similar to the TCHs, the DCCHs are bidirectional – i.e., they can be found in the uplink and downlink. They transmit the signaling information that is necessary during a connection. As the name implies, DCCHs are *dedicated* to one specific connection.

Standalone Dedicated Control CHannel (SDCCH)   After acceptance of a connection request, the SDCCH is responsible for further establishing this connection. The SDCCH ensures that the MS and the BS stay connected during the authentification process. After this process has been finished, a TCH is finally assigned for this connection via the SDCCH.

Slow Associated Control CHannel (SACCH)   Information regarding the properties of the radio link are transmitted via the SACCH. This information need not be transmitted very often, and therefore the channel is called *slow*. The MS informs the BS about the strength and quality of the signal received from serving BSs and neighboring BSs. The BS sends data about the power control and runtime of the signal from the MS to the BS. The latter is necessary for the *timing advance*, which will be explained later.

Fast Associated Control CHannel (FACCH)   The FACCH is used for HOs that are necessary for a short period of time; therefore, the channel has to be able to transmit at a higher rate than the SACCH. Transmitted information is similar to that sent by the SDCCH.

The SACCH is associated with either a TCH or a SDCCH; the FACCH is associated with a TCH.

24.4.2 Mapping Between Logical and Physical Channels

The signals of logical channels described above have to be transmitted via physical channels, which are represented by the timeslot number and the ARFCN. In order to better understand the mapping, we first have to realize that the time dimension is not only partitioned into periodically repeated frames of eight timeslots each, but that these frames and timeslots are the smallest units in the time grid. In fact, multiple frames are combined on different levels to make bigger frames (see Figure 24.7).

We have already seen above that eight timeslots with a duration of 577 μs each are combined as a frame. The duration of this frame, 4.61 ms, is the basic period of a GSM system. A total of 26 of these frames are combined as a *multiframe*, which has a duration of 120 ms. Furthermore, 51 of these multiframes are contained in one *superframe*, which has a length of 6.12 s. Finally, 2,048 of these superframes are combined into one *hyperframe*, which lasts 3 h and 28 min. The hyperframe is implemented mainly for cryptographic reasons, in order to guarantee privacy over the air interface. Therefore, encryption is applied to the payload data and the period of the encryption algorithm is exactly the length of one hyperframe.

Frames and multiframes

Figure 24.7   Structure of Global System for Mobile communications frames for traffic channels.
Adapted with permission from HP [1994] © Hewlett Packard.

Understanding the multiple frame structure enables us to discuss which timeslot contains which logical channel. Not all timeslots have to be used for the TCH, as the available data rate on the physical channel is $2 \cdot 57$ bits/4.615 ms = 24.7 kbit/s, while a full rate TCH requires only a 22.8-kbit/s data rate. Therefore, the remaining 1.9 kbit/s may be used for other logical channels.

SACCH   As discussed above, 26 frames are combined as a multiframe. Of these 26, only 24 frames are dedicated to the TCH. The 13th (and sometimes the 26th) frame are used by the SACCH. The 26th frame is only employed if two half-rate connections share one physical channel; otherwise the timeslot of the 26th frame is an *idle frame*. The transmission rate of the SACCH is 950 bit/s. The data transmitted via the SACCH is processed differently from the data in the TCH. The bits of four consecutive SACCH bursts are processed together. For this purpose, four multiframes might be combined into a (nameless) higher order frame of length 480 ms. These four SACCH bursts contain 456 bits associated with SACCH data and are used to transmit 184 actual data bits. The data bits are (i) first encoded with a (224, 184) block code, (ii) have four tail bits added, and (iii) then everything is encoded with the regular rate-1/2 convolutional encoder; this leads to the total of $2 \cdot 228$ bit = 456 bit.

FACCH   An FACCH does not have to be permanently available. It is only necessary in special situations – e.g., when a handover has to be performed. Therefore, no timeslots are *reserved* for the FACCH. Instead, normal TCH-related bursts of a connection are partly used for FACCH purposes in case this is required. The above-mentioned control bits (stealing bits) between the midamble and the datablocks of a burst indicate whether an FACCH is present in this burst or not – i.e., "steals" bits from the TCH. The 184 bits of an FACCH are encoded in the same way as SACCH bits. In order to transmit the resulting 456 bits via the normal TCH timeslots, eight consecutive frames are used: the even payload bits of the first four bursts and the odd bits of the second four bursts are replaced by bits from the FACCH.

Common Logical Channels The FACCH and SACCH use the physical channel of the associated connection. This is possible as the physical channel supports a slightly higher data rate than is necessary for one TCH connection. Therefore, it is possible to transmit signaling in timeslots belonging to the same physical channel. However, the other logical signaling channels are not associated with a TCH connection, either because they are required for *establishing* a connection or because they are used even in the absence of a TCH channel. Therefore, all these channels operate in the first burst of each frame of the so-called "BCCH carrier." This assignment strategy makes sure that one physical channel in each cell is permanently occupied. This leads, of course, to a loss of capacity, especially in cells that use only one carrier. However, there is one option to overcome this: if the cell is full, no new connections can be established. Therefore, no timeslots have to be reserved for signaling related to new connections, and also the first slot of the BCCH carrier can be used for a normal TCH channel.[8] Furthermore, the frames are combined as higher order frames in a different way. A total of 51 frames are combined into a multiframe, which has a duration of 235 ms. CCCHs are unidirectional, with the RACH being the only channel in the uplink, while several common channels exist in the downlink.

RACH The RACH is necessary only for the uplink. During each multiframe, 8 data bits, encoded into 36 bits, are transmitted via the RACH. These 36 bits are transmitted as an *access burst*. The structure of an access burst has to differ from normally transmitted bursts. At the time the MS requests a connection, it is not yet aware of the runtime of the signal from the MS to the BS. This runtime might be in the range from 0 to 100 μs where the maximal value is defined by the maximal cell range of 30 km. Therefore, a larger guard time is necessary to ensure that a random burst does not collide with other bursts in adjacent timeslots. After the connection is established, the BS informs the MS about the runtime and therefore the MS can reduce the size of the guard times by employing *timing advance*, which will be discussed later. A complete random access burst has the following structure. It starts with 8 tail bits, which are followed by 41 synchronization bits. Afterward, the 36 bits of encoded data and 3 additional tail bits are transmitted. This adds to a total of 88 bits and leaves a guard time of 100 μs at the end, which corresponds to 68.25 bits. As the RACH is the only unassociated control channel in the uplink, the timeslot numbered 0 may be used for random access burst in every frame.

Common Channels in the Downlink The other common channels – such as FCCH, SCH, BCCH, PCH, and AGCH – can only be found in the downlink and have a fixed order in the multiframe. Figure 24.8 illustrates this structure. Remember that only timeslot 0 in each frame carries a CCCH. Of the 51 frames in this multiframe, the last one is always idle. The remaining 50 frames are divided into blocks of 10 frames. Each of these blocks starts with a frame containing the FCCH. Afterward the SCH is transmitted during the next frame. The first block of frames contains four BCCHs (in frames 3–6) followed by four frames which contain the PCH or AGCH (frames 7–10). The other four blocks of 10 frames also start with the FCCH and SCH frames, and then consist either of PCH- or AGCH-carrying frames. The FCCH and the SCH employ bursts that have a special structure (this is discussed in the next section). As the MSs of neighboring cells continuously evaluate the signal strength of the first timeslot of the frames on the BCCH carrier, the BS always has to transmit some information during these timeslots, even when there is no connection request.

SDCCH The SDCCH may occupy a physical channel by itself, or – in case the common channels do not occupy all the available slots on the BCCH – it may be transmitted during the first timeslots on the BCCH. In the latter case, either four or eight SDCCHs share this physical channel.

---

[8] Nevertheless, this option is not implemented by most providers.

Figure 24.8 Mapping of broadcast channels (FCCH, SCH, and BCH) and common control channels to timeslots numbered 0 (compare [CME 20, 1994]).

24.5 Synchronization

Up to now, we have assumed that the BS and the MS are synchronized in time and frequency. However, only the BS is required by the standard to have a high-quality time and frequency reference. For the MS, such a reference would be too expensive. Thus, the MS synchronizes its frequency and time references with those of the BS. This is done in three steps: first, the MS tunes its carrier frequency to that of the BS. Next, the MS synchronizes its timing to the BS by using synchronization sequences. Finally, the timing of the MS is additionally shifted with respect to the timing of the BS to compensate for the runtime of the signal between the BS and MS (*timing advance*).

24.5.1 Frequency Synchronization

As mentioned before, the BS uses very precise rubidium clocks, or GPS (Global Positioning System) signals, as frequency references. Due to space and cost limitations, the oscillators at the MS are quartz oscillators, which have much lower precision. Fortunately, this is not a problem, since the BS can transmit its high-precision frequency reference periodically and the MS can adjust its local oscillator based on this received reference. Transmission of the reference frequency is done via the FCCH. As we discussed in the previous section, the FCCH is transmitted during timeslots with index 0 of roughly every tenth frame on the BCCH. An FCCH burst consists of 3 tail bits at the beginning, 142 all-zero bits in the middle, and 3 tail bits at the end. The usual guard period (length equivalent to 8.25 bits) is appended. It should be noted that it is not the carrier frequency that is transmitted as a reference, but rather the carrier modulated with a string of zeros. This equals a sinusoidal signal with a frequency that is the carrier frequency *offset* by the MSK modulation frequency. As this offset is completely deterministic, it does not change the principle underlying the synchronization process.

24.5.2 Time Synchronization

Time synchronization information is transmitted from the BS to the MS via the SCH. SCH bursts contain information regarding the current index of the hyperframe, superframe, and multiframe. This is not a lot of information, but has to be transmitted very reliably. This explains the relatively complex coding scheme on the SCH. The MS uses the transmitted reference numbers regarding the multiframes, etc. to set its internal counter. This internal counter is not only a time reference with respect to the timeslot and frame grid, but also serves as a time reference within a timeslot with a quarter-bit precision. This reference is initially adjusted by considering the start of SCH bursts received at the MS. The MS then transmits the RACH burst relative to this internal reference. Based on the reception of the RACH, the BS can estimate the roundtrip time between the BS and the MS and use this information for timing advance (described in the next section).

24.5.3 Timing Advance

GSM supports cell ranges of up to 30 km, so that propagation delay between the BS and the MS might be as big as 100 μs. Thus, the following situation might occur: consider user A being at a 30-km distance from the BS, and transmitting bursts in timeslot TS 3 of every frame. User B is located close to the BS and accesses timeslot TS 4. The propagation delay of user A is around 100 μs, whereas the propagation delay of user B is negligible. Therefore, if propagation delay is not compensated, the end of a burst from user A partly overlaps with the beginning of a burst from user B at the BS (this situation is illustrated in Figure 24.9).

Figure 24.9   Overlapping bursts assuming uncompensated propagation delay.

To overcome this problem, the propagation delay from MS to BS is estimated by the BS during the initial phase of establishing a connection. The result is transmitted to the MS, which then sends its bursts *advanced* (with respect to the regular timing structure) to ensure that the bursts *arrive* within the dedicated timeslots at the BS. As the access bursts are transmitted before the MS is aware of the propagation delay, it now becomes clear why they must have a bigger guard period than normal transmission bursts: the guard period must be big enough to accommodate the worst case propagation delay – i.e., an MS at the boundary of a cell with maximum size.

There are some very big cells in rural areas where propagation delays exceed foreseen timing advances. In these cases, it might be necessary to use only every second timeslot as otherwise timeslots from different users would collide at the BS. This implies a waste of capacity. However, as this might only occur in rural areas with big cell ranges and low subscriber densities, the actual loss for the provider is small.

24.5.4 Summary of Burst Structures

Finally, Figure 24.10 provides an overview of the different kinds of bursts and illustrates the functions of their bits.

Normal

3 start bits	58 data bits (encrypted)	26 training bits	58 data bits (encrypted)	3 stop bits	8.25 bits guard period

FCCH burst

3 start bits	142 zeros	3 stop bits	8.25 bits guard period

SCH burst

3 start bits	39 data bits (encrypted)	64 training bits	39 data bits (encrypted)	3 stop bits	8.25 bits guard period

RACH burst

8 start bits	41 synchronization bits	36 data bits (encrypted)	3 stop bits	68.25 bits extended guard period

Dummy burst

3 start bits	58 mixed bits	26 training bits	58 mixed bits	3 stop bits	8.25 bits guard period

Figure 24.10 Structure of timeslots in the Global System for Mobile communications. Reproduced with permission from Rappaport [1996] © IEEE.

24.6 Coding

To transmit speech via the physical GSM channel the "speech signals" have to be translated into digital signals. This process should maintain a certain speech quality while keeping the required data rate as low as possible (see also Chapter 15). Different forms of speech coding were considered for GSM, and finally a *Regular Pulse Excited with Long Term Prediction* (RPE-LTP) solution was chosen (see Chapter 15). The digitized speech that is obtained in such a way then has to be protected by FEC in order to remain intelligible when transmitted over typical cellular channels (uncoded Bit Error Rates (BERs) of $\sim 10^{-3}$ to 10^{-1}). Both block and convolutional codes are used for this purpose in GSM.

Thus, voice transmission in GSM represents a typical example of the paradox of speech communications. First, redundancy is removed from the source data stream during the speech-coding process, and then redundancy is added in the form of error-correcting coding before transmission. The reason for this approach is that the original redundancy of the speech signal is rather inefficient at ensuring intelligibility of speech when transmitted over wireless channels. In this section, we first describe voice encoding, and subsequently channel coding; these can be seen as important applications of the principles expounded in Chapters 15 and 14, respectively.

24.6.1 Voice Encoding

Like most voice encoders (also referred to as *vocoders*), the GSM vocoder is not a classical source-coding processes like, e.g., the *Huffman code*. Rather, GSM uses a lossy compression method, meaning that the original signal cannot be reconstructed perfectly, but that the compression and decompression procedures lead to a signal which is similar enough to the original one to allow comfortable voice communications. As GSM has evolved, so has the speech coder. For the first release of GSM, an RPE-LTP approach was used. The idea behind this approach is to consider the human voice as output from a time-varying filter bank which is excited periodically. Both parameters describing the filter bank and the excitation process are transmitted. Since the samples of a voice signal are correlated, any sample can be predicted approximately by linearly combining previous samples. Evidently, correlation reflects the redundancy of the voice signal. However, the correlation properties of the signal vary with time, therefore the filter bank has to be time varying as well.

Later on, an enhanced speech coder was introduced that improved speech quality without increasing the required data throughput. A more detailed description of GSM speech coding can be found in Appendix 24.B (see *www.wiley.com/go/molisch*), and the principles of speech coding in general are described in Chapter 15.

The data created by the vocoder are divided into different classes, which have different vulnerabilities to bit errors. By this we mean that the bits have different levels of importance for the *perceived* quality of the reconstructed signal. The bits in class 1a are important, as an error in these bits is perceived as a gross distortion of the signal. Therefore, they are protected by a convolutional code and additional block coding. The slightly less important bits of class 1b are protected just by a convolutional code, while the bits associated with class 2 are transmitted without further channel encoding.

Another method to reduce the data rate is *Voice Activity Detection* (VAD). It detects periods when the user is not speaking, and ceases transmission during these periods – this is *Discontinuous Transmission* (DTX). DTX increases the battery lifetime of the MS and reduces co-channel interference with other users.

24.6.2   Channel Encoding

Let us first give an overview of the encoding procedure. Figure 24.11 illustrates the channel coding applied to voice data. For every 20-ms voice signal there are 50 very important bits (class 1a). A block code adds 3 parity bits. This coding is not error *correcting*, but only allows *detection* of bit errors within these 50 bits. The 132 bits of class 1b are attached. After attaching 4 tail bits to determine the final state of the Viterbi decoder, a convolutional code with rate-1/2 is applied. This

Figure 24.11   Channel coding for voice data in the Global System for Mobile communications.

results in 378 bits which are transmitted together with the 78 bits of class 2. Thus, for every 20 ms of the voice signal, 456 bits have to be transmitted. In the following, the details of the different encoder blocks will be discussed.

Block Encoding

Block Encoding of Voice Data As discussed above, only class-1a bits of the voice data are encoded using a (53,50) block code. This is a very "weak" block code. It is only supposed to detect bit errors and cannot detect more than three bit errors within the 50 class 1a bits reliably. However, this is sufficient, since a block is completely discarded if an error is detected within the class-1a bits; the receiver then smoothes the resulting signal by "inventing" a block. Figure 24.12 shows the linear shift register representation of the block encoder. As the code is systematic, the 50 data bits pass through the encoder unchanged. However, each of them impacts the state of the shift register. The final state of the shift register determines the 3 parity bits which are attached to the 50 class-1a bits. Class 1a, 1b, and parity check bits are then reordered and interleaved. Finally, four all-zero tail bits are attached, which are needed for the convolutional decoder (see below).

Generator-Polynomial
$G_4(D) = D^3 + D + 1$

1 ... 50 CKL: SW closed
51 ... 53 CKL: SW open

Figure 24.12 Shift register structure for voice block encoding, Cla (53,50) systematic, cyclic block encoder. *In this figure:* CKL, Clock; SW, Switch.

Block Encoding of Signaling Data As mentioned in Section 24.4.1, the signaling information has to have stronger protection against bit errors than the voice data. While a bit error in voice-related data might lead to an unintelligible audio signal for 20 ms, a bit error in signaling bits can have a more severe impact – e.g., handover to a wrong cell and therefore loss of connection. Thus, higher redundancy is required. For most of the control channels, only 184 signal bits are transmitted within 20 ms (instead of 260 for speech). This allows better error correction. Signaling bits are encoded with a (224,184) Fire code. The Fire code is defined by the generator polynomial:

$$G(D) = D^{40} + D^{26} + D^{23} + D^{17} + D^3 + 1 \tag{24.1}$$

Fire codes are block codes which are particularly capable of correcting burst errors. Burst errors are defined as a series of bit errors, meaning that two or more consecutive bits are wrong; such error bursts occur, e.g., when Viterbi decoding fails (see Chapter 14). A total of 4 tail bits are attached to the resulting 224 bits. The result is fed into the convolutional encoder at code rate-1/2, which is the same as that used for class 1 of the voice signal. For selected logical signaling channels, such as RACH and SCH, different generator polynomials are used. The interested reader is referred to Steele and Hanzo [1999] and the GSM specifications.

Convolutional Encoding

Both the class-1 bits of the voice data and all of the signaling information are encoded with a convolutional coder at code rate-1/2 (see Section 14.3). The bits are fed into a 5-bit shift register. For each new input bit, two codebits are calculated according to the generator polynomials

$$\left. \begin{array}{l} G1(D) = 1 + D + D^3 + D^4 \\ G2(D) = 1 + D^3 + D^4 \end{array} \right\} \tag{24.2}$$

and transmitted. The 4 final tail bits attached to the input sequence ensure that the encoder terminates in the all-zero state at the end of each encoded block.

Interleaving

Due to the nature of fading channels, bit errors may occur in bursts in some transmission blocks – e.g., if those blocks were transmitted during a deep fade. Interleaving orders the bits in such a manner that the burst errors due to the channel are (hopefully) distributed evenly (see Section 14.7.1) Evidently, the more the interleaver distributes corrupted bits, the better. However, latency of the speech signal puts an upper limit on *interleaver depth:* In order to give acceptable speech quality, the delay of the signal should be less than 100 ms.

GSM interleaves the data of two blocks (henceforth called "a" and "b") in the following way: first, each of the blocks is divided into eight subblocks. Specifically, each bit receives an index $i \in \{0, \ldots, 455\}$, and the bits are sorted into subblocks with index $k \in \{0, \ldots, 8\}$ according to $k = i \bmod 8$. Each subblock of block "a" contributes one half of the bits in a transmission burst (114 bits). The other half is associated with subblocks of either a previous or a succeeding block "b." Figure 24.13 illustrates diagonal interleaving.

Frame number

$i+0$	$i+1$	$i+2$	$i+3$	$i+4$	$i+5$	$i+6$	$i+7$

0a	4b	1a	5b	2a	6b	3a	7b	4a	0b	5a	1b	6a	2b	7a	3b

114 Bits	114 Bits

Figure 24.13 Diagonal interleaving for traffic channel/slow associated control channel/fast associated control channel data.
Reproduced with permission from Rappaport [1996] © IEEE.

24.6.3 Cryptography

One of the most severe shortcomings of analog mobile communications was the ease with which it could be intercepted. Anybody with a frequency scanner was able to eavesdrop on phone conversations. This posed a threat – e.g., for business people dealing with confidential material. Furthermore, even political scandals have been known to develop as a consequence of eavesdropped conversations.

In a digital system, this problem can be solved by "standard" means: once the audio signal is represented by a bitstream, cryptographic procedures, which had long before been developed for military applications, can be easily applied. For GSM, intercepting a conversation requires a *man-in-the-middle attack*, which involves implementing a BTS, to which the target MS would log on, and forwarding the intercepted signal (to stop the victim noticing the attack) – an exceedingly cumbersome and costly approach. Law enforcement thus typically obtains the collaboration of the network providers, and intercepts conversations not over the air, but *after* the BTS.

Encryption of the transmitted signal is achieved by simply using an XOR operation on the bits on one hand, and a Pseudo Noise (PN)-sequence on the other hand. A PN-sequence is based on feedback linear shift registers and its periodicity is 3.5 hours. Thus, even knowing the sequence does not enable interception, as the listener has to know which part of the sequence is currently in use. The algorithm for encryption of the data, the A5 algorithm, and algorithms involved in authentication, the A3 and A8, were originally only disclosed to members of the MoU. However, they have been reverse engineered in recent years and successful attacks have been developed. Nevertheless, all these attacks involve a lot of effort and investment. Thus, the GSM air interface still provides the user with a high level of privacy.

24.6.4 Frequency Hopping

Slow frequency hopping is an optional feature in GSM, where the carrier frequency is changed for each transmission burst (see also Section 18.1). This helps to mitigate small-scale fading: if the carriers employed are separated by more than one coherence bandwidth of the channel (see Chapter 6), each frame is transmitted over a channel with independent fading realization.[9] As the data belonging to one payload (voice data) packet are interleaved over eight bursts, the probability that all of them are transmitted via bad channels is negligible. This makes it more likely that the packet can be reconstructed at the receiver. A similar effect occurs for (narrowband) interference. In either case (fading and interference), frequency hopping leads to an effective whitening of noise and interference.

The coherence bandwidth of GSM channels can vary from several hundred kHz to a few MHz (see Chapter 7). Given that an operator typically owns only a few MHz of spectrum, and only a subset of frequencies can be used in each cell (see Chapter 17), there can be correlation between the frequency channels used for the hopping. Still, frequency hopping provides some advantages even in this case: co-channel interference from other cells, in particular, is whitened.

In order for the receiver to follow the hopping pattern of the transmitter, both link ends have to be aware of the order in which carrier frequencies are to be used. The control sequence governing this pattern can specify hops over up to 64 different carriers, but it may also specify the "degenerated" case (no hopping) in which one frequency is used over and over.

The BS determines the actual frequency hopping sequence by selecting one of a set of predefined PN-sequences and matching this sequence to the available frequencies for the cell. Furthermore, it informs the MS about the hopping sequence as well as the phase of the sequence – i.e., when to start – during call setup (for details see Steele and Hanzo [1999]). Finally, we note that frequency hopping is *not* applied to the physical channels related to BCHs and CCCHs, as the MS is supposed to "find" them easily.

24.7 Equalizer

Since the symbol duration of GSM is shorter than typical channel delay spreads, ISI occurs, making it necessary to perform equalization (see Chapter 16). However, as GSM is an open standard, neither the structure nor the algorithms of the equalizer are specified. The signal structure just provides

---

[9] This is particularly important if the MS is not moving: without frequency hopping, it would "see" the same channel at all times; thus if it is in a fading dip, the Bit Error Rate (BER) would be very high.

the necessary "hooks" (means of implementation), such as a training sequence used to estimate the channel impulse response. The most important properties of the training sequence are as follows:

- The training sequence is 26 bits long.
- It is transmitted in the middle of a burst and, hence, is also called the *midamble* – it is preceded by 57 data bits, and followed by another 57 data bits.
- Eight different PN-sequences are defined for the midamble – different midambles may be used in different cells, and thus help to distinguish between those cells.

The eight PN-sequences are designed in such a way that their autocorrelation function has a peak of amplitude 26 at zero offset, surrounded by at least five zeros for positive and negative offsets. Therefore, the channel impulse response can be simply estimated by cross-correlating the received midamble with the sequence, as long as the channel impulse response is less than 5 symbol durations long. The cross-correlation thus represents a scaled version of the channel impulse response. This information is used to correct the ISI for all symbols within one burst.[10]

GSM uses a *mid*amble for training as it is supposed to support MS speeds of up to 250 km/h. At this speed, the MS covers roughly one-eighth of a wavelength during transmission of a burst (500 µs). The impulse response of the channel shows some variations over this distance. Were the training sequence transmitted at the beginning of the burst (*preamble*), the resulting channel estimate would no longer be sufficiently accurate at the end of the burst. Since training is transmitted in mid-burst, the estimate is still sufficiently accurate at both the start and the end of a burst.

As mentioned above, the GSM standard does not specify any particular equalizer design. Actually, the equalizer is one of the reasons that products from different manufacturers can differ in price and quality. However, most implemented equalizers are Viterbi equalizers. The assumed constraint length of the channel, which relates to the number of states of the trellis, reflects a tradeoff between the complexity and performance of a Viterbi equalizer. Constraint length is identical to the memory of the channel – in other words, the length of the channel impulse response in units of symbol durations. In Chapter 7 we saw that COST 207[11] channel models normally have impulse response lengths of up to 15 µs, which equals 4 symbol durations.[12] We also note that Viterbi equalization can be well combined with convolutional decoding.

We emphasize again that a delay-dispersive fading channel with an appropriate equalizer at the receiver leads to *lower* average bit error probabilities than a flat-fading channel. As the different versions of a symbol arriving at the receiver at different time instants propagate over different paths, their amplitudes undergo independent fading. In other words, delay dispersion leads to delay diversity (see also Chapter 13).

Table 24.1 summarizes the key parameters of GSM.

24.8 Circuit-Switched Data Transmission

When the GSM standard was originally drafted, voice communication was envisioned as the main application. Some data transmission – like the Short Message Service (SMS) and a point-to-point data transmission channel with a 9.6-kbit/s data rate – were already included, but were not considered sufficiently important to merit the introduction of much additional complexity. Thus, data transmission was handled in a circuit-switched mode, just like voice transmission.

---

[10] Note that the SCH and the RACH use longer training patterns. To simplify implementation, the same algorithm is normally used for equalization of all three different bursts.

[11] COST stands for European COoperation in the field of Scientific and Technical research.

[12] Note that the constraint length of a Viterbi equalizer can be required to be longer due to other effects – e.g., ISI due to the GMSK. Therefore, 4–6 is a practical value for constraint length.

Table 24.1 Key parameters of the Global System for Mobile communications

Parameter	Value
Frequency range	
GSM900	880–915 MHz (uplink)
	925–960 MHz (downlink)
GSM1800	1710–1785 MHz (uplink)
	1805–1880 MHz (downlink)
GSM1900	1850–1910 MHz (uplink – U.S.A.)
	1930–1990 MHz
	(downlink – U.S.A.)
Multiple access	FDMA/TDMA/FDD
Selection of physical channel	Fixed channel allocation/intracell
	handover/frequency hopping
Carrier distance	0.2 MHz
Modulation format	GMSK ($B_G T = 0.3$)
Effective frequency usage per duplex speech	50-kHz/channel
connection	
Gross bit rate on the air interface	271 kbit/s
Symbol duration	3.7 µs
Channels per carrier	8 full slots (13 kbit/s user data)
Frame duration	4.6 ms
Maximal RF transmission power at the MS	2 W
Voice encoding	13 kbit/s RPE-LTP
Diversity	Channel coding with interleaving
	Channel equalization
	Antenna diversity (optional)
	Frequency hopping (optional)
Maximal cell range	35 km
Power control	30-dB dynamics

In general, the circuit-switched data transmission modes of GSM have severe disadvantages. A main issue is the low data rate of less than 10 kbit/s.[13] Furthermore, the long time needed to set up a connection, as well as the relatively high costs of holding a connection, make it very unattractive, e.g., for Internet browsing. There was simply a significant mismatch between the low-data-rate connection-based services offered by GSM, and the new Web applications, which require high data volumes in bursts interrupted by long idle periods. Only SMS text messaging proved to be successful. For these reasons, packet-switched (also known as connectionless) transmission (see Section 17.4) was introduced later on.

24.9   Establishing a Connection and Handover

In this section, we discuss initial establishing of a connection, and the handover procedure, using the logical channels described in Section 24.4. Furthermore, we explore the kind of messages that

---

[13] The High Speed Circuit Switched Data (HSCSD) mode provides higher data rates based on circuit-switched transmission.

need to be exchanged during these processes. As a first step, we define various elements of a GSM system that are required for these functionalities.

24.9.1 Identity Numbers

An MS or a subscriber can be localized within the network by using identity numbers.[14] An active GSM MS has multiple identity numbers.

Mobile Station ISDN Number (MS ISDN)

The MS ISDN is the unique phone number of the subscriber in the public telephone network. The MS ISDN consists of *Country Code* (CC), the *National Destination Code* (NDC), which defines the regular GSM provider of the subscriber, and the subscriber number. The MS ISDN should not be longer than 15 digits.

International Mobile Subscriber Identity (IMSI)

The IMSI is another unique identification for the subscriber. In contrast to the MS ISDN, which is used as the phone number of the subscriber within the GSM network *and* the normal public phone network, the IMSI is only used for subscriber identification in the GSM network. It is used by the *Subscriber Identity Module* (SIM), which we explain later, the HLR, and the VLR. It consists again of three parts: the *Mobile Country Code* (MCC, three digits), the *Mobile Network Code* (MNC, two digits), and the *Mobile Subscriber Identification Number* (MSIN, up to ten digits).

Mobile Station Roaming Number (MSRN)

The MSRN is a temporary identification that is associated with a mobile if it is not in the area of its HLR. This number is then used for routing of connections. The number consists again of a CC, MNC, and a TMSI, which is given to the subscriber by the GSM network (s)he is roaming into.

International Mobile Station Equipment Identity (IMEI)

The IMEI is a means of identifying hardware – i.e., the actual mobile device. Let us note here that the three identity numbers described above are all either permanently or temporarily associated with the subscriber. In contrast, the IMEI identifies the actual MS used. It consists of 15 digits: six are used for the *Type Approval Code* (TAC), which is specified by a central GSM entity; two are used as the *Final Assembly Code* (FAC), which represents the manufacturer; and six are used as a *Serial Number* (SN), which identifies every MS uniquely for a given TAC and FAC.

24.9.2 Identification of a Mobile Subscriber

In analog wireless networks, every MS was uniquely identified by a single number that was permanently associated with it. All connections that were established from this MS were billed to its registered owner. GSM is more flexible in this respect. The subscriber is identified by his SIM, which is a plug-in chipcard roughly the size of a postage stamp. A GSM MS can only make and

---

[14] Note that we distinguish between a subscriber and the hardware (s)he is using.

receive calls when such a SIM is plugged in and activated.[15] All calls that are made from the MS are billed to the subscriber whose SIM is plugged in. Furthermore, the MS only receives calls going to the number of the SIM owner. This makes it possible for the subscriber to easily replace the MS, or even rent one for a short time.

As the SIM is of fundamental importance for billing procedures, it has to have several security mechanisms. The following information is saved on it:

- *Permanent security information*: this is defined when the subscriber signs a contract with the operator. It consists of the IMSI, the authentication key, and the access rights.
- *Temporary network information*: this includes the TMSI, location area, etc.
- *Information related to the user profile*: e.g., the subscriber can store his/her personal phonebook on the SIM – in this way the phonebook is always available, independent of the MS the subscriber uses.

The SIM can be locked by the user. It is unlocked by entering the *Personal Unblocking Key* (PUK). If a wrong code is entered ten times, the SIM is finally blocked and cannot be reactivated. Removing the SIM and then plugging it into the same or another MS does not reset the number of wrong trials. This blocking mechanism is an important security feature in case of theft.

The *Personal Identification Number* (PIN) serves a similar function as the PUK. The user may activate the PIN function, so that the SIM requests a four-digit key every time an MS is switched on. In contrast to the PUK, the PIN may be altered by the user. If a wrong PIN is entered three times, the SIM is locked and may be unlocked only by entering the PUK.

24.9.3 Examples for Establishment of a Connection

In the following, we give two examples for the steps that are performed when a connection is established. Both the user identification numbers and the different logical channels (see Section 24.4) play a fundamental role in this procedure.

If a subscriber wants to establish a connection from his MS, the following procedure is performed between the MS and the BTS to initialize the connection:

1. The MS requests an SDCCH from the BS by using the RACH.
2. The BS grants the MS access to an SDCCH via the AGCH.
3. The MS uses the SDCCH to send a request for connection to the MSC. This includes the following activities: the MS tells the MSC which number it wants to call. The authentication algorithm is performed; in this context it is evaluated if the MS is allowed to make a requested call (e.g., an international call). Furthermore, the MSC marks the MS as busy.
4. The MSC orders the BSC to associate a free TCH with the connection. The BTS and the MS are informed of the timeslot and carrier number of the TCH.
5. The MSC establishes a connection to the network to which the call should go – e.g., the PSTN. If the called subscriber is available and answers the call, the connection is established.

A call that is incoming from another network starts the following procedure:

1. A user of the public phone network calls a mobile subscriber, or more precisely, an MS ISDN. The network recognizes that the called number belongs to a GSM subscriber of a specific provider, since the NDC in the MS ISDN contains information about the network. The PSTN thus establishes a connection to a gateway MSC[16] of the GSM provider.

---

[15] Emergency calls can be made without a SIM.
[16] A gateway MSC is an MSC with a connection to the regular phone network.

2. The gateway MSC looks in the HLR for the subscriber's information and the routing information (the current location area of the subscriber).
3. The HLR translates the MS ISDN into the IMSI. If call forwarding is activated – e.g., to a voicemail box – the process is altered appropriately.[17]
4. If the MS is roaming, the HLR is aware of the MSC it is connected to, and sends a request for the MSRN to the MSC that is currently hosting the MS.
5. The hosting MSC sends the MSRN to the HLR. The gateway MSC can now access this information at the HLR.
6. As the MSRN contains an identification number of the hosting MSC, the gateway can now forward the call to the hosting MSC. Additional information – e.g., the caller ID – is included.
7. The hosting MSC is aware of the *location area* of the mobile. The location area is the area controlled by one BSC. The MSC contacts this BSC and requests it to page the MS.
8. The BSC sends a paging request to all the BTSs that cover the location area. These transmit the paging information via the BCH.
9. The called MS recognizes the paging information and sends its request for an SDCCH.
10. The BSC grants access to an SDCCH via the AGCH.
11. Establishment of the connection via the SDCCH follows the same steps as described in bullets 3 and 4 of the "MS-initiated call." If the mobile subscriber answers the incoming call, the connection is established.

24.9.4 Examples of Different Kinds of Handovers

A *handover* is defined as the procedure where an active MS switches the BTS with which it maintains a link; it is a vital part of mobility in cellular communications. Handover is performed when another BTS is capable of providing better link quality than the current one. In order to determine whether another BTS could provide better link quality, the MS monitors the signal strength of the BCH of neighboring BTSs. Since the BCH does not use power control, the MS measures the maximum signal power available from other BTSs. It transmits the results of these measurements to the BSC. Furthermore, the currently active BTS measures the quality of the uplink and sends this information to the BSC. Based on all this information, the BSC decides if and when to initiate a handover. Since the MS contributes to the handover decision, this procedure is called *Mobile Assisted Hand Over* (MAHO).

Let us now consider some more details in this procedure:

- The received signal strength from different BTSs is averaged over a few seconds (the exact values are selected by the network provider); this ensures that small-scale fading does not lead to a handover. Otherwise, an MS exposed to a similar signal strength from two BTSs would constantly switch from one BTS to the other by just moving over a small distance.
- Receive power is measured at 1-dB resolution in the range of $-103\,\mathrm{dBm}$ to $-41\,\mathrm{dBm}$. The lower bound reflects the sensitivity of GSM receivers – i.e., the minimum signal power required for communication.
- Furthermore, a handover is initialized when the necessary timing advance exceeds the specified limit of $235\,\mu s$. If the MS is so far away from the BTS that a bigger timing advance is necessary, a handover is made to a closer BTS.
- Even more importantly, a handover is initialized when the signal quality becomes too low due to interference.
- The BS transmits (via the BCCH) several parameters that support the handover procedure.

---

[17] This information is found in the HLR as well.

In the following, we describe three different types of HOs: the most simple involves only BTSs controlled by the same BSC. A more complex case arises if two different BSCs are connected to the same MSC. The most complex case involves different MSCs.

Case 24.1 – Handover between BTSs Belonging to the same BSC

The steps for this case are illustrated in Figure 24.14:

1. The BSC orders the new BTS to activate a new physical channel.
2. The BSC uses the FACCH of the link between the MS and the old BTS in order to transmit information about the carrier frequency and timeslot of the physical channel for the new BTS.
3. The MS switches to the new carrier frequency and timeslot and sends handover access bursts. These bursts are similar to RACH bursts: they are shorter than normal transmission bursts, as the necessary timing advance is unknown and has to be evaluated first by the new BTS.
4. After the new BTS has detected the handover bursts, it sends the necessary timing advance and power control information to the MS via the FACCH of the new channel.
5. The MS informs the BSC that the handover was successful.
6. The BSC requests the old BTS to switch off the old channel.

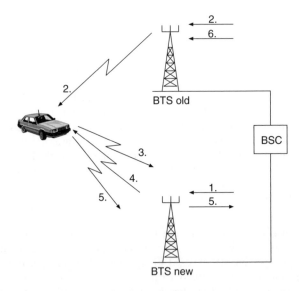

Figure 24.14   Handover between two base transceiver stations of the same base station controller.

Case 24.2 – Handover between Two BTSs that are Controlled by Different BSCs but the Same MSC

The steps for this case are illustrated in Figure 24.15.

1. The old BSC informs the MSC that a handover to a specific BTS is necessary.
2. The old MSC knows which BSC controls this BTS (the new BTS) and informs this new BSC about the upcoming handover.

Figure 24.15 Handover between two cells belonging to different base station controllers but the same mobile switching center.

3. The new BSC requests the new BTS to activate a physical channel.
4. The new BSC informs the MS about the carrier frequency and timeslot for the new link. This information goes via the MSC, the old BSC, and the old BTS, and is finally transmitted to the MS on the FACCH of the old BTS.
5. The MS switches to this new carrier frequency and timeslot and transmits access bursts (compare item 3 in Case 24.1).
6. After detecting handover bursts, the BTS transmits information regarding timing advance and power control to the MS via the FACCH.
7. The MS informs the old BSC about the successful handover (via the new BSC and the MSC).
8. The new BSC instructs the old BSC (via the MSC) to relinquish the connection to the MS.
9. The old BSC instructs the old BTS to deactivate the old physical channel.

Case 24.3 – Handover between Two BTSs which are Associated with Two Different MSCs

The steps for this case are illustrated in Figure 24.16:

1. The old BSC informs its own MSC (in the following called "MSC-A") about the necessary handover.
2. MSC-A recognizes that the requested handover involves a BTS associated with another MSC (in the following called "MSC-B") and contacts this MSC-B.
3. MSC-B associates a handover number with the process, so that it is able to reroute the connection. Subsequently, it informs the new BSC about the upcoming handover.
4. The new BSC orders the new BTS to activate a physical channel.
5. MSC-B gets the information about the carrier frequency and timeslot of the new physical channel and forwards this information to MSC-A. Furthermore, it informs MSC-A about the handover number of the connection.

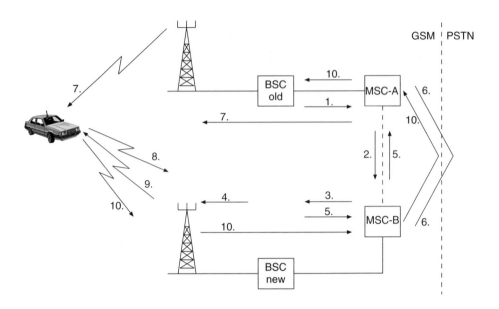

Figure 24.16   Handover between two cells belonging to two different mobile switching centers.

6. A connection between MSC-A and MSC-B is established.
7. MSC-A informs the MS about the carrier frequency and timeslot of the new physical channel. The information goes from MSC-A via the old BSC and the old BTS, whence it is transmitted to the MS via the FACCH.
8. As in Cases 24.1 and 24.2, the MS transmits handover bursts on the new physical channel.
9. After detecting the handover bursts, the new BTS instructs the MS about power control and timing advance.
10. The MS informs MSC-A about the successful handover; this information goes via the new link over the new BTS, the new BSC and MSC-B. From this time on, MSC-A forwards the connection to MSC-B. However, the connection is still maintained by MSC-A. MSC-A acts therefore as a so-called *anchor MSC*.
11. The old physical channel is deactivated.
12. After the connection ends, the new location area of the MS is established. Therefore, the VLR of MSC-B informs the HLR that the MS is now in its area and the HLR updates its entries regarding the location of the MS. Furthermore, the HLR request the VLR of MSC-A to delete all entries associated with the MS.

We see from these examples that the handover procedure depends on where in the network the switching centers are located.

24.10   Services and Billing

24.10.1   Available Services

In contrast to analog cellular networks, GSM offers a variety of services in addition to regular phone calls. Although providing those services is not a big effort for the network provider, they were a major motivation for customers to switch from analog systems to GSM or another digital mobile phone system. We thus briefly discuss in the following the services offered by GSM, distinguishing between (i) *teleservices*, (ii) *bearer services*, and (iii) *supplementary services*. Teleservices provide a connection between two communication partners, though they may use some additional devices to exchange information via this connection. Bearer services allow the subscriber to access another system. Thus, they provide a connection from the MS to an access point to another network (not a specific terminal in another network). Supplementary services support or manage both teleservices and bearer services.

Teleservices

- *Regular phone calls*: this is still the most common application of GSM.
- *Dual Tone Multi Frequency* (DTMF): this is a signaling method to allow the subscriber to control a device connected to a phoneline using the keypad of the MS. A typical example is the remote checking of an answering machine connected to a regular landline.
- *Emergency calls*: calls to emergency numbers have priority in GSM. If a cell is already full, an emergency call leads to interruption of another connection. Remember also that emergency calls can be made from any MS, even without a valid SIM.
- *FAX*: the protocol that is used to transmit *Commite' Consultatif International de Telegraphique et Telephonique (CCITT) group 3* faxes is incompatible with regular GSM connections. The classical fax protocol was developed to transmit a picture that was converted into a specific digital form over an analog phoneline. GSM, on the other hand, provides either voice connections with a voice-encoding function or data transmission channels for digital data. To send or receive a fax via GSM, adapters are thus required both at the GSM network end and at the MS. The *Terminal Adapter Function* (TAF) provides a general interface between GSM devices and other devices, and is combined with a special fax adapter.
- *SMS*: a short message consists of up to 160 alphanumerical characters. A short message may be transmitted from or to an MS. If it is transmitted to an MS that is switched off, the message is accepted by the GSM system and stored in a dedicated database (*SMS center*). The message is delivered once the MS is switched on and its location is known. This service has turned out to be staggeringly successful, with more than a trillion SMSs sent every year.
- *Cell broadcast*: a cell broadcast transmits a broadcast message of up to 93 alphanumeric characters to all the MSs within the range of a BTS. This may be used to transmit, e.g., news regarding the local (car) traffic situation.
- *Voicemail*: the voicemail service allows the subscriber to forward incoming calls to a service center of the GSM network that acts like an answering machine. After listening to a message from the subscriber the calling party may leave a voice message, which the subscriber can retrieve by connecting to this service center. Typically, a subscriber is notified about new messages via SMS.
- *Fax mail*: this is a service that allows the subscriber to forward incoming faxes to a special service center in the GSM network. The faxes can be retrieved by the subscriber by connecting to this center either with a GSM connection or from the PSTN.

Bearer Services

- *Connections to the PSTN*: this allows the user to connect, e.g., to modems connected to an analog landline.
- *Connections to the Integrated Services Digital Network (ISDN)*: all digital information can be transmitted over the digital network.
- *Connections to the packet-switched networks*: the user may also access packet-switched networks.

Supplementary Services

GSM enables the network provider to offer the subscriber a variety of additional, supplementary services:

- *Call forwarding*: the user can select under which conditions calls to his/her mobile number are forwarded to another number: (i) always; (ii) in case the MS cannot be reached; (iii) the user is making or receiving another call; or (iv) the user does not answer after a specified number of rings.
- *Blocking of outgoing calls*: the user or the provider may block outgoing calls under some of the following conditions: (i) all outgoing calls; (ii) all international calls; or (iii) all international calls, with the exception of those to the country of origin in case the subscriber is abroad.
- *Blocking of incoming calls*: this feature is of interest when the subscriber has to pay part of (or all of) the charges for an incoming call. The feature may be activated always or when the user is roaming out of the original network.
- *Advice of charges*: the user may be able to access an estimation of the call charges.
- *Call hold*: the user may put a connection on hold to make or receive another call and then continue with the first connection.
- *Call waiting*: during a call, the subscriber may be informed about another incoming call. He/she may either answer this call by putting the other call on hold or reject it. This feature is available for all circuit-switched connections except emergency calls.
- *Conference calls*: this feature enables connection to multiple subscribers simultaneously. It is only possible for normal voice communications.
- *Caller ID*: the phone number of the incoming call is displayed.
- *Closed groups*: subscribers in GSM, ISDN, and other networks may be defined as a specific user group. Members of this group can, e.g., be allowed to make calls only within the group.

24.10.2 Billing

In GSM, billing for the variety of different subscriber plans is not only an economics issue but also a technical challenge that involves the design of an *OSS*. In contrast to the regular public phone system, not all fees have to be paid by the party initiating the calls. Furthermore, accounting for supplementary services has to be done separately. To give an impression of the complexity of the accounting involved, we discuss one particular example here.[18]

Example 24.1   *Billing in the Global System for Mobile communications.*

The example involves the following communication parties:

- Subscriber A originates from Austria but is temporarily in Poland.
- Subscriber B is an English subscriber staying in France with a rented MS but his own English SIM.
- Subscriber C is Italian. He is on vacation and the option "If user does not answer, forward call to Subscriber B" is activated.
- Subscriber D is a subscriber to a U.S. service, but is currently in Mexico.

---

[18] Note that we base the discussion of this example partly on European billing procedures. This is fundamentally different from U.S. billing. In the U.S.A., mobile numbers are similar to landline numbers. Therefore, the calling party just pays the regular fees for a landline call, whereas the mobile subscriber pays the landline-to-mobile fees even for incoming calls.

Communication now follows the steps below:

1. Subscriber A calls subscriber C in Italy.
2. As subscriber C is not answering, the call is forwarded to subscriber B.
3. Subscriber B is in France and is currently speaking on his MS. Therefore, subscriber A activates the option "automatic call to busy MS." Thus, the MS automatically initiates a call the moment the other MS is no longer busy.
4. After subscriber B finishes his conversation, the MS of subscriber A initiates a connection to the MS of subscriber B.
5. This connection is first routed to England, where the HLR of subscriber B is located.
6. From there it gets forwarded to France, where subscriber B is right now.
7. During the conversation, subscriber B needs some information from subscriber D. Therefore, he initiates a "conference call" and calls subscriber D.
8. The call to subscriber D is first routed to the U.S.A. and from there to Mexico, where subscriber D is temporarily staying.

Now the question arises as to which subscriber is charged for which fees?

- Subscriber A has to pay the fees for a call from Poland to Italy. He has to pay both the "international call" charges, and the roaming fees (as he is not in his home country). Furthermore, he has to pay for the service "automatic call to busy MS."
- Subscriber B has to pay for a connection from England to France (for the incoming call), the charges for an international call (from France to the U.S.A.), and the roaming charges (initiating a call while being in a different network). Further, he has to pay for the "conference call" feature.
- Subscriber C has to pay for the connection from Italy to England and the "call forwarding" feature involved.
- Subscriber D has to pay for a "received call" in the U.S.A. (note that in the U.S.A., the called party pays for a received call the same way as for an active call), and the roaming fees from the U.S.A. to Mexico.

We see that for the same conversation different subscribers get charged different fees depending on their roaming. Subscribers do not have to be actively involved in the conversation to be charged (see subscriber C in the above example). This example gives a taste of the complexity of the billing software in the OSS.

24.11 Glossary for GSM

AB	Access Burst
AC	Administration Centre
ACCH	Associated Control CHannel
ACM	Address Complete Message
AGCH	Access Grant CHannel
ARFCN	Absolute Radio Frequency Channel Number
AUC	Authentication Center
BCC	Base station Color Code
BCCH	Broadcast Control CHannel
BCF	Base Control Function

BCH	Broadcast CHannel
Bm	Traffic channel for full-rate voice coder
BN	Bit Number
BNHO	Barring all outgoing calls except those to Home PLMN
BS	Base Station
BSC	Base Station Controller
BSI	Base Station Interface
BSIC	Base Station Identity Code
BSS	Base Station System
BSSAP	Base Station Application Part
BTS	Base Transceiver Station
CA	Cell Allocation
CBCH	Cell Broadcast CHannel
CC	Country Code
CCBS	Completion of Calls to Busy Subscribers
CCCH	Common Control CHannel
CCPE	Control Channel Protocol Entity
CI	Cell Identify
CM	Connection Management
CONP	Connect Number Identification Presentation
CUG	Closed User Group
DB	Dummy Burst
DCCH	Dedicated Control CHannel
DRM	Discontinuous Reception Mechanisms
DTAP	Direct Transfer Application Part
DTE	Data Terminal Equipment
DTMF	Dual Tone Multi Frequency (signalling)
DRX	Discontonuous Reception
DTX	Discontonuous Transmission Mechanisms
EIR	Equipment Identify Register
FB	Frequency correction Burst
FACCH	Fast ACCH
FACCH/F	Full-rate FACCH
FACCH/H	Half-rate FACCH
FCH	Frequency Correction Channel
FN	Frame Number
GMSC	Gateway Mobile Services Switching Centre
GSM	Global System for Mobile communications
HDLC	High Level Data Link Control
HLR	Home Location Register
HMSC	Home Mobile-services Switching Centre
HSN	Hop Sequence Number
IAM	Initial Address Message
ICB	Incoming Calls Barred
ID	Identification
IMEI	International Mobile station Equipment Identity
IMSI	International Mobile Subscriber Identity
ISDN	Integrated Services Digital Network
IWF	Inter Working Function
Kc	Cipher Key

Ki	Key used to calculate SRES
Kl	Location Key
Ks	Session Key
LAC	Location Area Code
LAI	Location Area Identify
LAP	hyphen;Dm Link Access Protocol on Dm Channel
LPC	Linear Prediction Coding (Voice Codec)
LR	Location Register
MA	Mobile Allocation
MACN	Mobile Allocation Channel Number
MAF	Mobile Additional Function
MAIO	Mobile Allocation Index Offset
MAP	Mobile Application Part
MCC	Mobile Country Code
ME	Maintenace Entity
MEF	Maintenace Entity Function
MIC	Mobile Interface Controller
MNC	Mobile Network Code
MS	Mobile Station
MSC	Mobile-services Switching Centre
MSCU	Mobile Station Control Unit
MS	ISDN Mobile Station ISDN Number
MSL	Main Signaling Link
MSRN	Mobile Station Roaming Number
MT	Mobile Terminal
MTP	Message Transfer Part
MUMS	Multi User Mobile Station
NB	Normal Burst
NBIN	A parameter in the hopping sequence
NCELL	Neighbouring (adjacent) Cell
NDC	National Destination Code
NF	Network Function
NM	Network Management
NMC	Network Management Centre
NMSI	National Mobile Station Identification number
NSAP	Network Service Access Point
NT	Network Termination
OACSU	Off Air Call Set Up
O&M	Operations & Maintenance
OCB	Outgoing Calls Barred
OMC	Operations & Maintenance Center
OS	Operating Systems
PAD	Packet Assembly/Disassambly facility
PCH	Paging CHannel
PIN	Personal Identification Number
PLMN	Public Land Mobile Network
PSPDN	Public Switched Public Data Network
PSTN	Public Switched Telephone Network
PTO	Public Telecommunications Operators
RA	Random Mode Request information field

RAB	Random Access Burst
RACH	Random Access Channel
RFC	Radio Frequency Channel
RFN	Reduced TDMA Frame Number
RLP	Radio Link Protocol
RNTABLE	Table of 128 integers in the hopping sequence
RPE	Regular Pulse Excitation (Voice Codec)
RXLEV	Received Signal Level
RXQUAL	Received Signal Quality
SABM	Set Asynchronous Balanced Mode
SACCH	Slow Associated Control Channel
SACCH/C4	Slow, SACCH/C4 Associated Control CHannel
SACCH/C8	Slow, SACCH/C8 Associated Control CHannel
SACCH/T	Slow, TCH Associated Control CHannel
SACCH/TF	Slow, TCH/F Associated Control CHannel
SACCH/TH	Slow, TCH/H Associated Control CHannel
SAP	Service Access Points
SAPI	Service Access Points Indicator
SB	Synchronization Burst
SCCP	Signalling Connection Control Part
SCH	Synchronisation CHannel
SCN	Sub Channel Number
SDCCH	Standalone Dedicated Control CHannel
SDCCH/4	Standalone Dedicated Control CHannel/4
SDCCH/8	Standalone Dedicated Control CHannel/8

24.12   Appendices

Please go to *www.wiley.com/go/molisch*

Further Reading

This current chapter is of course only a brief overview of GSM technology. Much more detailed information may be found in the GSM specifications. Note, however, that they encompass 5,000 pages and are also written as a technical specification and not as a textbook, and most engineers only read small sections of them. Another useful source of information is the monograph by Mouly and Pautet [1992]. A detailed GSM chapter can also be found in Steele and Hanzo [1999], Steele et al. [2001], and Schiller [2003]. GPRS is discussed in Cai and Goodman [1997]; Bates [2008]; GSM network aspects are discussed in Eberspaecher et al. [2009].

For updates and errata for this chapter, see wides.usc.edu/teaching/textbook

25

IS-95 and CDMA 2000

25.1 Historical Overview

Direct-sequence spread-spectrum communications has a long history, dating back to the middle of the 20th century (see Chapter 18). However, it was deemed to be too complex for commercial applications for a long while. In 1991, the U.S.-based company Qualcomm proposed a system that was adopted by the *Telecommunications Industry Association (U.S.)* (TIA) as *Interim Standard 95* (IS-95). This system became the first commercial Code Division Multiple Access (CDMA) system that achieved wide popularity. In the years after 1992, cellular operators in the U.S.A. started to switch from analog (Advanced Mobile Phone System (AMPS)) to digital communications. While the market remained fragmented, IS-95 was adopted by a considerable number of operators, and by 2005 was used by two of the four major operators in the U.S.A. It also obtained a dominant market position in South Korea.

The original IS-95 system did not fully exploit the flexibility inherent in CDMA systems; however, later refinements and modifications did make the system more flexible, and thus ready for data communications. In the late 1990s, the need for further enhancement of data communications capabilities became apparent. The new third-generation systems needed to be able to sustain high data rates, thus enabling audio- and videostreaming, Web browsing, etc. This would require higher data rates and flexible systems that could easily support multiple data rates with fine granularity. CDMA seemed well suited to this approach, and was chosen by all major manufacturers. However, no unique standard evolved. The IS-95 proponents (mostly U.S.-based) developed the so-called CDMA 2000 standard, which is backward-compatible with IS-95, and allows a seamless transition. Operators that were using the Global System for Mobile communications (GSM) standard for their second-generation system are generally opting for the Wideband Code Division Multiple Access (WCDMA) standard described in Chapter 26. CDMA 2000 and WCDMA are quite similar, but have enough differences to make them incompatible.

The main part of this chapter describes the original IS-95 standard. The Appendices (see *www.wiley.com/go/molisch*) then summarize the changes in the CDMA 2000 standard.

25.2 System Overview

IS-95 is a CDMA system with an additional Frequency Division Multiple Access (FDMA) component. The available frequency range is divided into frequency bands of 1.25 MHz; duplexing is done in the frequency domain. In the U.S.A., frequencies from 1850–1910 MHz are used for the uplink,

Wireless Communications, Second Edition   Andreas F. Molisch
© 2011 John Wiley & Sons, Ltd

and 1930–1990 MHz are used for the downlink band.[1] Within each band, traffic channels, control channels, and pilot channels are separated by the different codes (chip sequences) with which they are spread. IS-95 specifies two possible speech coder rates: 13.3 or 8.6 kbit/s. In both cases, coding increases the data rate to 28.8 kbit/s. The signal is then spread by a factor of 64, resulting in a chip rate of 1.2288 Mchip/s. Theoretically, each cell can sustain 64 speech users. In practice, this number is reduced to 12–18, due to imperfect power control, nonorthogonality of spreading codes, etc.

The downlink signals generated by one Base Station (BS) for different users are spread by different Walsh–Hadamard sequences (see Section 18.2.6), and thus orthogonal to each other. This puts an upper limit of 64 channels on each carrier. In the uplink, different users are separated by spreading codes that are not strictly orthogonal. Furthermore, interference from other cells reduces signal quality at the BS and Mobile Station (MS).

BSs use transmit powers between 8 and 50 W, depending on the coverage area required. MSs use peak powers of some 200 mW; accurate power control makes sure that all signals arriving at the BS have the same signal strength. Traffic channels and control channels are separated by different spreading codes. All BSs are synchronized, using signals from GPS (Global Positioning System) to obtain an accurate system time. This synchronization makes it much easier for the MS to detect signals from different BSs and manage the handover from cell to cell.

The requirements for the network and switching system, as well as operating support, servicing, and billing, are quite similar to those in GSM, and will not be repeated here. Similarly, billing considerations are the same as in GSM.

25.3 Air Interface

25.3.1 Frequency Bands and Duplexing

As mentioned in Section 25.2, IS-95 is a CDMA system with an additional FDMA component; it uses Frequency Domain Duplexing (FDD) for separation of uplink and downlink. In the U.S.A., IS-95 operates in the 1850–1990-MHz frequency band. This band, called the Personal Communication System (PCS) band, is divided into units of 50 kHz width, so that the center frequencies are related to the channel numbers n_{ch} as

$$f_c = (1850 + n_{ch} \cdot 0.05) \, \text{MHz} \quad \text{for the uplink} \tag{25.1}$$

$$f_c = (1930 + n_{ch} \cdot 0.05) \, \text{MHz} \quad \text{for the downlink} \tag{25.2}$$

where $n_{ch} = 1, \ldots, 1199$. An IS-95 system requires 1.25 MHz – i.e., 25 of these channels.

There is also a version operating in the 800-MHz band (historically speaking, this was the first band to be used for IS-95). Extensions to other bands are happening as they become available (see also Section 27.1).

25.3.2 Spreading and Modulation

IS-95 uses error correction coding, as well as different types of spreading codes, in order to spread the source data rate of 8.6 kbit/s (for rate-set-1) or 13.3 kbit/s (for rate-set-2) to a chip rate of 1.2288 Mchip/s. Encoding is usually done with standard convolutional encoders (between rate 1/3 and 3/4), with spreading is done with so-called "M-ary orthogonal keying" and/or multiplication by spreading sequences. More details are discussed in Section 25.4.

---

[1] The IS-95 uses the word "reverse link" for the uplink, and "forward link" for the downlink. In order to stay consistent with the notation in our book, we stick here with up- and downlink. However, note that many abbreviations of the IS-95 standard use "F" (for forward) and "R" (for reverse) to indicate whether a channel is used in downlink or uplink.

25.3.3 Power Control

Power control is one of the key aspects of a CDMA cellular system, and the accuracy of power control determines to a large extent the capacity of the system (see also Chapter 18). For this reason, IS-95 foresees a quite involved power control procedure that is intended to make sure that both large-scale fading and Small-Scale Fading (SSF) is compensated by power control. Note that there are two measures for the quality of power control: accuracy and speed. Accuracy indicates how much the received power deviates from the ideal level in a "steady state" where the received power changes slowly, or not at all. The speed of power control determines how quickly power control can adapt to changing channel conditions.

IS-95 foresees both open-loop and closed-loop mechanisms. For the open-loop mechanism, the MS observes the power it receives from downlink signals, reaches a conclusion about the current path loss, and adjusts its transmit power accordingly. This mechanism can only compensate for shadowing and average path loss, but not for SSF, because the SSF is different at the uplink and downlink frequencies (see Section 20.1.6). At the start of a call, in particular, open-loop power control is the only available method for controlling MS power.

The closed-loop mechanism uses two different feedbacks: inner-loop and outer-loop power control. In both loops, the BS observes the signal it receives from an MS, and then sends a command to that MS to adjust its power appropriately. In the inner loop, the BS observes the Signal-to-Interference-and-Noise Ratio (SINR), and as a consequence requests an adjustment of the transmit power of the MS. This is done at 1.25 ms intervals, and the command sent to the MS is to either increase or decrease the power by 1 dB. In the outer loop, the BS appraises the performance of the closed loop using the statistics of the frame quality of uplink transmission. If the frame error rate is too high, closed-loop power control is used to request transmission at a higher power; specifically, the SINR target is adjusted. This is done once at 20-ms intervals (frame).

Power control also exists for the downlink. This rather crude power control allows power to be adjusted only in a rather limited range (about ± 6 dB around the nominal value), and at a low speed (once per frame). It is based on the closed-loop scheme, where the MS measures the received signal quality, and requests an adjustment to BS transmit power. The MS sends the ratio of the number of bad frames and number of total frames to the BS in a *power measurement report message*. It is worth noting that power control in the downlink is *not* essential for functioning of the system. The different signals all arrive at the MS after having gone through the same channel, and thus suffer from the same amount of fading. Rather, downlink power control serves to minimize the overall power transmitted by the BS, and thus the interference to other cells.

25.3.4 Pilot Signal

Each BS sends out a pilot signal that the MS can use for timing acquisition, channel estimation, and to help with the handover process. The pilot signal always has the same form (a Pseudo Noise (PN)-sequence with 32,768 chips – i.e., 26.7 ms long). Pilots from different BSs are offset with respect to each other by 64 chips, which corresponds to a 52.08-μs offset. This is usually long enough that pilots from other BSs are not confused with long-delayed echoes of the desired pilot (though exceptions exist, see Chapter 7). By just correlating the received signal with the pilot PN-sequence, the MS can determine the signal strength of all the BSs in its surroundings (this is discussed more in Section 25.7).

25.4 Coding

Before a speech signal can be transmitted over the air interface, it first has to be digitized and encoded. IS-95 foresees the use of different speech coders (vocoders, see Chapter 15). They have different bit rates. As spreading and modulation should not be affected by these different rates, it also implies that error correction coding has to be different: it has to be designed in such a way that output from the channel coder is always 19.2 kbit/s for the downlink and 28.8 kbit/s for the uplink.

25.4.1 Speech Coders

IS-95 uses a number of different speech encoders. The original system foresaw a vocoder with 8.6 kbit/s, the IS-96A coder. However, it turned out to have poor speech quality: even in the absence of transmission errors, speech quality was unsatisfactory, and quickly degraded as the frame error rate of transmission increased. For this reason, the CDMA Development Group (CDG)-13 vocoder was introduced soon after. This coder (also known as *Qualcomm Code Excited Linear Prediction* (QCELP)) is actually a variable-rate speech coder that dynamically selects one of three to four available data rates (13.3, 6.2, 1, and possibly 2.7 kbit/s) for every 20-ms speech frame, depending on voice activity and energy in the speech signal. The encoder then determines the formant, pitch, and codebook parameters, which are required for Code Excited Linear Prediction (CELP) algorithms (see Chapter 15). The pitch and codebook parameters are determined in an "analysis by synthesis" approach (again, see Chapter 15), using an exhaustive search over all possible parameter values. This search is computationally intensive, and was – especially in the early days of IS-95 – one of the major complexity factors, especially for MSs. For the 13.3-kbit/s mode, each packet (representing 20 ms) consists of 32 bits for Linear Predictive voCoder (LPC) information, 4 pitch subframes with 11 bits each, and 4×4 codebook subframes with 12 bits each. The *Enhanced Variable Rate Coder* (EVRC) is based on very similar principles, but uses a smaller number of bits both while the voice user is active (170 bits per 20-ms interval), and during transmission pauses. It furthermore includes adaptive noise suppression, which enhances overall speech quality.

A key aspect of all vocoders is the variable data rate. People are usually silent for approximately 50% of the time (while they are listening to the person at the other end of the line). During that time, the data rate is reduced to about 1 kbit/s. As discussed in Chapter 18, this leads to a significant increase in total capacity.

25.4.2 Error Correction Coding

Error Correction Coding for 8.6 kbit/s in the Uplink

Forward error correction is different for 8.6 kbit/s and 13.3 kbit/s. However, it is identical for two existing vocoders that are based on 8.6-kbit/s output: the IS-96A vocoder and the EVRC vocoder. These vocoders are associated with rate-set-1, and thus have the following encoding steps:

1. Encoding starts with 172 bits for each 20-ms frame from the vocoder.
2. In a next step, 12 *Frame Quality Indicator* (FQI) bits are added. These bits act as parity check bits, and allow determination of whether the frame has arrived correctly or not.
3. Adding an 8-bit encoder tail brings the number of bits to 192.
4. These bits are then encoded with a rate-1/3 convolutional encoder with constraint length 9. The three generator polynomials are:

$$\left.\begin{array}{l} G1(D) = 1 + D^2 + D^3 + D^5 + D^6 + D^7 + D^8 \\ G2(D) = 1 + D + D^3 + D^4 + D^7 + D^8 \\ G3(D) = 1 + D + D^2 + D^3 + D^4 + D^5 + D^8 \end{array}\right\} \qquad (25.3)$$

This brings the bit rate up to 28.8 kbit/s.

Error Correction Coding for 13.3 kbit/s in the Uplink

For the CDG-13 coder, encoding steps are in principle similar, but different numerical values are used:

1. Encoding starts with 267 bits (including some unused bits) for each 20-ms frame.
2. A frame erasure bit is added.
3. A total of 12 FQI bits are added (again, to indicate whether the frame has arrived correctly).
4. Adding an 8-bit tail in order to help the Viterbi decoder. This brings the number of bits per 20-ms frame to 288.
5. These bits are then encoded with a rate-1/2 convolutional encoder with constraint length 9. The two generator polynomials are:

$$\left.\begin{array}{l} G1(D) = 1 + D + D^2 + D^3 + D^5 + D^7 + D^8 \\ G2(D) = 1 + D^2 + D^3 + D^4 + D^8 \end{array}\right\} \tag{25.4}$$

Error Correction Coding for 8.6 kbit/s in the Downlink

Error correction coding is somewhat different in the downlink. It uses the same combination of FQI bits and tail bits as the uplink, but then uses a rate-1/2 convolutional encoder to bring the bit rate to 19.2 kbit/s. The generator vectors are given by Eq. (25.4). Data with rate 19.2 kbit/s are then further processed as described in Section 25.3.4.

Error Correction Coding for 13.3 kbit/s in the Downlink

This mode uses the same encoding steps as the 13.3-kbit/s uplink procedure. However, this leads to a 28.8-kbit/s rate, while only 19.2 kbit/s can be transmitted in one downlink traffic channel. The output from the convolutional encoder is thus punctured, in order to yield the desired bit rate. We can also interpret this as encoding the vocoder output using a rate-3/4 convolutional code. Puncturing eliminates the third and fifth bit of each 6-bit symbol repetition block (see above). This corresponds to eliminating 2 bits created by $G2(D)$, while the bits from $G1(D)$ are completely transmitted.

Interleaving

Output from the convolutional encoder is sent through a block interleaver (see Section 14.7.1) of length 576 for rate set 2. More specifically, the interleaver has a matrix structure similar to the one outlined in Figure 14.25 with 32 rows and 18 columns. The data are written line by line – i.e., filling first the first column, then the second column, and so on – and read out orthogonally (see, again, Figure 14.25).

25.5   Spreading and Modulation

25.5.1   Long and Short Spreading Codes and Walsh Codes

IS-95 uses three types of spreading codes: long spreading codes, short spreading codes, and Walsh codes. These codes play different roles in the uplink and the downlink. In this section, we describe just the codes themselves. In the subsequent two sections, we describe how those codes are used in the uplink and downlink, respectively.

Walsh Codes

Walsh codes are strictly orthogonal codes that can be constructed systematically. As we saw in Section 18.2.6, we define the $n + 1$-order Walsh–Hadamard matrix $\mathbf{H}_{\text{had}}^{(n+1)}$ in terms of the nth order matrix:

$$\mathbf{H}_{\text{had}}^{(n+1)} = \begin{pmatrix} \mathbf{H}_{\text{had}}^{(n)} & \mathbf{H}_{\text{had}}^{(n)} \\ \mathbf{H}_{\text{had}}^{(n)} & \overline{\mathbf{H}}_{\text{had}}^{(n)} \end{pmatrix} \tag{25.5}$$

where $\overline{\mathbf{H}}$ is the modulo-2 complement of H. The recursive equation is initialized as

$$\mathbf{H}_{\text{had}}^{(1)} = \begin{pmatrix} 1 & 1 \\ 1 & -1 \end{pmatrix} \tag{25.6}$$

The Walsh codes in IS-95 are the columns of the complement of $\mathbf{H}_{\text{had}}^{(6)}$.

Short Spreading Codes

IS-95 also uses two spreading codes that are PN-sequences, generated with a shift register of length 15, and thus with a periodicity of $2^{15} - 1$. A single zero is inserted in the sequence, in order to increase periodicity to $2^{15} = 32,768$ chips, which corresponds to 26.7 ms. The generator polynomials of the sequence are

$$G_i(x) = x^{15} + x^{13} + x^9 + x^8 + x^7 + x^5 + 1 \tag{25.7}$$

$$G_q(x) = x^{15} + x^{12} + x^{11} + x^{10} + x^6 + x^5 + x^4 + x^3 + 1 \tag{25.8}$$

As we will see later on, each BS uses a time-shifted version of the short spreading code. The *spreading offset index* of the code indicates this timeshift.[2]

Long Spreading Codes

The third type of codes, called "long spreading codes," is also based on PN-sequences. For the long codes, the shift registers have length 42, so that periodicity is $2^{42} - 1$, corresponding to more than 40 days. The generator polynomial is:

$$G_1 = x^{42} + x^{35} + x^{33} + x^{31} + x^{27} + x^{26} + x^{25} + x^{22} + x^{21} + x^{19}$$
$$+ x^{18} + x^{17} + x^{16} + x^{10} + x^7 + x^6 + x^5 + x^3 + x^2 + x^1 + 1 \tag{25.9}$$

Output from the shift register is then modulo-2 added with the *long-code mask*. This long-code mask is different for different channels: for access channels (see Section 25.4), it is derived from the paging and access channel numbers and the BS identification. For traffic channels, it can either be derived from the *Electronic Serial Number* (ESN) (*public mask*) or from an encryption algorithm (*private mask*).

25.5.2 Spreading and Modulation in the Uplink

Modulation and coding in the uplink are achieved through a combination of steps. The starting point is a bitstream with 28.8 kbit/s, which is obtained by providing error correction coding to the vocoder signal (see Section 25.4):

---

[2] Index zero corresponds to a sequence that had 15 zeros, followed by a 1, at 0h 00 on January 6, 1980.

- The first step involves mapping the data sequence to Walsh code codewords. Remember that each Walsh code is 64 chips long. The transmitter then takes groups of 6 bits ($x_0, \ldots, x_5$), and maps them to one Walsh code symbol $\tilde{\mathbf{x}} = [\tilde{x}_0, \tilde{x}_1, \ldots, \tilde{x}_{63}]$, according to the rule:

$$j_{\text{Walsh}} = x_0 + 2x_1 + 4x_2 + 8x_3 + 16x_4 + 32x_5 \tag{25.10}$$

$$\tilde{x}_i = 1 + (\mathbf{H}_{\text{had}}^{(6)})_{i, j_{\text{Walsh}}} \tag{25.11}$$

This achieves spreading by a factor 64/6. The chip rate of output from the Walsh encoder is thus 307.2 kchip/s. This can also be seen as M-ary orthogonal modulation; in other words, each group of 6 bits is represented by one modulation symbol that is orthogonal to all other admissible modulation symbols. A key advantage of this technique is that it allows noncoherent demodulation.
- The next step involves spreading the Walsh encoder output to 1.2288 Mchip/s. This is achieved by multiplication of the Walsh code output by the long spreading code. Note that (in the uplink) it is the long spreading code that provides channelization, and thus allows different traffic channels and access channels to be distinguished. Output from the long spreading code is sent to the databurst randomizer, whose role will be described later.
- As a last step, the (spread) data stream is separated into in-phase and quadrature-phase components, each of which is separately multiplied by short spreading sequences. This multiplication does not result in a change of chip rate. The chip stream on the I- and Q-branch are then used to modulate the local oscillator with Offset Quadrature Amplitude Modulation (OQAM) (as discussed in Chapter 11).

Figure 25.1 shows a block diagram of uplink transmission. Figure 25.2 shows the data rates involved.

25.5.3 Databurst Randomization and Gating for the Uplink

Up to now we have considered the case when the output of the channel coder actually has a data rate of 28.8 kbit/s – i.e., a source rate of 14.4 or 9.6 kbit/s. However, depending on the source, data, a

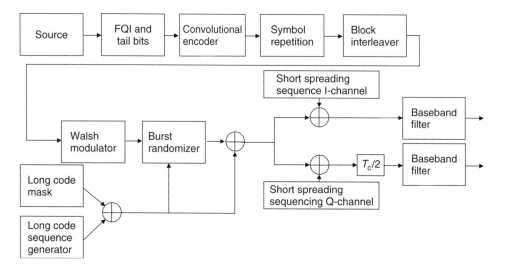

Figure 25.1 Block diagram of an IS-95 mobile station transmitter.

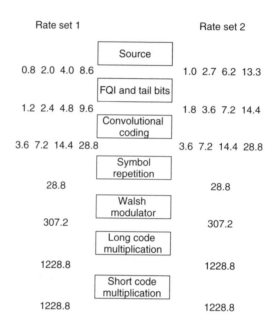

Figure 25.2 Data and chip rates at different interfaces in the uplink transmitter.

lower rate (14.4 kbit/s, 7.2 kbit/s, or 3.6 kbit/s) can also be the output of a convolutional encoder. In this case, encoded symbols are repeated (several times, if necessary) until a data rate of 28.8 kbit/s is achieved. It is these repeated data that are sent to the block interleaver for further processing.

However, it would waste resources to transmit all of these repeated data at full power. For the uplink, this problem is solved by gating off the transmitter part of the time. If, e.g., the coded data rate is 14.4 kbit/s (source data rate 7.2 kbit/s), then the transmitter is turned on only 1/2 of the time. As a consequence, average transmit power is only 1/2 of the full-data-rate case, and the interference seen by other users is only half as large.

Determination of the time for gating off the transmitter is actually quite complicated. The first issue is that gating has to be coordinated with the interleaver. For example, for the 7.2-kbit source data rate mode, each 1.25-ms-long group of output symbols is repeated once.[3] Gating thus eliminates one of these two symbol groups.

A decision about which of two symbol groups actually gets transmitted is determined by the long spreading sequence, according to the following algorithm:

- Consider the second-to-last 1.25-ms symbol group in the frame (20-ms period) immediately preceding the currently considered frame.
- Take the last 14 bits of the long spreading sequence used for spreading this symbol group, label them as $[b_0, b_1, \ldots, b_{13}]$.
- The gating of the symbol groups in the currently considered frame is now determined by these bits. Transmission is done of some of the 15 symbol groups contained in a 20-ms frame:
  - for the 14.4 (9.6) kbit/s source rate mode, always transmit;
  - for the 7.2 (4.8) kbit/s mode, transmit the first of the two identical groups if $b_i = 0$, $i = 0, \ldots, 7$, otherwise transmit the second frame;

---

[3] A symbol group of 1.25 ms is also often referred to as a *Power Control Group* (*PCG*), since fast power control can change every 1.25 ms.

     ○ for the 3.6 (2.4) kbit/s mode, transmit during $b_{2i} + 4i, i = 0, \ldots, 3$ if $b_{i+8} = 0$ or transmit during $2 + b_{2i+1} + 4i$ if $b_{i+8} = 1$;

     ○ for the 1.8 (1.2) kbit/s mode, transmit during $b_{4i} + 8i, i = 0, \ldots, 1$ if $b_{2i+8} = 0$ and $b_{i+12} = 0$, or transmit during $2 + b_{4i+1} + 8i$ if $b_{2i+8} = 1$ and $b_{i+12} = 0$, or transmit during $4 + b_{4i+2} + 8i$ if $b_{2i+9} = 0$ and $b_{i+12} = 1$, or transmit during $6 + b_{4i+3} + 8i$ if $b_{2i+9} = 1$ and $b_{i+12} = 1$.

Thus, the gating of sequences is pseudorandom, and different for each user (remember that the long spreading sequence is different for each user). Thus, in a system with low-rate users, interference with other users is "smeared out" – i.e., no user sees interference from all other users at the same time.

25.5.4 Spreading and Modulation in the Downlink

In the downlink, the different spreading codes are used in a very different way. The starting point is an encoded data stream with a rate of 19.2 kbit/s – i.e., lower by a factor of 1/3 than for the uplink. This bitstream is then scrambled and spread using the following steps:

- In a first step, the data stream is scrambled using the long spreading sequence. Remember that the long spreading sequence is defined to have a chip rate of 1.2288 Mchip/s. However, for downlink application, we do not want to use it for spreading, only for scrambling. We thus need to reduce the chip rate to 19.2 kchip/s. This is achieved by just using every 64th chip in the sequence. This decimated long-code sequence is the modulo-2 added to the data sequence.
- Next, data are spread by using Walsh sequences. In the downlink, Walsh sequences are used for channelization and spreading. Each traffic channel is assigned one Walsh sequence, with a chip rate of 1.2288 Mchip/s. This sequence is periodically repeated (with a period of 64 chips), and multiplied by the scrambled data sequence. We can interpret this alternatively as mapping each data bit either to the (user-specific) Walsh sequence or to its modulo-2 complement.
- Finally, output from the spreader is multiplied separately in the I- and Q-branch by the short spreading sequence. Note that in the downlink, the modulation format is Quadrature Amplitude Modulation (QAM), not OQAM.

Figure 25.3 shows a block diagram of the BS transmitter (downlink transmission principle), and Figure 25.4 shows the data rates involved.

In the downlink, there are also provisions for transmitting with data rates lower than 9.6 kbit/s or with a 14.4-kbit/s source rate. Also in this case, symbols are repeated to ensure maximum data rates, but the energy per transmitted symbol is reduced proportionally to the repetition factor, so a constant energy level is transmitted and a desired bit energy level is achieved. This treatment of the lower data rates in the downlink is different from the blanking used in the uplink.

25.5.5 Discussion

When confronted by different spreading schemes for uplink and downlink, the immediate question arises: "Why?" It would seem that a spreading scheme that is good for one direction should also prove to be advantageous for the other direction. However, we have to remember that there is an inherent asymmetry in a cellular system. Signals transmitted from the BS to the various MSs can be transmitted synchronously in a very easy manner – after all, the BS has control over when to transmit those signals. All signals then arrive at a given MS at the same time, having suffered the same attenuation and distortion by the channel.[4] Thus, if the signals are spread by perfectly

---

[4] Note that we are discussing here the different signals that arrive at a *given* MS. When considering the received signal at two different MSs, we find that these signals have different delays and different distortions.

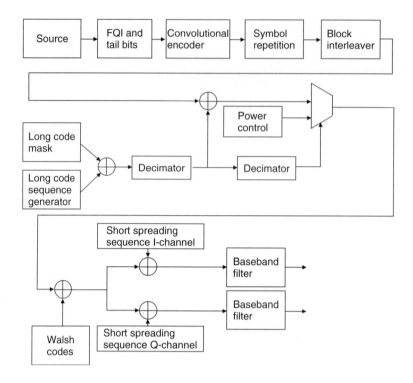

Figure 25.3 Block diagram for IS-95 downlink.

Figure 25.4 Data rates and chip rates in the downlink.

orthogonal codes at the transmitter, they can be perfectly separated at the receiver (this assumes nondistorting channels). It follows from this reasoning that (perfectly orthogonal) Walsh codes are used for spreading in the downlink. Since 64 users should be able to communicate within one cell, the spreading factor is 64, and the (coded) source data rate must not exceed 19.2 kbit/s. Short spreading sequences can be used for distinction between different BSs.

For the uplink, on the other hand, it is impossible for signals from different users to arrive at the same time. Thus, the use of Walsh codes for user separation is not possible. Walsh codes can have high cross-correlation when they have an arbitrary timeshift relative to each other. Rather, channelization is achieved by PN-sequences, which have low cross-correlation for arbitrary timeshifts. Walsh codes are used as part of the modulation scheme: since a 64-chip symbol is used to represent 6 bits, inherent redundancy and thus error-correcting capability are present.

25.6   Logical and Physical Channels

25.6.1   Traffic Channels

Traffic channels are the channels on which the voice data for each user are transmitted. We have already discussed them above, and here just reiterate the key data: there are two possible rate sets: rate-set-1, with a [9.6, 4.8, 2.4, 1.2]-kbit/s source data rate, and rate-set-2, with a [14.4, 7.2, 3.6, 1.8]-kbit/s source data rate. These source data – i.e., output from the vocoder – are possibly repeated, and then convolutionally encoded and interleaved.

The data are subsequently spread and modulated. The spreading and modulation operations are different in the uplink and the downlink, and are described in detail in Section 25.5.

A number of control messages are also transmitted on traffic channels. These include the following:

- For the uplink: Power Measurement Report, Pilot Strength Measurement, Handoff Completion, Long Code Transition Request, Long Code Transition Response, Data Burst, Request Analog Service.
- For the downlink: Neighbor List, Pilot Measurement Request, Handoff Direction, Long Code Transition Request, Long Code Transition Response, Data Burst, Analog Handoff Direction.

25.6.2   Access Channel

The access channel is a channel in the uplink that is used for signaling by MSs without a current call. Access channel messages include security messages (BS challenge, authentication challenge response), page response, origination, and registration.

The access channel uses a (source) data rate of 4.8 kbit/s. Spreading and modulation is very similar to uplink traffic channels with the same data rate. However, no gating is done, and all repeated symbols are actually transmitted. Since the access channel is spread (channelized) with a long spreading code, a considerable number of access channels can exist. As a matter of fact, up to 32 access channels exist for each paging channel (see below). An MS randomly chooses one of the active access channels before starting an access attempt.

A call initiated by the MS starts with a message on the access channel. The MS sets the initial power (based on the pilot power that it observes), and transmits a probe. If this probe is acknowledged before a timeout, then access was successful. If not, then the MS waits a random time, and then transmits the probe with increased power. This process is repeated until either access is successful, or the probe has reached the maximum admissible power; in the latter case, access is deemed to have failed.

25.6.3 Pilot Channels

The pilot channel allows the MS to acquire the timing for a specific BS, obtain the transfer function from BS to MS, and estimate the signal strength for all BSs in the region of interest. The pilot channel is similar to a downlink traffic channel, but shows some important peculiarities:

- It is not power controlled: the reason for this is that (i) it is used by many MSs (so it would not be clear which MS should determine the power control anyway), and (ii) it is used for estimating the attenuation of various links, which can only be done if transmit power is clearly defined and known to all MSs.
- It uses Walsh code 0 for transmission: this code is the all-zero code.
- It has higher transmit power than traffic channels: because of its importance, typically 20% of total BS power is assigned to the pilot channel.

The pilot channel is easy to demodulate, because it is just an all-zero sequence spread by the short spreading code. The only difference between pilots transmitted from different BSs is a temporal offset. After an MS has acquired the pilot, it can then more easily demodulate the synchronization channel (see next subsection), as the timing of that synchronization channel is locked to the pilot.

25.6.4 Synchronization Channel

The synchronization channel transmits information about system details that are required for the MS to synchronize itself to the network. Examples of such information include network identifier, PN-offset, long-code state, system time (from GPS), local time differential to system time, and the rate at which paging channels operate.

The synchronization channel transmits data at 1.2 kbit/s. After convolutional encoding with rate 1/2 and repetition, the data rate is transmitted. Note that this channel is not scrambled (no application of the long-code mask). Each frame of the synchronization channel is aligned at the start of the PN-sequence.

25.6.5 Paging Channel

The paging channel transmits system and call information from the BS to the MS. Several paging channels can exist within each cell; each of them is a 9.6-kbit/s channel. The information on the paging channel can include the following:

- Page message to indicate incoming call.
- System information and instructions:
  o handoff thresholds;
  o maximum number of unsuccessful access attempts;
  o list of surrounding cell PN-offsets;
  o channel assignment messages.
- Acknowledgments to access requests.

25.6.6 Power Control Subchannel

The power control subchannel provides signaling for compensation of SSF. IS-95 divides signals into *PCGs* of 1.25 ms duration. The BS estimates the Signal-to-Noise Ratio (SNR) for each user for each PCG. It then sends a power control command to the MS within two PCGs, and the MS reacts to it within 500 µs. This command signifies either an increase or a decrease by 1 dB, and thus requires 1 bit. Consequently, the data rate of the power control channel is 800 bit/s.

The power control subchannel is inserted into the traffic channel by simply replacing some of the traffic data symbols. Each PCG contains 24 modulation symbols; however, only the first 16 are candidates for replacement. The exact location is determined by the long-code mask: bits 20–23 of the long-code mask in the previous PCG determine which bit is replaced.

25.6.7 Mapping Logical Channels to Physical Channels

In the downlink, mapping is done in a rather straightforward way: different channels are assigned different Walsh codes for spreading. Specifically, the pilot channel uses Walsh-code-0, the paging channels use Walsh-code-1 through Walsh-code-7, the synchronization channel uses Walsh-code-32, and the traffic channels use all other Walsh codes.

In the uplink, only traffic channels and access channels exist. These are assigned different long spreading codes, and are thus also mapped to different coded channels in all cases.

25.7 Handover

One of the most important advantages of CDMA is "soft handover," which helps the performance of MSs especially at the cell edges (see Section 18.3.2). In this section, we discuss how IS-95 determines the available BSs so as to carry out soft handover, and how handover is actually carried out.

Each BS sends out the same PN-sequence as a pilot signal, just with an offset of 64 chips (see Section 25.3.4). This gives a total of 512 possible pilot signals. By observing these pilots, the MS always knows the signal strengths of all BSs in its environment. However, it is much too complicated and slow to monitor such a huge number of BSs – most of which are below the noise floor anyway. Thus, the MS only observes pilot signals within a given window.

The MS divides the available BSs into groups or *sets*. The *active set* contains pilots that are made available to a specific MS by the Mobile-services Switching Center (MSC); it can contain up to six pilots. The *candidate set* contains pilots that are strong enough to be demodulated, but are not contained in the active set. The *neighbor set* contains a list of BSs that are "nearby," but do not have sufficient strength to move into the candidate set. A list of the neighbor set is sent by the BS to the MS at regular intervals.

Monitoring of pilots and assignment of BSs to different sets are done the following way: when the MS finds a new pilot (not yet in the candidate set), it compares its strength with a parameter T_ADD. If the pilot is stronger than this threshold, the MS sends a Pilot Strength Measurement Message (PSMM) to the BS; the BS might then order the MS to add the pilot to the candidate set. The MS also monitors the strength of pilots in the candidate set, and if they become stronger than a threshold T_COMP, it moves the pilot from the candidate set to the active set. There are also mechanisms for dropping a pilot from the active or candidate set, if the strength falls below a certain threshold.

When the MS is in soft-handover mode, then it can communicate with members of the active set simultaneously. In the downlink, the MS just combines the signals from different BSs. In the uplink, the BSs in the active set determine which of them gets the best signal quality; this BS is then the one used for demodulation. The former case provides combining gain, while the latter provides selection gain.

25.8 Appendices

Please see companion website *www.wiley.com/go/molisch*

Further Reading

A summary of IS-95 can be found in Liberti and Rappaport [1999]. Additional information on CDMA 2000 can be found in the book by Garg [2000], Vanghi et al. [2004], Eternad [2004] and the papers by Tiedemann [2001] and Willenegger [2000]. The 1xEV-DO (EVolution-Data Optimized) mode is described in Sindhushayana and Black [2002] and Parry [2002]. The speech codec for CDMA 2000 is described in Jelinek et al. [2004].

For updates and errata for this chapter, see wides.usc.edu/teaching/textbook

26

WCDMA/UMTS

26.1 Historical Overview

This chapter is a short summary of the *Wideband Code Division Multiple Access* (WCDMA) standard for third generation (3G) cellular telephony. This standard is also part of a group of standards that are known as the *Universal Mobile Telecommunication System* (UMTS), *Third Generation Partnership Project* (3GPP), and *International Mobile Telecommunications* (IMT-2000). In this section, we will review the history of this standard and show its relationship to other third-generation standards.

The success of second-generation cellphones, especially the Global System for Mobile communications (GSM), motivated the development of a successor system. The International Telecommunications Union (ITU) announced goals for a standard for a 3G system, called IMT-2000:

- better spectral efficiency;
- higher peak data rates – namely, up to 2 Mbit/s indoor and 384 kbit/s outdoor. This should result in a choice of channels with a bandwidth of 5 MHz instead of 200 kHz;
- supporting multimedia applications, meaning the transmission of voice, arbitrary data, text, pictures, audio and video, which requires increased flexibility in the choice of data rates;
- backward compatibility to second-generation systems.

Europe started research activities toward this goal in the early 1990s, first under the auspices of European Union research programs, and then a more formal process within the European Telecommunications Standards Institute (ETSI). Based on the above list of requirements, different groups developed proposals, which ranged from Orthogonal Frequency Division Multiplexing (OFDM) solutions over broadband Time Division Multiple Access (TDMA) systems to Code Division Multiple Access (CDMA) protocols on the physical layer. During a final poll in January 1998 two drafts were selected: broadband CDMA – also known as *Frequency Division Duplexing* (FDD) mode – which is intended as the basic system, and T/CDMA with Joint Detection-Time/Code Division Multiple Access (JD-TCDMA) – also known as Time Division Duplexing (TDD) mode – which is to support high-data-rate applications. The FDD mode and TDD mode are often subsumed under the name WCDMA. FDD mode is the more important part, and will be at the center of our attention in this chapter. WCDMA is mostly used as an abbreviation for radio access technology, while the expression UMTS refers to the complete system, including the Core Network (CN).

These two systems were then included as the European proposal for the IMT-2000 family. A Japanese proposal for a WCDMA system was fairly similar to the European proposal, and thus

Wireless Communications, Second Edition   Andreas F. Molisch
© 2011 John Wiley & Sons, Ltd

merged. Still, it was not possible to reach agreement for a single 3G system within the ITU. Besides the Japanese/European proposal, the (mainly U.S.-favored) CDMA 2000 proposal (see Chapter 25) had strong support. Furthermore, Enhanced Data rates for GSM Evolution (EDGE) (see Chapter 24), Digital Enhanced Cordless Telecommunications (DECT) (see the Appendices at *www.wiley.com/go/molisch*), and Universal Wireless Communications (UWC) 136 (a further development of the American second-generation standard IS-136 (Interim Standard)) were advocated. In order to avoid a deadlock, the ITU decided to declare all of these proposals to be valid "modes" of IMT-2000. Of course, having five modes in a single standard is just as bad as having five different standards, but it was a face-saving measure for all participants. After 2000, the Chinese standard Time Division Synchronous Code Division Multiple Access (TD-SCDMA) has also received great attention, and is becoming an important part of 3G cellular developments. This standard, as well as WiMAX (Chapter 28) was later accepted into the IMT-2000 family.

The further development of the merged Japanese/European standard is being done by the 3GPP industry organization, whose members include ETSI, ARIB (Association of Radio Industries and Businesses, the Japanese standardization organization) and several other members. The original specifications are also known as "Release 99." Subsequent improvements were included in later releases – in particular High-Speed Packet Access (HSPA), which was first realized for the downlink, and then also the uplink. While there were advantages to having a broad base of companies contributing to the specifications, 3GPP faced some serious problems: the size of the specifications grew beyond reasonable limits, and became so complicated that it hampered implementation. The resulting high development costs and delays in time to market almost led to the death of the system, and it started to take off only around 2005. The exception was Japan, where implementation proceeded much faster than elsewhere. It is worth noting that Japan did not implement the full UMTS specifications, but rather a simplified system called "FOMA (Japanese version of the UMTS standard)." Development in Japan was also helped by the fact that second-generation Japanese Digital Cellular (JDC) reached its capacity limits earlier and the alternative Personal Communication Systems (PCSs) (compare with the DECT material on the companion website) could not meet demand completely.

UMTS development was also hampered by regulatory developments. The first UMTS-related licenses were granted first in 1999 in Finland and the other European countries followed in 2000. Particularly, the bidding process for the licenses in the United Kingdom and Germany received a lot of public attention, as several billion Euros were finally paid for each license. According to the regulatory authorities, the first UMTS networks should have started operating in 2002. However, the real start of operation has been delayed due to technical problems and the fact that the market for mobile high-data-rate applications had to be developed first. Only by the end of 2003 did the major mobile providers in Europe start to offer subscriptions to UMTS and the appropriate devices. But by 2009 these problems had been overcome, and UMTS has turned out to be extremely successful all over the world.

The remainder of this chapter describes mainly the physical layer of WCDMA/UMTS. Many of the networking functions are similar to GSM, and we simply refer the reader to Chapter 24 for these.

26.2   System Overview

26.2.1   Physical-Layer Overview

We first summarize the physical layer of WCDMA – i.e., communications between the Mobile Station (MS) and the Base Station (BS) via the air interface. The UMTS standard uses a number of unusual abbreviations. For example, the MS is called *User Equipment* (UE). In order to stay consistent with the notation in the remainder in this book, we will stick to MS. Similarly, the BS is called *Node-B* in the UMTS standard.

The WCDMA air interface uses CDMA for distinguishing between different users, and also between users and some control channels. We distinguish between spreading codes that are responsible for bandwidth expansion of the signals, and scrambling codes that are mainly used for distinguishing signals from different MSs and/or BSs. In addition, WCDMA also shows a timeslot structure: the time axis is divided into units of 10 ms, each of which is subdivided into slots of 0.67 ms. Depending on the location within a timeslot, a symbol might have different meanings.

WCDMA uses a number of logical channels for data and control information, which will be discussed in Section 26.4. They are then mapped to physical channels – i.e., channels that are distinguished by spreading codes, scrambling codes, and positions within the timeslot.

26.2.2   Network Structure

To discuss how a mobile network provider may introduce UMTS, we have to distinguish between the *MS*, the *Radio Access Network* (RAN), and the CN. The MS and the *UMTS Terrestrial Radio Access Network* (UTRAN) communicate with each other via the air interface, as discussed in the previous section. The UTRAN consists of multiple *Radio Network Subsystems* (RNSs), each of which contains several *Radio Network Controllers* (RNCs), each of which controls one or several BSs (Node-Bs).

The CN connects the different RNSs with each other and other networks, like ISDN and data packet networks. The CN can be based on an upgraded GSM CN or might be implemented as a completely new Internet Protocol (IP)-based network. Details about the different functional units of a GSM CN (mobile switching center, home location register, etc.) can be found in Chapter 24. The network functionalities for packet data are similar to the ones of the General Packet Radio Service (GPRS) (see Appendix 24.C. at *www.wiley.com/go/molisch.*).

Another way of looking at the UMTS architecture is to organize it in two domains:

1. *UE domain*, which consists of:
   o *User Service Identity Module* (USIM).
   o *Mobile Equipment* (ME) consisting of:
     – *Terminal Equipment* (TE);
     – *Terminal Adapter* (TA);
     – *Mobile Termination* (MT).
2. *Infrastructure domain*, which consists of:
   o The access network domain consisting of:
     – UTRAN.
   o The CN domain consisting of:
     – *Inter Working Unit* (IWU);
     – serving network;
     – transit network;
     – home network;
     – application network.

The physical properties of a link between two network entities and the signals transmitted via this link together with the functions of these signals are collectively referred to as an *interface*. Usually, an interface is standardized. Figure 26.1 shows the relevant interfaces in UMTS.

26.2.3   Hierarchical Cellular Structure

UMTS is intended to achieve global availability and thereby enable roaming worldwide. Therefore, the coverage area in UMTS is divided hierarchically into layers. The higher layers cover a bigger

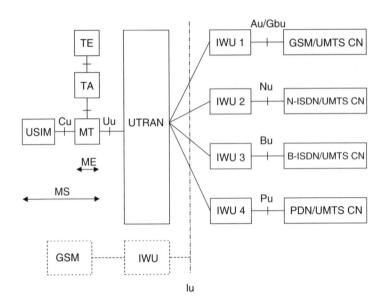

Figure 26.1   Interfaces in Universal Mobile Telecommunications System.

area than the lower layers. The highest layer achieves worldwide coverage by using satellites.[1]
The lower layers are the macrolayer, microlayer, and picolayer. They constitute the *UTRAN*. Each
layer consists of several cells. The lower the layer the smaller the cells. Thus, the macrolayer is
responsible for nationwide coverage with macrocells. Microcells are used for additional coverage
in the urban environment and picocells are employed in buildings or "hotspots" such as airports
or train stations. This concept is known and has been discussed and partly implemented for a
while. However, UMTS was supposed to cover all aspects of this concept worldwide and with the
initial rollout. In practice, however, it has been rather different: in the initial phase, only the big
cities were covered with a few cells, and large-area coverage is achieved by dual-mode devices
that can communicate with either WCDMA or EDGE/GPRS/GSM networks. By 2009, this rollout
deployment mode is still used in the U.S.A., while Japan and most of Europe have achieved
coverage of all areas with WCDMA.

26.2.4   Data Rates and Service Classes

The maximal data rate and the highest supported user velocity are different for each hierarchy layer.
The macrolayer supports at least 144 kbit/s at speeds up to 500 km/h. In the microlayer data rates
of 384 kbit/s at maximal speeds of 120 km/h are achievable. In picocells, maximal user speeds of
10 km/h and maximal data rates of 2 Mbit/s are supported. Figure 26.2 compares data rates and
maximal user velocity with other cellular (GSM) or wireless standards.

   Maximal Bit Error Rates (BERs) and transmission delays are grouped into sets out of which the
user may choose the following:

- *Conversational*: this class is mainly intended for speech services, similar to GSM. The delays
  for this type of service should be on the order of 100 ms or less; larger values are experienced
  as unpleasant interruptions by users. BERs should be on the order of 10^{-4} or less.

---

[1] Given the past difficulties of satellite cellular communications systems (e.g., IRIDIUM), it is questionable whether
this layer will actually be implemented.

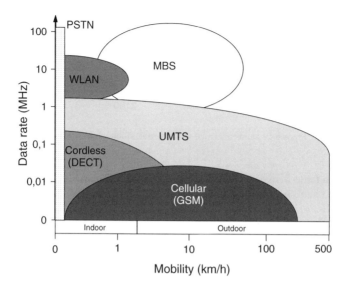

Figure 26.2 Data rates versus mobility for Universal Mobile Telecommunications System, Global System for Mobile communications, Digital Enhanced Cordless Telephone, wireless local area network mobile broadband systems and landline networks.

- *Streaming*: audio- and videostreaming are viewed as one important application of WCDMA. Larger delays (in excess of 100 ms) can be tolerated, as the receiver typically buffers several seconds of streaming material. BERs are typically smaller, as noise in the audio (music) signal is often considered to be more irksome than in a voice (telephone) conversation.
- *Interactive*: this category encompasses applications where the user requests data from a remote appliance. The most important category is Web browsing, but also database retrievals and interactive computer games fall into this category. Also for this category, there are upper limits to tolerable delay – the time between choosing a certain website and its actual appearance on the screen should not exceed a few seconds. BERs have to be lower, typically 10^{-6} or less.
- *Background class*: this category encompasses services where transmission delays are not critical. These services encompass email, Short Message Service (SMS), etc.

26.3 Air Interface

26.3.1 Frequency Bands and Duplexing

In most of the world, UMTS utilizes frequencies in ranges from 1,900 MHz to 2,025 MHz and from 2,110 MHz to 2,200 MHz (see Figure 26.3). Within these bands there is a dedicated subband for the *Mobile Satellite Service* (MSS).[2] The MSS uplink uses the band from 1,980 MHz to 2,010 MHz and the MSS downlink the band from 2,170 MHz to 2,200 MHz. The remaining parts of bands are split for the two modes of terrestrial operation, UMTS-TDD and UMTS-FDD. UMTS-FDD uses the band from 1,920 MHz to 1,980 MHz for the uplink and the band from 2,110 MHz to 2,170 MHz for the downlink. As the name indicates, TDD mode does not distinguish uplink and downlink by using different carrier frequencies but rather by accessing different timeslots on the same carrier. Therefore, this mode does not require symmetric frequency bands. It can simply use all of the remaining frequencies.

---

[2] For a detailed description of bands for 3G and fourth generation (4G) systems, see Chapter 27.

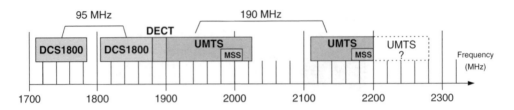

Figure 26.3 Frequency bands allocated to Universal Mobile Telecommunications System.

An exception to these frequency allocations is the U.S.A., where 3G phones are to be placed in PCS bands – i.e., uplink is in the band from 1,850 to 1,910 MHz, and downlink in the band from 1,930 to 1,990 MHz.

26.3.2 Time Domain Duplexing and Frequency Domain Duplexing Modes

FDD operating mode is intended for use in macrocells and microcells, whereas TDD operating mode is intended for use in picocells. In the TDD mode it is more difficult to handle large propagation delays between the MS and BS, as transmitting and receiving timeslots might overlap. Therefore, it can only be employed in picocells. However, TDD has the advantage that big asymmetries in throughput on the downlink and uplink can be easily supported. This is particularly important for applications like Web browsing for which the MS receives much more information than it transmits. For medium access, TDD mode employs JD-TCDMA and FDD mode WCDMA. We will not, however, consider the TDD mode further in this chapter.

26.3.3 Radio-Frequency-Related Aspects

Power Classes and Receiver Sensitivity

MSs: these are divided into four classes, according to their transmit power. The maximal powers are 33, 27, 24, and 21 dBm, corresponding to power classes 1, 2, 3, and 4. Power is measured before the antenna. Thus, the antenna characteristic does not have an impact on these limits. The receiver of the MS has to be so sensitive that given a received signal power of -117 dBm per 3.84-MHz channel a BER of 10^{-3} is still achievable with a 12.2-kbit/s data rate. Note that this specification includes the effects of forward error correction.

BS: there are no transmission powers specified for the BS. However, typical values for the transmit power are in the 10–40-W range. The receiver in the BS has to be so sensitive that given a received signal power of -121 dBm a BER of 10^{-3} for a 12.2-kbit/s is still feasible. Since noise power in a bandwidth of 12.2 kHz is -133 dBm, this is a quite challenging task.

The specifications also prescribe the resistance of the receiver to blocking – i.e., being able to function even in the presence of strong interferers. Processing gain can be used to decrease the effect of interference. However, if the Radio Frequency (RF) elements of the receiver have a limited dynamic range, then the interferer might drive the receive amplifier into saturation, resulting in a signal from which the desired part can no longer be recovered. Similarly, limits to intermodulation and other RF effects are specified [Richardson 2005].

Frequency Bands

The regular intercarrier spacing is 5 MHz. This is the nominal distance. In fact, the network provider might select the carrier distance to be any multiple of 200 kHz. Therefore, carriers are referred to using the *UTRA Absolute Radio Frequency Channel Number* (UARFCN) which refers to multiples of 200 kHz. The frequency deviation of the local oscillators of the MS is limited to 10^{-7} (0.1 parts per million), which equals roughly 200 Hz.

From a purely theoretical point of view, emissions outside the assigned 5-MHz band should be zero. The basis pulse for modulation is a raised cosine pulse with a roll-off factor of $\alpha = 0.22$. Given a chip rate of 3.84 Mchip/s the signal bandwidth is $(1 + \alpha)/T_C = 4.7$ MHz.[3] However, due to nonideal filter implementation, emissions outside this band do appear in practice. These emissions include both out-of-band emissions (defined as emissions a distance of 2.5 MHz to 12.5 MHz from the center frequency) and spurious emissions, which refer to emissions farther away from the center frequency.

Out-of-Band Emissions

There is a set of limitations for out-of-band emissions at a distance of 2.5 MHz to 12.5 MHz:

- **Spectrum emission mask**: Figure 26.4 shows the emission mask for a BS. For a given distance Δf to the center frequency, there is a limit for the maximum emitted power within a 30-kHz and 1-MHz bandwidth, respectively.
- **Adjacent Channel Leakage Ratio** (ACLR): this is a measure of how much power leaks from the desired band into the adjacent band. This ratio should be better than 45 dB and 50 dB for the first or second adjacent channel (5-MHz or 10-MHz carrier distance), respectively.

Spurious Emissions

Spurious emissions are emissions far away from the used channel ($\Delta f > 12.5$ MHz). They are due to harmonics emissions, intermodulation products, etc. The limits for these emissions are rather soft – e.g., -30 dBm for a 1-MHz bandwidth. However, for frequencies which are used by other mobile systems, such as GSM, DECT, Digital Cellular System (DCS1800), or UMTS Terrestrial Radio Access (UTRA-TDD), they are more severe. Especially when a UMTS BS is colocated with a GSM BS, emission limits of -98 dBm/100 kHz might apply, thus ensuring that UMTS signals do not interfere with GSM signals.

26.4 Physical and Logical Channels

26.4.1 Logical Channels

Similar to GSM, we distinguish between different logical channels in UMTS, which are mapped to physical channels. The logical channels in UMTS are sometimes referred to as *transport channels*. There are two kinds of transport channels: *common transport channels* and *Dedicated (transport) CHannels* (DCHs).

---

[3] The mismatch between signal bandwidth and nominal carrier distance can only be explained from a historic perspective. Originally, a chip rate of 4.096 Mchip/s was planned. With $\alpha = 0.22$ this results in a signal bandwidth of 5 MHz.

Frequency offset Δf from carrier (in MHz)

Figure 26.4   Spectrum mask for a base station in Wideband Code Divison Multiple Access.
Reproduced from [UMTS 1999] © 1999. 3GPP Technical Specification (TSs) and Technical Report (TR)s are the property of ARIB,
Alliance for Telecommunications Industry Solutions (ATIS), ETSI, China Communications Standards Association (CCSA),
Telecommunications Technology Association of Korea (TTA), and Telecommunication Technology Committee (TTC) who jointly
own the copyright in them. They are subject to further modifications and therefore provided "as is" for information purposes only.
Further use is strictly prohibited.

Common Channels

Common channels are relevant to all or at least a group of MSs in a cell. Thus, all of them receive
the information transmitted on these channels in the downlink and may access the channels in the
uplink. There are different kinds of common channels:

- *Broadcast CHannel* (BCH): the BCH is only found in the downlink. Both cell-specific and
  network-specific information is transmitted on it. For example, the BS uses this channel to
  inform all MSs in the cell about free access codes and available access channels. This channel
  has to be transmitted with relatively high power, as all MSs within the cell have to be able to
  receive it. Thus, on this channel neither power control nor smart antennas are implemented.
- *Paging CHannel* (PCH): this is also a channel that can be found only in the downlink. It is used
  to tell an MS about an incoming call. Since attenuation of the channel to the MS, as well as the
  location of the MS, is not known the PCH is transmitted with high power and without employing
  smart antennas. Depending on whether the current cell of the MS is known, paging information
  is either transmitted in only one cell or several cells.
- *Random Access CHannel* (RACH): the RACH is only used in the uplink. The MS uses it to
  initialize a connection to the BS. It can employ open-loop power control, but no smart antennas,
  as the BS must be able to receive signals on the RACH from every MS within the cell. As it is
  a *random* access channel, collisions might occur. Therefore, the structure of bursts in the RACH
  are different from that of other channels – this is described below in more detail.

- *Forward Access CHannel* (FACH): the FACH is used to transmit control information to a specific mobile. However, as the FACH is a common channel and therefore received by more than one MS, explicit addressing of the desired MS is required (*in-band identification*, UE-ID at the beginning of the packet). In contrast with this, a dedicated control channel has an implicit addressing of the desired MS: the MS is specified by the carrier and the spreading code used. The FACH can also transmit short user information packets. It can employ smart antennas, as information is transmitted to one specific, localized mobile.
- *Common Packet CHannel* (CPCH): the CPCH is an uplink channel, and can be interpreted as the counterpart of the FACH. It can transmit both control and information packets. If the FACH and the CPCH are used together closed-loop power control is possible.
- *Downlink Shared CHannel* (DSCH): the DSCH is a downlink channel similar to the FACH. It sends mainly control data, but also some traffic data, to multiple MSs. Explicit addressing has to be used on this channel, as the same CDMA code is used for all MSs. The reason for using the same code for multiple MSs lies in the limited amount of *short* spreading codes (see next section). Under normal circumstances, one spreading code would be permanently reserved for one MS, even if the traffic is bursty. The cell would thus quickly run out of codes (remember that there are very few codes that can be used for high-data-rate traffic). On the DSCH one short code is used for several mobiles and data are multiplexed in time. The DSCH supports fast power control, use of smart antennas, and rate adaptation from transmission frame to transmission frame. Note that the DSCH uses a different approach in HSDPA – namely, the use of multiple short codes simultaneously.

Dedicated Channels

Dedicated channels are present in both the uplink and the downlink. They are used to transmit both higher layer signaling and actual user data. As the position of the MS is known when a dedicated channel is in use, smart antennas, as well as fast power control and adaptation of the data rate on a frame-by-frame basis, can be used:

- *Dedicated (transport) channel (DCH)*: this is the only type of dedicated logical channel. MS addressing is inherent as for each MS a unique spreading code is used.

26.4.2 Physical Channels

WCDMA transmits control and user data on the same logical channel (in contrast to GSM) – namely, the DCH. However, for physical channels we distinguish between channels for control and user information. Combined transmission of both is called the *Coded Composite Traffic CHannel* (CCTrCH). Note that there are also some physical channels that are not related to a specific logical channel.

Uplink

In the uplink we find dedicated control and user data channels, which are transmitted simultaneously via I- and Q-code multiplexing (see Section 26.6):

- Pilot bits, *Transmit Power Control* (TPC), and *Feed Back Information* (FBI) are transmitted via the *Dedicated Physical Control CHannel* (DPCCH). Furthermore, the *Transport Format Combination Indicator* (TFCI) may be transmitted via the DPCCH (see also Section 26.5.2). The TFCI contains the instantaneous parameter of all data transport channels that are multiplexed on the DPDCH (see below). The spreading factor for the DPCCH is constant – namely, 256. Thus, ten control information bits are transmitted in each slot.

- The actual user data is transmitted via the *Dedicated Physical Data CHannel* (DPDCH). Spreading factors between 4 and 256 can be used. The DPDCH and DPCCH are transmitted at the same time and on the same carrier by using the I- and Q-branch, respectively.
- The *RACH* is not only a logical but also a physical channel (Physical RACH or PRACH). A *slotted ALOHA* approach is used for medium access (see Chapter 17). The burst structure of the RACH is completely different from that of the dedicated channel. Data packets may be transmitted via the PRACH, too. Packets can be transmitted either in pure PRACH mode, meaning that PRACH bursts as described in Section 26.6 are used, or in the *Uplink Common Packet CHannel* (UCPCH). The UCPCH is an extension of the PRACH. It is used in combination with a DPCCH in the downlink. Therefore, fast power control is possible.
- The *Physical Common Packet CHannel* (PCPCH) has a similar burst structure to the PRACH. Information is transmitted after a preamble. Initially, several access preambles are transmitted with ascending transmission power until the BS receives the necessary signal strength. After the BS acknowledges reception, another preamble is transmitted to detect eventual collisions with packets from other MSs that try to access the PCPCH at the same time. Before user data are transmitted, a power control preamble of length 0 or 8 timeslots can be transmitted. Only then are the actual data transmitted; the length of this transmission period is a multiple of the frame duration – i.e., 10 ms.

Downlink

Of course, the dedicated data and control channels, DPDCH and DPCCH, can be found in the downlink, too. However, they are multiplexed in a different manner, which is discussed in Section 26.6. The frame and slot timing are shown in Figure 26.5.

Furthermore, the downlink features the following common control channels:

- *Primary Common Control Physical CHannel* (P-CCPCH): bursts transmitted via the CCPCHs are similar to those of the DPCCH. However, there is no power control associated with the CCPCH and therefore the power control bits do not have to be transmitted. The P-CCPCH has an idle period in each frame; this idle period is 256 chips long. The P-CCPCH carries the broadcast channel, and it is thus critical that it can be demodulated by all MSs in the cell. Thus, it uses a constant, high spreading factor – namely, 256.
- *Secondary Common Control Physical CHannel* (S-CCPCH): the main difference between the primary and the secondary CCPCH is that the data rate and spreading factor are fixed in the former, whereas they are variable in the latter. The S-CCPCH carries the FACH and the PCH.
- *Synchronization CHannel* (SCH): the SCH does not relate to any logical channel. Its function is explained in Section 26.7. The SCH is multiplexed with the P-CCPCH onto one timeslot: it sends 256 chips for synchronization. During the above-mentioned idle period of a P-CCPCH burst, the SCH is transmitted by sending 256 chips for synchronization.
- *Common PIlot CHannel* (CPICH): this channel has a constant spreading factor of 256. The CPICH consists of a primary and a secondary CPICH. The primary CPICH serves as a reference for phase and amplitude for common channels in the entire cell. The primary CPICH is transmitted all over the cell and the secondary might be transmitted in selected directions. The primary and secondary pilot channels differ in the spreading and scrambling codes used: the primary CPICH always uses the primary scrambling code and a fixed channelization code, so that there is only one such code per cell.

CPICHs are particularly important for establishing a connection as during this period the pilots of the dedicated channels are not available. Furthermore, pilot channels provide an indication of

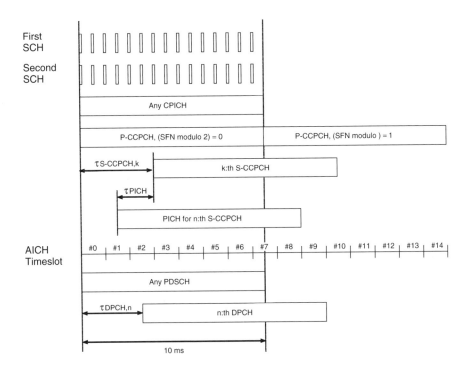

Figure 26.5 Frame and slot timing of downlink physical channels.

signal strength at the MS and are therefore important for handover procedures. The cell size of a BS can be varied by varying the transmit power of pilot channels. By reducing transmit power the area over which the BS provides the strongest signal is decreased. This reduces the traffic load of a BS.

- Apart from these channels, the downlink further features the *Physical Downlink Shared CHannel* (PDSCH), which carries the DSCH, the *Acquisition Indication CHannel* (AICH), which provides feedback about whether synchronization was successful or not, and the *Page Indication CHannel* (PICH), which supports paging.

Matching of Logical and Physical Channels

Figure 26.6 illustrates how logical channels, also referred to as transport channels, are matched to physical channels. Details regarding frame and slot timing can be found in the standard.

26.5 Speech Coding, Multiplexing, and Channel Coding

26.5.1 Speech Coder

The speech coder used in UMTS is an *Adaptive Multi Rate* (AMR) coder that also has a strong similarity to the enhanced speech coder used in GSM. AMR codecs are based on *Algebraic Code*

Transport Channels   **Physical Channels**

DCH ——————————— Dedicated Physical Data CHannel (DPDCH)

                                          Dedicated Physical Control CHannel (DPCCH)

RACH ——————————— Physical Random Access CHannel (PRACH)

CPCH ——————————— Physical Common Packet CHannel (PCPCH)

                                          Common Pilot CHannel (CPICH)

BCH ——————————— Primary Common Control Physical CHannel (P-CCPCH)

FACH ——————————— Secondary Common Control Physical CHannel (S-CCPCH)

PCH

                                          Synchronisation CHannel (SCH)

DSCH ——————————— Physical Downlink Shared CHannel (PDSCH)

                                          Acquisition Indication CHannel (AICH)

                                          Page Indication Channel (PICH)

Figure 26.6   Matching of physical and logical channels.
Reproduced from [UMTS 1999] © 1999. 3GPP TSs and TRs are the property of ARIB, ATIS, ETSI, CCSA, TTA, and TTC who jointly own the copyright in them. They are subject to further modifications and are therefore provided "as is" for information purposes only. Further use strictly prohibited.

Excited Linear Prediction (ACELP) (see Chapter 15). WCDMA-AMR contains eight different encoding modes, with source rates ranging from 4.75 to 12.2 kbit/s, as well as a "background noise" mode.[4]

26.5.2   *Multiplexing and Interleaving*

Multiplexing, coding, and interleaving are very complicated procedures that allow a high degree of flexibility. A data stream coming from upper layers has to be processed before it can be transmitted via the transport channels on the air interface. Transport channels are processed in blocks of 10-, 20-, 40- or 80-ms duration. We first discuss multiplexing and coding in the uplink. The block diagram in Figure 26.7a illustrates the order of the processes involved in multiplexing and coding:

- When processing a transport block, the first step is to append a *Cyclic Redundancy Check* (CRC) field. This field, which can be 8, 12, 16, or 24 bits long, is used for the purpose of error detection. It is calculated for each block of data for one transmission time interval from the code polynomials:

$$G(D) = D^8 + D^7 + D^4 + D^3 + D + 1 \quad \text{for 8-bit CRC} \tag{26.1}$$

$$G(D) = D^{12} + D^{11} + D^3 + D^2 + D + 1 \quad \text{for 12-bit CRC} \tag{26.2}$$

$$G(D) = D^{16} + D^{12} + D^5 + 1 \quad \text{for 16-bit CRC} \tag{26.3}$$

$$G(D) = D^{24} + D^{23} + D^6 + D^5 + D + 1 \quad \text{for 24-bit CRC} \tag{26.4}$$

and attached at the end of the block.

---

[4] A later standardized "wideband" AMR that results in even better speech quality uses nine modes with rates between 6.6 and 23.85 kbit/s.

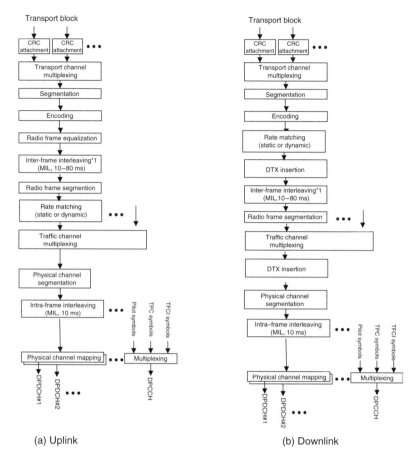

Figure 26.7 Multiplexing and coding.

- Afterwards the datablocks are concatenated or segmented into blocks that have suitable size for channel coding. The blocks should not be too small, because that increases the relative impact of the overhead (tail bits), and makes the performance of turbo codes worse. On the other hand, the blocks should not be too big; otherwise, decoding can become too complicated. For convolutional encoding the blocksize is typically 500 bits, for turbo codes approximately 5,000 bits. Tail bits are appended to help the decoder: 8 tail bits if convolutional encoding is used, and 4 tail bits for turbocoding.
- The blocks are then encoded with convolutional codes or turbo codes, details of which are discussed in the next subsection.
- The resulting encoded blocks then undergo *radio frame size equalization*. This makes sure that the amount of data is the same for each radio frame.
- In case the block spans more than one frame of length 10 ms, *interframe interleaving* is applied, which interleaves bits over the different frames of this block.
- If necessary, the blocks are then divided into 10-ms transmission blocks, a process called *radio frame segmentation*.

- The encoded block then undergoes *rate matching* – i.e., the rate of the block is then adapted to the desired rate by puncturing or selected bit repetition. Repetition is usually preferred, except for some special high-data-rate cases.
- If multiple data transport channels are transmitted, the resulting transmission blocks or frames are then time multiplexed. Each block is accompanied by a TFCI. This contains the rate information for the current block, and is therefore very important – if the TFCI is lost, then the whole frame is lost.
- The resulting data stream is then fed into a second interleaver, which interleaves the bits over one radio transmission frame, *intraframe interleaving*. In case multiple physical data channels are used, the transmission frames are then mapped to these multiple channels. Otherwise, a single dedicated physical channel is used.

The multiplexing operation in the downlink is slightly different, as outlined in Figure 26.7b, as the order of some of the steps is different. The main difference lies in the insertion of DTX (Discontinuous Transmission) bits, which indicate when to turn the transmission off. Depending on whether fixed or variable symbol positions are used, DTX indication bits are inserted at different points in the multiplexing/coding chain.

Coding

There are two modes of channel coding:

1. Convolutional codes are used with a coding rate of 1/2 for common channels and 1/3 for dedicated channels. Convolutional codes are mainly used for "normal" applications with a data rate of up to 32 kbit/s. The constraint length of the encoders is 9. Figure 26.8 shows the structure of the two encoders. The code polynomials for the rate-1/2 encoder are:

$$G1(D) = 1 + D^2 + D^3 + D^4 + D^8 \tag{26.5}$$

$$G2(D) = 1 + D + D^2 + D^3 + D^5 + D^7 + D^8 \tag{26.6}$$

and for the rate-1/3 encoder:

$$G1(D) = 1 + D^2 + D^3 + D^5 + D^6 + D^7 + D^8 \tag{26.7}$$

$$G2(D) = 1 + D + D^3 + D^4 + D^7 + D^8 \tag{26.8}$$

$$G3(D) = 1 + D + D^2 + D^5 + D^8 \tag{26.9}$$

2. Turbo codes are mainly used for high-data-rate (> 32 kbit/s) applications. The code rate is 1/3. A parallel concatenated code is used (see Chapter 14). Two recursive systematic convolutional encoders are employed (see Figure 26.9). The data stream is fed into the first one directly, and into the second one after passing an interleaver. Both encoders have a coding rate of 1/2. Thus, output is the original bit X or X' and the redundancy bits Y or Y', which are output from the recursive shift registers. However, as X equals X' only X, Y, and Y' are transmitted. Thus, the code rate of the turbo encoder is 1/3.

Table 26.1 provides an overview of the different coding modes used in different channels.

(a) Rate 1/2 convolutional encoder

(b) Rate 1/3 convolutional encoder

Figure 26.8   Structure of convolutional encoders.

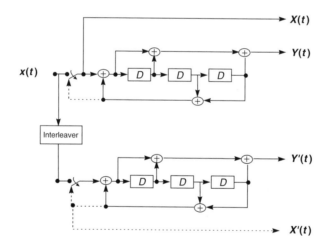

Figure 26.9   Structure of a turbo encoder.

26.6   Spreading and Modulation

26.6.1   *Frame Structure, Spreading Codes, and Walsh–Hadamard Codes*

WCDMA relies on CDMA for multiple access. However, transmission timing is still based on a hierarchical timeslot structure similar to GSM's: frames of duration $T_f = 10$ ms are divided into

Table 26.1 Forward error correction and different logical channels

Transport channel type	Coding scheme	Coding rate
BCH		
PCH		1/2
RACH	Convolutional code	
CPCH, DCH, DSCH, FACH		1/3, 1/2
	Turbo code	1/3

15 timeslots, each of which has a 12-bit-long *System Frame Number* (SFN). Each timeslot has a duration of 0.667 ms which equals 2,560 chips. The configuration of frames and timeslots is different for uplink and downlink.

WCDMA uses two types of code for spreading and multiple access: *channelization codes* and *scrambling codes* (compare also IS-95, Chapter 25). The former spread the signal by increasing the occupied bandwidth in accordance with the basic principle of CDMA. The latter do not lead to bandwidth expansion but help to distinguish between cells and/or users. In the following we discuss the codes and modulation for the different channels.

Channelization codes in WCDMA are *Orthogonal Variable Spreading Factor* (OVSF) codes (as discussed in Section 18.2.6, see also Section 25.5.1).

For scrambling codes, a long and a short code exist. Both are complex codes, and are derived from real-valued codes in accordance with the following expression:

$$C_{Scrambler}(k) = c_1(k) \cdot (1 + j \cdot (-1)^k \cdot c_2(2\lfloor k/2 \rfloor)) \tag{26.10}$$

Here, k is the chip index and c_1 and c_2 are real-valued codes.

For short codes c_1 and c_2 are two different members of the very large Kasami sets of length 256 (see Section 18.2.5). It is worth noting that the duration of the short code equals symbol duration only for spreading factor 256. Otherwise, a "short" code in WCDMA is not a short code in the sense of Chapter 18.

The long code is a Gold code, a combination of two Pseudo Noise (PN)-sequences that each have length $2^{25} - 1$. The I- and Q-part, c_1 and c_2 in Eq. (26.10), are versions of the same Gold sequence, shifted relative to each other. The codes are truncated to a length of 10 ms – i.e., one frame.

26.6.2 Uplink

Dedicated Channels

Figure 26.10 is a block diagram illustrating spreading and modulation for the uplink.

- Under normal circumstances, user data (DPDCH$_1$) and control data (DPCCH) are transmitted on the in-phase component and the control channel on the quadrature-phase component. First, channelization codes c_d and c_c are applied to the data and the control channel, respectively. As mentioned above these codes actually increase the signal bandwidth. Afterwards, the I- and Q-branch are treated as a complex signal. Processing the control and data channel like this is called I–Q multiplexing.
- If the data rate for the user is very high, then up to five additional data channels may be transmitted in parallel on the I-branch and Q-branch. These channels are then distinguished by applying different spreading codes, $c_{d,k}$ $k \in [1, \cdots, 3](!)$. The spreading factor is in this case 4.

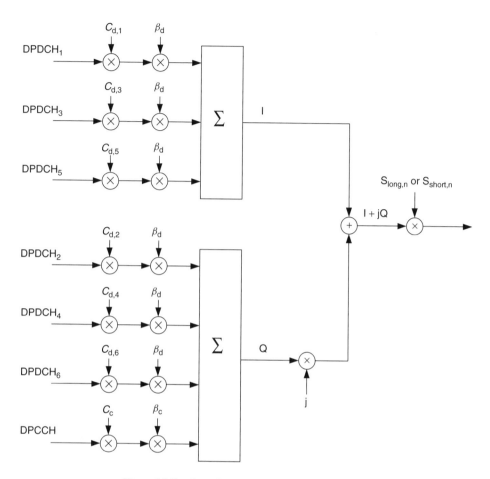

Figure 26.10 Spreading and modulation on the uplink.

Therefore, the total data rate is given by:

$$\frac{4 \text{ Mchip/s}}{4 \text{ chips/bit}} \cdot 6 \cdot \frac{1}{3} \tag{26.11}$$

The 4 chips/bit represent the spreading. The factor 6 relates to the maximal six transmitted channels and factor 1/3 relates to channel coding. Thus, the maximal achievable net user data rate is 2 Mbit/s.

- The transmit power of the control channel relative to the data channel(s) is determined by the ratio of spreading factors, β_c/β_d. The received noise power is proportional to the (unspread) bandwidth and thus the data rate. Consequently, the transmit power has to be the higher the lower the spreading factor in order to make sure that the Signal-to-Noise Ratio (SNR) at the receiver is the same for different data. The spreading factor of control data is always 256.
- Then the complex scrambling code $S_{long,n}$ or $S_{short,n}$ is applied and the signal is fed into the complex modulator. The complex signal usually shows a power imbalance between the I- and

Q-factor. In order to deal with this problem, spreading and scrambling codes are designed in such a way that the signal constellation is rotated by 90° between two subsequent chips.

- The resulting signal is modulated using Quadrature Amplitude Modulation (QAM) with Nyquist basis pulses. More precisely, the basic pulses are raised cosine pulses with a roll-off factor of $\alpha = 0.22$. Modulation precision is supposed to be better than 17%. Precision is measured by taking the ratio of the power of the difference signal between the actual and the ideal signal and the power of the ideal signal – *Error Vector Magnitude* (EVM).

I–Q multiplexing is used in the uplink in order to limit the crest factor. Even when there are no user data to be transmitted – e.g., in no-voice periods – the control channel is continuously active. Therefore, the RF amplifier in the MS does not have to switch on and off. This eases hardware requirements as the amplifier has to cover a smaller dynamic range. Furthermore, amplitude variations from slot to slot or burst to burst could lead to interference with audioprocessing equipment, like the microphone of the MS or the hearing aid of the user, as these activities would be in the audible frequency range.

We next turn to look in more detail at the control channel. It carries the pilot bits, the TPC bits, the FBI, and the TFCI. As the spreading factor is constantly 256, 10 bits of control information are transmitted per timeslot, see Figure 26.11.

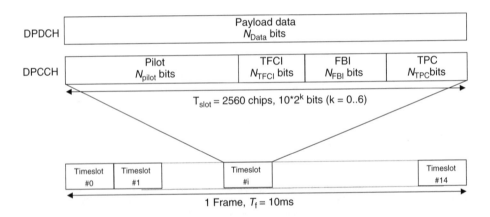

Figure 26.11   Frame structure on the uplink.

Random Access Channel

We next turn our attention to the burst structure of the RACH (see Figure 26.12). The start time for RACH transmission should be $t_0 + k \cdot 1.33$ ms, where t_0 refers to the start of a regular frame on the BCH. At the beginning of an access burst several preambles of length 4,096 chips are transmitted. The first preamble has power that is determined by "open-loop" power control – i.e., the MS measures the strength of the BCH, and from that concludes the amount of small-scale-averaged path gain; this (plus a safety margin) in turn determines transmit power. If the MS does not receive an acknowledgement of its access request on the AICH, it transmits the preamble again, with increased power. After receiving an acknowledgement, the actual access data (a field that is either 10 ms or 20 ms long) are transmitted.

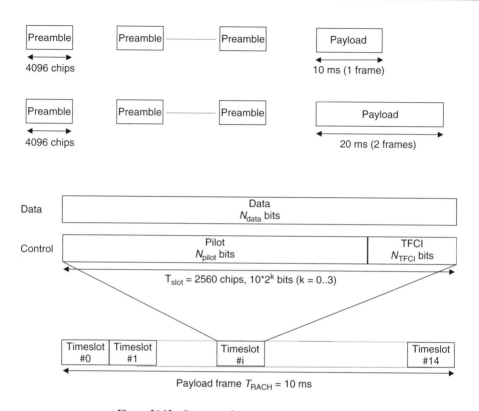

Figure 26.12 Structure of random access transmission.

The preamble contains one of a set of 16 predefined signature sequences, which are transmitted with a spreading factor of 256. The "data" containing message part of length 10 ms is similar to the DPCH divided into 15 timeslots. Within each of those timeslots 8 pilot bits and 2 TFCI bits are transmitted with a spreading factor of 256; this constitutes the control information for layer 1. Furthermore, a "data" message, usually containing control information for layer 2, is transmitted with a spreading factor between 32 and 256. Layer-1 control information and the "data" message are transmitted simultaneously by I–Q multiplexing (see Figure 26.13). Thus, the structure of the message part of the RACH is very similar to a frame of a dedicated physical channel. However, neither a TPC nor an FBI field are transmitted.

Physical Common Packet Channel

The burst structure of the PCPCH is quite similar to that of a transmission on the PRACH. One or several access preambles are transmitted initially with ascending transmission power until the received power at the BS is sufficient and the BS sends an acknowledgement to the MS. Afterwards, one preamble is transmitted whose sole purpose is the detection of collisions with other packets. Furthermore, power control information can be transmitted. Afterwards the actual data are transmitted. This message consists of one or several frames of 10 ms each, each of which is again divided into 15 timeslots.

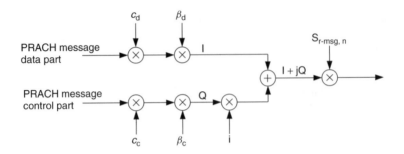

Figure 26.13 Modulation on the physical random access channel.

Base Station Processing

The BS has to process the received signal periodically in the following way [Holma and Toskala 2000]:

- It receives one frame, despreads it, and stores it with the sampling frequency given by the highest used data rate of this frame. Note that different data rates and spreading factors may be used during one frame.
- For each timeslot: (i) the channel impulse response is estimated using the pilots; (ii) the Signal-to-Interference Ratio (SIR) is estimated; (iii) power control information is transmitted to the MS; and (iv) power control information from the MS is decoded and transmission power is adapted accordingly.
- For every second or fourth timeslot: FBI bits are decoded.[5]
- For each 10-ms frame: TFCI information is decoded to get the decoding parameters for the DPDCH.
- For each interleaver period, which is either 10, 20, 40, or 80 ms: user data transmitted via the DPDCH is decoded.

Spreading Codes

The uplink uses OVSF codes for spreading. However, they are not used for channelization (distinguishing between users in the uplink). Therefore, different users can use the same spreading codes. As a matter of fact, assignment of spreading codes to the different channels of any MS is predefined. The DPCCH is always spread with the first code, the code with index 0, using spreading factor 256. A DPDCH channel is spread using the code with index $SF/4$, where SF is the spreading factor of the channel. In case multiple data channels are transmitted, the spreading factor is 4, when spreading codes with indices 1, 2, or 3 are used; one code is used for both a channel on the I-branch and a channel on the Q-branch. The spreading factor of a traffic channel may vary from frame to frame. There are additional code selection criteria for the PRACH and PCPCH.

Scrambling Codes

As mentioned above, signals from different users are distinguished by different scrambling codes. Both "short" and "long" codes may be used (see Chapter 18). There are no strict rules about when

---

[5] Note that two or four FBI blocks make up one complete FBI command, like an update of antenna weights.

to apply short or long codes. However, it is recommended to use short codes when the BS has multiuser detection capability (see Section 18.4). Short codes limit the computational complexity for multiuser detection, as the size of the cross-correlation matrices involved is smaller. In case no multiuser detection is implemented, long codes should be used as they provide better whitening of interference than short codes.

The BS assigns the specific scrambling code to the MS during connection setup with the *Access Grant Message*.

Codes for Random Access Channels

Random access channels use codes that are different from those in regular channels. These codes are specific to one BS, and two neighboring BSs should not use the same or similar codes. The preamble is transmitted first, which serves for synchronization and identification (see below). The preamble is designed in such a manner that it is very robust against initial uncertainties in frequency synchronization. Furthermore, RACH codes are only transmitted with a binary alphabet, which eases receiver design.

26.6.3 Downlink

Spreading and modulation of the data and control channels for the downlink is different from the uplink. In the downlink, data and control channels are *time-multiplexed* and the resulting single bitstream is then Quadrature-Phase Shift Keying (QPSK) modulated. We may interpret QPSK modulation of a single bitstream as a serial/parallel conversion into two streams on the I- and Q-branch. The resulting I-signal and Q-signal are spread separately using the same channelization code. The complex scrambling code is then applied to the resulting complex signal. Finally, the scrambled complex signal is fed into the complex modulator.

Figure 26.14 illustrates the frame and timeslot structure for the DDPCH. The first datablock containing bits is transmitted. Then the TPC field is sent, followed by the TFCI field. A second datafield is transmitted thereafter. The timeslot is concluded with the pilot field. A spreading factor between 4 and 512 can be used. The purpose and properties of the pilot, TFCI, and TPC fields are the same as for the uplink.

Figure 26.14 Frame and timeslot structure on the downlink.

The spreading factor for one user does not vary from frame to frame. It is chosen to accommodate the highest occurring data rate of this user; if a lower rate is required instantaneously, DTX is employed – i.e., transmission is blanked out for a while. The MS can be informed about the current data rate by the TFCI field.

The reasons that DTX can be employed in the downlink (in contrast to the uplink) are that (i) the appearance of audible interference resulting from on/off switching of transmit amplifiers is not an issue on the downlink; the common control and synchronization channels, like the BCH or the SCH, are transmitted continuously anyway; and (ii) high values of the crest factor cannot be avoided anyway as transmission of several CDMA signals in parallel always leads to high crest factors.

Mobile Station Processing

The MS receiver has to perform operations for the downlink similar to those of the BS receiver in the uplink. However, there are some differences [Holma and Toskala 2000]:

- The spreading factor of all channels apart of the DSCH is constant over time.
- There is no need for the FBI field.
- There is one CPICH in addition to the pilots of dedicated physical channels, which improves the channel estimate.
- Smart antennas can be used in the downlink.

Channelization Codes

The signals for different users are inherently synchronized in the downlink, as they come from the same transmitter – i.e., the BS (see also Chapter 25). Therefore, OVSF codes enable good separation of the signals for different users within one cell. The BS informs the MS about the code used during connection setup. The BS ensures that the same code (or a "mother code" in the OVSF tree) is only used once in the cell. If necessary, the code can be changed during a connection.

Scrambling Codes

Long codes for the downlink are the same as in the uplink; short codes are not used for the downlink. There are 512 primary scrambling codes,[6] which are divided into 64 groups of 8 codes each. Primary, secondary, and alternative codes are unambiguously related to each other. Each cell has exactly one primary scrambling code that is used on the CCPCH and CPICH. The other downlink channels might use the same primary or a related secondary code. However, use of more than one scrambling code only makes sense when other measures, like smart antennas, are used to increase cell capacity. Otherwise, one scrambling code in conjunction with spreading codes already achieves maximum user separation.

Overview

Table 26.2 provides an overview of the different codes used.

---

[6] Additionally, there are 512×25 secondary scrambling codes and 8,192 left and 8,192 right alternative scrambling codes for compressed mode.

Table 26.2 Overview of different channelization and scrambling codes

	Channelization codes (variable chip rate)		Scrambling codes (constant chip rate)			
	Uplink	Downlink	Uplink			Downlink
Separates	Channels (I/Q, DPDCH) of one user	Channels (DPDCH) and users	Users			Cells
	Assigned to connection	Assigned to connection and user	Assigned to user by BS			Assigned to cell
Reuse	Within the same cell	In all other cells	In other cells			In distant cells (code planning)
Selection	Fixed (given by the SF)	Variable	Variable			Fixed
Length of code	Short	Short	"Short"		Long	Long
Enables	Variable data rates	Different data rates	Multiuser detection			Cell search
Code family	OVSF (real)	OVSF (real)	Complex, based on VL Kasami		Complex, based on segments of Gold codes	Complex, based on segments of Gold codes
Codelength	4–256	4–512	256		38,400	38,400
Number of codes	Up to 256		> 10^6		≫	512 (8, 192)

High-Data-Rate Users

There are basically two possibilities to support high data rates in UMTS:

1. Transmission with a low spreading factor. This is the most straightforward way to increase the data rate. The lowest admissible spreading factor in WCDMA is 4; note that a spreading factor of 1 would lead to a pure Frequency Division Multiple Access (FDMA) system. The drawback of a low spreading factor is that InterSymbol Interference (ISI) becomes significant: with a spreading factor of 4, the symbol duration can be as low as 1 µs. Thus, the receiver has to cope not only with interchip interference, which is effectively dealt with by a Rake receiver structure, but also with severe ISI that needs to be eliminated by an equalizer (Chapter 16). This additional implementation complexity may increase the price of the unit, particularly for an MS.
2. The data stream can be serial/parallel converted and then transmitted using multiple codes, where each code has a sufficiently high spreading factor. This is called "multicode transmission." ISI is not a problem with this approach. However, two other problems appear: (i) the crest factor of a multicode signal is high (this is not a problem on the downlink, as multiple users have to be supported anyway); and (ii) efficiency is lowered as every additionally transmitted code leads to additional overhead.

26.7 Physical-Layer Procedures

26.7.1 Cell Search and Synchronization

The search for the signal of the strongest BS and synchronization with this BS has to be done in several steps.

In a first step, the MS *synchronizes with the timeslot* timing of the strongest signal it can observe. This is done by searching for the 256-chip primary synchronization code. Every BS periodically transmits this sequence, and it is identical for the whole system. The MS thus just correlates the incoming signal with this synchronization sequence to identify all available BSs. The output of this correlator shows several peaks at different delays, corresponding to the signals of different BSs and the various echoes (MPCs) of these signals. By selecting the highest peak, the MS achieves timeslot synchronization with the strongest signal. In other words, the MS is aware of the commencement of a timeslot in the strongest BS. However, it still does not know about frame timing, as it does not know, e.g., whether the first or the tenth timeslot is currently received.

Therefore, *frame synchronization* has to take place in the next step. This is done by observing the secondary synchronization channel; in the same step the codegroup of the primary scrambling code used by the BS is determined. As discussed earlier, the CCPCH and CPICH are transmitted in the downlink with one of 512 primary scrambling codes. These 512 primary codes are divided into 64 groups. The secondary synchronization channel transmits a sequence of 15 codes in 15 consecutive timeslots. Each code is a letter of the same alphabet. Thus, we may interpret the code sequence as a word with 15 letters. This codeword is repeated periodically. The MS can thus determine the group used and achieve frame synchronization by observing this codeword. Codewords are chosen in such a manner that a cyclic shift of one codeword never results in another codeword. Therefore, the MS can determine which codeword is transmitted (this indicates the used codegroup) and where the codeword starts (this indicates frame timing).

Let us illustrate this procedure with a simplified example. Assume, that the alphabet size used is five and three codewords of length 4 can be transmitted: (i) *abcd*; (ii) *aedc*; and (iii) *bcde*. Further, assume that *edca* is received. As (i) does not feature the letter *e* and (iii) does not feature the letter *a*, the received word has to be a cyclically shifted version of (ii). Once we know that (ii) is transmitted, we can easily see that frame timing has to be adjusted by one slot.

After the codegroup has been determined, the MS has to determine which code of this group is used on the CCPCH. Therefore, the CCPCH undertakes symbolwise correlation with all possible eight codes of that codegroup. Once this is achieved, the CCPCH can be properly demodulated.

26.7.2 Establishing a Connection

A connection initialized by the MS requires the following procedures:

1. The MS synchronizes itself with the strongest BS as described in the previous section.
2. The MS decodes the BCH and thereby acquires information regarding: (i) the spreading code of the preamble and the scrambling code for the message part of the PRACH; (ii) the codes available; (iii) the access timeslots available; (iv) possible spreading factors for messages; (v) the interference level on the uplink; and (vi) the transmit power of the CCPCH.
3. The MS selects a spreading code for the preamble and a scrambling code for messages.
4. The MS determines the spreading factor for the message.
5. Based on the measured signal strength of the CCPCH and information on the transmit power of the BS, the MS estimates the attenuation on the uplink. Based on the attenuation estimate and information regarding the uplink interference level, the MS estimates the necessary transmission power for the preamble.
6. The MS randomly selects an access timeslot and a signature from the available set.
7. The MS transmits the preamble. In case of successful acquisition the BS transmits an acknowledgement on the AICH.
8. In case the MS does not receive an acknowledgement from the BS it repeats the preamble with increased transmission power.

9. Once the BS indicates acquisition of the preamble, the MS starts to transmit the access message in the next available timeslot.
10. The MS waits for an access grant message from the network. If it does not receive this message within a predetermined time, it repeats the steps from step 5 on.

26.7.3 Power Control

Power control is an essential part of a CDMA system, as it is necessary to control mutual interference (see Chapter 18). Inner-loop power control in WCDMA, in particular, is supposed to adapt to *small-scale* fading for speeds up to 500 km/h. Therefore, the power control procedures in UMTS have to be rather fast. An update of transmit power occurs with every timeslot – i.e., every 0.667 ms. There is also outer-loop power control which continuously adjusts the target SIR for inner-loop power control (see also power control in IS-95, Chapter 25).

Uplink

The uplink uses a closed-loop procedure for power control. The BS estimates the power of the received signal and controls it by transmitting TPC instructions to the MS, which changes its transmit power accordingly. TPC bits are transmitted with the DPCCH, and contain instructions to increase or decrease power. The possible stepsizes are 1 dB or 2 dB with uncertainties of ± 0.5 dB or 1 dB. However, transmit power is not decreased below, for example, -44 dBm. Given a maximal transmission power of 33 dBm for class-1 units, the RF amplifier has to cover a dynamic range of 77 dB, which is a rather severe hardware requirement.

As for soft handover, the situation is rather complicated as the MS can receive different TPC bits from the different BSs involved. The UMTS standard specifies an algorithm that determines a means of obtaining a combined power control command from these different commands. The combined instruction is a function of weighted single-power-control commands. The weighting applied is proportional to reliability of the individual signals. Some particular rules apply for special situations. For example, for both compressed mode and the common packet channel a special procedure is implemented, as longer gaps between transmissions and TPC commands occur.

It is not possible to use closed-loop power control on the PRACH, as a dedicated channel has not yet been established between the MS and the BS. Therefore, open-loop power control has to be employed. The MS measures the average received power of the CCPCH over some time to average out small-scale fading effects. This is necessary because, due to the frequency duplex, small-scale fading on the uplink is unrelated to that on the downlink. In other words, it is not possible to determine from the instantaneous received power on the downlink, the level of instantaneous received power on the uplink. Average received power allows estimation of just the necessary *average* transmit power for the uplink. However, this is at least a good starting value.

Downlink

All signals on the downlink (to one MS) suffer from the same attenuation. Furthermore, downlink signals are orthogonal, due to the use of OVSF codes for channelization, so that the despreading operation results in a good SIR.[7] Therefore, power control on the downlink aims mainly at maintaining a good SNR.[8] A closed-loop is used for power control in the downlink. Each MS

---

[7] Note, however, that orthogonality can be destroyed in frequency-selective channels.
[8] This argumentation only considers one cell. Considering intercell interference, the SIR in one cell might be decreased by increasing the transmission power of all channels. However, this in turn decreases the SIR in all neighboring cells.

measures signal strength and quality and transmits TPC commands to the BS via the DPCCH. As long as at least one MS requests more transmission power the BS increases the power of all channels. Therefore, MSs at cell borders actually control the transmission power of the BS.

One problem is that the MS estimates the SNR *after* the Rake receiver. A cheap device, which employs only a few Rake fingers, captures less received signal energy than a more sophisticated device with more Rake fingers. Therefore, a cheap device tends to request more transmit power from the BS than an expensive one. Thus, the UMTS network levels out differences in the designs of units by ensuring good receive qualities independent of receiver design quality. Of course, this might limit the motivation of manufacturers to develop and produce good MSs.

Transmit Diversity

The downlink may implement the option of transmit diversity with two antennas. It distinguishes between closed-loop diversity, for which the transmitter requires Channel State Information (CSI), and open-loop diversity, for which the transmitter does not require any information about the channel. Three modes of diversity are specified:

- Orthogonal space–time block coding with two antennas – i.e., the Alamouti code (see Section 20.2).
- The transmit signal might be switched from one antenna to the other. This mode is only used for the SCH.
- Closed-loop transmit diversity: two antennas can transmit the same stream, by applying complex weights for each antenna. The pilot bits on signals from the two antennas are mutually orthogonal sequences. Therefore, the receiver can estimate the channel impulse responses of the two channels separately. Antenna weights are then determined such that signals from the two transmit antennas add up constructively at the MS (see Chapter 13). The computed weights are digitized, and transmitted to the BS via the FBI field of the DPCCH. This mode of diversity is optional for BSs, but MSs have to be able to support it. Support for more advanced multiple-antenna techniques was introduced in later releases.

26.7.4 Handover

Intrafrequency Handover

A connection handover between two BSs on the same carrier frequency is performed as a *soft handover*. The MS has a connection to both BSs during handover (see Chapter 18). Thus, signals from both BSs are used during this time and the Rake receiver processes them similarly to two paths of a multipath signal with two or more fingers. As BSs use different scrambling codes in WCDMA, the Rake receiver in the MS has to be able to apply different codes in each finger.

In preparation for soft handover, the MS has to acquire synchronization with other BSs. This synchronization process is similar to the one described above, apart from the fact that the MS has a priority list for codegroups. This list contains the codegroups used by the neighboring cells for handover, and is continuously updated.

Cell selection for soft handover can be made by different criteria, like signal strength after despreading or wideband power (Received Signal Strength Indication RSSI). No particular algorithms are specified in the standard. However, it is suggested to divide cells into the active cell and neighboring candidate cells that provide good signal strength.

A so-called *softer handover* is a special case of soft handover, in which the MS switches between two sectors of the same BS. Algorithms and processes are similar to those used for soft handover with the difference that only one BS is involved.

Interfrequency Handover

This kind of handover takes place in the following situations:

- Two BSs employ different carriers.
- The MS switches between hierarchy layers in the hierarchical cell structure.
- Handover to other providers or systems – e.g., GSM – is necessary.
- The MS switches from TDD mode to FDD mode or vice versa.

Interfrequency handover is a "hard" handover during which the connection between the MS and the old BS is first interrupted before the MS establishes a new connection with a new BS. There are two ways of measuring signal strength, etc. on other frequencies while the old connection is still active: (i) the MS might feature two receivers, so that one can measure on other frequencies while the first one still receives user data; and (ii) transmission is done in compressed mode, so that data that normally are transmitted within a 10-ms frame are compressed to 5 ms. The time that is freed up can then be used for measurements on other frequencies. Compression can be achieved, for example, by puncturing the data stream or reduction of the spreading factor.

26.7.5 Overload Control

The UTRAN has to ensure that it does not accept too many users or provide users with too high data rates, as then the system would be overloaded. Several means can be used for this [Holma and Toskala 2000]:

- No increase in transmit power in the downlink, even when one MS requests it. This of course implies lower signal quality in the downlink. If data packets are transmitted, the Automatic Repeat reQuest (ARQ) rate increases, which in turn lowers throughput. In case of voice transmission, audio quality is decreased.
- The target Signal-to-Interference-and-Noise Ratio (SINR) in the uplink can be decreased. This decreases the transmission power of MSs. It also results in decreased transmission quality for all subscribers.
- Throughput of the data packet channels can be lowered. This decreases the speed of data connections but maintains the quality of voice services.
- The output data rate of UMTS voice encoders can be decreased, as UMTS employs an encoder with a variable rate. This decreases audio quality, but has no impact on data applications.
- Some active links can be transferred to other frequencies.
- Regular phone calls can be handed over to the GSM network.
- Data connections or phone calls can be terminated.

26.8 Glossary for WCDMA

3GPP	Third Generation Partnership Project
ACELP	Algebraic Code Excited Linear Prediction
ACLR	Adjacent Channel Leakage Ratio
AD	Access Domain
AICH	Acquisition Indication Channel
AMR	Adaptive Multi Rate
AN	Access Network

ARQ	Automatic Repeat reQuest
ATM	Asynchronous Transfer Mode
BCH	Broadcast CHannel
CCPCH	Common Control Physical Channel
CCTrCH	Coded Composite Traffic CHannel
CDMA	Code Division Multiple Access
CN	Core Network
CND	Core Network Domain
CPCH	Common Packet CHannel
CPICH	Common PIlot CHannel
CRC	Cyclic Redundancy Check
DCH	Dedicated (transport) Channel
DPCCH	Dedicated Physical Control CHannel
DPDCH	Dedicated Physical Data CHannel
DSCH	Downlink Shared Channel
EVM	Error Vector Measurement
FACH	Forward Access CHannel
FBI	Feed Back Information
IMT	International Mobile Telecommunications
IP	Internet Protocol
ISI	Inter Symbol Interference
IWU	Inter Working Unit
JD-TCDMA	Joint Detection-Time/Code Division Multiple Access
ME	Mobile Equipment
MSS	Mobile Satellite Service
MT	Mobile Termination
Node-B	Base station
PCH	Paging Channel
P-CCPCH	Primary Common Control Physical CHannel
PCPCH	Physical Common Packet Channel
PDSCH	Physical Downlink Shared CHannel
PICH	Page Indication Channel
PRACH	Physical Random Access CHannel
QoS	Quality of Service
RACH	Random Access CHannel
RAN	Radio Access Network
RNC	Radio Network Controller
RNS	Radio Network Subsystem
SAP	Service Access Point
SAPI	Service Access Point Identifier
S-CCPCH	Secondary Common Control Physical CHannel
SCH	Synchronization CHannel
SFN	System Frame Number
SMS	Short Message Service
TA	Terminal Adapter
TE	Terminal Equipment
TFCI	Transmit Format Combination Indicator
TFI	Transport Format Indicator
TPC	Transmit Power Control
UARFCN	UTRA absolute radio frequency channel number

UE	User Equipment
UED	User Equipment Domain
UE-ID	User Equipment in-band IDentification
UMTS	Universal Mobile Telecommunications System
UCPCH	Uplink Common Packet CHannel
USIM	User Service Identity Module
UTRA	UMTS Terrestrial Radio Access
UTRAN	UMTS Terrestrial Radio Access Network
W-CDMA	Wideband CDMA

Further Reading

The most authoritative source for UMTS is, of course, the standard itself, whose most recent version can be found at *www.3gpp.org* However, this material is exceedingly difficult to read. Good summaries can be found in Holma and Toskala 2007] and Richardson [2005], as well as a number of other monographs. The High-Speed Packet Access (HSPA) part is described very nicely in Dahlman et al. [2008].

For updates and errata for this chapter, see wides.usc.edu/teaching/textbook

27

3GPP Long-Term Evolution

27.1 Introduction

27.1.1 History

In 2004, just as the first widespread rollout of Wideband Code Division Multiple Access (WCDMA) systems was happening, the Third Generation Partnership Project (3GPP) industry consortium started to work on fourth-generation (4G) systems. It was predicted at that time (and borne out by later developments) that the data rates and spectral efficiencies of WCDMA would not meet the demand of future applications; therefore, a new system had to be developed. In a somewhat bold move, it was decided to completely change both the air interface and the core network. The air interface was to move to Orthogonal Frequency Division Multiplexing (OFDM) as modulation, and Orthogonal Frequency Division Multiple Access (OFDMA), with (limited) support for Multiple Input Multiple Output system (MIMO) antenna technology. The core network was to evolve into a pure packet-switched network. The new standard became known as *3GPP Long-Term Evolution*, or simply *LTE*.

The development of LTE originally took place in parallel to the further evolution of WCDMA. Around 2007/2008, LTE started to take the center stage of the 3GPP meetings. The basic parameters of the air interface were soon agreed on, but the implementation details required an enormous effort to achieve compromises yet stay reasonably simple and self-consistent. At the time of this writing, Release 8 of the LTE specifications has been finalized. It provides for data rates up to 300 Mbit/s in the downlink (DL). Further improvements, in particular extending the use of MIMO for increasing the spectral efficiency, are foreseen for future releases. Release 10, also known as *LTE-Advanced* is intended to provide data rates up to 1 Gbit/s. LTE-Advanced will also be submitted to the International Telecommunications Union (ITU) as one of the candidates for International Mobile Telecommunications (IMT)-Advanced cellular systems (compare Section 26.1).

LTE has received strong support from the vast majority of cellphone and infrastructure manufacturers. Most notably, the *3GPP2* alliance (which had promoted the cdma2000 systems, a rival to 3GPP's WCDMA system) decided to terminate the development of its own incipient 4G standard; rather, its members are now participating in the development of LTE. Therefore, both the existing WCDMA and cdma2000 network operators will eventually migrate to LTE. While there are still two different flavors of LTE (a Frequency Domain Duplexing (FDD) and a Time Domain Duplexing (TDD) mode, where the latter is especially important for China), great effort has been made to make those two modes as similar as possible; and indeed the duplexing method is the only essential difference between them.

Wireless Communications, Second Edition   Andreas F. Molisch
© 2011 John Wiley & Sons, Ltd

One might wonder why there is such eagerness by the cellular industry to move to a new standard when the investments into 3G have not been amortized yet. The reasons are manifold, and include (i) need for improved data rates and spectral efficiency in particular in dense urban environments, (ii) for some operators, the possibility to "leapfrog" from 2G directly to 4G technology, (iii) the competition from WiMAX (Chapter 28), and (iv) the possibility of acquiring new spectrum in the name of getting it for a new system.

The current chapter describes LTE Release 8. An update, describing the modifications for LTE-Advanced, might be posted in the future on the companion website *www.wiley.com/go/molisch*.

27.1.2 Goals

LTE aims to achieve a *peak* data rate of 100 Mbit/s in the downlink and 50 Mbit/s in the uplink (UL), respectively, with a 20-MHz spectrum allocation for *each* of the downlink and uplink. Thus, the required spectral efficiency is 5 and 2.5 bit/s/Hz for the downlink and uplink, respectively. However, due to the wide range of applications and requirements, LTE defines a number of different types of Mobile Stations (MSs) that present a tradeoff between complexity and performance (see Table 27.1).

Table 27.1 Performance requirements for different MS classes

Category	1	2	3	4	5
Peak data rate DL (Mbit/s)	10	50	100	150	300
Max DL modulation	64 QAM	64 QAM	64 QAM	64 QAM	64 QAM
Peak data rate UL (Mbit/s)	5	25	50	50	75
Max UL modulation	16 QAM	16 QAM	16 QAM	16 QAM	64 QAM
Max number of layers for DL MIMO	1	2	2	2	4

For latency, the goals distinguish between

1. *Control-plane latency* (defined as the time for a handset to transition from various nonactive states to active states), which are between 50 and 100 ms, depending on the state in which the MS originally was. Furthermore, at least 400 active MSs per cell should be supported.
2. *User-plane latency* (defined as the time it takes to transmit a small Internet Protocol (IP) packet to the edge node of the *Radio Access Network*, RAN), which should not exceed 5 ms in a network with a single MS (i.e., no congestion problems).

For operation under realistic circumstances, LTE defined performance requirements relative to the performance of WCDMA systems (though that comparison baseline does not include some of the more advanced features of WCDMA; in particular, no Spatial Multiplexing (SM)). Generally, user throughput should improve 2–4 times. The system is intended to be optimized for low speeds (0–15 km/h), since the main usage, especially for data services, is expected to be for nomadic terminals. Slight performance degeneration is allowed for speeds up to 120 km/h, while for truly high-speed applications (up to 500 km/h), only basic connectivity needs to be retained. Future releases of LTE also include a strong support for *Multimedia Broadcast and Multicast Services* (MBMS) applications. A spectral efficiency of 1 bit/s/Hz is required – while that seems low compared to the peak data rate of normal (unicast) systems, it is worth remembering that sustaining a data rate to multiple users simultaneously is harder than for a single user: e.g., beamforming cannot be employed.

The transition from WCDMA/HSPA(High Speed Packet Access) (including legacy Global System for Mobile communication (GSM) systems) to LTE should be made as seamless as possible.

This implies that for a considerable number of years, the two systems will coexist, usually in the same frequency band. Transitions from one system to another will be frequently required, especially during the initial rollout of LTE, when only parts of the country will be covered by LTE Base Stations (BSs). Transition times for real-time applications should be less than 300 ms, and for nonreal-time applications should be 500 ms.

27.2 System Overview

27.2.1 Frequency Bands and Spectrum Flexibility

LTE can be operated in a variety of frequency bands that are assigned by national frequency regulators, based on the decisions of the World Radio Conference. This spectrum can be, in principle, used for any member of the IMT-2000 and IMT-Advanced family. It originally encompassed the frequency bands discussed in Section 26.3; later additional frequencies were assigned, which became available through the so-called "digital dividend," – i.e., spectrum that was freed up when TV was converted to digital transmission techniques that required less spectrum than the old analog techniques. Tables 27.2 and 27.3 show the bands available by the time of this writing (2009). However, not all bands are available in all countries. In Europe, band 1 is the same as that assigned for WCDMA, thus anticipating existing operators to migrate, without new frequency assignments, from WCDMA to LTE. Migration from GSM to LTE is anticipated by the creation of bands 3 and 8 (in Europe). Similarly, bands 2, 4, 10, cover the Personal Communication System (PCS) frequencies currently occupied by operators in the U.S.A., while band 5 covers the frequencies long used by U.S. operators for lower-frequency operation. Bands 6 and 9 cover the traditional Japanese operator frequencies. Bands that became available through the digital dividend include several blocks of spectrum in the 700-MHz range have been recently auctioned off (bands 12, 13, 14, 17) in the U.S.A. In Europe and Asia, current activity concentrates on the 2,300–2,700 MHz range (bands 7, 38, 40); further spectrum in the 3,400–3,600-MHz range will become available in the near future. Note that some newly available spectra are reserved exclusively for specific systems, while other frequencies might be used by the operators as they deem fit.

Table 27.2 Bands for FDD operation of LTE

Operating band	UL (MHz)	DL (MHz)	Bandwidth						
			1.4	3	5	10	15	20	
1	1920–1980	2110–2170			✓	✓	✓	✓	Europe, Asia
2	1850–1910	1930–1990	✓	✓	✓	✓	✓	✓	America
3	1710–1785	1805–1880	✓	✓	✓	✓	✓	✓	Europe, Asia
4	1710–1755	2110–2155	✓	✓	✓	✓	✓	✓	America
5	824–849	869–894	✓	✓	✓	✓			America
6	830–840	875–885			✓	✓			Japan
7	2500–2570	2620–2690			✓	✓	✓	✓	Europe, Asia
8	880–915	925–960	✓	✓	✓	✓			Europe, Asia
9	1750–1785	1845–1880			✓	✓	✓	✓	Japan
10	1710–1770	2110–2170			✓	✓	✓	✓	Americas
11	1428–1453	1476–1501			✓	✓	✓	✓	Japan
12	698–716	728–746	✓	✓	✓	✓			Americas
13	777–787	746–756	✓	✓	✓	✓			Americas
14	788–798	758–768	✓	✓	✓	✓			Americas
17	704–716	734–746	✓	✓	✓	✓			Americas

Table 27.3 Bands for TDD operation of LTE

Operating band	Band	Bandwidth						
		1.4	3	5	10	15	20	
33	1900–1920			✓	✓	✓	✓	Europe, Asia
34	2010–2025			✓	✓	✓		Europe, Asia
35	1850–1910	✓	✓	✓	✓	✓	✓	
36	1930–1990	✓	✓	✓	✓	✓	✓	
37	1910–1930			✓	✓	✓	✓	
38	2570–2620			✓	✓			Europe
39	1880–1920			✓	✓	✓	✓	China
40	2300–2400			✓	✓	✓		Europe, Asia

For migration from second- and third-generation systems to LTE, operators that own more than 5 MHz of spectrum can make a "soft transition" by first assigning a 5-MHz block to LTE, while retaining the remainder of the band for their legacy services. As more users switch to LTE, additional parts of the spectrum can be rededicated to use in LTE.

LTE can also be operated with various bandwidths. The most common bandwidths are anticipated to be 5 and 10 MHz, but lower bandwidths (1.4 and 3 MHz) as well as higher bandwidths (15 and 20 MHz) are also foreseen. When peak data rates are mentioned, they usually refer to usage in the 20-MHz spectrum. Due to the use of OFDM as modulation format, bandwidths can be adjusted with great ease by changing the number of subcarriers, without changing any of the other parameters in the system.

27.2.2 Network Structure

The network structure in LTE is quite simple in principle (and actually, simplified with respect to the GSM and WCDMA structure): there is only a single type of access point, namely, the eNodeB (or BS, in our notation).[1] Each BS can supply one or more cells, providing the following functionalities:

- air interface communications and PHYsical layer (PHY) functions;
- radio resource allocation/scheduling;
- retransmission control.

The X2 interface is the interface between different BSs. Information that is important for the coordination of transmissions in adjacent cells (e.g., for intercell interference reduction) can be exchanged on this interface. Each BS is connected by the S1 interface to the core network.

For LTE, a new core network, called System Architecture Evolution (SAE) or Enhanced Packet Core (EPC) was developed. It is based on packet-switched transmission. It consists of (i) a Mobility Management Entity (MME), (ii) the serving gateway (connecting the network to the RAN), and (iii) the packet data network gateway, which connects the network to the Internet. In addition, the Home Subscriber Server is defined as a separate entity. The structure is sketched in Figure 27.1. The core network fulfills, *inter alia*, the following functions:

- mobility management;

---

[1] Remember that in WCDMA, there is a distinction between nodeB and Radio Network Controller (RCN).

Figure 27.1 Network structure and interface definitions of LTE. *In this figure*: HSS, Home Subscriber Server; PDN, Public Data Network.

- subscriber management and charging;
- quality of service provisioning, and policy control of user data flows;
- connection to external networks.

27.2.3 Protocol Structure

The transmission protocol of LTE is divided into several layers. At the top, the *Packet Data Convergence Protocol* (PDCP) performs functions related to data integrity (like enciphering) and IP header compression. The PDCP hands its packets, called *Service Data Units* (SDUs) to the *Radio Link Control* (RLC). The RLC segments and/or concatenates the SDUs into packets that are more suitable for transmission over the radio channel, the *Protocol Data Units* (PDUs). Due to the large dynamic range of the transmission data rates, the size of a PDU can be adjusted dynamically, but in any case, a PDU can contain bits from one or more SDUs; conversely one SDU might be segmented, and its segments transmitted in multiple PDUs. The RLC also makes sure that at the receiver (RX), all PDUs arrive (and arrange for retransmission if they do not) and hand them to the PDCP in their correct order.

The *Medium Access Control* (MAC) handles the scheduling of the PDUs, as well as the Hybrid Automatic Repeat reQuest (HARQ) for retransmissions on the PHY.[2] Finally, the *PHY* handles all the processes of actually transmitting data over the air, including coding and modulation. Note that the PHY not only interfaces with the MAC layer (layer 2), but also the Radio Resource Control (RRC) of layer 3 (see Figure 27.2). MAC layer and PHY are at the center of this chapter.

27.2.4 PHY and MAC Layer Overview

Although the details of the PHY are quite intricate, the key features can be summarized in the following straightforward bullet points:

- In the downlink, LTE uses OFDM as modulation (Section 19.1–19.8); in the uplink, it uses OFDM precoded with a Discrete Fourier Transform (DFT). If the transmitter (TX) (i.e., the MS)

---

[2] see Section 27.5.2 for the relationship between HARQ and RLC retransmission.

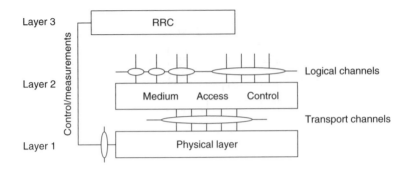

Figure 27.2 Protocol structure of LTE.

uses all available subcarriers, then this is identical to the single-carrier transmission with Cyclic
Prefix (CP), as described in Section 19.11. However, in LTE, an MS can also use just a subset
of subcarriers and apply the DFT precoding to this subset only.

- The multiple access format for both uplink and downlink is OFDMA combined with Time
  Division Multiple Access (TDMA). In other words, the spectral resources, as represented in
  the time/frequency plane, are assigned in a flexible manner to the different users. Furthermore,
  different users can have different data rates. The transmissions for a specific user are scheduled
  to happen in those frequency bands that offer the best propagation conditions, thus exploiting
  multiuser diversity.
- *Multicast/Broadcast over Single Frequency Network* (MBSFN), i.e., transmission of the same
  information from different BSs, can be realized in a straightforward manner using OFDM, as
  long as the runtime differences of the signals from the different BSs are less than the cyclic prefix.
- LTE provides means for intercell interference coordination, i.e., making sure that signals emitted
  in one cell do not interfere catastrophically with signals in the neighboring cells. Note that
  LTE, using essentially Frequency Division Multiple Access (FDMA)/TDMA for multiple access,
  is more dependent on intercell interference coordination than, e.g., WCDMA. The interference
  coordination used in LTE tends to be more sophisticated than the simple "frequency reuse"
  discussed in Section 17.6.1.
- Support for the multiple antennas, including receive diversity, various forms of transmit diversity,
  and spatial multiplexing (see Chapter 20).
- Adaptive modulation and coding, together with advanced coding schemes.
- FDD or TDD can be used for duplexing, depending on the assigned frequency bands. FDD and
  TDD modes are very similar (in contrast to WCDMA), though a number of signaling details and
  parameter choices differ. To keep the description compact, this chapter only considers the FDD
  mode unless otherwise stated. LTE also foresees half-duplexing, where transmission and reception
  for a specific terminal are distinguished both by different times *and* by different frequencies; this
  eases implementation of MSs without affecting spectral efficiency (see Section 17.5).

Generating the PHY signal for the downlink then consists of the following steps (see Figure 27.3):

- *Error correction encoding*: (see Section 27.3.6).
- *Scrambling of coded bits*: the bits of all transport channels are scrambled by multiplication with
  a Pseudo Noise (PN) (Gold) sequence. Note that – in contrast to WCDMA – it is the bits, and
  not the complex-valued symbols, that are scrambled.

Figure 27.3 Overview of the physical layer procedure.

Figure 27.4 Mapping of bit combinations onto symbols.

- *Modulation of scrambled bits to generate complex-valued modulation symbols*: the modulation formats used for data transmission are Quadrature-Phase Shift Keying (QPSK), 16-Quadrature Amplitude Modulation (QAM), and 64-QAM, with Gray mapping (i.e., signal points located next to each other are distinguished only by 1 bit); see Figure 27.4. The choice of the modulation format depends on the quality of the propagation channel: for better signal-to-interference- and Signal-to-Noise Ratio (SNR), higher order modulation can be employed.
- *Mapping of the modulation symbols onto the transmission layers*: LTE foresees multiple layers (roughly equivalent to "spatial streams" of Section 20.2) for the transmission with multiple antennas; see also Section 27.3.7.
- *Precoding of the symbols on each layer for transmission on the antenna ports*: this step is also related to multiple-antenna transmission.
- *Mapping of symbols* to Resource Elements (REs): assign which symbols are to be transmitted in which time/frequency resource (i.e., time and subcarrier). In the case of multiple transmit antennas, this mapping is done at each antenna port separately.
- *Generating the time domain OFDM signal*: (again, for each antenna port separately).

For the uplink, the steps are almost identical, except that:

- The assignment of symbols to time/frequency resources is different; only contiguous subcarriers can be used by one MS.
- Data scrambling is done with sequences that depend on the MS.
- The signals are DFT encoded before being sent to the OFDM modulator.

27.3 Physical Layer

27.3.1 Frames, Slots, and Symbols

In LTE, the time axis is divided into entities that play an important role in the transmission of different channels. These time entities have the following hierarchy (see Figure 27.5):

- The fundamental time unit of LTE transmission is a *radio frame*, which has a duration of 10 ms.
- Each radio frame is divided into 10 *subframes* (each being 1ms long). Subframes are the fundamental time unit for most LTE processing, like scheduling.
- Each subframe consists of two *slots*, which are each 0.5 ms long.
- Each slot consists of 7 (or 6) symbols.

Duration of the different units is often given in terms of the sampling time $T_s = 1/30,720,000$ s. Note that this "sampling time" is a bookkeeping unit; RXs are not obligated to actually sample at the corresponding rate. In particular, for bandwidths <15 MHz, a larger sampling time (lower sampling frequency) is feasible.

Figure 27.5 Structure of one slot in LTE.

Let us now turn to the details of a symbol. Since the modulation format is OFDM (regular OFDM for the downlink, and DFT-precoded OFDM in the uplink), multiple subcarriers are present. The regular spacing between the subcarriers is $\Delta f = 15$ kHz. The 15-kHz spacing of the subcarriers leads to an OFDM symbol duration (without cyclic prefix) of $67\,\mu s = 2048\,T_s$. One subcarrier, for the duration of 1 OFDM symbol, is called a *resource element*. We can fit 6 or 7 OFDM symbols into one slot, depending on the duration of the cyclic prefix. In the "regular" case, the duration of the cyclic prefix is 160 T_s in the first OFDM symbol and 144 T_s for the subsequent symbols. A long cyclic prefix is 512 T_s, so that only a total of 6 OFDM symbols fit into one slot. Such a long cyclic prefix is used in environments with large delay spread and/or for MB-SFN. To simplify the notation, the description in the remainder of this chapter assumes "normal-length" cyclic prefix unless otherwise stated.

Time/frequency resources are assigned to different users as integer multiples of a *Resource Block* (RB) (Figure 27.6). More precisely, an RB is 12 subcarriers (180 kHz) over the duration of one slot.[3] For the uplink, only contiguous RBs can be assigned to one MS. Furthermore, the number of RBs has to be decomposable into factors of 2, 3, and 5; this is done to ensure an efficient implementation: with that prescription, any of the necessary DFTs can be composed of radix-2, radix-3, and radix-5 butterfly structures.

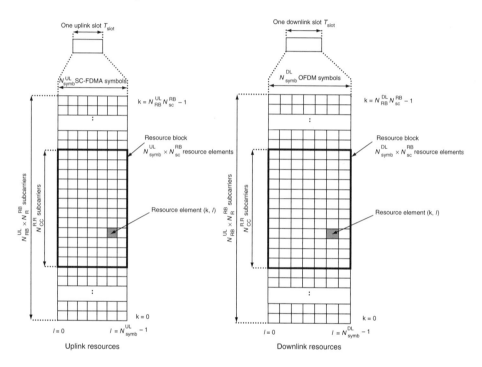

Figure 27.6 Resource blocks for uplink and downlink. *In this figure*: SC-FDMA, Single-Carrier Frequency Division Multiple Access.

In the TDD case, subframes can be assigned flexibly to uplink and downlink, with the exception of subframes 0 and 5, which are always used for the downlink, and subframe 2, which is always used for the uplink. For every transition from downlink to uplink, there is a guard interval, to avoid collisions between the packets "on the air" (see Section 17.5). Consequently, there are subframes that contain three distinct parts: a Downlink Pilot Time Slot (DwPTS), Uplink Pilot Time Slot (UpPTS), and a guard interval between them. Note that a guard interval is not necessary for a transition from uplink to downlink.

---

[3] A reduced subcarrier spacing of 7.5 kHz is also foreseen, together with an extended cyclic prefix. In that case, only 3 OFDM symbols fit into one slot; but 24 subcarriers are contained in one RB.

27.3.2 Modulation

Uplink and downlink use different modulation formats: while the downlink employs "classical" OFDM, the uplink uses a format that can be interpreted as single-carrier transmission (compare to Chapter 19), but in LTE can be better characterized as DFT-precoded OFDM (Figure 27.7).

The implementation of the downlink transmission is straightforward: the modulation is OFDM with 15-kHz subcarrier spacing. Data intended for different MSs are multiplexed onto the different RBs; each RB can employ a different modulation format. For each OFDM symbol, the overall signal is then subjected to an Inverse Fast Fourier Transform (IFFT) transformation; the cyclic prefix is prepended, and the signal is upconverted to passband for transmission (the more complicated case of multiantenna transmission is described in Section 27.3.7).

For the uplink, an MS has a number of contiguous subcarriers available for signaling. The MS maps the symbols onto the input of a DFT whose size equals the number of subcarriers. The output of this DFT is then mapped onto the subcarriers, which are processed like in the downlink (IFFT, cyclic prefix, upconversion to passband). The combination of the DFT with the IFFT inherent in the OFDM implementation results in a single-carrier signal with cyclic prefix, which can be effectively equalized by frequency domain equalization (see Section 19.11). The bandwidth of the signal corresponds to the number of subcarriers used in the transmission, multiplied with the subcarrier spacing of 15 kHz (i.e., the same as for the uplink).

Figure 27.7 Block diagram of DFT-precoded OFDM. *In this figure*: DAC, Digital to Analog Converter; P/S, Parallel to Serial; RF, Radio Frequency; S/P, Serial to Parallel.

27.3.3 Mapping to Physical Resources – Downlink

The complex symbols from the modulator (and possibly assigned to particular layers and antenna ports) have to be mapped onto the Physical Resource Blocks (PRBs), which are the basic units for the generation of the OFDM signal. In order to facilitate the implementation, the mapping proceeds in two steps: (i) map the symbols (in the sequence they occur) onto *Virtual Resource Blocks* (VRBs), (ii) map the VRBs onto the *PRBs*.

For the mapping of the symbols onto VRBs, the BS has to assign which VRBs are to be used for the signal intended for a particular MS. This assignment can either be contiguous, or noncontiguous. The following three types of allocation exist: types 0 and 1, which support noncontiguous allocation, and type 2, which only supports contiguous allocation:

- *Type 0*: it groups RBs (the size of the groups depends on the bandwidth), and then provides a bitmap to indicate which elements of that group are assigned to a particular MS. If there are 8 groups, and the bitmap reads 10011000, then groups 1, 4, 5 are used for the MS in question. The motivation for assigning RB groups (instead of RBs) is to reduce the size of the bitmap codeword.
- *Type 1*: here the RBs are grouped into interlaced subsets. A particular MS is then allocated to a particular subset. Within that subset, again a bitmap indicates the RBs that are actually used.
- *Type 2*: indicates simply the starting point, and the length of the block allocation. It is thus much shorter than types 0 and 1 that need bitmaps.

In a second step, the VRBs are mapped to PRBs. For this mapping, two different approaches are possible, corresponding to two choices of VRBs: localized, or distributed. A *localized* VRB with location i is directly mapped onto a PRB with location i; in other words, the mapping of symbols to PRBs is purely determined by the mapping of symbols to VRBs as described above. For *distributed* VRBs, we map RBs that are contiguous in the virtual domain onto noncontiguous PRBs. More precisely, the mapping proceeds in two steps (see Figure 27.8):

1. *RB-pair interleaving*: since each subframe (= scheduling interval) consists of two slots, scheduling involves the assignment of RB pairs. In RB-pair interleaving, RB pairs that are adjacent in the virtual domain are separated in the PRB domain.
2. *RB distribution*: the two elements of a VRB pair are mapped onto two different PRBs that are separated in the frequency domain by approximately half the system bandwidth.

The different types of resource allocation provide flexibility to adjust the allocation method to the amount of channel state information and the admissible overhead. If the BS has full channel state information, then a fully flexible approach employing type-0 allocation encoding and localized RBs allow full exploitation of the available multiuser diversity. Using localized RBs with contiguous assignment (type 2) has a smaller overhead, since the assigned RBs can be described in a very

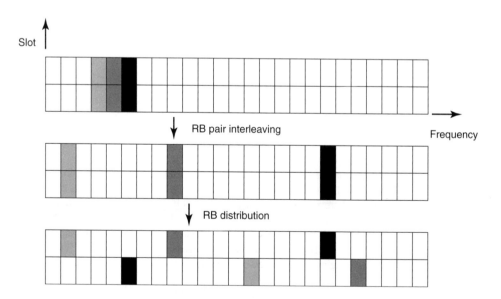

Figure 27.8 Mapping from VRBs to PRBs.
Reproduced from Dahlman et al. [2008] © Academic Press.

simple manner. However, it might not be optimum, particularly when many RBs are assigned to a particular MS, because some of the assigned resources might be on subcarriers that have bad channel quality for this particular MS.

The main reason for using distributed VRBs is to achieve frequency diversity, which is useful when scheduling according to the channel state is not feasible (e.g., in fast-changing environments). Of course, a distributed assignment can also be achieved by using type-0 or type-1 allocation with a suitable bitmap allocation. However, the use of the VRBs allows to provide frequency diversity with type-2 resource allocation notification (small overhead), i.e., assigning a *contiguous* block of *VRBs*.

27.3.4 Mapping to Physical Resources – Uplink

The mapping of symbols to the specific frequency resources in the uplink is somewhat simpler, since there are no distributed RBs. For one slot, a particular user can only occupy contiguous subcarriers. However, the mapping from VRB to PRB can change from slot to slot, thus realizing frequency hopping. The assignment can be done according to one of two methods: (i) use of a cell-specific hopping pattern, or (ii) explicit prescription of the hopping pattern.

For cell-specific hopping patterns, a part of the available cell bandwidth is subdivided into a number of subbands. During each new hop, the position of the RBs is (cyclically) shifted by the width of the subband multiplied with an integer number provided by the cell-specific random sequence. In addition, the RB might be "mirrored" (which effectively shifts its position within a subband), depending on whether a "mirror bit" is set.

It is also possible to explicitly prescribe by how many RB widths the RB in the second slot should hop compared to the first one. This information is transmitted in the scheduling grants.

27.3.5 Pilots or Reference Signals

Downlink

Reference Signals (RSs), or pilots in our notation, are used to estimate the channel. As outlined in Section 19.5.2, scattered pilots are used to provide sufficient sampling in the time and frequency domains, while limiting the amount of overhead and thus detrimental effects on the spectral efficiency.

For downlink channel estimation, it is usually sufficient to have a cell-wide pilot, i.e., a single pilot that is broadcast by the BS. From this broadcast signal, any MS can estimate the channel from the BS to that particular MS. Since the pilot can be heard by all MSs, it has to cover the whole cell bandwidth. The MS can use the channel estimates for (i) coherent demodulation, or (ii) channel quality estimation to be fed back to the BS (note that an MS can obtain channel estimates outside the RBs assigned to it).

The pilots are transmitted on the following locations within each RB: first OFDM symbol, $1 + i$ subcarrier; first OFDM symbol, $7 + i$ subcarrier; fifth OFDM symbol, $4 + i$ subcarrier, fifth OFDM symbol, $10 + i$ subcarrier, where $i = 0, \ldots 5$ is a frequency shift parameter specific to a cell, and all additions are to be taken modulo 12. Although it is not prescribed in the standard, it is beneficial for the system performance to orthogonalize the pilots of adjacent cells by using different frequency shifts. Of course there is still interference from data subcarriers in an adjacent cell, but that can be suppressed either by averaging (in channels with large coherence time/bandwidth) or power boosting of the pilot tones.

The complex symbols that are to be transmitted on the pilot locations are also cell specific. However, the sequence of pilot symbols is independent of the assigned bandwidth: it is computed assuming a 20-MHz bandwidth; if some subcarriers do not exist because the available bandwidth is smaller, those symbols are simply not transmitted. This allows an easier cell identification (ID) during the cell search procedure.

If multiple antenna elements are used by the BS, then the number of channels that need to be estimated is larger. The pilots for the different antennas are orthogonal, i.e., an RE carrying a pilot for one antenna element is completely empty (no pilot or data assigned on the other antenna elements). In the case of 2 antenna elements, the BS uses 2 different frequency shifts (offset by 3 subcarriers). In the case of 4 antenna elements, antenna elements 3 and 4 transmit on the same subcarriers as antenna elements 1 and 2 respectively, but on the second OFDM symbol (instead of the first and fifth), and thus have a lower density of pilots. Details are shown in Figure 27.9.

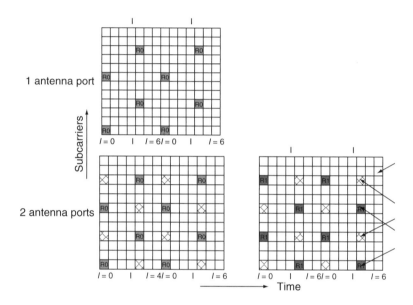

Figure 27.9 Pilots for the downlink.
From [3GPP LTE]. © 2009. 3GPP™ TSs and TRs are the property of ARIS, ATIS, CCSA, ETSI, TTA and TTC who jointly own the copyright in them. They are subject to further modifications and are therefore provided to you "as is" for information purposes only. Further use is strictly prohibited.

The cell-wide pilots cannot be used when the BS applies non-codebook-based beamforming, i.e., beamforming in which the MS does not know the beamforming coefficients. In that case, the pilots have to undergo the same beamforming as the user data, so that all the MS sees is an "effective" channel (the concatenation of beamformer and physical channel) that it can estimate from the pilot and use for coherent demodulation. Clearly, this effective channel is specific for one MS. The location (in the time/frequency plane) of those user-specific pilots is orthogonal to the locations of the cell-wide pilots, namely, on the fourth OFDM symbol (subcarriers 1,5,9) and 7th OFDM symbol (subcarriers 3,7,11).

In the case of MB-SFN, the cell-wide RSs cannot be applied either. Rather, a special MBSFN reference signal is defined. The mapping is shown in Figure 27.10; we see that the spacing in the frequency domain is closer than for the pilots we discussed above. The reason for this is that MB-SFN channels usually exhibit stronger frequency selectivity than unicast channels: in addition to multipath propagation, the MB-SFN channels also exhibit the effects of the different runtimes of the signals from the different BSs.

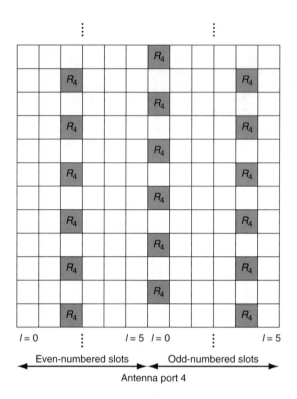

Uplink

The requirements for pilot signals in the uplink are significantly different from the pilot signals in the downlink. Firstly, *multiple* pilots need to be transmitted, since the BS has to learn the channel from multiple MSs to the BS. Secondly, the data that need to be demodulated are always well localized in the frequency domain. Thirdly, the existence of the pilot should not upset the single-carrier properties of the transmit signal. For these reasons, the pilots in the uplink show a very different structure from those of the downlink (Figure 27.11).

The main pilot is the *demodulation pilot* (demodulation reference signal), which is used to determine – with high accuracy – the channel transfer function in the specific RB(s) used by an MS for data transmission. In order to retain the single-carrier properties of the transmit signal, the pilot is not frequency multiplexed with the data signal. Rather, a complete OFDM symbol (over the bandwidth used by this particular MS) is assigned to the pilot. In particular, of the 7 OFDM symbols in one slot, the fourth symbol is reserved for demodulation pilot.

The exact form of the pilot is defined in the frequency domain (i.e., after having passed through the DFT precoder). It is based on Zadoff–Chu sequences, which are defined as

$$X^{(u)}(k) = e^{j\pi uk(k+1)/M_{ZC}} \qquad 0 \leq k \leq M_{ZC} \qquad (27.1)$$

where M_{ZC} is the length of the sequence, and u is the index of the sequence. Zadoff–Chu sequences have the remarkable property that they have constant amplitude in the time domain (important for the

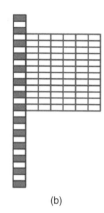

Figure 27.11 Structure of uplink demodulation pilot (a) and sounding pilot (b).

power amplifiers) and the frequency domain (so that the transmit SNR at all measured frequencies is the same); the autocorrelation function of such a sequence is a delta function. For this reason, it belongs to the class of Constant Amplitude Zero AutoCorrelation (CAZAC) sequences. The number of Chadoff–Zhu sequences is limited to the number of integers that are relative-prime to the sequence length – if M_{ZC} is a prime number, this number equals $M_{ZC} - 1$.

In order to circumvent the restrictions of having a small number of sequences, LTE uses the following two modifications of Zadoff–Chu sequences:

1. *Periodically extended sequences*: choosing M_{ZC} equal to the number of used subcarriers would result in a low number of available sequences, because the number of subcarriers is always an integer multiple of 12 (number of subcarriers in an RB). LTE thus uses for M_{ZC} the largest prime number that is smaller than the desired sequence length. For example, for a signal occupying 4 RBs (48 subcarriers), M_{ZC} is chosen as 47, so that there are 46 available sequences. Since the length of the sequences is not long enough to cover all available subcarriers, the sequence is simply cyclically extended (i.e., after the last element, we start again with the first, then the second, ...). Note that the minimum number of available sequences is 30: if the number of RBs is ≥ 3, then the above rule provides at least 30 sequences; for the case of 1 and 2 RBs, the standard tabulates a set of 30 different sequences that are to be used.

2. *Phase rotations*: in order to generate additional sequences, each (cyclically extended) Zhadoff–Chu sequence can be multiplied with a linear phase shift in the frequency domain, i.e.,

$$X'(k) = X(k)e^{j\alpha k} \tag{27.2}$$

where different shifts α lead to different sequences (this corresponds to a cyclic shift in the time domain). With a suitable choice of phase shifts, the resulting sequences can be made completely orthogonal to each other: when the phase shift is an integer multiple of $2\pi/12$, then orthogonality is obtained over 1 RB. The orthogonality can be destroyed by two effects: (i) frequency selectivity of the channel that is so strong that the channel varies significantly over the RB, and (ii) timing misalignment of the two considered sequences. Since signals within a cell are typically well time aligned, pilots based on the same Zadoff–Chu sequence, but with different phase shifts, can be used within one cell.

The different sequences are then assigned to cells, such that each cell has 1 periodically extended sequence[4] and can derive orthogonal sequences based on phase shifts. The sequences are assigned to cells based on cell ID (for the *Physical Uplink Control CHannel* PUCCH, see Section 27.4.10) or explicitly signaled (for *Physical Uplink Shared CHannel* (PUSCH), see Section 27.4.11). The standard also foresees the possibility of group hopping (i.e., the "fundamental" sequence for each cell is changed on a regular basis).

The demodulation pilot for the PUSCH is finally designed in the following way: in the first slot of a subframe, the pilot is the cyclically extended Zadoff–Chu sequence with a shift consisting of three terms: (i) a pseudorandom cell-specific shift, (ii) a deterministic, user-specific shift communicated in the Downlink Control Information (DCI), and (iii) an additional shift term communicated by the higher layers. The pilot in the second slot in a subframe is a shifted version of the first. For the PUCCH, somewhat different rules apply; for details see the standard.

A second type of pilot, the *sounding pilot* (sounding reference signal), is intended to enable efficient scheduling. The BS uses it to estimate the frequency domain channel response over the entire system bandwidth; based on this knowledge, it can then assign the users to the most suitable subcarriers. Therefore, the sounding pilot transmitted by each user typically occupies the entire system bandwidth. This large bandwidth can be achieved by (i) sending a DFT-OFDM symbol that spans the whole system bandwidth, or (ii) sending a sequence of frequency-hopped signals that each occupy a smaller bandwidth, but, taken together, occupy the whole bandwidth. The latter approach is more complicated, but has the advantage that the power-spectral density of the sounding pilot is higher. Various bandwidths (have to be multiples of 4 RBs) are admissible.

The sounding pilot is based on Zadoff–Chu sequences, having a shift $\alpha = 2\pi n_{SRS}/8$, where n_{SRS} is a parameter that can take on values between 0 and 7. That sequence is then mapped onto a frequency comb, i.e., every second subcarrier. The transmission always occurs in the last symbol of a subframe, but not in every subframe: the transmission interval can be chosen in a range from 2–160 ms. This accommodates a wide variety of coherence times of the channel. In order to avoid collisions between sounding pilot and payload (PUSCH) transmissions, the BS ensures that there are no PUSCH transmissions in the cell whenever *any* MS transmits a sounding pilot. Orthogonality between sounding pilots from different MSs is handled by a combination of two methods: (i) transmission on different frequency combs, and (ii) transmission of the sequences with different linear phase shifts (similar to the demodulation pilot). Note that the channel estimates that form the basis of scheduling need not be as accurate as those used for demodulation.

27.3.6 Coding

The encoding of the information is performed in several steps (see Figure 27.12).

Cyclic Redundancy Check

For each transport block, we compute a Cyclic Redundancy Check (CRC) of length 24 bits (or, 16 or 8 bits under some circumstances). It is calculated for each block of data for one transmission time interval from the code polynomials

$$G(D) = D^8 + D^7 + D^4 + D^3 + D + 1 \qquad \text{for 8-bit CRC} \qquad (27.3)$$

$$G(D) = D^{16} + D^{12} + D^5 + 1 \qquad \text{for 16-bit CRC} \qquad (27.4)$$

$$G(D) = D^{24} + D^{23} + D^6 + D^5 + D + 1 \qquad \text{for 24-bit CRC} \qquad (27.5)$$

---

[4] and 2 sequences of length 72 or longer.

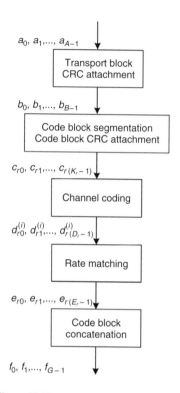

Figure 27.12   Encoding procedure in LTE.
From [3GPP LTE]. © 2009. 3GPP™ TSs and TRs are the property of ARIS, ATIS, CCSA, ETSI, TTA and TTC
who jointly own the copyright in them. They are subject to further modifications and are therefore provided
to you "as is" for information purposes only. Further use is strictly prohibited.

$$G(D) = D^{24} + D^{23} + D^{18} + D^{17} + D^{14} + D^{11} + D^{10} + D^7 + \ldots$$

$$+ D^6 + D^5 + D^4 + D^3 + D + 1 \quad \text{for 24-bit CRC (alternative)} \quad (27.6)$$

and attached at the end of the block.

If the transport block is too large, data are parsed into code blocks, where each block has a maximum length of 6144 bits, and – if necessary – filler bits are inserted such that each code block has a permissible length (only certain discrete values of code block length are allowed). Then for each code block, another CRC is computed.

Convolutional Codes

Convolutional codes are used in LTE only for the encoding of control information, not for the actual payload data. In particular, the standard defines a length-7 tail-biting convolutional code with the following code polynomials.

$$G1(D) = 1 + D^2 + D^3 + D^5 + D^6 \quad (27.7)$$

$$G2(D) = 1 + D + D^2 + D^3 + D^6 \quad (27.8)$$

$$G3(D) = 1 + D + D^2 + D^4 + D^6 \quad (27.9)$$

This code is used for the following information:

- Broadcast CHannel (BCH);
- DCI;
- UL control information.

Turbo Codes

Turbo codes are applied to the Uplink Shared CHannel (UL-SCH) and Downlink Shared CHannel (DL-SCH), Paging CHannel (PCH), and Multicast CHannel (MCH) (see Section 27.4). In contrast to WCDMA, there is no option to encode payload data with convolutional codes; only turbo codes are allowed. This is mainly due to the fact that turbo codes are now so well established, and RXs are sufficiently optimized, that there is no need for the slightly simpler (but worse-performing) convolutional codes anymore. The turbo encoder is the same as in WCDMA, with the exception of the interleaver.

Two recursive systematic convolutional encoders are employed (see Figure 27.13). The data stream is fed into the first one directly, and into the second one after passing an interleaver. Both encoders have a coding rate of 1/2. Thus, the output is the original bit X or X' and the redundancy bits Y or Y', which are the output of the recursive shift registers. However, as X equals X' only X, Y, and Y' are transmitted. Thus, the code rate of the turbo encoder is 1/3.

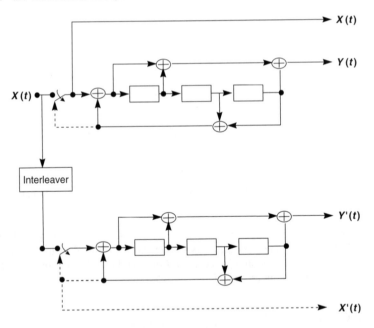

Figure 27.13   Structure of the turbo encoder.

In LTE, the interleaver is a Quadrature Permutation Polynomial (QPP) interleaver that moves bits from location

$$\left(f_1 i + f_2 i^2\right) \bmod K \tag{27.10}$$

to location i. Here, K is the blocksize, and f_1 and f_2 are tabulated constants that depend on K. After the encoding, a rate matching and code block concatenation is performed.

HARQ

LTE employs various forms of HARQ for the DL-SCH and UL-SCH (it is not useful for other channel types). When an RX receives a transport block, it attempts to decode and sends a single feedback bit (whose timing indicates which transport block it is associated with) to indicate whether a retransmission is necessary. More details are given in Section 27.5.2.

27.3.7 Multiple-Antenna Techniques

LTE makes extensive use of multiple-antenna techniques. Let us first establish some notations:

- *Codewords* are the parsed output of the decoders. In LTE, either one or two codewords are transmitted simultaneously. In the case of transmit diversity, the number of codewords is always 1.
- The codewords are mapped onto *layers*, which contain different symbols. Note that a codeword can be mapped onto several layers: the maximum number of layers is 4, while the maximum number of codewords is 2. If there were no ST coding, then the "layers" would correspond to the "spatial streams" discussed in Section 20.2. Each layer contains the same *number* of symbols.[5]
- The layers are finally mapped onto *antenna ports* by means of a precoding.

Transmit Diversity

For the case of 2 transmit antennas, LTE foresees Space Frequency Block Coding (SFBC), i.e., Alamouti encoding on adjacent subcarriers. This can be interpreted as mapping the symbols of a codeword onto the layers according to

$$A = \begin{bmatrix} s_1 & -s_2^* \\ s_2 & s_1^* \end{bmatrix} \tag{27.11}$$

For four antennas, the mapping of the codewords onto the four layers is done according to the following mapping:

$$A = \begin{bmatrix} s_1 & -s_2^* & & \\ s_2 & s_1^* & & \\ & & s_3 & -s_4^* \\ & & s_4 & s_3^* \end{bmatrix} \tag{27.12}$$

so that, e.g., the transmission vector on frequency 1 is $[s_1, 0, -s_2^*, 0]$. This scheme gives a rate of 1. The layers are then directly mapped onto the antenna ports.

Spatial Multiplexing

Closed-Loop Spatial Multiplexing   In closed-loop spatial multiplexing, the codewords are mapped onto the layers by a very simple procedure: if the number of codewords equals the number of layers, then each layer simply contains the symbols from one codeword. If there is one codeword and two layers, then the symbols of the codeword are alternatingly assigned to layer 1 and layer 2: if there are 2 codewords and 4 layers, then symbols from codeword 1 are alternatingly assigned to layers 1 and 2, while symbols from codeword 2 are alternatingly assigned to layers 3 and 4.

---

[5] If the number of layers is 3 and the number of codewords is 2, then 1 codeword has to have twice as many symbols as the other one.

Layers are then multiplied with a matrix W to provide the signals at the antenna ports. The precoding matrix is constituted from the elements of a codebook, and the MS feeds back to the BS the index of the codebook entry that it wants the BS to use. The codebooks are Fast Fourier Transform (FFT)-based for the case of 2 antenna ports, the entries are as follows:

Codebook index	Number of layers	
	1	2
0	$\frac{1}{\sqrt{2}}\begin{bmatrix} 1 \\ 1 \end{bmatrix}$	$\frac{1}{\sqrt{2}}\begin{bmatrix} 1 & 0 \\ 0 & 1 \end{bmatrix}$
1	$\frac{1}{\sqrt{2}}\begin{bmatrix} 1 \\ -1 \end{bmatrix}$	$\frac{1}{2}\begin{bmatrix} 1 & 1 \\ 1 & -1 \end{bmatrix}$
2	$\frac{1}{\sqrt{2}}\begin{bmatrix} 1 \\ j \end{bmatrix}$	$\frac{1}{2}\begin{bmatrix} 1 & 1 \\ j & -j \end{bmatrix}$
3	$\frac{1}{\sqrt{2}}\begin{bmatrix} 1 \\ -j \end{bmatrix}$	

In addition to the codebook-based beamforming, the BS also can do beamforming based on other criteria. In that case, the MS does not know the precoding vector. Instead, it has to estimate the "effective channel," i.e., the concatenation of precoding and propagation channel, from a user-specific pilot that is precoded in exactly the same way as the data; see also Section 27.3.5.

Open-Loop Spatial Multiplexing   In the case that a feedback of the desired beamforming is not available (e.g., because the channel changes too rapidly), an open-loop spatial multiplexing can be used. In that case, the layers are first multiplied by a (subcarrier-independent) matrix U, then a (subcarrier-dependent) diagonal matrix $D(i)$, and finally by a matrix W. For the case of 2 antenna ports and 2 layers,

$$W = \frac{1}{2}\begin{bmatrix} 1 & 1 \\ 1 & -1 \end{bmatrix} \quad U = \frac{1}{\sqrt{2}}\begin{bmatrix} 1 & 1 \\ 1 & \exp(-j2\pi/2) \end{bmatrix} \quad D(i) = \begin{bmatrix} 1 & 0 \\ 0 & \exp(-j2\pi/2) \end{bmatrix} \quad (27.13)$$

where the $D(i)$ effectively leads to a large cyclic shift of the considered block.

27.4   Logical and Physical Channels

27.4.1   Mapping of Data onto (Logical) Subchannels

LTE provides *logical channels* (which are defined by the type of information that they carry), that are mapped to *transport channels* and from there to *physical channels* (which are defined by their physical properties, i.e., time, subcarrier, etc.). The logical channels are similar to those in WCDMA, but repeated here in order to enable independent reading of the chapters:

- Traffic channels
  - *Dedicated Traffic CHannel (DTCH)*: it carries the user data for all ULs, as well as for those downlink data that are not multicast/broadcast.
  - *Multicast Traffic CHannel (MTCH)*: it carries the user data for multicast/broadcast downlink transmission.
- Control channels

○ *Broadcast Control CHannel (BCCH)*: it carries system information data that are broadcast to the MSs in a cell. Note the difference from the MTCH, which also broadcasts to MSs, but carries user data.
○ *Paging Control CHannel (PCCH)*: it pages MSs in multiple cells (i.e., when it is not known exactly in which cell the MS currently is located).
○ *Common Control CHannel (CCCH)*: it transmits control data for the Random Access (RA), i.e., when a connection is started.
○ *Dedicated Control CHannel (DCCH)*: it is used for the transmission of control information that relates to a specific MS (as opposed to the system information relevant for all MSs, which is broadcast in the BCCH).
○ *Multicast Control CHannel (MCCH)*: it carries the control information related to multicast/broadcast services.

These channels are mapped onto the following transport channels:

• *Broadcast Channel (BCH)*: it carries part of the BCCH (the remainder is on the DL-SCH described below). It has a fixed format, so that any MS can listen to it easily.
• *Paging Channel (PCH)*: it carries the PCCH.
• *Multicast Channel (MCH)*: it is used to support broadcast/multicast transmission. It has a semistatic scheduling and transport format.
• *DownLink Shared Channel (DL-SCH) and UpLink Shared Channel (UL-SCH)*: they carry the user data, as well as most of the control information (except the one already mentioned above).

The data on transport channels are organized into transport blocks; in each transmission time interval (usually a subframe), one transport block is transmitted. A transport format is associated with each transport block.

Finally, these transport channels are mapped onto physical channels; there are also physical channels that do not carry any transport channel, but are purely used for PHY functionality.

• Downlink
   ○ *Physical Broadcast CHannel (PBCH)*: it carries the BCH.
   ○ *Physical Downlink Shared CHannel (PDSCH)*: it carries the DL-SCH, i.e., user data, some control data for the downlink, as well as the PCH.
   ○ *Physical Multicast CHannel (PMCH)*: it carries the MCH, which contains the multicast payload, as well as some of the control information for multicast.
   ○ *Physical Downlink Control CHannel (PDCCH)*: it carries control information, such as scheduling that is required for reception of the PDSCH. This channel does not carry any transport channel.
   ○ *Physical Control Format Indicator CHannel (PCFICH)*: it carries control information about the PDCCH. This channel does not carry any transport channel.
   ○ *Physical HARQ Indicator CHannel (PHICH)*: it carries the feedback bits indicating whether a retransmission of transport blocks is necessary. This channel does not carry any transport channel.
   ○ *Synchronization Signal (SS)*.
• Uplink
   ○ *Physical Uplink Shared CHannel (PUSCH)*: it is the uplink counterpart to the PDSCH.
   ○ *Physical Uplink Control Channel (PUCCH)*: it carries mainly three types of information: (i) channel state feedback, (ii) resource requests (remember that the BS performs the scheduling, i.e., assigns all the resources also for the uplink; thus the MS must request resources when it has data to transmit), and (iii) HARQ feedback bits.

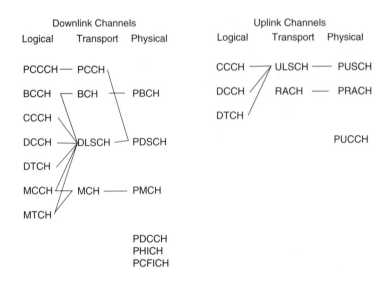

Figure 27.14 Mapping between logical, transport, and physical channels.

○ *Physical Random Access CHannel (PRACH)*: it is used for the random access, i.e., MS communicating to the BS before a connection with scheduling has been established.

Figure 27.14 summarizes the mapping between the channels.

27.4.2 Synchronization Signals

The *SS* carries information about the timing of the cell, as well as the *cell ID*. LTE actually provides two SSs, the Primary Synchronization Signal (PSS) and the Secondary Synchronization Signal (SSS). In contrast to other systems, these signals are not called "channels," but perform similar functions as, e.g., the synchronization channels in WCDMA. To understand the functionality of the SS, keep in mind that there are 504 cell IDs defined for LTE, which are divided into 168 ID groups.

The PSS is transmitted in the last symbol of the first slot of subframes 0 and 5 of every frame, extending over 72 subcarriers. The waveform transmitted during that slot is one of three allowed Zadoff–Chu sequences of length 63, extended with 5 zeroes at the lower and 5 zeroes at the upper band edge. Which of the three sequences is transmitted depends on the cell ID within a group (but note that the PSS does not provide the group ID; that is transmitted in the SSS). The MS can obtain the fine-resolution timing from the PSS (within one symbol, because the signal lasts only one symbol, and even within one OFDM sample, by correlation with the Zadoff–Chu sequence). However, there is still a timing ambiguity by multiples of 5 ms, i.e., the periodicity of the PSS signal; therefore, frame timing is not available.

The SSS signal is transmitted in the symbol directly before every PSS signal. The SSS carries information about the cell ID group: the signal also extends over 72 subcarriers. It is an interleaved combination of two *m*-sequences (see Section 18.2.6) of length 31. The resulting length-62 sequence is extended with zeroes at the band edges, just like for the PSS. Only 168 sequences are valid, and signify the cell ID group. In contrast to PSS, the signal transmitted in the first slot of a frame is different from the signal in the second slot: while it contains the same *m*-sequences, the interleaving in the second slot is different from that in the first slot. This allows the RX to acquire frame timing (i.e., ambiguity of timing is multiples of 10 ms), as well as the cell ID, from a single observation of an SSS.

The reason that both PSS and SSS use 72 subcarriers is that this is the smallest bandwidth permissible for an LTE system. At the time of the reception of those signals the MS does not know yet the actual system bandwidth in the considered cell; the SSs thus use the same bandwidth for all cells, namely the smallest possible.

Once a signal has acquired cell ID and frame timing, it can receive the pilot (remember downlink pilots depend on the cell ID), which are needed for the reception and decoding of the BCH.

27.4.3   Broadcast Channel

The BCH contains the *Master Information Block* (MIB), which contains critical information about the cell system information:

- System bandwidth in the cell.
- *PHICH configuration*: some bits describing the specific configuration of the PHICH channel (for details see the standard). Knowledge of this configuration is necessary to receive the control information.
- System frame number.

The baseband processing and transmission of the MIB bits on the BCH proceed in the following steps:

- Addition of a 16-bit CRC.
- Encoding with tail-biting rate 1/3 convolutional code.
- Scrambling.
- QPSK modulation.
- *Antenna mapping*: this is trivial if only a single antenna is used. If the BS has two antennas, then space-frequency block codes *must* be used. If the BS has four antennas, then the combination of antenna hopping and space-frequency block codes (see Section 27.3.7) *must* be used. This is done so that the MS can learn from the BCH how many antennas the BS has.
- *Demultiplexing*: the signal is mapped onto four consecutive frames; particularly onto the first subframe of each of those four frames. Within each such subframe, the signal is transmitted on the first 4 symbols of the second slot, over 72 subcarriers (the choice of 72 subcarriers is motivated the same way as for the PSS and SSS). Repetition coding is used to "fill up" this available resource, which is much larger than the number of bits that need to be transmitted.

Note that the BCH extends over 40 ms, while the timing acquired from the PSS and SSS has an ambiguity of multiples of 10 ms. The MS must therefore try to decode the BCH with four different timing shifts, and determine from the CRC which is the correct one. This decoding also provides the two least significant bits of the system frame number; for this reason, those two bits are not included in the MIB.

27.4.4   General Aspects of Control Channels Associated with a DL-SCH

There are three physical channels, namely PCFICH, PDCCH, and PHICH, associated with the DL-SCH. The control information contained by these channels is also called L1/L2 signaling, because it is relevant for both the PHY and the MAC layer. It is transmitted in the *control region* located at the beginning of each subframe; this control region occupies all the subcarriers (with the exception of the pilots) of the signal, and either (i) the first, (ii) the first two, or (iii) the first three OFDM symbols (or four in the case of very narrow bandwidth); the amount of occupied OFDM symbols can be changed from one subframe to the next. The reason for transmitting control information at the beginning of a subframe is that information contained in it is required for the decoding of the user data; in particular, if an MS sees from the scheduling information that none of the subsequent user data are intended for it, the MS can power down part of the receive circuitry.

The mapping of the control channels to particular REs is done in units of RE groups which consist of four REs each. This is done because the BS has up to 4 antenna elements that can be used for transmit diversity (note that spatial multiplexing is not used for control information).

27.4.5 Physical Control Format Indicator CHannel

The PCFICH carries the information of how many OFDM symbols are dedicated to the control region. Since the control regions can comprise up to three symbols, this information requires 2 bits. The reason for having a dynamical choice of the control region size is that the usage in a cell can vary dramatically. Sometimes, resources are taken up by a few high-data-rate users (in which case the control region, whose size is mainly determined by the resource allocation information, can be shorter). At other times, there are a lot of users (e.g., when most users voice users), which requires a longer control region.

If the PCFICH information is decoded incorrectly, all the remainder of a subframe is interpreted incorrectly. Therefore, the information uses a strong error correction, namely a rate 1/16 block code. The encoded bits are then scrambled with a scrambling sequence that depends on the subframe number and the cell ID, and thus reduces the effect of interference. Since the information is needed for the interpretation of the remainder of the subframe, it is always at the same location and modulation format (QPSK). In order to enhance frequency diversity, the 16 QPSK symbols are mapped onto 4 RE groups that are each spaced 1/4 of the available bandwidth apart; the starting subcarrier depends on the cell ID.

27.4.6 Physical HARQ Indicator CHannel

The PHICH carries the feedback information for the HARQ. Groups of eight feedback bits are formed in the following way: first, each feedback bit is repeated three times (repetition coding), and Binary Phase Shift Keying (BPSK) modulated. The resulting waveforms are multiplexed through a combination of In-phase – Quadrature phase (IQ) multiplexing and code multiplexing with length-4 orthogonal spreading sequences. The scrambled output of the multiplexer is mapped to three RE groups. The PHICH group number, IQ branch, and spreading sequence are related to the index of the first RB on which the data block that is to be acknowledged was originally transmitted.

27.4.7 Physical Downlink Control CHannel

The PDCCH occupies the largest part of the control zone. For each of the users, it carries the DCI, i.e., information that is required for decoding the payload information. In particular, it carries information about the resource allocation, i.e., which RBs in the data region are assigned to which MS (see also Section 27.3.3). The PDCCH also carries, for each user, the transport format, i.e., the modulation and coding scheme, power control information, and control information for spatial multiplexing.

There are seven different formats for encoding the DCI, reflecting the tradeoff between size of the control information and performance. They are as follows:

- *Format 1*: the fundamental DCI format, using Type 0 or 1 resource allocation notification. It does not assume multiple antennas at the BS.
- *Format 1A*: similar to Format 1, but more compact, since it employs Type 2 resource allocation notification.
- *Format 1B*: like Format 1A, but intended when the BS uses precoding for multiple antenna elements.
- *Format 1C*: extremely compact format with fixed modulation (QPSK); to be used only for special system messages.
- *Format 1D*: like format 1B, but with an additional power offset message.
- *Format 2*: using Type 0 or 1 resource allocation notification, this format is intended for systems with closed-loop spatial multiplexing.
- *Format 2A*: like Format 2, but for systems with open-loop spatial multiplexing.

In addition to the resource allocation information, the formats also contain some additional information (see also Table 27.4):

- *Modulation/coding*: those 5 bits indicate the modulation scheme and code rate used in the block. Of the possible 32 combinations, 29 are used to actually indicate combinations of modulation format and code rate; the remaining 3 can be used to indicate modulation format during retransmission.
- *HARQ process number*: this 3-bit message is the index of the HARQ process associated with the DCI. As explained in Section 27.5.2, several HARQ processes are active simultaneously.
- *New data indicator*: it indicates whether the data are a retransmission and thus need to be combined with previously received soft information for HARQ.
- *Redundancy version*: it indicates the type of redundancy used in the HARQ transmission.
- *Flag for 1A/0 differentiation*: this is a single bit indicating whether Format 1A or Format 0 (which is used for uplink scheduling grants, see below) is used.
- *Localized/distributed VRB*: flag indicating whether localized or distributed VRBs are used.
- *Gap value for VRB*: the spacing between the two parts of the VRB.
- *Transport blocksize index*: only used in the 1C Format, it indicates the size of the data block to be transmitted. Note that in the other formats, the number of transport blocks follows

Table 27.4 DCI in various DCI message formats

Field	1	1A	1B	1C	1D	2	2A
Modulation/coding	✓	✓	✓		✓	✓(2)	✓(2)
HARQ process number	✓	✓	✓		✓	✓	✓
New data indicator	✓	✓	✓		✓	✓(2)	✓(2)
Redundancy version	✓	✓	✓		✓	✓(2)	✓(2)
TPC command for PUCCH	✓	✓	✓		✓	✓	✓
Flag for 1A/0 differentiation		✓					
Localized/distributed VRB		✓	✓		✓		
Gap value for VRB				✓			
Transport blocksize index				✓			
TPMI for precoding			✓		✓	✓	
Downlink power offset					✓		
Downlink assignment index	✓	✓	✓		✓	✓	✓
Transport block to codeword swap flag						✓	✓
Open-loop precoding info							✓

TPMI, Transmitted Precoding Matrix Indicator.

implicitly from the number of allocated RBs and the code rate (the standard contains a table providing this mapping, since there are minor deviations from the nominal rates obtained by closed-form computations).

- *Precoding Matrix Indicator (PMI) for precoding*: the bits in this field contain the index of the precoding codebook for closed-loop multiple-antenna systems (see Section 27.3.7.).
- *DL assignment index*: it is related to the HARQ operation in the TDD mode; for details see the standard.
- *Transport block to codeword swap flag*: it indicates how the two transport blocks (in spatial multiplexing system) are mapped to the codewords.
- *Open-loop precoding info*: this field indicates the type of open-loop precoding (e.g., details of the cyclic delay diversity) used in the transmission with multiple-antenna systems.

Uplink Control Information

Also, control information related to the uplink is carried in the PDCCH. Remember that the control of all transmission parameters rests with the BS, while the MS can only make requests. Thus, the PDCCH carries uplink scheduling grants, which tell the MS when, and with which format, to transmit. Just like for the downlink, that includes information on which RBs to transmit, the transport format, power control, and spatial multiplexing information. This information for the uplink is transported in message of Format 0, which employs Type 2 resource allocation format.[6] Finally, Format 3 is used for the transmission of power control commands for PUCCH and PUSCH with 2-bit power adjustments.

Once all the control information is assembled, the actual transmit signal is obtained in the following way:

1. A CRC is computed. The CRC depends on the identification number of the MS for which the DCI is intended. Thus, each MS determines whether information is intended for it by checking the CRC: if the CRC checks, then the MS concludes that (i) the DCI is intended for it and (ii) was decoded correctly.
2. The CRC-protected DCI is encoded with a rate 1/3 tail-biting convolutional code (i.e., a convolutional code that ensures that starting state and end state in the trellis are identical).
3. After rate matching, multiple PDCCHs are multiplexed.
4. The resulting data are scrambled by a PN sequence (Gold sequence, see Section 18.2.6), and QPSK modulated. The QPSK symbols are grouped into groups of four symbols each. In contrast to the PCFICH, these symbols are not directly mapped onto RE groups, but first interleaved, and then subjected to a cell-specific cyclic shift. The purpose of these measures is to ensure (i) full exploitation of the frequency diversity, and (ii) avoiding consistent collisions between the same signals in neighboring cells.

27.4.8   Physical Random Access CHannel

The Random Access CHannel (RACH) is intended for signals from MSs that do not yet have resources assigned to them. Even though the MS knows the cell bandwidth by the time it uses the PRACH, it still eases the implementation to have the same PRACH in all systems. Therefore, the PRACH is transmitted on 72 subcarriers (just like the SS). In the time domain, the BS usually

---

[6] In addition to that allocation, Format 0 also contains the Flag for 1A/0 differentiation, a Hopping Flag (indicating whether frequency hopping is to be used), modulation/coding and redundancy version, new data indicator, Transmit Power Control (TPC) commands for the scheduled PUSCH, cyclic Shift for the demodulation pilot, Channel Quality Indicator (CQI) request (i.e., a request from the BS to the MS to provide it with the channel quality information), and a UL index employed only in the TDD mode.

reserves 1-ms long blocks for the PRACH during which no uplink payload data are scheduled, so that there is no possibility of collision. The periodicity of the PRACH intervals can be configured, e.g., depending on how many possible users are in the cell, and what the latency requirements for random access are.

The most common configuration, a 1-ms PRACH, consists of: (i) 0.1-ms cyclic prefix, (ii) 0.8-ms long Zadoff–Chu sequence, and (iii) 0.1-ms guard interval. Note the requirement for long guard intervals because when the PRACH is used, the necessary timing advance (see Section 24.5.3) has not yet been established. There are 64 different sequences that can be used in the "main part" of the PRACH: either distinguished by phase shifts or sequences with different index u (see Section 27.3.5). The advantage of using different phase shifts is that those sequences are truly orthogonal; however, the amount of phase shift is lower bounded by the runtime and delay dispersion of the signal in the cell. If the cell size is below 1.5 km, all 64 sequences can be produced by phase shifting of one basic sequence; otherwise, sequences with different index have to be used. Different MSs pick, at random, the sequence that they want to use for the PRACH. Due to the large number of sequences, it is unlikely (though not impossible) that two MSs pick the same sequence and try to use it in the same timeslot. If that occurs, later stages of the random access procedure resolve this collision (see Section 27.5.1).

The PRACH channel also foresees power ramping. In other words, if the first attempt at reaching the BS is not successful, the transmit power of the random access preamble is increased, and transmission is repeated. This approach is similar to the one in WCDMA, though it is not as important: usually a random access preamble is orthogonal to other random access preambles (due to the different sequences being used) and payload data (due to the time/frequency resource being dedicated exclusively to the PRACH).

27.4.9 General Aspects of Control Signals Associated with PUSCH

In LTE, there is little information that needs to be signaled in the uplink, because the BS has central control of almost all transmission parameters. The few pieces of information that need to be sent in uplink direction are as follows:

- *Scheduling requests*: indications that the MS wants to send data; the actual resource assignment is then sent by the BS in the DCI.
- *Channel state information*: this is essential for the BS to determine the appropriate transmission format and scheduling. The channel state information contains the following pieces:
  - *Rank Indicator (RI)*: this is the rank of the channel matrix, which in turn determines the maximum number of layers that can be transmitted in a meaningful way (see Chapter 20). Relevant for multiantenna systems only.
  - *PMI*: this is the index of the codebook entry that should be used for precoding by the BS (see Section 27.3.7). Due to the frequency selectivity of the channel a different setting could be optimal for different RBs. The standard allows to transmit either a different setting for each RB or a setting for a group of RBs; in the extreme case of "wideband" feedback, there is one setting for the whole available bandwidth. Clearly, the smaller the groups for which the settings are transmitted, the better the performance, but also the larger the overhead. The PMI is relevant for multiantenna systems only.
  - *CQI*: this index actually represents the modulation and coding scheme that should be used. This is information about the downlink direction, which is intended to help the BS in its task. While the BS can choose to ignore the recommendation and use a different precoder setting, it then has to signal explicitly to the MS which setting it is using.
- *HARQ ACKnowledgements (ACKs)*.

While the amount of information is small, the way that it is communicated is rather complex. The complexity is partly due to the fact that the information should be transmitted both when there is payload being sent for the particular MS (i.e., it has an active PUSCH) and in the case where there is no PUSCH; in the latter case, the information is sent on a separate channel called the *Physical Uplink Control CHannel*.

27.4.10 PUCCH

Since there are so few bits on the PUCCH, it would be difficult to transmit them in the "standard" uplink fashion and still retain good spectral efficiency. Rather, the uplink control information from multiple users is multiplexed onto the same RB by two measures:

1. The information from different MSs are used to modulate orthogonal sequences; those orthogonal sequences have the same form as uplink pilots, i.e., phase-shifted versions of cyclically extended Zadoff–Chu sequences (see Section 27.3.5). Typically, 6 different phase shifts, $2\pi n/6$, $n = 0, \ldots 5$ are used. Each sequence is multiplied with a BPSK or a QPSK signal, so that each sequence carries 1 or 2 bits of information. In order to randomize intercell interference, the phase shifts of the sequences are changed from symbol to symbol, where the index of the used phase is determined by a PN sequence whose initialization depends on the cell ID, as well as on the slot in which transmission takes place.
2. Block-wise spreading over the symbols in a slot is used to further enhance the multiplexing capacity. This block spreading is achieved by assigning different Walsh–Hadamard sequences (see Section 18.2.5) to different users, if the desired spreading factor is 2 or 4, and DFT sequences if the spreading factor is 3. The blockwise spreading is only used for PUCCH Formats 1/1a/1b (see below). For randomization of intercell interference, the assigned spreading sequence changes (in a pseudorandom way) from slot to slot.

The PUCCH signals are transmitted in the RBs at the lower and upper band edges of the system, so that they cannot disturb the contiguous assignment of RBs to the different users. From slot to

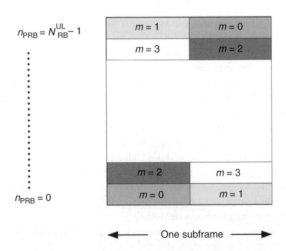

Figure 27.15   Resource assignment for the PUCCH.

Table 27.5 Modulation scheme for PUCCH

PUCCH format	Modulation scheme	Bits per subframe
1	OOK	N/A
1a	BPSK	1
1b	QPSK	2
2	QPSK	20
2a	QPSK+BPSK	21
2b	QPSK+BPSK	22

OOK, On Off Keying.

slot, the PUCCH hops between the lower and upper band edge, thus providing maximum frequency diversity (see Figure 27.15).

Depending on the amount of information that needs to be transmitted, the PUCCH uses different modulation formats (and possibly additional spreading). Table 27.5 gives an overview of the defined formats and associated modulation.

27.4.11 PUSCH

If there is an active PUSCH for a given MS, then the control data are time multiplexed with the payload data. The rules for multiplexing are motivated by the following concerns:

- The HARQ ACK has to be robust; it is therefore transmitted close to a pilot. Thus, there is no degradation of the channel estimate even for fast-varying channels. Similarly, the RI is sent close to the pilot.
- The rate matching should not depend on whether HARQ information is transmitted or not. Therefore, HARQ ACKs are punctured into the data stream.
- The modulation format is the same as that for the payload data, in order to ensure the single-carrier properties of the signaling. However, the code rate can be adapted to the channel state. Different codes are available for different pieces of information: repetition coding and simplex coding for ACKs and RIs, and Reed–Muller block codes or convolutional codes for CQI/PMI.

27.5 Physical Layer Procedures

27.5.1 Establishing a Connection

Scanning and Synchronization

The very first thing an MS has to do is to acquire the timing of the signals in the cell it is in; this is done via the SSs described in Section 27.4.2. Only after that can it learn about vital cell information (transmitted on the BCH), and perform the other functions it needs to do for communication (see Section 27.4.3).

After the acquisition of the timing and reception of the BCH, additional cell system information is communicated in the *System Information Blocks* (SIBs) transmitted via the DL-SCH, similar to the reception of payload data.

There are a number of different SIBs defined, depending on the type of information that is to be transmitted.

- *SIB 1*: it contains information about the access to the cell, information about cell selection, etc. It also contains a System Indicator (SI) window length, which is needed for the reception of all other SIBs.
- *SIB 2*: it contains configuration information valid for all MSs, like the configuration of common channels, pilot configuration, timers, etc.
- *SIB 3–8*: they contain information related to intersystem, interfrequency, and intrafrequency handover.
- *SIB 9*: it contains an identifier for the home BS.
- *SIB 10 and 11*: they contain information for earthquake and tsunami warning systems.

An SIB 1 is scheduled periodically every 80 ms, with a fixed timing. The repetition period for the other SIBs can be configured by the network operator. The MS has to search for the corresponding information within a time window whose width is communicated in SIB 1.

Random Access (RA)

If an MS wants to join the network, it has to let the BS know its request. Obviously, the MS does not have resources assigned to it at this time; it must therefore make a contention-based access (also known as random access): it is possible that its request might collide with requests from other MS. LTE specifies a procedure for this access, proceeding in four steps:

1. MS transmits an RA preamble (see Section 27.4.8), which allows the BS to compute the required timing advance. This is the only step using different PHY signaling. The subsequent steps are transmitted like normal data (except that HARQ is not used).
2. BS transmits an RA response: this contains (i) the index of the RA preamble for which the response is valid (note that at this time the BS does not know the ID of the MS yet, only the type of RA preamble that was used), (ii) the timing advance to be used by the MS, (iii) the resources to be used by the MS for the signaling in the subsequent step, and (iv) a temporary ID. Note that if several MSs used the same RA preamble, then the RA response is valid for all those MSs, and the resulting collision has to be resolved in the subsequent steps.
3. MS transmits Radio Resource Control (RRC) signaling information, which contains its ID. The details of the information depend on how much information the BS already has about this MS, e.g., whether the MS accesses network for the first time, or just reestablishes a link after the connection has been interrupted.
4. BS transmits a contention resolution message. As described above, ambiguities could have occurred if multiple MSs try to access the system with the same random access preamble. In the contention resolution message, the BS transmits explicitly the ID of the MS to which it assigns resources.

Paging

For paging, each MS is assigned (in the DL-SCH) a "paging interval," and a certain subframe in which a paging message might be transmitted. Thus, the MS needs to wake up from a sleep state only once per paging interval and listen whether there are data for it. The paging interval can be configured, representing a tradeoff between energy saving and latency.

27.5.2   Retransmissions and Reliability

Retransmission is an important part of LTE to ensure transmission quality. There are two components of the retransmission:

1. *Hybrid ARQ (HARQ)*: it provides for an integrated PHY/MAC approach for retransmission of data blocks that were not received successfully the first time, in such a way that the data from the multiple transmission can be combined. The retransmissions occur quickly.
2. *Radio Link Control (RLC)*: it is a higher layer retransmission protocol that arranges for the retransmissions of all the data blocks that fail even after HARQ. This mechanism is quite a bit slower, but will be invoked only rarely if HARQ is configured the right way. Being a "fallback" solution, it provides the extra reliability required for some applications (e.g., file transfer). The retransmission function can be switched off whenever it is not necessary/helpful (e.g., for voice calls, when the delay introduced by the protocol would be larger than what can be tolerated in phone conversations (see Chapter 15)).

The principle of HARQ is the same for uplink and downlink: (i) the TX sends a transport block, (ii) the RX tries to decode. If successful, it sends an ACK. Note that in the case of spatial multiplexing, two transport blocks can exist, so that the ACK has to have 2 bits, (iii) if the previous transmission was successful, the TX sends a new packet (and indicates this fact in the "new data" field); if the previous transmission was unsuccessful, the transport block is retransmitted (the "new data" field showing that it is a retransmission, and the "redundancy version" field indicating which type of retransmission is used).

However, there is one key differences between uplink and downlink: retransmission in the downlink might occur using arbitrary RBs. Retransmission in the uplink is fixed: it is always eight subframes after the first transmission attempt, and the RBs used for retransmission are the same as for the first transmission.[7]

Since there is a delay of eight frames between retransmissions, up to eight parallel HARQ processes must be active such that payload data can be transmitted during every frame. The principle is outlined in Figure 27.16.

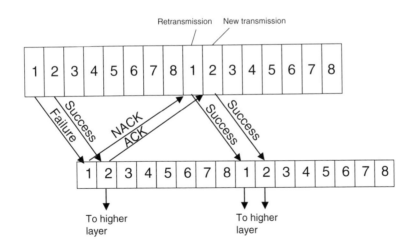

Figure 27.16 Scheduling for HARQ.

---

[7] Retransmission on a different RB can be achieved through a clever combination of the "new data indicator" and the scheduling grant, see Dahlmann et al. [2008, Section 19.1.1.2].

27.5.3   Scheduling

Scheduling in the context of LTE is the question at what time, and on which subcarriers, information for/of which MS is transmitted. Furthermore, it also involves the choice of the transport format, i.e., transport blocksize, modulation and coding scheme, and the multiple-antenna scheme. Under normal circumstances, the scheduling decision is transmitted afresh in every subframe, since it can change from subframe to subframe. However, in the case of voice calls (and other applications that have a low data rate but continuous data), semipersistent scheduling is used. Essentially, the BS tells the MS that (until further notice) it has the same resources allocated to it in every n-th frame; after that message, no further scheduling info needs to be transmitted to that MS, thus reducing the overhead.

All scheduling decisions are made by the serving BS (though that BS might use information it received from the MSs and from other BSs, in order to provide a better quality). This is true for both the downlink and the uplink. Note, however, that an MS that has multiple radio bearers to transmit for the uplink decides by itself which one to transmit on the resources assigned to it by the BS.

The standard does not define *how* the scheduling is to be done. In general, the scheduler will try to exploit the multiuser gain (Section 20.1.9) as much as possible, but also the interference to other cells, as well as the latency (due to backlog) for each MS must be taken into account.

27.5.4   Power Control

Power control in the uplink of LTE consists of an open-loop and a closed-loop power control. The open-loop mechanism establishes a baseline for the desired transmit power: the MS determines the downlink path loss from a pilot with known transmit power, and from that computes the necessary uplink power (including necessary margins in the process). A power control signal that is transmitted in the PDCCH then "fine-tunes" the transmit power. One-bit power control signals request the power to be changed by ± 1 dB; 2-bit signals select from the set $[-1, 0, 1, 3]$ dB. No downlink power control is specified, though a BS can adjust the power at will (since such adjustments do not contradict the standard).

While the mechanisms of power control are somewhat similar to that of WCDMA, their importance and motivation are very different. For WCDMA, power control is essential for proper functioning of the data transmission. In LTE (just like in GSM), power control is just used for increasing battery lifetime, and reduction of intercell interference.

27.5.5   Handover

Handovers in LTE are "hard handovers," in the sense that communication occurs between MS and one BS at one time, not to two simultaneously. On a more detailed level, the handover from a source BS to a target BS proceeds in the following three phases:

1. Handover preparation
   (a) The source BS configures the measurements the MS has to perform and report. Specifically, it sets the thresholds such that reports from the MS to the BS are required if certain measurement results (e.g., signal quality to neighboring BS) exceed those thresholds. Alternatively, the BS requires periodic reports.
   (b) MS sends its (periodic or aperiodic) measurement results.
   (c) BS makes a handover decision based on the measurement results (e.g., to a cell with larger path gain).

    (d) The source BS sends a handover request to the target BS, usually via the X2 interface (the interface between two BSs).

    (e) The target BS performs an admission control. If the target cell has no resources available, the connection might have to be terminated.

    (f) If it admits the handover, the target BS sends a "handover request acknowledgement" to the source BS.

2. Handover execution

    (a) The source BS sends a handover command to the MS, and at the same time starts to forward downlink packets (i.e., packets it receives from the network for this MS) to the target BS. Transmission of those packets by the target BS has to wait until the target BS can actually communicate to the MS (see below).

    (b) The source BS tells the target BS which packets were already acknowledged by the MS.

    (c) The MS synchronizes itself to the target BS via the RACH (a preliminary synchronization was already achieved during the cell identification process, when the cell did its measurements).

    (d) The target BS transmits the uplink resource allocation and timing advance to the MS.

    (e) The MS sends a "handover confirm" message to the target BS. From that time on, target BS and MS can communicate with each other.

3. Handover completion

    (a) The target BS sends a "path switch" message to the MME, (see Section 27.2.2), requesting that data for the MS are henceforth sent to the target BS.

    (b) The MME forwards this message to the serving gateway.

    (c) The serving gateway switches to the target BS the route the data for the MS have to take.

    (d) The serving gateway confirms the switch to the MME.

    (e) The MME confirms the "path switch" message to the target BS.

    (f) The target BS sends a message to the source BS, telling it to release the resources still reserved for the MS.

    (g) The source BS releases the resources.

27.6 Glossary for LTE

BCCH	Broadcast Control CHannel
BCH	Broadcast CHannel
CCCH	Common Control CHannel
CQI	Channel Quality Indicator
DCCH	Dedicated Control CHannel
DCI	Downlink Control Information
DL-SCH	Downlink Shared CHannel
DwPTS	Downlink Pilot Time Slot
DTCH	Dedicated Traffic CHannel
HSPA	High Speed Packet Access
LTE	Long-Term Evolution
MBMS	Multimedia Broadcast and Multicast Services
MB-SFN	Multicast/broadcast in a single-frequency network
MCCH	Multicast Control CHannel
MCH	Multicast CHannel
MIB	Master Information Block
MME	Mobility Management Entity
MTCH	Multicast Traffic CHannel

PBCH	Physical Broadcast CHannel
PCCH	Paging Control CHannel
PDCP	Packet Data Convergence Protocol
PCH	Paging CHannel
PCFICH	Physical Control Format Indicator CHannel
PDCCH	Physical Downlink Control CHannel
PDSCH	Physical Downlink Shared CHannel
PDU	Protocol Data Units
PHICH	Physical HARQ Indicator CHannel
PMCH	Physical Multicast CHannel
PMI	Precoding Matrix Indicator
PRACH	Physical Random Access CHannel
PRB	Physical Resource Block
PSS	Primary synchronization signal
PUCCH	Physical Uplink Control CHannel
PUSCH	Physical Uplink Shared CHannel
QPP	Quadrature Permutation Polynomial
RAN	Radio Access Network
RB	Resource Block
RE	Resource Element
RI	Rank Indicator
RLC	Radio Link Control
RRC	Radio Resource Control
RS	Reference Signal
SAE	System Architecture Evolution
SDU	Service Data Units
SI	System Indicator
SIB	system information block
SS	synchronization signal
SSS	Secondary synchronization signal
TPMI	Transmitted Precoding Matrix Indicator
UL-SCH	Uplink Shared CHannel
UpPTS	Uplinklink Pilot Time Slot
VRB	Virtual Resource Block

Further Reading

The official resource for studying the LTE standard is, of course, the standards documentation, which is available from *www.3gpp.org*. Of particular interest is the 36.2xx series. An excellent summary of the standard, which concentrates on the PHY and MAC layer, is Part II of Dahlman et al. [2008]. It describes not only *what* has to be implemented but also *why*. Another recent book summarizing the standard (with less detailed description of the specifications, and more results on system simulations) is Holma and Toskala [2009]. A number of good overview articles are in the April 2009 issue of the IEEE Communications Magazine. Furthermore, relay technologies for LTE are discussed in Yang et al. [2009], and femtocells in Golaup et al. [2009].

For updates and errata for this chapter, see wides.usc.edu/teaching/textbook

28

WiMAX/IEEE 802.16

28.1 Introduction

Worldwide Interoperability for Microwave Access (WiMAX) is a wireless communications standard for *Metropolitan Area Networks* (MANs), e.g., networks covering whole cities or even whole countries. Originally intended as a standard for Fixed Wireless Access (FWA) using millimeter wave bands (11–60 GHz), it concentrated more and more on providing mobility, and in its latest incarnation has become a competition to third- and fourth-generation cellular systems. Since 2005, it has received an enormous amount of media interest, though at the time of this writing (2009) the chances for commercial success are not quite clear yet.

28.1.1 History

As mentioned in Chapter 1, "last mile" access seemed a promising application for wireless technology in the 1990s, with an initial emphasis on voice communication. However, the deregulation of the telephone market in the U.S.A. and Europe eliminated the financial attractiveness for simply providing an alternative way of delivering voice. Broadband Internet access seemed like a more rewarding application, especially since at that time Digital Subscriber Line (DSL) and cable–modem connections were available only in very few places. Many of the *FWA* broadband systems developed during that time operated in the >10-GHz range, since ample bandwidth (BW) was available there. In order to overcome the market fragmentation of multiple proprietary systems, in 1999 the Institute of Electrical and Electronics Engineers (IEEE) established a standardization group *IEEE 802.16* for wireless MANs that established a standard for broadband FWA in the frequency range from 10 to 66 GHz. The standard, which was based on single-carrier modulation, was approved in 2001, but had little impact.

A fundamental problem of operation in the >10-GHz range is that it requires Line Of Sight (LOS) connection between transmitter (TX) and receiver (RX), so that not only the Base Station (BS) but also the antennas of the user equipment have to be located outdoors, usually above rooftop heights. For this reason, the IEEE developed standards for operation between 2 and 11 GHz. In this range, LOS is not necessary; however, ability to effectively deal with delay spread becomes vital. Based on these considerations, alternatives to the single-carrier modulation were developed, in particular Orthogonal Frequency Division Multiplexing (OFDM). Also, less bandwidth is available at those lower frequencies, so that spectral efficiency became a more important issue; as a

Wireless Communications, Second Edition   Andreas F. Molisch
© 2011 John Wiley & Sons, Ltd

consequence Multiple Input Multiple Output systems (MIMO) technology became an essential part of the standard. The new standard was consolidated, together with the >10-GHz standard, into *IEEE 802.16-2004*, also known as *fixed WiMAX* or *IEEE 802.16d*.

During the early 2000s, laptop computers had proliferated, and seemed prime candidates for direct communication with WiMAX BSs. As a consequence, it became necessary to support (limited) mobility in WiMAX. The *IEEE 802.16e* group developed a standard that fulfilled this requirement. In addition to mobility support, a number of further additions and modifications was made, e.g., it provides a more flexible Orthogonal Frequency Division Multiple Access (OFDMA). The standard, also known as *mobile WiMAX*, was approved in 2005 and in 2007 became part of the *International Telecommunications Union*'s (ITU) International Mobile Telecommunications 2000 (IMT-2000) family of standards. A further enhancement, which should provide full mobility as well as higher data rates and spectral efficiencies, is currently being developed by the *IEEE 802.16m* group and will be submitted to the *IMT-Advanced* family of standards.

The IEEE 802.16 standard specifies the PHYsical layer (PHY and *Medium Access Control* (MAC) layer, but does not ensure interoperability between standard-compliant devices. The reasons for this lack are (i) the 802.16 standard does not cover network layer and system architecture, (ii) the 802.16 standard is not sufficiently detailed to allow full interoperability in the PHY and MAC layer, and (iii) the 802.16 standard contains a bewildering number of options (e.g., allows five completely different modulation/multiple-access methods), so that building devices that can cover all those options in a cost-effective way becomes almost impossible. For these reasons, an industry alliance called the *WiMAX forum* has established further specifications that ensure interoperability. The problem of the multiple options is solved by the definitions of *WiMAX profiles*, which are essentially a subset of the 802.16 options that all WiMAX-compliant devices have to be able to fulfill. The WiMAX specifications also enable interoperability, and devices can be certified by a designated "certification lab" to be fully compliant to those specifications. In the remainder, we will use the somewhat sloppy terminology of applying "WiMAX" as synonymous with both "real" WiMAX specifications and IEEE 802.16 specifications.

28.1.2 *WiMAX versus Existing Cellular Systems*

Mobile WiMAX has become, de facto, a full-fledged cellular standard, and is in direct competition to Third Generation Partnership Project (3GPP), Code Division Multiple Access (CDMA) 2000. WiMAX has some important advantages in this battle:

- The WiMAX standard was originally intended for data communications; voice is more of an afterthought that is enabled by *Voice over Internet Protocol* (VoIP) communications. As the emphasis of future cellular systems will lie on data communications, this is beneficial for WiMAX.
- The modulation format and multiple-access format of WiMAX, i.e., MIMO/OFDM/OFDMA, is more suitable for high-data-rate communications.
- The standard is *much* simpler than the 3GPP standard, covering "only" 1,400 pages (compared to the 50,000 pages of 3GPP). As a consequence, the standard can be understood and realized even by relatively small companies, which increases the number of possible suppliers and drives down costs.[1]
- The system architecture is Internet Protocol (IP) based, so that no costly development and deployment of a separate backbone network is required (note, however, that the 3GPP network is an

---

[1] On the downside, many passages of the original text were incomplete, ambiguous, or contradictory. A "Cleaning up" project (release D2), which was published in 2009, significantly reduced those problems.

evolution of the Global System for Mobile communication (GSM) network; as a consequence, GSM operators migrating to a third-generation system prefer the GSM-style network).

- The standard is strongly supported by computer chip manufacturers, which will build WiMAX RXs into most laptops. This leads to an automatic customer base for WiMAX.

However, WiMAX is a yet unproven system with no established customer base, and tries to compete with a system that has strong support from the cellular industry and hundreds of millions of customers. At the time of this writing, only one major operator each in the U.S.A., Japan, and Korea, has committed to widespread WiMAX deployment.

Even more difficult to predict will be the competition between 3GPP-LTE (Long-Term Evolution) (see Chapter 27) and 802.16e. Those systems will use quite similar technology, so that market dynamics will outweigh technical considerations even more. Similarly, their extensions, LTE-Advanced and IEEE 802.16m, will compete head-to-head.

28.2   System Overview

28.2.1   Physical Layer Overview

WiMAX specifies a number of completely different PHYs. We will only treat the OFDMA standard, which is used for mobile communications in the 2–11 GHz band.

Let us briefly summarize this OFDMA PHY. Both Frequency Domain Duplexing (FDD) and Time Domain Duplexing (TDD) are specified, though TDD is by far the more common option. In TDD, the time axis is divided into 5 ms *frames*, which serve first the *downlink* (DL) and then the *uplink* (UL). In each of the uplink and downlink parts (*subframes*), OFDM symbols are transmitted, where different subcarriers, and different OFDM symbols, are assigned to different users. The subcarriers assigned to a user are either distributed over the available bandwidth to achieve high-frequency and high-interference diversity, or are all adjacent, so that Adaptive Modulation and Coding (AMC), based on *Channel State Information* (CSI) at the TX, can be more easily applied. The subframes are divided into *zones*, and during each zone, the nature of the assignment of subcarriers to users can be different. At the beginning of a frame, a notification MAP is sent out, describing which subcarriers are intended for which user.

Various types of multiple-antenna techniques can be used for the transmission in both uplink and downlink. Space–time block codes, spatial multiplexing, as well as antenna selection can be used to increase diversity and data rates.

28.2.2   Frequency Bands

WiMAX does not specify any particular operating frequency, but can work at any carrier frequency between 2 and 11 GHz. The spectrum assignments are left up to the national frequency regulators. The following bands have emerged as the most important:

- *1.9/2.1-GHz band*: in September 2007, WiMAX was made part of the IMT-2000 family of standards in the ITU. As a consequence, national regulators will probably allow the deployment of WiMAX in the frequency bands that are foreseen for third-generation cellular systems. In the U.S.A., this is the 1.9-GHz band, in most other countries, it is the frequency band between 1.9 and 2.2 GHz.
- *2.5-GHz band*: parts of the band between 2.5 and 2.7 GHz might be used for deployment in the U.S.A., Canada, Japan, Russia, parts of Central and South America, and possibly India. The 2.3–2.43 GHz band is used for the Wireless Broadband (WiBro) system, a variant of WiMAX, in Korea and Australia, and also the U.S.A.

- *3.5-GHz band*: parts of the band between 3.3 and 3.8 GHz have been assigned to FWA in large parts of Europe, Australia, Canada, South America, and Africa. The total available bandwidth varies from country to country.
- *5-GHz band*: unlicensed WiMAX systems could be deployed in the band between 5.25 and 5.85 GHz, which is assigned as *Unlicensed National Information Infrastructure* (U-NII) in the U.S.A., and with similar dedications in other countries. However, the unlicensed nature of the bands results in higher interference and low admissible transmit power. Essentially, WiMAX systems would go head-to-head with Wireless Fidelity (WiFi) in this frequency band (see Chapter 29).

Further discussions of the frequency allocations for IMT-Advanced systems can be found in Chapter 27.

28.2.3 MAC Layer Overview

The MAC layer consists of three main parts: (i) the *MAC Convergence Sublayer* (CS), (ii) the *MAC Common Part Sublayer*, and (iii) the *MAC Security Sublayer*.

The CS receives data packets, also known as *Service Data Units* (SDUs) from higher layers. The format of those SDUs can be different, according to the standard used in the networking layer. For example, they might be Transmission Control Protocol/Internet Protocol (TCP/IP) packets (from Internet traffic), or ATM (*Asynchronous Transfer Mode*) packets. The task of the CS is to reformat those packets into data units that are agnostic of the higher layers, and at the same time enable more efficient over-the-air transmission. For example, the CS provides packet header suppression: the SDUs contain a lot of redundant information in their headers, e.g., the IP addresses contained in every TCP/IP packet. The header suppression is based on the use of certain rules that allow the RX to reconstruct the header. The CS puts out MAC Packet Data Units (PDUs).

The Common Part Sublayer of the MAC provides the essential support functions for the over-the-air transmission of the information. It includes such functions as signaling the choice of the modulation/coding scheme, feedback of CSI, and bandwidth allocation. It also provides *fragmentation* and *packing*: it often happens that the SDUs (packets from a higher layer) are not of a size that can be easily transmitted in an "over-the-air" packet. If the packets are too large, the MAC fragments them into smaller parts, and adds information that tells the RX how to piece together those fragments. Similarly, if SDUs are too small (so that their transmission would increase the overhead), they can be packed together to be transmitted over the air.

WiMAX defines as a *service flow* a MAC transport service that is assigned to a specific user, and has a certain Quality of Service (QoS) requirement like latency, jitter, etc. It is characterized by the following parameters: service flow identifier (ID), *Connection Identifier* (CID), provisioned QoS parameters (the recommended QoS parameters), admitted QoS parameters (the QoS parameters actually allocated to the service flow), active QoS parameter set (QoS parameters that are being provided at a given time), and the authorization module.

28.2.4 Network Structure

The network architecture is outlined in Figure 28.1. It is not within the purview of the IEEE 802.16 standardization, but rather developed by the WiMAX Forum. A main goal of this architecture is to separate the over-the-air access (the *Access Service Network* (ASN) from the Internet access (*Connectivity Service Network* (CSN). It is a goal that the two networks can be owned by different providers.

Figure 28.1 WiMAX network structure and connection interfaces.

The ASN provides the over-the-air connectivity corresponding to the PHY and MAC layer as defined in the IEEE 802.16 standard, plus a number of related functions that are not defined in the standard (e.g., scheduling and resource management). It discovers which networks are available and connects the user to the preferred (permissible) CSN. Each BS is connected to an ASN Gateway, which has somewhat similar functionality as the BS controller in GSM (compare Chapter 24). WiMAX defines a number of different functional splits between BS and Gateway.

During network discovery, the Mobile Station (MS) discovers both the access service provider (which is characterized by a unique 24-bit "operator ID" in the DL-MAP) and the available *Network Service Provider* (NSP). The selection of the NSP and the functioning of the Communication Service Provider (CSP) have little to do with the wireless operation of WiMAX, and will thus not be described here any further. The situation is similar to WiFi (Chapter 29), where an access point just provides a wireless link that can be seen as a "cable replacement" for a wired Internet connection; getting access and making payments to the Internet service provider is independent of this operation.

28.3 Modulation and Coding

28.3.1 *Modulation*

OFDM

The modulation scheme in WiMAX is standard OFDM with cyclic prefix, as described in Chapter 19. The following parameters determine the settings:

- Bandwidth B.
- The number of used subcarriers N_{used}.
- The oversampling factor n_{samp}, which, together with the bandwidth and the number of used subcarriers, determines the subcarrier spacing:

$$\Delta f = \frac{n_{samp} B}{N_{FFT}} \qquad (28.1)$$

where N_{FFT} is the smallest power of 2 that is larger than N_{used}.[2] The purpose of the oversampling factor is to easily stretch or compress the spectrum, so that it can be effectively fit into a given spectral mask without changing other parameters like the number of used subcarriers. $n_{samp} = 28/25$ for all bandwidths that are integer multiples of 1.25, 1.5, 2, or 2.75 MHz; it is 8/7 for bandwidths that are integer multiples of 1.75 MHz and any other bandwidths not mentioned before. The useful symbol time is $1/\Delta f$.
- N_{cp}, the length of the cyclic prefix, which can be 1/4, 1/8, 1/16, or 1/32 of the useful symbol time.

The following Fast Fourier Transform (FFT) sizes are foreseen: 2048, 1024, 512, and 128. The FFT size of 256 is left out because it is used by another PHY transmission mode of WiMAX that is based on OFDM (not OFDMA!).

The subcarriers can be used for any of the following three purposes:

1. *Data subcarriers*: for carrying information.
2. *Pilot subcarriers*: for channel estimation and tracking.
3. *Null carriers*: containing no energy, used for guard bands in the frequency domain. Also the Direct Current (DC) subcarrier is not assigned any energy; this helps to avoid amplifier saturation; furthermore information on DC subcarriers would get lost in direct-conversion receivers.

The number of used subcarriers and null carriers depends on the subcarrier assignment scheme, and are discussed in Section 28.4.

Modulation Formats

The output of the channel encoder is mapped onto the modulation constellation, which can be Gray mapped 4-Quadrature Amplitude Modulation (QAM), 16-QAM, or 64-QAM (64-QAM is optional in the uplink). The constellations are normalized by multiplication with a factor c (as indicated in Figure 28.2 to have the same energy.

As a first step, we have to generate random numbers w_k that are the output of a shift register $(1 + x^9 + x^{11}$, see Figure 28.3). Those numbers are used to multiply all the modulation and pilot tones. For the data modulation, the k-th subcarrier (with k indexing the *physical* subcarrier), is multiplied with

$$2\left(\frac{1}{2} - w_k\right) \qquad (28.2)$$

before being modulated by the data.

Modulation and Coding Rates

WiMAX allows a number of different combinations of code rate and modulation formats. 4-QAM and 16-QAM can be combined with rate 1/2 and 3/4 codes, while 64-QAM can be combined with rate 1/2, 2/3, 3/4, and 5/6 codes.

---

[2] We simplify the notation here by writing the numerator as nB. Strictly speaking, it is $8000*\mathrm{floor}(n_{samp}B/8000)$.

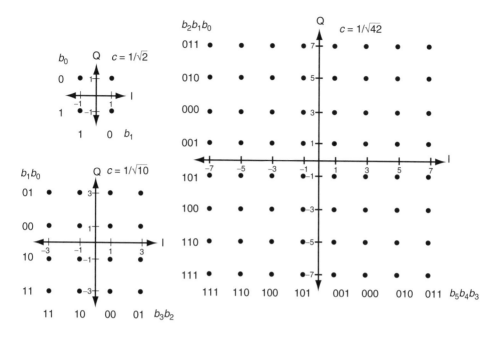

Figure 28.2 Modulation alphabets for WiMAX.

Reproduced with permission from [IEEE 802.16]. © IEEE.

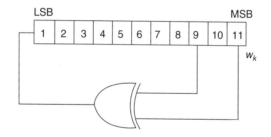

Figure 28.3 Shift register for generating random numbers for modulation.

Reproduced with permission from [IEEE 802.16]. © IEEE.

28.3.2 Coding

The WiMAX standard includes a number of different Forward Error Correction (FEC) schemes, as well as Automatic Repeat reQuest (ARQ). The different error correction schemes offer different tradeoffs between complexity and performance. However, only the simplest schemes, namely, the standard convolutional coding, and standard ARQ, are mandatory, while the other coding forms are optional. We will omit descriptions of block turbo codes and Low Density Parity Check (LDPC) codes, since they are not required in the WiMAX system profiles.

For all the FEC schemes, the source data are first randomized by modulo-2 addition to the output of a shift register $(1 + x^{14} + x^{15})$. Encoding is then done over so-called "FEC blocks," which consist of several subchannels (for a definition of subchannels, see Section 28.3.1.2). The FEC blocks are the output of segmentation.

Convolutional Codes

The mandatory FEC code in WiMAX is a tail-biting convolutional code with rate 1/2 and a constraint length of 7. The code polynomials are

$$G1(D) = 1 + D + D^2 + D^3 + D^6 \qquad (28.3)$$

$$G2(D) = 1 + D^2 + D^3 + D^5 + D^6 \qquad (28.4)$$

Higher code rates, in particular rate 2/3 and rate 3/4 codes, are derived from this code by puncturing.

In order to keep different code blocks independent of each other, the encoder must be initialized properly. Tail Biting Convolutional Codes (TBCC) use the last 6 bits of the code block to initialize the states of the encoder. This way the first and last state of the encoder are same – i.e., in the trellis representation, only those paths which start and end on a same node/state are the valid codewords. This overcomes the need to zero force the trellis to get back to zero state (compare Chapter 14).

After encoding, the data are interleaved in a two-step process. The first step makes sure that adjacent bits are not transmitted on adjacent subcarriers, which improves the frequency diversity. In a second step, adjacent bits are alternatingly mapped onto more or less important bits in the constellation.

Turbo Codes

One way to achieve higher performance in WiMAX is the use of turbo codes (see Section 14.6). A remarkable aspect of the turbo codes in WiMAX is that they are duobinary codes, i.e., the turbo encoding is not done on bits, but rather on 2-bit symbols. The resulting code has advantages compared to "normal" turbo codes in that it has larger minimum distance, better convergence, and less sensitivity to puncturing. On the downside, the decoder is more complex.

Figure 28.4 shows the structure of the turbo encoder. A two-bit symbol A/B is used as an input to recursive convolutional encoders in two incarnations: directly, and after going through an interleaver. The interleaver first flips bits within a symbol (it does that for every second symbol only); and then interleaves symbols. Each of the two convolutional encoders produces two parity bits, one for the normal order of the input symbol and one for the output of the interleaver. The resulting symbols are then interleaved (note that the "regular" WiMAX interleaver, as used, e.g., for the convolutional codes, is not used for turbo coding).

As mentioned above, the standard foresees LDPC codes (compare Section 14.7) only as optional modes.

HARQ

WiMAX foresees two optional types of Hybrid Automatic Repeat reQuest (HARQ), namely, chase combining, and incremental redundancy, as described in Appendix 14.A.[3] In chase combining (also called type-I HARQ), an FEC block is simply repeated, and the RX combines soft versions of the received signal. In *incremental redundancy* transmission (also called type-II HARQ), the TX sends multiple, differently encoded versions of the source data. The different encoding is achieved by changing the puncturing patterns. Which puncturing pattern is used is communicated in the beginning of an FEC block, by means of the SubPacket IDentity (SPID).

---

[3] In either case, a 16-bit cyclic redundancy check is transmitted with every HARQ packet, so that it can be easily determined whether a packet was properly received.

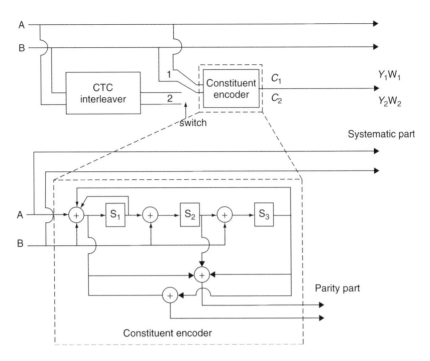

Figure 28.4   Structure of the duobinary turbo encoder in WiMAX.
Reproduced with permission from [IEEE 802.16]. © IEEE.

28.4   Logical and Physical Channels

WiMAX is based on OFDMA, and assigns data to particular subcarriers in particular OFDM symbols. One of the most important, and also most confusing, aspects of WiMAX is how this mapping is done – there are a number of different options, and all of them are rather complicated. The following first describes the basic principles and terminology, and only then goes into the details of the different allocations, even if that results in some redundancy of the description.

28.4.1   Frames and Zones

The largest fundamental unit in the time–frequency plane is a *frame*, which is 5-ms long,[4] and encompasses all subcarriers. It is divided into *subframes* that describe the time (in the TDD mode) for uplink and downlink. Each subframe is divided into *permutation zones*, during which one particular subcarrier permutation is used (we explain below what subcarrier permutations are). Figure 28.5 shows the subdivision of frames into subframes and permutation zones.

28.4.2   Details of Frame Structure

Figure 28.6 shows the detailed frame structure (with only the mandatory zones). It can be seen that each zone contains user data. It is important to know that the contiguous arrangement of the data

---

[4] In 802.16-2004, between 2 and 20 ms.

Figure 28.5 Zone structure of an OFDMA frame. *In this figure:* IE, Information Element.
Reproduced with permission from [IEEE 802.16]. © IEEE.

Figure 28.6 Structure of an OFDMA frame.
Reproduced with permission from [IEEE 802.16]. © IEEE.

in the time–frequency plane describes *logical* subchannels. In other words, a user might occupy logical channels 5 and 6; but those logical subchannels might be mapped to widely distributed (in the time–frequency plane) *physical* subcarriers, due to the subcarrier permutation that we discuss below in greater detail.

In addition to the data bursts for the separate users, a frame contains preambles and control signaling (*DL and UL-MAP*, *Frame Control Header* (FCH)). Also, these control signals undergo a subcarrier permutation. The FCH is strongly protected: it is sent with Quadrature-Phase Shift

Keying (QPSK) modulation with rate 1/2 convolutional code, and in addition repeated 4 times; the FCH information is sent on 4 subchannels with successive logical subchannel numbers. The FCH contains the DL frame prefix (which state which subchannels are used for the DL-MAP), and other details about the DL-MAP (length of the MAP, use of repetition coding, etc.). It is always sent on the lowest numbered subchannels after the DL frame preamble.

After the FCH, the DL-MAP and the UL-MAP are sent. They specify which data regions (OFDM symbols and subchannels) are being assigned to which user for uplink and downlink; note that the MAPs may be compressed.

We then have to distinguish between *subcarriers* and *subchannels*. A subchannel is a logical unit of data, which is mapped onto a *collection of physical subcarriers* (not necessarily contiguous); the exact number and arrangement of the subcarriers is defined as the *subcarrier permutation* (note that there are several different subcarrier permutation modes, Partial Use of SubCarriers (PUSC), (Full Usage of SubCarriers) (FUSC), and AMC, which are discussed below). A contiguous (in time and/or frequency) set of subchannels is known as a *data region*, and is always transmitted using the same *burst profile* (modulation format, code rate, type of code). The number of data carriers per subchannel is given in Table 28.1.

Table 28.1 Number of subchannels (from [Eklund et al. 2006])

Permutation/FFT size	128	512	1024	2048
PUSC – DL	3	15	30	60
PUSC – UL	4	17	35	70
PUSC – UL (optional permutation)	6	24	48	96
FUSC – DL (either permutation)	2	8	16	32
TUSC				
AMC	2	8	16	32

The smallest possible data allocation unit is called a *slot*. The size of a slot is defined in units of subchannels (in the frequency domain) and OFDM symbols (in the time domain). The numerical values for the size of a slot depend on the subcarrier permutation as well as on uplink/downlink; details are given in Table 28.2. Each user is assigned one or multiple slots for communicating its data. The details of which slots are associated with each user are communicated in the *DL-MAP* and *UL-MAP* fields.

Table 28.2 Size of slots in units of subchannels

	UL	DL
PUSC (either permutation)	1×3	1×2
FUSC (either permutation)		1×1
TUSC1/TUSC2		1×3

28.4.3 Mapping of Data onto (Logical) Subchannels

The (MAC) data are mapped onto a data region in the following way:

- For the downlink:
  - Segment the data into blocks sized to fit into one slot.
  - Number the slots in such a way that the lowest numbered slot occupies the lowest numbered subchannel in the lowest numbered OFDM symbol.

o Continue the mapping so that the OFDMA subchannel index is increased. When the edge of the data region is reached, continue the mapping from the lowest numbered OFDMA subchannel in the next available symbol.

- For the uplink, mapping occurs in a very similar manner. However, note that due to a difference in the definition of the data region, we tend to get allocations that are narrow in the frequency domain and wide across the symbols. This helps the MS to spend as much as power possible on a per-subcarrier basis, while keeping the total power transmitted still within limits.

28.4.4 Principles of Preamble and Pilots

Subcarrier Allocation – Why all those Different Modes?

A subcarrier permutation is given by a mathematical scheme defining how data symbols (with logical index) is mapped onto subcarriers.

The different permutations are suitable for different circumstances. Consider the following distinctions:

1. Distributed/adjacent data allocation: in PUSC, FUSC, subcarriers for one user are distributed, while for banded AMC they are adjacent. Distributed allocation is good if we are not aware of per-carrier channel quality while adjacent allocation is good if we know the frequency-dependent channel quality so that multiuser diversity can be exploited.
2. Reuse type: PUSC and AMC allow for partial reuse, while FUSC allows for full reuse.
3. Dedicated/Broadcast pilot allocations: we could have pilots attached to allocations/groups/subchannels, or pilots spread all over the bandwidth. In AMC, the pilots can be seen as part of the subchannel, in PUSC as part of the subchannel groups, while in FUSC they are global. Consequently AMC and PUSC could have pilots dedicated to a user and enable beamforming for that user over a set of subcarriers.

Preamble

The DL preamble that is sent by the BS at the beginning of each frame can be used for channel estimation, interference estimation, etc. It covers all subcarriers during one OFDM symbol. The subcarriers are divided into three interlaced groups: the first group contains the (physical) subcarrier $0, 3, 6, \ldots$; the second group the subcarriers $1, 4, 7, \ldots$ and the third group $2, 5, 8, \ldots$; guard tones are present. The preamble is sent with boosted power.

Pilot Symbol Placement

The pilot tones are modulated the same way as the preamble. The power of the pilot tones is boosted 2.5 dB above the power of the data tones.

For the FUSC and DL-PUSC, the pilot tones are allocated first; what remains are data subcarriers, which are divided into subchannels that are used exclusively for data, resulting in a set of common pilots.[5] For PUSC in the uplink, the set of used subcarriers is first partitioned into subchannels, and then the pilot subcarriers are allocated from within each subchannel, so that each subchannel contains its own set of pilot subcarriers. The reason for this is that in the uplink, common pilots do not make sense (the signals from the different users undergo different channels, even if they are at very similar frequencies).

---

[5] For the DL-PUSC, there is a set of common pilots in each so-called "major group," which is explained below.

28.4.5 PUSC

PUSC is the mandatory subcarrier allocation mode that has to be used in every frame. The basic principle of the PUSC is to divide all the subcarriers into groups, which can be used in different cell sectors. In other words, only a part of the subcarriers is used in each sector – hence the name. It is noteworthy that this concept, which decreases co-channel interference, is similar to frequency reuse in Frequency Division Multiple Access (FDMA) networks. It is used as the default mode because it is more robust than other schemes.

Downlink PUSC

For the downlink, a number of *clusters* consisting of 14 subcarriers each are formed. Within such a cluster, the pilots are arranged as depicted in Figure 28.7.

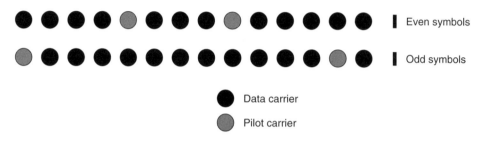

Figure 28.7   Pilot structure in the downlink PUSC.
Reproduced with permission from [IEEE 802.16]. © IEEE.

The clusters are then divided into to six *groups*; the assignment is done by first renumbering the clusters (using a pseudorandom numbering scheme), and then putting the first one-sixth of the (renumbered) clusters into group 1, the next one-sixth into group 2, and so on. A subchannel is then created by taking two clusters from the same group. PUSC can assign all, or just a subset, of the groups to a TX; in other words, different groups can be assigned to different sectors in a cellular system. By default, groups 0, 2, and 4 are assigned to sectors 1, 2, and 3. This is similar to an FDMA system with a frequency reuse of 1/3 (compare Section 17.6.4). Note, however, that PUSC has greater flexibility, since three groups can be distributed at will to the sectors.

Let us now go into the details of the subcarrier assignment (this description closely follows [IEEE 802.16]):

1. Divide the subcarriers into clusters (there are N_{clusters} of them) containing 14 adjacent subcarriers each. The number of clusters, N_{clusters}, varies with FFT sizes. For example, the 1024 FFT results in 60 clusters: $60 \times 14 = 840$ used subcarriers, plus the left guard tones (92), DC tone (1), and right guard tone (91) gives the total of 1024 carriers used in the FFT. The physical cluster numbers are simply assigned in sequence (the lowest frequencies get cluster number 0, followed by $1, 2, \ldots$). For the 128 FFT, there are only 6 clusters.
2. Renumber the physical clusters into logical clusters using the following formula:

$$LogicalClusterNumber = \begin{cases} RenumberingSequence(PhysicalCluster) & \text{First DL zone, or} \\ UseAllSCindicator = 0 \text{ in } STC\_DL\_Zone\_IE \\ RenumberingSequence((PhysicalCluster) & \text{otherwise} \\ + 13DL\_PermBase) \bmod N_{\text{clusters}} \end{cases}$$

$$(28.5)$$

The renumbering sequence is a random sequence specified in the standard. The DL permutation base is an integer parameter that can be set by the BS and communicated to the MSs in the DL_MAP field. The parameter *UseAllSCindicator* is communicated to the MS in the *STC_DL_Zone_IE* Information Element (essentially, a control signaling bit).

3. Allocate the logical clusters to groups. The allocation is different for different FFT sizes. For the 128 FFT, there are 3 major groups, with group 0 containing clusters 0–1, group 2 containing clusters 2–3, and group 4 containing clusters 4–5; assignment is similar for FFT size 512. For FFT size 1024, we divide the clusters into 6 major groups, group 0 includes clusters 0–11, group 1 includes clusters 12–19, group 2 includes clusters 20–31, group 3 has clusters 32–39, group 4 has clusters 40–51, and group 5 has clusters 52–59; note that groups 0, 2, 4 have a different number of clusters in them compared to groups 1, 3, 5; the assignments are similar for FFT size 2048.

An example for steps 1–3 for FFT size 128 is given in Figures 28.8 and 28.9.

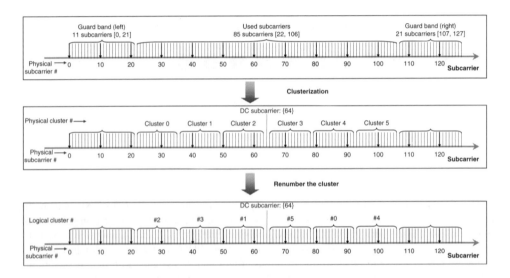

Figure 28.8   Clustering of subcarriers and assigning logical cluster numbers.

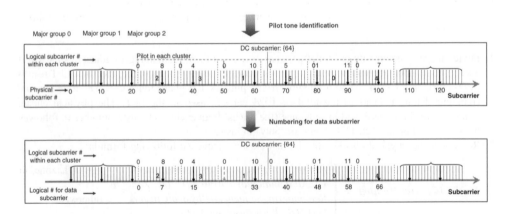

Figure 28.9   Numbering of data subcarriers From [Tao 2007].

4. Subcarriers are allocated to subchannels separately in each major group for each OFDMA symbol. We first allocate the pilot carriers within each cluster and then take all remaining data carriers within the symbol. The parameters vary with FFT sizes. After assigning the pilots, the remaining tones are now first assigned a continuous increasing index ("logical subcarrier index," in the notation of Eklund et al. [2006]). These logical subcarriers are assigned to logical subchannels. In particular, the k-th subcarrier of the s-th subchannel is assigned to the following logical subcarrier index:

$$Subcarrier(k, s) = N_{\text{subchannels}} \cdot n_k + \{p_s[n_k \bmod N_{\text{subchannels}}] + DL\_PermBase\} \bmod N_{\text{subchannels}}$$
$$(28.6)$$

where

- $Subcarrier(k, s)$: physical subcarrier index of subcarrier k in subchannel s;
- s: index number of a subchannel, $s \in \{0, \ldots, N_{\text{subchannels}} - 1\}$;
- $n_k = (k + 13s) \bmod N_{\text{subcarriers}}$;
- where k index of the subcarrier within a subchannel, $k \in \{0, \ldots, N_{\text{subcarriers}} - 1\}$;
- $p_s[j]$: series obtained by rotating a base sequence cyclically to the left s times;
- $IDcell$: an integer ranging from 0 to 31, which identifies the particular BS segment and is specified by the MAC layer (it is identical to IDCell in the first zone).

An example of this subcarrier assignment can be found in Figure 28.10.

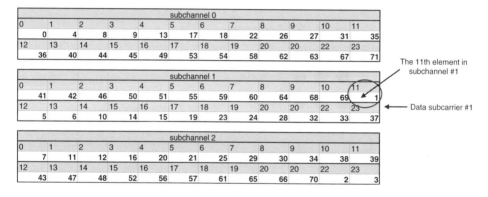

Figure 28.10   Assignment of data carriers to elements of subchannels. From [Tao 2007].

It is also useful to look at this problem in another way: if we are given a logical subchannel, what steps do we have to take to arrive at the associated physical assignment [Nyuami 2007]?

1. As a first step, we have to establish the division of all the available subcarriers into clusters, and the clusters into major groups. This is done according to steps 1–3 in the procedure above, see also Figures 28.8 and 28.9. To give another concrete example, consider "major group 3" with FFT size 1024. It contains (as mentioned in step 3 above) the logical clusters 32–39. From Eq. (28.5) and the renumbering sequence, we find that (for DL PermutationBase = 5), the associated physical clusters are 3, 6, 53, 20, 45, 57, 28, 19. This allows the definition of all the physical subcarriers that are associated with a particular major group.

2. We next assign the pilot tones. We know the index numbers of the physical clusters from the previous step, and we know from Figure 28.7 how the physical clusters are distributed within each group. We can thus easily determine the physical subcarrier indices of the pilot tones (note

again that the counting starts from the first nonguard subcarrier): 42, 54, 84, 96, 742, 754, . . .
for the odd symbols, and similarly for the even symbols.

3. Now we turn to the data subcarriers. There are 24 data subcarriers in a logical subchannel; note that $N_{\text{subchannel}} = 4$ in our example, because we have 8 physical clusters, giving 96 data carriers ($8 \times 14 = 112$ subcarriers, minus 16 pilots). The indices of the logical subcarriers (within the major group) associated with subcarrier $k = 0, \ldots 23$ are computed from Eq. (28.6). By considering now the mapping between physical and logical subcarriers, we arrive at the physical indices of the subcarriers that are to be used.

Uplink PUSC

In the uplink, the time–frequency plane is first divided into *tiles*, which consist of four contiguous subcarriers times three OFDM symbols. The tiles are then assigned in a pseudorandom way to groups, six of the tiles in a group create one subchannel. Note that the tiles contain pilot carriers, and it is the tiles that are assigned in a pseudorandom way to the subchannels. Thus, every subchannel carries the required pilot tones in them. In the standard PUSC mode, the corners of each tile are used as pilot tones (see Figure 28.11).

Figure 28.11   Pilot structure in the uplink PUSC within each tile.
Reproduced with permission from [IEEE 802.16]. © IEEE.

The procedure of mapping the subchannels to the physical carriers works in the following steps:

1. In the first step, the physical tiles are mapped to logical tiles according to

$$Tiles(s, n) = N_{\text{subchannels}} \cdot n + (Pt[(s + n) mod N_{\text{subchannels}}] + UL\_PermBase) mod N_{\text{subchannels}}$$

where
(a) $Tiles(s, n)$ is the physical tile index, increasing from smaller to larger subcarriers
(b) n is the tile index $0 \ldots 5$ in a subchannel (there are always 6 tiles in each subchannel, irrespective of the FFT size);
(c) Pt is the tile permutation, as given in tables in the standard;
(d) s is the subchannel number $s \in \{0 \ldots N_{\text{subchannels}} - 1\}$;
(e) UL_PermBase is an integer value assigned by the BS;
(f) $N_{\text{subchannels}}$ is the number of subchannels, which depends on the FFT size.
An example is given in Figure 28.12 and 28.13.

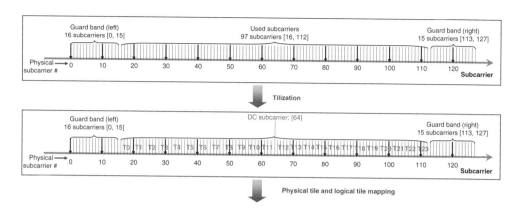

Figure 28.12   Division of subcarriers into tiles. From [Tao 2007].

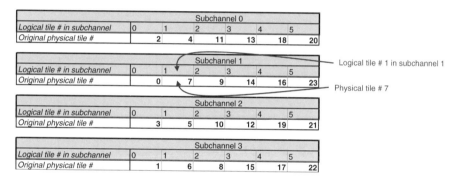

Figure 28.13   Mapping of logical to physical tiles. From [Tao 2007].

2. Within the tiles belonging to a slot, start to number the available physical units (the nonpilot subcarriers in the three available OFDM symbols) consecutively. Start at the earliest OFDM symbol, with the lowest subcarrier. Then increase the index along the frequency axis until you reach the highest subcarrier, then start to number the next OFDM symbol again starting from the lowest subcarrier (see Figure 28.14).

3. Then map the data onto the physical units according to the following equation

$$Subcarrier(n, s) = (n + 13 \cdot s) \mathrm{mod} N_{\mathrm{subcarriers}}$$

where

(a) *Subcarrier*(n, s) is the permutated subcarrier index corresponding to data subcarrier n is subchannel s;

(b) n is a running index $0 \ldots 47$, indicating the data constellation point within a subchannel;

(c) s is the subchannel number;

(d) $N_{\mathrm{subchannels}}$ is the number of subcarriers per slot.

For an example, see Figures 28.15 and 28.16.

A data rotation scheme is applied to the UL-PUSC: during each slot, the numbering of the subchannels is changed.

Figure 28.14 Numbering of subcarriers within tiles. From [Tao 2007].

Subchannel 0

	1	2	3	4	5	6	7	8	9	10	11	12	13	14	15	16	17	18	19	20	21	22	23	24
Original data subcarrier #							0	1	2	3	4	5	6	7	8	9	10	11						
New data subcarrier #							0	1	2	3	4	5	6	7	8	9	10	11						
Original data subcarrier #	12	13	14	15	16	17	18	19	20	21	22	23	24	25	26	27	28	29	30	31	32	33	34	35
New data subcarrier #	12	13	14	15	16	17	18	19	20	21	22	23	24	25	26	27	28	29	30	31	32	33	34	35
Original data subcarrier #							36	37	38	39	40	41	42	43	44	45	46	47						
New data subcarrier #							36	37	38	39	40	41	42	43	44	45	46	47						

Subchannel 1

	1	2	3	4	5	6	7	8	9	10	11	12	13	14	15	16	17	18	19	20	21	22	23	24
Original data subcarrier #							0	1	2	3	4	5	6	7	8	9	10	11						
New data subcarrier #							13	14	15	16	17	18	19	20	21	22	23	24						
Original data subcarrier #	12	13	14	15	16	17	18	19	20	21	22	23	24	25	26	27	28	29	30	31	32	33	34	35
New data subcarrier #	25	26	27	28	29	30	31	32	33	34	35	36	37	38	39	40	41	42	43	44	45	46	47	0
Original data subcarrier #							36	37	38	39	40	41	42	43	44	45	46	47						
New data subcarrier #							1	2	3	4	5	6	7	8	9	10	11	12						

Subchannel 2

	1	2	3	4	5	6	7	8	9	10	11	12	13	14	15	16	17	18	19	20	21	22	23	24
Original data subcarrier #							0	1	2	3	4	5	6	7	8	9	10	11						
New data subcarrier #							26	27	28	29	30	31	32	33	34	35	36	37						
Original data subcarrier #	12	13	14	15	16	17	18	19	20	21	22	23	24	25	26	27	28	29	30	31	32	33	34	35
New data subcarrier #	38	39	40	41	42	43	44	45	46	47	0	1	2	3	4	5	6	7	8	9	10	11	12	13
Original data subcarrier #							36	37	38	39	40	41	42	43	44	45	46	47						
New data subcarrier #							14	15	16	17	18	19	20	21	22	23	24	25						

Subchannel 3

	1	2	3	4	5	6	7	8	9	10	11	12	13	14	15	16	17	18	19	20	21	22	23	24
Original data subcarrier #							0	1	2	3	4	5	6	7	8	9	10	11						
New data subcarrier #							39	40	41	42	43	44	45	46	47	0	1	2						
Original data subcarrier #	12	13	14	15	16	17	18	19	20	21	22	23	24	25	26	27	28	29	30	31	32	33	34	35
New data subcarrier #	3	4	5	6	7	8	9	10	11	12	13	14	15	16	17	18	19	20	21	22	23	24	25	26
Original data subcarrier #							36	37	38	39	40	41	42	43	44	45	46	47						
New data subcarrier #							27	28	29	30	31	32	33	34	35	36	37	38						

Figure 28.15 Data subcarrier permutation. From [Tao 2007].

An optional PUSC mode uses a different tile size (three subcarriers by three OFDM symbols), and uses only the middle subcarrier/symbol entity as pilot, resulting in higher spectral efficiency, but worse channel estimation performance.

28.4.6 TUSC

The TUSC is a downlink permutation that is symmetrical with the UL-PUSC (remember that UL-PUSC and DL-PUSC use different subcarrier assignment, and thus cannot exploit reciprocity. TUSC 1 corresponds to the UL-PUSC, TUSC 2 to the optional UL-PUSC.

28.4.7 *FUSC*

FUSC, which exists only for the downlink, foresees only a single segment that contains all subcarrier groups. Each subchannel consists of 48 subcarriers that are distributed over the total available system bandwidth. The mapping of the subchannels to the subcarriers depends on the *cell ID*, and on a parameter called *PermutationBase* that the BS can set. The mapping furthermore changes with each OFDM symbol.

FUSC provides a very high degree of diversity. (i) During the transmission of one OFDM symbol, frequency diversity as well as interference diversity are provided by the fact that the data are mapped onto subcarriers that are distributed over the whole bandwidth. Although this gives not quite the same diversity as transmitting on all possible subcarriers, it still results in a quite high diversity gain. Furthermore, the transmission on a sparse set of subcarriers means that each user creates less worst case adjacent-cell interference (of course, the total amount of interference is identical to that created by an OFDM/Time Division Multiple Access (TDMA) system, but the interference in a worst case, i.e., during the time that a user close to the cell edge is transmitted to, is lower). (ii) Since the mapping changes from OFDM symbol to the next, an additional degree of interference diversity is provided. However, simulations have shown that the system starts to lose throughput if the load per cell is more than 1/3. This indicates that the efficiency is comparable to PUSC.

Let us now go into some more details. Consider first the assignment of the pilots (Figure 28.16): there are two sets of constant pilots and two sets of variable pilots; in the FUSC, all those pilots are used (the subdivision of the pilot into two sets is related to their use in the space–time coding mode; see below). The indices of the subcarriers belonging to the different sets are specified in Section 8.4.6.1.2.2 of the standard. The equation for the assignment of the variable pilots is given as

$$PilotsLocation = VariableSet\#x + 6 \cdot (FUSC\_SymbolNumber \bmod 2) \qquad (28.7)$$

After the pilots have been assigned, the data are mapped to the remaining physical subcarriers in a procedure similar to step 4 of the PUSC procedure. In other words, the available carriers (excluding guard carriers, DC carrier, and pilots) are numbered consecutively as "logical" subcarriers, and the data carried by the different subcarriers are assigned by Eq. (28.6) to those indices. Figures 28.17–28.19 give an example.

Figure 28.16 Final assignment of subcarriers. From [Tao 2007].

- Symbol number 0 (or any even number)

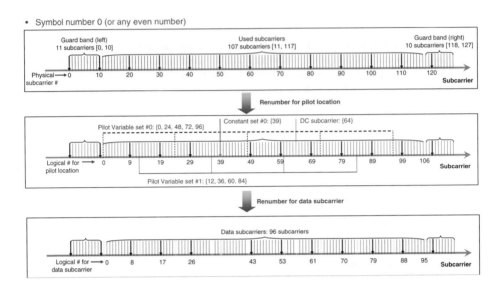

Figure 28.17 Insertion of pilot tones for even-numbered symbols in FUSC. From [Tao 2007].

- For IDcell = 0, DL_PermBase = [1 0], symbolNumber = 0, k in {0, ...,
  $N_{subcarriers}-1$} (i.e. {0, ..., 47}):
  - Subcarrier(k, 0) =

{ 1	2	5	6	9	10	13	14
17	18	21	22	25	26	29	30
33	34	37	38	41	42	45	46
49	50	53	54	57	58	61	62
65	66	69	70	73	74	77	78
81	82	85	86	89	90	93	94}

  -
  - Subcarrier(k, 1) =

{ 27	28	31	32	35	36	39	40
43	44	47	48	51	52	55	56
59	60	63	64	67	68	71	72
75	76	79	80	83	84	87	88
91	92	95	0	3	4	7	8
11	12	15	16	19	20	23	24}

Figure 28.18 Renumbering of subcarriers in the FUSC.

There is also another mode called "optional FUSC," which is distinguished by a different pilot assignment. Here, the subcarriers are placed eight subcarriers apart, and at each OFDM symbol, all the pilots are offset by three subcarriers. The pilots thus "cycle" through all subcarriers over a period of eight OFDM symbols. If BSs are synchronized, there is a danger of "catastrophic" collisions, i.e., if pilots of two adjacent cells collide, they collide at all times (in contrast to the standard FUSC, where each cell uses a different temporal change for the pilot assignment pattern so that interference at one time does not entail interference at another time).

28.4.8 AMC

In the *AMC* permutation (also known as *adjacent subcarrier permutation*), nine contiguous physical subcarriers (for the duration of one OFDM symbol) are grouped together in

Figure 28.19 Mapping of elements of logical channels to subcarriers. From [Tao 2007].

what is called a *bin*. The middle subcarrier of each bin is used as a pilot tone. A group of four rows of bins (i.e., bins at four frequencies) is called a *physical band*. An AMC subchannel consists of 6 adjacent (in either time or frequency) bins within one band; i.e., one bin over a period of six OFDM symbols, or three bins for two OFDM symbols, or of two adjacent bins over a period of three OFDM symbols, or one bin over six OFDM symbols.

Data from one subchannel are assigned to subcarrier/OFDM symbols according to a permutation equation. The mapping of subcarriers to subchannels does not change with time (at least not within one data region). There is one pilot tone in each bin, namely, the middle carrier. However, if AMC is used in connection with adaptive antennas, a more elaborate pilot pattern is used, as shown in Figure 28.20.

28.4.9 Channel Sounding

In order to enable closed-loop transmission scheme, the standard foresees that a Channel State Information at the Transmitter (CSIT)-capable MS transmits a *channel sounding* waveforms, so that the BS can determine the BS-to-MS channel response under the assumption of reciprocity. The sounding information is also useful for closed-loop multi-antenna transmission.

There is a special zone for channel sounding. In this zone, the MS transmits the sounding signal, which is a subsequence of a Golay sequence. The signal is transmitted in a part of the total available bandwidth. It can either use all available subcarriers in that band (in that case, different users employ different phase shifts of the Golay sequence to ensure separability of the users) or the signal is not transmitted on every subcarrier, but rather only on every D-th subcarrier. Possibly the MS also feeds back the (average) Signal-to-Interference-and-Noise Ratio (SINR) observed on the subcarriers; furthermore, the interference power per subcarrier can be indicated by assigning a transmit power to each pilot subcarrier that is proportional to the inverse of the interference power observed by the MS.

Furthermore, the MS can also directly feed back explicit channel information. It can transmit the downlink channel coefficients together with the uplink pilots, or just use a single additional symbol to feed back the received pilot coefficients.

Figure 28.20   Pilot pattern for Adaptive Antenna System (AAS) mode in the AMC zone.
Reproduced with permission from [IEEE 802.16]. © IEEE.

28.5   Multiple-Antenna Techniques

Multiple-antenna techniques are an important part of the WiMAX standard. It uses both space–time coding (Section 20.2.9) and spatial multiplexing (Section 20.2.8).

28.5.1   Space–Time Coding and Spatial Multiplexing

2 Transmit Antennas

For space–time coding with 2 transmit antennas, a conventional Alamouti scheme is used. When used together with the PUSC subcarrier assignment, then the pilot structure for the downlink changes to the one depicted in Figure 28.21. If FUSC is used, the pilots within the symbols are divided between the antennas. Antenna 0 uses VariableSet#0 and ConstantSet#0 for even symbols while antenna 1 uses VariableSet#1 and ConstantSet#1 for even symbols; the assignment is vice versa for the odd symbols.

Pilot and data mapping for the example of AMC subcarrier allocation is shown in Figure 28.22.

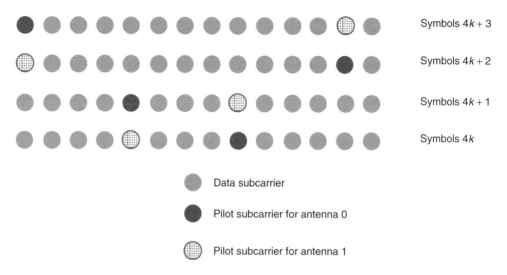

Figure 28.21 Cluster structure for FUSC with Alamouti code.
Reproduced with permission from [IEEE 802.16]. © IEEE.

Antenna #0

s_0	$-s_{16}^*$	s_{32}	$-s_{48}^*$	s_{64}	$-s_{80}^*$
		s_{33}	$-s_{49}^*$	s_{65}	$-s_{81}^*$
s_1	$-s_{17}^*$	s_{34}	$-s_{50}^*$	s_{66}	$-s_{82}^*$
s_2	$-s_{18}^*$	s_{35}	$-s_{51}^*$	s_{67}	$-s_{83}^*$
s_3	$-s_{19}^*$	s_{36}	$-s_{52}^*$		
s_4	$-s_{20}^*$	s_{37}	$-s_{53}^*$	s_{68}	$-s_{84}^*$
s_5	$-s_{21}^*$	s_{38}	$-s_{54}^*$	s_{69}	$-s_{85}^*$
s_6	$-s_{22}^*$			s_{70}	$-s_{86}^*$
s_7	$-s_{23}^*$	s_{39}	$-s_{55}^*$	s_{71}	$-s_{87}^*$
s_8	$-s_{24}^*$	s_{40}	$-s_{56}^*$	s_{72}	$-s_{88}^*$
		s_{41}	$-s_{57}^*$	s_{73}	$-s_{89}^*$
s_9	$-s_{25}^*$	s_{42}	$-s_{58}^*$	s_{74}	$-s_{90}^*$
s_{10}	$-s_{26}^*$	s_{43}	$-s_{59}^*$	s_{75}	$-s_{91}^*$
s_{11}	$-s_{27}^*$	s_{44}	$-s_{60}^*$		
s_{12}	$-s_{28}^*$	s_{45}	$-s_{61}^*$	s_{76}	$-s_{92}^*$
s_{13}	$-s_{29}^*$	s_{46}	$-s_{62}^*$	s_{77}	$-s_{93}^*$
s_{14}	$-s_{30}^*$			s_{78}	$-s_{94}^*$
s_{15}	$-s_{31}^*$	s_{47}	$-s_{63}^*$	s_{79}	$-s_{95}^*$

| s | Data subcarrier |
| Pilot / Null subcarrier |

Figure 28.22 Data mapping for Alamouti code with AMC permutation.
Reproduced with permission from [IEEE 802.16]. © IEEE.

The Alamouti encoding can also be used in the uplink. Again, the pilots are split between the 2 transmit antennas, though in a somewhat different fashion compared to the downlink.

Another possible transmission scheme for a two-antenna configuration is straightforward spatial multiplexing. Either symbol s_i is transmitted from the first antenna, and symbol s_{i+1} from the second antenna (*vertical encoding*), or two separate streams are transmitted from the two antennas (*horizontal encoding*) (see Figure 28.23).

Antenna #0 with vertical encoding

s_0	s_{32}	s_{64}	s_{96}	s_{128}	s_{160}
PILOT	NULL	s_{66}	s_{98}	s_{130}	s_{162}
s_2	s_{34}	s_{68}	s_{100}	s_{132}	s_{164}
s_4	s_{36}	s_{70}	s_{102}	s_{134}	s_{166}
s_6	s_{38}	s_{72}	s_{104}	PILOT	NULL
s_8	s_{40}	s_{74}	s_{106}	s_{136}	s_{168}
s_{10}	s_{42}	s_{76}	s_{108}	s_{138}	s_{170}
s_{12}	s_{44}	PILOT	NULL	s_{140}	s_{172}
s_{14}	s_{46}	s_{78}	s_{110}	s_{142}	s_{174}
s_{16}	s_{48}	s_{80}	s_{112}	s_{144}	s_{176}
PILOT	NULL	s_{82}	s_{114}	s_{146}	s_{178}
s_{18}	s_{50}	s_{84}	s_{116}	s_{148}	s_{180}
s_{20}	s_{52}	s_{86}	s_{118}	s_{150}	s_{182}
s_{22}	s_{54}	s_{88}	s_{120}	PILOT	NULL
s_{24}	s_{56}	s_{90}	s_{122}	s_{152}	s_{184}
s_{26}	s_{58}	s_{92}	s_{124}	s_{154}	s_{186}
s_{28}	s_{60}	PILOT	NULL	s_{156}	s_{188}
s_{30}	s_{62}	s_{94}	s_{126}	s_{158}	s_{190}

Antenna #0 with horizontal encoding

s_0^0	s_{16}^0	s_{32}^0	s_{48}^0	s_{64}^0	s_{80}^0
PILOT	NULL	s_{33}^0	s_{49}^0	s_{65}^0	s_{81}^0
s_1^0	s_{17}^0	s_{34}^0	s_{50}^0	s_{66}^0	s_{82}^0
s_2^0	s_{18}^0	s_{35}^0	s_{51}^0	s_{67}^0	s_{83}^0
s_3^0	s_{19}^0	s_{36}^0	s_{52}^0	PILOT	NULL
s_4^0	s_{20}^0	s_{37}^0	s_{53}^0	s_{68}^0	s_{84}^0
s_5^0	s_{21}^0	s_{38}^0	s_{54}^0	s_{69}^0	s_{85}^0
s_6^0	s_{22}^0	PILOT	NULL	s_{70}^0	s_{86}^0
s_7^0	s_{23}^0	s_{39}^0	s_{55}^0	s_{71}^0	s_{87}^0
s_8^0	s_{24}^0	s_{40}^0	s_{56}^0	s_{72}^0	s_{88}^0
PILOT	NULL	s_{41}^0	s_{57}^0	s_{73}^0	s_{89}^0
s_9^0	s_{25}^0	s_{42}^0	s_{58}^0	s_{74}^0	s_{90}^0
s_{10}^0	s_{26}^0	s_{43}^0	s_{59}^0	s_{75}^0	s_{91}^0
s_{11}^0	s_{27}^0	s_{44}^0	s_{60}^0	PILOT	NULL
s_{12}^0	s_{28}^0	s_{45}^0	s_{61}^0	s_{76}^0	s_{92}^0
s_{13}^0	s_{29}^0	s_{46}^0	s_{62}^0	s_{77}^0	s_{93}^0
s_{14}^0	s_{30}^0	PILOT	NULL	s_{78}^0	s_{94}^0
s_{15}^0	s_{31}^0	s_{47}^0	s_{63}^0	s_{79}^0	s_{95}^0

S — Data subcarrier   PILOT — Pilot subcarrier   NULL — Null subcarrier reserved for Art #1 Pilot

(s_x^0 : 0th stream, STC coded Symbol)

Figure 28.23   Data mapping for multiple transmit antennas with layered transmission.
Reproduced with permission from [IEEE 802.16]. © IEEE.

4 Transmit Antennas

For the case of 4 transmit antennas, a number of possible transmission schemes exist:

- *Combination of Alamouti code with beamforming*: An Alamouti-coded signal is transmitted from two sets containing two antennas each. Within each set, a phase shift is used between the elements to achieve beamforming gain in the desired direction. This gives a rate of 1.
- *Alamouti code with antenna grouping*: the BS transmits one Alamouti block (2 symbols) from transmit antennas 1 and 2, and the next Alamouti block from antennas 3 and 4 (the other antenna

elements do not transmit anything). Mathematically speaking, the STC encoding matrix is

$$A = \begin{bmatrix} s_1 & -s_2^* & & \\ s_2 & s_1^* & & \\ & & s_2 & -s_4^* \\ & & s_4 & s_3^* \end{bmatrix} \tag{28.8}$$

Again, this scheme gives a rate of 1. Note that different permutations of this matrix are possible.

$$A_1 = \begin{bmatrix} s_1 & -s_2^* & & \\ s_2 & s_1^* & & \\ & & s_2 & -s_4^* \\ & & s_4 & s_3^* \end{bmatrix} \quad A_2 = \begin{bmatrix} s_1 & -s_2^* & & \\ & & s_3 & -s_4^* \\ s_2 & s_1^* & & \\ & & s_4 & s_3^* \end{bmatrix} \quad A_3 = \begin{bmatrix} s_1 & -s_2^* & & \\ & & s_3 & -s_4^* \\ & & s_4 & s_3^* \\ s_2 & s_1^* & & \end{bmatrix}$$
$$\tag{28.9}$$

The mapping of subscript k to determine the matrix A_k is either given by a quasi-random assignment according to:

$$k = mod(floor((\text{logical\_data\_subcarrier\_number\_for\_first\_tone\_of\_code} - 1)/2), 3) + 1 \tag{28.10}$$

or chosen according to feedback from the MS.

- *Combination of spatial multiplexing with Alamouti code*: the first Alamouti block is transmitted from antennas 1 and 2, and simultaneously a second Alamouti block is sent from antennas 3 and 4. The next symbols are the sent via the antenna sets 1–3 and 2–4.

$$B = \begin{bmatrix} s_1 & -s_2^* & s_5 & -s_7^* \\ s_2 & s_1^* & s_6 & -s_8^* \\ s_3 & -s_4^* & s_7 & s_5^* \\ s_4 & s_3^* & s_8 & s_6^* \end{bmatrix} \tag{28.11}$$

This gives a rate of 2. Six possible permutations of this matrix are possible, and the permutation is chosen by an equation similar to Eq. (28.10) or according to feedback from the MS.

- *Pure spatial multiplexing*, with a precoding matrix

$$C = \begin{bmatrix} s_1 \\ s_2 \\ s_3 \\ s_4 \end{bmatrix} \tag{28.12}$$

- *Spatial multiplexing together with antenna selection*: one, two, or three antennas (and spatial streams) can be active; the power of the active antennas is boosted so that the *total* transmit power stays unchanged.

Note that transmission schemes are also defined for 3 BS antenna elements, but for the sake of brevity are not discussed here.

28.5.2 MIMO Precoding

If the TX has *CSI*, the overall performance of a MIMO system can be improved, as discussed in Section 20.2. The improvement is achieved by weighting the transmit signal with a matrix **T**, so that the signals $\tilde{s}$ that are actually transmitted are

$$\tilde{s} = Ts \tag{28.13}$$

where the dimension of the matrix $\mathbf{T}$ is $N_t \times N_{STC}$ and $\mathbf{s}$ is a vector containing the output of the space–time encoder with dimension N_{STC}. The precoding matrix $\mathbf{T}$ should be chosen according to some form of CSI obtained by reciprocity or feedback. Note that the antenna selection (bullet point 4 in the previous section) and antenna grouping (i.e., choosing the correct index k for the space–time coding matrices A_k or B_k) according to CSI is a form of MIMO precoding; we are just using specific forms of permutation matrices for $\mathbf{T}$.

For more elaborate kinds of precoding, the WiMAX standard defines a codebook of precoding matrices. The set of possible precoding matrices is indexed, and the RX only feeds back the index of the desired matrix. This reduces the feedback load compared to full feedback. There are two sets of codebooks, one with 8 entries, and one with 64. This allows to tradeoff the feedback overhead with the possible performance gain, see Section 20.2.11.

The TX (usually the BS) can also request the RX to feed back the explicit channel coefficients or can determine them through the channel sounding procedure; in those cases, the TX is free to choose any arbitrary precoding coefficients.

For the PUSC and FUSC operation, the same precoding matrix is used for all employed subcarriers. For the band-AMC operation, the BS might request either a common precoding matrix for all bands, or request feedback for the N best bands, where N is a programmable number.

The RX can either feed back long-term or short-term matrix indices. Long-term matrices, which are based on the average CSI, are less sensitive to fast movement, but give in general less performance improvement than short-term prediction.

28.5.3 SDMA and Soft Handover in MIMO

During the soft handover, the available transmit antennas of the BSs constitute an antenna pool whose elements can be used in the same fashion as described above.

28.6 Link Control

28.6.1 Establishing a Connection

Scanning and Synchronization

When an MS tries to enter a network, it first has to identify the possible operating frequency and synchronize itself to the network transmissions. This is achieved by first listening to the DL preamble on the possible operating frequencies. Typically, the MS will first listen on the frequencies it was last active on, and then – if not successful – start to scan other possible frequencies.

Once the MS hears a DL preamble, it can then establish timing, frequency synchronization and obtain the BS ID, listening to control messages such as FCH, DL-MAP, and UL-MAP. In particular, it then knows the parameters of the ranging channel (see below) that is needed for the initial ranging procedure.[6]

Initial Ranging

The initial ranging signal is sent on a ranging channel, which contains six or more adjacent logical subchannels, starting with the lowest subchannel, and using the UL-PUSC structure described above. The structure of the shift register producing the ranging signal is shown in Figure 28.24. The MS determines the power of the ranging method it should use from two parameters that the

---

[6] Note that ranging has little to do with establishing a distance between TX and RX, but rather refers to signaling to enter the network.

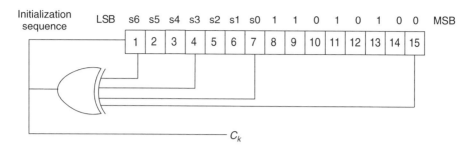

Figure 28.24 Shift register for generation of ranging code.
Reproduced with permission from [IEEE 802.16]. © IEEE.

BS broadcasts, as well as from the power the MS receives from the BS (this approach works well in a TDD system where channel reciprocity can be used).

By its very nature, the initial ranging is a contention-based multiple access, i.e., collisions can occur. A certain mitigation of the collisions is achieved by the fact that different MSs use different ranging sequences. However, it can still happen that ranging is unsuccessful; in this case a retransmission is done. The retransmission occurs at a suitable backoff delay, and with increased power; both measures are aimed to increase the probability that the retransmission is successful.

Parameter Assignment and Capability Exchange

After the BS has received the ranging request message from the MS, it sends a ranging response signal that contains the CID assigned to the MS (additional CIDs might be assigned later, as needed). The ranging response can also contain further fine-tuning of power levels and timing.

Subsequently, the MS sends a list of its capabilities, e.g., which modulation/coding formats it supports, which of the available multiantenna schemes it supports, etc.

After these steps are finished, the MS registers itself on the network by sending a registration requirement message on the network; it contains a code that allows the BS to check whether the message is authentic. Then, the MS (or BS) can initiate a service flow; for this purpose, it has to be first checked whether the MS is authorized to receive this specific type of service flow.

Periodic Ranging

An MS that is already synchronized to the BS will send out periodic ranging signals and bandwidth requests. The periodic ranging signals use outputs from the same shift register as the initial ranging, but the duration of the signal is either one or three OFDM symbols. The results of the periodic ranging are used to adjust synchronization and power.

28.6.2 Scheduling and Resource Request

The scheduling algorithm is not prescribed in WiMAX; rather it is left to the implementer. All decisions about resource allocations are done by the BS; the MS can only request certain resources.[7]

---

[7] Strictly speaking, the request does not come from an MS, but rather a CID, which is assigned to a specific service covering the MS. One MS might have several CIDs, e.g., streaming video and VoIP, in parallel. Each of the services then makes its own request for resources. However, the grant of resources by the BS is an aggregate

A resource request by the MS might be a stand-alone command, or it can be piggy-backed onto a generic MAC Packet Data Unit (PDU). The request is also often called "bandwidth request."

Requests can be expressed as *aggregate* or *incremental*. For an incremental request, the BS increases its currently perceived value of the required resources by a certain amount; for an aggregate request, the BS replaces its perception by the communicated amount.

WiMAX foresees *polling* in order to give the MSs the opportunity to request resources and/or additional CIDs. The polling might be unicast, i.e., each MS gets its own slot to make a request (the assignment is communicated in the UL-MAP of the downlink subframe). If unicast polling would lead to too much overhead, a multicast or broadcast polling can be done, where several MSs are assigned the same resource for making their requests. In order to minimize collisions, only the MSs that actually have a request do transmit. Furthermore, collisions are resolved by a backoff algorithm (i.e., if the request does not reach the BS, the MS assumes that it was lost, and increases the possible backoff window from which the actual transmission time is chosen at random).

28.6.3 QoS

Guaranteeing a certain quality of service is essential for the many Internet services, including video streaming, VoIP, etc. WiMAX defines the following scheduling services:

- *Best effort*: this is the lowest level service that provides no guarantees for the service quality. Data are sent whenever resources are available (after all other services have been satisfied). An MS may use only the contention-based polling to request resources.
- *Real-time polling*: in this scheme, the BS provides unicast polling for the MS to request resources. The opportunities for requests have to be frequent enough that the latency requirements of the service of the MS can be satisfied. Typical applications for this service are streaming services that generate variable size packets, like Motion Picture Experts Group (MPEG) movies. In a modified version of this service (extended real-time polling), the polling opportunities can be used either for resource requests or for transmission of uplink data.
- *Non-real-time polling services*: are similar, but rely on contention-based polling instead of unicast polling (there might be unicast polling opportunities, but only at very large intervals; the main resource requests occur in a contention-based fashion).
- *Unsolicited grants*: in this service the BS assigns resources to an MS without the need for the MS to request the resources, thus eliminating all the overhead of the polling. This type of service makes sense if the MS creates fixed-size packets on a periodic basis, e.g., in VoIP.

28.6.4 Power Control

The WiMAX standard requires that the MS reports the maximum available power and the normalized transmitted power. The BS can use that information for assigning modulation and coding schemes, and also channel allocation. The values are quantized in 0.5-dB steps, ranging from -64 to $+63.5$ dBm for maximum available power, and -84 to 43.5 dBm for currently used power. Note that the standard does not specify the details of the power control algorithm to be used, only the reporting requirements.

In addition to the closed-loop power control, an open-loop power control is specified in the standard.

---

grant to the MS (i.e., encompassing the service for all the MSs). The uplink scheduler in the MS then assigns those resources to the different services according to their needs.

It is noteworthy that when an MS changes the number of used subcarriers, it keeps the *power-spectral density*, not the *total power*, constant.

28.6.5   Handover and Mobility Support

As in any mobile cellular system, handover between cells is a critical functionality for mobile WiMAX (note that for fixed WiMAX systems, it is less important, though situations can occur where the channel changes due to moving interacting objects, necessitating a handover). It requires to first identify a new cell that the MS should be handed over to (called *scanning* in WiMAX), and then the actual handover, which can be either a *hard handover* (break the connection to the old BS before sending user data to the new one) or a *soft handover* (retain simultaneous connection to the old and the new cell for some time).

Scanning and Association

Either the MS or the BS (possibly via a higher layer) can request a handover and initiate a scanning procedure. During the scanning interval, the MS listens to other, adjacent cells, and measures the received field strengths and SINRs. It can then optionally associate with (i.e., establish a preliminary connection with) one or more of the scanned cells. In *association level 0*, the MS initiates ranging in the same way as it would during network association, namely, by transmitting a ranging signal during the contention period. In *association level 1*, the currently active BS coordinates with the BSs in the newly associated cells, so that the MS gets a contention-free ranging slot. *Association level 2* also does this; in addition the ranging responses from the new BSs are fed back via the backbone to the old BS, which then aggregates them and sends them to the MS.

Hard Handover

After the scanning, either the MS or the BS might decide to move the connection to a different cell. Appropriate signaling messages are sent from either the MS or the BS. The MS then synchronizes to the new BS by listening to the frame preamble, and decoding the FCH, DL-MAP, and other control messages. The MS then performs ranging with the new BS (this step can be simplified by methods similar to the association stage). After the successful connection to the new BS has been established, the connection to the old BS is terminated. In a network optimized hard handover, discovery of the new BS and handover negotiation is done in parallel or before the MS handover procedure.

Soft Handover

The *soft handover*, also known as *Macro-Diversity HandOver* (MDHO) helps to improve the quality of a connection near the cell edge. Its principle is the same as the one discussed for CDMA in Chapter 18. During the soft handover, the MS keeps connection to several BSs, which are called the "diversity set." One of those BSs can have a privileged role as *anchor BS*. In one implementation, the anchor BS communicates the resource allocations of *all* the BSs of the diversity set to the MS; in another implementation the MS has to listen to the DL-MAP and UL-MAP of each BS separately.

For the uplink, each BS separately decodes the received signals, and selects the best one (again, this is similar to CDMA systems, see, e.g., Chapter 25). For the downlink, the BS can either receive the signals from the multiple BSs with multiple antennas, and do a combination in baseband (note that – in contrast to CDMA signals – no distinction of the signals by means of different spreading codes is possible). Alternatively, all BS signals can be completely identical and synchronized, and can thus be added up in the Radio Frequency (RF) domain by the MS.

28.6.6 *Power Saving*

Power saving is very important for WiMAX, even though it is only defined as optional in the standard. Most of the operations required for the operation, like FFTs with large number of points, are very energy consuming, and the short battery lifetime of WiMAX handsets is a major obstacle to commercial success. For this reason, *sleep mode* and *idle mode* are important.

In order to initiate a *sleep mode*, an active MS (i.e., an MS with at least one active CID) negotiates with the BS a time window in which the MS will not receive or send any data. The length of the window depends on the power saving class (there are a total of three such classes). During the window, the BS either buffers or discards all data intended for the sleeping MS. After the sleep window, a listen window follows, during which the MS resynchronizes itself.

The *idle mode* is intended to save battery power by avoiding needless signaling related to intercell handovers of inactive MSs. To achieve this, the BSs of the system are grouped into "paging groups." An MS in idle mode now only needs to notify the network when it moves from one paging group to the next (instead of having a handover every time it moves from one cell to the next);[8] the changing of the paging group is done by a "location update" procedure. This greatly decreases the battery drain of the MS, and at the same time saves spectral resources.

28.7 Glossary for WiMAX

AAA	Authentication, Authorization, and Accounting
AAS	Adaptive Antenna System
AMC	Adaptive Modulation and Coding
ASN	Access Services Network
ATM	Asynchronous Transfer Mode
BS	Base Station
CID	Connection Identifier
CS	Convergence Sublayer
CSN	Core Services Network
FA	Foreign Agent
FCH	Frame Control Header
FUSC	Full Use of Subcarriers
HA	Home Agent
HO	HandOver
IE	Information Element
MAP	Maximum A Posteriori
MDHO	Macro-Diversity HandOver
NAP	Network Access Provider
NSP	Network Service Provider
PUSC	Partial Use of Subcarriers
RS	Relay Station
MS	Mobile Station
SDU	Service Data Unit

---

[8] There is a tradeoff in the size of the paging groups: if the groups are too large, then a lot of system-wide resources have to be spent when there is incoming traffic for a (previously idle) MS, because the MS has to be paged in many cells. If the groups are too small, the overhead for signaling to change paging groups becomes significant.

SPID	SubPacket IDentity
TUSC	Tiled Use of Subcarriers
WiBro	Wireless Broadband
WiMAX	Worldwide Interoperability for Microwave Access

Further Reading

The authoritative source for WiMAX is the (consolidated) IEEE standard [IEEE 802.16-2009]; however, as mentioned previously, it is not easy to read. It is augmented by a book of some of the participants of the standard [Eklund et al. 2006]. In addition to the IEEE documents, also the WiMAX system profiles [WiMAX 1.5] are essential reading for actual implementers of the standard.

The most readable introduction to the standard is the textbook by Andrews et al. [2007], which nicely summarizes the standard and also provides a useful introduction to the underlying technology. Also the book by Nuaymi [2007] gives a good summary of the standard, while assuming more knowledge of the readers about fundamental technologies. More recent developments are reviewed in the June 2009 issue of the IEEE Communications Magazine. The multi-hop relay extension of WiMAX, called 802.16j, is described in Hart et al. [2009].

For updates and errata for this chapter, see wides.usc.edu/teaching/textbook

29

Wireless Local Area Networks

29.1 Introduction

29.1.1 History

In the late 1990s, wired fast Internet connections became widespread both in office buildings and in private residences. For companies, a fast intranet as well as fast connections to the Internet became a necessity. Furthermore, private consumers became frustrated with the long download times of dialup connections for elaborate webpages, music, etc., as connection speeds were limited to 56 kbit/s. They therefore opted for cable connections (several Mbit/s) or Digital Subscriber Lines (DSLs) (up to 1 Mbit/s in the U.S.A., and more than 20 Mbit/s in Japan) for their computer connections. At the same time, laptop computers started to be widely used in the workplace. This combination of factors spurred demand for wireless data connections – from the laptop to the nearest wired Ethernet port – that could match the speed of wired connections.

In the following years, two rival standards were developed. The ETSI (European Telecommunications Standards Institute) started to develop the HIPERLAN (*HIgh PERformance Local Area Network*) standard, while the IEEE (*Institute of Electrical and Electronics Engineers*) established the 802.11 group – the number 802 refers to all standards of the IEEE dealing with local and metropolitan area networks, and the suffix 11 was assigned for Wireless Local Area Networks (WLANs). In subsequent years, the 802.11 standard gained widespread acceptance, while HIPERLAN became essentially extinct.

Actually, it is not correct to speak of *the* 802.11 standard. 802.11 encompasses a number of different standards (see Table 29.1), which are not all interoperable. To understand the terminology, we first have to summarize the history of the standard. The "original" 802.11 standard was intended to provide data rates of 1 and 2 Mbit/s; since it operated in the 2.45-GHz ISM (*Industrial, Scientific, and Medical*) band, the frequency regulator in the U.S.A. – the Federal Communications Commission (FCC) – required that spectrum-spreading techniques be used. For this reason, the original 802.11 standard defined two modes: (i) frequency hopping, and (ii) direct-sequence spreading; these two modes were incompatible with each other.

It soon became obvious that higher data rates were demanded by users. Two subgroups were formed: 802.11a, which investigated Orthogonal Frequency Division Multiplexing (OFDM)-based schemes, and 802.11b, which attempted to retain the direct-sequence approach. 802.11b became popular first; it defined a standard that allowed an 11-Mbit/s data rate in a 20-MHz channel. Though the scheme was no longer really spread spectrum, the FCC approved its use. The standard was later adopted by an industry group called WiFi (Wireless Fidelity), which was formed to ensure true

Wireless Communications, Second Edition   Andreas F. Molisch
© 2011 John Wiley & Sons, Ltd

Table 29.1 The IEEE 802.11 standards and their main focus

Standards	Scope
802.11 (original)	Define a WLAN standard that includes both MAC and PHY functions
802.11a	Define a high-speed (up to 54 Mbps) PHY supplement in the 5-GHz band
802.11b	Define a high-speed (up to 11 Mbps) PHY extension in the 2.4-GHz band
802.11d	Operation in additional regulatory domains
802.11e	Enhance the original 802.11 MAC to support QoS (applies to 802.11a/b/g)
802.11f	Define a recommended practice for interaccess point protocol (applies to 802.11a/b/g)
802.11g	Define a higher rate (up to 54 Mbps) PHY extension in the 2.4-GHz band
802.11h	Define MAC functions that allow 802.11a products to meet European regulatory requirements
802.11i	Enhance 802.11 MAC to provide improvement in security (applies to 802.11a/b/g)
802.11j	Enhance the current 802.11 MAC and 802.11a PHY to operate in Japanese 4.9-GHz and 5-GHz bands
802.11n	Enhance the 802.11a and 802.11 PHY to operate at data rates up to 600 Mbit/s
802.11p	Modify the 802.11a standard for car-to-car communications
802.11s	Provide a protocol for autoconfiguring mesh networks
802.11w	Provide data integrity and authentication

interoperability between all WiFi-certified products.[1] WiFi gained widespread market acceptance after 2000. The data rate of 11 Mbit/s still was not sufficient for many applications – especially in light of the fact that actual throughput in practical situations is closer to 3–5 Mbit/s. For this reason, the work of the 802.11a group became of greater interest. 802.11a specified an alternative PHYsical layer (PHY) that uses OFDM and higher order modulation alphabets, allowing up to 54-Mbit/s data transfer rate (again, this rate is nominal, and true throughput is lower by about a factor of 2). This mode also works in a different frequency band (above 5 GHz), which is less "crowded" – i.e., has to deal with fewer interferers. The 802.11g group uses the same modulation format in the 2.45-GHz ISM band, and has by now become the most popular standard. Further modifications of the 802.11a standard are provided by the 802.11h and 802.11j standards, which adapt it to European and Japanese regulations, respectively. The 802.11n standard, which provides higher throughput by using Multiple Input Multiple Output system (MIMO) as well as possibly using larger bandwidth, was approved in 2009.

In addition, the original MAC (Medium Access Control) has been amended: the 802.11e standard provides modifications to the MAC that allow us to better ensure certain levels for *Quality of Service* (QoS). Additionally, a number of further subgroups of the 802.11 standardization group have been formed, all dealing with "amendments" and "additions" to the original standard. Realistically speaking, though, an 802.11a device, using the 802.11e MAC, bears no resemblance to the original 802.11 standard.

---

[1] Not all 802.11b products are completely interoperable.

Due to the multitude of 802.11 standards, we only present the most important. In the following, we give an overview of 802.11a as well as 802.11n PHYs, and the 802.11 MAC layer. More details can be found, as always, in the official standards publications [*www.802wirelessworld.com*] and the multitude of books that have been published on that topic. An excellent summary of the earlier versions of the standard can be found in O'Hara and Petrick [2005], and of the 802.11n standard in Perahia and Stacey [2008].

29.1.2 Applications

The main markets for WLANs are as follows:

- Wireless networks in office buildings and private homes, to allow unhindered Internet access from anywhere within a building. Access points (equivalent to base stations in cellular systems) for WLANs need to follow the specifications, but also allow a certain amount of vendor leeway. For example, multiple antennas can be used at the access point, without leading to incompatibilities. As far as "clients" (equivalent to mobile stations in cellular systems) are concerned, WLAN cards have turned into a mass market with very little distinction between products from different vendors, and are often built into laptops. As a consequence, research tends to focus on methods for production cost reduction (implementation with smaller chip area, low-cost semiconductor technology), while research on access points includes a broader field of topics.
- "Hotspots" – i.e., wireless access points that allow the public to connect to the Internet. These hotspots are often set up in coffee shops, hotels, airports, etc. Several providers also have a "nationwide" or even "continentwide" network of hotspots, so that subscribers can log in at many different locations.[2] However, it must be stressed that coverage from these networks is *much* lower than for cellular networks. For this reason, research is ongoing on how to seamlessly integrate WLANs with cellular networks or large-area networks.

29.1.3 Relationship between the Medium Access Control Layer and the PHY

Before going into details of the MAC and the PHY, we first have to establish some notation used by the 802.11 community. The data payload received from upper layers is attached to headers and trailers at both the MAC and PHY before it gets transmitted on the air. For example, each *MAC Service Data Unit* (MSDU) received from the *Logic Link Control* (LLC) layer is appended with a 30-byte-long MAC header and a 4-byte-long *Frame Check Sequence* (FCS) trailer to form the *MAC Protocol Data Unit* (MPDU). The same MPDU, once handed over to the PHY, is then called the *Physical Layer Service Data Unit* (PSDU). And, then a *Physical Layer Convergence Procedure* (PLCP) preamble and header, and proper tail bits and pad bits are attached to the PSDU to finally generate the *Physical Layer Protocol Data Unit* (PPDU) for transmission. The relationships among MSDU, MPDU, PSDU, and PPDU are illustrated in Figure 29.1.

From just this brief paragraph, the reader will have seen that – as with most standards – the alphabet soup of the numerous acronyms is a major hurdle to understanding this standard. For this reason, there is a list of acronyms and their meaning in the frontmatter of this book (see p. xxxi) and a separate list of abbreviations for some chapters in the appendices.

---

[2] In most cases, users are charged a per-minute fee or a flat rate for 24 hours of usage. Nationwide networks often have an option for monthly or annual subscriptions.

Figure 29.1 Relations among the MAC service data unit, MAC protocol data unit, physical layer service data unit, and physical layer protocol data unit.

Reproduced with permission from IEEE 802.11 © IEEE.

29.2   802.11a/g – Orthogonal Frequency Division Multiplexing-Based Local Area Networks

In an attempt to attain higher data rates, the 802.11 Working Group (WG) published their 802.11a standard defining a PHY for high-speed data communications based on OFDM in 1999. The standard is defined for the 5-GHz band, where more bandwidth is available, and less interference is present. However, the achievable range is not as good as in the 2.45-GHz band. Therefore, the same PHY, but working in the 2.45-GHz band, was introduced as 802.11g standard. It is currently the dominant version of WLAN standards. Its main properties are the following (see also Table 29.2):

- use of the 5.15–5.825 GHz band (in the U.S.A.) for 11a and 2.4-2.27 GHz for the 11g standard;
- 20-MHz channel spacing;
- data rates include 6, 9, 12, 18, 24, 36, 48, and 54 Mbps, where support of 6, 12, and 24 Mbps is mandatory;
- OFDM with 64 subcarriers, out of which 52 are user modulated with Binary or Quadrature-Phase Shift Keying (BPSK/QPSK), 16-Quadrature Amplitude Modulation (16-QAM), or 64-QAM;
- forward error correction, using convolutional coding with coding rates of 1/2, 2/3, or 3/4 as Forward Error Correction (FEC) coding.

Table 29.2   Important parameters of the 802.11a PHY layer

Information data rate	6, 9, 12, 18, 24, 36, 48, 54 Mbit/s
Modulation	BPSK, OPSK, 16-QAM, 64-QAM
FEC	$K = 7$ convolutional code
Coding rate	1/2, 2/3, 3/4
Number of subcarriers	52
OFDM symbol duration	4 μs
Guard interval	0.8 μs
Occupied bandwidth	16.6 MHz

29.2.1   Frequency Bands

In the U.S.A., the frequency bands 5.15–5.25, 5.25–5.35, and 5.725–5.825 GHz, called the *Unlicensed National Information Infrastructure* (U-NII) bands, are used for 802.11a. These channels are numbered, starting every 5 MHz, according to the formula:

$$\text{Channel center frequency} = 5,000 + 5 \times n_{\text{ch}}(MHz) \tag{29.1}$$

where $n_{\text{ch}} = 0, 1, \ldots, 200$. The transmit powers in the 5.15–5.25, 5.25–5.35, and 5.725–5.825–GHz bands are limited to 40, 200, and 800 mW, respectively.

Obviously, each 20-MHz channel used by 802.11a occupies four channels in the U-NII band. Recommended channel usage in the U.S.A. is given in Table 29.3. In Japan, the assigned carrier frequencies are slightly lower.

The band plan for 802.11a in the U.S.A. is also given in Figure 29.2.

Table 29.3   Frequency assignment for 802.11a in the U.S.A

Bands (GHz)	Allowed power	Channel numbers (n_{ch})	Channel center frequency (MHz)
U-NII lower band (5.15–5.25)	40 mW (2.5 mW/MHz)	36	5,180
		40	5,200
		44	5,220
		48	5,240
U-NII middle band (5.25–5.35)	200 mW (12.5 mW/MHz)	52	5,260
		56	5,280
		60	5,300
		64	5,320
U-NII upper band (5.725–5.825)	800 mW (50 mW/MHz)	149	5,745
		153	5,765
		157	5,785
		161	5,805

29.2.2   Modulation and Coding

802.11a uses OFDM as its modulation format, enabling high data rates. The principles of OFDM were described in Chapter 19, so here we simply analyze details specific to 802.11a. A typical block diagram of a transceiver is shown in Figure 29.3.

In 802.11a, OFDM with 64 subcarriers is specified. Powers of 2 are habitually used as numbers for OFDM subcarriers, as they allow the most efficient implementation via Fast Fourier Transforms (FFTs). However, only 52 of the 64 subcarriers are actually used (modulated and transmitted), while the other 12 subcarriers are null-carriers that do not carry any useful information; useful carriers are indexed from -26 to 26, without a Direct Current (DC) component. Of these 52 subcarriers, 4 are used as pilots – namely, subcarriers number $-21, -7, 7, 21$. The pilot should be BPSK-modulated by a pseudorandom sequence to prevent generation of spectral lines.

The other 48 subcarriers carry the PSDU data. BPSK, QPSK, 16-QAM, or 64-QAM are all admissible modulation alphabets, depending on the channel state. Note, however, that the standard does *not* foresee truly adaptive modulation in the sense that the modulation alphabet can differ from subcarrier to subcarrier. Rather, the system uses an average "transmission quality" criterion

Figure 29.2 802.11a channel plan.

Reproduced with permission from IEEE 802.11 © IEEE.

Figure 29.3 Block diagram of a 802.11a transceiver.

Reproduced with permission from IEEE 802.11 © IEEE. HPA, High Power Amplifier; LNA, Low Noise Amplifier.

to adapt the data rate to the current channel state. Rate adaptation is achieved by modifying either the modulation alphabet or the rate of the error correction code (see below), or both.

The duration of an OFDM symbol is $4\,\mu s$, including a cyclic prefix of $0.8\,\mu s$. This is sufficient to accommodate the maximum excess delay of most indoor propagation channels, including factory halls and other challenging environments.

For FEC, 802.11a uses a convolutional encoder with coding rates 1/2, 2/3, or 3/4, depending on the desired data rate. The generator vectors are $G1 = 133$ and $G2 = 171$ (in octal notation), for the rate-1/2 coder shown in Figure 29.4. Higher rates are derived from this "mother code" by puncturing.

All encoded data bits are interleaved by a block interleaver with a blocksize equal to the number of bits in a single OFDM symbol. The interleaver works in two steps (permutations). The first permutation ensures that adjacent coded bits are mapped onto nonadjacent subcarriers.

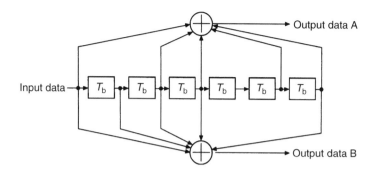

Figure 29.4 Convolutional encoder (K = 7).
Reproduced with permission from IEEE 802.11 © IEEE.

The second ensures that adjacent coded bits are mapped alternately onto both less and more significant bits of the constellation and, thereby, long runs of low-reliability bits are avoided.

Table 29.4 summarizes the rates that can be achieved with different combinations of alphabets and coding rates, as well as the OFDM modulation parameters.

Table 29.4 Data rates in 802.11a

Data rate (Mbit/s)	Modulation	Coding rate	Coded bits per subcarrier	Coded bits per OFDM symbol	Data bits per OFDM symbol
6	BPSK	1/2	1	48	24
9	BPSK	3/4	1	48	36
12	QPSK	1/2	2	96	48
18	QPSK	3/4	2	96	72
24	16-QAM	1/2	4	192	96
36	16-QAM	3/4	4	192	144
48	64-QAM	2/3	6	288	192
54	64-QAM	3/4	6	288	216

29.2.3 Headers

For transmission, a preamble and a PLCP header are prepended to the encoded PSDU data[3] that are received from the MAC layer, creating a PPDU. At the receiver (RX), the PLCP preamble and header are used to aid demodulation and data delivery. A PPDU frame format is shown in Figure 29.5.

The PLCP header is transmitted in the SIGNAL field of the PPDU. It incorporates the RATE field, a LENGTH field, a TAIL field, and so on:

- *RATE (4 bits)*: indicates transmission data rate.
- *LENGTH (12 bits)*: indicates the number of octets in the PSDU.
- *Parity (1 bit)*: parity check.
- *Reserved (1 bit)*: future use.

---

[3] Plus pilot tones.

Figure 29.5 PHY-protocol-data-unit frame format.
Reproduced with permission from IEEE 802.11 © IEEE.

- *TAIL (6 bits)*: convolutional-coding tail.
- *SERVICE (16 bits)*: initialization of the scrambler.

29.2.4 Synchronization and Channel Estimation

Synchronization is achieved by means of the PLCP preamble field. It consists of 10 short symbols and 2 long symbols (Figure 29.6).

Figure 29.6 Orthogonal frequency division multiplexing training structure.
Reproduced with permission from IEEE 802.11 © IEEE.

The training sequence starts out with 10 short symbols of duration $0.8\,\mu s$ that allow the RX to detect the signal, adjust the Automatic Gain Control (AGC), and perform a coarse-frequency offset estimation. These short symbols consist of just 12 subcarriers, which are modulated by elements of the following sequence:

$$S_{-26,26} = \sqrt{(13/6)}\{0, 0, 1 + j, 0, 0, 0, -1 - j, 0, 0, 0, 1 + j, 0, 0, 0, -1 - j, 0, 0, 0,$$
$$-1 - j, 0, 0, 0, 1 + j, 0, 0, 0, 0, 0, 0, 0, -1 - j, 0, 0, 0, -1 - j, 0, 0, 0, 1$$
$$+j, 0, 0, 0, 1 + j, 0, 0, 0, 1 + j, 0, 0, 0, 1 + j, 0, 0\} \tag{29.2}$$

Multiplication by a factor of $\sqrt{(13/6)}$ is done so as to normalize the average power of the resulting OFDM symbol, which utilizes 12 of the 52 subcarriers.

These symbols are followed by 2 long training symbols that serve for both channel estimation and finer frequency offset estimation, preceded by a Guard Interval (GI). A long OFDM training symbol consists of 53 subcarriers (including a 0 value at DC), which are modulated by elements of sequence L, given by

$$L_{-26,26} = \{1, 1, -1, -1, 1, 1, -1, 1, -1, 1, 1, 1, 1, 1, 1, -1, -1, 1, 1, -1, 1, -1, 1, 1, 1, 1, 0, 1,$$
$$-1, -1, 1, 1, -1, 1, -1, 1, -1, -1, -1, -1, -1, 1, 1, -1, -1, 1, -1, 1, -1, 1, 1, 1, 1\} \tag{29.3}$$

The PLCP preamble is followed by the SIGNAL and DATA fields. The total training length is $16\,\mu s$. The dashed boundaries in the figure denote repetitions due to periodicity of the inverse Fourier transform.

Table 29.5 summarizes the most important parameters for 802.11a mode.

Table 29.5   Parameters of 802.11a

Parameter	Value
Number of data subcarriers	48
Number of pilot subcarriers	4
Subcarrier spacing	0.3125 MHz
IFFT/FFT period	$3.2\,\mu s$
Preamble duration	$16\,\mu s$
Duration of OFDM symbol	$4.0\,\mu s$
Guard interval for signal symbol	$0.8\,\mu s$
Guard interval for training symbol	$1.6\,\mu s$
Short training sequence duration	$8\,\mu s$
Long training sequence duration	$8\,\mu s$

29.3   IEEE 802.11n

29.3.1   Overview

The 802.11n standard offers up to (nominal) 600 Mbit/s bit rate. These high data rates, as well as improved reliability, are necessitated by a number of new applications: (i) wireless computer networks require higher data transfer rates between various computers at home, and (due to the emergence of fiber-to-the-home) transfer rates from the computer to the wired Internet port at users' homes, (ii) Audio and Video (AV) applications, e.g., transfer of videos from laptops, hard-disk video recorders, and DVD players to TVs, and (iii) Voice over Internet Protocol (VoIP) applications, which require lower data rates, but high reliability.

The 802.11n group was established in 2002. In September 2004, a number of different technical proposals were presented, which were subsequently consolidated into two proposals supported by a major industry alliances each: TGnSync, and WWise. After more than a year of negotiations and fights, a compromise draft proposal was approved by 802.11n in January 2006. Since then, the draft was in the process of revisions and corrections, and a final version of the 11n standard was approved in 2009. But even before then, several companies already sold "pre-n" products that followed the 11n draft standard, and remain compatible with the final standard..

802.11n achieves high data rates mainly by two methods: use of multiple-antenna techniques (see Chapter 20), and increase of the available bandwidth from 20 to 40 MHz. The generic structure of an 11n transceiver is shown in Figure 29.7. The source data stream is first scrambled and then (only for data rates >300 Mbit/s) divided into two parallel data streams, to reduce the processing speed requirements of the encoder/decoder. A number of different codecs are available: binary convolutional codes are the default solution, while Low Density Parity Check

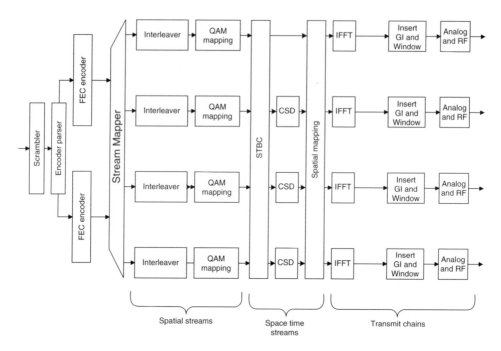

Figure 29.7 Block diagram of an IEEE 802.11n transmitter.
Reproduced with permission from [IEEE P802.11n] © IEEE.

(LDPC) codes are an option for high-performance transmission (for more details see Section 14.7). The thus-encoded bits are divided into a number of spatial streams (compare Section 20.2), which are to be transmitted in parallel from the antennas. Each of the spatial streams is then interleaved, mapped onto complex modulation symbols, and grouped into OFDM symbols. Next, the spatial streams can be modulated by Alamouti codes and/or cyclic shift diversity (for details see Section 29.4.3). Next, the "spatial mapping" distributes the spatial streams onto the modulation/upconversion chains. In each of the chains, the symbols are submitted to an Inverse Fast Fourier Transformation (IFFT), a guard interval is inserted, and the signal is upconverted to the passband; this part of the processing is identical to 802.11a.

29.3.2 Modulation and Coding

For a single spatial stream, the modulation and coding schemes are very similar to those of 802.11a. The modulation formats are BPSK, QPSK, 16-QAM, and 64-QAM. When convolutional coding is used, code rates 1/2, 2/3, and 3/4 are the same as for 802.11a; an additional code rate of 5/6 (for higher throughput in very good channel conditions) was introduced as well. This results in a total of 8 Modulation and Coding Scheme (MCS) schemes with data rates from 6.5 to 65 Mbit/s. When multiple antennas are present, the transmitter (TX) can either use the same MCS for all spatial streams (this makes sense if the TX has no channel state information), or it can have different MCSs for different streams. A total of 32 MCSs are defined for the case of equal modulation (the 8 "fundamental" MCSs equally used on 1, 2, 3, or 4 spatial streams). For unequal modulation, a further 44 MCSs (different combinations of the existing MCSs on the various spatial streams) are defined, though they are not mandatory.

802.11n also introduces the concept of a short GI: the system can adaptively decide whether the length of the cyclic prefix is the "normal" 800 ns or shortened to 400 ns; the latter is used to increase the spectral efficiency in environments where the delay spread is so small that a short cyclic prefix is sufficient.

The 802.11n standard also introduces LDPC encoding (compare Section 14.7), which achieves extremely low error probability at the price of higher decoding complexity. The parity-check matrices can be partitioned into square subblocks (submatrices), which are either cyclic permutations of the identity matrix, or all-zero matrices. Twelve different codes are defined, which are all based on the same codestructure. The codeword sizes and submatrix sizes are 648 (27), 1296 (54), and 1944 (81).

29.3.3 Multiple-Antenna Techniques

The key to the 802.11n standard is the use of multiple-antenna techniques. The standard foresees a number of different techniques, in particular (i) spatial multiplexing, (ii) space–time block coding, (iii) eigenbeamforming, and (iv) antenna selection. The basics of most of those techniques are outlined in Section 20.2; here, we only deal with the specific implementation in 802.11n.

Space–Time Coding

Space–time block codes can be used in an 802.11n system to increase the robustness of the system. In particular, Alamouti codes are used, and can be combined with spatial multiplexing. If there are two transmit Radio Frequency (RF) chains, then one spatial stream is mapped onto those chains (and from there to the antennas) by means of the standard Alamouti code. If there are three transmit antennas and two spatial streams, then one stream is mapped to two RF chains by means of Alamouti encoding, and one stream is mapped directly to the remaining chain. For four transmit chains, either three streams (where one of them is Alamouti encoded), or two streams (with each of them Alamouti encoded) can be used. Different modulation schemes can be used on the different streams; this is motivated by the fact that Alamouti-encoded streams are more robust and can sustain a higher modulation scheme than nonencoded schemes.

Another way of achieving transmit diversity is the use of "Cyclic Shift Diversity" (CSD). This method, which is somewhat similar to delay diversity as described in Chapter 13, introduces a different delay for each signal. In contrast to conventional delay diversity, where signals are linearly delayed, in CSD the OFDM symbols are *cyclically* shifted. In other words, this means that the signal on the k-th subcarrier is shifted by $\exp[-j2\pi k\Delta_F\tau_i]$, where k indexes the subcarrier frequency, Δ_F is the spacing of the subcarriers, and τ_i is the cyclic shift applied to the i−th signal. The cyclic shifts are 0, −400, −200, and −600 ns for the first, second, third, and fourth spatial stream, respectively.

Spatial Multiplexing and Beamforming

The dividing of the original data stream results in a total number N_{SS} spatial streams, which can be smaller than, or equal to, the number of available RF chains N_{RF} for the upconversion. In any case, linear combinations of the spatial streams are assigned to the RF chains; these combinations are described by means of the so-called "spatial mapping" matrix $\mathbf{Q}$, so that for each time instant the vector of spatial-stream vectors $\mathbf{x}$ is mapped onto the signals for the RF chains $\mathbf{y}$ as $\mathbf{y} = \mathbf{Qx}$. The following possibilities are defined in the standard:

- *Direct mapping*: this method is used if $N_{SS} = N_{RF}$. In the simplest case, $\mathbf{Q}$ is either an identity matrix, or it is a diagonal matrix in which the elements perform CSD, so that $Q_{i,i} =$

$\exp[-j2\pi k\Delta_F\tau_i]$. The CSD serves to avoid inadvertent beamforming when similar signals occur in the different spatial streams.

- *Spatial mapping for the case* $N_{SS} = N_{RF}$: in this case, $\mathbf{Q}$ is the product of a CSD matrix with a square matrix with orthogonal columns, such as a Fourier matrix or a Hadamard matrix.
- *Spatial matrix for the case* $N_{SS} < N_{RF}$: in this case, some of the spatial streams are duplicated (so that the total number of streams becomes equal to N_{RF}). All the streams are then power adjusted (so that the total power stays constant) and mapped onto the transmit RF chains by means of a CSD matrix.
- *Beamforming steering matrix*: any matrix that improves the overall Bit Error Rate (BER) can be used as matrix $\mathbf{Q}$. Realistically speaking, the matrix is based on channel state information at the TX (see Section 29.3.6). In particular, if the TX knows the instantaneous channel transfer matrix $\mathbf{H}$, it can perform an eigenvalue decomposition of $\mathbf{H}$ on each subcarrier, precode with the right singular matrix (see Section 20.2.5), and possibly weight the streams according to the waterfilling rules.

Antenna Selection

There are situations where the number of available antenna elements is larger than the number of RF chains – either because of cost reasons, or because the maximum number of RF chains foreseen in the 802.11n standard is 4. In those cases, antenna selection allows improvement of the system performance.

The available RF chains are connected to the "instantaneously best" antenna elements via electronic switches. However, in order to determine the "best" antenna elements, the complete channel (from each transmit to each RX antenna element) has to be sounded. This is achieved in two or more subsequent packets. For example, for transmit antenna selection, the first packet is transmitted from the first subset of antennas, the next packet from a second subset of antennas, and so on. After all subsets have been tried out, the RX sends an information to the TX about which subset to use. Receive antenna selection works analogously.

29.3.4   20-MHz and 40-MHz Channels

802.11n allows the use of either 20 MHz or 40 MHz bandwidth, see Figure 29.8. In the former case, 802.11n uses more subcarriers (56) than 802.11a (52). For 40-MHz bandwidth, 114 subcarriers are used. Subcarriers are added in the middle and in the place of the DC subcarrier. The phases of signals in the upper channel are rotated by $+90^\circ$ in reference to the lower channel.

Figure 29.8   20-MHz and 40-MHz subcarriers.

29.3.5 Headers and Preambles

Another important point is the backward compatibility to 802.11a/g. Networks with either 11a/g or 11n access point, and a mixture of 11a/g and 11n client have to work. Many of the details of the 11n standard, in particular the design of preamble, can only be understood in the light of the requirement for backward compatibility.

There are three types of PLCP preambles (i.e., the part of the preamble that is used for synchronization and channel estimation, compare Section 29.2.4), see Figure 29.9. The first type is the legacy preamble, which is identical to the 802.11a preamble; this is to be used if only legacy (802.11a) devices are used at a given time.

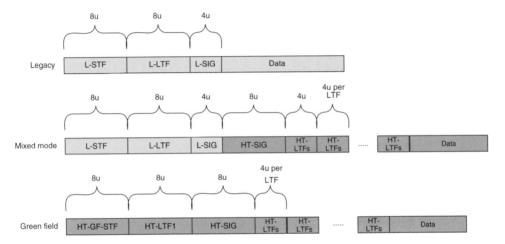

Figure 29.9 Types of preambles in IEEE 802.11n. *In this figure:* HT-GF-STF, High Throughput – GreenField – Short Training Field; L-SIG, Legacy SIGnalling field; L-STF, Legacy Short Training Field.

If both 802.11a and 802.11n devices are present in a Local Area Network (LAN), then the mixed-mode preamble has to be used. It starts with the same fields as the legacy preamble, which are then followed by the High Throughput-SIGnal field (HT-SIG), see Figure 29.10. This contains information about a number of MIMO parameters, bandwidth allocations, etc., that are unique to 11n devices. In particular, it contains information about:

- modulation and coding scheme;
- bandwidth indication (20 or 40 MHz);
- *smoothing*: indicates whether frequency-domain smoothing is recommended as part of channel estimation;
- not-sounding (i.e., whether the current PPDU is a sounding PPDU, see Section 29.3.6);
- aggregation (whether the data portion of the packet participates is part of a data aggregation transmission);
- Space Time Block Code (STBC) (indication of space–time coding);
- FEC encoding (convolutional or LDPC);
- short GI: whether long or short cyclic prefix are used;
- number of extension spatial streams;
- Cyclic Redundancy Check (CRC): error detection for HT-SIG;
- tail bits for terminating convolutional code.

Figure 29.10 High-throughput signaling field.

The HT-SIG is encoded with a rate 1/2 convolutional code, and BPSK modulated, to ensure high robustness in the transmission.

If there are only 11n devices in a LAN, then the greenfield preamble can be used, which omits all the "legacy" fields, and thus provides a shorter and more efficient preamble.

29.3.6 Channel Estimation

The MIMO channel between TX and RX is estimated from the High Throughput – Long Training Field (HT-LTFs). If the TX provides training for the exactly available spatial streams, then the preamble uses exactly N_{SS} training symbols (except for the case of three spatial streams, which requires four training symbols). If the TX is providing more training fields than is required for the current number of spatial streams, more spatial dimensions can be estimated, which enables, e.g., beamforming for eigenvalue decomposition. In this case, the PPDU is called a *sounding PPDU*. Thus, the HT long training field portion has one or two parts. The first part consists of N_{SS} Long Training Fields (LTFs) (known as Data-HT-LTFs) that are necessary for demodulation of the HT-Data portion of the PPDU. The optional second part (the one occurring only in sounding PPDUs) consists of the HT-LTFs that may be used to sound extra spatial dimensions. Figure 29.11 shows an example.

Another way of sounding all spatial dimensions is to use a PPDU that contains no payload data. The preamble of such a *Null Data Packet* then contains only data-HT-LTFs, but the (nominal) N_{SS} is chosen such that all required spatial dimensions can be sounded.

The HT-LTFs provide the Channel State Information (CSI) that is required for the reception of the signals, but can also be used at the TX to enable an appropriate precoding (i.e., creation of a suitable matrix **Q** for each subcarrier). The CSI at the TX can either be obtained through explicit feedback, or from the principle of reciprocity.

Figure 29.11 Example for sounding PPDU with extension HT-LTFs.

- In the case of explicit feedback, the RX determines either the effective channel matrix $\mathbf{H}_{\text{eff}}$ (which is the product of the matrix $\mathbf{Q}$ with the channel matrix $\mathbf{H}$) or the beamforming matrix. The real and imaginary parts of the channel matrix coefficients are quantized to 4, 5, 6, or 8 bits. This can result in rather large channel matrices that need to be fed back; therefore, a compressed feedback is foreseen as well.
- In implicit feedback, the TX employs the channel reciprocity to obtain knowledge about the channel. Since all transmissions are at the same frequency, and the channels are changing only slowly, the channel matrix is approximately the same for uplink and the downlink. However, this is true only for the actual propagation channel (from TX antenna connector to RX antenna connector), while the upconversion/downconversion RF chains are *not* necessarily reciprocal. The 802.11n standard thus foresees a procedure for calculating a set of calibration matrices that can be applied at the transmit side of a STAtion (STA i.e., an access point or client), to correct the amplitude and phase differences between the transmit and receive chains in the STA. The procedure, which has to be performed only at very large intervals, involves essentially determining channel coefficients implicitly as well as explicitly; the difference between the results can then be used to establish the calibration matrices.

29.4 Packet Transmission in 802.11 Wireless Local Area Networks

There are nine MAC services specified by IEEE 802.11. These include distribution, integration, association, reassociation, disassociation, authentication, deauthentication, privacy, and MSDU delivery. Six of the services are used to support MSDU delivery between STAs (used as a generic expression for 802.11 devices, both access points and clients). Three of the services are used to control 802.11 WLAN access and confidentiality. Each of the services is supported by one or more MAC frame types. The IEEE 802.11 MAC uses three types of messages: *data, management*, and *control*. Some of the services are supported by MAC management messages and some by MAC data messages. All messages gain access to the WM (Wireless Medium) via the IEEE 802.11 MAC medium access method which includes both contention-based and contention-free channel access methods: *Distributed Coordination Function* (DCF) and *Point Coordination Function* (PCF). In the following, the 802.11 MAC functions and services will be described.

29.4.1 General Medium Access Control Structure

The 802.11 MAC uses a temporal superframe structure with *Contention Period* (CP)[4] and *Contention Free Period* (CFP) alternately as shown in Figure 29.12. Superframes are separated by

---

[4] Note that it is only in this section that we use the abbreviation CP for contention period (and not for cyclic prefix). Since we are talking about the MAC layer only, no confusion can arise.

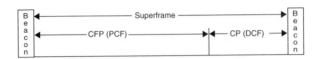

Figure 29.12   Superframe structure of 802.11.
Reproduced with permission from IEEE 802.11 © IEEE.

periodic management frames, the so-called "beacon frames." During the CP, DCF is used for channel access, while PCF is used for channel access during the CFP.

The 802.11 MAC uses different interframe gaps, denoted as *Inter Frame Spaces* (IFSs), in order to control medium access – i.e., to give STAs in specific cases a higher or lower priority. These IFSs are (in the order shortest to longest):

- *Short Inter Frame Space* (SIFS);
- *Priority Inter Frame Space* (PIFS);
- *Distributed Inter Frame Space* (DIFS);
- *Extended Inter Frame Space* (EIFS).

Their actual values depend on PHY parameters.

29.4.2   Frame Formats

The MAC frame format comprises a set of fields that contain various types of control information as well as the actual frame body, all of which occur in a fixed order in all frames. Figure 29.13 depicts the general MAC frame format. The fields Address 2, Address 3, Sequence Control, Address 4, and Frame Body are only present in certain frame types.

Figure 29.13   Medium-access-control frame format (a typical MPDU).
Reproduced with permission from O'Hara and Petrick [2005] © IEEE.

When the MSDUs handed down to the MAC become too large, it becomes difficult to transmit them in one block: obviously, the probability of a block error – i.e., that one of the bits in the block is in error – increases with duration of the block.[5] As each block error might lead to the necessity of retransmission, this is highly undesirable. Thus, MSDUs have to be fragmented in order to increase transmission reliability. This fragmentation is done when an MSDU size exceeds the fragmentation threshold. In this case, the MSDU will be broken into multiple fragments with an MPDU size equal to a fragmentation threshold, and a special field (the "more_fragments" field) is set to 1 in all but the last fragment. The receiving STA acknowledges each fragment individually. The channel is not released until the complete MSDU has been transmitted successfully or until

---

[5] Note that, in general, this effect may be offset by the fact that larger blocks allow the use of better codes, like highly efficient LDPC codes. However, for convolutional codes this is not relevant.

a nonacknowledgment has been received for a fragment. In the latter case, the source STA will recontend for the channel following the normal rules and retransmit the nonacknowledged fragment, as well as all the subsequent ones.

29.4.3 Packet Radio Multiple Access

Carrier Sense Multiple Access

The DCF employs *Carrier Sense Multiple Access with Collision Avoidance* (CSMA/CA), as described in Chapter 17, plus a random backoff mechanism. Support for DCF is mandatory for all STAs. In DCF mode, each STA checks whether the channel is idle before attempting to transmit. If the channel has been sensed idle for a DIFS period, transmission can begin immediately. If the channel is determined to be busy, the STA will defer until the end of the current transmission. After the end of the current transmission, the STA will select a random number called a "backoff timer," in the range between 0 and a *Contention Window*(CW). This is the time the WM has to be free before the STA might try to transmit again. The size of the CW increases (up to a limit) every time a transmission has to be deferred. If transmission is not successful, the STA thinks that a collision has occurred. Also in this case, the CW is doubled, and a new backoff procedure starts again. The process will continue until transmission is not successful (or discarded). The basic access method and backoff procedure are shown in Figures 29.14 and 29.15, respectively.

Figure 29.14   Basic access method.
Reproduced with permission from IEEE 802.11 © IEEE.

Physical and virtual carrier-sense functions are used to determine the state of the channel. When either function indicates a busy channel, the channel should be considered busy; otherwise, it should be considered idle. A physical carrier-sense mechanism is provided by the PHY. A virtual carrier-sense mechanism is provided by the MAC. This mechanism is referred to as the *Network Allocation Vector* (NAV). The NAV maintains a prediction of future traffic on the medium based on duration information that is announced in the DURATION field in the transmitted frames.

Polling

PCF is an optional medium access mode for 802.11. It provides contention-free frame transfer, based on polling (see Chapter 17). The *Point Coordinator* (PC) resides in the BS (access point). All STAs inherently obey the medium access rules of the PCF and set their NAV at the beginning of each CFP. The PCF relies on the PC to perform polling, and enables polled STAs to transmit without contending for the channel. When polled by the PC, an STA transmits only one MPDU, which can be to any destination. If the transmitted dataframe is not in turn acknowledged, the STA

Figure 29.15 Timing backoff procedure.

Reproduced with permission from IEEE 802.11 © IEEE.

does not retransmit the frame unless it is polled again by the PC, or it decides to retransmit during the CP. An example of PCF frame transfer is given in Figure 29.16. At the beginning of each CFP, the PC senses and makes sure the channel is idle for one PIFS before sending the beacon frame. All STAs adjust their NAVs according to the broadcast CFP duration value in the beacon. After one SIFS time of the beacon, the PC may send out a *Contention-free Poll* (CF-Poll), data, or data plus a CF-Poll. Each polled STA can get a chance to transmit to another STA or respond to the PC after one SIFS with an acknowledgment (plus possibly data).

Figure 29.16 Point-coordination-function frame transfer.

Reproduced with permission from IEEE 802.11 © IEEE.

As discussed above, the DCF and PCF coexist in a manner that permits both to operate concurrently.[6] The two access methods alternate, with a CFP followed by a CP. Since the PCF is built on top of the DCF, there are no conflicts between DCF and PCF when they coexist in the system. All STAs will inherently obey the medium access rules of the PCF. STAs will stay silent during CFP unless they are polled.

---

[6] Within the same Basic Service Set (BSS).

29.5 Alternative Wireless Local Area Networks and Future Developments

As we mentioned in the introduction, there is a multitude of "802.11-named" standards, most of which have not gained widespread popularity. Among these standards, the original 802.11 standard with its 1-Mbit/s direct-sequence spreading mode is a typical example. Furthermore, the frequency-hopping mode of this standard never gained popularity. Finally, the standard also defined a mode for infrared communications between computers; this application never gained significant popularity as well. While the 802.11b standard was enormously popular in the mid-2000s, it has in the meantime lost much ground to the 802.11g standard.

In all the discussions above, we have concentrated on WLANs that have one access point, plus a number of clients that connect to this access point. The 802.11 group of standards also establishes modes for peer-to-peer communications. This approach has not gathered widespread popularity either.

We have also mentioned the HIPERLAN standards developed by ETSI. The HIPERLAN II standard, in particular, bears considerable similarity to the 802.11a PHY, though the MAC is based on Time Division Multiple Access (TDMA) instead of CSMA. While a number of research papers have been published on this standard, it has not gained practical relevance, and even its previous proponents have switched to using 802.11a.

802.11 has also started activities on standards that can provide even higher throughput than the 11n standard. One of these envisioned high-throughput standards works at carrier frequencies around 60 GHz, where a very large bandwidth (approximately 7 GHz) is available. Due to the high carrier frequency, attenuation is strong, and the scheme works best in line-of-sight situations, or at least for TX and RX being in the same room. Another scheme, which works in the usual microwave regime, exploits multiple-antenna techniques that are more advanced than those in 11n and/or have more antenna elements, to achieve a higher throughput.

29.6 Glossary for WLAN

AC	Access Category
AIFS	Arbitration Inter Frame Spacing
AP	Access Point
CAP	Controlled Access Period
CCK	Complementary Code Keying
CFB	Contention Free Burst
CFP	Contention Free Period
CF-Poll	Contention-free Poll
CP	Contention Period
CSMA/CA	Carrier Sense Multiple Access with Collision Avoidance
CW	Contention Window
DCF	Distributed Coordination Function
DIFS	Distributed Inter Frame Space
DLP	Direct Link Protocol
EDCA	Enhanced Distributed Channel Access
EIFS	Extended Inter Frame Space
FCS	Frame Check Sequence
HCCA HCF	(Hybrid Coordination Function) Controlled Channel Access
HC	Hybrid Coordinator
HCF	Hybrid Coordination Function

HIPERLAN	High PERformance Local Area Network
IEEE	Institute of Electrical and Electronic Engineers
IFS	Inter Frame Space
ISM	Industrial, Scientific, and Medical
MBOA	Multi Band OFDM Alliance
MPDU	MAC Protocol Data Unit
MSDU	MAC Service Data Unit
NAV	Network Allocation Vector
PAN	Personal Area Network
PC	Point Coordinator
PCF	Point Coordination Function
PIFS	Priority Inter Frame Space
PLCP	Physical Layer Convergence Procedure
PPDU	Physical Layer Protocol Data Unit
PSDU	Physical Layer Service Data Unit
QAP	QoS Access Point
QoS	Quality of Service
QSTA	QoS STAtion
SIFS	Short Infer Frame Space
STA	STAtion
TC	Traffic Category
TS	Traffic Stream
TXOP	Transmission Opportunity
TSPEC	Traffic SPECifications
U-NII	Unlicensed National Information Structure
UP	User Priority

Further Reading

The official standards documents for the 802.11 standard can be found online at *www.802wirelessworld.com* Excellent summaries of the older 802.11 versions (802.11, 11b, 11a/g) and the recent 802.11n standard are given in O'Hara and Petrick [2005], and Perahia and Stacey [2008], respectively. A historically interesting comparison of IEEE 802 and HIPERLAN is found in Doufexi et al. [2002]. The principles of the WiMedia–MBOA (Multi Band OFDM Alliance) specifications are described in Siriwongpairat and Liu [2007]. A comparison between WLAN and Wireless Personal Area Network (WPAN) standards is given in Cooklev [2004].

For updates and errata for this chapter, see wides.usc.edu/teaching/textbook

30

Exercises

Peter Almers, Ove Edfors, Hao Feng, Fredrik Floren, Anders Johanson, Johan Karedal, Buon Kiong Lau, Christian Mehlführer, Andreas F. Molisch, Jan Plasberg, Barbara Resch, Jonas Samuelson, Junyang Shen, Andre Stranne, Fredrik Tufvesson, Anthony Vetro, and Shurjeel Wyne

30.1 Chapter 1: Applications and Requirements of Wireless Services

1. What was the timespan between:
   (a) invention of the cellular principle and deployment of the first widespread cellular networks;
   (b) start of the specification process of the GSM system and its widespread deployment;
   (c) specification of the IEEE 802.11b (WiFi) system and its widespread deployment?
2. Which of the following systems cannot transmit in both directions (duplex or semiduplex): (i) cellphone, (ii) cordless phone, (iii) pager, (iv) trunking radio, (v) TV broadcast system?
3. Assuming that speech can be digitized with 10 kbit/s, compare the difference in the number of bits for a 10-s voice message with a 128-letter pager message.
4. Are there conceptual differences between (i) wireless PABX and cellular systems, (ii) paging systems and wireless LANs, (iii) cellular systems with closed user groups and trunking radio?
5. What are the main problems in sending very high data rates from an MS to a BS that is far away?
6. Name the factors that influence the market penetration of wireless devices.

30.2 Chapter 2: Technical Challenges of Wireless Communications

1. Does the carrier frequency of a system have an impact on (i) small-scale fading, (ii) shadowing? When moving over a distance x, will variations in the received signal power be greater for low frequencies or high frequencies? Why?
2. Consider a scenario where there is a direct path from BS to MS, while other multipath components are reflected from a nearby mountain range. The distance between the BS and MS is 10 km, and the distance between the BS and mountain range, as well as the MS and mountain range, is 14 km. The direct path and reflected components should arrive at the RX within 0.1 times the symbol duration, to avoid heavy ISI. What is the required symbol rate?
3. Why are low carrier frequencies problematic for satellite TV? What are the problems at very high frequencies?
4. In what frequency ranges can cellphones be found? What are the advantages and drawbacks?
5. Name two advantages of power control.

30.3   Chapter 3: Noise- and Interference-Limited Systems

1. Consider an RX that consists (in this sequence) of the following components: (i) an antenna connector and feedline with an attenuation of 1.5 dB; (ii) a low-noise amplifier with a noise figure of 4 dB and a gain of 10 dB, and a unit gain mixer with a noise figure of 1 dB. What is the noise figure of the RX?

2. Consider a system with 0.1-mW transmit power, unit gain for the transmit and receive antennas, operating at 50-MHz carrier frequency with 100-kHz bandwidth. The system operates in a suburban environment. What is the receive SNR at a 100-m distance, assuming free-space propagation? How does the SNR change when changing the carrier frequency to 500 MHz and 5 GHz? Why does the 5-GHz system show a significantly lower SNR (assume the RX noise figure is 5 dB independent of frequency)?

3. Consider a GSM uplink. The MS has 100-mW transmit power, and the sensitivity of the BS RX is -105 dBm. The distance between the BS and MS is 500 m. The propagation law follows the free-space law up to a distance of $d_{break} = 50$ m, and for larger distances the receive power is similar to $(d/d_{break})^{-4.2}$. Transmit antenna gain is -7 dB; the receive antenna gain is 9 dB. Compute the available fading margin.

4. Consider a wireless LAN system with the following system specifications:
   $f_c = 5$ GHz
   $B = 20$ MHz
   $G_{TX} = 2$ dB
   $G_{RX} = 2$ dB
   Fading margin = 16 dB
   Path loss = 90 dB
   $P_{TX} = 20$ dBm
   TX losses: 3 dB
   Required SNR: 5 dB
   What is the maximum admissible RF noise figure?

5. Consider an environment with propagation exponent $n = 4$. The fading margin between the median and 10% decile, as well as between the median and 90% decile is 10 dB each. Consider a system that needs an 8-dB SIR for proper operation. How far do serving and interfering BSs have to be apart so that the MS has sufficient SIR 90% of the time at the cell boundary? Make a worst case estimate.

30.4   Chapter 4: Propagation Mechanisms

1. Antenna gain is usually given in relation to an isotropic antenna (radiating/receiving equally in all directions). It can be shown that the effective area of such an antenna is $A_{iso} = \lambda^2/4\pi$. Compute the antenna gain G_{par} of a circular parabolic antenna as a function of its radius r, where the effective area is $A_e = 0.55A$ and A is the physical area of the opening.

2. When communicating with a geostationary satellite from Earth, the distance between TX and RX is approximately 35,000 km. Assume that Friis' law for free-space loss is applicable (ignore any effects from the atmosphere) and that stations have parabolic antennas with gains 60 dB (Earth) and 20 dB (satellite), respectively, at the 11-GHz carrier frequency used.
   (a) Draw the link budget between transmitted power P_{TX} and received power P_{RX}.
   (b) If the satellite RX requires a minimum received power of -120 dBm, what transmit power is required at the Earth station antenna?

3. A system operating at 1 GHz with two 15-m-diameter parabolic antennas at a 90-m distance are to be designed.

(a) Can Friis' law be used to calculate the received power?

(b) Calculate the link budget from transmitting antenna input to receiving antenna output assuming that Friis' law *is* valid. Compare P_{TX} and P_{RX} and comment on the result.

(c) Determine the Rayleigh distance as a function of antenna gain G_{par} for a circular parabolic antenna as in Problem 1.

4. A TX is located 20 m from a 58-m-high brick wall ($\varepsilon_r = 4$), while an RX is located on the other side, 60 m away from the wall. The wall is 10 cm thick and can be regarded as lossless. Let both antennas be of height 1.4 m and using a center frequency of 900 MHz.

(a) Considering TE waves, determine the field strength at the RX caused by transmission through the wall, $E_{through}$.

(b) The wall can be regarded as a semi-infinite thin screen. Determine the field strength at the RX caused by diffraction over the wall, E_{diff}.

(c) Determine the ratio of the magnitudes of the two field strengths.

5. Show that for a wave propagating from medium 1 to medium 2, the reflection coefficients for TE and TM can be written as

$$
\left.\begin{aligned}
\rho_{TM} &= \frac{\varepsilon_r \cos \Theta_e - \sqrt{\varepsilon_r - \sin^2 \Theta_e}}{\varepsilon_r \cos \Theta_e + \sqrt{\varepsilon_r - \sin^2 \Theta_e}} \\[2mm]
\rho_{TE} &= \frac{\cos \Theta_e - \sqrt{\varepsilon_r - \sin^2 \Theta_e}}{\cos \Theta_e + \sqrt{\varepsilon_r - \sin^2 \Theta_e}}
\end{aligned}\right\}
\tag{30.1}
$$

if medium 1 is air and medium 2 is lossless with a dielectric constant ε_r.

6. Waves are propagating from the air toward a lossless material with a dielectric constant ε_r.

(a) Determine an expression for the angle that results in totally transmitted waves – i.e., $|\rho_{TM}| = 0$. Is there a corresponding angle for TE waves?

(b) Assuming that $\varepsilon_r = 4.44$, plot the magnitude for TE and TM reflection coefficients. Determine the angle for which $|\rho_{TM}| = 0$.

7. For propagation over a perfectly conducting ground plane, the magnitude of total received field E_{tot} is given by

$$
|E_{tot}(d)| = E(1\text{m}) \frac{1}{d} 2 \frac{h_{TX} h_{RX}}{d} \frac{2\pi}{\lambda}
\tag{30.2}
$$

Show that if transmit power is P_{TX} and transmit and receive antenna gains are G_{TX} and G_{RX}, respectively, the received power P_{RX} is given by

$$
P_{RX}(d) = P_{TX} G_{TX} G_{RX} \left(\frac{h_{TX} h_{RX}}{d^2} \right)^2
\tag{30.3}
$$

8. Assume that we have a BS with a 6-dB antenna gain and an MS with antenna gain of 2 dB, at heights 10 m and 1.5 m, respectively, operating in an environment where the ground plane can be treated as perfectly conducting. The lengths of the two antennas are 0.5 m and 15 cm, respectively. The BS transmits with a maximum power of 40 W and the mobile with a power of 0.1 W. The center frequency of the links (duplex) are both at 900 MHz, even if in practice they are separated by a small duplex distance (frequency difference).

(a) Assuming that Eq. (4.24) holds, calculate how much received power is available at the output of the receive antenna (BS antenna and MS antenna, respectively), as a function of distance d.

(b) Plot the received powers for all valid distances d – i.e., where Eq. (4.24) holds *and* the far field condition of the antennas is fulfilled.

9. The following system is used to give an estimate of the exposure to electromagnetic waves from a BS and an MS: consider communication between the BS and MS separated by a distance d in the 900-MHz band. The ground plane is treated as perfectly conducting, with antenna heights $h_{BS} = 10\,\text{m}$ and $h_{MS} = 1.5\,\text{m}$. The antenna gains are $G_{BS} = 6\,\text{dB}$ and $G_{MS} = 2\,\text{dB}$, respectively. On a straight line between the BS and MS, at a distance of 3 m from the MS and a height $h_{ref} = 1.5\text{m}$, there is a Reference Antenna (RA), picking up signals from both, which can be used to measure exposure. The RA has a gain G_{ref}. Assume that Eq. (4.24) can be used to describe transmission between the BS and MS as well as between the BS and RA, although transmission between the MS and RA is better described by Friis' law due to the short distance.

   (a) Assume that the BS has a 10-dB lower requirement on the received power available at the antenna output and determine expressions for (as functions of distance d and RX sensitivity level $P_{RX,MS}^{min}$ of the MS):

       (i) required transmit power by the BS, $P_{TX,BS}$;
       (ii) required transmit power by the MS, $P_{RX,BS}$;
       (iii) received power in the reference antenna from the BS, $P_{RX,ref}^{BS}$;
       (iv) received power in the reference antenna from the MS, $P_{RX,ref}^{MS}$.

   (b) Use the expressions from (a) to determine the difference in dB between $P_{RX,ref}^{MS}$ and $P_{RX,ref}^{BS}$ as a function of d. Plot the result for $d = 50 - 5{,}000\,\text{m}$.

10. Communication is to take place from one side of a building to the other as depicted in Figure 30.1, using 2-m-tall antennas. Convert the building into a series of semi-infinite screens and determine the field strength at the receive antenna caused by diffraction using Bullington's method for (a) $f = 900\,\text{MHz}$, (b) $f = 1{,}800\,\text{MHz}$, and (c) $f = 2.4\,\text{GHz}$.

11. Derive the reflection and transmission coefficient for TE and TM waves. *Hint*: use the continuity conditions for parallel and perpendicular fields at the interface.

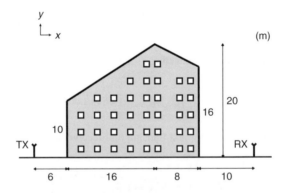

Figure 30.1 The geometry for Problem 10.

30.5 Chapter 5: Statistical Description of the Wireless Channel

1. A mobile RX traveling at 25 m/s receives the following two multipath components:

$$E_1(t) = 0.1\cos[2\pi \cdot 2 \cdot 10^9 t - 2\pi \upsilon_{max}\cos(\gamma_1)t + 0]\text{V} \cdot \text{m}^{-1}$$
$$E_2(t) = 0.2\cos[2\pi \cdot 2 \cdot 10^9 t - 2\pi \upsilon_{max}\cos(\gamma_2)t + 0]\text{V} \cdot \text{m}^{-1} \tag{30.4}$$

Assume that the RX moves directly opposite to the direction of propagation of the first component and exactly in the direction of propagation of the second component. Compute the power per unit area at the receive antenna at time instants $t = 0, 0.1$, and 0.2s. Compute the average power per unit area received over this interval.

Hint: The power per unit area is given by the magnitude of the average Poynting vector, $S_{avg} = \frac{1}{2} \cdot \frac{E^2}{Z}$, where averaging is performed over one period of the electric field. E is the amplitude of the total electric field and $Z_0 = 377\,\Omega$ for air.

2. Show that the Rice distribution with a large Rice factor, $K_r = \frac{A^2}{2\sigma^2}$, can be approximated by a Gaussian distribution with mean value A.

3. A mobile communication system is to be designed to the following specifications: the instantaneous received amplitude r must not drop below 10% of a specified level r_{min} at the cell boundary (maximum distance from BS) for 90% of the time. The signal experiences both small-scale Rayleigh fading and large-scale log-normal fading, with $\sigma_F = 6\,$dB. Find the required fading margin for the system to work.

4. A radio system is usually specified in such a way that an RX should be able to handle a certain amount of Doppler spread in the received signal, without losing too much in performance. Assume that only the mobile RX is moving and that the maximal Doppler spread is measured as twice the maximal Doppler shift. Further, assume that you are designing a mobile communication system that should be able to operate at both 900 MHz and 1,800 MHz.
   (a) If you aim at making the system capable of communicating when the terminal is moving at 200 km/h, which maximal Doppler spread should it be able to handle?
   (b) If you design the system to be able to operate at 200 km/h when using the 900-MHz band, at what maximal speed can you communicate if the 1,800-MHz band is used (assuming the same Doppler spread is the limitation)?

5. Assume that, at a certain distance, we have a deterministic propagation loss of 127 dB and large-scale fading, which is log-normally distributed with $\sigma_F = 7\,$dB.
   (a) How large is the outage probability (due to large-scale fading) at that particular distance, if our system is designed to handle a maximal propagation loss of 135 dB?
   (b) Which of the following alternatives can be used to lower the outage probability of our system, and why are they/are they not possible to use?
      (i) Increase the transmit power.
      (ii) Decrease the deterministic path loss.
      (iii) Change the antennas.
      (iv) Lower the σ_F.
      (v) Build a better RX.

6. Assume a Rayleigh-fading environment in which the sensitivity level of an RX is given by a signal amplitude r_{min}, then the probability of outage is $P_{out} = \Pr\{r \leq r_{min}\} = cdf(r_{min})$.
   (a) Considering that the squared signal amplitude r^2 is proportional to the instantaneous received power C through the relation $C = K \cdot r^2$, where K is a positive constant, determine the expression for the probability of outage expressed in terms of RX sensitivity level C_{min} and mean received power $\overline{C}$.
   (b) The fading margin (level of protection against fading) is expressed as

$$M = \frac{\overline{C}}{C_{min}} \tag{30.5}$$

   or (in dB)

$$M_{dB} = \overline{C}_{dB} - C_{min\ dB} \tag{30.6}$$

   Determine a closed-form expression for the required fading margin, in dB, as a function of outage probability P_{out}.

7. For an outage probability P_{out} of up to 5% in a Rayleigh-fading environment, we assume that we can use the approximation:

$$P_{\text{out}} = \Pr\{r \leq r_{\text{min}}\} \approx \frac{r_{\text{min}}^2}{2\sigma^2} \tag{30.7}$$

where r_{min} is the sensitivity level of the RX.
   (a) Determine a closed-form expression for the required fading margin, in dB, as a function of outage probability P_{out}, when the above approximation is used. Denote the resulting approximation of the fading margin as $\tilde{M}_{dB}$.
   (b) Determine the largest error between the approximate $\tilde{M}_{dB}$ and the exact M_{dB} values for outage probabilities $P_{\text{out}} \leq 5\%$.
8. In some cases, the "mean" of the received signal amplitude r is expressed in terms other than $\overline{r^2}$ – i.e., terms that are proportional to the mean received power. A common value to be found in propagation measurements is the median value r_{50} – i.e., the value showing the signal is below 50% of the time, and above 50% of the time.
   Now, assume that we have small-scale fading described by the Rayleigh distribution.
   (a) Derive the expression for $cdf(r)$ when given the median value r_{50} of the received signal.
   (b) Derive the expressions for the required fading margins in dB (for a certain P_{out}), expressed both in relation to $\overline{r^2}$ (the mean power) and in relation to r_{50}^2. Call these fading margins $M_{\text{mean}|\text{dB}}(P_{\text{out}})$ and $M_{\text{median}|\text{dB}}(P_{\text{out}})$.
   (c) Compare the expressions for $M_{\text{mean}|\text{dB}}(P_{\text{out}})$ and $M_{\text{median}|\text{dB}}(P_{\text{out}})$ and try to find a simple relation between them.
9. For Rayleigh fading, derive expressions for the level crossing rate $N_r(r_{\text{min}})$ and the $\text{ADF}(r_{\text{min}})$ in terms of the fading margin $M = \overline{r^2}/r_{\text{min}}^2$ rather than parameters Ω_0 and Ω_2 (compare Eqs. (5.49) and (5.50)).
10. Assume that we are going to design a wireless system with maximal distance $d_{\text{max}} = 5\,\text{km}$ between the BS and MS. The BS antenna is at a 20-m elevation and the MS antenna is at 1.5-m elevation. The carrier frequency of choice is 450 MHz and the environment consists of rather flat and open terrain. For this particular situation, we find a propagation model called Egli's model, which says that the propagation loss is

$$\Delta L_{|\text{dB}} = 10 \log \left(\frac{f_{|\text{MHz}}^2}{1,600} \right) \tag{30.8}$$

in addition to what is predicted by the theoretical model for propagation over a ground plane. The model is only valid if the distance is smaller than the radio horizon:

$$d_h \approx 4100 \left(\sqrt{h_{\text{BS}|\text{m}}} + \sqrt{h_{\text{MS}|\text{m}}} \right)_{|\text{m}} \tag{30.9}$$

The calculated value of propagation loss is a *median* value. The narrowband link is subjected to both large-scale log-normal fading, with $\sigma_F = 5\,\text{dB}$, and small-scale Rayleigh fading. In addition, there is an RX sensitivity level of $C_{\text{min}|\text{dBm}} = -122$ for the MS (at the antenna output). Both the BS and MS are equipped with gain 2.15-dB $\lambda/2$ dipoles.
   (a) Draw a link budget for the downlink – i.e., for the link from the BS to MS. Start with the input power $P_{TX|\text{dB}}$ to the antenna and follow the budget through to the received median power $C_{\text{median}|\text{dB}}$, at the MS antenna output. Then, between $C_{\text{median}|\text{dB}}$ and $C_{\text{min}|\text{dB}}$ there will be a fading margin $M_{|\text{dB}}$.
   (b) The system is considered operational (has coverage) if the instantaneous received power is not below $C_{\text{min}|\text{dB}}$ more than 5% of the time. Calculate the fading margin needed as a consequence of small-scale fading.

(c) We wish for the system to have a 95% boundary coverage – i.e., to be operational at 95% of the locations at the maximal distance d_{max}. Calculate the fading margin needed as a consequence of large-scale fading to fulfill this requirement.

(d) Calculate the required transmit power $P_{TX|dB}$, by adding the fading margins obtained in (b) and (c) to a total fading margin $M_{|dB}$, which is to be inserted in the link budget. Remember that the propagation model gives a median value and make the necessary compensation, if necessary.

   Note: Adding the two fading margins from (b) and (c) does not give the lowest possible fading margin. In fact the system is slightly overdimensioned, but it is much simpler than combining the statistics of the two fading characteristics (giving the Suzuki distribution).

11. Let us consider a simple interference-limited system, where there are two transmitters, TX A and TX B, both with antenna heights 30 m, at a distance of 40 km. They both transmit with the same power, use the same omnidirectional $\lambda/2$ dipole antennas, and use the same carrier frequency, 900 MHz. TX A is transmitting to RX A located at a distance d in the direction of TX B. Transmission from TX B is interfering with the reception of RX A, which requires an average (small-scale averaged) Carrier-to-Interference ratio (C/I) of $(C/I)_{min} = 7$ dB. Incoming signals to RX A (wanted and interfering) are both subject to independent 9-dB log-normal large-scale fading. The propagation exponent in the environment we are studying is $\eta = 3.6$ – i.e., the received power decreases as $d^{-\eta}$.

   (a) Determine the fading margin required to give a 99% probability that (C/I) is not below $(C/I)_{min}$.

   (b) Using the fading margin from (a), determine the maximal distance d_{max} between TX A and RX A.

   (c) Can you, by studying the equations, give a quick answer to what happens to the maximal distance d_{max} (as defined above) if TX A and TX B were located at 20 km from each other?

30.6   Chapter 6: Wideband and Directional Channel Characterization

1. Explain the difference between *spreading function* and *scattering function*.
2. Give examples of situations where the information contained in the time frequency correlation function $R_H(\Delta t, \Delta f)$ can be useful.
3. Assume that the measured impulse response of Figure 6.6 can be approximated as stationary. Furthermore, approximate the PDP as consisting of two clusters, each having exponential decay on a linear scale; i.e.,

$$P(\tau) = \begin{cases} a_1 e^{-b_1 \tau} & 0 \leq \tau \leq 20\,\mu s \\ a_2 e^{55.10^{-6} - b_2 \tau} & 55\,\mu s \leq \tau \leq 65\,\mu s \\ 0 & \text{elsewhere} \end{cases} \tag{30.10}$$

First, find the coefficients a_i and b_i and then calculate the time-integrated power, the average mean delay, and the average root mean square (rms) delay spread, all given in Eqs. (6.37)–(6.39). If the coherence bandwidth is approximated by

$$B_{coh} \approx \frac{1}{2\pi S_\tau} \tag{30.11}$$

where S_τ is the rms delay spread, would you characterize the channel as flat- or frequency-selective for a system bandwidth of 100 kHz? Use Figure 6.4 to justify your answer.

4. As described in Section 6.5.5, the GSM system has an equalizer of 4 symbol durations, corresponding to 16 µs – i.e., multipath components arriving within that window can be processed

by the RX. Calculate the interference quotient for a window of duration $16\,\mu s$ and starting at $t_0 = 0$ for the data measured in Figure 6.6 using the approximation in Problem 3. Perform this calculation first for the whole PDP as specified in Problem 3 and then again while ignoring all components arriving after $20\,\mu s$. Compare the two results.

5. The coherence time T_c gives a measure of how long a channel can be considered to be constant, and can be approximated as the inverse of the Doppler spread. Obtain an estimate of the coherence time in Figure 6.7.

6. In wireless systems, one has to segment the transmitted stream of symbols into blocks, also called *frames*. In every frame, one usually also inserts symbols that are known by the RX, so-called *pilot symbols*. In this manner, the RX can estimate the current value of the channel, and thus coherent detection can be performed. Therefore, let the RX in every frame be informed of the channel gain for the first symbol in the frame, and assume that the RX then believes that this value is valid for the whole frame. Using the definition of coherence time in Problem 5 and assuming Jakes' Doppler spectrum, estimate the maximum speed of the RX for which the assumption that the channel is constant during the frame is still valid. Let the framelength be 4.6 ms.

7. In a CDMA system, the signal is spread over a large bandwidth by multiplying the transmitted symbol by a sequence of short pulses, also called *chips*. The system bandwidth is thus determined by the duration of a chip. If the chip duration is $0.26\,\mu s$ and the maximum excess delay is $1.3\,\mu s$, into how many delay bins do the multipath components fall? If the maximum excess delay is 100 ns, is the CDMA system wideband or narrowband?

30.7 Chapter 7: Channel Models

1. Give a physical interpretation of the log-normal distribution (is it realistic?).
2. Assume that we are calculating propagation loss in a medium-size city, where a BS antenna is located at a height of $h_b = 40$ m, the MS at $h_m = 2$ m. The carrier frequency of the transmission taking place is $f = 900$ MHz and the distance between the two is $d = 2$ km.
   (a) Calculate the predicted propagation loss L_{Oku} between isotropic antennas by using the formula for free-space attenuation, in combination with Okumura's measurements (Figures 7.12–7.13).
   (b) Calculate the predicted propagation loss L_{O-H} between isotropic antennas by using parameterization of Okumura's measurements provided by Hata – i.e., use the Okumura–Hata model.
   (c) Compare the results. If there are differences, where do you think they come from and are they significant?
3. Assume that we are calculating propagation loss at $f_0 = 1,800$ MHz in a medium-size city environment with equally spaced buildings of height $h_{Roof} = 20$ m, at a building-to-building distance of $b = 30$ m and $w = 10$-m-wide streets. The propagation loss we are interested in is between a BS at height h_b and an MS at height $h_m = 1.8$ m, with a distance of $d = 800$ m between the two. The MS is located on a street which is oriented at an angle $\varphi = 90°$ relative to the direction of incidence (direction of incoming wave). For this purpose, the COST 231–Walfish–Ikegami model is a suitable tool.
   (a) Check that the given parameters are within the validity range of the model.
   (b) Calculate the propagation loss when the BS antenna is located 3 m above the rooftops – i.e., when $h_b = 23$ m.
   (c) Calculate the propagation loss when the BS antenna is located 1 m below the rooftops – i.e., when $h_b = 19$ m, and comment on the difference from (b).
4. When evaluating GSM systems, a wideband model is required since the system is designed in such a way that different delays in the channel are experienced by the RX. The COST 207 model is created for GSM evaluations:

(a) Draw PDPs for the tap delay line implementations RA, TU, BU, and HT using Tables 7.3–7.6 in the text. Use the dB scale for power and the same scale for the delay axis of all four plots so that you can compare the four PDPs.

(b) Convert the delays into path lengths, using $d_i = c\tau_i$, where c is the speed of light, and make notations of these path lengths in the graphs drawn in (b). Using these path lengths, try to interpret the distribution of power in the four scenarios (from where does the power come?).

5. For the COST 207 channel models (created for GSM evaluations),

(a) Find the rms delay spread of the COST 207 environments RA, TU.

(b) What is the coherence bandwidth of the RA and TU channels?

(c) Could two different function PDPs have the same rms delay spread?

6. The correlation between the elements of an antenna array are dependent on angular power distribution, element response, and element spacing. An antenna array consists of two omnidirectional elements separated by distance d and placed in a channel with azimuthal power spectrum $f_\phi(\phi)$.

(a) Derive an expression for correlation.

(b) [MATLAB] Plot the correlation coefficient between two elements for different antenna spacings for 100 MPCs with a uniform angular distribution $(0, 2\pi)$.

30.8 Chapter 8: Channel Sounding

1. Assume a simple direct RF pulse channel sounder that generates a signal which is a sequence of narrow probing pulses. Each pulse has duration $t_{on} = 50$ ns, and the pulse repetition period is $20\,\mu$s. Determine the following:

(a) minimum time delay that can be measured by the system;

(b) maximum time delay that can be measured unambiguously.

2. As the wireless propagation channel may be treated as a linear system, why are m-sequences (PN-sequences) used for channel sounding? *Hint*: elaborate on the auto- and cross-correlation properties of white noise, and discuss its similarities to PN-sequences.

3. A maximal length sequence (m-sequence) is generated from a Linear Feedback Shift Register (LFSR) with certain allowed connections between the memory elements and the modulo-2 adder. In Figure 30.2 an LFSR is shown with $m = 3$ memory elements, connected so as to generate an m-sequence. Determine the following:

(a) the period M_c of the m-sequence;

(b) the complete m-sequence $\{C_m\}$, given that the memory is initialized with $a_{k-1} = 0$, $a_{k-2} = 0$, $a_{k-3} = 1$.

Figure 30.2 A maximal LFSR sequence generator.

4. Autocorrelation refers to the degree of correspondence between a sequence and a shifted copy of itself. If the ±1 sequence $\{\acute{C}_m\}$ is defined as $\acute{C}_m = 1 - 2C_m$, where $\{C_m\} \in (0, 1)$ is an m-sequence, then the ACF $R_{\acute{C}_m}(\tau)$, defined as

$$R_{\acute{C}_m}(\tau) = \frac{1}{M_c} \sum_{m=1}^{M} \acute{C}_m \acute{C}_{m+\tau} = \begin{cases} 1, & \tau = 0, \pm M_c, \pm 2M_c, \ldots \\ -\frac{1}{M_c}, & \text{otherwise} \end{cases} \qquad (30.12)$$

is also periodic. If the code waveform $p(t)$ is the square-wave equivalent of sequence $\{\acute{C}_m\}$ with pulse duration T_c then determine the ACF for all values of τ.

(a) Plot the ACF for $T_c = 1$, $M_c = 7$, and $-8 \leq \tau \leq 8$.

(b) Neglecting the effects of system noise, what is the dynamic range over which the system can detect received signals?

5. In an STDCC system, the maximum measurable Doppler shift is given by

$$\upsilon_{\max} = \frac{1}{2K_{\mathrm{scal}}M_c T_c} \tag{30.13}$$

Given that $K_{\mathrm{scal}} = 5{,}000$, $M_c = 31$, $T_c = 0.1\,\mu\mathrm{s}$, and the carrier frequency is 900 MHz, determine the following:

(a) The maximum permissible velocity of the MS.

(b) The longest time delay that can be measured.

(c) If we increase the m-sequence length to $M_c = 63$, what happens to the above quantities?

6. Plane waves from two separate sources impinge on a uniform linear array as shown in Figure 30.3. The array consists of four isotropic antenna elements with interelement spacing of $\frac{\lambda}{2}$ where λ is the carrier wavelength. The covariance matrix of array outputs is provided for two different cases. Determine the angles of arrival of the two waves in each case, using conventional (Fourier–Bartlett) beamforming.

(a)

$$\mathbf{R}_{rr} = \begin{bmatrix} 1.2703 & 0.3559 + 0.6675i & 0.4215 - 0.1392i & 0.9818 + 0.7086i \\ 0.3559 - 0.6675i & 1.3252 & 0.3489 + 0.6493i & 0.4805 - 0.1864i \\ 0.4215 + 0.1392i & 0.3489 - 0.6493i & 1.2475 & 0.3314 + 0.5787i \\ 0.9818 - 0.7086i & 0.4805 + 0.1864i & 0.3314 - 0.5787i & 1.2344 \end{bmatrix} \tag{30.14}$$

(b)

$$\mathbf{R}_{rr} = \begin{bmatrix} 0.9198 & 0.2125 + 0.7644i & -0.5295 + 0.1202i & 0.1355 - 0.5488i \\ 0.2125 - 0.7644i & 0.8957 & 0.1661 + 0.7244i & -0.5104 + 0.0526i \\ -0.5295 - 0.1202i & 0.1661 - 0.7244i & 0.8323 & 0.1444 + 0.6853i \\ 0.1355 + 0.5488i & -0.5104 - 0.0526i & 0.1444 - 0.6853i & 0.8283 \end{bmatrix} \tag{30.15}$$

Figure 30.3 Plane waves S_1 and S_2 impinging on a four-element ULA. The direction of propagation is indicated by arrows perpendicular to the wavefronts.

7. Repeat the estimation problem discussed in Problem 6 using the ESPRIT algorithm as outlined in Appendix 8.A (see *www.wiley.com/go/molisch*).

8. Let M samples of the impulse response of a frequency flat channel be measured during a time interval t_{meas} – the channel is assumed to be time invariant during this interval. Furthermore, it is assumed that the measurement noise in these samples is iid with zero-mean and variance σ^2. Prove that averaging over M samples of the impulse response enhances the SNR by a factor M.

30.9   Chapter 9: Antennas

1. A BS for GSM900 MHz has output power of 10 W; the cell is in a rural environment, with a decay exponent of 3.2 for distances larger than 300 m. The cell coverage radius is 37 km when measured by the MS in the absence of a user. Calculate how much smaller the coverage radius becomes for the median user in Figure 9.1 when she/he places the cellphone in close proximity to the head.
2. Sketch possible antenna placements on a PDA for a 2.4-GHz antenna for WiFi. Assume that the antenna is a patch antenna with effective permittivity $\varepsilon_r = 2.5$. Take care to minimize the influence of the user's hand.
3. Consider a helical antenna with $d = 5$ mm operating at 1.9 GHz. Calculate the diameter D such that polarization becomes circular.
4. Consider an antenna with a pattern:

$$G(\phi, \theta) = \sin^n(\theta/\theta_0)\cos(\theta/\theta_0) \tag{30.16}$$

where $\theta_0 = \pi/1.5$.
(a) What is the 3-dB beamwidth?
(b) What is the 10-dB beamwidth?
(c) What is the directivity?
(d) Compute the numerical values of (i), (ii), (iii) for $\theta_0 = \pi/1.5, n = 5$.
5. Compute the input impedance of a half-wavelength slot antenna; assume uniform current distribution. *Hint*: the input impedance of antenna structure A and its complement B (a structure that has metal whenever structure A has air, and has air whenever structure A has metal) are related by

$$Z_A Z_B = \frac{Z_0^2}{4} \tag{30.17}$$

where Z_0 is the free-space impedance $Z_0 = 377\,\Omega$.
6. Let the transmit antenna be a vertical $\lambda/2$ dipole, and the receive antenna a vertical $\lambda/20$ dipole.
(a) What are the radiation resistances at TX and RX?
(b) Assuming ohmic losses due to $R_{ohmic} = 10\,\Omega$, what is the radiation efficiency?
7. Consider an antenna array with N elements where all elements have the same amplitude, and a phase shift:

$$\Delta = -\frac{2\pi}{\lambda}d_a \tag{30.18}$$

(a) What is the gain in the endfire direction?
(b) What is the gain in the endfire direction if

$$\Delta = -\left[\frac{2\pi}{\lambda}d_a + \frac{\pi}{N}\right] \tag{30.19}$$

30.10   Chapter 10: Structure of a Wireless Communication Link

1. A directional coupler has a directivity of 20 dB, an insertion attenuation of 1 dB, and a coupling attenuation of 20 dB. The reflection coefficients of all ports (with respect to the nominal impedance of $Z_0 = 50\,\Omega$) are 0.12, and the phase shift in the main path is 0, in the coupled path $\pi/2$, and 0 between the decoupled paths.
   (a) Compute the S-matrix of this directional coupler.
   (b) Is the directional coupler passive, reciprocal, symmetrical?
2. The input impedance of an antenna $Z = 70 - 85j\,\Omega$ is transformed by putting it in series with an inductance L. The antenna operates at 250 MHz.
   (a) What is the transformed impedance if $L = 85\,$nH?
   (b) Which value does L have to take on to make the transformed impedance real?
3. Consider the concatenation of two amplifiers with gains G_1, G_2 and noise figures F_1, F_2. Show that in order to minimize total noise figure, the amplifier with the smaller value of

$$M = \frac{F - 1}{1 - 1/G} \tag{30.20}$$

   has to be placed first.
4. For the measurement of received power for AGCs, it is common to use diodes, assuming that the output (diode current) is proportional to the square of the input (voltage). However, true characteristics are better described by an exponential function:

$$i_D \exp(U/U_t) - 1 \tag{30.21}$$

   where the voltage U_t is a constant. This function can be approximated by a quadratic function only as long as the voltage is low.
   (a) Show that the diode current contains a DC component.
   (b) What spectral components are created by the diode? How can they be eliminated?
   (c) How large is the maximum diode current if the error (due to the difference between ideal quadratic and exponential behavior) in measured power is to remain below 5%?
5. An amplifier has cubic characteristics:

$$U_{out} = a_0 + a_1 U_{in} + a_2 U_{in}^2 + a_3 U_{in}^3 \tag{30.22}$$

   Let there be two sinusoidal input signals (with frequencies f_1, f_2) at the input, both with a power of $-6\,$dBm. At output, we measure desired signals (at f_1, f_2) with 20-dBm power, and third-order intermodulation products (at $2f_2 - f_1$ and $2f_1 - f_2$) with a power of $-10\,$dBm. What is the intercept point? In other words, at what level of the input signal do intermodulation products become as large as desired signals?
6. A sawtooth received signal is to be quantized by a linear ADC (linear spacing of quantization steps).
   (a) What is the minimum number of quantization steps if quantization noise (variance between ideal and quantized received signal) has to stay below 10, 20, 30 dB?
   (b) How does the result change when – due to a badly working AGC – the peak amplitude of the received signal is only half as large as the maximum amplitude that the ADC could quantize?

30.11   Chapter 11: Modulation Formats

1. Both GSM and DECT use GMSK, but with different Gaussian filters ($B_G T = 0.3$ in GSM, $B_G T = 0.5$ in DECT). What are the advantages of having a larger bandwidth time product? Why is the lower one used in GSM?

2. Derive the smallest frequency separation for orthogonal binary frequency shift keying.
3. In the EDGE standard (high-speed data in GSM networks), $\frac{3\pi}{8}$-shifted 8-PSK is used. Sketch transitions in the signal space diagram for this modulation format. What is the ratio between average value and minimum value of the envelope? Why is it that $\frac{\pi}{4}$-shifted 8-PSK cannot be used?
4. MSK can be interpreted as offset QAM with specific pulse shapes. Show that MSK has a constant envelope using this interpretation.
5. Consider two functions:

$$f(t) = \begin{cases} 1 & 0 < t < T/2 \\ -2 & T/2 < t < T \end{cases} \tag{30.23}$$

and

$$g(t) = 1 - (t/T) \tag{30.24}$$

for $0 < t < T$.
   (a) Find a set of expansion functions, using Gram–Schmidt orthogonalization.
   (b) Find points in the signal space diagram for $f(t)$, $g(t)$, and the function that is unity for $0 < t < T$.
6. For the bit sequence of Figure 11.6, plot $p_D(t)$ for *differentially encoded* BPSK.
7. Relate the mean signal energy of 64-QAM to the distance between points in the signal space diagram.
8. A system should transmit as high a data rate as possible within a 1-MHz bandwidth, where out-of-band emissions of -50 dBm are admissible. The transmit power used is 20 W. Is it better to use MSK or BPSK with root-raised cosine filters with $\alpha = 0.35$? *Note:* this question only concentrates on spectral efficiency, and avoids other considerations like the peak-to-average ratio of the signal.

30.12   Chapter 12: Demodulation

1. Consider a point-to-point radio link between two highly directional antennas in a stationary environment. The antennas have antenna gains of 30 dB, distance attenuation is 150 dB, and the RX has a noise figure of 7 dB. The symbol rate is 20 Msymb/s and Nyquist signaling is used. It can be assumed that the radio link can be treated as an AWGN channel without fading. How much transmit power is required (disregarding power losses at TX and RX ends) for a maximum BER of 10^{-5}:
   (a) When using coherently detected BPSK, FSK, differentially detected BPSK, or noncoherently detected FSK?
   (b) Derive the exact bit and symbol error probability expressions for coherently detected Gray-coded QPSK. Start by showing that the QPSK signal can be viewed as two antipodal signals in quadrature.
   (c) What is the required transmit power if Gray-coded QPSK is used?
   (d) What is the penalty in increased BER for using differential detection of Gray-coded QPSK in (c)?
2. Use the full union-bound method for upper bounding the BER for higher order modulation methods.
   (a) Upper bound the BER for Gray-coded QPSK by using the full union bound.
   (b) In which range of E_b/N_0 is the difference between the upper bound in (a) and the exact expression less than 10^{-5}?
3. Consider the 8-Amplitude Modulation Phase Modulation (8-AMPM) modulation format shown in Figure 30.4. What is the average symbol energy expressed in d_{min} (all signal alternatives assumed to be equally probable)? What is the nearest neighbor union bound on the BER?

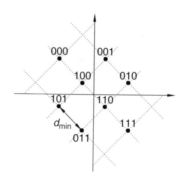

Figure 30.4   8-AMPM constellation.

4. Use the nearest neighbor union bound method to calculate approximate BERs for higher order modulation methods.
   (a) Use the nearest neighbor union bound to calculate the approximate upper bound on BER for 8-PSK as shown in Figure 30.5.
   (b) Use the nearest neighbor union bound to calculate the approximate upper bound on BER for Gray-coded 8-PSK as shown in Figure 30.6. How large is the approximate gain from Gray coding at a BER of 10^{-5}?

Figure 30.5   Constellation diagram for 8PSK.

Figure 30.6   Grey-coded 8PSK.

5. Obtain simulated BER results for Gray-coded QPSK and 8-PSK by creating a MATLAB™ program. Run simulations for E_b/N_0 in the range from 0 to 15 dB. In which region does the nearest neighbor union bound on BER (see Problem 3) appear to be tight?

6. Calculate the nearest neighbor union bound for Gray-coded 16-QAM. Assuming that the BER must not exceed 10^{-5}, what are the useful ranges of E_b/N_0 for adaptive switching between the two modulation schemes QPSK and 16-QAM for maximum achievable data rate?
7. Assume that M-ary orthogonal signaling is used. Based on the union bound, how small does E_b/N_0 need to be to allow for completely error-free transmissions as $M \to \infty$?
8. Consider the point-to-point radio link introduced in Problem 1. By how much must transmit power be increased to maintain the maximum BER of 10^{-5} if the channel is flat-Rayleigh-fading:
    (a) When are coherently detected BPSK and FSK, and DPSK and noncoherent FSK used?
    (b) What is the required increase in transmit power if the channel is Ricean with $K_r = 10$ and DPSK is used? What are the expected results for $K_r \to 0$?
9. Consider a mobile radio link with a carrier frequency of $f_c = 1,200$ MHz, a bit rate of 3 kbit/s, and a required maximum BER of 10^{-4}. The modulation format is MSK with differential detection. A maximum transmit power of 10 dBm (EIRP) is used with 5-dB gain antennas and RXs with noise figures of 9 dB.
    (a) Assuming that the channel is flat-Rayleigh-fading and that BS and MS heights are 40 and 3 m, respectively, what is the achievable cell radius in a suburban environment according to the Okumura–Hata path loss model?
    (b) Assume that the channel is frequency-dispersive Rayleigh fading and characterized by a classical Jakes' Doppler spectrum. What is the maximum mobile terminal speed for an irreducible BER due to frequency dispersion of 10^{-5}?
10. Consider a mobile radio system using MSK with a bit rate of 100 kbit/s. The system is used for transmitting IP packets of up to 1,000 bytes. The packet error rate must not exceed 10^{-3} (without the use of an ARQ scheme).
    (a) What is the maximum allowed average delay spread of the mobile radio channel?
    (b) What are typical values of average delay spread in indoor, urban, and rural environments for mobile communication systems?

30.13   Chapter 13: Diversity

1. To illustrate the impact of diversity, we look at the average BER for BPSK and MRC, given approximately by Eq. (13.35). Assume that the average SNR is 20 dB.
    (a) Calculate the average BER for $N_r = 1$ receive antennas.
    (b) Calculate the average BER for $N_r = 3$ receive antennas.
    (c) Calculate the SNR that would be required in a one-antenna system in order to achieve the same BER as a three-antenna system at 20 dB.
2. Assume a situation where we have the possibility of adding one or two extra antennas, with uncorrelated fading, to an RX operating in a Rayleigh-fading environment. The two diversity schemes under consideration are RSSI-driven selection diversity or MRC diversity.
    (a) Assume that the important performance requirement is that the instantaneous $\bar{E}_b/N_0$ cannot be below some fixed value more than 1% of the time (1% outage). Determine the required fading margin for one, two, and three antennas with independent Rayleigh fading and the corresponding "diversity gains" for the two- and three-antenna cases when using RSSI-driven selection diversity and MRC diversity.
    (b) Assume that the important performance requirement is that the average BER is 10^{-3} (when using BPSK). Determine the required average $\bar{E}_b/N_0$ for one, two, and three antennas with independent Rayleigh fading and the corresponding "diversity gains" for the two- and three-antenna cases when using RSSI-driven selection diversity and MRC diversity.
    (c) Perform (a) and (b) again, but with 10% outage in (b) and an average BER of 10^{-2} in (c). Compare the "diversity gains" with the ones obtained earlier and comment on the differences!

3. Let an RX be connected to two antennas, for which the SNRs are independent and exponentially distributed using the same average SNR. RSSI-driven selection diversity is employed and the outage probability is P_{out}. We are interested in the fading margin.
   (a) Derive an expression in terms of P_{out} for the fading margin when only one antenna is used.
   (b) Derive an expression in terms of P_{out} for the fading margin when both antennas are used.
   (c) Use the two results above to calculate the diversity gain for an outage probability of 1%.

4. In a wideband CDMA system, the Rake RX can take advantage of the multipath diversity arising from the delay dispersion of the channel. Under certain assumptions, the Rake RX acts as a maximal ratio combiner where branches correspond to the delay bins of the CDMA system. Assume a rectangular PDP and that the number of Rake fingers is equal to the number of resolvable multipaths. Furthermore, assume that each MPC is Rayleigh fading. We require that the instantaneous BER exceeds 10^{-3} only 1% of the time when using BPSK on a channel with an average SNR of 15 dB. How many resolvable multipaths must the channel consist of in order for the BER requirement to be fulfilled?

5. In order to reduce the complexity, a hybrid selection MRC scheme can be used instead of full MRC. If we have five antennas but only use the three strongest, what is the loss in terms of mean SNR compared with full MRC?

6. Consider a scenario where the TX is equipped with two antennas and the RX only one. The TX sends two symbols in the following fashion (see Figure 30.7). In the first symbol interval, symbol s_1 is the TX from the first antenna and symbol s_2 from the second antenna. In the second symbol interval s_2^* is transmitted from the first antenna and $-s_1^*$ from the second antenna. The (complex-valued) attenuation between the first transmit antenna and the receive antenna is h_1, and the corresponding attenuation for the second transmit antenna is h_2. The attenuations are assumed to be constant over both signaling intervals. At the RX, AWGN is added; n_1 in the first symbol interval and n_2 in the second (see also Section 20.2).
   (a) Derive the output from the receive antenna for both symbol intervals. Let the output for the first interval be r_1, and r_2 for the second.
   (b) Let the following operation be performed at the receiving end:

$$\left.\begin{array}{l} \hat{s}_1 = h_1^* r_1 - h_2 r_2^* \\ \hat{s}_2 = h_2^* r_1 + h_1 r_2^* \end{array}\right\} \tag{30.25}$$

What is achieved by this operation?

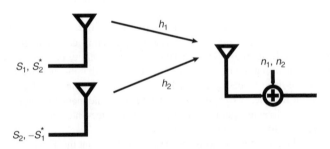

Figure 30.7 Principle of Alamouti codes.

7. Consider an N_r-branch antenna diversity system using the MRC rule. All branches are subject to Rayleigh fading and the fading at one branch is independent of fading at all other branches. Derive the pdf of the SNR at the output of the combiner for the following cases:

(a) The average SNR per branch is $\bar{\gamma}$, for all branches.

(b) The average SNR for the ith branch is $\bar{\gamma}_i$. Let the $\bar{\gamma}_i$'s be distinct.

8. In CDMA systems, each symbol is spread over a large bandwidth by multiplying the symbol using a spreading sequence. If s is the transmitted symbol and $\xi(t)$ is the spreading sequence, the received baseband signal can be written as

$$r(t) = \sum_{m=1}^{M} \alpha_m \xi(t - \tau_m)s + n(t) \tag{30.26}$$

where M is the number of resolvable multipaths, α_m is the complex gain of the mth multipath, τ_m the delay of the mth multipath, and $n(t)$ is an AWGN process. Assume that gains $\{\alpha_m\}$ and delays $\{\tau_m\}$ can be estimated perfectly, and that the spreading sequence has perfect autocorrelation properties. Let there be one matched filter per multipath and a device that combines the outputs of the matched filters. Derive an expression for the SNR at the output of the combiner.

9. As described in Section 13.5.2 in the text, the RX in a switched diversity system with two antennas takes as its input the signal from one antenna as long as the SNR is above some threshold. When the SNR falls below the threshold, the RX switches to the second antenna, regardless of the SNR at the second antenna. If the antennas fade independently and according to the same distribution, it can be shown that the cdf of the SNR at the RX is

$$cdf_\gamma(\gamma) = \begin{cases} \Pr(\gamma_1 \leq \gamma_t \text{ and } \gamma_2 \leq \gamma), & \text{for } \gamma < \gamma_t \\ \Pr(\gamma_t \leq \gamma_1 \leq \gamma \text{ or } [\gamma_1 \leq \gamma_t \text{ and } \gamma_2 \leq \gamma]), & \text{for } \gamma \geq \gamma_t \end{cases} \tag{30.27}$$

where γ is the SNR after the switching device (i.e., the SNR at the RX), γ_1 is the SNR of the first antenna, γ_2 is the SNR of the second antenna, and γ_t is the switching threshold.

(a) For Rayleigh fading where both antennas have the same mean SNR, give the cdf and pdf of γ.

(b) If we use mean SNR as our performance measure, what is the optimum switching threshold and resulting mean SNR? What is the gain in dB of using switched diversity compared with that of a single antenna? Compare this gain with that of maximal ratio combining and selection diversity.

(c) If, instead, our performance measure is average BER, what is the optimum switching threshold for binary noncoherent FSK? What is the average BER for an SNR of 15 dB? Compare this with the case of a single antenna. Remember that the BER for binary noncoherent FSK is

$$\text{BER} = \frac{1}{2}\exp\left(-\frac{\gamma}{2}\right) \tag{30.28}$$

10. In maximal ratio combining each branch is weighted with the complex conjugate of that branch's complex fading gain. However, in practice the RX must somehow estimate fading gains in order to multiply received signals by them. Assume that this estimation is based on pilot symbol insertion, in which case the weights become subject to a complex Gaussian error. If an N_r-branch diversity system with Rayleigh fading and a mean SNR Γ on each branch is used, the pdf of the output SNR becomes [Tomiuk et al. 1999]

$$pdf_\gamma(\gamma) = \frac{(1-\rho^2)^{N_r-1}e^{-\gamma/\bar{\gamma}}}{\bar{\gamma}} \sum_{n=0}^{N_r-1} \binom{N_r-1}{n} \left[\frac{\rho^2\gamma}{(1-\rho^2)\bar{\gamma}}\right]^n \frac{1}{n!} \tag{30.29}$$

where ρ^2 is the normalized correlation coefficient between fading gain on a branch and its estimate – i.e., the weight.

(a) Show that the pdf above can be written as a weighted sum of N_r ideal-maximal-ratio SNR pdfs.

(b) What happens when fading gains and weights become fully uncorrelated?

(c) What happens when fading gains and weights become fully correlated?

(d) The average error rate for an ideal N_r-branch MRC channel, for a certain modulation scheme, is denoted $P_e(\bar{\gamma}, N_r)$. Using the result from (a), give a general expression for the average error rate for N_r-branch MRC with imperfect weights.

(e) For many schemes, the average error rate for ideal MRC for large mean SNRs can be approximated as

$$\tilde{P}_e(\bar{\gamma}, N_r) = \frac{C(N_r)}{\bar{\gamma}^{N_r}} \qquad (30.30)$$

where $C(s)$ is a constant specific to the modulation scheme. Using the result from (d), find out what happens to the average error rate for N_r-branch MRC with imperfect weights when the mean SNR grows very large and $\rho < 1$.

11. Consider an N_r-branch diversity system. Let the signal on the kth branch be $\tilde{s}_k = s_k e^{-j\phi_k}$, noise power on each branch be N_0, and noise be independent between the branches. Each branch is phase adjusted to zero phase, weighted with α_k, and then the branches are combined. Give expressions for the branch SNR, the combined SNR, and then derive the weights α_k that maximize the combined SNR. With optimal weights, what is the combined SNR expressed in terms of the branch SNRs?

30.14   Chapter 14: Channel Coding

1. Let us consider a linear cyclic (7, 3) block code with generator polynomial $G(x) = x^4 + x^3 + x^2 + 1$.

   (a) Encode the message $U(x) = x^2 + 1$ systematically, using $G(x)$.

   (b) Calculate the syndrome $S(x)$ when we have received (probably corrupted) $R(x) = x^6 + x^5 + x^4 + x + 1$.

   (c) A close inspection reveals that $G(x)$ can be factored as $G(x) = (x + 1)T(x)$, where $T(x) = x^3 + x + 1$ is a primitive polynomial. This implies, we claim, that the code can correct all single errors and all double errors, where the errors are located next to each other. Describe how you would verify this claim.

2. We have a binary systematic linear cyclic (7, 4) code. The codeword $X(x) = x^4 + x^2 + x$ corresponds to the message $U(x) = x$. Can we, with just this information, calculate the codewords for all messages? If so, describe in detail how it is done and which code properties you use.

3. Show that for a cyclic (N, K) code with generator polynomial $G(x)$ there is only one codeword with degree $N-K$ and that codeword is the generator polynomial itself. In a cyclic block code, a cyclic shift of a codeword results in another valid codeword.

4. The polynomial $x^{15} + 1$ can be factored into irreducible polynomials as

$$x^{15} + 1 = (x^4 + x^3 + 1)(x^4 + x^3 + x^2 + x + 1)$$

$$\cdot (x^4 + x + 1)(x^2 + x + 1)(x + 1)$$

Using this information, list all generator polynomials generating binary cyclic (15, 8) codes.

5. Assume that we have a linear (7, 4) code, where the codewords corresponding to messages $\mathbf{u} = [1000], [0100], [0010],$ and $[0001]$ are given as

Message					Codeword						
1	0	0	0	→	1	1	0	1	0	0	0
0	1	0	0	→	0	1	1	0	1	0	0
0	0	1	0	→	0	0	1	1	0	1	0
0	0	0	1	→	0	0	0	1	1	0	1

   (a) Determine all codewords in this code.

(b) Determine the minimum distance, d_{min}, and how many errors, t, the code can correct.

(c) The code above is not in systematic form. Calculate the generator matrix **G** for the corresponding systematic code.

(d) Determine the parity check matrix **H**, such that $\mathbf{HG}^T = 0$.

(e) Is this code cyclic? If so, determine its generator polynomial.

6. We have a linear systematic (8, 4) block code, with the following generator matrix:

$$\mathbf{G} = \begin{bmatrix} 1 & 0 & 0 & 0 & 1 & 1 & 0 & 1 \\ 0 & 1 & 0 & 0 & 0 & 1 & 1 & 1 \\ 0 & 0 & 1 & 0 & 1 & 1 & 1 & 0 \\ 0 & 0 & 0 & 1 & 1 & 0 & 1 & 1 \end{bmatrix}$$

(a) Determine the codeword corresponding to the message $\mathbf{u} = [1011]$, the parity matrix **H**, and calculate the syndrome when the word $\mathbf{y} = [010111111]$ is received.

(b) By removing the fifth column of **G**, we get a new generator matrix $\mathbf{G}^*$, which generates a (7, 4) code. The new code is, in addition to its linear property, also cyclic. Determine the generator polynomial by inspection of $\mathbf{G}^*$. It can be done by observing a property of cyclic codes – which?

7. Show that the following inequality (the Singleton bound) is always fulfilled for a linear (N, K) block code:

$$d_{min} \leq N - K + 1$$

8. Show that the Hamming bound

$$2^{N-K} \geq \sum_{i=0}^{t} \binom{N}{i}$$

needs to be fulfilled if we want syndrome decoding to be able to correct t errors using a linear binary (N, K) code. *Note*: A code that meets the Hamming bound with equality is called a *perfect code*.

9. Show that Hamming codes are perfect $t = 1$ error-correcting codes.

10. Consider the convolutional encoder in Figure 14.3.

(a) If binary antipodal transmission is used, we can represent coding and modulation together in a trellis where 1s and 0s are replaced by +1s and −1s instead. Draw a new version of the trellis stage in Figure 14.5a, using this new representation.

(b) A signal is transmitted over an AWGN channel and the following (soft) values are received:

$$-1.1; 0.9; -0.1 \quad -0.2; -0.7; -0.6 \quad 1.1; -0.1; -1.4 \quad -0.9; -1.6; 0.2$$
$$-1.2; 1.0; 0.3 \quad \quad 1.4; 0.6; -0.1 \quad \quad -1.3; -0.3; 0.7$$

If these values were detected as binary digits, before decoding, we would get the binary sequence in Figure 14.5b. This time, however, we are going to use soft Viterbi decoding with squared Euclidean metric. Perform soft decoding such that it corresponds to the hard decoding performed in Figure 14.5d–f. After the final step, are there the same survivors as for the hard-decoding case?

11. Block codes on fading channels: the text indicates (the expression is stated for a special case in Section 14.8.2) that the BER of a t-error correcting (N, K) block code over a properly interleaved Rayleigh-fading channel with hard decoding is proportional to

$$\sum_{i=t+1}^{N} K_i \left(\frac{1}{2 + 2\overline{\gamma}_B} \right)^i \left(1 - \frac{1}{2 + 2\overline{\gamma}_B} \right)^{N-i}$$

where the K_is are constants and $\overline{\gamma}_B$ is the average SNR. The text also states that in general a code with minimum distance d_{min} achieves a diversity order of $\left\lceil \frac{d_{min}-1}{2} \right\rceil + 1$. Prove this statement using the expression above for proportionality of the BER.

30.15   Chapter 15: Speech Coding

1. Explain the main drawbacks to lossless speech coders.
2. Describe the three basic types of speech coders.
3. What are the most relevant spectral characteristics of speech?
4. In this problem, we study the spectra of vowels. Based on the formant frequencies for the vowels /iy/ (270, 2290, and 3010 Hz) and /aa/, (730, 1090, 2240 Hz),
   (a) Sketch the spectral envelopes (the magnitude transfer function of the vocal tract filter). The excitation signal is a series of delta pulses spaced 1/80 Hz apart.
   (b) Sketch the pole positions in the complex plane.
   (c) Indicate the tongue position for each vowel.
   (d) Synthesize the vowels artificially using MATLAB.
5. A glottal pulse can be modeled by

$$g(n) = \begin{cases} 1 + \cos(\pi \frac{n}{T}) & n = 0, \ldots, T - 1 \\ 0 & \text{otherwise} \end{cases}$$

   (a) Sketch $g(n)$.
   (b) Calculate the Discrete-time Fourier Transform (DtFT) of $g(n)$.
6. Here we study a popular method for speech time-scale modification: Waveform Similarity Over-Lap and Add (WSOLA). In WSOLA, frames of length N samples are extracted from the original speech waveform and then these frames are overlapped 50% and added to form the output. Here we study a special case of WSOLA. To formalize we index each frame with i, and denote the frame length N. If we extract nonoverlapping frames, the extraction point (index of the first sample in the frame) is iN. The extracted frames are overlapped 50% and added (see Figure 30.8).
   (a) Calculate (sketch) the output for the input signal in Figure 30.9. Use a frame length $N = 8$ and calculate the output for 6 extracted frames.
   (b) To improve the quality of the output, waveform similarity matching can be performed. This changes the position of the frames that are extracted slightly; the extracted frames are still overlapped 50 % and added. The nominal extraction point for frame i is iN but the actual point is $iN + k_i^*$, where

$$k_i^* = \text{argmax}_{k_i} \hat{R}(k_i), k_i \in [-\Delta, \Delta]$$

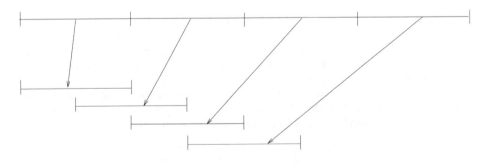

Figure 30.8   Principle of WSOLA.

Figure 30.9 The signal in Problem 15.6.

and

$$\hat{R}(k_i) = \sum_{n=0}^{N-1} x_{nc}(n)x_{ex}(n+k_i)$$

is a cross-correlation. $x_{nc}(n) = x(n+iN-N/2+k^*_{i-1})$, $n = 0, \ldots, N-1$ is the natural continuation, i.e., the waveform that continues after the previously extracted frame. $x_{ex}(n+k_i) = x(n+k_i+iN)$ is the extracted frame. WSOLA tries to extract a frame that correlates well with the natural continuation.

(c) For the input signal in Figure 30.9, calculate (sketch) the output. Use $\Delta = 2$ and, as before, $N = 8$. (Hint: The first frame ($i = 0$) is just left where it is, and $k^*_0 = 0$.)

7. If we define the short-time spectrum of a signal in terms of its short-time Fourier transform as

$$S_m(e^{j\omega}) = |X_m(e^{j\omega})|^2$$

and we define the short-time autocorrelation of the signal as

$$R_m(k) = \sum_{n=-\infty}^{\infty} x_m(n)x_m(n+k)$$

where $x_m(n) = x(n)w(n-m)$, then show that for

$$X_m(e^{j\omega}) = \sum_{n=-\infty}^{\infty} x(n)w(n-m)e^{-j\omega n}$$

$R_m(k)$ and $S_m(e^{j\omega})$ are related as a normal (long-time) Fourier transform pair. In other words, show that $S_m(e^{j\omega})$ is the (long-time) Fourier transform of $R_m(k)$, and vice versa.

8. To illustrate the effect of the window position, consider the periodic unit sample sequence

$$x(n) = \sum_{l=-\infty}^{\infty} \delta(n-lP)$$

and a triangle analysis window $w(n)$ of length P. Calculate the DtFT and the Discrete Fourier Transform (DFT) of the short-time section $x_m(n) = x(n)w(n-m)$.

9. An AR (Auto Regressive) sequence $\{x_n\}$ has been formed by a stationary white random sequence u_n passing through a second order all-pole filter (purely recursive filter) with the difference equation

$$x(n) + c_1 x(n-1) + c_2 x(n-2) = u(n)$$

(a) If $x(n)$ has unity variance ($R_{xx}(0) = 1$), calculate $R_{xx}(1)$, $R_{xx}(2)$ and $R_{xx}(3)$.

(b) Use the Levinson – Durbin algorithm (see [Haykin 1991] or [http://ccrma.stanford.edu/~jos/lattice/Levinson_Durbin_algorithm.html]) to calculate the predictor coefficients $a^{(p)}$ for $j = 1, \ldots, p$ and the prediction error $V^{(p)}$ for $p = 3$.

10. A sequence $\{x_n\}$ has been formed as an MA (Moving Average) sequence by using a stationary white random sequence u_n according to

$$x(n) = u(n) + u(n - 1)$$

with zero-mean and unity variance.

(a) Determine the autocorrelations $R_{XX}(0)$, $R_{XX}(1)$ and $R_{XX}(2)$.

(b) Determine the optimum predictor coefficients $a_1^{(2)}$ and $a_2^{(2)}$ and the associated error measure $\alpha_2^{(2)}$ of a second order MEV predictor ($p = 2$).

(c) In an application, $x(n)$ is to be predicted using only a first-order predictor ($p = 1$). Determine the predictor coefficient $a_1^{(l)}$ and the associated error measure $\alpha^{(l)}$ in this case.

11. Considering narrowband speech representation ($f_s = 8000$ Hz), human beings are known to produce typically 1 formant per kHz, resulting in 4 formants (or poles) to code. With this knowledge, we would normally choose a linear prediction filter of order 8, which is capable of modeling all the formants. Despite that, typically 10 filter coefficients are used to describe the LP filters. Give an explanation.

12. To make linear prediction analysis adapt to the quasi-stationary speech source, the autocorrelations from stochastic linear prediction can be replaced by short-time estimates of the ACF, calculated over samples inside a window starting at sample index m

$$R_m(k) = \sum_{n=-\infty}^{\infty} x_m(n)x_m(n + k)$$

where $x_m(n) = x(n)w(n - m)$. In this problem, we study if this approach is optimal when processing short blocks.

To this end we define, just like in the stochastic case, the "predictor" $\hat{x}_m(n) = \sum_{k=1}^{p} a_k x_m(n - k)$, and the prediction error $e_m(n) = x_m(n) - \hat{x}_m(n)$. We replace the expected squared error with the energy of the prediction error sequence

$$\epsilon = \sum_{n=-\infty}^{\infty} e_m(n)^2$$

and minimize ϵ with respect to $a_l, \ldots, a_p$. Show that the coefficients that minimize ϵ are $\mathbf{a} = \mathbf{R}^{-l}\mathbf{r}$, where $\mathbf{R}$ is a Toeplitz matrix with the first row being $(R_m(0), R_m(1), \ldots, R_m(p - 1))$, and $\check{r} = (R_m(1), R_m(2), \ldots, R_m(p))^{\mathrm{T}}$, i.e., our ad hoc scheme of replacing the stochastic correlations with the short-time dito makes sense.

For simplicity you can set $m = 0$.

Note that we know all the samples in the frame $x_m(n)$ and prediction is perhaps a misleading name. A more appropriate term is least square fitting. The name used in the literature is "the autocorrelation method."

13. Consider a stochastic MA signal

$$x(n) = u(n) + 0.5u(n - 1)$$

where $u(n)$ is iid, zero-mean with variance $\sigma_U^2 = 0.77$.

(a) Calculate $R_x(k)$.

(b) Calculate the predictor coefficients for a predictor of order 1,2,3 (e.g., using the Levinson – Durbin algorithm).

(c) Consider next the design of a linear predictor. What is the minimum prediction error variance we can achieve (allowing arbitrarily high predictor order)?

14. A speech signal $x(t)$ is applied to a quantizer. Assume that a sample of the input signal has a density function as follows:

$$p_X(x) = \begin{cases} k.e^{-|x|} & -4 < x < 4 \\ 0 & \text{otherwise} \end{cases}$$

(a) Compute the constant k.

(b) Compute the stepsize Δ in a 4-level midrise quantizer (a midrise quantizer has an output that is $\lfloor x/Q \rfloor$, so that, e.g., any input value between 0 and Δ is represented by the quantized value $\Delta/2$). Choose the stepsize as the smallest possible value that does not result in overload distortion.

(c) Determine the variance σ_Q^2 of the quantizer error. The assumption that $p_X(x)$ is constant in each interval is not permitted.

(d) Determine the (SNR) at the output of the quantizer.

15. Long-term statistics for the amplitude of a speech signal are commonly assumed to be Laplace distributed. The Laplace probability density function (with zero mean) is given by

$$p_X(x) = \frac{1}{\sqrt{2\sigma_X^2}} : e^{-\sqrt{\frac{2}{\sigma_X^2}}|x|}$$

The variance is $\sigma_X^2 = 0.02$ and the overload level $x_{\max} = 1$. Before quantization, a Laplace-distributed random variable X is compressed by a compression function $f(x)$, leading to the transformed random variable C at compressor output. The compressor is of μ-law type

$$\text{output}(x) = x_{\max} \frac{\log\left[1 + \mu \frac{|x|}{x_{\max}}\right]}{\log(1+\mu)} \sin(x) \tag{30.31}$$

with $\mu = 100$. Compute the probability density function $p_C(c)$ of the signal at compressor output. Find an expression for the probability $P(c \leq C \leq c + dc)$ in terms of $p_C(c)$ and $p_X(x)$.

16. Consider split VQ of LP parameter vectors. We wish to encode 10-dimensional vectors of line spectral frequencies. We use a 2-split, where the first split is six dimensional, and the second is four dimensional. A total of 24 bits is used to quantize a whole vector. How should the bits be allocated between the splits to minimize computational complexity?

17. Analyze the complexity of 2-stage VQ. Assume the first stage has K bits, and the second stage has $24 - K$ bits. Measure the computational complexity in terms of the total number of codevectors that need to be tested during encoding.

(a) Sketch the complexity as a function of K. Use a logarithmic scale for the complexity!

(b) What value of K minimizes the computational complexity?

(c) What value of K minimizes the expected distortion?

30.16 Chapter 16: Equalizers

1. To mitigate the effects of multipath propagation, we can use an equalizer at the RX. A simple example of an equalizer is the linear zero-forcing equalizer. Noise enhancement is, however, one of the drawbacks to this type of equalizer. Explain the mechanism behind noise enhancement and name an equalizer type where this is less pronounced.

2. State the main advantage and disadvantage of blind equalization and name three approaches to designing blind equalizers.
3. The "Wiener–Hopf" equation was given by $\mathbf{R} \mathbf{e}_{opt} = \mathbf{p}$. For real-valued white noise n_m with zero mean and variance σ_n^2, calculate the correlation matrix $\mathbf{R} = E\{\mathbf{u}^*\mathbf{u}^T\}$ for the following received signals:
   (a) $u_m = a \sin(\omega m) + n_m$;
   (b) $u_m = b u_{m-1} + n_m$; $a, b \in R, b \neq \pm 1$.
4. Consider a particular channel with the following parameters:

$$\mathbf{R} = \begin{bmatrix} 1 & 0.576 & 0.213 \\ 0.576 & 1 & 0.322 \\ 0.213 & 0.322 & 1 \end{bmatrix}, \quad \mathbf{p} = \begin{bmatrix} 0 & 0.278 & 0.345 \end{bmatrix}, \quad \text{and} \quad \sigma_S^2 = 0.7$$

   Write the MSE equation for this channel in terms of real-valued equalizer coefficients.
5. An infinite-length ZF equalizer can completely eliminate ISI, as long as the channel transfer function is finite in the transform domain – i.e., Eq. (16.40). Here, we investigate the effect of using a finite-length equalizer to mitigate ISI.
   (a) Design a five-tap ZF equalizer for the channel transfer function described in Table 30.1 – i.e., an equalizer that forces the impulse response to be 0 for $i = -2, -1, 1, 2$ and 1 for $i = 0$.
   Hint: A 5×5 matrix inversion is involved.
   (b) Find the output of the above equalizer and comment on the results.

Table 30.1   Channel transfer function

n	f_n
−4	0
−3	0.1
−2	−0.02
−1	0.2
0	1
1	−0.1
2	0.05
3	0.01
4	0

6. When transmitting 2-Amplitude Shift Keying (2-ASK) (with alternatives −1 and +1 representing "0" and "1," respectively) over an AWGN channel, ISI is experienced as a discrete time equivalent channel $F(z) = 1 + 0.5z^{-1}$ (see Figure 30.10). When transmitting over this channel, we assume that the initial channel state is −1, and the following noisy sequence is received when transmitting 5 consecutive bits. Transmission continues after this, but we only have the following information at this stage:

$$0.66 \quad 1.59 \quad -0.59 \quad 0.86 \quad -0.79$$

   (a) What would the equalizer filter be if we apply a ZF linear equalizer?
   (b) What is the memory of this channel?
   (c) Draw one trellis stage with states, input symbols, and output symbols shown.
   (d) Draw a full trellis for this case and apply the Viterbi algorithm to find the maximum-likelihood sequence estimate of the transmitted 5-bit sequence.

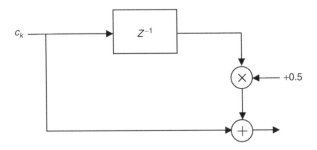

Figure 30.10   Block diagram of the channel $F(z) = 1 + 0.5z^{-1}$.

7. In general, the MSE equation is a quadratic function of equalizer weights. It is always positive, convex, and forms a hyperparabolic surface. For a two-tap equalizer, the MSE equation takes the form:

$$Ae_1^2 + Be_1e_2 + Ce_2^2 + De_1 + Ee_2 + F$$

with $A, B, C, D, E \in \mathbb{R}$.
For the following data, make a contour plot of the hyperbolic surface formed by the MSE equation:

$$\mathbf{R} = \begin{bmatrix} 1 & 0.651 \\ 0.651 & 1 \end{bmatrix}$$

$$\mathbf{p} = \begin{bmatrix} 0.288 & 0.113 \end{bmatrix}^T, \quad \text{and} \quad \sigma_S^2 = 0.3$$

8. As stated in the text, the choice of the μ parameter strongly influences the performance of the LMS algorithm. It is possible to study the convergence behavior in isolation by assuming perfect knowledge of $\mathbf{R}$ and $\mathbf{p}$.
   (a) For the data provided in Exercise 16.7, plot the convergence of the LMS algorithm for $\mu = 0.1/\lambda_{\max}, 0.5/\lambda_{\max}, 2/\lambda_{\max}$ with initial value $\mathbf{e} = \begin{bmatrix} 1 & 1 \end{bmatrix}^T$ and make comparisons between the results:

$$R = \begin{bmatrix} 1 & 0.651 \\ 0.651 & 1 \end{bmatrix}$$

$$\mathbf{p} = \begin{bmatrix} 0.288 & 0.113 \end{bmatrix}^T, \quad \text{and} \quad \sigma_S^2 = 0.3$$

   Hint: For the real-valued problem, the gradient of the MSE equation is given by $\frac{\partial}{\partial e_n}\text{MSE} = \nabla_n = -2\mathbf{p} + 2\mathbf{R}e_n$.
   (b) Plot the convergence paths of the equalizer coefficients for all three cases.

30.17   Chapter 17: Multiple Access and the Cellular Principle

1. An analog cellular system has 250 duplex channels available (250 channels in each direction). To obtain acceptable transmission quality, the relation between reuse distance (D) and cell radius (R) has to be at least $D/R = 7$. The cell structure is designed with a cell radius of $R = 2$ km. During a busy hour, the traffic per subscriber is on average one call of 2-minute duration. The network setup is modeled as an Erlang-B loss system with the blocking probability limited to 3%.

(a) Calculate the following:
  (i) The maximal number of subscribers per cell.
  (ii) The capacity of the network in Erlangs/km^2. (Assume that the cell area is $A_{cell} = \pi R^2$.)
(b) The analog system above is modernized for digital transmission. As a consequence, the channel separation has to be doubled – i.e., only 125 duplex are now available. However, digital transmission is less sensitive to interference and acceptable quality is obtained for $D/R = 4$. How is the capacity of the network affected by this modernization (in terms of Erlangs/km^2)?
(c) To increase the capacity of the network in B, the cells are made smaller, with a radius of only $R = 1$ km. How much is capacity increased (in terms of Erlangs/km^2) and how many more BSs are required to cover the same area?

2. A system specifies a blocking level to be less than 5% for 120 users each with an activity level of 10%. When a user is blocked, it is assumed to be cleared immediately – i.e., the system is an Erlang-B system. Assume two scenarios: (i) one operator and (ii) three operators. How many channels are needed for the two scenarios?

3. A system specifies a blocking level to be less than 5% for 120 users, each with an activity level of 10%. When a user is blocked, it is assumed to be placed in an infinite queue – i.e., the system is an Erlang-C system. Assume two scenarios: (i) one operator and (ii) three operators.
  (a) How many channels are needed for the two scenarios (compare with the previous problem)?
  (b) What is the average waiting time if the average call duration is 5 min?

4. TDMA requires a temporal guard interval.
  (a) The cell radius of a mobile system is specified as 3,000 m and the longest impulse response in the cell is measured as 10 µs. What is the minimum temporal guard interval needed to avoid overlapping transmissions?
  (b) How is the temporal guard interval reduced in GSM?

5. Consider a cellular system of hexagonal structure and a reuse distance D (distance to the closest co-channel BS), a cell radius of R, and a propagation exponent η. It is assumed that all six co-channel BSs transmit independent signals with the same power as the BS in the studied cell.
  (a) Show that the downlink-carrier-to-interference ratio is bounded by

$$\left(\frac{C}{I}\right) > \frac{1}{6}\left(\frac{R}{D-R}\right)^{-\eta} \qquad (30.32)$$

  (b) Figure 30.11 shows an illustration of the worst case uplink scenario, where communication from MS 0 to BS 0 is affected by interference from other co-channel mobiles in first-tier co-channel cells. The worst interference scenario is when co-channel mobiles (MS 1 to MS 6) communicating with their respective BSs (BS 1 to BS 6) are at their respective cell boundaries, in the direction of BS 0. Calculate the C/I at BS 0, given that all mobiles transmit with the same power, and compare your expression with the one above.

6. Consider a system with a requirement for successful demodulation such that $SNIR = 7$ dB operating in an environment with 10-dB SNR. The received power at the MS is equal for both the signal of interest and the interferer. However, the interferer is suppressed by 10 dB from, e.g., spreading gain. Compute the maximum effective throughput in a slotted ALOHA system.

7. Consider a cellular system with TDD; let the cell radius be 1.5 km, and the duplexing time (time that the system is either in uplink or downlink) be 1 ms. What is the worst-case loss of spectral efficiency due to the time-of-flight of a signal through the cell?

8. Consider the system layout of Figure 30.11, and let the distance between two BSs be four times the cell radius; the mean path loss follows a d^{-4} law. Assume that each link suffers not only from path loss, but also from shadowing with standard deviation $\sigma = 6$ dB. Derive the mean interference power, and the cdf of the interference power.

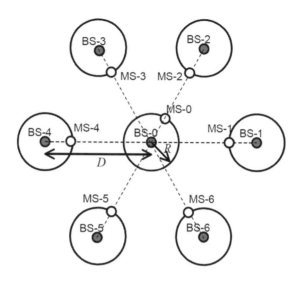

Figure 30.11 Cell structure.

9. Consider a cellular packet radio system where the MSs constantly sense the channel; if they detect that the channel is free, they wait a random time before sending a packet (CSMA). Let this random wait time t_{wait} be distributed as $\exp(-t_{wait}/\tau_{wait})$. Due to the finite runtime t_{run} of the signals in the cell, collisions can occur. Let the runtime be approximated as a random variable uniformly distributed between 0 and $\tau_{run} = 1$. What is the probability of a collision if two users are trying to access the channel?

30.18   Chapter 18: Spread Spectrum Systems

1. Two MSs are communicating with the same BS. The information to be sent by the two MSs is

$$s_1 = \begin{bmatrix} 1 & -1 & 1 & 1 \end{bmatrix} \tag{30.33}$$

$$s_2 = \begin{bmatrix} -1 & 1 & -1 & 1 \end{bmatrix} \tag{30.34}$$

spread by spreading sequences c_1 and c_2. The radio channels between the MSs and BS are described by impulse responses h_1 and h_2.
   (a) Plot the received signal before and after despreading for a $h_1 = h_2 = [1]$ using PN spreading sequences c_1 and c_2 of length 4, 16, and 128. Find the information sent in the despreaded sequence.
   (b) Plot the received signal before and after despreading for $h_1 = h_2 = \begin{bmatrix} 1 & 0.5 & 0.1 \end{bmatrix}$ using PN-sequences c_1 and c_2 of length 4, 16, and 128. Find the information sent in the despreaded sequence.
   (c) Plot the received signal before and after despreading for $h_1 = \begin{bmatrix} 1 & 0.5 & 0.1 \end{bmatrix}$, $h_2 = \begin{bmatrix} 1 & 0 & 0.5 \end{bmatrix}$ using PN-sequences c_1 and c_2 of length 4, 16, and 128. Find the information sent in the despreaded sequence.
   (d) Repeat (a), (b), and (c) for Hadamard sequences of length 16.
2. Consider a frequency-hopping system with four possible carrier frequencies, and hopping sequence 1, 2, 3, 4. What are hopping sequences of length four that have only one collision with this sequence (for arbitrary integer shifts between sequences)?

3. Consider a system with frequency hopping plus simple code (7, 4 Hamming code) with generator matrix:

$$
\mathbf{G} = \begin{bmatrix} 1 & 0 & 0 & 0 & 1 & 1 & 0 \\ 0 & 1 & 0 & 0 & 1 & 0 & 1 \\ 0 & 0 & 1 & 0 & 0 & 1 & 1 \\ 0 & 0 & 0 & 1 & 1 & 1 & 1 \end{bmatrix}
\tag{30.35}
$$

Assume BPSK modulation, and compute the BER with and without frequency hopping. An interleaver maps every symbol to alternating frequencies (seven are available). The coherence time is larger than the symbol duration. Every frequency is independently Rayleigh fading. Let the RX use hard decoding. Plot the BER as a function of average SNR (use MATLAB).

4. Consider an MFEP with ELP impulse response:

$$
f_{\mathrm{M}}(t) = \begin{cases} s^*(T_{\mathrm{s}} - t), & 0 < t < T_{\mathrm{s}} \\ 0, & \text{otherwise} \end{cases}
\tag{30.36}
$$

where $s(t)$ denotes the ELP-transmitted signal having energy $2E_{\mathrm{s}}$, and T_{s} is the symbol duration. Show that for a slowly varying WSSUS channel, the correlation function of the MFEP output is

$$
R_y(t_1, t_2) = \begin{cases} \int_{-\infty}^{+\infty} P_{\mathrm{h}}(0, \tau) \tilde{R}_{\mathrm{s}}^*(t_1 - T_{\mathrm{s}} - \tau) \tilde{R}_{\mathrm{s}}(t_2 - T_{\mathrm{s}} - \tau) \, d\tau & \text{for } |t_2 - t_1| < \frac{2}{B_{\mathrm{s}}} \\ 0 & \text{otherwise} \end{cases}
\tag{30.37}
$$

where

$$
\tilde{R}_{\mathrm{s}}(t - T_{\mathrm{s}} - \tau) \triangleq \int_0^{+\infty} s(t - \alpha - \tau) f_{\mathrm{M}}(\alpha) \, d\alpha
\tag{30.38}
$$

What is the autocorrelation of the noise?

5. Consider a channel with three taps that are Nakagami-m-fading, and have mean powers 0.6, 0.3, 0.1, and m-factors of 5, 2, and 1.
   (a) What is the diversity order – i.e., the slope of the BER versus SNR curves at high SNRs – when maximum ratio combining is applied?
   (b) Give a closed-form equation for the average BER of BPSK in such a channel.
   (c) Plot the BER as a function of average SNR, and compare it with pure Rayleigh fading (other parameters identical).

6. A CDMA handset is at the boundary of the cell, and thus in soft handover. It operates in a rich multipath environment (and thus sees a large number of resolvable MPCs). The shadowing standard deviation is $\sigma_{\mathrm{F}} = 5\,\mathrm{dB}$ and mean received SNR is $8\,\mathrm{dB}$. What is the probability of outage if the handset requires a 4-dB SNR to operate? *Hint*: for computing the distribution of the sum of log-normally distributed variables, convert to natural logarithms, and match the first and second moments of the desired approximation and of the given sum of powers.

7. Consider a CDMA system, operating at 1,800 MHz, with a cell size of 1 km, circularly shaped cells, and users distributed uniformly in the cell area. The path loss model is the following: free space up to a distance of 100 m, and $n = 4$ beyond that. In addition, Rayleigh fading is superimposed; neglect shadow fading. Let power control ensure that the signal received at the desired BS is constant at $-90\,\mathrm{dBm}$. Simulate the average power received from these handsets at the *neighboring* BS (use MATLAB).

8. Consider a CDMA system with three users, each of which is spread using a PN-sequence. The spreading sequences are generated from the shift register of Figure 30.12, where the initializations are [100], [110], and [101], respectively. Let the three TXs send out sequences $\begin{bmatrix} 1 & -1 & 1 & 1 \end{bmatrix}, \begin{bmatrix} -1 & -1 & -1 & 1 \end{bmatrix}, \begin{bmatrix} -1 & 1 & -1 & 1 \end{bmatrix}$. The gains from the TXs to the RXs $h_{1,j}$

Figure 30.12 A maximal LFSR sequence generator.

are 1, 0.6, and 11.3 and let there be no power control. Consider the noise sequence (generate 21 noise samples with variance 0.3). Compute the received signal both before and after despreading. What bit sequences are detected for the three users? How do the results change when perfect power control is implemented? Use MATLAB for the simulation.

9. MATLAB: Consider again the CDMA signal from Problem 8. Write a MATLAB program that performs ZF multiuser detection, and one that performs serial interference cancellation. What are the detected signals in the two cases?

10. Derive the power-spectral density of TH-IR with 2-PPM and short spreading sequences – i.e., the duration of the spreading sequence is equal to the symbol duration.

11. Consider a CDMA system with a target SINR of 6 dB. At the cell boundary, the SNR is 9 dB. The spreading factor is 64; the orthogonality factor is 0.4. How many users can be served per cell (disregard adjacent cell interference).

12. Consider the downlink of a CDMA system with 4 Mchip/s chip rate. The cell contains 4 data users, with (coded) data rates of 500, 500, 250, and 125 kbit/s. How many speech users (with coded data rate of 15.6 kbit/s can be served, assuming that downlink spreading codes should be completely orthogonal?

13. Soft handover always leads to better reliability of a link, but does not necessarily increase cellular capacity. Under what circumstances can soft handover decrease the capacity in the (i) uplink, and (ii) downlink?

30.19 Chapter 19: Orthogonal Frequency Division Multiplexing (OFDM)

1. Consider an eight-tone OFDM system transmitting a time domain baseband signal $s[n] = \{1\ 4\ 3\ 2\ 1\ 3\ 1\ 2\}$ over a channel with an impulse response of $h[n] = \{2\ 0\ 2\ -1\}$.
   (a) Draw the block diagram of a baseband OFDM system.
   (b) What is the minimum required length of the cyclic prefix?
   (c) Plot the signal vector for
      (i) $s[n]$;
      (ii) $s_C[n] = s[n]$ with cyclic prefix;
      (iii) $y_C[n] =$ after passing through the channel $h[n]$;
      (iv) $y[n] =$ after cyclic prefix deletion;
      (v) $Y[k] =$ after DFT;
      (vi) $\hat{S}[k] = Y[k]/H[k]$;
      (vii) $\hat{s}[n] =$ IDFT of $\hat{S}[k]$.

2. Consider an OFDM system with mean output power normalized to unity. Let the system have a power amplifier that amplifies with a cutoff characteristic – i.e., amplifies linearly only between amplitude levels $-A_0$ and A_0, and otherwise emits levels $-A_0$ and A_0, respectively. How much larger than unity must A_0 be so that the probability of cutoff is less than (a) 10%, (b) 1%, (c) 0.1%?

3. Show that in the presence of insufficient cyclic prefix and ISI, the resulting signal can be described as

$$\mathbf{Y}^{(i)} = \mathbf{Y}^{(i,i)} + \mathbf{Y}^{(i,i-1)} = \mathbf{H}^{(i,i)} \cdot \mathbf{X}^{(i)} + \mathbf{H}^{(i,i-1)} \cdot \mathbf{X}^{(i-1)} \qquad (30.39)$$

where $\mathbf{Y}^{(i,i-1)}$ is the ISI term and $\mathbf{Y}^{(i,i)}$ contains the desired data disturbed by ICI. Derive $\mathbf{H}^{(i,i)}$ as described in Eq. (19.19), and obtain equations for the ISI matrix $\mathbf{H}^{(i,i-1)}$.

4. Consider an eight-point OFDM system with Walsh–Hadamard spreading, operating in a channel with impulse response $h = [0.5 + 0.2j, \quad -0.6 + 0.1j, \quad 0.2 - 0.25j]$ and $\sigma_n^2 = 0.1$. Assume that the system prepends a four-point cyclic prefix:
   (a) Assuming data vector $[10110001]$ and BPSK modulation, sketch
      (i) the data vector;
      (ii) the Walsh – Hadamard (WH) – transformed signal;
      (iii) the transmit signal (after Fourier transform and prepending of the cyclic prefix);
      (iv) the received signal (disregarding noise);
      (v) the received signal after Fourier transformation and ZF equalization;
      (vi) the received signal after Walsh–Hadamard transformation.
   (b) What is the noise enhancement for an ZF RX?
   (c) Simulate the BER of the system using ZF equalization? Use MATLAB for the simulations.

5. Consider an OFDM system with a 128-point FFT where each OFDM symbol is 128-μs-long. It operates in a slowly time-variant (i.e., negligible Doppler), frequency-selective channel. The PDP of the channel is $P_h = \exp(-\tau/16\,\mu s)$, and the average SNR is 8 dB. Compute the duration of the cyclic prefix that maximizes the SINR at the RX.

6. Consider a channel with $\sigma_n^2 = 1$, $\alpha_n^2 = 1, 0.1, 0.01$, and total power $\sum P_n = 100$. Compute the capacity when using waterfilling. What (approximate) capacity can be achieved if the TX can only use BPSK?

7. Consider an OFDM system with coding across the tones, where the code is a block code with Hamming distance $d_H = 7$.
   (a) If all tones that carry bits of this code fade independently, what is the diversity order that can be achieved?
   (b) In a channel with a 5-μs rms delay spread and exponential PDP, what spacing between these tones is necessary so that fading is independent? Approximate "independent" as "correlation coefficient less than 0.3."

8. Consider an FDMA system, using raised-cosine pulses ($\alpha = 0.35$) on each carrier, and using BPSK modulation.
   (a) What is the spectral efficiency if signals on carriers are to be completely orthogonal?
   (b) What is the spectral efficiency of an OFDM system with BPSK if the number of carriers is very large (i.e., no guard bands required)?

9. Consider an OFDM system with a nonlinear power amplifier, with a backoff of 0 dB, resulting in an interference power to adjacent users according to Figure 19.13. Let the system require 30 dB SIR when operating in an uncoded mode. By coding with a rate 5/6 code, the SIR requirement can be reduced to 25 dB. Does the coding improve the possible spectral efficiency?

10. Consider a channel with $\sigma_n^2 = 1$, $\alpha_n^2 = 1, 0.3, 0.1$. Compare the capacity that can be achieved with (i) waterfilling and (ii) giving equal power to the channels above the waterfilling threshold, while giving zero to the channels below the threshold. Give the results for the cases $\sum P_n = 2, 10, 50$.

30.20 Chapter 20: Multiantenna Systems

1. State three advantages of using a smart antenna system, as opposed to the conventional single-antenna system.

2. Consider the uplink of a CDMA system with spreading factor $M_C = 128$, and $SIR_{threshold} = 6\,dB$. How many users could be served in the system if the BS had only a single antenna? Now, let the BS have $N_r = 2, 4$, or 8 antenna elements. How many users can be served according to the simplified equations of Section 20.1.1? By how much is that number reduced if the angular power spectrum of the desired user is Laplacian with $APS(\phi) = (6/\pi)\exp(-|\phi - \phi_0|/(\pi/12))$, for $\phi_0 = \pi/2$?

3. A MIMO system can be used for three different purposes, one of which is groundbreaking and has contributed most to its popularity. List all three and explain its most popular use in greater detail.

4. For the following realization of a channel for a 3×3 MIMO system, calculate the channel capacity with and without channel knowledge at the TX for a mean SNR per receive branch ranging from $0\,dB$ to $30\,dB$, in steps of $5\,dB$. Comment on the results:

$$\mathbf{H} = \begin{bmatrix} -0.0688 - j1.1472 & -0.9618 - j0.2878 & -0.4980 + j0.5124 \\ -0.5991 - j1.0372 & 0.5142 + j0.4967 & 0.6176 + j0.9287 \\ 0.2119 + j0.4111 & 1.1687 + j0.5871 & 0.9027 + j0.4813 \end{bmatrix}$$

5. Section 20.2.1 states that the number of possible data streams for spatial multiplexing is limited by the number of transmit/receive antennas (N_t, N_r) and the number of significant scatterers N_S. Using an appropriate channel model, demonstrate the limitation due to N_S for the case of a 4×4 MIMO system. The four-element arrays are spaced apart by 2λ.

6. The Laplacian function:

$$f(\theta) = \frac{1}{\sqrt{2}\sigma_S}e^{-\sqrt{2}|\theta|/\sigma_S}, \quad \theta \in (-\pi, \pi] \qquad (30.40)$$

has been proposed as a suitable power angular spectrum at the BS. On the other hand, due to the richness of scatterers in its surroundings, a uniform angular distribution is often assumed for the MS; that is,

$$f(\theta) = \frac{1}{2\pi}, \quad \theta \in (-\pi, \pi] \qquad (30.41)$$

Using the Kronecker model, study the effect of angular spread (at the BS) on the 10% outage capacity of the channel for different array spacings (at the BS) for the case of a 4×4 MIMO system. The SNR is assumed to be $20\,dB$ and uniform linear arrays are used. Arrays at the MS are spaced apart by $\lambda/2$.

7. Consider a downlink scenario where a 3×3 MIMO system is used to boost the throughput of an indoor wireless LAN. Uniform linear arrays (spaced apart by $\lambda/2$) are used at both the user's MS (a PDA) and a wall-mounted BS. Calculate the additional transmit power needed (in percent) to compensate for the loss of expected (or mean) channel capacity when the user moves from an LOS region into an NLOS region (see Figure 30.13).

   In the LOS region, the user receives 6 mW of power in the LOS path and 3 mW in all the other paths combined. Noise power at the RX is constant at 1 mW. Here we assume that LOS consists of a single path and the statistics of NLOS components do not vary when the RX moves from the LOS to the NLOS scenario. Heavy scattering in the environment ensures that NLOS paths follow a Rayleigh distribution. For simplicity, path loss is neglected and equal power transmission is assumed. Repeat the problem for 5% outage capacity to be maintained. Comment on the difference from the expected capacity case.

8. Generate channel capacity cdfs for an iid MIMO channel with $N_R = 4$ and $N_T = [1, \ldots, 8]$. The channel matrix is normalized as $\|\mathbf{H}\|_F^2 = N_r N_t$.
   (a) What is the 10%, 50%, and 90% outage capacity gain compared with a SISO system, assuming that no CSI is available at the TX?
   (b) Comment on the gain from using more transmit antennas than receive antennas?

Figure 30.13   User moving from LOS to NLOS region.

9. Use the Kronecker model and plot the 10% outage capacity as a function of $r = 0, 0.1, \ldots, 1$ for marginal correlation matrices of

$$\mathbf{R}_R = \mathbf{R}_T = \begin{bmatrix} 1 & r & r^2 \\ r & 1 & r \\ r^2 & r & 1 \end{bmatrix} \tag{30.42}$$

10. Show that

$$\sum_{k=1}^{M} \log\left(1 + \frac{\bar{\gamma}}{N_t}\sigma_k^2\right) = \log\det\left(\mathbf{I}_{N_r} + \frac{\bar{\gamma}}{N_t}\mathbf{H}\mathbf{H}^\dagger\right) \tag{30.43}$$

where σ_k, $\bar{\gamma}$, N_t, and M are the kth singular value of $\mathbf{H}$, the average received SNR, the number of transmit elements, and the number of nonzero singular values of $\mathbf{H}$, respectively.

11. Derive the capacity of a MIMO system with ZF receivers. First derive an exact description for a deterministic (fading) channel. Then use a high-SNR approximation, and assume $N_r = N_r$ to estimate the capacity distribution in a flat-fading channel.

12. What is the capacity of a VBLAST system with successive interference-cancellation reception? Compare with the capacity of ZF receivers (see previous example), and give an intuitive explanation of the differences.

13. Keyhole channels do not show Rayleigh amplitude statistics of the transfer function matrix entries.
    Derive the correct amplitude statistics for a perfect keyhole.

30.21   Chapter 21: Cognitive Radio

1. List three reasons why TV spectrum bands are desirable for the implementation of cognitive radio.

2. State the reason for following two mechanisms of the cognitive radio:
   (a) Why does the cognitive radio need to sense spectrum bands before accessing them?
   (b) Why does the cognitive radio need to periodically sense spectrum bands?

3. Consider a primary system in which primary users work on a spectrum band with SNR γ_p, and the spectrum bandwidth is W. The probability that primary users use the whole spectrum band is p, and the probability that the primary users use none of them is $1 - p$. The cognitive radio users access this spectrum band only if it is not used by primary users. When only cognitive radio users use this band, the SNR is γ_s. The background noise is with constant power density.
   (a) Assuming the cognitive radio users can perfectly sense the occupancy state of this spectrum band, give the sum Shannon capacity of primary and cognitive systems.
   (b) Assume the cognitive radio users have probability α ($\alpha < 1$) to detect the existence of primary users, and have probability b ($b < 1$) to detect the absence of primary users. Give the Shannon capacity of primary and cognitive systems.

4. Derive the expression for the minimum number of samples N that can achieve given probability of detection $P_d(P_d = 1 - P_{md})$ and probability of false alarm P_f. Show that in low SNR region $(|h|^2/\sigma_n^2 \ll 1)$, N scales as $O(1/SNR^2)$.

5. In Problem 4, we have derived that in low SNR region, N scales as $O(1/SNR^2)$. For matched filter RXs, in low SNR region, N scales as $O(1/SNR)$, which means that matched filter has better performance. However, compared with matched filter, energy detection is always a more practical sensing scheme for the cognitive radio to detect the existence of primary users. State the reasons for this.

6. Consider a more general multinode detection scheme, in which there are M secondary users. The secondary system makes the final decision based on the number of secondary users claiming the existence of primary users, and let Λ be the number of this kind of secondary users, the decision rule is formulated as follows:

$$\begin{cases} \Lambda > K, \text{ decide primary users exist} \\ \Lambda \leq K, \text{ decide primary users does not exist} \end{cases}$$

where $K = 1, 2, \ldots, M - 1$. Assume every secondary user has the same probability of false alarm (P_f) and probability of missed detection (P_m). Give the expressions for the final probability of false alarm ($P_{f,network}$) and probability of missed detection ($P_{m,network}$).

7. The energy detection is a commonly used method in spectrum sensing. During each sensing period, $2N$ sample number is obtained. r_n is the received sampling value, $n = 1, 2, \ldots, 2N$. The decision statistics is

$$y = \sum_{i=1}^{2N} |r_n|^2$$

the probability density function of y is

$$\begin{cases} f_Y(y|H_0) = \frac{1}{2^N \Gamma(N)} y^{N-1} \exp\left(-\frac{y}{2}\right), & H_0 \text{ (primary user is absent)} \\ f_Y(y|H_1) = \frac{1}{2}\left(\frac{y}{2\gamma}\right)^{\frac{N-1}{2}} \exp\left(-\frac{2\gamma+y}{2}\right) I_{N-1}(\sqrt{2\gamma y}), & H_1 \text{ (primary use is present)} \end{cases}$$

where the γ is the SNR, $\Gamma(\cdot)$ is the gamma function, and $I_v(\cdot)$ is vth-order modified Bessel function of the first kind as follows:

$$I_{N-1}(\sqrt{2\gamma\theta}) = \left(\frac{\gamma\theta}{2}\right)^{\frac{N-1}{2}} \sum_{k=0}^{\infty} \frac{\left(\frac{\gamma\theta}{2}\right)^k}{k!\Gamma(N+k)}$$

Then, the probability of false alarm P_f and probability of detection P_d ($P_d = 1 - P_{md}$) can be evaluated as follows:

$$\begin{cases} P_f = P\{Y > \theta | H_0\} = \int_\theta^\infty f_Y(y|H_0) dy = \frac{\Gamma\left(N, \frac{\lambda}{2}\right)}{\Gamma(N)} \\ P_d = P\{Y > \theta | H_1\} = \int_\theta^\infty f_Y(y|H_1) dy = Q_N(\sqrt{2\gamma}, \sqrt{\theta}) \end{cases}$$

where θ is the threshold of decision. $\Gamma(\cdot, \cdot)$ is the incomplete gamma function, and $Q_N(\cdot, \cdot)$ is the generalized Marcum Q-function. Prove that given N and γ, P_d is a concave function of P_f (a is a concave function of b if the second derivative of a over b is negative).

8. Consider another simple example of optimization problem in Eq. (21.17). There are k TXs with TX power $P_1, P_2, \ldots, P_k$. There are also k RXs, and RX i is designed to receive the signal from

TX i ($i = 1, 2, \ldots, k$). The power received at RX j, from TX i, is given by $G_{ji} P_i (G_{ji} > 0)$, where G_{ji} is the channel gain from TX i to RX j. Then, the SINR at RX i is given by

$$S_i = \frac{G_{ii} P_i}{\sigma_i^2 + \sum_{m \neq i} G_{im} P_m}$$

where σ_i^2 is the background noise power at RX i. We require that SINR at any RX is above a threshold S_{min}:

$$\frac{G_{ii} P_i}{\sigma_i^2 + \sum_{m \neq i} G_{im} P_m} \geq S_{min}, \qquad i = 1, 2, \ldots, k$$

We also have limits on the TX power:

$$P_i^{min} \leq P_i \leq P_i^{max}, \qquad i = 1, 2, \ldots, k$$

The problem of minimizing the total TX power is formulated as follows:

$$\text{minimize} \qquad P_1 + P_2 + \cdots + P_n$$

subject to $\qquad P_i^{min} \leq P_i \leq P_i^{max}, \qquad i = 1, 2, \ldots, k$

$$\frac{G_{ii} P_i}{\sigma_i^2 + \sum_{m \neq i} G_{im} P_m} \geq S_{min}, i = 1, 2, \ldots, k$$

Convert this problem into a linear programming one, which can be solved very easily. Note that linear programs are problems that can be formulated as follows:

$$\text{Minimize } \mathbf{c}^T \mathbf{x}$$

$$\text{Subject to } \mathbf{A}\mathbf{x} \leq \mathbf{b}$$

where $\mathbf{x}$ is the vector of variables (to be determined), and $\mathbf{b}$ are vectors of (known) coefficients and $\mathbf{A}$ is a (known) matrix of coefficients.

30.22 Chapter 22: Relaying, Multi-Hop, and Cooperative Communications

1. Consider a relay system with the source, relay, and destination at location (0,0), (0,500), and (0,1000), respectively. The operating frequency is 1 GHz. Assume equal transmit power at source and relay of 1 W; let the considered bandwidth be 10 MHz. If the path gain is determined by free space law with superimposed Nakagami fading ($m = 2$) for both links, determine the pdf of the transmission rate from source to destination for MDF. Similarly, derive the SNR pdf for DDF.

2. To analyze the impact of feedback, consider a system with 1-bit feedback from the destination to the source and relay. The protocol is as follows:
   Both destination and relay listen to the message transmitted from the source in the first time slot, during which the destination gets the CSI of source destination link; after the first time slot, the destination compares CSI of source destination link and CSI of relay destination link (priorily obtained by destination through training sequence), and selects the better one to work in the second timeslot by sending back 1-bit feedback to destination and relay. The relay performs in the AF mode.
   (a) Derive the pdf of the achievable data rate for source, relay, and destination all being on one line.

(b) Determine the diversity order.

(c) Determine the average data rate, and the 10% outage rate, and compare it to the no-feedback case.

The system settings are the same as Exercise 1.

3. Prove the optimum power distribution for decode and forward with repetition coding shown as Eq. (22.2).

4. Show that

$$SER \leq \frac{(M-1)P_n^2}{M^2} \cdot \frac{MbP_s\sigma_{s,r}^2 + (M-1)bP_r\sigma_{r,d}^2 + (2M-1)P_n}{\left(P_n + bP_s\sigma_{s,d}^2\right)\left(P_n + bP_s\sigma_{s,r}^2\right)\left(P_n + bP_r\sigma_{r,d}^2\right)}$$

is an upper bound for the error probability of uncoded M-PSK in a DDF protocol. Where $b = \sin^2(\pi/M)$; $\sigma_{s,r}^2, \sigma_{r,d}^2, \sigma_{s,d}^2$ are the variances of the channel coefficients $h_{s,r}, h_{r,d}, h_{s,d}$, respectively. Assume that the channel coefficients are modeled as zero-mean complex Gaussian random variables. (Hint: First consider the SER formula for uncoded M-PSK in the single link case shown as Eq. (12.66) in Chapter 12; then analyze the error events for relay case under DDF mode and get the closed-form of SER expression; finally, do appropriate approximation and manipulation on the closed-form to get the upper bound expression).

5. Consider a system with one source, three relays, and one destination. Relay selection scheme is used in this network, and all channels are subject to Rayleigh fading with the same characteristics.

(a) Derive the outage probability for fixed transmit power, and derive the power that is necessary to achieve <1% outage.

(b) Suppose the transmitting power at source and relay is variable. The power allocation scheme is shown as Eq. (22.2). Derive the outage probability. (Hint: $\int_0^\infty \exp\left[-\left(ax + \frac{b}{x}\right)\right] dx = \frac{2\sqrt{b}\, Bessel\, K\left(1, 2\sqrt{ab}\right)}{\sqrt{a}}$, where $Bessel\, K\,(\cdot,\cdot)$ is modified second kind Bessel function). What is the overall power expenditure to guarantee <1% outage probability in this scheme?

6. Consider a system with one source, two relays, and one destination. Alamouti coding is adopted when both relays successfully received the message during the first time slot.

(a) Derive the outage probability of this system, given that the source and two relays have equal transmission power P.

(b) What happens if relays are much closer to source than to destination?

7. Prove Eqs. (22.30) and (22.31) in a coded cooperation scheme. Hint: consider the four cases depicted in Figure 22.7, and analyze the corresponding outage events for each user; assume a high SNR regime, i.e. the mean SNR at each link is very high and therefore, you can use equivalent Taylor's series representing exponential terms to simplify the derivation; coded cooperation has diversity order 2, so higher order 1/SNR terms in Taylor's series can be written as $O\left(\frac{1}{\bar{\gamma}^3}\right)$, where $\bar{\gamma}$ is very large).

8. Matlab exercise: write a program to find the optimum route from node 1 to node 13 in the network topology shown as Figure 30.14 using Dijkstra algorithm.

9. Matlab exercise: As is shown in Figure 30.15, three source nodes transmit data to one destination through three orthogonal channels with carrier frequencies 900-MHz, 1000-MHz, and 1100-MHz, respectively; the channels have equal bandwidth 10-MHz; suppose the path gain is determined by free-space law without small scale fading, and the traveling distances are $d_1 = 3000\,\text{m}, d_2 = 2000\,\text{m}$ and $d_3 = 5000\,\text{m}$; packets arrive to the source nodes satisfy Poisson processes with mean rates $\bar{\lambda}_1 = 8/9\ bits/slot, \bar{\lambda}_2 = 10/9\ bits/slot, \bar{\lambda}_3 = 5/9\ bits/slot$, where each timeslot is $0.1\mu s$; the service delay of the destination node can be assumed zero.

(a) Suppose the control parameter V is set to 10^5, and no extra constraint is placed on the amount of power allocated to each node. Using backpressure algorithm, write a Matlab program to get the power allocated to each source node averaged over 10^4 timeslots to reach the least amount of total power, while maintaining the network stable.

(b) Plot a figure showing how the time averaged total power evolves with V value.

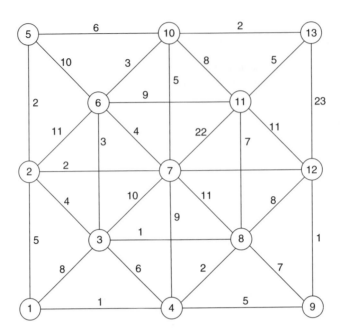

Figure 30.14 Problem setup for the Dijkstra algorithm in Exercise 30.22.8.

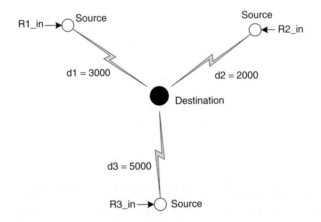

Figure 30.15 Problem setup for backpressure algorithm in Exercise 30.22.9.

30.23   Chapter 23: Video Coding

1. Explain the reasons for using the YCbCr color space for video coding.
2. Consider a Markov-1 process with $\rho = 0.95$. Compute the transform coding gain for the 4-point DCT and 8-point DCT.
3. Show that the variance of quantization error for a uniformly distributed source is $\Delta^2/12$, where Δ is the quantization step size. Hint: model the quantization error as a uniform random variable in the range $[-\Delta/2, \Delta/2]$.

4. Consider an alphabet containing 5 symbols, $x_i, i = 1 \ldots 5$, with probabilities: $p(x_1) = 0.35$, $p(x_2) = 0.2$, $p(x_3) = 0.15$, $p(x_4) = 0.15$, $p(x_5) = 0.15$. Construct a Huffman code and compute the average length of the code.

5. Explain the reasons for error propagation when the compressed video bitstream has experienced bit error or packet loss.

6. Consider a compressed video that is subject to unequal error protection. Should greater protection be given to I-pictures or P-pictures? Explain the reason why.

7. Why is UDP preferred over TCP for video streaming?

30.24 Chapter 24: GSM – Global System for Mobile Communications

1. Consider a GSM operator that has licenses for both 900- and 1,800-MHz bands. How should they be used in the buildup of the network?

2. Which is true: the following combinations of devices *must* be bought from the same vendor: (i) BTS–MS, (ii) BTS–BSC, (iii) BSC–MSC?

3. Assuming that directions of arrival are uniformly distributed at the MS, how large is the correlation coefficient (for a GSM1800 system) between the channel in the middle and the end of the burst when the MS moves at 250 km/h? How large is the correlation coefficient between the channels at the beginning and end of the burst?

4. Consider a GSM system at 1,900 MHz operating in a TU environment. What is the correlation coefficient between two channels that are separated by (i) one carrier frequency, (ii) one auctioned frequency block (5 MHz), or (iii) one duplex frequency.

5. Explain the difference between the fast and slow associated control channel. When are they used?

6. Explain the difference between block FEC for voice and control data. Why are different schemes used?

7. What is the common property of all midambles? Why are different midambles defined?

8. A user has repeatedly entered the wrong PIN code into his phone, so that it is now blocked. Is there a possibility to unblock the phone?

9. Consider the following billing problem: subscriber A is Swedish but temporarily based in Denmark. Subscriber B is in Finland. Subscriber C is in France, but forwards all calls to subscriber D in England. A calls B, and wants to conference in subscriber C. As C has call forwarding on, the call goes to D, and that subscriber is then active in the phone conference. Who pays which fees?

10. Compare the required SNR to achieve 10^{-2} BER of uncoded GMSK and 8-PSK in an AWGN channel. Note that GMSK is used for GSM data transmission, while 8-PSK is used in EDGE.

11. While the use of GSM phones on airplanes is currently illegal, it is still interesting to hypothetically analyze the possibility of communicating with ground BSs during flight.
    (a) Establish a link budget assuming that the plane is flying at 10 km altitude, over an area where the cell radii are 30 km. Assume furthermore a 5 dB penetration loss through the aircraft hull, and a −10 dB antenna gain of the BS antenna in the direction of the airplane.
    (b) Discuss whether (and if yes, how much) fading margin is required.
    (c) Using typical airplane speeds, compute how often handovers between cells would occur. How does that change in an area where the cell radius is 1 km? How would that influence the quality of the link?

12. Discuss the effectiveness of the error-correction coding of GSM in a static channel (user and IOs are stationary) with and without frequency hopping. What conclusions can we draw for system design?

13. Both HSCSD, GPRS, and EDGE aim at increasing the rate of data transmission in GSM. What are the main differences, and what are the peak data rates that can be achieved with those schemes?
14. What information is contained in the HLR, and at what occasions is that information updated?

30.25   Chapter 25: IS-95 and CDMA 2000

1. What are the spreading factors for the uplink and the downlink for rate set 1 and rate set 2 in IS-95?
2. What is the maximum speed of the MS that the power control of IS-95 can follow in the (i) 800-MHz band and (ii) the 1,900-MHz band?
3. What is the typical percentage of total power transmitted from a BS that a pilot in IS-95 is using? If that percentage is halved, by what percentage could the cell size be increased? How would that affect capacity?
4. What code rates are used for convolutional codes in IS-95? Where are they applied?
5. How are physical channels separated in IS-95? Compare this approach with GSM.
6. How is the transmission of low data rates handled in IS-95 in the uplink and downlink. Comment on the difference.
7. How is frequency diversity obtained in the multicarrier mode of cdma2000 in the uplink and downlink? What are the advantages and drawbacks of the two approaches?
8. Consider the power control in IS-95. Assume 20 users in a cell, each of which has power control with $a \pm 1$ dB error. What is the variance of the total interference power? How much must the SIR margin be increased to guarantee a 95% outage probability?
9. How are pilot tones from different cells distinguished in IS-95?
10. Describe the spreading and modulation methods for uplink and downlink of IS-95.

30.26   Chapter 26: WCDMA/UMTS

1. What service classes are defined in UMTS? For what purposes are they used, and how does that affect admissible BERs and delays?
2. Two operators are building up systems in two adjacent 5-MHz bands. Let an MS in operator A's system be at the cell boundary, operating at 3 dB above sensitivity level. What is the minimum required path loss from the BS of operator B in order to still function? Assume that all antennas have omniradiation patterns.
3. Compare the way pilot signals are transmitted in WCDMA/UMTS to the method of IS-95.
4. What is the maximum data rate in an UMTS uplink? How is that achieved?
5. How often is feedback information transmitted? Assuming radiation incident only in the horizontal plane, and a uniformly distributed azimuthal power spectrum, what is the maximum velocity admissible so that envelope correlation between channel-state observation and its use is higher than 0.9? How does this result change when radiation is incident isotropically (both in azimuth and elevation)?
6. Explain how data channels and control channels are multiplexed in the uplink and downlink of UMTS.
Why are different methods used?
7. Describe the role of channelization codes and scrambling codes in UMTS. Which types of codes are being used?
8. Establish the link budget for the uplink of a 12.2 kbit/s data rate channel in a vehicular (fast-moving) environment, with soft handover. Compute the results assuming the following values: antenna gain at BS and MS is 18 and -3 dB, respectively, and extra loss through the car is

8 dB. The BS noise figure is 5 dB, interference margin is 3 dB, required E_b/N_0 is 5 dB, cable loss at BS is 2 dB, shadowing fading margin is 7.2 dB, no fast fading margin is required. For the path loss, assume a simple breakpoint model with $d_{break} = 100$ m, $n = 2$ for $d < d_{break}$, and $n = 4$ for $d > d_{break}$. What is the range that can be achieved?

9. Describe the steps that BS and MS take for the processing of data.
10. How is cell search performed in UMTS?

30.27   Chapter 27: 3GPP Long Term Evolution

1. Describe the difference between service data units and protocol data units. How is the mapping from one to the other done?
2. Draw a block diagram of an LTE downlink/uplink transmitter.
3. Assuming a latency requirement of 1 ms, a system bandwidth of 10 MHz, single antenna port transmission, regular cyclic prefix length, and rate 1/3 coded 16-QAM transmission, what is the minimum data rate that can be transmitted without having to send "empty" bits? (Note: this aims at using the fact that data have to be transmitted in units of resource blocks.)
4. Onto what subcarriers are the virtual resource blocks 1 and 2 mapped for:
   (a) localized resource blocks, and
   (b) distributed resource blocks.
   Assume 5 MHz channel bandwidth corresponding to 25 resource blocks and a transmission with regular cyclic prefix length. For a detailed definition of the resource block mapping refer to 3GPP TS 36.211, Section 6.2.3.
5. MATLAB: write a Matlab program for the estimation of channel coefficients from the demodulation resource symbols, using LMMSE and LS estimation (see Chapter 19). Evaluate the channel estimation accuracy in a tapped-delay line channel with Rayleigh fading, where the taps are at 0 ns, 100 ns, and 300 ns, with strength 1, 0.4, 0.2.
6. Consider a system with 10 MHz bandwidth, 1 MHz coherence bandwidth (assume block fading), and 10 users that require resource blocks of size 1 MHz each. Compute the probability of "optimum-assignment collisions", i.e., that the same subcarriers would be optimum for different users.
7. MATLAB: Consider an LTE system with 20 MHz operating at a carrier frequency of 2.1 GHz. Coherence bandwidth is 1 MHz (assume block fading, that is, divided the full bandwidth into 1 MHz blocks that are fading independently). Obtain, by simulation, the pdf of the loss of SNR compared to "ideal" sounding for a user driving at (i) 30 km/h and (ii) 120 km/h. In other words, how much SNR does the user lose because it has outdated SNR and therefore suboptimum assignment of its frequency?
8. MATLAB: Generate 1) a (cyclically-extended) Zadoff–Chu sequence without phase rotation and 2) a (cyclically-extended) Zadoff–Chu sequence with phase rotation $\alpha = 2\pi/12$. For these two sequences simulate the pdf of the loss of orthogonality within one resource block due to frequency selectivity in a channel with exponential PDP with decay time constant (i) 100 ns, (ii) 1 μs, (iii) 10 μs?
9. Which physical downlink channels have no equivalent transport or logical channels?
10. How is the modulation and coding scheme indicated in LTE? What combinations of code rates and modulation formats are possible? (Hint: 3GPP TS 36.213.)
11. Assume a Poisson probability for users to try and join the network. What is the rate of requests required (within the 1 ms PRACH) so that the probability of collision becomes 0.1?
12. Describe the steps of the handover procedure.
13. When employing a regular cyclic prefix and 10 MHz system bandwidth, what is the maximum time dispersion that does not lead to intersymbol interference?

30.28   Chapter 28: WiMAX/IEEE 802.16

1. For a system with 10 MHz bandwidth and FFT size 1024, compute for downlink PUSC:
   (a) the subcarrier spacing,
   (b) the possible durations of the cyclic prefix,
   (c) the number of null carriers, and
   (d) the number of pilot symbols per OFDM symbol.
2. What is a "zone"? Which zones are mandatory, and which are optional, in uplink and downlink subframes?
3. How many pilot tones are in each cluster of downlink PUSC? What is the resulting loss of spectral efficiency?
4. For FFT size 1024, to which physical subcarriers is logical subchannel 17 mapped to in PUSC? Assume DL_PermBase = 5 and odd OFDM symbol numbers. For the definition of the basic permutation sequence, the renumbering sequence, and the detailed specification of the PUSC symbol structure refer to the WiMAX standard Section 8.4.6.1.2.1.
5. For FFT size 1024, to which physical subcarriers is logical subchannel 12 mapped to in FUSC? Assume DL_PermBase = 5 and FUSC_SymbolNumber = 0. For a detailed specification of the FUSC symbol structure refer to the WiMAX standard Section 8.4.6.1.2.2.
6. Show the assignment of pilot tones in uplink PUSC tiles. How large is the loss in spectral efficiency due to the presence of pilot tones?
7. Which physical subcarriers/OFDM symbols is subchannel 2 mapped onto in uplink PUSC with FFT size 512? Assume UL_PermBase = 5. For the definition of the tile permutation sequence refer to the WiMAX standard Section 8.4.6.2.
8. Consider the transmission of a WiMAX signal in a flat-fading channel without coding. Operating mode is QPSK, mean SNR is 10 dB. Packets of size 128 bits are transmitted without coding. In Scenario 1, a single-antenna transmission with HARQ is used; in Scenario 2 Alamouti-encoded transmission with ARQ is used. Analytically derive and compare the relative spectral efficiency of the schemes.

   Hint: $\int_0^\infty \frac{\gamma^{(L-1)}}{(L-1)!\bar{\gamma}^L} \exp\left(-\frac{\gamma}{\bar{\gamma}}\right) Q\left(\sqrt{2\gamma}\right) d\gamma = \left[\frac{1}{2}(1-\mu)\right]^L \sum_{k=0}^{L-1} \binom{L-1+k}{k} \left[\frac{1}{2}(1+\mu)\right]^k$

   with $\mu = \sqrt{\frac{\bar{\gamma}}{1+\bar{\gamma}}}$
9. What are the main steps in establishing a connection between MS and BS?

30.29   Chapter 29: Wireless Local Area Networks

1. In 802.11a, what is the loss of spectral efficiency due to (i) not all subcarriers carrying data, (ii) cyclic prefix, (iii) training sequence and signaling field (assume that 16 OFDM symbols are transmitted).
2. What are the maximum data rates of 802.11, 802.11b, and 802.11a.
3. What are the mechanisms that 802.11a uses to achieve high maximum throughput?
4. Estimate the frequency diversity order that an 802.11a system can achieve in a typical office environment (delay spread 50 ns) and a large open space (delay spread 250 ns). What is the impact of coding?
5. The 802.11n standard foresees an optional shortening of the guard interval from 800 ns to 400 ns. What are the advantages and drawbacks? Up to which channel delay spread should the shortened guard interval be used? Assume uncoded transmission and an exponential power delay profile, and an average SNR of 10 dB.
6. Consider the effect of narrowband interference on an 802.11a system.
   (a) What is the effective bandwidth of a Bluetooth signal, which uses GMSK with $B_G T = 0.5$?

(b) How many tones are effectively blocked if we assume that the power spectral density in the Bluetooth signal is equal to the 802.11 signal and a 20 dB SINR is required per tone?

(c) How many tones are blocked if the Bluetooth signal is 20 dB stronger than the 802.11 signal?

7. Consider the problem is covering a whole city with WiFi. Assume that transmission occurs at the maximum admissible power, and that operation at 11 Mbit/s requires an SNR of 5 dB to keep the number of required retransmissions low.

(a) Assume first a network in the 2.45- GHz range that only aims to cover outdoor locations. Using the Walfish – Ikegami model (Appendix 7B), establish a link budget for a city in the U.S.A. Use the following parameters for the Walfish – Ikegami model: BS antenna height $h_{BS} = 12.5$ m, building height 12 m, building-to-building distance 50 m, street width 25 m, MS antenna height 1.5 m, orientation 30 degree for all paths; correction factor according to "metropolitan center" environment. Assume an NLOS situation.

(b) Assuming a wall penetration loss of 10 dB, how much is the coverage distance reduced?

(c) For a city with a size of 10 km², give the number of access points required for covering the outdoor-only case, and the outdoor-plus-indoor case. Assuming costs of 1,000 $ for each access point, what is the total cost of setting up the network? For a MAC efficiency of 50 %, estimate the number of users that can be covered with an average data rate of 300 kbit/s.

References

3GPP LTE Third-Generation Partnership Project, TS 36.201, 36.211, 36.212, on www.3gpp.org (2010).

Abdi et al. 2000 A. Abdi, K. Wills, H. A. Barger, M. S. Alouini, and M. Kaveh, "Comparison of the level crossing rate and average fade duration of Rayleigh, Rice and Nakagami fading models with mobile channel data", *Proceedings of VTC Fall 2000*, pp. 1850–1857 (2000).

Abramowitz and Stegun 1965 M. Abramowitz and I. A. Stegun, *Handbook of Mathematical Functions*, National Bureau of Standards, Washington (1965).

Abramson 1970 N. Abramson, "The ALOHA system – Another alternative for computer communications", *Proceedings of Fall 1970 AFIPS Computer Conference* (1970).

Acharya and Yates 2007 J. Acharya and R. D. Yates, "A framework for dynamic spectrum sharing between cognitive radios", *IEEE Int. Conf. Commun.*, 5166–5171 (2007).

Adachi and Ohno 1991 F. Adachi and K. Ohno, "BER performance of QDPSK with postdetection diversity reception in mobile radio channels", *IEEE Trans. Veh. Technol.*, 40, 237–249 (1991).

Adachi and Parsons 1989 F. Adachi and J. D. Parsons, "Error rate performance of digital FM mobile radio with postdetection diversity", *IEEE Trans. Commun.*, 37, 200–210 (1989).

Akino et al. 2009 T. Koike-Akino, A. F. Molisch, P. Orlik, Z. Tao, and T. Kuze, Unified analysis of linear block precoding for distributed antenna systems, *IEEE Globecom* (2009).

Akyildiz et al. 2006 I. F. Akyildiz, W. Y. Lee, M. C. Vuran, and S. Mohanty, "Next generation/dynamic spectrum access/cognitive radio wireless networks: A survey", *Comput. Netw.*, 50(13), 2127–2159 (2006).

Alamouti 1998 S. M. Alamouti, "A simple transmit diversity technique for wireless communications", *IEEE J. Sel. Area. Commun.*, 16, 1451–1458 (1998).

Almers et al. 2003 P. Almers, F. Tufvesson, and A. F. Molisch, "Measurement of keyhole effect in wireless multiple-input – multiple-output (MIMO) channels", *IEEE Commun. Lett.*, 7, 373–375 (2003).

Almers et al. 2007 P. Almers, E. Bonek, A. Burr, et al., "Survey of channel and radio propagation models for wireless MIMO systems", *Eurasip J. Wireless Commun. Networking*, Article ID 19070, 19 (2007).

Alouini and Goldsmith 1999 M. S. Alouini and A. J. Goldsmith, "Area spectral efficiency of cellular mobile radio systems", *IEEE Trans. Veh. Technol.*, 48, 1047–1066 (1999).

Andersen 1991 J. B. Andersen, "Propagation parameters and bit errors for a fading channel", *Proceedings of Commsphere '91*, paper 8.1 (1991).

Andersen 1997 J. B. Andersen, "UTD multiple-edge transition zone diffraction", *IEEE Trans. Antennas Propagat.*, 45, 1093–1097 (1997).

Andersen 2000 J. B. Andersen, "Antenna arrays in mobile communications: Gain, diversity, and channel capacity", *IEEE Antennas Propagat. Mag.*, 42, 12–16 (2000).

Andersen 2002 J. B. Andersen, *Power Distributions Revisited*, COST 273 TD(02)004 (2002).

Andersen and Hansen 1977 J. B. Andersen and F. Hansen, "Antennas for VHF/UHF personal radio: A theoretical and experimental study of characteristics and performance", *IEEE Trans. Veh. Technol.*, VT-26, 349–357 (1977).

Andersen et al. 1990 J. B. Andersen, S. L. Lauritzen, and C. Thommesen, "Distribution of phase derivatives in mobile communications", *Proc. IEE, Part H*, 137, 197–204 (1990).

Andersen et al. 1995 J. B. Andersen, T. S. Rappaport, and S. Yoshida, "Propagation measurements and models for wireless communications channels", *IEEE Commun. Mag.*, 33(1), 42–49 (1995).

Anderson 2003 H. R. Anderson, *Fixed Broadband Wireless System Design*, Wiley (2003).

Anderson 2005 J. B. Anderson, *Digital Transmission Engineering*, 2nd edition, Prentice-Hall (2005).

Anderson et al. 1986 J. B. Anderson, T. Aulin, and C. E. Sundberg, *Digital Phase Modulation*, Plenum (1986).

Andrews et al. 2001 M. R. Andrews, P. P. Mitra, and R. de Carvalho, "Tripling the capacity of wireless communications using electromagnetic polarization", *Nature*, 409, 316–318 (2001).

Andrews et al. 2007 J. G. Andrews, A. Ghosh, and R. Muhamed, *Fundamentals of WiMAX: Understanding Broadband Wireless Networking*, Prentice Hall (2007).

Annamalai et al. 2000 A. Annamalai, C. Tellambura, and V. K. Bhargava, "A general method for calculating error probabilities over fading channels", *Proceedings of the International Conference in Communication 2000*, pp. 36–40 (2000).

Annavajjala et al. 2007 R. Annavajjala, P. C. Cosman, and L. B. Milstein, "Statistical channel knowledge-based optimum power allocation for relaying protocols in the high SNR regime", *IEEE J. Sel. Area. Commun.*, 25, 292–305 (2007).

Arai et al. 1988 Y. Arai, T. Agui, and M. Nakajima, "A fast DCT-SQ scheme for images", *Trans. IEICE*, E71, 1095 (1988).

Ariyavisitakul 2000 S. L. Ariyavisitakul, "Turbo space-time processing to improve wireless channel capacity", *IEEE Trans. Commun.*, 48, 1347–1359 (2000).

Ashtiani et al. 2003 F. Ashtiani, J. A. Salehi, and M. R. Aref, "Mobility modeling and analytical solution for spatial traffic distribution in wireless multimedia networks", *IEEE J. Sel. Area. Commun.*, 21, 1699–1709 (2003).

Asplund et al. 2006 H. Asplund, A. A. Glazunov, A. F. Molisch, K. I. Pedersen, and M. Steinbauer, "The COST 259 directional channel model – II. Macrocells", *IEEE Trans. Wireless Commun.*, 5, 3434–3450 (2006).

Atal and Remde 1982 B. Atal and J. Remde, "A new model of LPC excitation for producing natural-sounding speech at low bit rates", *Proceedings of the IEEE International Conference on Acoustics, Speech, and Signal Processing, ICASSP'82'*, Paris, pp. 614–617 (1982).

Atal and Schroeder 1984 B. S. Atal and M. R. Schroeder, "Stochastic coding of speech signals at very low bit rates", *Proceedings of the IEEE International Conference in Communication ICC'84'*, Amsterdam (The Netherlands), pp. 1610–1613 (1984).

Ayadi et al. 2002 J. Ayadi, A. A. Hutter, and J. Farserotu, "On the multiple input multiple output capacity of Rician channels", *Proc. Int. Symp. Wireless Personal Multimedia Communications 2002*; 402–406 (2002).

Badsberg et al. 1995 M. Badsberg, J. Bach Andersen, and P. Mogensen, *Exploitation of the Terrain Profile in the Hata Model*, COST 231 TD(95)9 (1995).

Bahai et al. 2004 A. R. S. Bahai, B. R. Saltzberg, and M. Ergen, *Multi-Carrier Digital Communications: Theory and Applications of OFDM*, 2nd edition, Springer (2004).

Bahl et al. 1974 L. R. Bahl, J. Cock, F. Jelink, and J. Raviv, "Optimal decoding of linear codes for minimum symbol error rate", *IEEE Trans. Inform. Theory*, 20, 248–287 (1974).

Balanis 2005 C. A. Balanis, *Antenna Theory: Analysis and Design*, 3rd edition, Wiley (2005).

Barclay 2002 L. W. Barclay, *Propagation of Radiowaves*, 2nd edition, IET Press (2002).

Barry et al. 2003 J. R. Barry, D. G. Messerschmidt, and E. A. Lee, *Digital Communications*, 3rd edition, Kluwer (2003).

Bass and Fuks 1979 F. G. Bass and I. M. Fuks, *Wave Scattering from Statistically Rough Surfaces*, Pergamon (1979).

Bates 2008 R. J. B. Bates, *GPRS: General Packet Radio Service*, McGraw Hill (2008).

Beckman and Lindmark 2007 C. Beckman and B. Lindmark, "The evolution of base station antennas for mobile communications", *2007 International Conference on Electromagnetics in Advanced Applications*, pp. 85–92 (2007).

van de Beek et al. 1995 J.-J. van de Beek, O. Edfors, M. Sandell, S. K. Wilson, and P. O. Börjesson, "On channel estimation in OFDM systems", *Proceedings of the IEEE Vehicle Technology Conference*, Vol. 2, pp. 815–819, Chicago, IL, July (1995).

van de Beek et al. 1999 J.-J. van de Beek, P. O. Borjesson, M.-L. Boucheret, et al., "A time and frequency synchronization scheme for multiuser OFDM", *IEEE J. Sel. Area. Commun.*, 17, 1900–1914 (1999).

Belfiore and Park 1977 C. A. Belfiore and J. H. Park, "Decision feedback equalization", *Proc. IEEE*, 67, 1143–1156 (1977).

Bello 1963 P. A. Bello, "Characterization of randomly time-variant linear channels", *IEEE Trans. Commun.*, 11, 360–393 (1963).

Bello and Nelin 1963 P. Bello and B. D. Nelin, "The effect of frequency selective fading on the binary error probabilities of incoherent and differentially coherent matched filter receivers", *IEEE Trans. Commun.*, 11, 170–186 (1963).

Benedetto and Biglieri 1999 S. Benedetto and E. Biglieri, *Principles of Digital Transmission: With Wireless Applications*, Kluwer (1999).

diBenedetto et al. 2005 M. G. diBenedetto, T. Kaiser, A. F. Molisch, I. Oppermann, C. Politano, and D. Porcino, *UWB. Communication Systems–A Comprehensive Overview*, EURASIP Publishing (2005).

diBenedetto et al. 2006 M. G. diBenedetto, T. Kaiser, A. F. Molisch, I. Oppermann, C. Politano, and D. Porcino (eds), *UWB. Communication Systems – A Comprehensive Overview*, Hindawi Publishing (2006).

Benvenuto et al. 2010 N. Benvenuto, R. Dinis, D. Falconer, and S. Tomasin, "Single carrier modulation with nonlinear frequency domain equalization: An idea whose time has come – again", *Proc. IEEE*, 98, 69–96 (2010).

Berger 1971 T. Berger, *Rate Distortion Theory: A Mathematical Basis for Data Compression*, Prentice-Hall, Englewood Cliffs (1971).

Bergljung 1994 C. Bergljung, *Diffraction of Electromagnetic Waves by Dielectric Wedges*, Ph.D. thesis, Lund Institute of Technology, Lund, Sweden (1994).

Berrou et al. 1993 C. Berrou, A. Glavieux, and P. Thitimajshima, "Near Shannon limit error-correcting coding and decoding: Turbo-codes", *Proceedings of the IEEE International Conference on Communications, ICC '93* (1993).

Bertoni 2000 H. L. Bertoni, *Radio Propagation for Modern Wireless Systems*, Prentice-Hall (2000).

Bettstetter et al. 1999 C. Bettstetter, H. J. Voegel, and J. Eberspecher, "GSM phase 2+ general packet radio service GPRS: Architecture, protocols, and air interface", *IEEE Commun. Surveys, Third Quarter 1999*, 2(3) (1999).

Bi et al. 2001 Q. Bi, G. L. Zysman, and H. Menkes, "Wireless mobile communications at the start of the 21st century", *IEEE Commun. Mag.*, 39(1), 110–116 (2001).

Biglieri et al. 1991 E. Biglieri, *Introduction to Trellis-Coded Modulation with Applications*, MacMillan, New York (1991).

Biglieri et al. 2007 E. Biglieri, R. Calderbank, A. Constantinides, A. Goldsmith, A. Paulraj, and H. V. Poor, *MIMO Wireless Communications*, Cambridge University Press (2007).

Blanz and Jung 1998 J. J. Blanz and P. Jung, "A flexibly configurable spatial model for mobile radio channels", *IEEE Trans. Commun.*, 46, 367–371 (1998).

Blaunstein 1999 N. Blaunstein, *Radio Propagation in Cellular Networks*, Artech House (1999).

Bletsas et al. 2006 A. Bletsas, A. Khisti, D. P. Reed, and A. Lippman, "A simple cooperative diversity method based on network path selection", *IEEE J. Sel. Area. Commun.*, 24, 659–672 (2006).

Bletsas et al. 2007 A. Bletsas, H. Shin, and M. Z. Win "Cooperative communications with outage-optimal opportunistic relaying", *IEEE Trans. Wireless Commun.*, 6, 3450–3460 (2007).

Boelcskei et al. 2008 H. Boelcskei, D. Gesbert, C. B. Papadias, and A.-J. van der Veen (eds), *Space-Time Wireless Systems: From Array Processing to MIMO Communications*, Cambridge University Press (2008).

Bottomley et al. 2000 G. E. Bottomley, T. Ottosson, and Y. P. E. Wang, "A generalized RAKE receiver for interference suppression", *IEEE J. Sel. Area. Commun.*, 18, 1536–1545 (2000).

Boudreau et al. 2009 G. Boudreau, J. Panicker, N. Guo, R. Chang, N. Wang, and S. Vrzic, "Interference coordination and cancellation for 4G networks", *IEEE Comm. Mag.*, April, 74–81 (2009).

Boukerche et al. 2009 A. Boukerche, M. Z. Ahmad, D. Turgut, and B. Turgut, "A taxonomy of routing protocols for mobile ad-hoc networks", in A. Boukerche (ed.), *Algorithms and Protocols for Wireless and Mobile Ad-hoc Networks*, Wiley (2009).

Bowman 1987 J. J. Bowman (ed.), *Electromagnetic and Acoustic Scattering by Simple Shapes*, Hemisphere, New York (1987).

Boyd and Vandenberghe 2004 S. Boyd and L. Vandenberghe, *Convex Optimization*, Cambridge University Press (2004).

Braun and Dersch 1991 W. R. Braun and U. Dersch, "A physical mobile radio channel model", *IEEE Trans. Veh. Technol.*, 40, 472–482 (1991).

Brennan and Cullen 1998 C. Brennan and P. Cullen, "Tabulated interaction method for UHF terrain propagation problems", *IEEE Trans. Antennas Propagat.*, 46, 881, 1998.

Buehler 1994 H. Buehler, *Estimation of Radio Channel Time Dispersion for Mobile Radio Network Planning, Dissertation*, Technical University Vienna (1994).

Burr 2001 A. Burr, *Modulation and Coding for Wireless Communications*, Prentice Hall (2001).

Cadambe and Jafar 2008 V. R. Cadambe and S. A. Jafar, "Interference alignment and spatial degrees of freedom for the K user interference channel", *IEEE Int. Conf. Commun.*, 971–975 (2008).

Cai and Giannakis 2003 X. Cai and G. B. Giannakis, "Bounding performance and suppressing intercarrier interference in wireless mobile OFDM", *IEEE Trans. Commun.*, 51, 2047–2056 (2003).

Cai and Goodman 1997 J. Cai and D. J. Goodman, "General packet radio service in GSM", *IEEE Commun. Mag.*, 35(10), 122–131 (1997).

Calcev et al. 2007 G. Calcev, D. Chizhik, B. Goransson, et al., "A wideband spatial channel model for system-wide simulations", *IEEE Trans. Veh. Technol.*, 56, 389–403 (2007).

Callaway et al. 2002 E. Callaway, P. Gorday, L. Hester, et al., "Home networking with IEEE 802.15.4: A developing standard for low-rate wireless personal area networks", *IEEE Commun. Mag.*, 40(8), 70–77 (2002).

Cardieri and Rappaport 2001 P. Cardieri and T. Rappaport, "Statistical analysis of co-channel interference in wireless communications systems", *Wireless Commun. Mobile Comput.*, 1, 111–121 (2001).

Catreux et al. 2001 S. Catreux, P. F. Driessen, and L. J. Greenstein, "Attainable throughput of an interference-limited multiple-input multiple-output cellular system", *IEEE Trans. Commun.*, 48, 1307–1311 (2001).

Chan 1992 G. K. Chan, "Effects of sectorization on the spectrum efficiency of cellular radio systems", *IEEE Trans. Veh. Technol.*, 41, 217–225 (1992).

Chandrasekhar et al. 2008 V. Chandrasekhar, J. Andrews, and A. Gatherer, "Femtocell networks: A survey", *IEEE Commun. Mag.*, 46(9), 59–67 (2008).

Chang 1966 R. W. Chang, "Synthesis of band-limited orthogonal signals for multichannel data transmission", *Bell Sys. Techn. J.*, 45, 1775–1796 (1966).

Chatschik 2001 B. Chatschik, "An overview of the Bluetooth wireless technology", *IEEE Commun. Mag.*, 39(12), 86–94 (2001).

Chen and Chuang 1998 Y. Chen and J. C. I. Chuang, "The effects of time-delay spread on unequalized TCM in a portable radio environment", *IEEE Trans. Veh. Technol.*, 46, 375–380 (1998).

Chen and Luk 2009 C. N. Chen and K. M. Luk, *Antennas for Base Stations in Wireless Communications*, McGraw Hill (2009).

Chen et al. 2005 J. Chen, L. Jia, X. Liu, G. Noubir, and R. Sundaram, "Minimum energy accumulative routing in wireless networks", *IEEE INFOCOM*, 2005, 1875–1886 (2005).

Chennakeshu and Saulnier 1993 S. Chennakeshu and G. J. Saulnier, "Differential detection of Pi/4-shifted-DQPSK for digital cellular radio", *IEEE Trans. Veh. Technol.*, 42, 46–57 (1993).

Chiang 2005 M. Chiang, "Balancing transport and physical layers in wireless multihop networks: Jointly optimal congestion control and power control", *IEEE J. Sel. Area. Commun.*, 23, 104–116 (2005).

Choi et al. 2001 Y.-S. Choi, P. J. Voltz, and F. Cassara, "On channel estimation and detection for multicarrier signals in fast and frequency selective Rayleigh fading channel", *IEEE Trans. Commun.*, 49, 1375–1387 (2001).

Chollet et al. 2005 G. Chollet, A. Esposito, M. Faundez-Zanuy, and M. Marinaro, *Nonlinear Speech Modeling and Applications*, Vol. 3445, *Lecture Notes in Computer Science*, Springer-Verlag, Berlin, New York (2005).

Chrisanthopoulou and Tsoukatos 2007 M. P. Chrisanthopoulou and K. P. Tsoukatos, "Joint beamforming and power control for CDMA uplink throughput maximization", *18th IEEE International Symposium on Personal, Indoor and Mobile Radio Communication*, Athens (2007).

Chuah et al. 2002 C. N. Chuah, D. Tse, J. M. Kahn, and R. Valenzuela, "Capacity scaling in MIMO wireless systems under correlated fading", *IEEE Trans. Inform. Theory*, 48, 637–650 (2002).

Chuang 1987 J. Chuang, "The effects of time delay spread on portable radio communications channels with digital modulation", *IEEE J. Sel. Area. Commun.*, 5, 879–888 (1987).

Cimini 1985 L. J. Cimini, "Analysis and simulation of a digital mobile channel using orthogonal frequency division multiplexing", *IEEE Trans. Commun.*, 33, 665–675 (1985).

Clark 1998 M. V. Clark, "Adaptive frequency-domain equalization and diversity combining for broadband wireless communications", *IEEE J. Sel. Area. Commun.*, 16, 1385–1395 (1998).

Clarke 1968 R. Clarke, "A statistical theory of mobile radio reception", *Bell System Techn. J.*, 47, 957–1000 (1968).

CME 20 1994 Ericsson, *Course Handouts for Course CME 20 1994*, Vienna, Austria (1994).

Collin 1985 R. E. Collin, *Antennas and Radiowave Propagation*, McGraw Hill (1985).

Collin 1991 R. E. Collin, *Field Theory of Guided Waves*, IEEE Press, Piscataway (1991).

Cooklev 2004 T. Cooklev, *Wireless Communication Standards: A Study of IEEE 802.11, 802.15, and 802.16*, IEEE Press (2004).

COST 231 E. Damosso and L. Correira, *Digital Mobile Radio - The View of COST 231'*, European Union, Luxemburg (1999).

Coursey 1999 C. C. Coursey, *Understanding Digital PCS: The TDMA Standard*, Artech (1999).

Cover and El Gamal 1979 T. Cover and A. El Gamal, "Capacity theorems for the relay channel", *IEEE Trans. Inform. Theory*, 25, 572–584 (1979).

Cover and Thomas 2006 T. M. Cover and J. A. Thomas, *Elements of Information Theory*, 2nd edition, Wiley (2006).

Cox 1972 D. C. Cox, "Delay-doppler characteristics of multipath propagation at 910 MHz in a suburban mobile radio environment", *IEEE Trans. Antennas Propagat.*, 20, 625–635 (1972).

Cramer et al. 2002 R. J. Cramer, R. A. Scholtz, and M. Z. Win, "Evaluation of an ultra-wide-band propagation channel", *IEEE Trans. Antennas Propagat.*, 50, 541–550 (2002).

Crohn et al. 1993 I. Crohn, G. Schultes, R. Gahleitner, and E. Bonek, "Irreducible error performance of a digital portable communication system in a controlled time-dispersion indoor channel", *IEEE J. Sel. Area. Commun.*, 11, 1024–1033 (1993).

Cruz and Santhanam 2003 R. L. Cruz and A. V. Santhanam, "Optimal routing, link scheduling and power control in multihop wireless networks", *IEEE Infocom.*, 702–711 (2003).

Cullen et al. 1993 P. J. Cullen, P. C. Fannin, and A. Molina, "Wide-band measurement and analysis techniques for the mobile radio channel", *IEEE Trans. Veh. Technol.*, 42, 589–603 (1993).

Dahlman et al. 2008 E. Dahlman, S. Parkvall, J. Skold, and P. Beming, *3G Evolution: HSPA and LTE for Mobile Broadband*, Academic Press (2008).

Dam et al. 1999 H. Dam, M. Berg, R. Bormann, et al., "Functional test of adaptive antenna base stations for GSM", *3rd EPMCC European Personal Mobile Communications Conference*, Paris (1999).

Damosso and Correia 1999 E. Damosso and L. Correia (eds), *Digital Mobile Communications – The View of COST 231*, Commission of the European Union (1999).

Davey 1999 M. C. Davey, *Error Correction Using Low-Density Parity Check Codes*, Ph.D. thesis, Cambridge University (1999).

David and Benkner 1997 K. David and T. Benkner, *Digital Mobile Radio Systems*, Teubner (1996) [in German].

David and Nagaraja 2003 H. A. David and H. N. Nagaraja, *Order Statistics*, Wiley (2003).

Deller et al. 2000 J. R. Deller, Jr., J. H. Hansen, and J. G. Proakis, *Discrete-Time Processing of Speech Signals*, IEEE Press, New York (2000).

Devroye et al. 2007 N. Devroye, P. Mitran, M. Sharif, S. Ghassemzadeh, and V. Tarokh, "Information theoretic analysis of cognitive radio systems," in V. Bhargava and E. Hossain (eds), *Cognitive Wireless Communications*, Springer (2007).

Devroye and Tarokh 2007 N. Devroye and V. Tarokh, "Fundamental limits of cognitive radio networks", in F. H. P. Fitzek and M. Katz (eds), *Cognitive Wireless Networks: Concepts, Methodologies and Vision*, Springer (2007).

Deygout 1966 J. Deygout, "Multiple knife edge diffraction of microwaves", *IEEE Trans. Antennas Propagat.*, 14, 480–489 (1966).

Dietrich et al. 2001 C. B. Dietrich, K. Dietze, J. R. Nealy, and W. L. Stutzman, "Spatial, polarization, and pattern diversity for wireless handheld terminals", *IEEE Trans. Antennas Propagat.*, 49, 1271–1281 (2001).

Diggavi et al. 2004 S. N. Diggavi, N. Al-Dhahir, A. Stamoulis, and A. R. Calderbank, "Great expectations: The value of spatial diversity in wireless networks", *Proc. of IEEE*, 92, 219–270 (2004).

Dinan and Jabbari 1998 E. H. Dinan and B. Jabbari, "Spreading codes for direct sequence CDMA and wideband CDMA cellular networks", *IEEE Commun. Mag.*, 36(9), 48–54 (1998).

Divsalar and Simon 1990 D. Divsalar and M. K. Simon, "Multiple-symbol differential detection of MPSK", *IEEE Trans. Commun.*, 38, 300–308 (1990).

Dixon 1994 R. C. Dixon, *Spread Spectrum Systems with Commercial Applications*, Wiley (1994).

Doufexi et al. 2002 A. Doufexi, S. Armour, M. Butler, et al., "A comparison of the HIPERLAN/2 and IEEE 802.11a wireless LAN standards", *IEEE Commun. Mag.*, 40(5), 172–180 (2002).

Draper et al. 2008 S. C. Draper, L. Liu, A. F. Molisch, and J. S. Yedidia, "Routing in cooperative networks with mutual-information accumulation", *Proceedings of the International Conference in Communication* (2008).

Dubois-Ferriere 2006 H. Dubois-Ferriere, "Anypath Routing", Ph.D. thesis, Ecole Polytechnique Federale de Lausanne, Switzerland (2006).

Duel-Hallen et al. 1995 A. Duel-Hallen, J. Holtzman, and Z. Zvonar, "Multiuser detection for CDMA systems", *IEEE Pers. Commun. Mag.*, 2(2), 46–58 (1995).

Dunlop et al. 1999 J. Dunlop, D. Girma, and J. Irvine, *Digital Mobile Communications and the TETRA System*, Wiley (1999).

Durgin 2003 G. Durgin, *Space-Time Wireless Channels*, Cambridge University Press (2003).

Eberspaecher et al. 2009 J. Eberspaecher, H. J. Voegel, C. Bettstetter, and C. Hartmann, *GSM Architecture, Protocols, and Services*, 3rd edition, Wiley (2009).

Edfors et al. 1998 O. Edfors, M. Sandell, J. J. van de Beek, S. K. Wilson, and P. O. Borjesson, "OFDM channel estimation by singular value decomposition", *IEEE Trans. Commun.*, 46, 931–939 (1998).

Edfors et al. 2000 O. Edfors, M. Sandell, J. J. van de Beek, S. K. Wilson, and P. O. Borjesson, "Analysis of DFT-based channel estimators for OFDM", *Wireless Pers. Commun.*, 12, 55–70 (2000).

Eklund et al. 2006 C. Eklund, R. B. Marks, S. Ponnuswamy, K. L. Stanwood, and N. J. M. Van Waes, *Wireless-MAN: Inside the IEEE 802.16 Standard for Wireless Metropolitan Area Networks*, IEEE Press (2006).

Epstein and Peterson 1953 J. Epstein and D. W. Peterson, "An experimental study of wave propagation at 850 MC", *Proc. IEEE*, 41, 595–611 (1953).

Erätuuli and Bonek 1997 P. Erätuuli and E. Bonek, "Diversity arrangements for internal handset antennas", *8th IEEE International Symposium on Personal, Indoor and Mobile Radio Communications (PIMRC'97)*, Helsinki, Finland, pp. 589–593 (1997).

Erceg et al. 2004 V. Erceg et al., *TGn Channel Models*, IEEE document 802.11-03/940r4, on *www.802wirelessworld.com*, May (2004).

Ertel et al. 1998 R. B. Ertel, P. Cardieri, K. W. Sowerby, T. S. Rappaport, and J. H. Reed, "Overview of spatial channel models for antenna array communication systems", *IEEE Personal Communications*, 5(1), 10–22 (1998).

Eternad 2004 K. Eternad, *CDMA2000 Evolution: System Concepts and Design Principles*, Wiley (2004).

ETSI 1992 ETSI 300 175-1, *Radio Equipment and Systems (RES); Digital European Cordless Telecommunications (DECT) Common Interface Part 1: Overview, Part 2: Physical Layer ETSI*, Oktober (1992).

Fabregas et al. 2008 A. G. Fabregas, A. Martinez, and G. Caire, "Bit-interleaved coded modulation", *Found. Trends Commun. Inform. Theory*, 5, 1–2 (2008).

Failii 1989 E. Failii (ed.), *Digital Land Mobile Radio Communications – COST 207*, European Union, Brussels, Belgium (1989).

Falconer et al. 1995 D. D. Falconer, F. Adachi, and B. Gudmundson, "Time division multiple access methods for wireless personal communications", *IEEE Commun. Mag.*, 33(1), 50–57 (1995).

Falconer et al. 2003 D. Falconer, S. L. Ariyavisitakul, A. Benyamin-Seeyar, and B. Eidson, "Frequency domain equalization for single-carrier broadband wireless systems", *IEEE Commun. Mag.*, 40(4), 58–66 (2002).

Fant 1970 G. Fant, *Acoustic Theory of Speech Production*, Mouton, The Hague (1970).

Featherstone and Molkdar 2002 W. Featherstone and D. Molkdar, "Capacity benefits of GPRS coding schemes CS-3 and CS-4", *3G Mobile Communications Technologies*, 287–291 (2002).

Feldbauer et al. 2005 C. Feldbauer, G. Kubin, and W. B. Kleijn, "Anthropomorphic coding of speech and audio: A model inversion approach", *EURASIP J. Appl. Signal Process.*, 9, 1334–1349 (2005).

Felhauer et al. 1993 T. Felhauer, P. W. Baier, W. König, and W. Mohr, "Optimized wideband system for unbiased mobile radio channel sounding with periodic spread spectrum signals", *IEICE Trans. Commun.*, E76-B, 1016–1029 (1993).

Felsen and Marcuvitz 1973 L. B. Felsen and V. Marcuvitz, *Radiation and Scattering of Waves*, Prentice-Hall (1973).

Fleury 1990 B. Fleury, *Charakterisierung von Mobil- und Richtfunkkanälen mit Schwach Stationären Fluktuationen und Unkorrelierter Streuung (WSSUS)*, Dissertation, ETH Zuerich, Switzerland (1990).

Fleury 1996 B. H. Fleury, "An uncertainty relation for WSS processes and its application to WSSUS systems", *IEEE Trans. Commun.*, 44, 1632–1634 (1996).

Fleury 2000 B. H. Fleury, "First- and second-order characterization of direction dispersion and space selectivity in the radio channel", *IEEE Trans. Inform. Theory*, 46, 2027–2044 (2000).

Fleury et al. 1999 B. H. Fleury, M. Tschudin, R. Heddergott, D. Dahlhaus, and K. I. Pedersen, "Channel parameter estimation in mobile radio environments using the SAGE algorithm", *IEEE JSAC*, 17, 434–450 (1999).

Fontan and Espineira 2008 F. P, Fontan and P. M. Espineira, *Modelling the Wireless Propagation Channel: A Simulation Approach with MATLAB*, Wiley (2008).

Foschini and Gans 1998 G. J. Foschini and M. J. Gans, "On limits of wireless communications in a fading environment when using multiple antennas", *Wireless Personal Commun.*, 6, 311–335 (1998).

Foschini et al. 2003 G. J. Foschini, D. Chizhik, M. J. Gans, C. Papadias, and R. A. Valenzuela, "Analysis and performance of some basic space-time architectures", *IEEE J. Sel. Area. Commun.*, 21, 303–320 (2003).

Fragouli et al. 2005 C. Fragouli, J. Y. Le Boudec, and J. Widmer, *Network Coding: An Instant Primer*, EPFL LCA-REPORT-2005-010 *http://infoscience.epfl.ch/record/58339*.

Frullone et al. 1996 M. Frullone, G. Riva, P. Grazioso, and G. Falciasecca, "Advanced planning criteria for cellular systems", *IEEE Pers. Commun.*, 3(6), 10–15 (1996).

Fuhl 1994 J. Fuhl, Diploma thesis, Technical University Vienna, Austria (1994).

Fuhl et al. 1998 J. Fuhl, A. F. Molisch, and E. Bonek, "Unified channel model for mobile radio systems with smart antennas", *IEE Proc. Radar, Sonar Navigation*, 145, 32–41 (1998).

Fujimoto 2008 K. Fujimoto, *Mobile Antenna Systems Handbook*, 3rd edition, Artech (2008).

Fujimoto et al. 1987 K. Fujimoto, "A review of research on small antennas", *J. Inst. Electron. Inform. Commun. Eng.*, 70, 830–838 (1987).

Gahleitner 1993 R. Gahleitner, *Radio Wave Propagation in and into Urban Buildings*, Ph.D. thesis, Technical University, Vienna (1993).

Gallagher 1961 R. Gallagher, *Low Density Parity Check Codes*, Ph.D. thesis, Massachusetts Institute of Technology (1961).

Gallagher 2008 R. Gallagher, *Principles of Digital Communication*, Cambridge University Press (2008).

Garg 2000 V. K. Garg, *IS-95 CDMA & cdma2000: Cellular/PCS Systems Implementation*, Prentice-Hall (2000).

Gay and Benesty 2000 S. L. Gay and J. Benesty, *Acoustic Signal Processing for Telecommunication*, Kluwer Academic Publishers (2000).

Georgiadis et al. 2006 L. Georgiadis, M. Neely, and L. Tassiulas, "Resource allocation and cross layer control in wireless networks", *Found. Trends Networking*, 1, 1–144 (2006).

Gersho and Gray 1992 A. Gersho and R. M. Gray, *Vector Quantization and Signal Compression*, Kluwer Academic Publishers (1992).

Gesbert et al. 2002 D. Gesbert, H. Boelcskei, and A. Paulraj, "Outdoor MIMO wireless channels: Models and performance prediction", *IEEE Trans. Commun.*, 50(12), 1926–1935 (2002).

Gesbert et al. 2003 D. Gesbert, M. Shafi, D. S. Shiu, P. J. Smith, and A. Naguib, "From theory to practice: An overview of MIMO space-time coded wireless systems", *IEEE J. Sel. Area. Commun.*, 21, 281–302 (2003).

Gesbert et al. 2007 D. Gesbert, M. Kountouris, R. W. Heath, C. B. Chae, and T. Saelzer, "Shifting the MIMO paradigm", *IEEE Signal Process. Mag.*, 24(9), 36–46 (2007).

Ghavami et al. 2006 M. Ghavami, L. Michael, and R. Kohno, *Ultra Wideband Signals and Systems in Communication Engineering*, Wiley (2006).

Giannakis and Halford 1997 G. B. Giannakis and S. D. Halford, "Blind fractionally spaced equalization of noisy FIR channels: Direct and adaptive solutions", *IEEE Trans. Signal Process.*, 45, 2277–2292 (1997).

Gibson et al. 1998 J. D. Gibson, T. Berger, T. Lookabaugh, R. Baker, and D. Lindbergh, *Digital Compression for Multimedia: Principles & Standards*, Morgan Kaufman (1998).

Gilhousen et al. 1991 K. S. Gilhousen, I. M. Jacobs, R. Padovani, A. J. Viterbi, L. A. Weaver, and C. E. Wheatley, "On the capacity of a cellular CDMA system", *IEEE Trans. Veh. Technol.*, 40, 303–312 (1991).

Gitlin and Weinstein 1981 R. D. Gitlin and S. B. Weinstein, "Fractionally-spaced equalization: An improved digital transversal equalizer", *Bell System Tech. J.*, 60, 275–296 (1981).

Glassner 1989 A. S. Glassner, *An Introduction to Ray Tracing*, Morgan Kaufmann (1989).

Glisic and Vucetic 1997 S. Glisic and B. Vucetic, *Spread Spectrum CDMA Systems for Wireless Communications*, Artech House, London (1997).

Godara 1997 L. C. Godara, "Applications of antenna arrays to mobile communications. I. Performance improvement, feasibility, and system considerations", *Proceedings of IEEE 85, 1031-1060 and Application of Antenna Arrays to Mobile Communications. II. Beam-forming and Direction-of-arrival Considerations, Proceedings of IEEE 85"*, pp. 1195–1245 (1997).

Godara 2001 L. C. Godara, *Handbook of Antennas in Wireless Communications*, CRC Press, Boca Raton (2001).

Godard 1980 D. N. Godard, "Self-recovering equalization and carrier tracking in two-dimensional data communication systems", *IEEE Trans. Commun.*, 28, 1867–1875 (1980).

Goiser 1998 A. Goiser, *Handbuch der Spread-Spectrum Technik*, Springer, Wien (1998).

Goiser et al. 2000 A. Goiser, M. Z. Win, G. Chrisikos, and S. Glisic, "Code division multiple access", part 5 of A. F. Molisch (ed.), *Wireless Wideband Digital Communications*, Prentice-Hall, USA (2000).

Golaup et al. 2009 A. Golaup, M. Mustapha, and L. B. Patanapongpibul, "Femtocell access control strategy in UMTS and LTE", *IEEE Commun. Mag.*, 47(9), 117–123 (2009).

Goldsmith et al. 2003 A. Goldsmith, S. A. Jafar, N. Jindal, and S. Vishwanath, "Capacity limits of MIMO channels", *IEEE J. Select. Area. Commun.*, 21, 684–702 (2003).

Golomb and Gong 2005 S. Golomb and G. Gong, *Signal Design for Good Correlation: For Wireless Communication, Cryptography, and Radar*, Cambridge University Press (2005).

Gonzalez 1984 G. Gonzalez, *Microwave Transistor Amplifiers: Analysis and Design*, Prentice Hall (1984).

Gonzalez and Woods 2008 R. C. Gonzalez and R. E. Woods, *Digital Image Processing*, 3rd edition, Prentice Hall (2008).

Goodman et al. 1989 D. J. Goodman, R. A. Valenzuela, K. T. Gayliard, and B. Ramamurthi, "Packet reservation multiple access for local wireless communications", *IEEE Trans. Commun.*, 37, 885–890 (1989).

Gorokhov 1998 A. Gorokhov, "On the performance of the Viterbi equalizer in the presence of channel estimation errors", *IEEE Signal Process. Lett.*, 5, 321–324 (1998).

Gray 1989 R. M. Gray, *Source Coding Theory*, Kluwer Academic Publishers (1989).

Greenstein et al. 1997 L. J. Greenstein, V. Erceg, Y. S. Yeh, and M. V. Clark, "A new path-gain/delay-spread propagation model for digital cellular channels", *IEEE Trans. Veh. Technol.*, 46, 477–485 (1997).

Gross and Harris 1998 D. Gross and C. M. Harris, *Fundamentals of Queuing Theory*, 3rd edition, Wiley (1998).

Gupta and Kumar 2000 P. Gupta and P. R. Kumar, "The capacity of wireless networks", *IEEE Trans. Inform. Theory*, 46, 388–404 (2000).

Haardt and Nossek 1995 M. Haardt and J. A. Nossek, "Unitary ESPRIT: How to obtain increased estimation accuracy with a reduced computational burden", *IEEE Trans. Signal Process.*, 43, 1232–1242 (1995).

Haensler and Schmidt 2004 E. Haensler and G. Schmidt (eds), *Acoustic Echo and Noise Control*, John Wiley & Sons (2004).

Hagenauer and Hoeher 1989 J. Hagenauer and P. Hoeher, "A Viterbi algorithm with soft-decision outputs and its applications", *IEEE Globecom*, 1680–1686 (1989).

Han et al. 2007 Z. Han, Z. Ji, and K. J. R. Liu, "Non-cooperative resource competition game by virtual referee in multi-cell OFDMA networks", *IEEE J. Sel. Area. Commun.*, 25, 1079–1090 (2007).

Han and Poor 2009 Z. Han and V. H. Poor, "Impact of cooperative transmission on network routing", in Y. Zhang, H. H. Chen, and M. Guizani (eds), *Cooperative Wireless Communications*, CRC Press (2009).

Hansen 1998 R. C. Hansen, *Phased Array Antennas*, Wiley (1998).

Hanzo et al. 2000 L. Hanzo, W. Webb, and T. Keller, *Single- and Multi-Carrier Quadrature Amplitude Modulation: Principles and Applications for Personal Communications, WLANs and Broadcasting*, Wiley (2000).

Hanzo et al. 2001 L. Hanzo, F. C. A. Somerville, and J. P. Woodard, *Voice Compression and Communications*, IEEE Press Wiley Interscience, New York (2001).

Hanzo et al. 2003 L. Hanzo, M. Muenster, B. J. Choi, and T. Keller, *OFDM and MC-CDMA for Broadband Multi-User Communications, WLANs and Broadcasting*, Wiley (2003).

Harrington 1993 R. F. Harrington, *Field Computation by Moment Method*, Wiley/IEEE Press (1993).

Harryson et al. 2010 F. Harryson, J. Medbo, A. F. Molisch, A. Johansson, and F. Tufvesson, "Efficient experimental evaluation of a MIMO handset with user influence", *IEEE Trans. Wireless Commun.*, 9, 853–863 (2010).

Hart et al. 2009 M. Hart, Z. J. Tao, and Y. Zhou, *IEEE 802.16j Multi-hop Relay*, Wiley (2009).

Hashemi 1979 H. Hashemi, "Simulation of the urban radio propagation channel", *IEEE Trans. Veh. Technol.*, 28, 213–225 (1979).

Hashemi 1993 H. Hashemi, "Impulse response modeling of indoor radio propagation channels", *IEEE J. Sel. Area. Commun.*, 11, 943 (1993).

Haslett 2008 C. Haslett, *Essentials of Radio Wave Propagation*, Cambridge University Press (2008).

Hata 1980 M. Hata, "Empirical formula for propagation loss in land mobile radio services", *IEEE Trans. Veh. Technol.*, 29, 317–325 (1980).

Haykin 1991 S. Haykin, *Adaptive Filter Theory*, Prentice Hall, Englewood Cliffs (1991).

Haykin 2005 S. Haykin, "Cognitive radio: Brain-empowered wireless communications", *IEEE J. Sel. Area. Commun.*, 23, 201–220 (2005).

Heath 2011 R. W. Heath, Jr., *Advanced MIMO Communications*, Cambridge University Press (2011).

Heavens 1965 O. S. Heavens, *Optical Properties of Thin Film Solids*, Dover, New York (1965).

Hewlett-Packard 1994 Hewlett-Packard, *Schulungsunterlagen GSM*, Hewlett-Packard, German (1994).

Hirade et al. 1979 K. Hirade, M. Ishizuka, F. Adachi, and K. Ohtani, "Error-rate performance of digital FM with differential detection in land mobile radio channels", *IEEE Trans. Veh. Technol.*, 28, 204–212 (1979).

Hirasawa and Haneishi 1991 K. Hirasawa and M. Haneishi (eds), *Analysis, Design, and Measurements of Small and Low-Profile Antennas*, Artech House (1991).

Hoeher 1992 P. Hoeher, "A statistical discrete-time model for the WSSUS multipath channel", *IEEE Trans. Veh. Technol.*, 41, 461–468 (1992).

Holma and Toskala 2007 H. Holma and A. Toskala (eds), *WCDMA for UMTS: Radio Access for Third Generation Mobile Communications*, 4th edition, Wiley (2007).

Holma and Toskola 2009 H. Holma and A. Toskola (eds), *LTE for UMTS – OFDMA and SC-FDMA Based Radio Access*, Wiley (2009).

Holtzman and Jalloul 1994 J. M. Holtzman and L. M. Jalloul, "Rayleigh fading effect reduction with wideband DS/CDMA signals", *IEEE Trans. Commun.*, 42, 1012–1016 (1994).

Hong et al. 2007 Y. W. Hong, W. J. Huang, F. H. Chiu, and C.-C. Jay Kuo, "Cooperative communications in resource-constrained wireless networks", *IEEE Signal Proc. Mag.*, 5, 47–57 (2007).

Honig 2009 M. L. Honig (ed.), *Advances in Multiuser Detection*, Wiley (2009).

Hoppe et al. 2003 R. Hoppe, P. Wertz, F. M. Landstorfer, and G. Woelfle, "Advanced ray optical wave propagation modelling for urban and indoor scenarios including wideband properties", *Eur. Trans. Telecommun.*, 14, 61–69 (2003).

Hossain and Barghava 2008 E. Hossain and V. K. Barghava (eds), *Cognitive Wireless Communication Networks*, Springer (2008).

Hottinen et al. 2003 A. Hottinen, O. Tirkkonen, and R. Wichman, *Multi-antenna Transceiver Techniques for 3G and Beyond*, Wiley (2003).

Huang et al. 2001 X. Huang, A. Acero, and H. W. Hon, *Spoken Language Processing*, Prentice-Hall (2001).

Huffman 1952 D. A. Huffman, "A method for the construction of minimum redundancy codes", *Proceedings of IRE 40*, pp. 1098–1101 (1952).

Hunter and Nosratinia 2006 T. Hunter and A. Nosratinia, "Diversity through Coded Cooperation", *IEEE Transactions on Wireless Communications*, 5(2), 1–7, February (2006).

Hunter et al. 2006 T. E. Hunter, S. Sanayei, and A. Nosratinia, "Outage analysis of coded cooperation", *IEEE Trans. Inform. Theory*, 52, 375–391 (2006).

Hwang et al. 2009 T. Hwang, C. Yang, G. Wu, S. Li, and Y. G. Li, "OFDM and its wireless applications: A survey", *IEEE Trans. Veh. Technol.*, 58, 1673–1694 (2009).

IEEE P802.11n IEEE P802.11n™/D7.0 Draft STANDARD for Information Technology – Telecommunications and information exchange between systems – Local and metropolitan area networks – Specific requirements

IEEE 802.11 Institute of Electrical and Electronics Engineers, *Standard 802.11*, particularly the following documents: IEEE std 802.11-1999, Part 11: Wireless LAN Medium Access Control (MAC) and Physical Layer (PHY) specifications" (1999); IEEE 802.11e draft/D9.0, Part 11:Wireless Medium Access Control (MAC) and physical layer (PHY) specifications: *Medium Access Control (MAC) Quality of Service (QoS) Enhancements*, (2004); IEEE std 802.11a-1999, Part 11: Wireless LAN Medium Access Control (MAC) and Physical Layer (PHY) specifications: *High-Speed Physical Layer in the 5GHZ Band* (1999); IEEE std 802.11b-1999, Part 11: Wireless LAN Medium Access Control (MAC) and Physical Layer (PHY) specifications: *Higher-Speed Physical Layer Extension in the 2.4GHz Band* (1999).

IEEE 802.16-2009 IEEE 802.16 standardization group, *IEEE Standard for Local and Metropolitan Area Networks – Part 16: Air Interface for Broadband Wireless Access Systems*, at *http://wirelessman.org/pubs/80216Rev2.html* (2009).

Ikegami et al. 1984 F. Ikegami, S. Yoshida, T. Takeuchi, and M. Umehira, "Propagation factors controlling mean field strength on urban streets", *IEEE Trans. Antennas Propagat.*, 32, 822–829 (1984).

Intanagonwiwat et al. 2003 C. Intanagonwiwat, R. Govindan, D. Estrin, J. Heidemann, and F. Silva, "Directed diffusion for wireless sensor networking", *IEEE/ACM Trans. Networking*, 11, 2–16 (2003).

Itakura and Saito 1968 F. Itakura and S. Saito "Analysis synthesis telephony based on the maximum likelihood principle", *Proceedings of the 6th International Congress on Acoustics*, Tokyo, Japan, pp. C17–C20 (1968).

ITU 1993 International Telecommunications Union, *Video Codec for Audiovisual Services at px64 Kbit/s*, Recommendation H.261 (1993).

ITU 1996 International Telecommunications Union, *Video Coding for Low Bit Rate Communication*, ITU-T Recommendation H.263 (1996).

ITU 1997 International Telecommunications Union, *Guidelines for Evaluation of Radio Transmission Technologies for IMT-2000*, Recommendation ITU-R M.1225 (1997).

ITU 1998 International Telecommunications Union, *Studio Encoding Parameters of Digital Television for Standard 4:3 and Wide-screen 16:9 Aspect Ratios*, ITU-R, Recommendation BT.601-5 (1998).

ITU 2008 International Telecommunications Union, *Guidelines for Evaluation of Radio Interface Technologies for IMT-Advanced*, Recommendation ITU-R M.2135 (2008).

ITU 2009 International Telecommunications Union, *Information Technology – Coding of Audio-visual Objects – Part 10: Advanced Video Coding (AVC)*, 5th edition, ITU-T Recommendation H.264 *backslash*mid$ ISO/IEC 14496-10:2009 (2009).

Jafarkhani 2005 H. Jafarkhani, *Space-Time Coding: Theory and Practice*, Cambridge University Press (2005).

Jain 1989 A. K. Jain, *Fundamentals of Digital Image Processing*, Prentice Hall (1989).

Jajszczyk and Wagrowski 2005 A. Jajszczyk and M. Wagrowski, "OFDM for wireless communication systems", *IEEE Commun. Mag.*, 43(9), 18–20 (2005).

Jakes 1974 W. C. Jakes, *Microwave Mobile Communications*, IEEE Press (reprint) (1974).

Jamali and Le-Ngoc 1991 S. Jamali and T. Le-Ngoc, "A new 4-state 8PSK TCM scheme for fast fading, shadowed mobile radio channels", *IEEE Trans. Veh. Technol.*, 40, 216–222 (1991).

Jayant and Noll 1984 N. S. Jayant and P. Noll, *Digital Coding of Waveforms – Principles and Applications to Speech and Video*, Prentice-Hall, Englewood Cliffs (1984).

Jelinek et al. 2004 M. Jelinek, R. Salami, S. Ahmadi, B. Bessetle, P. Gournay, and C. Laflamme, "On the architecture of the cdma2000/spl reg/variable-rate multimode wideband (VMR-WB) speech coding standard", *Proceedings of IEEE International Conference in Acoustics, Speech, and Signal Processing*, I-281-4 (2004).

Ji and Liu 2007 Z. Ji and K. J. R. Liu, "Dynamic spectrum sharing: A game theoretical overview", *IEEE Commun. Mag.*, 45(5), 88–95 (2007).

Jiang and Hanzo 2007 M. Jiang and L. Hanzo, "Multiuser MIMO-OFDM for next-generation wireless systems", *Proc. IEEE*, 95, 1430–1469 (2007).

Jiang and Wu 2008 T. Jiang and Y. Wu, "Peak-to-average power ratio reduction techniques for OFDM signals", *IEEE Trans. Broadcast.*, 54, 257–268 (2008).

Johannesson and Zigangirov 1999 R. Johannesson and K. S. Zigangirov, *Fundamentals of Convolutional Coding*, Wiley - IEEE Press (1999).

Johnson et al. 2001 D. B. Johnson, D. A. Maltz, and J. Broch. "DSR: The dynamic source routing protocol for multi-hop wireless ad hoc networks", in C. E. Perkins (ed.), *Ad Hoc Networking*, Chapter 5, pp. 139–172, Addison-Wesley (2001).

Jovocic and Visvanath 2009 A. Jovocic and P. Viswanath, *Cognitive Radio: An Information-Theoretic Perspective*, IEEE Trans. Information Theory 55, 3945–3958 (2009).

Jurafsky and Martin 2000 D. Jurafsky and J. H. Martin, *Speech and Language Processing*, Prentice-Hall, Upper Saddle River (2000).

Kattenbach 1997 R. Kattenbach, *Characterisierung zeitvarianter Indoor Mobilfunkkanäle mittels ihrer System- und Korrelationsfunktionen*, Dissertation an der Universitä t GhK Kassel, publiziert beim Shaker-Verlag, Aachen (1997).

Kattenbach 2002 R. Kattenbach, "Statistical modeling of small-scale fading in directional radio channels", *IEEE J. Sel. Area. Commun.*, 20, 584–592 (2002).

Katzela and Naghshineh 2000 I. Katzela and M. Naghshineh, "Channel assignment schemes for cellular mobile telecommunication systems: A comprehensive survey", *IEEE Commun. Surveys Tutorials*, 3(2), 10–31 (2000).

Keller 1962 J. B. Keller, "Geometrical theory of diffraction", *J. Opt. Soc. Am.*, 2, 116–130 (1962).

Keller and Hanzo 2000 T. Keller and L. Hanzo, "Adaptive multicarrier modulation: A convenient framework for time-frequency processing in wireless communications", *Proc. IEEE*, 88, 611–640 (2000).

Kermoal et al. 2002 J. P. Kermoal, L. Schumacher, K. I. Pedersen, P. E. Mogensen, and F. Frederiksen, "A stochastic MIMO radio channel model with experimental validation", *IEEE J. Sel. Area. Commun.*, 20, 1211 (2002).

Khandani et al. 2003 A. Khandani, J. Abounadi, E. Modiano, and L. Zhang, "Cooperative routing in wireless networks", *Allerton Conference on Communications, Control and Computing*, October (2003).

Khun-Jush et al. 2002 J. Khun-Jush, P. Schramm, G. Malmgren, and J. Torsner, "HiperLAN2: Broadband wireless communications at 5 GHz", *IEEE Commun. Mag.*, 40(6), 130–136 (2002).

Kim et al. 1999 Y. H. Kim, I. Song, H. G. Kim, T. Chang, and H. M. Kim, "Performance analysis of a coded OFDM system in time-varying multipath Rayleigh fading channels", *IEEE Trans. Veh. Technol.*, 48, 1610–1615 (1999).

Kivekäs et al. 2004 O. Kivekäs, J. Ollikainen, T. Lehtiniemi, and P. Vainikainen, Member, "Bandwidth, SAR, and efficiency of internal mobile phone antennas", *IEEE Trans. Electromagn. Comp.*, 46, 71–76 (2004).

Kleijn 2005 W. B. Kleijn, *Information Theory and Source Coding*, Royal Institute of Technology (KTH), unpublished course notes (2005).

Kleijn and Granzow 1991 W. B. Kleijn and W. Granzow, "Methods for waveform interpolation in speech coding", *Digit. Signal Process.*, 1, 215–230 (1991).

Kleijn and Paliwal 1995 W. B. Kleijn and K. K. Paliwal, *Speech Coding and Synthesis*, Elsevier (1995).

Kleinrock and Tobagi 1975 L. Kleinrock and F. Tobagi, "Packet switching in radio channels: Part I – carrier sense multiple access modes and their throughput-delay characteristics", *IEEE Trans. Commun.*, 23, 1400–1416 (1975).

Klemenschits and Bonek 1994 T. Klemenschits and E. Bonek, "Radio coverage of road tunnels at 900 and 1800 MHz by discrete antennas", *Proceedings of PIMRC 1994*, pp. 411–415 (1994).

Kobayashi and Caire 2006 M. Kobayashi and G. Caire, "An iterative water-filling algorithm for maximum weighted sum-rate of Gaussian MIMO-BC", *IEEE J. Sel. Area. Commun.*, 24, 1640–1646 (2006).

Kohno et al. 1995 R. Kohno, R. Meidan, and L. B. Milstein, "Spread spectrum access methods for wireless communications", *IEEE Commun. Mag.*, 33(1), 58–67 (1995).

Kondoz 2004 A. M. Kondoz, *Digital Speech: Coding for Low Bit Rate Communication Systems*, 2nd edition, Wiley (2004).

Kouyoumjian and Pathak 1974 R. G. Kouyoumjian and P. H. Pathak, "A uniform geometrical theory of diffraction for an edge in a perfectly conducting surface", *Proc. IEEE*, 62, 1448–61 (1974).

Kozek 1997 W. Kozek, *Matched Weyl-Heisenberg Expansions of Nonstationary Environments*, Dissertation, Technical University Vienna, Austria (1997).

Kozek and Molisch 1998 W. Kozek and A. F. Molisch, "Nonorthogonal pulseshapes for multicarrier communications in doubly dispersive channels", *IEEE J. Sel. Area. Commun.*, 16, 1579–1589 (1998).

Kramer et al. 2005 G. Kramer, M. Gastpar, and P. Gupta, "Cooperative strategies and capacity theorems for relay networks", *IEEE Trans. Inform. Theory*, 51, 3037–3063 (2005).

Kramer et al. 2007 G. Kramer, I. Maric, and R. D. Yates, "Cooperative communications", *Found. Trends Networking*, 1(3–4), 1–167 (2007).

Kraus and Marhefka 2002 J. D. Kraus and R. J. Marhefka, *Antennas: For All Applications*, 3rd edition, McGrawHill (2002).

Kreuzgruber et al. 1993 P. Kreuzgruber, P. Unterberger, and R. Gahleitner, "A ray splitting model for indoor propagation associated with complex geometries", *Proceedings of 43rd IEEE Vehicular Technology Conference*, Secaucus, NJ, pp. 227–230 (1993).

Krim and Viberg 1996 H. Krim and M. Viberg, "Two decades of array signal processing - the parametric approach", *IEEE Signal Proc. Mag.*, 13(4), 67–94 (1996).

Kubin 1995 G. Kubin, "Nonlinear processing of speech", in W. B. Kleijn and K. K. Paliwal (eds), *Speech Coding and Synthesis*, pp. 557–610, Elsevier (1995).

Kuchar et al. 1997 A. Kuchar, J. Fuhl, and E. Bonek, "Spectral efficiency enhancement and power control of smart antenna system", *EPMCC 97*, Bonn, Germany, September 30–October 2 (1997).

Kuchar et al. 2002a A. Kuchar, J. P. Rossi, and E. Bonek, "Directional macro-cell channel characterization from urban measurements", *IEEE Trans. Antennas Propagat.*, 48, 137–146 (2002a).

Kuchar et al. 2002b A. Kuchar, M. Taferner, and M. Tangemann, "A real-time DOA-based smart antenna processor", *IEEE Trans. Veh. Technol.*, 51, 1279–1293 (2002b).

Kunz and Luebbers 1993 K. S. Kunz and R. J. Luebbers, *The Finite Difference Time Domain Method for Electromagnetics*, CRC Press (1993).

Laneman and Wornell 2003 J. N. Laneman and G. W. Wornell, "Distributed space-time-coded protocols for exploiting cooperative diversity in wireless networks", *IEEE Trans. Inform. Theory*, 49, 2415–2425 (2003).

Laneman et al. 2004 J. N. Laneman, D. N. C. Tse, and G. W. Wornell, "Cooperative diversity in wireless networks: Efficient protocols and outage behavior", *IEEE Trans. Inform. Theory*, 50, 3062–3080 (2004).

Larsen and Aarts 2004 E. R. Larsen and R. M. Aarts, *Audio Bandwidth Extension: Application of Psychoacoustics, Signal Processing and Loudspeaker Design*, Wiley (2004).

Larsson and Stoica 2008 E. G. Larsson and P. Stoica, *Space-Time Block Coding for Wireless Communications*, Cambridge University Press (2008).

Latief and Zhang 2007 K. B. Letaief and W. Zhang, "Cooperative spectrum sensing", in E. Hossein and V. K. Barghava (eds.), *Cognitive Wireless Communication Networks* Springer (2007).

Laufer et al. 2009 R. Laufer, H. Dubois-Ferriere, and L. Kleinrock, "Multirate anypath routing in wireless mesh networks", *IEEE INFOCOM*, 37–45 (2009).

Laurent 1986 P. A. Laurent, "Exact and approximate construction of digital phase modulations by superposition of amplitude modulated pulses", *IEEE Trans. Commun.*, 34, 150–160 (1986).

Laurila 2000 J. Laurila, *Semi-Blind Detection of Co-Channel Signals in Mobile Communications*, Dissertation TU Wien, Wien (2000).

Laurila et al. 1998 J. Laurila, A. F. Molisch, and E. Bonek, "Influence of the scatter distribution on power delay profiles and azimuthal power spectra of mobile radio channels", *Proc. ISSSTA'98*, 267–271 (1998).

Lawton and McGeehan 1994 M. C. Lawton and J. P. McGeehan, "The application of a deterministic ray launching algorithm for the prediction of radio channel characteristics in small-cell environments", *IEEE Trans. Veh. Technol.*, 43, 955–969 (1994).

Lee 1973 W. C. Y. Lee, "Effects on correlations between two mobile base-station antennas", *IEEE Trans. Commun.*, 21, 1214–1224 (1973).

Lee 1982 W. C. Y. Lee, *Mobile Communications Engineering*, McGraw Hill, New York (1982).

Lee 1986 W. C. Y. Lee, *Mobile Communications Design Fundamentals*, Sams, Indianapolis (1986).

Lee 1995 W. C. Y. Lee, *Mobile Cellular Telecommunications: Analog and Digital Systems*, McGraw Hill (1995).

Letaief and Zhang 2007 K. B. Letaief and W. Zhang, "Cooperative spectrum sensing", in E. Hossein and V. K. Barghava (eds), *Cognitive Wireless Communication Networks*, Springer (2007).

Li and Miller 1998 J. S. Lee and L. E. Miller, *CDMA Systems Engineering Handbook*, Artech House, London (1998).

Li and Stuber 2006 Y. G. Li and G. L. Stuber, *Orthogonal Frequency Division Multiplexing for Wireless Communications*, Springer (2006).

Li et al. 1997 J. Li, J. F. Wagen, and E. Lachat, "ITU model for multi-knife-edge diffraction", *Microwaves, Antennas Propagat., IEE Proc.*, 143, 539–541 (1997).

Li et al. 1998 Y. G. Li, L. C. Cimini, and N. R. Sollenberger, "Robust channel estimation for OFDM systems with rapid dispersive fading channels", *IEEE Trans. Commun.*, 46, 902–915 (1998).

Li et al. 1999 Y. G. Li, N. Seshadri, and S. Ariyavisitakul, "Channel estimation for OFDM systems with transmitter diversity in mobile wireless channels", *IEEE J. Sel. Area. Commun.*, 17(3), 461–471 (1999).

Liberti and Rappaport 1996 J. C. Liberti and T. S. Rappaport, "A geometrically based model for line of sight multipath radio channels", *Proceedings of IEEE Vehicular Technology Conference*, pp. 844–848 (1996).

Liberti and Rappaport 1999 J. C. Liberti and T. S. Rappaport, *Smart Antennas for Wireless Communications: IS-95 and Third Generation CDMA Applications*, Prentice-Hall (1999).

Liebenow and Kuhlmann 1993 U. Liebenow and P. Kuhlmann, "Determination of scattering surfaces in hilly terrain", *COST 231, TD (93)* 119, (1993).

Lin 2003 J. C. Lin, "Safety standards for human exposure to radio frequency radiation and their biological rationale", *IEEE Microwave Mag.*, 4(4), 22–26 (2003).

Lin and Costello 2004 S. Lin and D. J. Costello, *Error Control Coding*, 2nd edition, Prentice Hall (2004).

Liu et al. 1996 H. Liu, G. Xu, L. Tong, T. Kailath, "Recent developments in blind channel equalization: From cyclostationarity to subspaces", *Signal Process.*, 50, 83–99 (1996).

Liu et al. 2009 K. J. R. Liu, A. K. Sadek, W. Su, and A. Kwasinski, *Cooperative Communications and Networking*, Cambridge University Press (2009).

Lo 1999 T. K. Y. Lo, "Maximum ratio transmission", *IEEE Trans. Commun.*, 47, 1458–1461 (1999).

Loeliger 2004 H. A. Loeliger, "An introduction to factor graphs", *IEEE Signal Process. Mag.*, January, 28–41 (2004).

Loncar et al. 2002 M. Loncar, R. Müller, T. Abe, J. Wehinger, and C. Mecklenbräuker, "Iterative equalizer using soft-decoder feedback for MIMO systems in frequency-selective fading", *Proceedings of URSI General Assembly 2002* (2002).

Lott and Teneketzis 2006 C. Lott and D. Teneketzis, "Stochastic routing in ad-hoc networks", *IEEE Trans. Automat. Control*, 51, 52–70 (2006).

Love et al. 2003 D. J. Love, R. W. Heath, and S. Strohmer, "Grassmannian beamforming for multiple-input multiple-output wireless systems", *IEEE Trans. Inform. Theory*, 49, 2735–2747 (2003).

Love et al. 2008 D. J. Love, R. W. Heath, V. K. N. Lau, D. Gesbert, B. D. Rao, and M. Andrews, "An overview of limited feedback in wireless communication systems", *IEEE J. Sel. Area. Commun.*, 26, 1341–1365 (2008).

Lozano et al. 2008 A. Lozano, A. M. Tulino, and S. Verdu, "Multiantenna capacity myths and reality", in H. Boelcskei, D. Gesbert, C. B. Papadias, and A.-J. van der Veen (eds), *Space-Time Wireless Systems*, Cambridge University Press (2008).

Lucky et al. 1968 R. W. Lucky, J. Salz, and E. J. Weldon Jr., *Principles of Data Communication*, McGraw Hill (1968).

Lustman and Porrat 2010 Y. Lustmann and D. Porrat, Hebrew University, Jerusalem, private communication (2010).

MacAulay and Quatieri 1986 R. J. McAulay and T. F. Quatieri, "Speech analysis/synthesis based on a sinusoidal representation", *IEEE Trans. Acoustics, Speech, Signal Process.*, 34, 744–754 (1986).

MacKay 2002 D. J. C. MacKay, *Information Theory, Inference & Learning Algorithms*, Cambridge University Press (2002).

MacKay and Neal 1997 D. J. C. MacKay and R. M. Neal, "Near Shannon limit performance of low density parity check codes", *Electron. Lett.*, 33, 457–458 (1997).

Madan et al. 2009 R. Madan, N. B. Mehta, A. F. Molisch, and J. Zhang, "Energy-efficient decentralized control of cooperative wireless networks with fading", *IEEE Trans. Automat. Control*, 54, 512–527 (2009).

Mailloux 1994 R. J. Mailloux, *Phased Array Antenna Handbook*, Artech House (1994).

Malvar et al. 2003 H. Malvar, A. Hallapuro, M. Karczewicz, and L. Kerofsky, "Low-complexity transform and quantization in H.264/AVC", *IEEE Trans. Circ. Syst. Video Technol.*, 13, 598–603 (2003).

Manholm et al. 2003 L. Manholm, M. Johansson, and S. Petersson, Antennas with Electrical Beamtilt for WCDMA: Simulations and Implementation, *Swedish National Conference on Antennas* (2003).

Marcuse 1991 D. Marcuse, *Theory of Dielectric Optical Waveguides*, 2nd edition, Academic Press, Boston (1991).

Mardia et al. 1979 K. V. Mardia, J. T. Kent, and J. M. Bibby, *Multivariate Analysis*, Academic Press, London (1979).

Maric and Titlebaum 1992 S. V. Maric and E. L. Titlebaum, "A class of frequency hop codes with nearly ideal characteristics for use in multiple-access spread-spectrum communications and radar and sonar systems", *IEEE Trans. Commun.*, 40, 1442–1447 (1992).

Maric and Yates 2004 I. Maric and R. D. Yates, "Cooperative multihop broadcast for wireless networks", *IEEE J. Sel. Area. Commun.*, 22, 1080–1088 (2004).

Martirosyan et al. [2008] A. Martirosyan, A. Boukerche, and R. W. N. Pazzi, "A taxonomy of cluster-based routing protocols for wireless sensor networks", *The International Symposium on Parallel Architectures, Algorithms, and Networks* (2008).

Marzetta and Hochwald 1999 T. L. Marzetta and B. M. Hochwald, "Capacity of a mobile multiple-antenna communication link in Rayleigh at fading", *IEEE Trans. Inform. Theory*, 45, 139–157 (1999).

Matz 2003 G. Matz, "Characterization of non-WSSUS fading dispersive channels", *Proceedings of ICC '03*, pp. 2480–2484 (2003).

Matz and Hlawatsch 1998 G. Matz and F. Hlawatsch, "Time-frequency transfer function calculus (symbolic calculus) of linear time-varying systems (linear operators) based on a generalized underspread theory", *J. Math. Phys. (Special Issue on Wavelet and Time-Frequency Analysis)*, 39, 4041–4070 (1998).

Matz et al. 2002 G. Matz, A. F. Molisch, F. Hlawatsch, M. Steinbauer, and I. Gaspard, "On the systematic measurement errors of correlative mobile radio channel sounders", *IEEE Trans. Commun.*, 50, 808–821 (2002).

May and Rohling 2000 T. May and H. Rohling, "Orthogonal frequency division multiple access", part 4 of A. F. Molisch (ed.), *Wideband Wireless Digital Communications*, Prentice-Hall, U.S.A (2000).

Mayr 1996 B. Mayr, *Modulationsangepasste Codierung*, Lecture notes, TU Vienna (1996).

McEliece 2004 R. McEliece, *The Theory of Information and Coding*, Student edition, Cambridge University Press (2004).

McNamara et al. 1990 D. A. McNamara, C. W. I. Pistorius, and J. A. G. Malherbe, *Introduction to the Uniform Geometrical Theory of Diffraction*, Artech House, Boston, MA (1990).

Mehta et al. 2007 N. B. Mehta, J. Wu, A. F. Molisch, and J. Zhang, "Approximating a sum of random variables with a lognormal distribution", *IEEE Trans. Wireless Commun.*, 6, 2690–2699 (2007).

Mengali and D'Andrea 1997 U. Mengali and A. N. D'Andrea, *Synchronization Techniques for Digital Receivers*, Plenum (1997).

van der Meulen 1971 E. van der Meulen, "Three-terminal communication channels", *Adv. Appl. Prob.*, 3, 120–154 (1971).

Meurling and Jeans 1994 J. Meurling and R. Jeans, *The Mobile Phone Book*, ISBN 0-9524031-02 published by Communications Week International, London (1994).

Meyr and Ascheid 1990 H. Meyr and G. Ascheid, *Digital Communication Receivers, Phase-, Frequency-Locked Loops, and Amplitude Control*, Wiley (1990).

Meyr et al. 1997 H. Meyr, M. Moeneclaeye, and S. A. Fechtel, *Digital Communication Receivers, Vol. 2: Synchronization, Channel Estimation, and Signal Processing*, Wiley (1997).

Milstein 1988 L. B. Milstein, "Interference rejection techniques in spread spectrum communications", *Proc. IEEE*, 76, 657–671 (1988).

Moeller et al. 2010 S. Moeller, A. Sridharan, B. Krishnamachari, and O. Gnawali, "Routing without routes: the backpressure collection protocol", *9th ACM/IEEE Intl. Conf. Information Processing in Sensor Networks* (2010).

Molisch 2000 A. F. Molisch (ed.), *Wideband Wireless Digital Communications*, Prentice-Hall (2000).

Molisch 2001 A. F. Molisch, *A System Proposal for Wireless LANS with MIMO*, AT&T Research Labs Internal Report (2001).

Molisch 2002 A. F. Molisch, *Modeling of Directional Mobile Radio Channels*, Radio Science Bulletin, No. 302, September 2002, pp. 16–26 (2002).

Molisch 2004 A. F. Molisch, "A generic model for the MIMO wireless propagation channel", *IEEE Proc. Signal Proc.*, 52, 61–71 (2004).

Molisch 2005 A. F. Molisch, "Ultrawideband propagation channels – theory, measurement and models", *IEEE Trans. Veh. Technol., Invited*, 54, 1528–1545 (2005).

Molisch 2009 A. F. Molisch, "Ultrawideband propagation channels", *Proc. IEEE, Special Issue on UWB*, 97, 353–371 (2009).

Molisch and Hofstetter 2006 A. F. Molisch and H. Hofstetter, "The COST 273 MIMO channel model", in L. Correia (ed.), *Mobile Broadband Multimedia Networks*, Academic Press (2006).

Molisch and Steinbauer 1999 A. F. Molisch and M. Steinbauer, "Condensed parameters for characterizing wideband mobile radio channels", *Int. J. Wireless Inform. Net.*, 6, 133–154 (1999).

Molisch and Tufvesson 2004 A. F. Molisch and F. Tufvesson, "Multipath propagation models for broadband wireless systems", in M. Ibnkahla (ed.), *Digital Signal Processing for Wireless Communications Handbook*, Chapter 2, pp. 2.1–2.43, CRC Press (2004).

Molisch and Tufvesson 2005 A. F. Molisch and F. Tufvesson, "MIMO Channel capacity and measurements", in T. Kaiser (ed.), *Smart Antennas in Europe – State of the Art"*, EURASIP Publishing (2005).

Molisch and Win 2004 A. F. Molisch and M. Z. Win, "MIMO systems with antenna selection", *IEEE Microwave Mag.*, March, 46–56 (2004).

Molisch et al. 1995 A. F. Molisch, J. Fuhl, and E. Bonek, "Pattern distortion of mobile radio base station antennas by antenna masts and roofs", *Proceedings of the 25th European Microwave Conference*, Bologna, pp. 71–76 (1995).

Molisch et al. 1996 A. F. Molisch, J. Fuhl, and P. Proksch, "Error floor of MSK modulation in a mobile-radio channel with two independently-fading paths", *IEEE Trans. Veh. Technol.*, 45, 303–309 (1996).

Molisch et al. 1998 A. F. Molisch, H. Novak, and E. Bonek, "The DECT radio link", *(invited) Telektronikk*, 94, 45–53 (1998).

Molisch et al. 2002 A. F. Molisch, M. Steinbauer, M. Toeltsch, E. Bonek, and R. Thoma, "Capacity of MIMO systems based on measured wireless channels", *IEEE JSAC*, 20, 561–569 (2002).

Molisch et al. 2003a A. F. Molisch, J. R. Foerster and M. Pendergrass, "Channel models for ultrawideband Personal Area Networks", *IEEE Personal Communications Magazine*, 10, 14–21 (2003a).

Molisch et al. 2003b A. F. Molisch, A. Kuchar, J. Laurila, K. Hugl, and R. Schmalenberger, "Geometry-based directional model for mobile radio channels–principles and implementation", *European Trans. Telecommun.*, 14, 351–359 (2003b).

Molisch et al. 2005 A. F. Molisch, Y. G. Li, Y. P. Nakache, et al., "A low-cost time-hopping impulse radio system for high data rate transmission,", *EURASIP J. Appl. Signal Process.*, special issue on UWB (invited), 3, 397–412 (2005).

Molisch et al. 2006a A. F. Molisch, D. Cassioli, C. C. Chong, et al., "A comprehensive model for ultrawideband propagation channels", *IEEE Trans. Antennas Propagat.*, 54, special issue on wireless propagation, 3151–3166 (2006a).

Molisch et al. 2006b A. F. Molisch, H. Asplund, R. Heddergott, M. Steinbauer, and T. Zwick, "The COST 259 directional channel model – I. Overview and methodology," *IEEE Trans. Wireless Comm.*, 5, 3421–3433 (2006b).

Molisch et al. 2007 A. F. Molisch, N. B. Mehta, J. Yedidia, and J. Zhang, "Cooperative relay networks with mutual-information accumulation", *IEEE Trans. Wireless Commun.*, 6, 4108–4119 (2007).

Molisch et al. 2007a A. F. Molisch, M. Toeltsch, and S. Vermani, "Iterative methods for cancellation of intercarrier interference in OFDM systems", *IEEE Trans. Veh. Technol.*, 56, 2158 DH 2167 (2007).

Molisch et al. 2009 A. F. Molisch, F. Tufvesson, J. Karedal, and C. Mecklenbrauker, "Propagation aspects of vehicle-to-vehicle communications", *IEEE Wireless Commun. Mag.*, 16(6), 12 DH 22 (2009).

Molnar et al. 1996 B. G. Molnar, I. Frigyes, Z. Bodnar, and Z. Herczku, "The WSSUS channel model: Comments and a generalisation", *Proceedings of Globecom 1996*, 158–162 (1996).

Moon 2005 T. K. Moon, *Error Correction Coding: Mathematical Methods and Algorithms*, Wiley (2005).

Moroney and Cullen 1995 D. Moroney and P. Cullen, "A fast integral equation approach to UHF coverage estimation", in E. delRe (ed.), *Mobile and Personal Communications*, Elsevier Press, Amsterdam, The Netherlands (1995).

Moshavi 1996 S. Moshavi, "Multi-user detection for DS-CDMA Communications," *IEEE Commun. Mag.*, October, 124–136 (1996).

Motley and Keenan 1988 A. J. Motley and J. P. Keenan, "Personal communication radio coverage in buildings at 900 MHz and 1700 MHz", *Electron. Lett.*, 24, 763–764 (1988).

Mouly and Pautet 1992 M. Mouly and M. B. Pautet, *The GSM System for Mobile Communications*, self-publishing (1992).

MPEG 1 ISO/IEC 11172-2:1993, *Information Technology – Coding of Moving Pictures and Associated Audio for Digital Storage Media at up to about 1.5 Mbit/s – Part 2: Video*. (1993).

MPEG 2 International Telecommunications Union, *Information Technology - Generic Coding of Moving Pictures and Associated Audio Information - Part 2: Video*, 2nd edition, ITU-T Recommendation H.262 and ISO/IEC 13818-2:2000 (2000).

MPEG 4 ISO/IEC 14496-2:2004, *Information Technology – Coding of Audio-Visual Objects – Part 2: Visual*, 3rd edition (2004).

Mudumbai et al. 2009 R. Mudumbai, D. R. Brown, U. Madhow, and H. V. Poor, "Distributed transmit beamforming: Challenges and recent progress", *IEEE Commun. Mag.*, 47, 102–110 (2009).

Muirhead 1982 R. J. Muirhead, *Aspects of Multivariate Statistical Theory*, Wiley (1982).

Muquet et al. 2002 B. Muquet, Z. Wang, G. B. Giannakis, M. de Courville, and P. Duhamel, "Cyclic prefixing or zero padding for wireless multicarrier transmissions", *IEEE Trans. Commun.*, 50, 2136–2148 (2002).

Murota and Hirade 1981 K. Murota and K. Hirade, "GMSK modulation for digital mobile radio telephony", *IEEE Trans. Commun.*, 29, 1044–1050 (1981).

Nabar et al. 2004 R. U. Nabar, H. Bolcskei, and F. W. Kneubuhler, "Fading relay channels: Performance limits and space-time signal design", *IEEE J. Sel. Area. Commun.*, 22, 1099–1109 (2004).

Nakagami 1960 M. Nakagami, "The M-distribution: A general of intensity distribution of rapid fading", in W. Hoffman (ed.), *Statistical Methods of Radio Wave Propagation*, Pergamon Press (1960).

Namislo 1984 N. Namislo, "Analysis of mobile radio slotted ALOHA networks", *IEEE J. Sel. Area. Commun.*, 2, 583–588 (1984).

Narayanan et al. 2004 R. M. Narayanan, K. Atanassov, V. Stoiljkovic, and G. R. Kadambi, "Polarization diversity measurements and analysis for antenna configurations at 1800 MHz", *IEEE Trans. Antennas Propagat.*, 52, 1795–1810 (2004).

Nazer and Gastpar 2008 B. Nazer and M. Gastpar, "Compute-and-forward: A novel strategy for cooperative networks", *42nd Asilomar Conference on Signals, Systems and Computers*, pp. 69–73 (2008).

Nazer et al. 2009 B. Nazer, S. A. Jafar, M. Gastpar, and S. Vishwanath, "Ergodic interference alignment", *IEEE Int. Symp. Inform. Theory*, November/December 1769–1773 (2009).

Necker 2008 M. C. Necker, "Interference coordination in cellular OFDMA networks", *IEEE Network*, Nov/Dec., 12–19 (2008).

Neely 2006 M. J. Neely, "Energy optimal control for time varying wireless networks", *IEEE Trans. Inform. Theory*, 52, 2915–2934 (2006).

Neely and Urgaonkar 2009 M. J. Neely and R. Urgaonkar, "Optimal backpressure routing in wireless networks with multi-receiver diversity", *Ad Hoc Networks (Elsevier)*, 7, 862–881 (2009).

Neubauer et al. 2001 Th. Neubauer, H. Jaeger, J. Fuhl, and E. Bonek, "Measurement of the background noise floor in the UMTS FDD uplink band", *European Personal Mobile Communications Conference* (2001).

Nilsson et al. 1997 R. Nilsson, O. Edfors, M. Sandell, and P. O. Börjesson, "An analysis of two-dimensional pilot-symbol assisted modulation for OFDM", *Proceedings IEEE International Conference on Personal Wireless Communications*, pp. 71–74, Bombay, India, December 1997.

Noerpel et al. 1996 A. R. Noerpel, Y. B. Lin, and H. Sherry, "PACS: Personal communications system: A tutorial", *IEEE Pers. Commun.*, June, 32–43 (1996).

Norklit and Andersen 1998 O. Norklit and J. B. Andersen, "Diffuse channel model and experimental results for array antennas in mobile environments," *IEEE Trans. Antennas Propagat.*, 46, 834 (1998).

Nosratinia et al. 2004 A. Nosratinia, A. T. E. Hunter, and A. Hedayat, "Cooperative communication in wireless networks", *IEEE Commun. Mag.*, 42(10), 74–80 (2004).

Nuaymi 2007 L. Nuaymi, *WiMAX: Technology for Broadband Wireless Access*, Wiley (2007).

O'Hara and Petrick 2005 B. O'Hara and A. Petrick, *The IEEE 802.11 Handbook: A Designer's Companion*, 2nd edition, IEEE Standards Publications (2005).

O'Shaughnessy 2000 D. O'Shaughnessy, *Speech Communication: Human and Machine*, 2nd edition, IEEE Press (2000).

Oehrvik 1994 S. O. Oehrvik, *Radio School*, Ericsson, Stockholm (1994).

Oestges and Clerckx 2007 C. Oestges and B. Clerckx, *MIMO Wireless Communications: From Real-World Propagation to Space-Time Code Design*, Academic Press (2007).

Ogawa et al. 2001 K. Ogawa, T. Matsuyoshi, and K. Monma, "An analysis of the performance of a handset diversity antenna influenced by head, hand, and shoulder effects at 900 MHz.II. Correlation characteristics", *IEEE Trans. Veh. Technol.*, 50, 845–853 (2001).

Ogawa and Matsuyoshi 2001 K. Ogawa and T. Matsuyoshi, "An analysis of the performance of a handset diversity antenna influenced by head, hand, and shoulder effects at 900 MHz.I. Effective gain characteristics", *IEEE Trans. Veh. Technol.*, 50, 830–844 (2001).

Ogilvie 1991 J. A. Ogilvie, *Theory of Wave Scattering from Random Rough Surfaces*, IOP Publishing (1991).

Okumura et al. 1968 Y. Okumura, E. Ohmori, T. Kawano, and K. Fukuda, "Field strength and its variability in VHF and UHF land mobile services", *Rev. Elec. Commun. Lab.*, 16, 825–873 (1968).

Oppenheim and Schafer 2009 A. V. Oppenheim and R. W. Schafer, *Discrete-Time Signal Processing*, 3rd edition, Prentice Hall, Englewood Cliffs (2009).

Ozgur et al. 2007 A. Ozgur, O. Leveque, and D. N. C. Tse, "Hierarchical cooperation achieves optimal capacity scaling in ad hoc networks", *IEEE Trans. Inform. Theory*, 53, 3549–3572 (2007).

Paetzold 2002 M. Paetzold, *Mobile Fading Channels: Modelling, Analysis, & Simulation*, Wiley (2002); 2nd edition to appear (2010).

Pajusco 1998 P. Pajusco, "Experimental characterization of D.O.A at the base station in rural and urban area", *Proceedings of the IEEE VTC'98*, pp. 993–998 (1998).

Papoulis 1985 A. Papoulis, "Predictable processes and Wold's decomposition: A review", *IEEE Trans. Acoustics, Speech, Signal Process. ASSP*, 33, 933 (1985).

Papoulis 1991 A. Papoulis, *Probability, Random Variables, and Stochastic Processes*, 3rd edition, McGraw-Hill, New York (1991).

Parry 2002 R. Parry, "CDMA 2000, 1xEV", *IEEE Potentials*, October/November, 10–13 (2002).

Parsons 1992 J. D. Parsons, *The Mobile Radio Channel*, Wiley, New York (1992).

Parsons et al. 1991 J. D. Parsons, D. A. Demery, and A. M. D. Turkmani, "Sounding techniques for wideband mobile radio channels: A review", *Proc. Inst. Elect.Eng. – I*, 138, 437–446 (1991).

Paulraj and Papadias 1997 A. J. Paulraj and C. B. Papadias, "Space-time processing for wireless transmissions,", *IEEE Pers. Commun.*, 14(5), 49–83 (1997).

Paulraj et al. 2003 A. Paulraj, D. Gore, and R. Nabar, *Multiple Antenna Systems*, Cambridge University Press (2003).

Pawula et al. 1982 R. F. Pawula, S. O. Rice, and J. H. Roberts, "Distribution of the phase angle between two vectors perturbed by Gaussian noise", *IEEE Trans. Commun.*, 30, 1828–1841 (1982).

Pedersen et al. 1997 K. Pedersen, P. E. Mogensen, and B. Fleury, "Power azimuth spectrum in outdoor environments", *IEEE Electronics Lett.*, 33, 1583–1584 (1997).

Pedersen et al. 1998 G. F. Pedersen, J. O. Nielsen, K. Olensen, and I. Z. Kovács, *Measured Variation in Performance of Handheld Antennas for a Large Number of Test Persons*, COST259 TD(98)025 (1998).

Peel 2003 C. B. Peel, "On dirty-paper coding", *IEEE Signal Process. Mag.*, 20(3), 112–113 (2003).

Perahia and Stacey 2008 E. Perahia and R. Stacey, *Next Generation Wireless LANs: Throughput, Robustness, and Reliability in 802.11n*, Cambridge University Press (2008).

Perkins 2001 C. E. Perkins, *Ad Hoc Networking*, Addison-Wesley (2001).

Perkins and Royer 1999 C. E. Perkins and E. M. Royer, "Ad hoc on-demand distance vector routing", *Proceedings of 2nd IEEE Workshop on Mobile Computing Systems and Applications*, New Orleans, LA, February 1999, pp. 90–100 (1999).

Peterson and Davie 2003 L. L. Peterson and B. S. Davie, *Computer Networks: A Systems Approach*, 3rd edition, Academic Press (2003).

Petrus et al. 2002 P. Petrus, J. H. Reed, and T. S. Rappaport, "Geometrical-based statistical macrocell channel model for mobile environments", *IEEE Trans. Commun.*, 50, 495 (2002).

van der Plassche 2003 R. van der Plassche, *Cmos Integrated Analog-To-Digital and Digital-To-Analog Converters*, 2nd edition, Kluwer (2003).

Plenge 1997 C. Plenge, "Leistungsbewertung öffentlicher DECT-Systeme", *Dissertation, RWTH Aachen* (1997) [in German].

Polydoros and Weber 1984 A. Polydoros and C. L. Weber, "A unified approach to serial search spread-spectrum code acquisition-part 1: General theory", *IEEE Trans. Commun.*, 32, 542–549; "Part II: Matched filter receiver", *IEEE Trans. Commun.*, 32, 550–560 (1984).

Poor 2001 H. V. Poor, "Turbo multiuser detection: A primer", *J. Commun. Networks*, 3, 196–201 (2001).

Poor 2004 H. V. Poor, "Iterative multiuser detection", *IEEE Signal Process. Mag.*, 21, 81–88 (2004).

Pozar 2000 D. M. Pozar, *Microwave and RF Design of Wireless Systems*, Wiley, New York (2000).

Proakis 1968 J. G. Proakis, "On the probability of error for multichannel reception of binary signals", *IEEE Trans. Commun.*, 16, 68–71 (1968).

Proakis 1991 J. G. Proakis, "Adaptive equalization for TDMA digital mobile radio", *IEEE Trans. Veh. Technol.*, 40, 333–341 (1991).

Proakis 2005 J. G. Proakis and M. Salehi, *Digital Communications*, 5th edition, McGraw Hill, New York (2005).

Qiu 2002 R. C. Qiu, "A study of the ultra-wideband wireless propagation channel and optimum UWB receiver design", *IEEE J. Sel. Area. Commun.*, 20, 1628–1637 (2002).

Qiu 2004 R. C. Qiu, "A generalized time domain multipath channel and its application in ultra-wideband (UWB) wireless optimal receiver design – Part II: Physics-based system analysis", *IEEE Trans. Wireless Commun.*, 3, 2312–2324 (2004).

Quan et al. 2008 Z. Quan, S. Ciu, A. H. Sayed, and H. V. Poor, "Wideband spectrum sensing in cognitive radio networks", *Proceedings of the IEEE International Conference on Communication* (2008).

Quatieri 2002 T. F. Quatieri, *Discrete-Time Speech Signal Processing*, Prentice-Hall, Upper Saddle River (2002).

Rabiner 1994 L. Rabiner, "Applications of voice processing to telecommunications", *Proc. IEEE*, 82, 199–228 (1994).

Rabiner and Schafer 1978 L. R. Rabiner and R. W. Schafer, *Digital Processing of Speech Signals*, Prentice-Hall, Englewood Cliffs (1978).

Raleigh and Cioffi 1998 G. Raleigh and J. M. Cioffi, "Spatial-temporal coding for wireless communications", *IEEE Trans. Commun.*, 46, 357–366 (1998).

Ramachandran et al. 2004 I. Ramachandran, Y. P. Nakache, P. Orlik, J. Zhang, and A. F. Molisch, "Symbol spreading for ultrawideband systems based on multiband OFDM", *Proceedings of the Personal, Indoor, Mobile Radio Symposium*, 1204–1209 (2004).

Ramo et al. 1967 S. Ramo, J. R. Whinnery, and T. van Duzer, *Fields and Waves in Communication Electronics*, Wiley, New York (1967).

Rao and Yip 1990 K. R. Rao and P. Yip, *Discrete Cosine Transform: Algorithms, Advantages, Applications*, Academic Press (1990).

Rappaport 1996 T. S. Rappaport, *Wireless Communications - Principles and Practice*, 2nd edition 2001, IEEE Press, Piscataway, NJ (1996).

Rappaport 1998 T. S. Rappaport (ed.), *Smart Antennas: Adaptive Arrays, Algorithms, & Wireless Position Location*, IEEE Press (1998).

Rasinger et al. 1990 J. Rasinger, A. L. Scholtz, and E. Bonek, "A new enhanced-bandwidth internal antenna for portable communication systems", *Proceedings of the 40th IEEE Vehicular Technology Conference*, Orlando, pp. 7–12 (1990).

Razavi 1997 B. Razavi, *RF Microelectronics*, Prentice-Hall (1997).

Reed 2005 J. H. Reed, *An Introduction to Ultra Wideband Communication Systems*, Prentice-Hall (2005).

Reznik et al. 2007 Y. A. Reznik, A. T. Hinds, C. Zhang, L. Yu, and Z. Ni, "Efficient fixed-point approximations of the 8x8 inverse discrete cosine transform", *Proceedings of the SPIE 6696*, pp. 1–17 (2007).

Rice 1947 S. O. Rice, "Statistical properties of a sine wave plus random noise", *Bell System Techn. J.*, 27, 109–157 (1947).

Rice 2008 M. D. Rice, *Digital Communications: A Discrete-Time Approach*, Prentice-Hall, 2008.

Richardson 2005 A. Richardson, *WCDMA Design Handbook*, Cambridge University Press (2005).

Richardson and Urbanke 2008 T. Richardson and R. Urbanke, *Modern Coding Theory*, Cambridge University Press (2008).

Richardson et al. 2001 T. J. Richardson, M. A. Shokrollahi, and R. L. Urbanke, "Design of capacity-approaching irregular low-density parity-check codes", *IEEE Trans. Inform. Theory*, 47, 619–637 (2001).

Richter 2006 A. Richter, "The contribution of distributed diffuse scattering in radio channels to channel capacity: Estimation and modelling", *Proceedings of Asilomar Conference on Signals, Systems, and Computers*, Pacific Grove, CA, USA, October (2006).

Rohling 2005 H. Rohling, "OFDM-A flexible and adaptive air interface for a 4G communication system", XI International Symposium of Radio Science URSI 2005, Poznan, April (2005).

Rokhlin 1990 V. Rokhlin, Rapid solution of integral equations of scattering theory in two dimensions, *J. Comp. Phys.*, 96, 414, 1990.

Roy et al. 1986 R. Roy, A. Paulraj, and T. Kailath, "ESPRIT - A subspace rotation approach to estimation of parameters of cisoids in noise", *IEEE Trans. Acoustics, Speech, Signal Process.*, 34, 1340–1342 (1986).

Roy et al. 2004 S. Roy, J. R. Foerster, V. S. Somayazulu, and D. G. Leeper, "Ultrawideband radio design: The promise of high-speed, short-range wireless connectivity", *Proc. IEEE*, 92, 295–311 (2004).

Roy and Fortier 2004 S. Roy and P. Fortier, "A closed-form analysis of fading envelope correlation across a wideband basestation array", *IEEE Trans. Wireless Commun.*, 3, 1502–1507 (2004).

Royer and Toh 1999 E. M. Royer and C. K. Toh, "A review of current routing protocols for ad hoc mobile wireless networks", *IEEE Pers. Commun.*, 6(2), 46–55 (1999).

Rubin 1979 I. Rubin, "Message delays in FDMA and TDMA communication channels", *IEEE Trans. Commun.*, 27, 769–777 (1979).

Sadek et al. 2007 M. Sadek, A. Tarighat, and A. H. Sayed, "A leakage-based precoding scheme for downlink multi-user MIMO channels", *IEEE Trans. Wireless Commun.*, 6, 1711–1721 (2007).

Saleh and Valenzuela 1987 A. Saleh and R. A. Valenzuela, "A statistical model for indoor multipath propagation", *IEEE J. Sel. Area. Commun.*, 5, 128 (1987).

Salmi et al. 2009 J. Salmi, A. Richter, and V. Koivunen, "Detection and tracking of MIMO propagation path parameters using state-space approach", *IEEE Trans. Signal Process.*, 57, 1538–1550 (2009).

de Santo and Brown 1986 J. A. de Santo and G. S. Brown, *Progress in Optics*, Vol. 23, edited by E. Wolf, North-Holland (1986).

Sari et al. 2000 H. Sari, F. Vanhaverbeke, and M. Moeneclaey, "Extending the capacity of multiple access channels", *IEEE Commun. Mag.*, 38(1), 74–82 (2000).

Sato 1975 Y. Sato, "A method of self recovering equalization for multilevel amplitude modulation", *IEEE Trans. Commun.*, 23, 679–682 (1975).

Sayeed 2002 A. M. Sayeed, "Deconstructing multiantenna fading channels", *IEEE Trans. Signal Process.*, 50, 2563 (2002).

Sayre 2001 C. W. Sayre, *Complete Wireless Design*, McGraw Hill (2001).

Scaglione et al. 2006 A. Scaglione, D. L. Goeckel, and N. J. Laneman, "Cooperative communications in mobile ad-hoc networks", *IEEE Signal Process. Mag.*, 2006, 18–29 (2006).

Schiller 2003 J. Schiller, *Mobile Communications*, 2nd edition, Addison-Wesley (2003).

Schlegel and Perez 2003 C. B. Schlegel and L. C. Perez, *Trellis and Turbo Coding*, Wiley (2003).

Schmidl and Cox 1997 T. M. Schmidl and D. C. Cox, "Robust frequency and timing synchronization for OFDM", *IEEE Trans. Commun.*, 45, 1613–1621 (1997).

Schmidt 1986 R. Schmidt, "Multiple emitter location and signal parameters estimation", *IEEE Trans. Antennas Propagat.*, 34, 276–280 (1986).

Schniter 2004 P. Schniter, "Low-complexity equalization of OFDM in doubly selective channels", *IEEE Trans. Signal Process.*, 52, 1002–1011 (2004).

Schroeder and Atal 1985 M. Schroeder and B. Atal, "Code-excited linear prediction (CELP): High-quality speech at very low bit rates", *Proceedings of the IEEE International Conference Acoustics, Speech, Signal Process., ICASSP'85'*, pp. 937–940 (1985).

Schroeder et al. 1979 M. R. Schroeder, B. S. Atal, and J. L. Hall, "Optimizing digital speech coders by exploiting masking properties of the human ear", *J. Acoust. Soc. Am.*, 66, 1647–1652 (1979).

Scholtz 1982 R. A. Scholtz, "The origins of spread spectrum communications", *IEEE Trans. Commun.*, 30, 822–854 (1982).

Schwarz et al. 2007 H. Schwarz, D. Marpe, and T. Wiegand, "Overview of the scalable video coding extension of the H.264/AVC standard", *IEEE Trans. Circ. Syst. Video Technol.*, 17, 1103–1120 (2007).

Sendonaris et al. 2003 A. Sendonaris, E. Erkip, and B. Aazhang, "User cooperation diversity–Part I: System description and user cooperation diversity–Part II: Implementation aspects and performance analysis", *IEEE Trans. Commun.*, 51, 1927–1948 (2003).

Shafi et al. 2006 M. Shafi, M. Zhang, A. L. Moustakas, et al. "Polarized MIMO channels in 3D: Models, measurements and mutual information", *IEEE J. Sel. Areas Commun.*, 24, 514–527 (2006).

Shannon 1948 C. E. Shannon, "A mathematical theory of communication", *Bell System Tech. J.*, 27, 379–423 and 623–656 (1948).

Shannon 1949 C. E. Shannon, "Communication in the presence of noise", *Proc. IRE*, 37, 10–21 (1949).

Shannon 1959 C. E. Shannon, "Coding theorems for a discrete source with a fidelity criterion", *IRE National Convention Record*, 4, 142–163 (1959).

Shen et al. 2006 S. Shen, M. Guizani, R. C. Qiu, and T. L. Ngog, *Ultra-Wideband Wireless Communications and Networks*, Wiley (2006).

Shi and Sun 2000 Y. Q. Shi and H. Sun, *Image and Video Compression for Multimedia Engineering: Fundamentals, Algorithms and Standards*, 2nd edition, CRC Press (2000).

Shi et al. 2007 S. Shi, M. Schubert, and H. Boche, "Downlink MMSE transceiver optimization for multiuser MIMO systems: Duality and sum-MSE minimization", *IEEE Trans. Signal Process.*, 55, 5436–5446 (2007).

Shiu et al. 2000 D. Shiu, G. J. Foschini, M. J. Gans, and J. M. Kahn, "Fading correlation and its effect on the capacity of multielement antenna systems", *IEEE Trans. Commun.*, 48, 502–513 (2000).

Simon et al. 1994 M. K. Simon, J. K. Omura, R. A. Scholtz, and B. K. Levitt, *Spread Spectrum Communications Handbook*, Revised edition, McGraw Hill (1994).

Simon and Alouini 2004 M. K. Simon and M. S. Alouini, *Digital Communications Over Fading Channels*, 2nd edition, Wiley (2004).

Sindhushayana and Black 2002 N. T. Sindhushayana and P. J. Black, "Forward link coding and modulation for CDMA2000 1XEV-DO (IS-856)", *Proceedings fo the IEEE PIMRC 2002*, p. 1839–1846, 2002.

Singh et al. 2010 S. Singh, N. B. Mehta, and A. F. Molisch, "Moment-matched lognormal modeling of uplink interference with power control and cell selection", *IEEE Trans. Wireless Commun.* 9, 932–938 (2010).

Siriwongpairat and Liu 2007 W. P. Siriwongpairat and K. J. R. Liu, *Ultra-Wideband Communications Systems: Multiband OFDM Approach*, Wiley (2007).

Sklar 1997 B. Sklar, "A primer on turbo code concepts", *IEEE Commun. Mag.*, 35, 94–102 (1997).

Sklar 2001 B. Sklar, *Digital Communications - Fundamentals and Applications*, 2nd edition, Prentice Hall (2001).

Sklar and Harris 2004 B. Sklar and F. J. Harris, "The ABCs of linear block codes", *IEEE Signal Process. Mag.*, 21(4), 14–35 (2004).

Spencer et al. 2004a Q. H. Spencer, C. B. Peel, A. L. Swindlehurst, and M. Haardt, "An introduction to the multi-user MIMO downlink", *IEEE Commun. Mag.*, 42(10), 60–67 (2004a).

Spencer et al. 2004b Q. H. Spencer, A. L. Swindlehurst and M. Haardt, "Zero-forcing methods for downlink spatial multiplexing in multiuser MIMO channels", *IEEE Trans. Signal Processing*, 52, 461–471 (2004b).

Spencer et al. 2006 Q. H. Spencer, J. W. Wallace, C. B. Peel, et al., "Performance of multi-user spatial multiplexing with measured channel data", in G. Tsoulos (ed.), *MIMO System Technology for Wireless Communications*, CRC Press (2006).

Speth et al. 1999 M. Speth, S. A. Fechtel, G. Fock, and H. Meyr, "Optimum receiver design for wireless broadband systems using OFDM. I", *IEEE Trans. Commun.*, 47, 1668–1677 (1999); "II A case study", *IEEE Trans. Commun.*, 49, 571–578 (2001).

Spyropoulos et al. 2005 T. Spyropoulos, K. Psounis, and C. S. Raghavendra, "Spray and wait: An efficient routing scheme for intermittently connected mobile networks", *Proceedings of the 2005 ACM SIGCOMM Workshop on Delay-tolerant Networking*, pp. 252–259 (2005).

Stankovic et al. 2006 V. Stankovic, A. Host-Madsen, and Z. Xiong, "Cooperative diversity for wireless ad hoc networks", *IEEE Signal Process. Mag.*, 23(5), 37–49 (2006).

Steele and Hanzo 1999 R. Steele and L. Hanzo, *Mobile Communications*, 2nd edition, Wiley (1999).

Steele et al. 2001 R. Steele, C. C. Lee, and P. Gould, *GSM, cdmaOne and 3G Systems*, John Wiley & Sons (2001).

Steendam and Moenclaey 1999 H. Steendam and M. Moeneclaey, "Analysis and optimization of the performance of OFDM on frequency-selective time-selective fading channels", *IEEE Trans. Commun.*, 47, 1811–1819 (1999).

Stein 1964 S. Stein, "Unified analysis of certain coherent and noncoherent binary communications systems", *IEEE Trans. Inform. Theory*, 11, 239–246 (1964).

Steinbauer and Molisch 2001 M. Steinbauer and A. F. Molisch (eds), "Spatial channel models", in L. Correia (ed.), *Wireless Flexible Personalized Communications*, Wiley (2001).

Stojmenovic 2002 I. Stojmenovic, "Position-based routing in ad-hoc networks", *IEEE Commun. Mag.*, July, 128–134 (2002).

Stojnic et al. 2006 M. Stojnic, H. Vikalo, and B. Hassibi, "Rate maximization in multi-antenna broadcast channels with linear preprocessing", *IEEE Trans. Wireless Commun.*, 5, 2338–2342 (2006).

Strang 1988 G. Strang, *Linear Algebra and its Application*, 3rd edition, Harcourt Brace Jovanovich, San Diego (1988).

Stueber 1996 G. Stueber, *Principles of Mobile Communication*, Kluwer (1996), second edition 2001.

Stueber et al. 2004 G. L. Stuber, J. R. Barry, S. W. McLaughlin, et al., "Broadband MIMO-OFDM wireless communications", *Proc. IEEE*, 92, 271–294 (2004).

Stutzman and Thiele 1997 W. L. Stutzman and G. A. Thiele, *Antenna Theory and Design*, 2nd edition, Wiley, New York (1997).

Sun and Reibman 2001 M. T. Sun and A. R. Reibman (eds), *Compressed Video Over Networks*, Marcel Dekker (2001).

Suzuki 1977 H. Suzuki, "A statistical model for urban radio propagation", *IEEE Trans. Commun.*, 25, 673–680 (1977).

Suzuki 1982 H. Suzuki, "Canonic receiver analysis for M-ary angle modulations in Rayleigh fading environment", *IEEE Trans. Veh. Technol.*, 31, 7–14 (1982).

Swarts et al. 1998 F. Swarts, P. van Rooyan, I. Oppermann, and M. P. Lötter, *CDMA Techniques for Third Generation Mobile Systems*, Kluwer (1998).

Sweeney 2002 P. Sweeney, *Error Control Coding: From Theory to Practice*, Wiley (2002).

Sayed and Kailath 2001 A. H. Sayed and T. Kailath, "A survey of spectral factorization methods", *J. Num. Linear Algebra Appl.*, 8, 467–496 (2001).

Taga 1990 T. Taga, "Analysis for mean effective gain of mobile antennas in land mobile radio environments", *IEEE Trans. Veh. Technol.*, 39, 117–131 (1990).

Taga 1993 T. Taga, "Characteristics of space-diversity branch using parallel dipole antennas in mobile radio communications", *Electron. Commun. Jpn.*, 76(Pt. 1), 55–65 (1993).

Tao 2007 J. Tao, "IEEE 802.16e OFDMA PHY", Mitsubishi Electric Research Labs, internal report (2007).

Tarokh et al. 1998 V. Tarokh, N. Seshadri, and A. R. Calderbank, "Space-time coding for high data rate wireless communication: Performance criterion and code construction", *IEEE Trans. Inform. Theory*, 44, 744–765 (1998).

Tarokh et al. 1999 V. Tarokh, H. Jafarkhani, and A. R. Calderbank, "Space–time block codes from orthogonal designs", *IEEE Trans. Inform. Theory*, 45, 1456–1467 (1999).

Tekalp 1995 A. Murat Tekalp, *Digital Video Processing*, Prentice Hall (1995).

Telatar 1999 I. E. Telatar, "Capacity of multi-antenna Gaussian channels", *European Trans. Telecomm.*, 10, 585–595 (1999).

Thjung and Chai 1999 T. T. Tjhung and C. C. Chai, "Fade statistics in Nakagami-lognormal channels", *IEEE Trans. Commun.*, 47, 1769–1772 (1999).

Thomae et al. 2000 R. S. Thomae, D. Hampicke, A. Richter, et al., Identification of time-variant directional mobile radio channels, *IEEE Trans. Instrum. Meas.*, 49, 357–364 (2000).

Thomae et al. 2005 R. S. Thomae, M. Landmann, A. Richter, U. Trautwein, "Multidimensional high-resolution channel sounding", in *Smart Antennas in Europe – State-of-the-Art*, EURASIP Book Series, p. 27, Hindawi Publishing Corporation (2005).

Tiedeman 2001 E. Tiedemann, "CDMA2000 1X: New capabilities for CDMA networks", *IEEE Veh. Technol. Soc. News*, 48(4), 4–12 (2001).

Tobagi 1980 F. Tobagi, "Multiaccess protocols in packet communication systems", *IEEE Trans. Commun.*, 28, 468–488 (1980).

Tomiuk et al. 1999 B. R. Tomiuk, N. C. Beaulieu, and A. A. Abu-Dayya, "General forms for maximal ratio diversitywith weighting errors", *IEEE Transactions on Communications*, 47(4), April, 488–492 (1999).

Tong et al. 1994 L. Tong, G. Xu, and T. Kailath, "Blind identification and equalization based on second-order statistics: A time domain approach", *IEEE Trans. Inform. Theory*, 40, 340–349 (1994).

Tong et al. 1995 L. Tong, G. Xu, B. Hassibi and T. Kailath, "Blind identification and equalization based on second-order statistics: A frequency domain approach", *IEEE Trans. Inform. Theory*, 41, 329–334 (1995).

Tse and Visvanath 2005 D. Tse and P. Visvanath, *Fundamentals of Wireless Communications*, Cambridge University Press (2005).

Tsoulos 2001 G. V. Tsoulos, *Adaptive Antennas for Wireless Communications*, IEEE Press (2001).

Tsoulos 2006 G. Tsoulos (ed.), *MIMO Antenna Technology for Wireless Communications*, CRC press (2006).

Turin et al. 1972 G. L. Turin, F. D. Clapp, T. L. Johnston, S. B. Fine, and D. Lavry, "A statistical model of urbran multipath propagation", *IEEE Trans. Veh. Technol.*, 21, 1–9 (1972).

Tuttlebee 1997 W. H. W. Tuttlebee (ed.), *Cordless Telecommunications Worldwide*, Springer, London (1997).

UMTS 1999 Third Generation Partnership Project, 3GPP TS 25.104, 25.211, 25.212, 25.213 (1999).

Ungerboeck 1976 G. Ungerboeck, "Fractional tap-spacing equalizer and consequences for clock recovery in data modems", *IEEE Trans. Commun.*, 24, 856–864 (1976).

Ungerboeck 1982 G. Ungerboeck, "Channel coding with multilevel/phase signals", *IEEE Trans. Inform. Theory*, 28, 55–67 (1982).

Valenti 1999 M. C. Valenti, *Iterative Detection and Decoding for Wireless Communications*, Ph.D. thesis, Virginia Tech University (1999).

Valenzuela 1993 R. A. Valenzuela, "A ray tracing approach to predicting indoor wireless transmission", *Proceedings of the VTC 1993*, pp. 214–218 (1993).

Vanderveen and Paulraj 1996 A. J. Vanderveen and A. Paulraj, "An analytical constant modulus algorithm", *IEEE Trans. Signal Process.*, 44, 1136–1155 (1996).

Vanderveen et al. 1997 M. C. Vanderveen, C. B. Papadias, and A. Paulraj, "Joint angle and delay estimation (JADE) for multipath signals arriving at an antenna array", *IEEE Commun. Lett.*, 1, 12–14 (1997).

Vanghi et al. 2004 V. Vanghi, A. Damnjanovic, and B. Vojcic, *The CDMA2000 System for Mobile Communications*, Prentice-Hall PTR (2004).

Varshney and Kumar 1991 P. Varshney and S. Kumar, "Performance of GMSK in a land mobile radio channel", *IEEE Trans. Veh. Technol.*, 40, 607–614 (1991).

Vary et al. 1998 P. Vary and U. Heute and W. Hess, *Digitale Sprachsignalverarbeitung (Digital Speech Signal Processing, in German)*, B.G. Teubner, Stuttgart (1998).

Vaseghi 2000 S. V. Vaseghi, *Advanced Digital Signal Processing and Noise Reduction*, 2nd edition, John Wiley & Sons (2000).

Vaughan and Andersen 2003 R. Vaughan and J. B. Andersen, *Channels, Propagation and Antennas for Mobile Communications*, IEE Press (2003).

Verdu 1998 S. Verdu, *Multiuser Detection*, Cambridge University Press, Cambridge (1998).

Viswanath et al. 2002 P. Viswanath, C. N. C. Tse, and R. Laroia, "Opportunistic beamforming using dumb antennas", *IEEE Trans. Inform. Theory*, 48, 1277–1294 (2002).

Viterbi 1967 A. Viterbi, "Error bounds for convolutional codes and an asymptotically optimum decoding algorithm", *IEEE Trans. Inform. Theory*, 13, 260–269 (1967).

Viterbi 1995 A. J. Viterbi, *CDMA – Principles of Spread Spectrum Communication*, Addison-Wesley Wireless Communications Series (1995).

Vitetta et al. 2000 G. Vitetta, B. Hart, A. Mammela, and D. Taylor, "Equalization techniques for single-carrier, unspread modulation", in A. F. Molisch (ed.), *Wideband Wireless Digital Communications*, Prentice-Hall (2000).

Waldschmidt et al. 2004 C. Waldschmidt, S. Schulteis, and W. Wiesbeck, "Complete RF system model for analysis of compact MIMO arrays", *IEEE Trans. Veh. Technol.*, 53, 579–586 (2004).

Walfish and Bertoni 1988 J. Walfish and H. L. Bertoni, "A theoretical model of UHF propagation in urban environments", *IEEE Trans. Antennas Propagat.*, 822–829 (1988).

Wallace and Jensen 2004 J. W. Wallace and M. A. Jensen, "Mutual coupling in MIMO wireless systems: A rigorous network theory analysis", *IEEE Trans. Wireless Commun.*, 3, 1317–1325 (2004).

Wang 2002 Y. Wang, J. Ostermann, and Y. Q. Zhang, *Video Processing and Communications*, Prentice Hall (2002).

Wang and Giannakis 2000 Z. Wang and G. B. Giannakis, "Wireless multicarrier communications", *IEEE Signal Process. Mag.*, 17(3), 29–48 (2000).

Wang and Poor 2003 X. Wang and H. V. Poor, *Wireless Communication Systems: Advanced Techniques for Signal Reception*, Prentice-Hall (2003).

Wang and Zhu 1998 Y. Wang and Q. F. Zhu, "Error control and concealment for video communication – A review", *Proc. IEEE*, 85, 974–997 (1998).

Wang et al. 2007 Y. K. Wang, I. Bouazizi, M. Hannuksela, and I. Curcio, "Mobile video applications and standards", *Proceedings of the ACM Multimedia, Workshop on Mobile Video*, Augsburg, Germany, September (2007).

de Weck 1992 J. P. de Weck, *Real-time Characterization of Wideband Mobile Radio Channels*, Dissertation an der TU Wien, Wien, Österreich (1992).

Weichselberger et al. 2006 W. Weichselberger, M. Herdin, H. Ö zcelik, and E. Bonek, "A stochastic MIMO channel model with joint correlation of both link ends", *IEEE Trans. Wireless Commun.*, 5, 90–100 (2006).

Weinstein and Ebert 1971 S. Weinstein and P. Ebert, "Data transmission by frequency-division multiplexing using the discrete Fourier transform", *IEEE Trans. Commun.*, 19, 628–634 (1971).

Weisstein 2004 E. W. Weisstein. *Lindeberg-Feller Central Limit Theorem*. From MathWorld–A Wolfram Web Resource, *http://mathworld.wolfram.com/Lindeberg-FellerCentralLimitTheorem.html* (2004).

Wellens et al. 2007 M. Wellens, J. Wu, and P. Mahonen, "Evaluation of spectrum occupancy in indoor and outdoor scenario in the context of cognitive radio", *2nd International Conference in Cognitive Radio Oriented Wireless Networks and Communications*, 420–427 (2007).

Wiegand et al. 2003 T. Wiegand, G. J. Sullivan, G. Bjontegaard, and A. Luthra, "Overview of the H.264/AVC video coding standard", *IEEE Trans. Circ. Syst. Video Technol.*, 13, 560–576 (2003).

Willenegger 2000 S. Willenegger, "CDMA2000 physical layer: An overview", *IEEE J. Commun. Netw.*, 2(1), 5–17 (2000).

Wilson 1996 S. G. Wilson, *Digital Modulation and Coding*, Prentice Hall, Upper Saddle River (1996).

Wimax 1.5 Wimax Forum, "Wimax Forum Network Architecture Release 1.5", at *http://www.wimaxforum.org/resources/documents/technical/release*.

Win and Chrisikos 2000 M. Z. Win and G. Chrisikos, "Impact of spreading bandwidth and selection diversity order on selective Rake reception," in A. F. Molisch (ed.), *Wideband Digital Communications*, pp. 424–454, Prentice Hall (2000).

Win and Scholtz 1998 M. Z. Win and R. A. Scholtz, "Impulse radio: How it works", *IEEE Commun. Lett.*, 2, 36–38 (1998).

Win and Scholtz 2000 M. Z. Win and R. A. Scholtz, "Ultra -wide bandwidth time-hopping spread-spectrum impulse radio for wireless multiple-access communications", *IEEE Trans. Commun.*, 48, 679–691 (2000).

Win and Winters 1999 M. Z. Win and J. H. Winters, "Analysis of hybrid selection/maximal-ratio combining of diversity branches with unequal SNR in Rayleigh fading", *Proceedings of the VTC'99*, Spring, pp. 215–220 (1999).

WINNER 2007 IST-WINNER D1.1.2 P. Kyösti, J. Meinilä, L. Hentilä, et al., *WINNER II Channel Models*, ver 1.1, September 2007. Available: *https://www.ist-winner.org/WINNER2-Deliverables/D1.1.2v1.1.pdf* (2007).

Winters 1984 J. H. Winters, "Optimum combining in digital mobile radio with cochannel interference", *IEEE JSAC*, 2, 528–539 (1984).

Winters 1987 J. H. Winters, "On the capacity of radio communications systems with diversity in Rayleigh fading environments", *IEEE J. Sel. Area. Commun.* (1987).

Winters 1994 J. H. Winters, "The diversity gain of transmit diversity in wireless systems with Rayleigh fading", *Proceedings of the IEEE International Conference in Communications*, pp. 1121–1125 (1994).

Witrisal et al. 2009 K. Witrisal, G. Leus, G. Janssen, et al., "Noncoherent ultra-wideband systems", *IEEE Signal Process. Mag.*, 26(4), 48–66 (2009).

Wittneben 1993 A. Wittneben, "A new bandwidth efficient transmit antenna modulation diversity scheme for linear digital modulation", *Proceedings of the IEEE International Conference on Communication*, pp. 1630–1634 (1993).

Woerner et al. 1994 B. D. Woerner, J. H. Reed, and T. S. Rappaport, "Simulation issues for future wireless modems", *IEEE Commun. Mag.*, 32(7), 42–53 (1994).

Wolf 1978 J. K. Wolf, "Efficient maximum-likelihood decoding of linear block codes using a trellis", *IEEE Trans. Inform. Theory*, 24, 76–80 (1978).

Wong et al. 1999 C. Y. Wong, R. S. Cheng, K. B. Lataief, and R. D. Murch, "Multiuser OFDM with adaptive subcarrier, bit, and power allocation", *IEEE J. Sel. Area. Commun.*, 17, 1747–1758 (1999).

Wozencraft and Jacobs 1965 J. M. Wozencraft and I. M. Jacobs, *Principles of Communication Engineering*, Wiley, New York (1965).

Wymeersch 2007 H. Wymeersch, *Iterative Receiver Design*, Cambridge University Press (2007).

Xiao and Hu 2008 Y. Xiao and F. Hu (eds), *Cognitive Radio Networks*, CRC Press (2008).

Xiong 2006 F. Xiong, *Digital Modulation Techniques*, 2nd edition, Artech (2006).

Xu et al. 2000 Z. Xu, A. N. Akansu, and S. Tekinay, "Cochannel interference computation and asymptotic performance analysis in TDMA/FDMA systems with interference adaptive dynamic channel allocation", *IEEE Trans. Veh. Technol.*, 49, 711–723 (2000).

Xu et al. 2002 H. Xu, D. Chizhik, H. Huang, and R. A. Valenzuela, "A wave-based wideband MIMO channel modeling technique", *Proc. 13th IEEE Int. Symp. Personal, Indoor Mobile Radio Commun.*, 4, 1626 (2002).

Yang and Hanzo 2003 L. L. Yang and L. Hanzo, "Multicarrier DS-CDMA: A multiple access scheme for ubiquitous broadband wireless communications", *IEEE Commun. Mag.*, 41(10), 116–124 (2003).

Yang et al. 2009 Y. Yang, H. Hu, J. Xu, and G. Mao, "Relay technologies for WiMax and LTE-advanced mobile systems", *IEEE Commun. Mag.*, 47(10), 100–105 (2009).

Yeung 2006 R. W. Yeung, *A First Course in Information Theory*, Springer (2006).

Ying and Anderson 2003 Z. Ying and J. Anderson, "Multi band, multi antenna system for advanced mobile phone", *Swedish National Conference on Antennas* (2003).

Yongacoglu et al. 1988 A. Yongacoglu, D. Makrakis, and K. Feher, "Differential detection of GMSK using decision feedback", *IEEE Trans. Commun.*, 36, 641–649 (1988).

Yu et al. 1997 C. C. Yu, D. Morton, C. Stumpf, R. G. White, J. E. Wilkes, and M. Ulema, "Low-tier wireless local loop radio systems – Part 1 and 2, *IEEE Commun. Mag.*, March, 84–98 (1997).

Yu et al. 2002 W. Yu, G. Ginis, and J. M. Cioffi, "Distributed multiuser power control for digital subscriber lines", *IEEE J. Sel. Areas Commun.*, 20, 1105–1115 (2002).

Yu and Ottersten 2002 K. Yu and B. Ottersten, "Models for MIMO propagation channels: A review", *Wireless Commun. Mobile Comput.*, 2, 653 (2002).

Yu et al. 2004 W. Yu, W. Rhee, S. Boyd, and J. M. Cioffi, "Iterative water-filling for Gaussian vector multiple-access channels", *IEEE Trans. Inform. Theory*, 50, 145–152 (2004).

Zhang and Dai 2004 H. Zhang and H. Dai, "Cochannel interference mitigation and cooperative processing in downlink multicell multiuser MIMO networks", *Eur. J. Wireless Commun. Networking*, 222–235, 2004.

Zhang and Lu 2006 D. Zhang and J. Lu, *Joint Transceiver Design Using Linear Processing for Downlink Multiuser MIMO Systems*, 2006 Asia-Pacific Conference on Communications (2006).

Zhang et al. 2008 H. Zhang, X. Zhou, and T. Chen, "Ultra-wideband cognitive radio for dynamic spectrum accessing networks", in Y. Xiao and F. Hu (eds), *Cognitive Radio Networks*, CRC Press (2008).

Zhao and Sadler 2007 Q. Zhao and B. M. Sadler, "A survey of dynamic spectrum access", *IEEE Signal Process. Mag.*, 24(3), 79–89 (2007).

Zhao et al. 2007 Q. Zhao, B. Krishnamachari, and K. Liu, "Low-complexity approaches to spectrum opportunity tracking", *2nd International Conference Cognitive Radio Oriented Wireless Networks and Communications*, pp. 27–35 (2007).

Zheng and Tse 2003 L. Zheng and D. L. C. Tse, "Diversity and multiplexing: A fundamental tradeoff in multiple-antenna channels", *IEEE Trans. Inform. Theory*, 49, 1073–1096 (2003).

Ziemer et al. 1995 R. E. Ziemer, R. L. Peterson and D. E. Borth, *Introduction to Spread Spectrum Communications*, Prentice-Hall (1995).

Zienkiewicz and Taylor 2000 O. C. Zienkiewicz and R. L. Taylor, *Finite Element Method: Volume 1- The Basis*, Butterworth Heinemann (2000).

Index